Microwave Electronics

Microwave Electronics

Measurement and Materials Characterization

L. F. Chen, C. K. Ong and C. P. Neo
National University of Singapore

V. V. Varadan and V. K. Varadan
Pennsylvania State University, USA

John Wiley & Sons, Ltd

Other Wiley Editorial Offices

John Wiley & Sons Inc., 111 River Street, Hoboken, NJ 07030, USA

Jossey-Bass, 989 Market Street, San Francisco, CA 94103-1741, USA

Wiley-VCH Verlag GmbH, Boschstr. 12, D-69469 Weinheim, Germany

John Wiley & Sons Australia Ltd, 33 Park Road, Milton, Queensland 4064, Australia

John Wiley & Sons (Asia) Pte Ltd, 2 Clementi Loop #02-01, Jin Xing Distripark, Singapore 129809

John Wiley & Sons Canada Ltd, 22 Worcester Road, Etobicoke, Ontario, Canada M9W 1L1

Wiley also publishes its books in a variety of electronic formats. Some content that appears
in print may not be available in electronic books.

British Library Cataloguing in Publication Data

A catalogue record for this book is available from the British Library

ISBN 0-470-84492-2

Typeset in 10/12pt Times by Laserwords Private Limited, Chennai, India
Printed and bound in Great Britain by Antony Rowe Ltd, Chippenham, Wiltshire
This book is printed on acid-free paper responsibly manufactured from sustainable forestry
in which at least two trees are planted for each one used for paper production.

Contents

Preface   xi

1   Electromagnetic Properties of Materials   1
     1.1   Materials Research and Engineering at Microwave Frequencies   1
     1.2   Physics for Electromagnetic Materials   2
            1.2.1   Microscopic scale   2
            1.2.2   Macroscopic scale   6
     1.3   General Properties of Electromagnetic Materials   11
            1.3.1   Dielectric materials   11
            1.3.2   Semiconductors   16
            1.3.3   Conductors   17
            1.3.4   Magnetic materials   19
            1.3.5   Metamaterials   24
            1.3.6   Other descriptions of electromagnetic materials   28
     1.4   Intrinsic Properties and Extrinsic Performances of Materials   32
            1.4.1   Intrinsic properties   32
            1.4.2   Extrinsic performances   32
            References   34

2   Microwave Theory and Techniques for Materials Characterization   37
     2.1   Overview of the Microwave Methods for the Characterization of Electromagnetic Materials 37
            2.1.1   Nonresonant methods   38
            2.1.2   Resonant methods   40
     2.2   Microwave Propagation   42
            2.2.1   Transmission-line theory   42
            2.2.2   Transmission Smith charts   51
            2.2.3   Guided transmission lines   56
            2.2.4   Surface-wave transmission lines   73
            2.2.5   Free space   83
     2.3   Microwave Resonance   87
            2.3.1   Introduction   87
            2.3.2   Coaxial resonators   93
            2.3.3   Planar-circuit resonators   95
            2.3.4   Waveguide resonators   97
            2.3.5   Dielectric resonators   103
            2.3.6   Open resonators   115
     2.4   Microwave Network   119
            2.4.1   Concept of microwave network   119
            2.4.2   Impedance matrix and admittance matrix   119

	2.4.3	Scattering parameters	120
	2.4.4	Conversions between different network parameters	121
	2.4.5	Basics of network analyzer	121
	2.4.6	Measurement of reflection and transmission properties	126
	2.4.7	Measurement of resonant properties	134
		References	139

3	Reflection Methods	142
3.1	Introduction	142
	3.1.1	Open-circuited reflection	142
	3.1.2	Short-circuited reflection	143
3.2	Coaxial-line Reflection Method	144
	3.2.1	Open-ended apertures	145
	3.2.2	Coaxial probes terminated into layered materials	151
	3.2.3	Coaxial-line-excited monopole probes	154
	3.2.4	Coaxial lines open into circular waveguides	157
	3.2.5	Shielded coaxial lines	158
	3.2.6	Dielectric-filled cavity adapted to the end of a coaxial line	160
3.3	Free-space Reflection Method	161
	3.3.1	Requirements for free-space measurements	161
	3.3.2	Short-circuited reflection method	162
	3.3.3	Movable metal-backing method	162
	3.3.4	Bistatic reflection method	164
3.4	Measurement of Both Permittivity and Permeability Using Reflection Methods	164
	3.4.1	Two-thickness method	164
	3.4.2	Different-position method	165
	3.4.3	Combination method	166
	3.4.4	Different backing method	167
	3.4.5	Frequency-variation method	167
	3.4.6	Time-domain method	168
3.5	Surface Impedance Measurement	168
3.6	Near-field Scanning Probe	170
		References	172

4	Transmission/Reflection Methods	175
4.1	Theory for Transmission/reflection Methods	175
	4.1.1	Working principle for transmission/reflection methods	175
	4.1.2	Nicolson–Ross–Weir (NRW) algorithm	177
	4.1.3	Precision model for permittivity determination	178
	4.1.4	Effective parameter method	179
	4.1.5	Nonlinear least-squares solution	180
4.2	Coaxial Air-line Method	182
	4.2.1	Coaxial air lines with different diameters	182
	4.2.2	Measurement uncertainties	183
	4.2.3	Enlarged coaxial line	185
4.3	Hollow Metallic Waveguide Method	187
	4.3.1	Waveguides with different working bands	187
	4.3.2	Uncertainty analysis	187
	4.3.3	Cylindrical rod in rectangular waveguide	189
4.4	Surface Waveguide Method	190

	4.4.1	Circular dielectric waveguide	190
	4.4.2	Rectangular dielectric waveguide	192
4.5	Free-space Method		195
	4.5.1	Calculation algorithm	195
	4.5.2	Free-space TRL calibration	197
	4.5.3	Uncertainty analysis	198
	4.5.4	High-temperature measurement	199
4.6	Modifications on Transmission/reflection Methods		200
	4.6.1	Coaxial discontinuity	200
	4.6.2	Cylindrical cavity between transmission lines	200
	4.6.3	Dual-probe method	201
	4.6.4	Dual-line probe method	201
	4.6.5	Antenna probe method	202
4.7	Transmission/reflection Methods for Complex Conductivity Measurement		203
	References		205
5	Resonator Methods		208
5.1	Introduction		208
5.2	Dielectric Resonator Methods		208
	5.2.1	Courtney resonators	209
	5.2.2	Cohn resonators	214
	5.2.3	Circular-radial resonators	216
	5.2.4	Sheet resonators	219
	5.2.5	Dielectric resonators in closed metal shields	222
5.3	Coaxial Surface-wave Resonator Methods		227
	5.3.1	Coaxial surface-wave resonators	228
	5.3.2	Open coaxial surface-wave resonator	228
	5.3.3	Closed coaxial surface-wave resonator	229
5.4	Split-resonator Method		231
	5.4.1	Split-cylinder-cavity method	231
	5.4.2	Split-coaxial-resonator method	233
	5.4.3	Split-dielectric-resonator method	236
	5.4.4	Open resonator method	238
5.5	Dielectric Resonator Methods for Surface-impedance Measurement		242
	5.5.1	Measurement of surface resistance	242
	5.5.2	Measurement of surface impedance	243
	References		247
6	Resonant-perturbation Methods		250
6.1	Resonant Perturbation		250
	6.1.1	Basic theory	250
	6.1.2	Cavity-shape perturbation	252
	6.1.3	Material perturbation	253
	6.1.4	Wall-impedance perturbation	255
6.2	Cavity-perturbation Method		256
	6.2.1	Measurement of permittivity and permeability	256
	6.2.2	Resonant properties of sample-loaded cavities	258
	6.2.3	Modification of cavity-perturbation method	261
	6.2.4	Extracavity-perturbation method	265
6.3	Dielectric Resonator Perturbation Method		267
6.4	Measurement of Surface Impedance		268

	6.4.1	Surface resistance and surface reactance	268
	6.4.2	Measurement of surface resistance	269
	6.4.3	Measurement of surface reactance	275
6.5	Near-field Microwave Microscope		278
	6.5.1	Basic working principle	278
	6.5.2	Tip-coaxial resonator	279
	6.5.3	Open-ended coaxial resonator	280
	6.5.4	Metallic waveguide cavity	284
	6.5.5	Dielectric resonator	284
	References		286
7	**Planar-circuit Methods**		288
7.1	Introduction		288
	7.1.1	Nonresonant methods	288
	7.1.2	Resonant methods	290
7.2	Stripline Methods		291
	7.2.1	Nonresonant methods	291
	7.2.2	Resonant methods	292
7.3	Microstrip Methods		297
	7.3.1	Nonresonant methods	298
	7.3.2	Resonant methods	300
7.4	Coplanar-line Methods		309
	7.4.1	Nonresonant methods	309
	7.4.2	Resonant methods	311
7.5	Permeance Meters for Magnetic Thin Films		311
	7.5.1	Working principle	312
	7.5.2	Two-coil method	312
	7.5.3	Single-coil method	314
	7.5.4	Electrical impedance method	315
7.6	Planar Near-field Microwave Microscopes		317
	7.6.1	Working principle	317
	7.6.2	Electric and magnetic dipole probes	318
	7.6.3	Probes made from different types of planar transmission lines	319
	References		320
8	**Measurement of Permittivity and Permeability Tensors**		323
8.1	Introduction		323
	8.1.1	Anisotropic dielectric materials	323
	8.1.2	Anisotropic magnetic materials	325
8.2	Measurement of Permittivity Tensors		326
	8.2.1	Nonresonant methods	327
	8.2.2	Resonator methods	333
	8.2.3	Resonant-perturbation method	336
8.3	Measurement of Permeability Tensors		340
	8.3.1	Nonresonant methods	340
	8.3.2	Faraday rotation methods	345
	8.3.3	Resonator methods	351
	8.3.4	Resonant-perturbation methods	355
8.4	Measurement of Ferromagnetic Resonance		370
	8.4.1	Origin of ferromagnetic resonance	370
	8.4.2	Measurement principle	371

	8.4.3	Cavity methods	373
	8.4.4	Waveguide methods	374
	8.4.5	Planar-circuit methods	376
	References		379

9	Measurement of Ferroelectric Materials		382
	9.1	Introduction	382
		9.1.1	Perovskite structure	383
		9.1.2	Hysteresis curve	383
		9.1.3	Temperature dependence	383
		9.1.4	Electric field dependence	385
	9.2	Nonresonant Methods	385
		9.2.1	Reflection methods	385
		9.2.2	Transmission/reflection method	386
	9.3	Resonant Methods	386
		9.3.1	Dielectric resonator method	386
		9.3.2	Cavity-perturbation method	389
		9.3.3	Near-field microwave microscope method	390
	9.4	Planar-circuit Methods	390
		9.4.1	Coplanar waveguide method	390
		9.4.2	Coplanar resonator method	394
		9.4.3	Capacitor method	394
		9.4.4	Influence of biasing schemes	404
	9.5	Responding Time of Ferroelectric Thin Films	405
	9.6	Nonlinear Behavior and Power-Handling Capability of Ferroelectric Films	407
		9.6.1	Pulsed signal method	407
		9.6.2	Intermodulation method	409
	References		412

10	Microwave Measurement of Chiral Materials		414
	10.1	Introduction	414
	10.2	Free-space Method	415
		10.2.1	Sample preparation	416
		10.2.2	Experimental procedure	416
		10.2.3	Calibration	417
		10.2.4	Time-domain measurement	430
		10.2.5	Computation of ε, μ, and β of the chiral composite samples	434
		10.2.6	Experimental results for chiral composites	440
	10.3	Waveguide Method	452
		10.3.1	Sample preparation	452
		10.3.2	Experimental procedure	452
		10.3.3	Computation of ε, μ, and ξ of the chiral composite samples	453
		10.3.4	Experimental results for chiral composites	454
	10.4	Concluding Remarks	458
	References		458

11	Measurement of Microwave Electrical Transport Properties		460
	11.1	Hall Effect and Electrical Transport Properties of Materials	460
		11.1.1	Direct current Hall effect	461
		11.1.2	Alternate current Hall effect	461
		11.1.3	Microwave Hall effect	461

11.2		Nonresonant Methods for the Measurement of Microwave Hall Effect	464
	11.2.1	Faraday rotation	464
	11.2.2	Transmission method	465
	11.2.3	Reflection method	469
	11.2.4	Turnstile-junction method	473
11.3		Resonant Methods for the Measurement of the Microwave Hall Effect	475
	11.3.1	Coupling between two orthogonal resonant modes	475
	11.3.2	Hall effect of materials in MHE cavity	476
	11.3.3	Hall effect of endplate of MHE cavity	482
	11.3.4	Dielectric MHE resonator	484
	11.3.5	Planar MHE resonator	486
11.4		Microwave Electrical Transport Properties of Magnetic Materials	486
	11.4.1	Ordinary and extraordinary Hall effect	486
	11.4.2	Bimodal cavity method	487
	11.4.3	Bimodal dielectric probe method	489
	References		489
12	Measurement of Dielectric Properties of Materials at High Temperatures		492
12.1		Introduction	492
	12.1.1	Dielectric properties of materials at high temperatures	492
	12.1.2	Problems in measurements at high temperatures	494
	12.1.3	Overviews of the methods for measurements at high temperatures	496
12.2		Coaxial-line Methods	497
	12.2.1	Measurement of permittivity using open-ended coaxial probe	498
	12.2.2	Problems related to high-temperature measurements	498
	12.2.3	Correction of phase shift	500
	12.2.4	Spring-loaded coaxial probe	502
	12.2.5	Metallized ceramic coaxial probe	502
12.3		Waveguide Methods	503
	12.3.1	Open-ended waveguide method	503
	12.3.2	Dual-waveguide method	504
12.4		Free-space Methods	506
	12.4.1	Computation of ε_r^*	507
12.5		Cavity-Perturbation Methods	510
	12.5.1	Cavity-perturbation methods for high-temperature measurements	510
	12.5.2	TE_{10n} mode rectangular cavity	512
	12.5.3	TM mode cylindrical cavity	514
12.6		Dielectric-loaded Cavity Method	520
	12.6.1	Coaxial reentrant cavity	520
	12.6.2	Open-resonator method	523
	12.6.3	Oscillation method	524
	References		528
Index			531

Preface

Microwave materials have been widely used in a variety of applications ranging from communication devices to military satellite services, and the study of materials properties at microwave frequencies and the development of functional microwave materials have always been among the most active areas in solid-state physics, materials science, and electrical and electronic engineering. In recent years, the increasing requirements for the development of high-speed, high-frequency circuits and systems require complete understanding of the properties of materials functioning at microwave frequencies. All these aspects make the characterization of materials properties an important field in microwave electronics.

Characterization of materials properties at microwave frequencies has a long history, dating from the early 1950s. In past decades, dramatic advances have been made in this field, and a great deal of new measurement methods and techniques have been developed and applied. There is a clear need to have a practical reference text to assist practicing professionals in research and industry. However, we realize the lack of good reference books dealing with this field. Though some chapters, reviews, and books have been published in the past, these materials usually deal with only one or several topics in this field, and a book containing a comprehensive coverage of up-to-date measurement methodologies is not available. Therefore, most of the research and development activities in this field are based primarily on the information scattered throughout numerous reports and journals, and it always takes a great deal of time and effort to collect the information related to on-going projects from the voluminous literature. Furthermore, because of the paucity of comprehensive textbooks, the training in this field is usually not systematic, and this is undesirable for further progress and development in this field.

This book deals with the microwave methods applied to materials property characterization, and it provides an in-depth coverage of both established and emerging techniques in materials characterization. It also represents the most comprehensive treatment of microwave methods for materials property characterization that has appeared in book form to date. Although this book is expected to be most useful to those engineers actively engaged in designing materials property–characterization methods, it should also be of considerable value to engineers in other disciplines, such as industrial engineers, bioengineers, and materials scientists, who wish to understand the capabilities and limitations of microwave measurement methods that they use. Meanwhile, this book also satisfies the requirement for up-to-date texts at graduate and senior undergraduate levels on the subjects in materials characterization.

Among this book's most outstanding features is its comprehensive coverage. This book discusses almost all aspects of the microwave theory and techniques for the characterization of the electromagnetic properties of materials at microwave frequencies. In this book, the materials under characterization may be dielectrics, semiconductors, conductors, magnetic materials, and artificial materials; the electromagnetic properties to be characterized mainly include permittivity, permeability, chirality, mobility, and surface impedance.

The two introductory chapters, Chapter 1 and Chapter 2, are intended to acquaint the readers with the basis for the research and engineering of electromagnetic materials from the materials and microwave fundamentals respectively. As general knowledge of electromagnetic properties of materials is helpful for understanding measurement results and correcting possible errors, Chapter 1 introduces the general properties of various electromagnetic materials and their underlying physics. After making a brief review on the methods for materials properties characterization, Chapter 2 provides a summary of the basic microwave theory and techniques, based on which the methods for materials characterization are developed. This summary is mainly intended for reference rather than for tutorial purposes, although some of the important aspects of microwave theory are treated at a greater length. References are cited to permit readers to further study the topics they are interested in.

Chapters 3 to 8 deal with the measurements of the permittivity and permeability of low-conductivity materials and the surface impedance of high-conductivity materials. Two types of nonresonant methods, reflection method and transmission/reflection method, are discussed in Chapters 3 and 4 respectively; two types of resonant methods, resonator method and resonant-perturbation method, are discussed in Chapters 5 and 6 respectively. In the methods discussed in Chapters 3 to 6, the transmission lines used are mainly coaxial-line, waveguide, and free-space, while Chapter 7 is concerned with the measurement methods developed from planar transmission lines, including stripline, microstrip-, and coplanar line. The methods discussed in Chapters 3 to 7 are suitable for isotropic materials, which have scalar or complex permittivity and permeability. The permittivity of anisotropic dielectric materials is a tensor parameter, and magnetic materials usually have tensor permeability under an external dc magnetic field. Chapter 8 deals with the measurement of permittivity and permeability tensors.

Ferroelectric materials are a special category of dielectric materials often used in microwave electronics for developing electrically tunable devices. Chapter 9 discusses the characterization of ferroelectric materials, and the topics covered include the techniques for studying the temperature dependence and electric field dependence of dielectric properties.

In recent years, the research on artificial materials has been active. Chapter 10 deals with a special type of artificial materials: chiral materials. After introducing the concept and basic characteristics of chiral materials, the methods for chirality measurements and the possible applications of chiral materials are discussed.

The electrical transport properties at microwave frequencies are important for the development of high-speed electronic circuits. Chapter 11 discusses the microwave Hall effect techniques for the measurement of the electrical transport properties of low-conductivity, high-conductivity, and magnetic materials.

The measurement of materials properties at high temperatures is often required in industry, scientific research, and biological and medical applications. In principle, most of the methods discussed in this book can be extended to high-temperature measurements. Chapter 12 concentrates on the measurement of the dielectric properties of materials at high temperatures, and the techniques for solving the problems in high-temperature measurements can also be applied for the measurement of other materials property parameters at high temperatures.

In this book, each chapter is written as a self-contained unit, so that readers can quickly get comprehensive information related to their research interests or on-going projects. To provide a broad treatment of various topics, we condensed mountains of literature into readable accounts within a text of reasonable size. Many references have been included for the benefit of the readers who wish to pursue a given topic in greater depth or refer to the original papers.

It is clear that the principle of a method for materials characterization is more important than the techniques required for implementing this method. If we understand the fundamental principle underlying a measurement method, we can always find a suitable way to realize this method. Although the advances in technology may significantly change the techniques for implementing a measurement method, they cannot greatly influence the measurement principle. In writing this book, we tried to present the fundamental principles behind various designs so that readers can understand the process of applying fundamental concepts to arrive at actual designs using different techniques and approaches. We believe that an engineer with a sound knowledge of the basic concepts and fundamental principles for materials property characterization and the ability apply to his knowledge toward design objectives, is the engineer who is most likely to make full use of the existing methods, and develop original methods to fulfill ever-rising measurement requirements.

We would like to indicate that this text is a compilation of the work of many people. We cannot be held responsible for the designs described that are still under patent. It is also difficult to always give proper credits to those who are the originators of new concepts and the inventors of new methods. The names we give to some measurement methods may not fit the intentions of the inventors or may not accurately reflect the most characteristic features of these methods. We hope that there are not too many such errors and will appreciate it if the readers could bring the errors they discover to our attention.

There are many people to whom we owe many thanks for helping us prepare this book. However, space dictates that only a few of them can receive formal acknowledgements. But this should not be taken as a disparagement of those whose contributions remain anonymous. Our foremost appreciation goes to Mr. Quek Gim Pew, Deputy Chief Executive (Technology), Singapore Defence Science & Technology Agency, Mr. Quek Tong Boon, Chief Executive Officer, Singapore DSO National Laboratories, and Professor Lim Hock, Director, Temasek Laboratories, National University of Singapore, for their encouragement and support along the way. We are grateful to Pennsylvania State University and HVS Technologies for giving us permission to include the HVS Free Space Unit and the data in this book. We really appreciate the valuable help and cooperation from Dr. Li Zheng-Wen, Dr. Rao Xuesong, and Mr. Tan Chin Yaw. We are very grateful to the staff of John Wiley & Sons for their helpful efforts and cheerful professionalism during this project.

L. F. Chen
C. K. Ong
C. P. Neo
V. V. Varadan
V. K. Varadan

1

Electromagnetic Properties of Materials

This chapter starts with the introduction of the materials research and engineering at microwave frequencies, with emphasis laid on the significance and applications of the study of the electromagnetic properties of materials. The fundamental physics that governs the interactions between materials and electromagnetic fields is then discussed at both microscopic and macroscopic scales. Subsequently, we analyze the general properties of typical electromagnetic materials, including dielectric materials, semiconductors, conductors, magnetic materials, and artificial materials. Afterward, we discuss the intrinsic properties and extrinsic performances of electromagnetic materials.

1.1 MATERIALS RESEARCH AND ENGINEERING AT MICROWAVE FREQUENCIES

While technology decides how electromagnetic materials can be utilized, science attempts to decipher why materials behave as they do. The responses of materials to electromagnetic fields are closely determined by the displacement of their free and bounded electrons by electric fields and the orientation of their atomic moments by magnetic fields. The deep understanding and full utilization of electromagnetic materials have come from decoding the interactions between materials and electromagnetic fields by using both theoretical and experimental strategies.

This book mainly deals with the methodology for the characterization of electromagnetic materials for microwave electronics, and also discusses

the applications of techniques for materials property characterization in various fields of sciences and engineering. The importance of the research on the electromagnetic properties of materials at microwave frequencies can be understood in the aspects that follow.

Firstly, though it is an old field in physics, the study of electromagnetic properties of materials at microwave frequencies is full of academic importance (Solymar and Walsh 1998; Kittel 1997; Von Hippel 1995a,b; Jiles 1994; Robert 1988), especially for magnetic materials (Jiles 1998; Smit 1971) and superconductors (Tinkham 1996) and ferroelectrics (Lines and Glass 1977). The knowledge gained from microwave measurements contributes to our information about both the macroscopic and the microscopic properties of materials, so microwave techniques have been important for materials property research. Though magnetic materials are widely used in various fields, the research of magnetic materials lags far behind their applications, and this, to some extent, hinders us from making full application of magnetic materials. Until now, the electromagnetic properties of magnetic properties at microwave frequencies have not been fully investigated yet, and this is one of the main obstacles for the development of microwave magnetoelectrics. Besides, one of the most promising applications of superconductors is microwave electronics. A lot of effort has been put in the study of the microwave properties of superconductors, while many areas are yet to be explored. Meanwhile, as ferroelectric materials have great application potential in developing smart electromagnetic materials, structures, and

Microwave Electronics: Measurement and Materials Characterization   L. F. Chen, C. K. Ong, C. P. Neo, V. V. Varadan and V. K. Varadan
© 2004 John Wiley & Sons, Ltd   ISBN: 0-470-84492-2

devices in recent years, microwave ferroelectricity is under intensive investigation.

Secondly, microwave communications are playing more and more important roles in military, industrial, and civilian life, and microwave engineering requires precise knowledge of the electromagnetic properties of materials at microwave frequencies (Ramo *et al.* 1994). Since World War II, a lot of resources have been put into electromagnetic signature control, and microwave absorbers are widely used in reducing the radar cross sections (RCSs) of vehicles. The study of electromagnetic properties of materials and the ability of tailoring the electromagnetic properties of composite materials are very important for the design and development of radar absorbing materials and other functional electromagnetic materials and structures (Knott *et al.* 1993).

Thirdly, as the clock speeds of electronic devices are approaching microwave frequencies, it becomes indispensable to study the microwave electronic properties of materials used in electronic components, circuits, and packaging. The development of electronic components working at microwave frequencies needs the electrical transport properties at microwave frequencies, such as Hall mobility and carrier density; and the development of electronic circuits working at microwave frequencies requires accurate constitutive properties of materials, such as permittivity and permeability. Meanwhile, the electromagnetic interference (EMI) should be taken into serious consideration in the design of circuit and packaging, and special materials are needed to ensure electromagnetic compatibility (EMC) (Montrose 1999).

Fourthly, the study of electromagnetic properties of materials is important for various fields of science and technology. The principle of microwave remote sensing is based on the reflection and scattering of different objects to microwave signals, and the reflection and scattering properties of an object are mainly determined by the electromagnetic properties of the object. Besides, the conclusions of the research of electromagnetic materials are helpful for agriculture, food engineering, medical treatments, and bioengineering (Thuery and Grant 1992).

Finally, as the electromagnetic properties of materials are related to other macroscopic or microscopic properties of the materials, we can obtain information about the microscopic or macroscopic properties we are interested in from the electromagnetic properties of the materials. In materials research and engineering, microwave techniques for the characterization of materials properties are widely used in monitoring the fabrication procedure and nondestructive testing of samples and products (Zoughi 2000; Nyfors and Vainikainen 1989).

This chapter aims to provide basic knowledge for understanding the results from microwave measurements. We will give a general introduction on electromagnetic materials at microscopic and macroscopic scales and will discuss the parameters describing the electromagnetic properties of materials, the classification of electromagnetic materials, and general properties of typical electromagnetic materials. Further discussions on various topics can be found in later chapters or the references cited.

1.2　PHYSICS FOR ELECTROMAGNETIC MATERIALS

In physics and materials sciences, electromagnetic materials are studied at both the microscopic and the macroscopic scale (Von Hippel 1995a,b). At the microscopic scale, the energy bands for electrons and magnetic moments of the atoms and molecules in materials are investigated, while at the macroscopic level, we study the overall responses of macroscopic materials to external electromagnetic fields.

1.2.1　Microscopic scale

In the microscopic scale, the electrical properties of a material are mainly determined by the electron energy bands of the material. According to the energy gap between the valence band and the conduction band, materials can be classified into insulators, semiconductors, and conductors. Owing to its electron spin and electron orbits around the nucleus, an atom has a magnetic moment. According to the responses of magnetic moments to magnetic field, materials can be generally classified into diamagnetic materials, paramagnetic materials, and ordered magnetic materials.

1.2.1.1 Electron energy bands

According to Bohr's model, an atom is characterized by its discrete energy levels. When atoms are brought together to constitute a solid, the discrete levels combine to form energy bands and the occupancy of electrons in a band is dictated by Fermi-dirac statistics. Figure 1.1 shows the relationship between energy bands and atomic separation. When the atoms get closer, the energy bands broaden, and usually the outer band broadens more than the inner one. For some elements, for example lithium, when the atomic separation is reduced, the bands may broaden sufficiently for neighboring bands to merge, forming a broader

band. While for some elements, for example carbon, the merged broadband may further split into separate bands at closer atomic separation.

The highest energy band containing occupied energy levels at 0 K in a solid is called the *valence band*. The valence band may be completely filled or only partially filled with electrons. The electrons in the valence band are bonded to their nuclei. The conduction band is the energy band above the valence energy band, and contains vacant energy levels at 0 K. The electrons in the conduction band are called *free electrons*, which are free to move. Usually, there is a forbidden gap between the valence band and the conduction band, and the availability of free electrons in the conduction band mainly depends on the forbidden gap energy. If the forbidden gap is large, it is possible that no free electrons are available, and such a material is called an *insulator*. For a material with a small forbidden energy gap, the availability of free electron in the conduction band permits some electron conduction, and such a material is a semiconductor. In a conductor, the conduction and valence bands may overlap, permitting abundant free electrons to be available at any ambient temperature, thus giving high electrical conductivity. The energy bands for insulator, semiconductor, and good conductor are shown schematically in Figure 1.2.

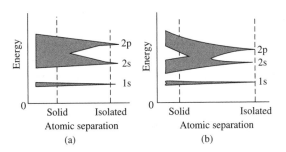

Figure 1.1 The relationships between energy bands and atomic separation. (a) Energy bands of lithium and (b) energy bands of carbon. (Bolton 1992) Source: Bolton, W. (1992), *Electrical and Magnetic Properties of Materials*, Longman Scientific & Technical, Harlow

Insulators

For most of the insulators, the forbidden gap between their valence and conduction energy bands

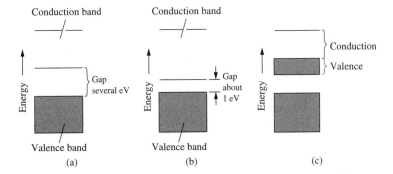

Figure 1.2 Energy bands for different types of materials. (a) Insulator, (b) semiconductor, and (c) good conductor. (Bolton 1992). Modified from Bolton, W. (1992), *Electrical and Magnetic Properties of Materials*, Longman Scientific & Technical, Harlow

is larger than 5 eV. Usually, we assume that an insulate is nonmagnetic, and under this assumption, insulators are called *dielectrics*. Diamond, a form of carbon, is a typical example of a dielectric. Carbon has two electrons in the 1s shell, two in the 2s shell, and two in the 2p shell. In a diamond, the bonding between carbon atoms is achieved by covalent bonds with electrons shared between neighboring atoms, and each atom has a share in eight 2p electrons (Bolton 1992). So all the electrons are tightly held between the atoms by this covalent bonding. As shown in Figure 1.1(b), the consequence of this bonding is that diamond has a full valence band with a substantial forbidden gap between the valence band and the conduction band. But it should be noted that, graphite, another form of carbon, is not a dielectric, but a conductor. This is because all the electrons in the graphite structure are not locked up in covalent bonds and some of them are available for conduction. So the energy bands are related to not only the atom structures but also the ways in which atoms are combined.

Semiconductors

The energy gap between the valence and conduction bands of a semiconductor is about 1 eV. Germanium and silicon are typical examples of semiconductors. Each germanium or silicon atom has four valence electrons, and the atoms are held together by covalent bonds. Each atom shares electrons with each of four neighbors, so all the electrons are locked up in bonds. So there is a gap between a full valence band and the conduction band. However, unlike insulators, the gap is relatively small. At room temperature, some of the valence electrons can break free from the bonds and have sufficient energy to jump over the forbidden gap, arriving at the conduction band. The density of the free electrons for most of the semiconductors is in the range of 10^{16} to 10^{19} per m^3.

Conductors

For a conductor, there is no energy gap between the valence gap and conduction band. For a good conductor, the density of free electrons is on the order of 10^{28} m^3. Lithium is a typical example of a conductor. It has two electrons in the 1s shell and one in the 2s shell. The energy bands of such elements are of the form shown in Figure 1.1(a). The 2s and 2p bands merge, forming a large band that is only partially occupied, and under an electric field, electrons can easily move into vacant energy levels.

In the category of conductors, superconductors have attracted much research interest. In a normal conductor, individual electrons are scattered by impurities and phonons. However, for superconductors, the electrons are paired with those of opposite spins and opposite wave vectors, forming Cooper pairs, which are bonded together by exchanging phonons. In the Bardeen–Cooper–Schrieffer (BCS) theory, these Cooper pairs are not scattered by the normal mechanisms. A superconducting gap is found in superconductors and the size of the gap is in the microwave frequency range, so study of superconductors at microwave frequencies is important for the understanding of superconductivity and application of superconductors.

1.2.1.2 Magnetic moments

An electron orbiting a nucleus is equivalent to a current in a single-turn coil, so an atom has a magnetic dipole moment. Meanwhile, an electron also spins. By considering the electron to be a small charged sphere, the rotation of the charge on the surface of the sphere is also like a single-turn current loop and also produces a magnetic moment (Bolton 1992). The magnetic properties of a material are mainly determined by its magnetic moments that result from the orbiting and spinning of electrons. According to the responses of the magnetic moments of the atoms in a material to an external magnetic field, materials can be generally classified into diamagnetic, paramagnetic, and ordered magnetic materials.

Diamagnetic materials

The electrons in a diamagnetic material are all paired up with spins antiparallel, so there is no net magnetic moment on their atoms. When an external magnetic field is applied, the orbits of the electrons change, resulting in a net magnetic moment in the direction opposite to the applied magnetic field. It should be noted that all materials have diamagnetism since all materials have orbiting electrons. However, for diamagnetic materials, the spin of the electrons does not contribute to the magnetism; while for paramagnetic and ferromagnetic materials, the effects of the magnetic dipole moments that result from the spinning of electrons are much greater than the diamagnetic effect.

Paramagnetic materials

The atoms in a paramagnetic material have net magnetic moments due to the unpaired electron spinning in the atoms. When there is no external magnetic field, these individual moments are randomly aligned, so the material does not show macroscopic magnetism. When an external magnetic field is applied, the magnetic moments are slightly aligned along the direction of the external magnetic field. If the applied magnetic field is removed, the alignment vanishes immediately. So a paramagnetic material is weakly magnetic only in the presence of an external magnetic field. The arrangement of magnetic moments in a paramagnetic material is shown in Figure 1.3(a). Aluminum and platinum are typical paramagnetic materials.

Ordered magnetic materials

In ordered magnetic materials, the magnetic moments are arranged in certain orders. According to the ways in which magnetic moments are arranged, ordered magnetic materials fall into several subcategories, mainly including ferromagnetic, antiferromagnetic, and ferrimagnetic (Bolton 1992; Wohlfarth 1980). Figure 1.3 shows the arrangements of magnetic moments in paramagnetic, ferromagnetic, antiferromagnetic, and ferrimagnetic materials, respectively.

As shown in Figure 1.3(b), the atoms in a ferromagnetic material are bonded together in such a way that the dipoles in neighboring atoms are all in the same direction. The coupling between atoms of ferromagnetic materials, which results in the

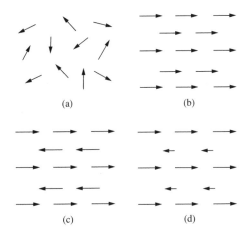

Figure 1.3 Arrangements of magnetic moments in various magnetic materials. (a) Paramagnetic, (b) ferromagnetic, (c) antiferromagnetic, and (d) ferrimagnetic materials. Modified from Bolton, W. (1992). *Electrical and Magnetic Properties of Materials*, Longman Scientific & Technical, Harlow

ordered arrangement of magnetic dipoles shown in Figure 1.3(b), is quite different from the coupling between atoms of paramagnetic materials, which results in the random arrangement of magnetic dipoles shown in Figure 1.3(a). Iron, cobalt, and nickel are typical ferromagnetic materials.

As shown in Figure 1.3(c), in an antiferromagnetic material, half of the magnetic dipoles align themselves in one direction and the other half of the magnetic moments align themselves in exactly the opposite direction if the dipoles are of the same size and cancel each other out. Manganese, manganese oxide, and chromium are typical antiferromagnetic materials. However, as shown in Figure 1.3(d), for a ferrimagnetic material, also called *ferrite*, the magnetic dipoles have different sizes and they do not cancel each other. Magnetite (Fe_3O_4), nickel ferrite ($NiFe_2O_4$), and barium ferrite ($BaFe_{12}O_{19}$) are typical ferrites.

Generally speaking, the dipoles in a ferromagnetic or ferrimagnetic material may not all be arranged in the same direction. Within a domain, all the dipoles are arranged in its easy-magnetization direction, but different domains may have different directions of arrangement. Owing to the random orientations of the domains, the material does not have macroscopic magnetism without an external magnetic field.

The crystalline imperfections in a magnetic material have significant effects on the magnetization of the material (Robert 1988). For an ideal magnetic material, for example monocrystalline iron without any imperfections, when a magnetic field *H* is applied, due to the condition of minimum energy, the sizes of the domains in *H* direction increase, while the sizes of other domains decrease. Along with the increase of the magnetic field, the structures of the domains change successively, and finally a single domain in *H* direction is obtained. In this ideal case, the displacement of domain walls is free. When the magnetic field *H* is removed, the material returns to its initial state; so the magnetization process is reversible.

Owing to the inevitable crystalline imperfections, the magnetization process becomes complicated. Figure 1.4(a) shows the arrangement of domains in a ferromagnetic material when no external magnetic field is applied. The domain

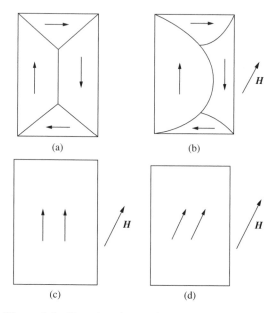

(a)   (b)

(c)   (d)

Figure 1.4 Domains in a ferromagnetic material. (a) Arrangement of domains when no external magnetic field is applied, (b) arrangement of domains when a weak magnetic field is applied, (c) arrangement of domains when a medium magnetic field is applied, and (d) arrangement of domains when a strong magnetic field is applied. Modified from Robert, P. (1988). *Electrical and Magnetic Properties of Materials*, Artech House, Norwood

walls are pinned by crystalline imperfections. As shown in Figure 1.4(b), when an external magnetic field *H* is applied, the domains whose orientations are near the direction of the external magnetic field grow in size, while the sizes of the neighboring domains wrongly directed decrease. When the magnetic field is very weak, the domain walls behave like elastic membranes, and the changes of the domains are reversible. When the magnetic field increases, the pressure on the domain walls causes the pinning points to give way, and the domain walls move by a series of jumps. Once a jump of domain wall happens, the magnetization process becomes irreversible. As shown in Figure 1.4(c), when the magnetic field *H* reaches a certain level, all the magnetic moments are arranged parallel to the easy magnetization direction nearest to the direction of the external magnetic field *H*. If the external magnetic field *H* increases further, the magnetic moments are aligned along *H* direction, deviating from the easy magnetization direction, as shown in Figure 1.4(d). In this state, the material shows its greatest magnetization, and the material is magnetically saturated.

In a polycrystalline magnetic material, the magnetization process in each grain is similar to that in a monocrystalline material as discussed above. However, due to the magnetostatic and magnetostrictions occurring between neighboring grains, the overall magnetization of the material becomes quite complicated. The grain structures are important to the overall magnetization of a polycrystalline magnetic material. The magnetization process of magnetic materials is further discussed in Section 1.3.4.1.

It is important to note that for an ordered magnetic material, there is a special temperature called *Curie temperature* (T_c). If the temperature is below the Curie temperature, the material is in a magnetically ordered phase. If the temperature is higher than the Curie temperature, the material will be in a paramagnetic phase. The Curie temperature for iron is 770 °C, for nickel 358 °C, and for cobalt 1115 °C.

1.2.2 Macroscopic scale

The interactions between a macroscopic material and electromagnetic fields can be generally described by Maxwell's equations:

$$\nabla \cdot \boldsymbol{D} = \rho \tag{1.1}$$

$$\nabla \cdot \boldsymbol{B} = 0 \tag{1.2}$$

$$\nabla \times \boldsymbol{H} = \partial \boldsymbol{D}/\partial t + \boldsymbol{J} \tag{1.3}$$

$$\nabla \times \boldsymbol{E} = -\partial \boldsymbol{B}/\partial t \tag{1.4}$$

with the following constitutive relations:

$$\boldsymbol{D} = \varepsilon \boldsymbol{E} = (\varepsilon' - \mathrm{j}\varepsilon'')\boldsymbol{E} \tag{1.5}$$

$$\boldsymbol{B} = \mu \boldsymbol{H} = (\mu' - \mathrm{j}\mu'')\boldsymbol{H} \tag{1.6}$$

$$\boldsymbol{J} = \sigma \boldsymbol{E} \tag{1.7}$$

where $\boldsymbol{H}$ is the magnetic field strength vector; $\boldsymbol{E}$, the electric field strength vector; $\boldsymbol{B}$, the magnetic flux density vector; $\boldsymbol{D}$, the electric displacement vector; $\boldsymbol{J}$, the current density vector; ρ, the charge density; $\varepsilon = \varepsilon' - \mathrm{j}\varepsilon''$, the complex permittivity of the material; $\mu = \mu' - \mathrm{j}\mu''$, the complex permeability of the material; and σ, the conductivity of the material. Equations (1.1) to (1.7) indicate that the responses of an electromagnetic material to electromagnetic fields are determined essentially by three constitutive parameters, namely permittivity ε, permeability μ, and conductivity σ. These parameters also determine the spatial extent to which the electromagnetic field can penetrate into the material at a given frequency.

In the following, we discuss the parameters describing two general categories of materials: low-conductivity materials and high-conductivity materials.

1.2.2.1 Parameters describing low-conductivity materials

Electromagnetic waves can propagate in a low-conductivity material, so both the surface and inner parts of the material respond to the electromagnetic wave. There are two types of parameters describing the electromagnetic properties of low-conductivity materials: constitutive parameters and propagation parameters.

Constitutive parameters

The constitutive parameters defined in Eqs. (1.5) to (1.7) are often used to describe the electromagnetic

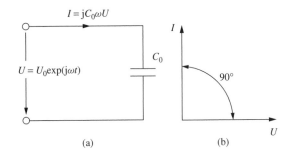

Figure 1.5 The current in a circuit with a capacitor. (a) Circuit layout and (b) complex plane showing current and voltage

properties of low-conductivity materials. As the value of conductivity σ is small, we concentrate on permittivity and permeability. In a general case, both permittivity and permeability are complex numbers, and the imaginary part of permittivity is related to the conductivity of the material. In the following discussion, we analogize microwave signals to ac signals, and distributed capacitor and inductor to lumped capacitor and inductor (Von Hippel 1995b).

Consider the circuit shown in Figure 1.5(a). The vacuum capacitor with capacitance C_0 is connected to an ac voltage source $U = U_0\exp(\mathrm{j}\omega t)$. The charge storage in the capacitor is $Q = C_0U$, and the current I flowing in the circuit is

$$I = \frac{\mathrm{d}Q}{\mathrm{d}t} = \frac{\mathrm{d}}{\mathrm{d}t}(C_0U_0\mathrm{e}^{\mathrm{j}\omega t}) = \mathrm{j}C_0\,\omega U \tag{1.8}$$

So, in the complex plane shown in Figure 1.5(b), the current I leads the voltage U by a phase angle of $90°$.

Now, we insert a dielectric material into the capacitor and the equivalent circuit is shown in Figure 1.6(a). The total current consists of two parts, the charging current (I_c) and loss current (I_l):

$$I = I_c + I_l = \mathrm{j}C\omega U + GU = (\mathrm{j}C\omega + G)U \tag{1.9}$$

where C is the capacitance of the capacitor loaded with the dielectric material and G is the conductance of the dielectric material. The loss current is in phase with the source voltage U. In the complex plane shown in Figure 1.6(b), the charging current I_c leads the loss current I_l by a phase angle of $90°$, and the total current I leads

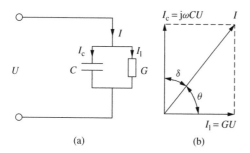

Figure 1.6   The relationships between charging current and loss current. (a) Equivalent circuit and (b) complex plane showing charging current and loss current

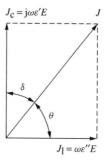

Figure 1.7   Complex plane showing the charging current density and loss current density

the source voltage U with an angle θ less than $90°$. The phase angle between I_c and I is often called *loss angle* δ.

We may alternatively use complex permittivity $\varepsilon = \varepsilon' - j\varepsilon''$ to describe the effect of dielectric material. After a dielectric material is inserted into the capacitor, the capacitance C of the capacitor becomes

$$C = \frac{\varepsilon C_0}{\varepsilon_0} = (\varepsilon' - j\varepsilon'')\frac{C_0}{\varepsilon_0} \tag{1.10}$$

And the charging current is

$$I = j\omega(\varepsilon' - j\varepsilon'')\frac{C_0}{\varepsilon_0}U = (j\omega\varepsilon' + \omega\varepsilon'')\frac{C_0}{\varepsilon_0}U \tag{1.11}$$

Therefore, as shown in Figure 1.6, the current density J transverse to the capacitor under the applied field strength E becomes

$$J = (j\omega\varepsilon' + \omega\varepsilon'')E = \varepsilon\frac{dE}{dt} \tag{1.12}$$

The product of angular frequency and loss factor is equivalent to a dielectric conductivity: $\sigma = \omega\varepsilon''$. This dielectric conductivity sums over all the dissipative effects of the material. It may represent an actual conductivity caused by migrating charge carriers and it may also refer to an energy loss associated with the dispersion of ε', for example, the friction accompanying the orientation of dipoles. The latter part of dielectric conductivity will be discussed in detail in Section 1.3.1.

According to Figure 1.7, we define two parameters describing the energy dissipation of a dielectric material. The dielectric loss tangent is given by

$$\tan\delta_e = \varepsilon''/\varepsilon', \tag{1.13}$$

and the dielectric power factor is given by

$$\cos\theta_e = \varepsilon''/\sqrt{(\varepsilon')^2 + (\varepsilon'')^2} \tag{1.14}$$

Equations (1.13) and (1.14) show that for a small loss angle δ_e, $\cos\theta \approx \tan\delta_e$.

In microwave electronics, we often use relative permittivity, which is a dimensionless quantity, defined by

$$\varepsilon_r = \frac{\varepsilon}{\varepsilon_0} = \frac{\varepsilon' - j\varepsilon''}{\varepsilon_0} = \varepsilon_r' - j\varepsilon_r'' = \varepsilon_r'(1 - j\tan\delta_e) \tag{1.15}$$

where ε is complex permittivity,
ε_r is relative complex permittivity,
$\varepsilon_0 = 8.854 \times 10^{-12}$ F/m is the permittivity of free space,
ε_r' is the real part of relative complex permittivity,
ε_r'' is the imaginary part of relative complex permittivity,
$\tan\delta_e$ is dielectric loss tangent, and
δ_e is dielectric loss angle.

Now, let us consider the magnetic response of low-conductivity material. According to the Faraday's inductance law

$$U = L\frac{dI}{dt}, \tag{1.16}$$

we can get the magnetization current I_m:

$$I_m = -j\frac{U}{\omega L_0} \tag{1.17}$$

where U is the magnetization voltage, L_0 is the inductance of an empty inductor, and ω is the angular frequency. If we introduce an

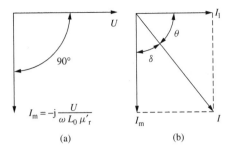

Figure 1.8 The magnetization current in a complex plane. (a) Relationship between magnetization current and voltage and (b) relationship between magnetization current and loss current

ideal, lossless magnetic material with relative permeability μ'_r, the magnetization field becomes

$$I_m = -j\frac{U}{\omega L_0 \mu'_r} \qquad (1.18)$$

In the complex plane shown in Figure 1.8(a), the magnetization current I_m lags the voltage U by 90° for no loss of magnetic materials. As shown in Figure 1.8(b), an actual magnetic material has magnetic loss, and the magnetic loss current I_l caused by energy dissipation during the magnetization cycle is in phase with U. By introducing a complex permeability $\mu = \mu' - j\mu''$ and a complex relative permeability $\mu_r = \mu'_r - j\mu''_r$ in complete analogy to the dielectric case, we obtain the total magnetization current

$$I = I_m + I_l = \frac{U}{j\omega L_0 \mu_r} = -\frac{jU(\mu' + j\mu'')}{\omega(L_0/\mu_0)(\mu'^2 + \mu''^2)} \qquad (1.19)$$

Similar to the dielectric case, according to Figure 1.8, we can also define two parameters describing magnetic materials: the magnetic loss tangent given by

$$\tan \delta_m = \mu''/\mu', \qquad (1.20)$$

and the power factor given by

$$\cos \theta_m = \mu''/\sqrt{(\mu')^2 + (\mu'')^2}. \qquad (1.21)$$

In microwave electronics, relative permeability is often used, which is a dimensionless quantity given by

$$\mu_r = \frac{\mu}{\mu_0} = \frac{\mu' - j\mu''}{\mu_0}$$
$$= \mu'_r - j\mu''_r = \mu'_r(1 - j\tan\delta_m) \qquad (1.22)$$

where μ is complex complex permeability,
μ_r is relative complex permeability,
$\mu_0 = 4\pi \times 10^{-7}$ H/m is the permeability of free space,
μ'_r is the real part of relative complex permeability,
μ''_r is the imaginary part of the relative complex permeability,
$\tan \delta_m$ is the magnetic loss tangent, and
δ_m is the magnetic loss angle.

In summary, the macroscopic electric and magnetic behavior of a low-conductivity material is mainly determined by the two complex parameters: permittivity (ε) and permeability (μ). Permittivity describes the interaction of a material with the electric field applied on it, while permeability describes the interaction of a material with magnetic field applied on it. Both the electric and magnetic fields interact with materials in two ways: energy storage and energy dissipation. Energy storage describes the lossless portion of the exchange of energy between the field and the material, and energy dissipation occurs when electromagnetic energy is absorbed by the material. So both permittivity and permeability are expressed as complex numbers to describe the storage (real part) and dissipation (imaginary part) effects of each.

Besides the permittivity and permeability, another parameter, quality factor, is often used to describe an electromagnetic material:

$$Q_e = \frac{\varepsilon'_r}{\varepsilon''_r} = \frac{1}{\tan \delta_e} \qquad (1.23)$$

$$Q_m = \frac{\mu'_r}{\mu''_r} = \frac{1}{\tan \delta_m} \qquad (1.24)$$

On the basis of the dielectric quality factor Q_e and magnetic quality factor Q_m, we can get the total quality factor Q of the material:

$$\frac{1}{Q} = \frac{1}{Q_e} + \frac{1}{Q_m} \qquad (1.25)$$

Propagation parameters

The propagation of electromagnetic waves in a medium is determined by the characteristic wave impedance η of the medium and the wave velocity v in the medium. The characteristic wave impedance η is also called the *intrinsic impedance* of the medium. When a single wave propagates with velocity v in the Z-positive direction, the characteristic impedance η is defined as the ratio of total electric field to total magnetic field at a Z-plane. The wave impedance and velocity can be calculated from the permittivity and permeability of the medium:

$$\eta = \sqrt{\frac{\mu}{\varepsilon}} \qquad (1.26)$$

$$v = \frac{1}{\sqrt{\mu\varepsilon}} \qquad (1.27)$$

From Eqs. (1.26) and (1.27), we can calculate the wave impedance of free space, $\eta_0 = (\mu_0/\varepsilon_0)^{1/2} = 376.7\ \Omega$, and the wave velocity in free space, $c = (\mu_0\varepsilon_0)^{-1/2} = 2.998 \times 10^8$ m/s. Expressing permittivity and permeability as complex quantities leads to a complex number for the wave velocity (v), where the imaginary portion is a mathematical convenience for expressing loss.

Sometimes, it is more convenient to use the complex propagation coefficient γ to describe the propagation of electromagnetic waves in a medium:

$$\gamma = \alpha + j\beta = j\omega\sqrt{\mu\varepsilon} = j\frac{\omega}{c}\sqrt{\mu_r\varepsilon_r} = j\frac{\omega}{c}n \qquad (1.28)$$

where n is the complex index of refraction, where ω is the angular frequency, α is the attenuation coefficient, $\beta = 2\pi/\lambda$ is the phase change coefficient, and λ is the operating wavelength in the medium.

1.2.2.2 Parameters describing high-conductivity materials

For a high-conductivity material, for example a metal, Eq. (1.28) for the complex propagation constant γ should be modified as

$$\gamma = \alpha + j\beta = j\omega\sqrt{\mu\varepsilon}\sqrt{1 - j\frac{\sigma}{\omega\varepsilon}} \qquad (1.29)$$

For a high-conductivity material, we assume $\sigma \gg \omega\varepsilon$, which means that the conducting current is much larger than the displacement current. So, Eq. (1.29) can be approximated by ignoring the displacement current term:

$$\gamma = \alpha + j\beta = j\omega\sqrt{\mu\varepsilon}\sqrt{\frac{\sigma}{j\omega\varepsilon}} = (1 + j)\sqrt{\frac{\omega\mu\sigma}{2}} \qquad (1.30)$$

We define the skin depth:

$$\delta_s = \frac{1}{\alpha} = \sqrt{\frac{2}{\omega\mu\sigma}} \qquad (1.31)$$

The physics meaning of skin depth is that, in a high-conductivity material, the fields decay by an amount e^{-1} in a distance of a skin depth δ_s. At microwave frequencies, the skin depth δ_s is a very small distance. For example, the skin depth of a metal at microwave frequencies is usually on the order of 10^{-7} m.

Because of the skin effect, the utility and behavior of high-conductivity materials at microwave frequencies are mainly determined by their surface impedance Z_s:

$$Z_s = R_s + jX_s = \frac{E_t}{H_t} = (1 + j)\sqrt{\frac{\mu\omega}{2\sigma}} \qquad (1.32)$$

where H_t is the tangential magnetic field, E_t is the tangential electric field, R_s is the surface resistance, and X_s is the surface reactance. For normal conductors, σ is a real number. According to Eq. (1.32), the surface resistance R_s and the surface reactance X_s are equal and they are proportional to $\omega^{1/2}$ for normal metals:

$$R_s = X_s = \sqrt{\frac{\mu\omega}{2\sigma}} \qquad (1.33)$$

1.2.2.3 Classification of electromagnetic materials

Materials can be classified according to their macroscopic parameters. According to conductivity, materials can be classified as insulators, semiconductors, and conductors. Meanwhile, materials can also be classified according to their permeability values. General properties of typical types of materials are discussed in Section 1.3.

When classifying materials according to their macroscopic parameters, it should be noted that we use the terms insulator, semiconductor, conductor, and magnetic material to indicate the dominant responses of different types of materials. All materials have some response to magnetic fields but, except for ferromagnetic and ferrimagnetic types, their responses are usually very small, and their permeability values differ from μ_0 by a negligible fraction. Most of the ferromagnetic materials are highly conductive, but we call them *magnetic materials*, as their magnetic properties are the most significant in their applications. For superconductors, the Meissner effect shows that they are a kind of very special magnetic materials, but in microwave electronics, people are more interested in their surface impedance.

Insulators

Insulators have very low conductivity, usually in the range of 10^{-12} to 10^{-20} $(\Omega m)^{-1}$. Often, we assume insulators are nonmagnetic, so they are actually dielectrics. In theoretical analysis of dielectric materials, an ideal model, perfect dielectric, is often used, representing a material whose imaginary part of permittivity is assumed to be zero: $\varepsilon'' = 0$.

Semiconductors

The conductivity of a semiconductor is higher than that of a dielectric but lower than that of a conductor. Usually, the conductivities of semiconductors at room temperature are in the range of 10^{-7} to 10^4 $(\Omega m)^{-1}$.

Conductors

Conductors have very high conductivity, usually in the range of 10^4 to 10^8 $(\Omega m)^{-1}$. Metals are typical conductors. There are two types of special conductors: perfect conductors and superconductors. A perfect conductor is a theoretical model that has infinite conductivity at any frequencies. Superconductors have very special electromagnetic properties. For dc electric fields, their conductivity is virtually infinite; but for high-frequency electromagnetic fields, they have complex conductivities.

Magnetic materials

All materials respond to external magnetic fields, so in a broad sense, all materials are magnetic materials. According to their permeability values, materials generally fall into three categories: diamagnetic ($\mu < \mu_0$), paramagnetic ($\mu \geq \mu_0$), and highly magnetic materials mainly including ferromagnetic and ferrimagnetic materials. The permeability values of highly magnetic materials, especially ferromagnetic materials, are much larger than μ_0.

1.3 GENERAL PROPERTIES OF ELECTROMAGNETIC MATERIALS

Here, we discuss the general properties of typical electromagnetic materials, including dielectric materials, semiconductors, conductors, magnetic materials, and artificial materials. The knowledge of general properties of electromagnetic materials is helpful for understanding the measurement results and correcting the possible errors one may meet in materials characterization. In the final part of this section, we will discuss other descriptions of electromagnetic materials, which are important for the design and applications of electromagnetic materials.

1.3.1 Dielectric materials

Figure 1.9 qualitatively shows a typical behavior of permittivity (ε' and ε'') as a function of frequency. The permittivity of a material is related to a variety of physical phenomena. Ionic conduction, dipolar relaxation, atomic polarization, and electronic polarization are the main mechanisms that contribute to the permittivity of a dielectric material. In the low frequency range, ε'' is dominated by the influence of ion conductivity. The variation of permittivity in the microwave range is mainly caused by dipolar relaxation, and the absorption peaks in the infrared region and above is mainly due to atomic and electronic polarizations.

1.3.1.1 Electronic and atomic polarizations

Electronic polarization occurs in neutral atoms when an electric field displaces the nucleus with respect to the surrounding electrons. Atomic polarization occurs when adjacent positive and negative

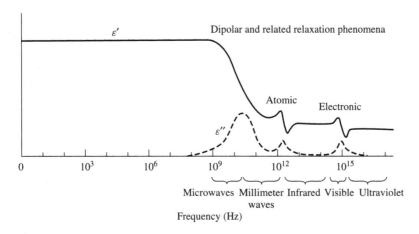

Figure 1.9   Frequency dependence of permittivity for a hypothetical dielectric (Ramo *et al.* 1994). Source: Ramo, S. Whinnery, J. R and Van Duzer, T. (1994). *Fields and Waves in Communication Electronics*, 3rd edition, John Wiley & Sons, Inc., New York

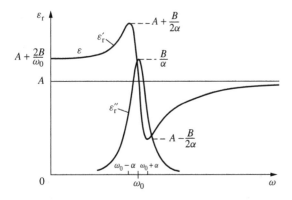

Figure 1.10   The behavior of permittivity due to electronic or atomic polarization. Reprinted with permission from *Industrial Microwave Sensors*, by Nyfors, E. and Vainikainen, P., Artech House Inc., Norwood, MA, USA, www.artechhouse.com

ions stretch under an applied electric field. Actually, electronic and atomic polarizations are of similar nature. Figure 1.10 shows the behavior of permittivity in the vicinity of the resonant frequency ω_0. In the figure, A is the contribution of higher resonance to ε_r' at the present frequency range, and $2B/\omega_0$ is the contribution of the present resonance to lower frequencies. For many dry solids, these are the dominant polarization mechanisms determining the permittivity at microwave frequencies, although the actual resonance occurs at a much higher frequency. If only these two polarizations are present, the materials are almost lossless at microwave frequencies.

In the following discussion, we focus on electronic polarization, and the conclusions for electronic polarization can be extended to atomic polarization. When an external electric field is applied to neutral atoms, the electron cloud of the atoms will be distorted, resulting in the electronic polarization. In a classical model, it is similar to a spring-mass resonant system. Owing to the small mass of the electron cloud, the resonant frequency of electronic polarization is at the infrared region or the visible light region. Usually, there are several different resonant frequencies corresponding to different electron orbits and other quantum-mechanical effects. For a material with s different oscillators, its permittivity is given by (Nyfors and Vainikainen 1989)

$$\varepsilon_r = 1 + \sum_s \frac{(n_s e^2)/(\varepsilon_0 m_s)}{\omega_s^2 - \omega^2 + j\omega^2 \alpha_s} \qquad (1.34)$$

where n_s is the number of electrons per volume with resonant frequency ω_s, e is the charge of electron, m_s is the mass of electron, ω is the operating angular frequency, and α_s is the damping factor.

As microwave frequencies are far below the lowest resonant frequency of electronic polarization, the permittivity due to electronic polarization is almost independent of the frequency and temperature (Nyfors and Vainikainen 1989):

$$\varepsilon_r = 1 + \sum_s \frac{N_s e^2}{\varepsilon_0 m_s \omega_s^2} \tag{1.35}$$

Eq. (1.35) indicates that the permittivity ε_r is a real number. However, in actual materials, small and constant losses are often associated with this type of polarization in the microwave range.

1.3.1.2 Dipolar polarization

In spite of their different origins, various types of polarizations at microwave and millimeter-wave ranges can be described in a similar qualitative way. In most cases, the Debye equations can be applied, although they were firstly derived for the special case of dipolar relaxation. According to Debye theory, the complex permittivity of a dielectric can be expressed as (Robert 1988)

$$\varepsilon_r = \varepsilon_{r\infty} + \frac{\varepsilon_{r0} - \varepsilon_{r\infty}}{1 + j\beta} \tag{1.36}$$

with

$$\varepsilon_{r\infty} = \lim_{\omega \to \infty} \varepsilon_r \tag{1.37}$$

$$\varepsilon_{r0} = \lim_{\omega \to 0} \varepsilon_r \tag{1.38}$$

$$\beta = \frac{\varepsilon_{r0} + 2}{\varepsilon_{r\infty} + 2} \omega\tau \tag{1.39}$$

where τ is the relaxation time and ω is the operating angular frequency. Equation (1.36) indicates that the dielectric permittivity due to Debye relaxation is mainly determined by three parameters, ε_{r0}, $\varepsilon_{r\infty}$, and τ. At sufficiently high frequencies, as the period of electric field E is much smaller than the relaxation time of the permanent dipoles, the orientations of the dipoles are not influenced by electric field E and remain random, so the permittivity at infinite frequency $\varepsilon_{r\infty}$ is a real number. As ε_∞ is mainly due to electronic and atomic polarization, it is independent of the temperature. As at sufficiently low frequencies there is no phase difference between the polarization P and electric field E, ε_{r0} is a real number. But the static permittivity ε_{r0} decreases with increasing temperature because of the increasing disorder, and the relaxation time τ

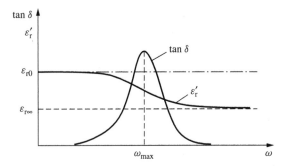

Figure 1.11 The frequency dependence of the complex permittivity according to the Debye relation (Robert 1988). Reprinted with permission from *Electrical and Magnetic Properties of Materials* by Robert, P., Artech House Inc., Norwood, MA, USA, www.artechhouse.com

is inversely proportional to temperature as all the movements become faster at higher temperatures.

From Eq. (1.36), we can get the real and imaginary parts of the permittivity and the dielectric loss tangent:

$$\varepsilon_r' = \varepsilon_{r\infty} + \frac{\varepsilon_{r0} - \varepsilon_{r\infty}}{1 + \beta^2} \tag{1.40}$$

$$\varepsilon_r'' = \frac{\varepsilon_{r0} - \varepsilon_{r\infty}}{1 + \beta^2} \beta \tag{1.41}$$

$$\tan \delta_e = \frac{\varepsilon_{r0} - \varepsilon_{r\infty}}{\varepsilon_{r0} + \varepsilon_{r\infty}\beta^2} \beta \tag{1.42}$$

Figure 1.11 shows the variation of complex permittivity as a function of frequency. At the frequency

$$\omega_{max} = \frac{1}{\tau} \cdot \sqrt{\frac{\varepsilon_{r0}}{\varepsilon_{r\infty}} \cdot \frac{\varepsilon_{r\infty} + 2}{\varepsilon_{r0} + 2}}, \tag{1.43}$$

the dielectric loss tangent reaches its maximum value (Robert 1988)

$$\tan \delta_{max} = \frac{1}{2} \cdot \frac{\varepsilon_{r0} - \varepsilon_{r\infty}}{\sqrt{\varepsilon_{r0}\varepsilon_{r\infty}}} \tag{1.44}$$

The permittivity as a function of frequency is often presented as a two-dimensional diagram, Cole–Cole diagram. We rewrite Eq. (1.36) as

$$\varepsilon_r' - \varepsilon_{r\infty} - j\varepsilon_r'' = \frac{\varepsilon_{r0} - \varepsilon_{r\infty}}{1 + j\beta} \tag{1.45}$$

As the moduli of both sides of Eq. (1.45) should be equal, we have

$$(\varepsilon_r' - \varepsilon_{r\infty})^2 + (\varepsilon_r'')^2 = \frac{(\varepsilon_{r0} - \varepsilon_{r\infty})^2}{1 + \beta^2} \tag{1.46}$$

After eliminating the term β^2 using Eq. (1.40), we get (Robert 1988)

$$(\varepsilon_r' - \varepsilon_{r\infty})^2 + (\varepsilon_r'')^2 = (\varepsilon_r' - \varepsilon_{r\infty})(\varepsilon_{r0} - \varepsilon_{r\infty}) \tag{1.47}$$

Eq. (1.47) represents a circle with its center on the ε_r' axis. Only the points at the top half of this circle have physical meaning as all the materials have nonnegative value of imaginary part of permittivity. The top half of the circle is called *Cole–Cole diagram*, as shown in Figure 1.12.

The relaxation time τ can be determined from the Cole–Cole diagram. According to Eqs. (1.40) and (1.41), we can get

$$\varepsilon_r'' = \beta(\varepsilon_r' - \varepsilon_{r\infty}) \tag{1.48}$$

$$\varepsilon_r'' = -(1/\beta)(\varepsilon_r' - \varepsilon_{r0}) \tag{1.49}$$

As shown in Figure 1.12, for a given operating frequency, the β value can be obtained from the slope of a line pass through the point corresponding to the operating frequency and the point corresponding to ε_{r0} or $\varepsilon_{r\infty}$. After obtaining the β value, the relaxation time τ can be calculated from β according to Eq. (1.39).

In some cases, the relaxation phenomenon may be caused by different sources, and the dielectric material has a relaxation-time spectrum. For example, a moist material contains water molecules bound with different strength. Depending on the moisture and the strength of binding

water, the material exhibits a distribution of relaxation frequencies. Often an empirical constant, a, is introduced and Eq. (1.36) is modified into the following form (Robert 1988):

$$\varepsilon_r = \varepsilon_{r\infty} + \frac{\varepsilon_{r0} - \varepsilon_{r\infty}}{1 + (j\beta_a)^{1-a}} \tag{1.50}$$

where a is related to the distribution of β values, and β_a denotes the most possible β value. The constant a is in the range $0 \leqslant a < 1$. When $a = 0$, Eq. (1.50) becomes Eq. (1.36), and in this case, there is only single relaxation time. When the value of a increases, the relaxation time is distributed over a broader range.

If we separate the real and imaginary parts of Eq. (1.50) and then eliminate β_a, we can find that the $\varepsilon_r''(\varepsilon_r')$ curve is also a circle passing through the points ε_{r0} and $\varepsilon_{r\infty}$, as shown in Figure 1.13. The center of the circle is below the ε_r' axis with a distance d given by

$$d = \frac{\varepsilon_{r0} - \varepsilon_{r\infty}}{2} \tan\theta \tag{1.51}$$

where θ is the angle between the ε_r' axis and the line connecting the circle center and the point $\varepsilon_{r\infty}$:

$$\theta = a\frac{\pi}{2} \tag{1.52}$$

Similar to Figure 1.12, only the points above the ε_r' axis have physical meaning. Equations (1.51) and (1.52) indicate that the empirical constant a can be calculated from the value of d or θ.

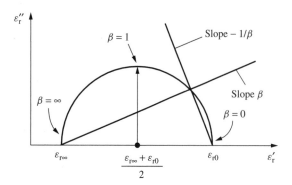

Figure 1.12   The Cole–Cole presentation for a single relaxation time (Robert 1988). Reprinted with permission from *Electrical and Magnetic Properties of Materials* by Robert, P., Artech House Inc., Norwood, MA, USA, www.artechhouse.com

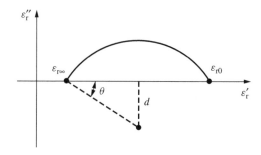

Figure 1.13   Cole–Cole diagram for a relaxation-time spectrum. Reprinted with permission from *Electrical and Magnetic Properties of Materials* by Robert, P., Artech House Inc., Norwood, MA, USA, www.artechhouse.com

1.3.1.3 Ionic conductivity

Usually, ionic conductivity only introduces losses into a material. As discussed earlier, the dielectric loss of a material can be expressed as a function of both dielectric loss (ε_{rd}'') and conductivity (σ):

$$\varepsilon_r'' = \varepsilon_{rd}'' + \frac{\sigma}{\omega\varepsilon_0} \qquad (1.53)$$

The overall conductivity of a material may consist of many components due to different conduction mechanisms, and ionic conductivity is usually the most common one in moist materials. At low frequencies, ε_r'' is dominated by the influence of electrolytic conduction caused by free ions in the presence of a solvent, for example water. As indicated by Eq. (1.53), the effect of ionic conductivity is inversely proportional to operating frequency.

1.3.1.4 Ferroelectricity

Most of the dielectric materials are paraelectric. As shown in Figure 1.14(a), the polarization of a paraelectric material is linear. Besides, the ions in paraelectric materials return to their original positions once the external electric field is removed; so the ionic displacements in paraelectric materials are reversible.

Ferroelectric materials are a subgroup of pyroelectric materials that are a subgroup of piezoelectric materials. For ferroelectric materials, the response of polarization versus electric field is nonlinear. As shown in Figure 1.14(b), ferroelectric materials display a hysteresis effect of polarization with an applied field. The hysteresis loop is caused by the existence of permanent electric dipoles in the material. When the external electric field is initially increased from the point 0, the polarization increases as more of the dipoles are lined up. When the field is strong enough, all dipoles are lined up with the field, so the material is in a saturation state. If the applied electric field decreases from the saturation point, the polarization also decreases. However, when the external electric field reaches zero, the polarization does not reach zero. The polarization at zero field is called the *remanent polarization*. When the direction of the electric field is reversed, the polarization decreases. When the reversed field reaches a certain value, called the *coercive field*, the polarization becomes zero. By further increasing the field in this reverse direction, the reverse saturation can be reached. When the field is decreased from the saturation point, the sequence just reverses itself.

For a ferroelectric material, there exists a particular temperature called the *Curie temperature*. Ferroelectricity can be maintained only below the Curie temperature. When the temperature is higher than the Curie temperature, a ferroelectric material is in its paraelectric state.

Ferroelectric materials are very interesting scientifically. There are rich physics phenomena near

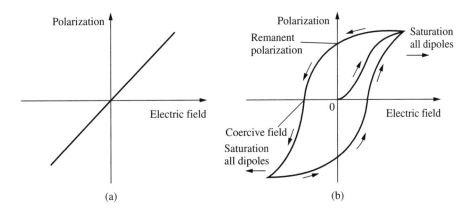

Figure 1.14 Polarization of dielectric properties. (a) Polarization of linear dielectric and (b) typical hysteresis loop for ferroelectric materials. Modified from Bolton, W. (1992). *Electrical and Magnetic Properties of Materials*, Longman Scientific & Technical, Harlow

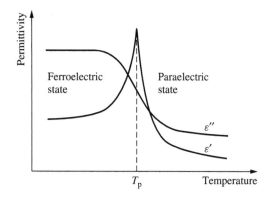

Figure 1.15   Schematic view of the temperature dependence of a ferroelectric material near its Curie temperature

the Curie temperature. As shown in Figure 1.15, the permittivity of a ferroelectric material changes greatly with temperature near the Curie temperature. Dielectric constant increases sharply to a high value just below the Curie point and then steeply drops just above the Curie point. For example, barium titanate has a relative permittivity on the order of 2000 at about room temperature, with a sharp increase to about 7000 at the Curie temperature of 120 °C. The dielectric loss decreases quickly when the material changes from ferroelectric state to paraelectric state. Furthermore, for a ferroelectric material near its Curie temperature, its dielectric constant is sensitive to the external electric field.

Ferroelectric materials have application potentials in various fields, including miniature capacitors, electrically tunable capacitors and electrically tunable phase-shifters. Further discussions on ferroelectric materials can be found in Chapter 9.

1.3.2 Semiconductors

There are two general categories of semiconductors: intrinsic and extrinsic semiconductors. An intrinsic semiconductor is also called a *pure semiconductor* or an *undoped semiconductor*. The band structure shown in Figure 1.2(b) is that of an intrinsic semiconductor. In an intrinsic semiconductor, there are the same numbers of electrons as holes. Intrinsic semiconductors usually have high resistivity, and they are often used as the starting materials for fabricating extrinsic semiconductors.

Silicon and germanium are typical intrinsic semiconductors.

An extrinsic semiconductor is obtained by adding a very small amount of impurities to an intrinsic semiconductor, and this procedure is called *doping*. If the impurities have a higher number of valence electrons than that of the host, the resulting extrinsic semiconductor is called *type n*, indicating that the majority of the mobile charges are negative (electrons). Usually the host is silicon or germanium with four valence electrons, and phosphorus, arsenic, and antimony with five valence electrons are often used as dopants in type n semiconductors. Another type of extrinsic semiconductor is obtained by doping an intrinsic semiconductor using impurities with a number of valence electrons less than that of the host. Boron, aluminum, gallium, and indium with three valence electrons are often used for this purpose. The resulted extrinsic semiconductor is called *type p*, indicating that the majority of the charge carriers are positive (holes).

Both the free charge carriers and bounded electrons in ions in the crystalline lattice have contributions to the dielectric permittivity $\varepsilon = \varepsilon' - j\varepsilon''$ (Ramo *et al.* 1994):

$$\varepsilon' = \varepsilon_1 - \frac{n_e e^2}{m(v^2 + \omega^2)} \qquad (1.54)$$

$$\varepsilon'' = \frac{n_e e^2 v}{\omega m(v^2 + \omega^2)} \qquad (1.55)$$

where ε_1 is related to the effects of the bound electrons to the positive background, n_e is the density of the charge carriers, v is the collision frequency, ω is the circular frequency, m is the mass of the electron, and $(n_e e^2/mv)$ equals the low frequency conductivity σ.

At microwave frequency ($\omega^2 \ll v^2$), for semiconductors with low to moderate doping, whose conductivity is usually not higher than 1 S/m, the second term of Eq. (1.54) is negligible. So the permittivity can be approximated as

$$\varepsilon = \varepsilon_1 - j\frac{\sigma}{\omega} \qquad (1.56)$$

Besides the permittivity discussed above, the electrical transport properties of semiconductors, including Hall mobility, carrier density, and conductivity are important parameters in the development of electronic components. Discussions on electrical transport properties can be found in Chapter 11.

1.3.3 Conductors

Conductors have high conductivity. If the conductivity is not very high, the concept of permittivity is still applicable, and the value of permittivity can be approximately calculated from Eqs. (1.54) and (1.55). For good conductors with very high conductivity, we usually use penetration depth and surface resistance to describe the properties of conductors. As the general properties of normal conductors have been discussed earlier, here we focus on two special types of conductors: perfect conductors and superconductors. It should be noted that perfect conductor is only a theoretical model, and no perfect conductor physically exists.

A perfect conductor refers to a material within which there is no electric field at any frequency. Maxwell equations ensure that there is also no time-varying magnetic field in a perfect conductor. However, a strictly static magnetic field should be unaffected by the conductivity of any value, including infinite conductivity. Similar to an ideal perfect conductor, a superconductor excludes time-varying electromagnetic fields. Furthermore, the Meissner's effect shows that constant magnetic fields, including strictly static magnetic fields, are also excluded from the interior of a superconductor. From the London theory and the Maxwell's equations, we have

$$B = B_0 e^{-z/\lambda_L} \qquad (1.57)$$

with the London penetration depth given by

$$\lambda_L = \left(\frac{m}{\mu n_e e^2} \right)^{\frac{1}{2}} \qquad (1.58)$$

where B is the magnetic field in the depth z, B_0 is the magnetic field at the surface $z = 0$, m is the mass of an electron, μ is permeability, n_e is the density of the electron, and e is the electric charge of an electron. So an important difference between a superconductor and a perfect conductor is that, for a superconductor, Eq. (1.57) applies for both time-varying magnetic field and static magnetic field; while for a perfect conductor, Eq. (1.57) only applies for time-varying magnetic fields.

For a superconductor, there exists a critical temperature T_c. When the temperature is lower than T_c, the material is in superconducting state, and at T_c, the material undergoes a transition from normal state into superconducting state. A material with low T_c is called a *low-temperature superconducting* (LTS) material, while a material with high T_c is called a *high-temperature superconducting* (HTS) material. LTS materials are metallic elements, compounds, or alloys, and their critical temperatures are usually below about 24 K. HTS materials are complex oxides and their critical temperature may be higher than 100 K. HTS materials are of immediate interest for microwave applications because of their very low surface resistance at microwave frequency at temperatures that can be readily achieved by immersion in liquid nitrogen or with cryocoolers. In contrast to metallic superconductors, HTS materials are usually anisotropic, exhibiting strongest superconductive behavior in preferred planes. When these materials are used in planar microwave structures, for example, thin-film transmission lines or resonators, these preferred planes are formed parallel to the surface to facilitate current flow in the required direction (Lancaster 1997; Ramo *et al.* 1994).

The generally accepted mechanism for superconductivity of most LTS materials is phonon-mediated coupling of electrons with opposite spin. The paired electrons, called *Cooper pairs*, travel through the superconductor without being scattered. The BCS theory describes the electron pairing process, and it explains the general behavior of LTS materials very well. However, despite the enormous efforts so far, there is no theory that can explain all aspects of high-temperature superconductivity. Fortunately, an understanding of the microscopic theory of superconductivity in HTS materials is not required for the design of microwave devices (Lancaster 1997; Shen 1994). In the following, we discuss some phenomenological theories based on the London equations and the two-fluid model. We will introduce some commonly accepted theories for explaining the responses of superconductors to electromagnetic fields, and our discussion will be focused on the penetration depth, surface impedance, and complex conductivity of superconductors.

1.3.3.1 Penetration depth

The two-fluid model is often used in analyzing superconductors, and it is based on the assumption that there are two kinds of fluids in a superconductor: a superconductive current with a carrier density n_s and a normal current with a carrier density n_n, yielding a total carrier density $n = n_s + n_n$. At temperatures below the transition temperature T_c, the equilibrium fractions of the normal and the superconducting electrons vary with the absolute temperature T:

$$\frac{n_n}{n} = \left(\frac{T}{T_c}\right)^4 \tag{1.59}$$

$$\frac{n_s}{n} = 1 - \left(\frac{T}{T_c}\right)^4 \tag{1.60}$$

From Eqs. (1.59) and (1.60), we can get the relationship between the penetration depth λ_L and temperature T:

$$\lambda_L(T) = \lambda_L(0)\left[1 - \left(\frac{T}{T_c}\right)^4\right]^{-\frac{1}{2}} \tag{1.61}$$

with

$$\lambda_L(0) = \sqrt{\frac{m_s}{\mu n q_s^2}}. \tag{1.62}$$

where m_s and q_s are the effective mass and electrical charge of the superconductive carriers. Eq. (1.62) indicates that the penetration depth has a minimum value of penetration depth $\lambda_L(0)$ at $T = 0\,\text{K}$.

1.3.3.2 Surface impedance and complex conductivity

The surface impedance is defined as the characteristic impedance seen by a plane wave incident perpendicularly upon a flat surface of a conductor. According to Eqs. (1.32) and (1.33), the surface impedance of normal conductors, such as silver, copper, or gold, can be calculated from their conductivity σ. For a normal conductor, the value of its conductivity σ is a real number, and the surface resistance R_s and the surface reactance X_s are

equal, and they are proportional to the square root of the operating frequency $\omega^{1/2}$.

If we want to calculate the impedance of a superconductor using Eq. (1.32), the concept of complex conductivity should be introduced. According to the two-fluid model, there are two types of currents: a superconducting current with volume density J_s and a normal current with volume density J_n. Correspondingly, the conductivity σ also consists of two components: superconducting conductivity σ_s and normal conductivity σ_n, respectively. The total conductivity of a superconductor is given by $\sigma = \sigma_s + \sigma_n$.

The superconducting conductivity σ_s is purely imaginary and does not contribute to the loss:

$$\sigma_s = \frac{1}{j\omega\mu\lambda_L^2} \tag{1.63}$$

While the normal conductivity σ_n contains both real and imaginary components and the real part contributes to the loss:

$$\sigma_n = \sigma_{n1} - j\sigma_{n2} = \left(\frac{n_n q_n^2}{m_n}\right)\frac{\tau}{1 + j\omega\tau}$$

$$= \left(\frac{n_n q_n^2 \tau}{m_n}\right)\frac{1 - j\omega\tau}{1 + (\omega\tau)^2} \tag{1.64}$$

where q_n is the electrical charge for the normal carriers, τ is the relaxation time for electron scattering, and m_n is the effective mass of the normal carriers. Therefore, the total conductivity σ of a superconductor is then obtained:

$$\sigma = \sigma_n + \sigma_s = \left(\frac{n_n q_n^2 \tau}{m_n}\right)\frac{1}{1 + (\omega\tau)^2}$$

$$- j\left(\frac{n_n q_n^2 \tau}{m_n}\right)\frac{\omega\tau}{1 + (\omega\tau)^2} - j\frac{1}{\omega\mu\lambda_L^2} \tag{1.65}$$

At microwave frequencies ($\omega\tau \ll 1$), Eq. (1.65) can be simplified as

$$\sigma = \sigma_1 - j\sigma_2 = \frac{n_n q_n^2 \tau}{m_n} - j\frac{1}{\omega\mu\lambda_L^2} \tag{1.66}$$

where σ_1 and σ_2 are the real and imaginary components of the complex conductivity. The real part of complex conductivity represents the loss due to the normal carriers, whereas its imaginary part represents the kinetic energy of the superconductive carriers.

From Eqs. (1.32), (1.33) and Eq. (1.66), we can calculate the surface impedance of a superconductor:

$$Z_s = R_s + jX_s$$

$$= \sqrt{\frac{j\omega\mu}{\sigma_1 - j\sigma_2}} = j\sqrt{\frac{\omega\mu}{\sigma_2}}\left(1 + j\frac{\sigma_1}{\sigma_2}\right)^{-\frac{1}{2}} \quad (1.67)$$

As usually $\sigma_1 \ll \sigma_2$, Eq. (1.67) can be simplified as

$$Z_s = R_s + jX_s$$

$$= \sqrt{\frac{\omega\mu}{\sigma_2}}\left(\frac{\sigma_1}{2\sigma_2} + j\right)$$

$$= \frac{\omega^2\mu^2\lambda_L^3 n_n q_n^2 \tau}{2m_n} + j\omega\mu\lambda_L \quad (1.68)$$

$$R_s = \frac{1}{2}\omega^2\mu^2\lambda_L^3\sigma_N\left(\frac{n_n}{n}\right) \quad (1.69)$$

$$X_s = \omega\mu\lambda_L \quad (1.70)$$

where σ_N is the conductivity of the superconductor in its normal state:

$$\sigma_N = \frac{nq_n^2\tau}{m_n} \quad (1.71)$$

$$\sigma_n = \frac{n_n q_n^2 \tau}{m_n} = \sigma_N \frac{n_n}{n} = \sigma_N\left(\frac{T}{T_c}\right)^4 \quad (1.72)$$

According to Eqs. (1.67) to (1.72), the two-fluid model leads to the prediction that the surface resistance R_s is proportional to ω^2 for superconductors, which is quite different from the $\omega^{1/2}$ frequency dependence for normal conductors.

1.3.4 Magnetic materials

As the penetration depth of metals at microwave frequencies is on the order of a few microns, the interior of a metallic magnetic material does not respond to a microwave magnetic field. So, metallic magnetic materials are seldom used as magnetic materials at microwave frequencies. Here, we concentrate on magnetic materials with low conductivity.

The frequency dependence of magnetic materials is quite complicated (Smit 1971; Fuller 1987), and some of the underlying mechanisms have not been fully understood. Figure 1.16 shows the typical magnetic spectrum of a magnetic material.

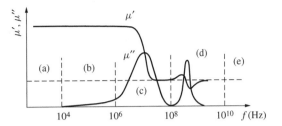

Figure 1.16 Frequency dependence of permeability for a hypothetical ferromagnetic material

At different frequency ranges, different physics phenomena dominate. In the low frequency range ($f < 10^4$ Hz), μ' and μ'' almost do not change with frequency. In the intermediate frequency range ($10^4 < f < 10^6$ Hz), μ' and μ'' change a little, and for some materials, μ'' may have a maximum value. In the high-frequency range ($10^6 < f < 10^8$ Hz), μ' decreases greatly, while μ'' increase quickly. In the ultrahigh frequency range ($10^8 < f < 10^{10}$ Hz), ferromagnetic resonance usually occurs. In the extremely high frequency range ($f > 10^{10}$ Hz), the magnetic properties have not been fully investigated yet.

1.3.4.1 Magnetization and hysteresis loop

Figure 1.17 shows the typical relationship between the magnetic flux density B in a magnetic material and the magnetic field strength H. As discussed in

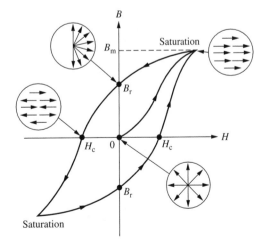

Figure 1.17 The hysteresis loop for a magnetic material

Section 1.2.1.2, at the starting point 0, the domains are randomly orientated, so the net magnetic flux density is zero. The magnetic flux density B increases with the increase of the magnetic field strength H, as the domains close to the direction of the magnetic field grow. This continues until all the domains are in the same direction with the magnetic field H and the material is thus saturated. At the saturation state, the flux density reaches its maximum value B_m. When the magnetic field strength is reduced to zero, the domains in the material turn to their easy-magnetization directions close to the direction of the magnetic field H, and the material retains a remanence flux density B_r. If we reverse the direction of the magnetic field, the domains grow in the reverse direction. When the numbers of the domains in the H direction and opposite the H direction are equal, that is, the flux density becomes zero, the value of the applied magnetic field is called *coercive field* H_c. Further increase in the strength of the magnetic field in the reverse direction results in further growth of the domains in the reverse direction until saturation in the reverse direction is achieved. When this field is reduced to zero, and then reversed back to the initial direction, we can get a closed hysteresis loop of the magnetic material.

In most cases, magnetic materials are anisotropic for magnetization. For a hexagonal ferrite, there exists an easy-magnetization direction and a hard-magnetization direction. As shown in Figure 1.18, in the easy-magnetization direction, saturation can

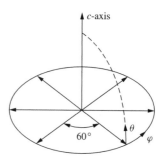

Figure 1.19 Preferential directions for a ferroxplana material (Smit 1971). Source: Smit, J. (editor), (1971), *Magnetic Properties of Materials*, McGraw-Hill, New York

be easily achieved, while in the hard-magnetization direction, high magnetic field is required for saturation. The magnetic field H_a corresponding to the cross point of the two magnetization curves is called *anisotropic field*.

There are two typical types of anisotropies of magnetic materials: axis anisotropy and plane anisotropy for a hexagonal structure. Figure 1.19 shows the potential directions for a ferroxplana material. If the easy-magnetization direction is along the *c*-axis, the material has uniaxial anisotropy, usually described by the anisotropic field H_a. If the easy-magnetization direction is in the *c*-plane, the material has planar anisotropy. Planar anisotropy is usually described by the anisotropic fields H_θ and H_φ, where H_θ is the magnetic field required for turning a domain in one preferential magnetization direction in the *c*-plane to another preferential magnetization direction in the *c*-plane through the hard-magnetization *c*-axis, and H_φ is the magnetic field required for turning a domain in one preferential magnetization direction in the *c*-plane to another preferential magnetization direction in the *c*-plane within the easy-magnetization plane.

The coercive field H_c is an important parameter in describing the properties of a magnetic material. The value of coercive field H_c is mainly governed by two magnetization phenomena: rotation of domain and movement of domain wall. It is related to intrinsic magnetic properties, such as anisotropic field and domain-wall energy, and it is also related to the microstructures of the material, such as grain size and domain-wall thickness.

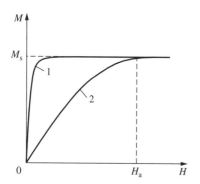

Figure 1.18 Magnetization curves for an anisotropic magnetic material. Curve 1 is the magnetization in the easy-magnetization direction and Curve 2 is the magnetization in the hard-magnetization direction

Besides, the amount and distribution of impurities in the material also affects the value of the coercive field H_c.

1.3.4.2 Definitions of scalar permeability

As the relationship between the magnetic flux density B and the magnetic field strength H is nonlinear, the permeability is not a constant but varies with the magnetic field strength. Usually, it is not necessary to have a complete knowledge of the magnetic field dependence of permeability. In the mathematical treatment of general applications, the relative permeability is simply a number denoted by the symbol μ_r, but for different cases, permeability has different physical meaning. On the basis of the hysteresis loop shown in Figure 1.20, we can distinguish four definitions of scalar permeability often used in materials research (Robert 1988).

The initial relative permeability is defined as

$$\mu_{ri} = \frac{1}{\mu_0} \lim_{H \to 0} \frac{B}{H} \qquad (1.73)$$

It is applicable to a specimen that has never been subject to irreversible polarization. It is a

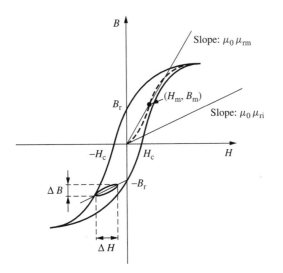

Figure 1.20 Definitions of four scalar permeabilities (Robert 1988). Reprinted with permission from *Electrical and Magnetic Properties of Materials*, by Robert, P., Artech House Inc., Norwood, MA, USA, www.artechhouse.com

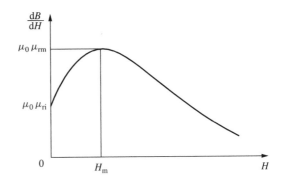

Figure 1.21 The dependence of permeability on magnetic field

theoretical value corresponding to a zero field, and in a strict meaning, it cannot be directly measured. Usually, the initial relative permeability is determined by extrapolation. In practice, μ_{ri} is often given as the relative permeability measured in a weak field lying between 100 and 200 A/m.

Figure 1.21 shows the relationship between (dB/dH) and H corresponding to the dashed line in Figure 1.20. The (dB/dH) value point at $H = 0$ equals the initial permeability discussed above. At the point H_m, which satisfies

$$\frac{d^2 B}{dH^2} = 0, \qquad (1.74)$$

the value of (dB/dH) reaches its maximum value, which is defined as maximum permeability $(\mu_0 \mu_{rm})$, as shown in Figures 1.20 and 1.21. The value of μ_{rm} can be taken as a good approximation of the relative permeability for a low-frequency alternating field with amplitude H_m.

Now, we consider the case when an alternating field H_2 is superimposed on a steady field H_1 parallel to H_2. If $H_2 \gg H_1$, the hysteresis loop is simply translated without substantial deformation. If $H_2 \ll H_1$, there will be an eccentric local loop, which is always contained within the main cycle. In the presence of a superimposed steady field H_1, the differential relative permeability $u_{r\Delta}$ is defined by

$$\mu_{r\Delta} = \frac{1}{\mu_0} \frac{\Delta B}{\Delta H} \qquad (1.75)$$

where ΔH is the amplitude of the alternating field and ΔB is the corresponding variation of the magnetic induction. The reversible relative permeability u_{rr} is the value of the differential relative permeability for an alternating field tending to zero

$$\mu_{rr} = \frac{1}{\mu_0} \lim_{\Delta H \to 0} \frac{\Delta B}{\Delta H} \qquad (1.76)$$

1.3.4.3 Soft and hard magnetic materials

According to the values of their coercive fields, magnetic materials can be classified into soft and hard magnetic materials. Figure 1.22(a) shows a typical hysteresis loop of a soft magnetic material. The term *soft* is applied to a magnetic material that has a low coercive field, so only a small magnetic field strength is required to demagnetize or reverse the direction of the magnetic flux in the material. Usually, soft magnetic material has high permeability. The area enclosed by the hysteresis loop is usually small, so little energy is lost in the magnetization cycle. In a microscopic scale, the domains in a soft magnetic material can easily grow and rotate. Soft magnetic materials are widely used for electrical applications, such as transformer cores. Figure 1.22(b) shows a typical hysteresis curve for a hard magnetic material. A hard magnetic material has a high coercive field, so it is difficult to demagnetize it. The permeability of a hard magnetic material is usually small. Besides, a hard magnetic material usually has a large area enclosed by the hysteresis loop. Hard magnetic materials are often used as permanent magnets.

It should be emphasized that the coercive field H_c is the criteria for the classification of soft and

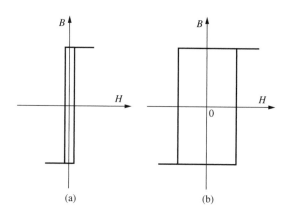

(a)   (b)

Figure 1.23   Rectangular hysteresis loops. (a) Soft magnetic material and (b) hard magnetic material

hard magnetic materials. Generally speaking, the coercive field of a soft magnetic material is less than ten oersted, while that of a hard magnetic material is larger than several hundred oersted. It should be noted that remanence flux density B_r is not a criteria for the classification of soft and hard magnetic materials. A magnetic material with rectangular hysteresis loop has a relatively high value of B_r, but high value of B_r does not mean high value of H_c. As shown in Figure 1.23(a) and (b), both soft and hard magnetic materials can have rectangular hysteresis loops.

For a material with rectangular hysteresis loop, when the magnetizing field is removed, the flux density almost remains unchanged, so that the remanence flux density is virtually the same as the saturation one. This means that, once the material is magnetized, it retains most of the flux density when the magnetizing field is switched off. These materials are often used in magnetic recording.

1.3.4.4 Magnetic resonance

Magnetic resonance is an important loss mechanism of magnetic materials, and should be taken into full consideration in the application of magnetic materials. For most of the magnetic materials, the energy dissipation at microwave frequencies is related to natural resonance and wall resonance.

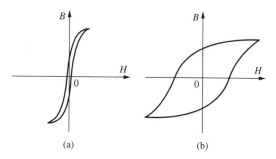

(a)   (b)

Figure 1.22   Hysteresis loops. (a) Soft magnetic materials and (b) hard magnetic materials. Source: Bolton, W. (1992), *Electrical and Magnetic Properties of Materials*, Longman Scientific & Technical, Harlow

Natural resonance

As shown in Figure 1.24, under a dc magnetic field H and ac magnetic field h, the magnetic moment M

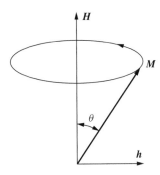

Figure 1.24 Precession of magnetic moment

makes a precession around the dc magnetic field H, and the ac magnetic field h provides the energy to compensate the energy dissipation of the precession. This is the origin of ferromagnetic resonance, and can be described by the Gilbert equation:

$$\frac{dM}{dt} = -\gamma M \times H + \frac{\lambda}{M} M \times \frac{dM}{dt} \qquad (1.77)$$

where $\gamma = 2.8\,\text{MHz/Oe}$ is the gyromagnetic ratio and λ is the damping coefficient. The dc magnetic field H includes external dc magnetic field H_0, anisotropic field H_a, demagnetization field H_d, and so on. If $H_0 = 0$, the ferromagnetic resonance is usually called *natural resonance*. In the following text, we concentrate on natural resonance of

ferrites and ferromagnetic resonance under the application of external dc magnetic field will be discussed in Chapter 8.

The resonance frequency f_r of a natural resonance is mainly determined by the anisotropic field of material. For a material with uniaxial anisotropy, the resonance frequency is given by

$$f_r = \gamma H_a \qquad (1.78)$$

For a material with planar anisotropic anisotropy, the resonance frequency is given by

$$f_r = \gamma (H_\theta \cdot H_\varphi)^{1/2} \qquad (1.79)$$

There are two typical types of resonances: Lorentzian type and Debye type. It should be indicated that, in actual materials, natural resonance may be in a type between the Lorentzian one and the Debye one. The Lorentzian type occurs when λ is much smaller than one, and it is also called *resonant type*. From Eq. (1.77), we can get

$$\mu_r = 1 + \frac{\chi_0}{1 - (f/f_r)^2 + \text{j}(2\lambda f/f_r)} \qquad (1.80)$$

where χ_0 is the static susceptibility of the material, f_r is the resonance frequency, and f is the operation frequency. Figure 1.25(a) shows a typical permeability spectrum of a resonance with Lorentzian type.

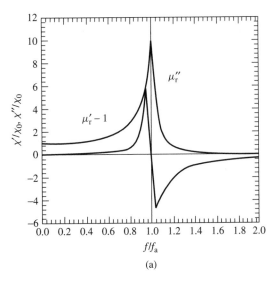

(a)

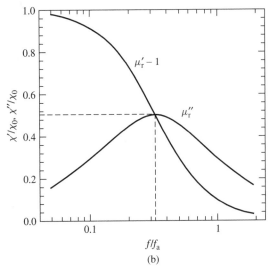

(b)

Figure 1.25 Two types of permeability spectrums. (a) Lorentzian type. The results are calculated based on Eq. (1.80) with $\lambda = 0.1$ and $f_a = f_r$. (b) Debye type. The results are calculated based on Eq. (1.81) with $f_a = f_r/\lambda$

The Debye type occurs when λ is much larger than one. The Debye type is also called *relaxation type*. From Eq. (1.77), we can get

$$\mu_r = 1 + \frac{A}{1 + j(\lambda f/f_r)} \qquad (1.81)$$

Figure 1.25(b) shows a typical permeability spectrum of Debye type.

The Snoek limit describes the relationship between the resonant frequency and permeability. For a material with uniaxial anisotropy, we have

$$f_r \cdot (\mu_r - 1) = \frac{2}{3}\gamma M_s \qquad (1.82)$$

where M_s is the saturated magnetization. For a material with a given resonance frequency, higher saturated magnetization corresponds to higher permeability. For a material with planar anisotropy, the Snoek limit is in the form of

$$f_r \cdot (\mu_r - 1) = \frac{1}{2}\gamma M_s \cdot \left(\frac{H_\theta}{H_\varphi}\right)^{1/2} \qquad (1.83)$$

Eq. (1.83) indicates that planar anisotropy provides more flexibility for the design of materials with expected resonant frequency and permeability.

Wall resonance

If a dc magnetic field H is applied to a magnetic material, the domains in the directions close to the direction of the magnetic field grow, while the domains in the directions close to the opposite directions of the magnetic field shrink. The growth and shrink of domains are actually the movements of the domain wall. If an ac magnetic field h is applied, the domain wall will vibrate around its equilibrium position, as shown in Figure 1.26. When the frequency of the ac magnetic field is equal to the frequency of the wall vibration, resonance occurs, and such a resonance is usually called *wall resonance*. Rado proposed a relationship between the resonance frequency f_0 and relative permeability μ_r (Rado 1953):

$$f_r \cdot (\mu_r - 1)^{1/2} = 2\gamma M_s \cdot \left(\frac{2\delta}{D}\right)^{1/2} \qquad (1.84)$$

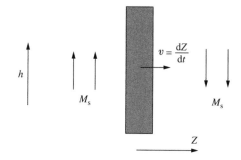

Figure 1.26   Mechanism of wall resonance

where δ and D are the thickness and the width of the domain wall respectively. M_s is the magnetization within a domain and it equals the saturated magnetization of the material.

The movement of domain wall is similar to a forced harmonic movement. So the wall resonance can be described using spring equation:

$$m_w \frac{d^2 Z}{dt^2} + \beta \frac{dZ}{dt} + \alpha Z = 2M_s h e^{j\omega t} \qquad (1.85)$$

where m_w is the effective mass of the domain wall, β is the damping coefficient, α is the elastic coefficient, and h is the amplitude of the microwave magnetic field. For a Lorentzian-type resonance, we have

$$\mu_r = 1 + \frac{A}{1 - (f/f_\beta)^2 + j(f/f_\tau)} \qquad (1.86)$$

where the intrinsic vibration frequency f_β is given by

$$f_\beta = (\alpha/m_w)^{1/2} \qquad (1.87)$$

and the relaxation frequency f_τ is given by

$$f_\tau = \alpha/\beta \qquad (1.88)$$

For most of the wall resonance, $f_\beta \gg f_\tau$, Eq. (1.86) becomes

$$\mu_r = 1 + \frac{A}{1 + j(f/f_\tau)} \qquad (1.89)$$

Eq. (1.89) represents a Debye-type resonance.

1.3.5 Metamaterials

Electromagnetic metamaterials are artificial structures with unique or superior electromagnetic properties. The special properties of metamaterials come from the inclusion of artificially fabricated, extrinsic, low-dimensional inhomogeneities. The development of metamaterials includes the design of unit cells that have dimensions commensurate with small-scale physics and the assembly of the unit cells into bulk materials exhibiting desired electromagnetic properties. In recent years, the research on electromagnetic metamaterials is very active for their applications in developing functional electromagnetic materials. In the following, we discuss three examples of metamaterials: chiral materials, left-handed materials, and photonic band-gap materials.

1.3.5.1 Chiral materials

Chiral materials have received considerable attention during recent years (Jaggard *et al.* 1979; Mariotte *et al.* 1995; Theron and Cloete 1996; Hui and Edward 1996) and might have a variety of potential applications in the field of microwaves, such as microwave absorbers, microwave antennae, and devices (Varadan *et al.* 1987; Lindell and Sihvola 1995). (Lakhtakia *et al.* 1989) has given a fairly complete set of references on the subject. (Bokut and Federov 1960; Jaggard *et al.* 1979; Silverman 1986; Lakhtakia *et al.* 1986) have studied the reflection and refraction of plane waves at planar interfaces involving chiral media. The possibility of designing broadband antireflection coatings with chiral materials was addressed by (Varadan *et al.* 1987). These researchers have shown that the introduction of chirality radically alters in scattering and absorption characteristics. In these papers, the authors have used assumed values of chirality parameter, permittivity, and permeability in their numerical results.

(Winkler 1956; Tinoco and Freeman 1960) have studied the rotation and absorption of electromagnetic waves in dielectric materials containing a distribution of large helices. Direct and quantitative measurements are made possible with the recent advances in microwave components and measurement techniques. Urry and Krivacic (1970) have measured the complex, frequency dependent values of $(n_L - n_R)$ for suspensions of optically active molecules, where n_L and n_R are the refractive indices for left- circularly polarized

(LCP) and right- circularly polarized (RCP) waves. LCP and RCP waves propagate with different velocities and attenuation in a chiral medium. Still, these differential measurements are unable to characterize completely the chiral medium. More recently, (Guire *et al.* 1990) has studied experimentally the normal incidence reflection of linearly polarized waves of metal-backed chiral composite samples at microwave frequencies. The beginning of a systematic experiment work came from (Umari *et al.* 1991) when they reported measurements of axial ratio, dichroism, and rotation of microwaves transmitted through chiral samples. However, in order to characterize completely the chiral composites, the chirality parameter, permittivity, and permeability have to be determined.

The chirality parameter, permittivity, and permeability can be determined from inversion of three measured scattering parameters. The new chirality parameter can be obtained only with the substitution of new sets of constitutive equations (Ro 1991; Sun *et al.* 1998),

$$\boldsymbol{D} = \varepsilon \boldsymbol{E} + \beta \varepsilon \nabla \times \boldsymbol{E} \text{ and} \qquad (1.90)$$

$$\boldsymbol{B} = \mu \boldsymbol{H} + \beta \mu \nabla \times \boldsymbol{H}, \text{ or} \qquad (1.91)$$

$$\boldsymbol{D} = \varepsilon \boldsymbol{E} + \mathrm{i}\xi \boldsymbol{B} \text{ and} \qquad (1.92)$$

$$\boldsymbol{H} = \mathrm{i}\xi \boldsymbol{E} + \boldsymbol{B}/\mu. \qquad (1.93)$$

Here, ε and μ are the usual permittivity and permeability respectively, while β and ξ are the chirality parameter that results from the handedness or lack of inversion symmetry in the microstructure of the medium. The values of chirality parameter, permittivity, and permeability vary with frequency, volume concentration of the inclusions, geometry and size of the inclusion, and the electromagnetic properties of the host medium. Further discussion on chiral materials can be found in Chapter 10.

1.3.5.2 Left-handed materials

A left-handed material is a material whose permeability and permittivity are simultaneously negative. It should be noted that the term "left handed" does not refer to either chirality or symmetry breaking. These other phenomena are often referred to as "left handed", but are distinct from the effects that we are discussing in left-handed materials.

All the normal materials are "right handed", which means that the relationship between the fields and the direction of wave vector follows the "right-hand rule". If the fingers of the right hand represent the electric field of the wave, and if the fingers curl around to the base of the right hand, representing the magnetic field, then the outstretched thumb indicates the direction of the flow of the wave energy. However, for a left-handed material, the relationship between the fields and the direction of wave vector follows the "left-hand rule".

Left-handed materials were first envisioned in the 1960s by Russian physicist Victor Veselago of the Lebedev Physics Institute. He predicted that when light passed through a material with both a negative dielectric permittivity and a negative magnetic permeability, novel optical phenomena would occur, including reversed Cherenkov radiation, reversed Doppler shift, and reversed Snell effect. Cherenkov radiation is the light emitted when a charged particle passes through a medium, under certain conditions. In a normal material, the emitted light is in the forward direction, while in a left-handed material, light is emitted in a reversed direction. In a left-handed material, light waves are expected to exhibit a reversed Doppler effect. The light from a source coming toward you would be reddened while the light from a receding source would be blue shifted.

The Snell effect would also be reversed at the interface between a left-handed material and a normal material. For example, light that enters a left-handed material from a normal material will undergo refraction, but opposite to what is usually observed. The apparent reversal comes about because a left-handed material has a negative index of refraction. Using a negative refractive index in Snell's law provides the correct description of refraction at the interface between left- and right-handed materials. As a further consequence of the negative index of refraction, lenses made from left-handed materials will produce unusual optics. As shown in Figure 1.27, a flat plate of left-handed material can focus radiation from a point source back to a point. Furthermore, the plate can amplify the evanescent waves from the source and thus the sub-wavelength details of the source can be restored at the image (Pendry 2000; Rao and Ong

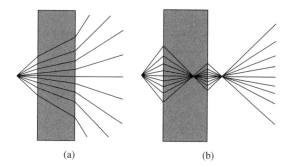

(a)   (b)

Figure 1.27   Effects of flat plates. (a) Flat plate made from a normal material and (b) flat plate made from a left-handed material

2003a, 2003b). Therefore, such a plate can work as a *superlens*.

Left-handed materials do not exist naturally. In Veselago's day, no actual left-handed materials were known. In the 1990s, John Pendry of Imperial College discussed how negative-permittivity materials could be built from rows of wires (Pendry *et al.* 1996) and negative-permeability materials from arrays of tiny resonant rings (Pendry *et al.* 1999). In 2000, David Smith and his colleagues constructed an actual material with both a negative permittivity and a negative permeability at microwave frequencies (Smith *et al.* 2000). An example of a left-handed material is shown in Figure 1.28. The raw materials used, copper wires and copper rings, do not have unusual properties of their own and indeed are nonmagnetic. But when

Figure 1.28   A left-handed material made from wires and rings. This picture is obtained from the homepage for Dr David R. Smith (http://physics.ucsd.edu/~drs/index.html)

incoming microwaves fall upon alternating rows of the rings and wires, a resonant reaction between the light and the whole of the ring-and-wire array sets up tiny induced currents, making the whole structure "left handed". The dimensions, geometric details, and relative positioning of the wires and the rings strongly influence the properties of the left-handed material.

However, the surprising optical properties of left-handed materials have been thrown into doubt by physicists. Some researchers said that the claims that left-handed materials could act as perfect lenses violate the principle of energy conservation (Garcia and Nieto-Vesperinas 2002). Meanwhile, some researchers indicated that "negative refraction" in left-handed materials would breach the fundamental limit of the speed of light (Valanju *et al.* 2002). But other researchers in the field defended their claims on left-handed materials. The debate should generate some light, and stimulate better experiments, which would benefit the understanding and utilization of this type of metamaterials. If the negative refraction and perfect lensing of left-handed materials can be proven, left-handed materials could have a wide range of applications including high-density data storage and high-resolution optical lithography in the semiconductor industry.

Finally, it should be indicated that many researchers in this field object to the term "left handed," which often refers to the structures exhibiting chirality. New descriptive terms have been introduced to refer to materials with simultaneously negative permittivity and permeability. "Backward wave materials" is used to signify the characteristic that materials with negative permittivity and permeability reverse the phase and group velocities. "Materials with negative refractive index" emphasizes the reversed Snell effect. And "double negative materials" is a quick and easy way to indicate that both the permittivity and permeability of the material are negative.

1.3.5.3 Photonic band-gap materials

A photonic band-gap (PBG) material, also called *photonic crystal*, is a material structure whose refraction index varies periodically in space. The periodicity of the refraction index may be in one dimension, two dimensions, or three dimensions. The name is applied since the electromagnetic waves with certain wavelengths cannot propagate in such a structure. The general properties of a PBG structure are usually described by the relationship between circular frequency and wave vector, usually called *wave dispersion*. The wave dispersion in a PBG structure is analogous to the band dispersion (electron energy versus wave vector) of electrons in a semiconductor. Figure 1.29(a) schematically shows a three-dimensional PBG structure, which is an array of dielectric spheres

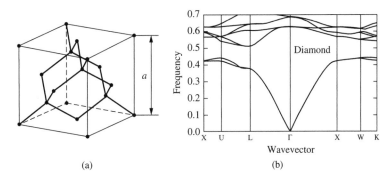

(a)   (b)

Figure 1.29 A PBG structure formed by dielectric spheres arranged in a diamond lattice. The background is air. The refraction index of the spheres is 3.6 and the filling ratio of the spheres is 34 %. The frequency is given in unit of c/a, where a is the cubic constant of the diamond lattice and c is the velocity of light in vacuum. (a) Schematic illustration of the three-dimensional structure and (b) wave dispersion of the PBG structure. Theoretical results are from (Ho *et al.* 1990). Source: Ho, K. M., Chan, C. T., and Soukoulis, C. (1990). "Existence of a photonic gap in periodic dielectric structures", *Phys. Rev. Lett.* **65**(25), 3152–3155. © 2003 The American Physical Society

surrounded by vacuum. The photonic band of the structure is shown in Figure 1.29(b).

The origin of the band gap stems from the very nature of wave propagation in periodic structures. When a wave propagates in a periodic structure, a series of refraction and reflection processes occur. The incident wave and the reflected wave interfere and may reinforce or cancel one another out according to their phase differences. If the wavelength of the incident wave is of the same scale as the period of the structure, very strong interference happens and perfect cancellation may be achieved. As a result, the wave is attenuated and cannot propagate through the periodic structure. In a broad sense, the electronic band gaps of semiconductors, where electron waves propagate in periodic electronic potentials, also fall into this category. Owing to the similarity of PBGs and the electronic band gaps, PBG materials for electromagnetic waves can be treated as semiconductors for photons.

The first PBG phenomenon was observed by Yablonovitch and Gmitter in an artificial microstructure at microwave frequency (Yablonovitch and Gmitter 1989). The microstructure was a dielectric material with about 8000 spherical air "atoms". The air "atoms" were arranged in a face-centered-cubic (fcc) lattice. Thereafter, many other structures and material combinations were designed and fabricated with superior PBG characteristics and greater manufacturability.

PBG materials are of great technological and theoretical importance because their stop-band and pass-band frequency characteristics can be used to mold the flow of electromagnetic waves (Joannopoulos *et al.* 1995). Extensive applications have been achieved using the concept of PBG in various fields, especially in optoelectronics and optical communication systems. The PBG is the basis of most applications of PBG materials, and it is characterized by a strong reflection of electromagnetic waves over a certain frequency range and high transmission outside this range. The center frequency, depth, and width of the band gap can be tailored by modifying the geometry and arrangement of units and the intrinsic properties of the constituent materials.

It should be noted that PBG structures also exist in the nature. The sparking gem opal, colorful wings of butterflies, and the hairs of a wormlike creature called the *sea mouse* have typical PBG structures, and their lattice spacing is exactly right to diffract visible light. It should also be noted that, although "photonic" refers to light, the principle of the band gap applies to all the waves in a similar way, no matter whether they are electromagnetic or elastic, transverse or longitude, vector or scalar (Brillouin 1953).

1.3.6 Other descriptions of electromagnetic materials

Besides the microscopic and macroscopic parameters discussed above, in materials research and engineering, some other macroscopic properties are often used to describe materials.

1.3.6.1 Linear and nonlinear materials

Linear materials respond linearly with externally applied electric and magnetic fields. In weak field ranges, most of the materials show linear responses to applied fields. In the characterization of materials' electromagnetic properties, usually weak fields are used, and we assume that the materials under study are linear and that the applied electric and magnetic fields do not affect the properties of the materials under test.

However, some materials easily show nonlinear properties. One typical type of nonlinear material is ferrite. As discussed earlier, owing to the nonlinear relationship between B and H, if different strength of magnetic field H is applied, different value of permeability can be obtained. High-temperature superconducting thin films also easily show nonlinear properties. In the characterization of HTS thin films and the development microwave devices using HTS thin films, it should be kept in mind that the surface impedance of HTS thin films are dependent on the microwave power.

1.3.6.2 Isotropic and anisotropic materials

The macroscopic properties of an isotropic material are the same in all orientations, so they can be represented by scalars or complex numbers. However, the macroscopic properties of an anisotropic material have orientation dependency, and they are usually represented by tensors or matrixes. Some crystals are anisotropic because of their crystalline structures. More discussion on anisotropic materials can be found in Chapter 8, and further discussion on this topic can be found in (Kong 1990).

1.3.6.3 Monolithic and composite materials

According to the number of constituents, materials can be classified into monolithic or composite materials. A monolithic material has a single constituent. While a composite material has several constituents, and usually one of the constituents is called *host medium*, the others are called *inclusions* or *fillers*. The properties of a composite material are related to the properties and fractions of the constituents, so the electromagnetic properties of composites can be tailored by varying the properties and fractions of the constituents. The study of the electromagnetic properties of composite materials has attracted much attention, with the aim of developing composites with expected electromagnetic properties.

The prediction of the properties of a composite from those of the constituents of the composite is a long-standing problem for theoretical and experimental physics. The mixing laws relating the macroscopic electromagnetic properties of composite materials to those of their individual constituents have been a subject of enquiry since the end of the nineteenth century. The ability to treat a composite with single effective permittivity and effective permeability is essential to work in many fields, for example, remote sensing, industrial and medical applications of microwaves, materials science, and electrical engineering.

The mean-field method and effective-medium method are two traditional approaches in predicting the properties of composite materials (Banhegyi 1994). In the mean-field method, we calculate the upper and lower limits of properties representing the parallel and perpendicular arrangements of the constituents. A practical method is to approximate the composite structure by elements of ellipsoidal shape, and various techniques are available to calculate the composite permittivity. For isotropic composites, closer limits can be calculated and, depending on morphological knowledge, more

sophisticated limits are possible. In an effective-medium method, we assume the presence of an imaginary effective medium, whose properties are calculated using general physical principles, such as average fields, potential continuity, average polarizability, and so on. Detailed discussion on effective-medium theory can be found in (Choy 1999).

To achieve more accurate prediction, numerical methods are often used in predicting the properties of composite materials. Numerical computation of the effective dielectric constant of discrete random media is important for practical applications such as geophysical exploration, artificial dielectrics, and so on. In such dielectrics, a propagating electromagnetic wave undergoes dispersion and absorption. Some materials are naturally absorptive owing to viscosity, whereas inhomogeneous media exhibit absorption due to geometric dispersion or multiple scattering. The scattering characteristics of the individual particles (or the inclusions) in the composite could be described by a transition or T-matrix and the frequency-dependent dielectric properties of the composite are calculated using multiple scattering theory and appropriate correlation functions between the particles (Varadan and Varadan 1979; Bringi *et al.* 1983; Varadan *et al.* 1984; Varadan and Varadan 1985).

More discussions on monolithic and composite materials can be found in (Sihvola 1999; Neelakanta 1995; Priou 1992; Van Beek 1967). In the following, we concentrate on the dielectric permittivity of composite materials, and we mainly discuss dielectric–dielectric composites and dielectric-conductor composites.

Dielectric–dielectric composites

The host media of composite materials are usually dielectric materials, and if the inclusions are also dielectric materials, such composites are called *dielectric–dielectric composites*. The shapes and structures of the inclusions affect the overall properties of the composites.

A composite with spherical inclusions is the simplest and a very important case. Consider a mixture with a host medium of permittivity ε_0 containing n inclusions in unit volume, with each of the inclusions having polarizability α. The permittivity ε_0 of the host medium can take any value, including complex ones. The effective permittivity ε_{eff} of a composite is defined as the ratio between the average electric displacement $\boldsymbol{D}$ and the average electric field $\boldsymbol{E}$: $\boldsymbol{D} = \varepsilon_{\text{eff}}\boldsymbol{E}$. The electric displacement $\boldsymbol{D}$ depends on the polarization $\boldsymbol{P}$ in the material, $\boldsymbol{D} = \varepsilon_0\boldsymbol{E} + \boldsymbol{P}$, and the polarization can be calculated from the dipole moments $\boldsymbol{p}$ of the n inclusions, $\boldsymbol{P} = n\boldsymbol{p}$. This treatment assumes that the dipole moments are the same for all inclusions. If the inclusions are of different polarizabilities, the polarization has to be summed by weighting each dipole moment with its number density, and the overall polarization thus consists of a sum or an integral over all the individual inclusions.

The dipole moment $\boldsymbol{p}$ depends on the polarizability and the exciting field $\boldsymbol{E}_{\text{e}}$: $\boldsymbol{p} = \alpha\boldsymbol{E}_{\text{e}}$. For spherical inclusions, the exciting field $\boldsymbol{E}_{\text{e}}$ is: $\boldsymbol{E}_{\text{e}} = \boldsymbol{E} + \boldsymbol{P}/(3\varepsilon_0)$. From the above equations, the effective permittivity can be calculated as a function of the dipole moment density $n\alpha$:

$$\varepsilon_{\text{eff}} = \varepsilon_0 + 3\varepsilon_0\frac{n\alpha}{3\varepsilon_0 - n\alpha} \qquad (1.94)$$

Equation (1.94) can also be written in the form of the Clausius–Mossotti formulas

$$\frac{\varepsilon_{\text{eff}} - \varepsilon_0}{\varepsilon_{\text{eff}} + 2\varepsilon_0} = \frac{n\alpha}{3\varepsilon_0} \qquad (1.95)$$

If the composite contains inclusions with different polarization, for example N types of spheres with different permittivities, Eq. (1.95) should be modified into (Sihvola 1989a; Sihvola and Lindell 1989b):

$$\frac{\varepsilon_{\text{eff}} - \varepsilon_0}{\varepsilon_{\text{eff}} + 2\varepsilon_0} = \sum_{i=1}^{N} \frac{n_i\alpha_i}{3\varepsilon_0} \qquad (1.96)$$

Let the permittivity of the background medium be ε_0, that of the inclusions be ε_1, and the volume fraction of the inclusions be f_1. The polarizability of this kind of inclusions depends on the ratio between the inside and the outside fields when the inclusions are in a static field. According to Sihvola(1989a) and Sihvola and Lindell (1989b), the polarizability of a spherical inclusion with radius a_1 is

$$\alpha = 4\pi\varepsilon_0 a_1^3 \frac{\varepsilon_1 - \varepsilon_0}{\varepsilon_1 + 2\varepsilon_0} \qquad (1.97)$$

So the effective permittivity of this mixture is

$$\frac{\varepsilon_{\text{eff}} - \varepsilon_0}{\varepsilon_{\text{eff}} + 2\varepsilon_0} = f_1\frac{\varepsilon_1 - \varepsilon_0}{\varepsilon_1 + 2\varepsilon_0} \qquad (1.98)$$

This formula is known as the Rayleigh's formula.

The success of a mixture formula for a composite relies on the accuracy in the modeling of its real microstructure details. Besides the Rayleigh's formula, several other formulas have been derived using different approximations of the microstructural details of the composite. Several other popular formulas for the effective permittivity ε_{eff} of two-phase nonpolar dielectric mixtures with host medium of permittivity ε_0 and spherical inclusion of permittivity ε_1 with volume fraction f_1 are listed in the following:

Looyenga's formula:

$$\varepsilon_{\text{eff}}^{\frac{1}{3}} = f_1\varepsilon_1^{\frac{1}{3}} + (1 - f_1)\varepsilon_0^{\frac{1}{3}} \qquad (1.99)$$

Beer's formula:

$$\varepsilon_{\text{eff}}^{\frac{1}{2}} = f_1\varepsilon_1^{\frac{1}{2}} + (1 - f_1)\varepsilon_0^{\frac{1}{2}} \qquad (1.100)$$

Lichtenecher's formula:

$$\ln\varepsilon_{\text{eff}} = f_1\ln\varepsilon_1 + (1 - f_1)\ln\varepsilon_0 \qquad (1.101)$$

In the above formulas (1.98–1.101), the interparticle actions between the inclusions are neglected. The above formulas can be extended to multiphase composites, but they are not applicable to composites with layered inclusions, because they ignore the interactions between the different layers in an inclusion. The properties of composites with layered spherical inclusions are discussed in (Sihvola and Lindell 1989b).

Dielectric-conductor composites

In a dielectric-conductor composite, the host medium is a dielectric material, while the inclusions are conductors. Such composite materials have extensive electrical and electromagnetic applications, such as antistatic materials, electromagnetic shields, and radar absorbers. Here we do not consider the frequency dependence of the electromagnetic properties of dielectric-conductor composites, and only consider the static limit ($\omega \to 0$). Discussions on the frequency dependence of the properties of such

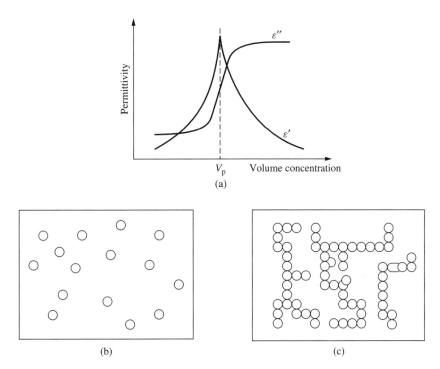

Figure 1.30 Percolation in a dielectric-conductor composite. (a) Change of static permittivity near the percolation threshold, (b) the case when the volume concentration of the fillers is less than the percolation threshold (V_p), and (c) the case when the volume concentration of the fillers is close to the percolation threshold (V_p). In (b) and (c), circle denotes inclusions, otherwise the host medium

composites can be found in (Potschke *et al.* 2003) and the references given therein.

For a dielectric-conductor composite, there exists a phenomenon called *percolation*. When the volume concentration of the conductive inclusions approaches the percolation threshold, the dielectric composite becomes conductive. One can observe a significant change in permittivity of the composites filled with conductive inclusions when it percolates. As shown in Figure 1.30(a), near the percolation threshold, the real part of permittivity of the composite increases quickly along with the increase of the volume concentration of the conductive inclusions and reaches its maximum value at the percolation threshold; while the imaginary part of permittivity monotonically increases with the increase of the volume concentration of the conductive inclusions. The origin of percolation phenomenon is the connection of the conductive inclusions. Figures 1.30(b) and (c) show distributions of the conductive inclusions in the

host medium when the volume concentration of the inclusions is less than and close to the percolation threshold respectively.

The location of the percolation threshold and the concentration dependence of permittivity and conductivity around the threshold depend on the properties of the host medium, conductive inclusions, and the morphology of the composite. Because of the rich physics phenomena near the percolation threshold, in percolation research, constructing models of permittivity or conductivity near the percolation threshold is of great theoretical importance and application meaning.

It should be emphasized that the geometry of the inclusions plays an important role in determining the percolation threshold and the electromagnetic properties of a dielectric-conductor composite. The general geometry of an inclusion is elliptic sphere, which, at special conditions, can be disk, sphere, and needle. In recent years, composites with fiber inclusions have attracted great attentions (Lagarkov *et al.* 1998). Fiber can be taken as a very thin and long needle. The mechanical and electrical performance of polymer materials may be greatly improved by adding carbon or metal fibers, and the resulted fiber-reinforced composites have a wide range of practical applications due to their unique mechanical, chemical, and physical properties. Fiber-filled composites present more possibilities of tailoring the dielectric properties. For example, high values of dielectric constant can be obtained at a low concentration of fiber inclusions, and composites filled with metal fibers possess pronounced microwave dielectric dispersion, which are very important for the development microwave absorbing materials.

Finally, it should be indicated that percolation phenomena also exists in many other systems, for example, the superconductivity of metal-superconductor composites and leakage of fluids through porous media.

1.4 INTRINSIC PROPERTIES AND EXTRINSIC PERFORMANCES OF MATERIALS

The properties of materials can be generally classified into two categories: intrinsic and extrinsic. Intrinsic properties of a material are independent of the size of the material. If the electromagnetic properties of a material are related to the geometrical structures and sizes, such properties are extrinsic.

1.4.1 Intrinsic properties

Most of the electromagnetic properties discussed above in this chapter are intrinsic, as they are governed by their respective underlying mechanisms, not by their geometries. This book concentrates on the characterization of intrinsic properties of materials. Here, we make a brief summary of the intrinsic properties of materials often studied in physics, materials sciences, and microwave electronics.

The parameters describing the intrinsic electromagnetic properties of materials generally include constitutive parameters, propagation parameters, and electrical transport properties. The constitutive parameter for low-conductivity materials mainly includes permittivity, permeability, conductivity, and chirality. Electromagnetic waves can propagate within low-conductivity materials, and the propagation parameters for low-conductivity materials mainly include wave impedance, propagation constant, and index of refraction. For conductors and superconductors, the main propagation parameters are skin depth and surface impedance. For semiconductors, the intrinsic properties are usually described by their electrical transport properties, including Hall mobility, conductivity, and carrier density.

1.4.2 Extrinsic performances

As the performances of electromagnetic materials and structures depend on their geometries, the performance-related properties are usually extrinsic. The design of functional materials and structures is to realize the desired extrinsic performances based on the intrinsic properties of the raw materials to be used. The performance-related properties could be monitored along the manufacturing procedures to ensure that the final products have the specified extrinsic performances.

There are varieties of extrinsic performances, and it is difficult to make a systematic classification. In the following, we discuss the characteristic impedance of a transmission line, the reflectivity of a Dallenbach layer, and the resonance of a dielectric resonator.

1.4.2.1 Characteristic impedance of a transmission line

The characteristic impedance of a transmission line should be taken into serious consideration in the design of high-speed circuits. Figure 1.31

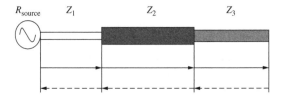

Figure 1.31 Impedance discontinuities in a transmission line. The solid arrows represent the transmission signals and the dashed arrows represent the reflection signals

shows a transmission line connected to a source, and how the voltage and current signals from the source interact with the transmission line as they propagate along the transmission line. If the characteristic impedance of the transmission line changes, either by a geometry change or a material change, some of the signals will be reflected. At the interface between two segments of transmission lines with characteristic impedances Z_1 and Z_2 respectively, the reflectivity is given by

$$\Gamma = \frac{V_{\text{reflected}}}{V_{\text{incident}}} = \frac{Z_2 - Z_1}{Z_2 + Z_1}. \qquad (1.102)$$

At each interface, there will be a series of reflections. Therefore, the impedance discontinuities distort signal, decrease the signal integrity, increase standing waves, increase rise-time degradation, and require longer setting time. The way to solving these problems is to use the same characteristic impedance throughout the transmission line, including traces and connectors. By considering manufacturability, cost, noise sensitivity, and power dissipation, the optimum Z_0 value for most systems is in the range of 50 to 80 Ω.

Microstrip is the most widely used transmission line in microwave electronics. As shown in Figure 1.32, the structural parameters of a microstrip line are the width W of the microstrip and the thickness d of the dielectric substrate. The characteristic impedance of a microstrip depends on the dielectric constant of the substrate and the structural parameters. For a thin substrate case ($d/W < 1$), the characteristic impedance of a microstrip line is given by (Pozar 1998):

$$Z_0 = \frac{120\pi}{\sqrt{\varepsilon_e}[W/d + 1.393 + 0.667 \ln(W/d + 1.444)]} \qquad (1.103)$$

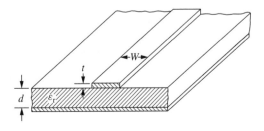

Figure 1.32 Geometry of a microstrip line

where the effective dielectric constant ε_e is given by

$$\varepsilon_e = \frac{\varepsilon_r + 1}{2} + \frac{\varepsilon_r - 1}{2} \frac{1}{\sqrt{1 + 12d/W}} \qquad (1.104)$$

If we want to fabricate a microstrip transmission line of 50 Ω from a substrate with dielectric thickness of 5 mils and dielectric constant of 5, the line width should be around 9 mils. If the line width varies by ± 1 mil, the characteristic impedance would vary by about $\pm 10\%$. If the dielectric constant varies by $\pm 10\%$, the characteristic impedance would vary by about $\pm 5\%$. Therefore, the line width is more sensitive to the characteristic impedance than the dielectric constant of the substrate. More discussions on microstrip line and other types of transmission lines will be made in Chapter 2.

1.4.2.2 Reflectivity of a Dallenbach layer

As shown in Figure 1.33, a Dallenbach layer is a homogeneous layer backed by a metal plate and the dissipation of microwave energy is made throughout the layer (Knott *et al.* 1993). The reflectivity of a Dallenbach layer is dependent on its thickness, and it is a typical example of extrinsic performance. In the design of a Dallenbach layer, we should consider the interferences between the reflections at different interfaces. When a microwave signal is incident to the layer, part of the signal is reflected at the interface between the free space and the material layer, and this part of signal is called the *first reflection*. Part

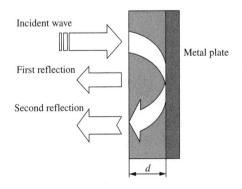

Figure 1.33 The structure and working principle of a Dallenbach layer

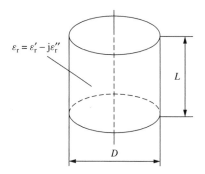

Figure 1.34 Configuration of a cylindrical dielectric resonator

of signal propagates into the material layer, and this part of signal is reflected at the interface between the material layer and the metal plate, and part of the signal, called the *second reflection*, comes out of the layer again. If the first reflection and second reflection have the same amplitude but are in opposite phase, they cancel each other, and so no actual reflection occurs. To ensure this cancellation, the thickness of the layer should be the quarter wavelength of the microwave signal in the material layer, and the layer should have proper impedance and loss factor.

1.4.2.3 Resonance of a dielectric resonator

The resonance of a dielectric resonator is another example of extrinsic performance. Figure 1.34 shows an isolated cylindrical dielectric resonator, with diameter D and length L. The resonant properties, including the resonant frequency and quality factor, of the dielectric resonator are determined by the dielectric permittivity of the dielectric materials and the geometrical parameters including diameter D and length L. More discussion on the resonant properties of dielectric resonators can be found in Chapter 2.

REFERENCES

Banhegyi, G. (1994). "Dielectric spectroscopy", in *Characterization of Composite Materials*, H. Ishida, Ed., Butterworth-Heinemann, Boston.

Van Beek, L. K. H. (1967). "Dielectric behavior of heterogeneous system", *Progress in Dielectrics*, **7**, 67–114.

Bokut, B. V. and Federov, F. I. (1960). "Reflection and refraction of light in an optically isotropic active media," *Optika i Spektroskopiya.*, **9**, 334–336.

Bolton, W. (1992). *Electrical and Magnetic Properties of Materials*, Longman Scientific & Technical, Harlow.

Brillouin, L. (1953). *Wave Propagation in Periodic Structures*, Dover Publications, New York.

Bringi, V. N. Varadan, V. K. and Varadan, V. V (1983). "Average dielectric properties of discrete random media using multiple scattering theory", *IEEE Transactions on Antennas and Propagation*, **AP-31**, 371–375.

Choy, T. C. (1999). *Effective Medium Theory: Principles and Applications*, Oxford University Press, New York.

Fuller, A. J. B. (1987). *Ferrites at Microwave Frequencies*, Peter Peregrinus Ltd., London.

Garcia, N. and Nieto-Vesperinas, M. (2002). "Left-handed materials do not make a perfect lens", *Physical Review Letters*, **88** (20), art. no. 207403, 1–4.

Guire, T. Varadan, V. V. and Varadan, V. K. (1990). "Influence of chirality on the reflection of em waves by planar dielectric slabs," *IEEE Transactions on Electromagnetic Compatibility*, **32**, 300–304.

Ho, K. M. Chan, C. T. and Soukoulis, C. (1990). "Existence of a photonic gap in periodic dielectric structures", *Physical Review Letters*, **65** (25), 3152–3155.

Hui, H. T. and Edward, K. N. (1996). "Modal expansion of dyadic Green's functions of the cylindrical chirowaveguide", *IEEE Microwave and Guided Wave Letters*, **6**, 360–362.

Jaggard, D. L. Mickelson, A. R. and Papas, C. H. (1979). "On electromagnetic waves in chiral media", *Applied Physics*, **18**, 211–216.

Jiles, D. (1994). *Introduction to the Electronic Properties of Materials*, Chapman & Hall, London.

Jiles, D. (1998). *Introduction to Magnetism and Magnetic Materials*, 2nd edition, Chapman & Hall, London.

Joannopoulos, J. D. Meade, R. D. and Winn, J. N. (1995). *Photonic Crystals: Molding the Flow of Light*, Princeton University Press, Princeton.

Kittel, C. (1997). *Introduction to Solid State Physics*, 7th edition, John Wiley & Sons, New York.

Knott, E. F. Shaeffer, J. F. and Tuley, M. T. (1993). *Radar Cross Section*, 2nd edition, Artech House, Boston.

Kong, J. A. (1990). *Electromagnetic Wave Theory*, 2nd edition, John Wiley & Sons, New York.

Lakhtakia, A. Varadan, V. V. and Varadan, V. K. (1986). "A Parametric study of microwave reflection characteristics of a planar achiral-chiral interface," *IEEE Transactions on Electromagnetic Compatibility*, **28**, 90–95.

Lakhtakia, A. Varadan, V. K. and Varadan, V. V. (1989). *Time-Harmonic Electromagnetic Fields in Chiral Media*, Lect. Note Ser. 35, Springer-Verlag, New York.

Lancaster, M. J. (1997). *Passive Microwave Device Applications of High-temperature Superconductors*, Cambridge University Press, Cambridge.

Lagarkov, A. N. Matytsin, S. M. Rozanov, K. N. and Sarychev, A. K. (1998). "Dielectric properties of fiber-filled composites", *Journal of Applied Physics*, **84**, 3806–3814.

Lindell, I. V. and Sihvola, A. H. (1995). "Plane-wave reflection from uniaxial chiral interface and its application to polarization transformation," *IEEE Transactions on Antennas and Propagation*, **43**, 1397–1404.

Lines, M. E. and Glass, A. M. (1977). *Principles and Applications of Ferroelectrics and Related Materials*, Clarendon Press, Oxford.

Mariotte, F. Guerin, F. Bannelier, P. and Bourgeade, A. (1995). "Numerical computations of the electromagnetic field scattered by complex chiral bodies", *Journal of Electromagnetic Waves and Applications*, **9**, 1459–1485.

Montrose, M. I. (1999). *EMC and the Printed Circuit Board: Design, Theory, and Layout made Simple*, IEEE Press, New York.

Neelakanta, P. S. (1995). *Handbook of Electromagnetic Materials, Monolithic and Composite Versions and their Applications*, CRC Press, Boca Raton.

Nyfors, E. and Vainikainen, P. (1989). *Industrial Microwave Sensors*, Artech House, Norwood.

Pendry, J. B. Holden, A. J. Stewart, W. J. and Youngs, I. (1996). "Extremely low frequency plasmons in metallic mesostructures", *Physical Review Letters*, **76** (25), 4773–4776.

Pendry, J. B. Holden, A. J. Robbins, D. J. and Stewart, W. J. (1999). "Magnetism from conductors and enhanced nonlinear phenomena", *IEEE Transactions on Microwave Theory and Techniques*, **47** (11), 2075–2084.

Pendry, J. B. (2000). "Negative refraction makes a perfect lens", *Physical Review Letters*, **85** (18), 3966–3969.

Potschke, P. Dudkin, S. M. and Alig, I. (2003). "Dielectric spectroscopy on melt processed polycarbonate – multiwalled carbon nanotube composites", *Polymer*, **44**, 5023–5030.

Pozar, D. M. (1998). *Microwave Engineering*, 2nd edition, John Wiley & Sons, New York.

Priou, A. Ed. (1992). *Dielectric Properties of Heterogeneous Materials*, Elsevier, New York.

Rado, G. T. (1953). "Magnetic spectra of ferrites", *Review of Modern Physics*, **25** (1), 81–89.

Ramo, S. Whinnery, J. R. and Van Duzer, T. (1994). *Fields and Waves in Communication Electronics*, 3rd edition, John Wiley & Sons, New York.

Rao, X. S. and Ong, C. K. (2003a). "Amplification of evanescent waves in a lossy left-handed material slab", *Physical Review B*, **68** (11), art. no. 113103, 1–4.

Rao, X. S. and Ong, C. K. (2003b). "Subwavelength imaging by a left-handed material superlens", *Physical Review E*, **68** (6), art. no. 067601, 1–3.

Ro, R. (1991). "Determination of the electromagnetic properties of chiral composites, using normal incidence measurements," Ph.D. Thesis, The Pennsylvania State University, Department of Engineering Science and Mechanics.

Robert, P. (1988). *Electrical and Magnetic Properties of Materials*, Artech House, Norwood.

Shen, Z. Y. (1994). *High-temperature Superconducting Microwave Circuits*, Artech House, Boston.

Sihvola, A. H. (1989a). "Self-consistency aspects of dielectric mixing theories", *IEEE Transactions on Geoscience and Remote Sensing*, **27**, 403–415.

Sihvola, A. H. and Lindell, I. V. (1989b). "Polarizability and effective permittivity of layered and continuously inhomogeneous dielectric spheres", *Journal of Electromagnetic Waves and Applications*, **3**, 37–60.

Sihvola, A. (1999). *Electromagnetic Mixing Formulas and Applications*, Institution of Electrical Engineers, London.

Silverman, M. P. (1986). "Reflection and refraction at the surface of a chiral medium: comparison of gyrotropic constitutive relations invariant and noninvariant under a duality transformation", *Journal of the Optical Society of America*, **3**, 830–837.

Smit, J. Ed. (1971). *Magnetic Properties of Materials*, McGraw-Hill, New York.

Smith, D. R. Padilla, W. J. Vier, D. C. Nemat-Nasser, S. C. and Schultz, S. (2000). "Composite medium with simultaneously negative permeability and permittivity", *Physical Review Letters*, **84** (18), 4184–4187.

Solymar, L. and Walsh, D. (1998). *Electrical Properties of Materials*, 6th edition, Oxford University Press, Oxford.

Sun, G. C. Yao, K. L. Liu, Z. L. and Huang, Q. L. (1998). "A study on measuring the electromagnetic parameters of chiral materials," *Journal of Physics D: Applied Physics*, **31**, 2109–2111.

Theron, I. P. and Cloete, J. H. (1996). "The optical activity of an artificial non-magnetic uniaxial chiral crystal at microwave frequencies," *Journal of Electromagnetic Waves and Applications*, **10**, 539–561.

Thuery, J. and Grant, E. H. Eds. (1992). *Microwaves: Industrial, Scientific and Medical Applications*, Artech House, Boston.

Tinkham, M. (1996). *Introduction to Superconductivity*, 2nd edition, McGraw-Hill, Singapore.

Tinoco, I. and Freeman, M. P. (1960). "The optical activity of oriented copper helices, II. Experimental," *Journal of Physical Chemistry*, **61**, 1196–2000.

Umari, M. Varadan, V. V. and Varadan, V. K. (1991). "Rotation and dichroism associated with microwave propagation in chiral composite samples," *Radio Science*, **26**, 1327–1334.

Urry, D. W. and Krivacic, J. (1970). "Differential scatter of left and right circularly polarized light by optically active particulate system," *Proceedings of the National Academy of Sciences of the United States of America*, **65**, 845–852.

Valanju, P. M. Walser, R. M. and Valanju, A. P. (2002). "Wave refraction in negative-index media: Always positive and very inhomogeneous", *Physical Review Letters*, **88** (18), art. no. 187401, 1–4.

Varadan, V. K. and Varadan, V. V. (1979). *Acoustic, Electromagnetic and Elastic Wave Scattering – Focus on the T-matrix Approach*, Pergamon Press, New York.

Varadan, V. V. and Varadan, V. K (1985). *Multiple Scattering of Waves in Random Media and Random Rough Surfaces*, Pennsylvania State University Press, Pennsylvania.

Varadan, V. V. Ma, Y and Varadan, V. K (1984). "Anisotropic dielectric properties of media containing nonspherical scatterers", *IEEE Transactions on Antennas and Propagation*, **AP-33**, 886–890.

Varadan, V. K. Varadan, V. V. and Lakhtakia, A. (1987). "On the possibility of designing anti-reflection coatings using chiral composites," *Journal of Wave-Material Interaction*, **2**, 71–81.

Von Hippel, A. R. (1995a). *Dielectrics and Waves*, Artech House, Boston.

Von Hippel, A. R. Ed. (1995b). *Dielectric Materials and Applications*, Artech House, Boston.

Winkler, M. H. (1956). "An experimental investigation of some models for optical activity," *Journal of Physical Chemistry*, **60**, 1656–1659.

Wohlfarth, E. P. (1980). *Ferromagnetic Materials: a Handbook on the Properties of Magnetically Ordered Substances*, Vol. 2, North Holland, Amsterdam.

Yablonovitch, E. and Gmitter, T. J. (1989). "Photonic band structure: the face-centered-cubic case", *Physical Review Letters*, **63** (18), 1950–1953.

Zoughi, R. (2000). *Microwave Non-Destructive Testing and Evaluation*, Kluwer Academic Publishers, Dordrecht.

2

Microwave Theory and Techniques for Materials Characterization

This chapter discusses the basic microwave theory and techniques for the characterization of electromagnetic materials. The methods for materials properties characterization generally fall into nonresonant methods and resonant methods; and, correspondingly, we mainly discuss two microwave phenomena: microwave propagation based on which the nonresonant methods are developed and microwave resonance based on which the resonant methods are developed. In our discussion, both the field approach and the line approach are used in analyzing electromagnetic structures. In the final part of this chapter, we introduce the concept of microwave network and discuss the experimental techniques for characterizing propagation and resonance networks.

2.1 OVERVIEW OF THE MICROWAVE METHODS FOR THE CHARACTERIZATION OF ELECTROMAGNETIC MATERIALS

There have been many extensive review papers (Afsar *et al.* 1986; Baker–Jarvis *et al.* 1993; Guillon 1995; Krupka and Weil 1998; Weil 1995; Zaki and Wang 1995) on the microwave methods for materials property characterization, and there are also several monographs on special measurement methods and for special purposes (Musil and Zacek 1986; Nyfors and Vainikainen 1989; Zoughi 2000). In this section, we focus on the basic principles for the measurement of the permittivity and permeability of low conductivity materials and the

surface impedance of high-conductivity materials. We do not discuss the detailed structures of fixtures and detailed algorithms for the calculation of materials properties, which will be discussed in Chapters 3 to 8. The characterization of chiral materials will be discussed in Chapter 10 and the measurement of microwave electrical transport properties will be discussed in Chapter 11.

The microwave methods for materials characterization generally fall into nonresonant methods and resonant methods. Nonresonant methods are often used to get a general knowledge of electromagnetic properties over a frequency range, while resonant methods are used to get accurate knowledge of dielectric properties at single frequency or several discrete frequencies. Nonresonant methods and resonant methods are often used in combination. By modifying the general knowledge of materials properties over a certain frequency range obtained from nonresonant methods with the accurate knowledge of materials properties at several discrete frequencies obtained from resonant methods, accurate knowledge of materials properties over a frequency range can be obtained.

In the following, we discuss the working principles of various measurement methods. It should be noted that the examples given below are used only to illustrate the basic configurations of the measurement methods, and they are not necessarily the configurations with highest accuracy and sensitivities. As will be discussed in later chapters, there are many practical considerations in the development of actual measurement methods.

Microwave Electronics: Measurement and Materials Characterization   L. F. Chen, C. K. Ong, C. P. Neo, V. V. Varadan and V. K. Varadan
© 2004 John Wiley & Sons, Ltd   ISBN: 0-470-84492-2

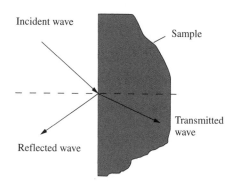

Figure 2.1   Boundary condition for material characterization using a nonresonant method

2.1.1 Nonresonant methods

In nonresonant methods, the properties of materials are fundamentally deduced from their impedance and the wave velocities in the materials. As shown in Figure 2.1, when an electromagnetic wave propagates from one material to another (from free space to sample), both the characteristic wave impedance and the wave velocity change, resulting in a partial reflection of the electromagnetic wave from the interface between the two materials. Measurements of the reflection from such an interface and the transmission through the interface can provide information for the deduction of permittivity and permeability relationships between the two materials.

Nonresonant methods mainly include reflection methods and transmission/reflection methods. In a reflection method, the materials properties are calculated on the basis of the reflection from the sample, and in a transmission/reflection method, the material properties are calculated on the basis of the reflection from the sample and the transmission through the sample.

Nonresonant methods require a means of directing the electromagnetic energy toward a material, and then collecting what is reflected from the material, and/or what is transmitted through the material. In principle, all types of transmission lines can be used to carry the wave for nonresonant methods, such as coaxial line, hollow metallic waveguide, dielectric waveguide, planar transmission line, and free space. In this section, we use hollow metallic waveguide or coaxial line as examples.

2.1.1.1 Reflection methods

In reflection methods, electromagnetic waves are directed to a sample under study, and the properties of the material sample are deduced from the reflection coefficient at a defined reference plane. Usually, a reflection method can only measure one parameter, either permittivity or permeability.

Two types of reflections are often used in materials property characterization: open-circuit reflection and short-circuit reflection, and the corresponding methods are called *open-reflection method* and *shorted reflection method*. As coaxial lines can cover broad frequency bands, coaxial lines are often used in developing measurement fixtures for reflection methods. Detailed discussions on reflection methods can be found in Chapter 3.

Open-reflection method

Figure 2.2 shows the basic measurement configuration of an open – reflection method. In actual applications, the outer conductor at the open end is usually fabricated into a flange to provide suitable capacitance and ensure the repeatability of sample loading (Li and Chen 1995; Stuchly and Stuchy 1980), and the measurement fixture is usually called the *coaxial dielectric probe*. This method assumes that materials under measurement are nonmagnetic, and that interactions of the electromagnetic field with the noncontacting boundaries of the sample are not sensed by the probe. To satisfy the second assumption, the thickness of

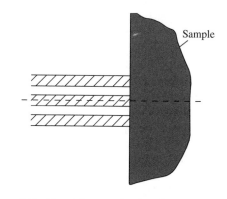

Figure 2.2   Coaxial open-circuit reflection

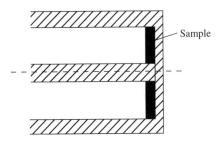

Figure 2.3 Coaxial short-circuit reflection

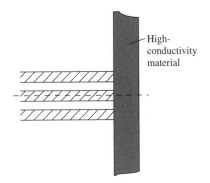

Figure 2.4 Reflection method for the measurement of surface impedance of high-conductivity materials

the sample should be much larger than the diameter of the aperture of the open-ended coaxial line, and, meanwhile, the material should have enough loss.

Shorted reflection method

Figure 2.3 shows a coaxial short-circuit reflection. In this method, the sample under study is usually electrically short, and this method is often used to measure magnetic permeability (Guillon 1995; Fannis *et al.* 1995). In this method, the permittivity of the sample is not sensitive to the measurement results, and in the calculation of permeability, the permittivity is often assumed to be ε_0.

Reflection method for surface impedance measurement

Besides their applications in the measurement of permittivity and permeability of low-conductivity materials, reflection methods are also used for the measurement of the surface impedance of high-conductivity materials. As shown in Figure 2.4, the high-conductivity material under study contacts the open end of a coaxial line. Microwave radiation can propagate into some extent of the high-conductivity material. The complex surface impedance of the sample can be extracted from the complex reflection coefficient. As the penetration depth of high-conductivity materials is small, microwave cannot go deep into the sample. This method does not require a very thick sample, but the thickness of the sample should be several times larger than the penetration depth. Because there are electrical currents flowing between the inner and the outer conductors of the coaxial line through the sample, this method requires

that both the inner and the outer conductors of the open end of the coaxial line have good electrical contact with the sample. This method has been used for the measurement of the surface resistance of high-temperature superconducting thin films (Booth *et al.* 1994).

2.1.1.2 Transmission/reflection methods

In a transmission/reflection method, the material under test is inserted in a piece of transmission line, and the properties of the material are deduced on the basis of the reflection from the material and the transmission through the material. This method is widely used in the measurement of the permittivity and permeability of low conductivity materials, and it can also be used in the measurement of the surface impedance of high-conductivity materials. Detailed discussions on transmission/reflection methods can be found in Chapter 4.

Permittivity and permeability measurement

Transmission/reflection method can measure both permittivity and permeability of low conductivity materials. Figure 2.5 shows the configuration of coaxial transmission/reflection method. The characteristic impedance of the piece of transmission line loaded with the sample is different from that of the transmission line without the sample, and such difference results in special transmission and reflection properties at the interfaces. The permittivity and permeability of the sample

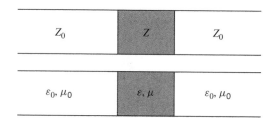

Figure 2.5   Coaxial transmission/reflection method

are derived from the reflection and transmission coefficients of the sample-loaded cell (Weir 1974; Nicolson and Rose 1970).

Surface impedance measurement

Reflection/transmission method can also be used for the measurement of surface impedance of high-conductivity thin films. As shown in Figure 2.6, the thin film under study forms a quasi-short circuit in a waveguide transmission structure. From the ratio of the transmitted power to the incident power and the phase shift across the thin film, the surface impedance of the thin film can be deduced. However, this method is only suitable for thin films whose thickness is less than the penetration depth of the sample, and requires a measurement system with very high dynamic range. This method has been used to study the microwave surface impedance of superconducting extrathin films (Dew–Hughes 1997; Wu and Qian 1997).

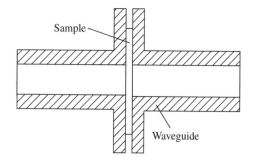

Figure 2.6   Transmission/reflection method for surface impedance measurements

2.1.2 Resonant methods

Resonant methods usually have higher accuracies and sensitivities than nonresonant methods, and they are most suitable for low-loss samples. Resonant methods generally include the resonator method and the resonant-perturbation method. The resonator method is based on the fact that the resonant frequency and quality factor of a dielectric resonator with given dimensions are determined by its permittivity and permeability. This method is usually used to measure low-loss dielectrics whose permeability is μ_0. The resonant-perturbation method is based on resonant-perturbation theory. For a resonator with given electromagnetic boundaries, when part of the electromagnetic boundary condition is changed by introducing a sample, its resonant frequency and quality factor will also be changed. From the changes of the resonant frequency and quality factor, the properties of the sample can be derived.

2.1.2.1 Resonator method

This method is often called *dielectric resonator method*. It can be used to measure the permittivity of dielectric materials and the surface resistance of conducting materials. More detailed discussions on the resonator method can be found in Chapter 5.

Measurement of dielectric permittivity

In a resonator method for dielectric property measurement, the dielectric sample under measurement serves as a resonator in the measurement circuit, and dielectric constant and loss tangent of the sample are determined from its resonant frequency and quality factor (Kobayashi and Tanaka 1980). Figure 2.7 shows the configuration often used in the dielectric resonator method. In this configuration, the sample is sandwiched between two conducting plates, and the resonant properties of this configuration are mainly determined by the properties of the dielectric cylinder and the two pieces of the conducting plates. In the measurement of the dielectric properties of the dielectric cylinder, we assume the properties of the conducting plates are known. The TE_{011} mode is often selected as the working mode, as this mode does

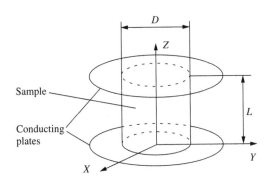

Figure 2.7 A dielectric cylinder sandwiched between two conducting plates

not have a transverse electric field between the sample and the conducting plates. Therefore, a small gap between the sample and the plates does not greatly affect the measurement results. This method can be used to measure high dielectric constant (Cohn and Kelly 1966), low loss (Krupka *et al.* 1994), and anisotropic materials (Geyer and Krupka 1995).

Surface resistance measurement

The configuration shown in Figure 2.7 can also be used for the measurement of the surface resistance of conductors. If the dielectric properties of the dielectric cylinder are known, from the quality factor of the whole resonant structure, we can calculate the surface resistance of the conducting plates.

2.1.2.2 Resonant-perturbation method

When a sample is introduced into a resonator, the resonant frequency and quality factor of the resonator will be changed, and the electromagnetic properties of the sample can be derived from the changes of the resonant frequency and quality factor of the resonator. Generally speaking, there are three types of resonant perturbations: cavity shape perturbation, wall-loss perturbation, and material perturbation. Cavity shape perturbation is often used to adjust the resonant frequency of a cavity. In the wall-loss perturbation method, part of the cavity wall is replaced by the sample under study, and the resonant frequency and quality

factor of the cavity are changed subsequently. The wall-loss perturbation method is usually used to measure the surface resistance of conductors. In the material perturbation method, the introduction of the material into a cavity causes changes in the resonant frequency and quality factor of the cavity. The material perturbation method is also called the *cavity-perturbation method*, and is suitable for measuring low-loss materials. More detailed discussions on resonant-perturbation methods can be found in Chapter 6.

Permittivity and permeability measurement

In the cavity perturbation method, the sample under study is introduced into an antinode of the electric field or magnetic field, depending on whether permittivity or permeability is being measured. As shown in Figure 2.8, if the sample under study is introduced into place *A* with maximum dielectric field and minimum magnetic field, the dielectric properties of the sample can be characterized; if the sample is inserted into place B with maximum electric field and minimum magnetic field, the magnetic properties of the sample can be characterized.

Surface resistance measurement

As shown in Figure 2.9, in this method, the end wall of a hollow metallic cavity is replaced by the

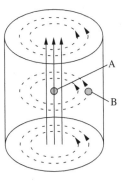

Figure 2.8 Cylindrical cavity (TM$_{010}$ mode) for measurement of materials properties using resonant-perturbation method. Position A is for permittivity measurement and position B is for permeability measurement

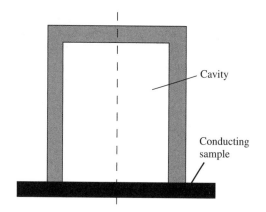

Figure 2.9 Cavity perturbation method for the measurement of surface resistance of conductors

conducting sample under test. With the knowledge of the geometry and the resonant mode of the cavity, the surface resistance of the conducting sample can be derived from the change of the quality factor due to the replacement of the end wall.

For better understanding and application of various methods for materials property characterization, in the following two sections, we discuss two important microwave phenomena: microwave propagation, which is the basis for nonresonant methods, and microwave resonance, which is the basis for resonant methods. In the last section of this chapter, we discuss microwave network, which is often used in microwave theoretical analysis and experimental measurement.

Field approach and line approach are often used in microwave theory and engineering. In the field approach, we analyze the distributions of the electric and magnetic fields. In the line approach, we use equivalent circuits to represent microwave structures. In the following discussions, both the field and the line approaches are used.

2.2 MICROWAVE PROPAGATION

As discussed above, there are two general types of methods for materials property characterization: nonresonant methods and resonant methods, and the microwave phenomena related to these two

types of methods are microwave propagation and microwave resonance. In this section, we discuss microwave propagation, and microwave resonance will be discussed in Section 2.3.

In this section, we start with transmission-line theory, and then we introduce transmission Smith Charts, which are powerful tools in microwave theory and engineering. Afterward, we discuss three categories of transmission lines widely used in materials property characterization: guided transmission lines, surface-wave transmission lines, and free space.

2.2.1 Transmission-line theory

In this part, line approach will be used to analyze transmission structures. We use equivalent circuits to represent transmission structures, and discuss the propagation of equivalent voltage and current along transmission structures.

2.2.1.1 General properties of transmission structures

We consider a general cylindrical metallic transmission structure whose cross sections do not change in z-direction. As shown in Figure 2.10, there are two types of metallic transmission lines, one is a single hollow metallic tube and the other consists of two or more conductors.

The propagation of an electromagnetic wave along a transmission structure can be analyzed using Maxwell's equations. From Eqs. (1.1)–(1.4), we can get the wave equations for electric field E

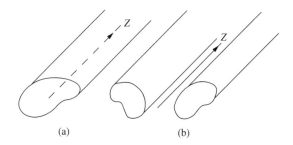

Figure 2.10 Two types of cylindrical metallic transmission lines. (a) Hollow tube and (b) two metal bars

and magnetic field H:

$$\nabla^2 E + k^2 E = 0 \qquad (2.1)$$

$$\nabla^2 H + k^2 H = 0, \qquad (2.2)$$

where $k = 2\pi/\lambda$ is the wave number and λ is the wavelength.

In the transmission structure, the electromagnetic fields can be decomposed into the transverse components (E_T and H_T) and the axial components (E_z and H_z):

$$E = E_T + E_z \qquad (2.3)$$

$$H = H_T + H_z \qquad (2.4)$$

We can also revolve the ∇ operator into the transverse part ∇_T and axial part ∇_Z:

$$\nabla = \nabla_T + \nabla_Z \qquad (2.5)$$

with

$$\nabla_Z = \hat{z}\frac{\partial}{\partial z} \qquad (2.6)$$

$$\nabla_T = \hat{x}\frac{\partial}{\partial x} + \hat{y}\frac{\partial}{\partial y} \qquad \text{(rectangular coordinates)} \qquad (2.7)$$

$$\nabla_T = \hat{r}\frac{\partial}{\partial r} + \hat{\phi}\frac{1}{r}\frac{\partial}{\partial \phi} \qquad \text{(cylindrical coordinates)} \qquad (2.8)$$

where $\hat{z}$, $\hat{x}$, $\hat{y}$, $\hat{r}$, and $\hat{\phi}$ are the unit vectors along their corresponding axes.

From Eqs. (2.1)–(2.8), we can get the propagation equations for metallic waveguides

$$(k^2 - k_z^2)H_T = -j\omega\varepsilon\hat{z} \times \nabla_T E_z - jk_z \nabla_T H_z \qquad (2.9)$$

$$(k^2 - k_z^2)E_T = -j\omega\mu\hat{z} \times \nabla_T H_z - jk_z \nabla_T E_z, \qquad (2.10)$$

where k_z is the wave number in Z-direction. Equations (2.9) and (2.10) show the relationships between the transverse and axial fields. If E_z and H_z are known, other components of electromagnetic waves can be calculated based on E_z and H_z.

There are three types of electromagnetic waves with special E_z and H_z. If $E_z = 0$, the electromagnetic wave is called *transverse electric (TE) wave*,

and such a wave is also called H wave. If $H_z = 0$, the electromagnetic wave is called *transverse magnetic (TM) wave*, and such a wave is also called E wave. If $E_z = 0$ and $H_z = 0$, the electromagnetic wave is called *transverse electromagnetic (TEM) wave*. In the following, we discuss the general properties of TE, TM, and TEM waves.

TE wave

For TE wave ($E_z = 0$), from Eqs. (2.9) and (2.10), we have

$$H_T = -\frac{jk_z}{k^2 - k_z^2}\nabla_T H_z \qquad (2.11)$$

$$E_T = \frac{j\omega\mu}{k^2 - k_z^2}\hat{z} \times \nabla_T H_z = -\frac{\omega\mu}{k_z}\hat{z} \times H_T \qquad (2.12)$$

Equation (2.12) indicates that E_T, H_T, and $\hat{z}$ are perpendicular to each other. If we define the wave impedance of TE wave as

$$\eta_{TE} = \frac{\omega\mu}{k_z}, \qquad (2.13)$$

Eq. (2.12) can be rewritten as

$$E_T = -\eta_{TE}\hat{z} \times H_T \qquad (2.14)$$

The wave impedance for TE wave can be rewritten as

$$\eta_{TE} = \frac{\omega\sqrt{\varepsilon\mu}}{k_z}\sqrt{\frac{\mu}{\varepsilon}} = \frac{k}{k_z}\eta \qquad (2.15)$$

with the wave impedance of plane wave

$$\eta = \sqrt{\frac{\mu}{\varepsilon}} \qquad (2.16)$$

and the wave number

$$k = \omega\sqrt{\varepsilon\mu} \qquad (2.17)$$

TM wave

For TM wave ($H_z = 0$), from Eqs. (2.9) and (2.10), we have

$$E_T = -\frac{jk_z}{k^2 - k_z^2}\nabla_T E_z \qquad (2.18)$$

$$H_T = -\frac{j\omega\mu}{k^2 - k_z^2}\hat{z} \times \nabla_T E_z = \frac{\omega\varepsilon}{k_z}\hat{z} \times E_T \qquad (2.19)$$

Equation (2.19) also indicates that E_T, H_T, and $\hat{z}$ are perpendicular to each other. Also, we define the wave impedance of TM wave:

$$\eta_{TM} = \frac{k_z}{\omega \varepsilon} \tag{2.20}$$

Similarly, the wave impedance for TM wave can be rewritten as

$$\eta_{TM} = \frac{k_z}{k} \eta \tag{2.21}$$

TEM wave

For TEM wave ($E_z = H_z = 0$), from Eqs. (2.9) and (2.10), we have

$$(k^2 - k_z^2)H_T = 0 \tag{2.22}$$

$$(k^2 - k_z^2)E_T = 0 \tag{2.23}$$

As $E_T \neq 0$ and $H_T \neq 0$, from Eqs. (2.22) and (2.23), we have

$$k = k_z \tag{2.24}$$

Equation (2.24) indicates that the wave number of TEM wave propagating in the z-direction is equal to the wave number of that of a plane wave. Also, the wave impedance for TEM wave is equal to that for plane wave:

$$\eta_{TEM} = \eta. \tag{2.25}$$

Also we have

$$E_T = -\eta_{TEM}\hat{z} \times H_T \tag{2.26}$$

Equation (2.26) shows that E_T, H_T, and $\hat{z}$ are perpendicular to each other.

Meanwhile, it can be proven that (Pozar 1998), for TEM wave, its field E_T at a cross section can be expressed by a scalar potential Φ:

$$E_T = -\nabla_T \Phi \tag{2.27}$$

We can further get that for a source-free transmission line

$$\nabla_T^2 \Phi = 0 \tag{2.28}$$

Equation (2.28) indicates that if a TEM wave can propagate in a transmission line, the static field can also be established there and vice versa.

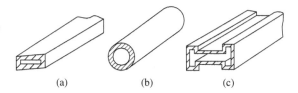

Figure 2.11   Transmission structures for TE and TM waves. (a) Rectangular waveguide, (b) circular waveguide, and (c) ridged waveguide

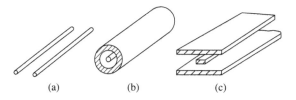

Figure 2.12   Transmission structures for TEM waves. (a) Two parallel lines, (b) coaxial line, and (c) stripline

Static electric field cannot be built within a hollow tube, and TEM wave also cannot propagate in a hollow tube. But TE wave and TM wave could propagate in hollow tubes, and some hollow tube examples are shown in Figure 2.11. TEM waves can only propagate in a transmission structure with at least two conductors, and some examples of TEM transmission lines are shown in Figure 2.12.

2.2.1.2 Propagation equations

The above discussion indicates that, in the cross section of a TEM transmission line, the distribution of the electromagnetic field is the same as that of the static field, so we can introduce the concepts of equivalent voltage, equivalent current, and equivalent impedance. Therefore, we can use the "line" method to analyze the TEM transmission line. However, to ensure that this method is applicable, the geometrical length l of the transmission line should be comparable to the wavelength λ, which means $l/\lambda \geq 1$, and the current and voltage distribution along the transmission line are not uniform. In engineering, such a transmission line is called *long line*.

It should be indicated that in the following discussion, we use TEM transmission line. Actually, TE transmission line and TM transmission line can

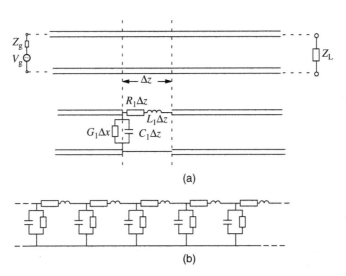

Figure 2.13   Equivalent of a transmission line. (a) Equivalent circuit for a short line element and (b) equivalent circuit for a long line. In the figure, Z_g and V_g are the impedance and driving voltage of the generator, and Z_L is the impedance of the load

also be analyzed in this way, provided suitable equivalent current and equivalent voltage are introduced.

Equivalent circuit of transmission line

Owing to the distribution properties of a transmission line, the voltage and current on a long line are functions of both time and positions. The distributions of the voltage and current are mainly determined by the shape, dimension, and the properties of the conductors and dielectrics. In following discussion, we assume that the cross section of the transmission structure does not change along with its axis.

We divide a long line into many short line elements with length Δz ($\Delta z \ll \lambda$), and represent the short line elements with effective parameters: $R_1 \Delta z$, $L_1 \Delta z$, $G_1 \Delta z$, and $C_1 \Delta z$, with R_1, L_1, G_1 and C_1 representing the resistance, inductance, conductance and capacitance of the line element respectively. The transmission line can therefore be represented by an equivalent circuit, as shown in Figure 2.13.

Telegrapher equations

Consider a short line element with length Δz and starting point at z, as shown in Figure 2.14. Let the

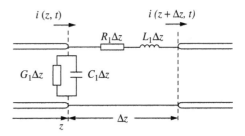

Figure 2.14   A short transmission line with length Δz at position z

voltage and current at the two ends of the element be: $v(z, t)$, $i(z, t)$ and $v(z + \Delta z, t)$, $i(z + \Delta z, t)$. According to the Kirchhoff's law, we have

$$-\Delta v(z, t) = v(z + \Delta z, t) - v(z, t)$$

$$= R_1 \Delta z \cdot i(z, t) + L_1 \Delta z \frac{\partial i(z, t)}{\partial t} \quad (2.29)$$

$$-\Delta i(z, t) = i(z + \Delta z, t) - i(z, t)$$

$$= G_1 \Delta z \cdot v(z, t) + C_1 \Delta z \frac{\partial v(z, t)}{\partial t} \quad (2.30)$$

Letting $\Delta z \to 0$, we can get the famous Telegrapher equations

$$-\frac{\partial v(z, t)}{\partial z} = R_1 i(z, t) + L_1 \frac{\partial i(z, t)}{\partial t} \quad (2.31)$$

$$-\frac{\partial i(z,t)}{\partial z} = G_1 v(z,t) + C_1 \frac{\partial v(z,t)}{\partial t} \quad (2.32)$$

The above Telegrapher equations can be rewritten as

$$-\frac{dV(z)}{dz} = Z_1 I(z) \quad (2.33)$$

$$-\frac{dI(z)}{dz} = Y_1 V(z) \quad (2.34)$$

with

$$v(z,t) = V(z)e^{j\omega t} \quad (2.35)$$

$$i(z,t) = I(z)e^{j\omega t} \quad (2.36)$$

$$Z_1 = R_1 + j\omega L_1 \quad (2.37)$$

$$Y_1 = G_1 + j\omega C_1. \quad (2.38)$$

Equations (2.33) and (2.34) indicate that, in the transmission line, the change of voltage is due to the series impedance Z_1 and the change of current is due to the parallel admittance Y_1.

Characteristic parameters of a transmission line

Following propagation equations can be obtained from Eqs. (2.33) and (2.34):

$$\frac{d^2 V}{dz^2} = \gamma^2 V \quad (2.39)$$

$$\frac{d^2 I}{dz^2} = \gamma^2 I \quad (2.40)$$

with the transmission constant given by

$$\gamma = \sqrt{Y_1 Z_1} = \alpha + j\beta \quad (2.41)$$

The general solutions for Eqs. (2.39) and (2.40) are

$$V = V_{0+}e^{-\gamma z} + V_{0-}e^{\gamma z} \quad (2.42)$$

$$I = I_{0+}e^{-\gamma z} + I_{0-}e^{\gamma z}, \quad (2.43)$$

where $V_{0\pm}$ and $I_{0\pm}$ are constants that should be defined by the boundary conditions. Equations (2.42) and (2.43) indicate that there may exist waves propagating along the $+z$ direction and the $-z$ direction, respectively. The phase velocity is given by

$$v_p = \frac{\omega}{\beta} \quad (2.44)$$

From Eqs. (2.33) and (2.37), we can get

$$I(z) = \frac{1}{Z_c} V_{0+}e^{-\gamma z} + \left(-\frac{1}{Z_c}\right) V_{0-}e^{\gamma z}, \quad (2.45)$$

where Z_c is the characteristic impedance of the transmission line:

$$Z_c = \frac{Z_1}{\gamma} = \sqrt{\frac{Z_1}{Y_1}} = \frac{1}{Y_c}, \quad (2.46)$$

where Y_c is the characteristic admittance of the transmission line.

By rewriting Eq. (2.45) as

$$I(z) = I_{0+}e^{-\gamma z} + I_{0-}e^{\gamma z} = \frac{V_{0+}}{Z_c}e^{-\gamma z} + \frac{V_{0-}}{-Z_c}e^{\gamma z} \quad (2.47)$$

we can get

$$\frac{V_{0+}}{I_{0+}} = Z_c \quad (2.48)$$

$$\frac{V_{0-}}{I_{0-}} = -Z_c \quad (2.49)$$

The negative sign in Eq. (2.49) is totally due to the definitions of voltage, current, and the direction of z, and it does not indicate negative impedance.

In summary, Eqs. (2.42) and (2.45) are the solutions of the Telegrapher Equations, and they indicate that there are voltage waves and current waves propagating in $+Z$ direction and $-Z$ direction. As the propagation constant γ is a complex number, these waves attenuate along the transmission line. From Eq. (2.46), we have

$$\gamma = \alpha + j\beta = \sqrt{Y_1 Z_1}$$
$$= \sqrt{(R_1 + j\omega L_1)(G_1 + j\omega C_1)} \quad (2.50)$$

with

$$\alpha = \sqrt{\frac{1}{2}\left(\sqrt{(R_1^2 + \omega^2 L_1^2)(G_1^2 + \omega^2 C_1^2)} - (\omega^2 L_1 C_1 - R_1 G_1)\right)} \quad (2.51)$$

$$\beta = \sqrt{\frac{1}{2}\left(\sqrt{(R_1^2 + \omega^2 L_1^2)(G_1^2 + \omega^2 C_1^2)} + (\omega^2 L_1 C_1 - R_1 G_1)\right)} \quad (2.52)$$

According to Eq. (2.46), the characteristic impedance of the transmission line is given by

$$Z_c = \frac{\sqrt{R_1 + j\omega L_1}}{\sqrt{G_1 + j\omega C_1}} = \sqrt{\frac{L_1}{C_1}} \sqrt{\frac{1 - j(R_1/\omega L_1)}{1 - j(G_1/\omega C_1)}} \quad (2.53)$$

In the following, we discuss three typical conditions: high loss, low loss, and no loss. In microwave electronics, most of the actual conditions can be approximated as these conditions.

High-loss transmission line

For a transmission line with high loss, we assume

$$\omega L_1 \ll R_1 \quad (2.54)$$

$$\omega C_1 \ll G_1 \quad (2.55)$$

From Eqs. (2.50)–(2.53), we can get

$$\alpha \approx \sqrt{R_1 G_1} \quad (2.56)$$

$$\beta \approx 0 \quad (2.57)$$

$$Z_c \approx \sqrt{\frac{R_1}{G_1}} \quad (2.58)$$

So the electromagnetic waves attenuate quickly, and cannot propagate in such a transmission line. Matched load can be taken as such a kind of transmission line.

Low-loss transmission line

For a low-loss transmission line, we assume

$$\omega L_1 \gg R_1 \quad (2.59)$$

$$\omega C_1 \gg G_1 \quad (2.60)$$

From Eqs. (2.50)–(2.53), we can get

$$\alpha \approx \frac{1}{2}\left(R_1\sqrt{\frac{C_1}{L_1}} + G_1\sqrt{\frac{L_1}{C_1}}\right) \quad (2.61)$$

$$\beta \approx \omega\sqrt{L_1 C_1} \quad (2.62)$$

Equation (2.62) indicates that β is almost independent of R_1 and G_1. The characteristic impedance

Z_c becomes

$$Z_c = \sqrt{\frac{L_1}{C_1}}\left(1 - \frac{j}{2}\left(\frac{R_1}{\omega L_1} - \frac{G_1}{\omega C_1}\right)\right) \approx \sqrt{\frac{L_1}{C_1}} \quad (2.63)$$

No-loss transmission line

This is an ideal case. At microwave frequencies, no actual transmission lines can be strictly no loss. However, if the transmission line is made of good conductors and low-loss dielectric, and the transmission line is not very long, we can neglect its loss. For a no-loss transmission line,

$$R_1 = 0, \quad (2.64)$$

$$G_1 = 0. \quad (2.65)$$

From Eqs. (2.50)–(2.53), we can get

$$\alpha = 0 \quad (2.66)$$

$$\beta = \omega\sqrt{L_1 C_1} \quad (2.67)$$

$$Z_c = \sqrt{\frac{L_1}{C_1}} \quad (2.68)$$

In this condition, Eqs. (2.42) and (2.43) become

$$V(z) = V_{0+}e^{-j\beta z} + V_{0-}e^{j\beta z} \quad (2.69)$$

$$I(z) = I_{0+}e^{-\beta z} + I_{0-}e^{j\beta z}$$
$$= \frac{V_{0+}}{Z_c}e^{-j\beta z} + \frac{V_{0-}}{-Z_c}e^{j\beta z} \quad (2.70)$$

Equations (2.69) and (2.70) indicate that both the voltage and current waves can propagate along $+Z$ direction and $-Z$ direction. As the characteristic impedance is a real number, the current and voltage are in phase. This is the most common situation in microwave engineering, and is usually called *lossless transmission line* or *ideal transmission line*.

2.2.1.3 Reflection and impedance

As discussed above, in a uniform transmission line, both the voltage wave and current wave have two components propagating along $+z$ direction

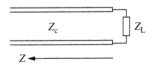

Figure 2.15   A transmission line connected to a load with impedance Z_L

and $-z$ direction, and these two components can be called *incident wave* and *reflection wave*. In nonresonant methods for materials property characterization, the sample under study is loaded to a transmission line. We will discuss how the load to a transmission line affects the relationship between the incident wave and reflected wave.

Voltage reflection coefficient

As shown in Figure 2.15, a load with impedance Z_L is connected to a piece of transmission line with length l. In analyzing the reflection properties, the origin of the axis is chosen at the place of load, and the positive direction of the axis is from the load to the generator, while the positive direction of current is still from the generator to the load. The relationship between the voltage and the current is determined by the loading impedance:

$$Z_L = \frac{V_L}{I_L} \tag{2.71}$$

The voltage reflection coefficient represents the voltage ratio between the reflected voltage V_- and incident voltage V_+:

$$\Gamma = \frac{V_-}{V_+} = \frac{V_{0-}e^{-j\beta z}}{V_{0+}e^{+j\beta z}} = \frac{V_{0-}}{V_{0+}}e^{-j2\beta z} \tag{2.72}$$

Equation (2.72) indicates that the reflection coefficient is related to the position along the z-axis. As the origin of the axis is chosen at the position of the load, the reflection at the load is: $\Gamma_L = V_{0-}/V_{0+}$. The reflection coefficient at a position (z) is

$$\Gamma = \Gamma_L e^{-j2k_z z} \tag{2.73}$$

It is clear that the amplitude of the reflection coefficient does not change along a uniform transmission line.

Input impedance

On the basis of the definition of reflection coefficient (Γ), the total voltage $V(z)$ and total current $I(z)$ along a transmission line can be expressed as

$$V(z) = V_+(1 + \Gamma) \tag{2.74}$$

$$I(z) = I_+(1 - \Gamma) \tag{2.75}$$

The relationships between the total voltage $V(z)$ and total current $I(z)$ are described by the input impedance Z_i and the input admittance Y_i:

$$Z_i = \frac{1}{Y_i} = \frac{V}{I} = \frac{V_+(1 + \Gamma)}{I_+(1 - \Gamma)} \tag{2.76}$$

Sometimes, we use normalized impedance:

$$\overline{z} = \frac{Z_i}{Z_c}, \tag{2.77}$$

where Z_c is the characteristic impedance of the transmission line, and normalized admittance:

$$\overline{y} = \frac{Y_i}{Y_c}, \tag{2.78}$$

where Y_c is the characteristic admittance of the transmission line. The relationships between the reflection, impedance, and admittance are listed in Table 2.1.

The reflection coefficient Γ is periodical with period $\lambda_g/2$, and so are the impedance and

Table 2.1   Relationships between reflection coefficient and impedance and admittance

	Unnormalized	Normalized
Relationship between impedance and reflection	$Z_i = Z_c\dfrac{1 + \Gamma}{1 - \Gamma}$	$\overline{z} = \dfrac{1 + \Gamma}{1 - \Gamma}$
	$\Gamma = \dfrac{Z_i - Z_c}{Z_i + Z_c}$	$\Gamma = \dfrac{\overline{z} - 1}{\overline{z} + 1}$
Relationship between admittance and reflection	$Y_i = Y_c\dfrac{1 - \Gamma}{1 + \Gamma}$	$\overline{y} = \dfrac{1 - \Gamma}{1 + \Gamma}$
	$\Gamma = \dfrac{Y_c - Y_i}{Y_c + Y_i}$	$\Gamma = \dfrac{1 - \overline{y}}{1 + \overline{y}}$

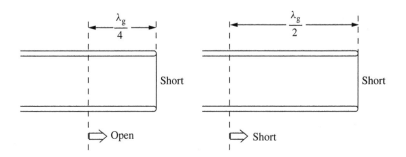

Figure 2.16 Impedance inversion along a transmission line. λ_g is the wavelength along the transmission line

admittance. We can find the inversion property of the impedance:

$$\overline{z}(z) = \frac{1}{\overline{z}(z + \lambda_g/4)}, \tag{2.79}$$

where $\overline{z}(z)$ is the normalized impedance at position z, and $\overline{z}(z + \lambda_g/4)$ is the normalized impedance at position $(z + \lambda_g/4)$. As shown in Figure 2.16, if a transmission line is loaded with a short, it is open, seen at the position $\lambda_g/4$ away from the short, and is short again seen at the position $\lambda_g/2$ away from the short.

Other expressions for input impedance

If we write the total voltage and total current in following forms

$$V(z) = (V_{0+} + V_{0-})\cos k_z z + j(V_{0+} - V_{0-})\sin k_z z, \tag{2.80}$$

$$I(z) = \frac{V_{0+} - V_{0-}}{Z_c}\cos k_z z + j\frac{V_{0+} + V_{0-}}{Z_c}\sin k_z z, \tag{2.81}$$

the input impedance is given by

$$Z_i(z) = \frac{(V_{0+} + V_{0-})\cos k_z z + j(V_{0+} - V_{0-})\sin k_z z}{[(V_{0+} - V_{0-})/Z_c]\cos k_z z + j[(V_{0+} + V_{0-})/Z_c]\sin k_z z} \tag{2.82}$$

As the input impedance at $z = 0$ is the impedance of the load Z_L, we have

$$Z_L = Z_c\frac{V_{0+} + V_{0-}}{V_{0+} - V_{0-}} \tag{2.83}$$

From Eqs. (2.82) and (2.83), we can get

$$Z_i = Z_c\frac{Z_L + jZ_c\tan k_z z}{Z_c + jZ_L\tan k_z z} \tag{2.84}$$

$$\overline{z}_i = \frac{\overline{z}_L + j\tan k_z z}{1 + j\overline{z}_L\tan k_z z} \tag{2.85}$$

$$Y_i = Y_c\frac{Y_L + jY_c\tan k_z z}{Y_c + jY_L\tan k_z z} \tag{2.86}$$

$$\overline{y}_i = \frac{\overline{y}_L + j\tan k_z z}{1 + j\overline{y}_L\tan k_z z} \tag{2.87}$$

The above equations indicate that if the load is purely reactive, the input impedance at any position is also purely reactive. We consider two special cases. For a short loading ($\overline{z}_L = 0$), we have

$$\overline{z}_i = j\tan k_z z \tag{2.88}$$

For an open loading ($\overline{z}_L = j\infty$), we have

$$\overline{z}_i = -j\frac{1}{\tan k_z z} \tag{2.89}$$

2.2.1.4 Typical working states of transmission lines

The distributions of current and voltage are determined by the properties of the loading. Here, we discuss three typical working states of transmission lines: pure travelling wave, pure standing wave, and mixed wave.

Pure travelling wave

In this state, there is no reflection wave: $\Gamma = 0$, $V(z) = V_+$, $Z_i = Z_c$, and $\overline{z}_i = 1$. The total voltage is the incident voltage, and the input impedance at any position in the transmission line equals the characteristic impedance of the transmission line. The ratio between the voltage and current equals to Z_c, and the current and voltage are in phase.

As the input impedance at any cross section at the transmission line equals to Z_c, the load impedance is also Z_c, and such a load is called a *matching load*. In microwave electronics, the matching loads for coaxial line and waveguide are not an actual resistance, but a kind of material or structure which can absorb all the energy from the generator.

Pure standing wave

In this state, the load does not absorb any energy, all the energy is reflected: $|\Gamma| = 1$. When the load is short ($Z_L = 0$), open ($Z_L = \infty$), or purely reactive ($Z_L = jX$), the transmission line is in pure standing-wave state.

Figure 2.17 shows the standing wave in a transmission line terminated by a short. In this state, $Z_L = 0$ and $\Gamma_L = -1$. According to Eqs. (2.80) and (2.81), we have

$$V(z) = j2V_{0+} \sin\left(\frac{2\pi}{\lambda}z\right) \quad (2.90)$$

$$I(z) = \frac{2V_{0+}}{Z_c} \cos\left(\frac{2\pi}{\lambda}z\right) \quad (2.91)$$

As shown in Figure 2.17(b), the phase difference between the voltage and current is 90°. Figure 2.17(c) shows the distributions of the amplitudes of the voltage and current along the transmission line.

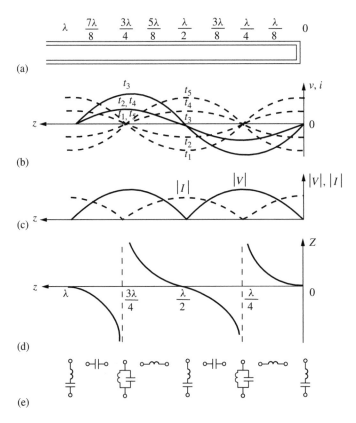

Figure 2.17 Standing wave of a shorted transmission line. (a) Short terminated transmission line, (b) instant distributions of voltage (full line) and current (dashed line), (c) amplitudes of voltage and current, (d) the impedance distribution along the transmission line, and (e) lumped equivalent circuits at typical positions

Equation (2.88) indicates that the impedance of the transmission line shorted at the end is sure reactive. As shown in Figure 2.17(d), the impedance periodically appears as inductive reactance and capacitive reactance along z-direction with period $\lambda/2$. Figure 2.17(e) shows the equivalent lumped-element circuits at several typical positions. A piece of shorted transmission line with length less than $\lambda/4$ is equivalent to a lumped inductance, and a piece of shorted transmission line with length equal to $\lambda/4$ is equivalent to a lumped parallel LC resonant circuit. A piece of shorted transmission line with length more than $\lambda/4$ but less than $\lambda/2$ is equivalent to a lumped capacitor, and a piece of shorted transmission line with length equal to $\lambda/2$ is equivalent to a lumped series LC resonant circuit.

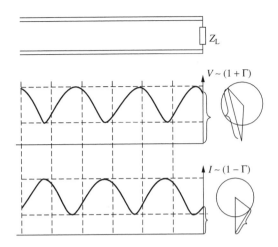

Figure 2.19 The voltage and current distributions along a loaded transmission line at the mixed state

Mixed wave

In most of the transmission lines, some of the energy is absorbed by the load and some of the energy is reflected. This state is a mixture of pure travelling wave and pure standing wave. The current and voltage along the line are

$$V = V_+(1 + \Gamma_L e^{-j2k_z z}) \tag{2.92}$$

$$I = \frac{V_+}{Z_c}(1 - \Gamma_L e^{-j2k_z z}), \tag{2.93}$$

where Γ_L is the reflection coefficient at the loading, and $\Gamma = \Gamma_L \exp(-j2k_z z)$ is the reflection coefficient at position z. As Γ_L is a constant, the voltage and current can be drawn using vector rotation method, as shown in Figure 2.18.

Figure 2.19 indicates that the distributions of voltage and current along a transmission line in mixed state are periodical with period $\lambda/2$. The voltage V is in the range of $V_+(1 + |\Gamma|)$ to $V_+(1 - |\Gamma|)$, while current is in the range of $I_+(1 + |\Gamma|)$ to $I_+(1 - |\Gamma|)$, with $I_+ = V_+/Z_c$.

2.2.2 Transmission Smith charts

In microwave engineering, it is often required to transform the impedance to different positions at the transmission line. Such transformation is usually complicated (Eqs. (2.83)–(2.87)), while the calculation of reflection coefficient is relatively simple (Eqs. (2.72) and (2.73)). As there are corresponding relationships between the impedance and reflection coefficient, it is helpful to draw charts to represent the corresponding relationships, so that the transformation between impedance and reflection can be easily conducted.

Actually, there are two types of charts, square chart and circle chart, for this purpose. In a square chart, the magnitude and phase of the reflection coefficient are plotted in rectangular normalized impedance coordinates, while in a circle chart, the real part and imaginary part of the normalized impedance are plotted on polar coordinates of the reflection coefficient. The equal radius curve represents the magnitude of reflection, while equal angle line represents the phase of reflection. As

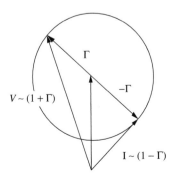

Figure 2.18 Vector forms of the voltage and current

the reflection coefficient cannot be larger than unity, the reflection coefficient is always within the unit circle of the polar coordinate. In fact, these two kinds of charts indicate the transformation relationships between the impedance complex plane and reflection complex plane. Usually, the second chart is more often used and is often called *Smith chart*.

Smith chart provides great convenience for transmission-line calculation. In the following discussion, we concentrate on Smith chart. There are two kinds of Smith charts: impedance Smith chart and admittance Smith chart. We will discuss the impedance Smith chart first, and the admittance Smith chart can then be transformed from the impedance Smith chart.

2.2.2.1 Impedance Smith chart

The impedance Smith chart is formed by plotting the impedance on the reflection polar plane. Figure 2.20 shows the Smith chart and square chart in the impedance plane. As $r \geq 0$ and $\Gamma \leq 1$, the transformation between the square chart and Smith chart is a transformation between the right half plane in the impedance plane and the unit circle area in the reflection plane.

The relationship between the reflection coefficient Γ and the normalized impedance is:

$$\bar{z} = \frac{1 + \Gamma}{1 - \Gamma} \qquad (2.94)$$

As $\bar{z} = r + jx$ and $\Gamma = u + jv$, where $r = R/Z_c$ is the normalized resistance and $x = X/Z_c$ is the normalized reactance, Eq. (2.94) can be rewritten as

$$r + jx = \frac{1 + (u + jv)}{1 - (u + jv)} = \frac{1 - u^2 - v^2 + j2v}{(1 - u)^2 + v^2} \qquad (2.95)$$

As the real parts and the imaginary parts at the two sides of Eq. (2.95) should be equal, respectively, we have

$$r = \frac{1 - (u^2 - v^2)}{(1 - u)^2 + v^2} \qquad (2.96)$$

$$x = \frac{2v}{(1 - u)^2 + v^2} \qquad (2.97)$$

The impedance Smith chart is based on Eqs. (2.96) and (2.97). In the following, we discuss several sets of special lines in the impedance Smith chart, including r circles, x circles, ρ circles, and θ lines.

r circles

We rewrite Eq. (2.96) into the following form:

$$\left(u - \frac{r}{1 + r}\right)^2 + v^2 = \left(\frac{1}{1 + r}\right)^2 \qquad (2.98)$$

Equation (2.98) represents a circle in the reflection plane with center $(r/(r + 1), 0)$ and radius

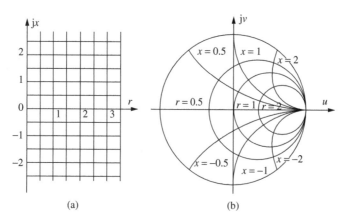

(a)   (b)

Figure 2.20   Normalized impedance charts. (a) Chart on impedance plane and (b) Smith chart (chart on reflection plane)

$1/(r + 1)$. Different r values correspond to different circles which form a series of tangent circles with the tangent point $(1, 0)$. When $r = 0$, the circle center is at the original point and the radius is 1, representing a pure reactive state. When $r \to \infty$, the center is at $(1, 0)$ and the radius becomes zero, so the circle becomes a point $(1, 0)$. Figure 2.21(a) shows a series of r circles.

x circles

Equation (2.97) can be modified into the following form:

$$(u - 1)^2 + \left(v - \frac{1}{x}\right)^2 = \left(\frac{1}{x}\right)^2 \qquad (2.99)$$

Equation (2.99) represents a circle in the reflection plane with center $(1, 1/x)$ and radius $1/x$. Different x values correspond to different circles which form a series of tangent circles with the tangent point $(1, 0)$. The center for the circle corresponding to $x = 0$ is at $(1, \infty)$, and its radius is ∞. The real axis of the impedance plane represents a pure resistive state $(x = 0)$. A circle with $x > 0$ is on the upper half, representing an inductive reactance state. A circle with $x < 0$ is at the lower half, representing a capacitive reactance state. For the cases $x = \pm\infty$, the center of the circle is at $(1, 0)$ and the radius is 0, so the circles become a point $(1, 0)$. Figure 2.21(b) shows a series of x circles.

ρ circles

The standing-wave coefficient ρ can be calculated from the reflection coefficient Γ:

$$\rho = \frac{1 + |\Gamma|}{1 - |\Gamma|} \qquad (2.100)$$

The points with the same standing-wave coefficient ρ form a circle whose center is the origin of the reflection plane. Different ρ values represent a series of circles having the same center. When $r > 1$, $\rho = r$ and when $r < 1$, $\rho = (1/r)$. A series of ρ circles are shown in Figure 2.22(c).

θ lines

According to the expression: $\Gamma = \Gamma' + j\Gamma'' = |\Gamma|e^{j\theta}$, it is clear that $\theta = \tan^{-1}(\Gamma''/\Gamma')$. As shown in Figure 2.22(c), the θ lines are a series of straight lines passing through the center. Usually, θ lines are not shown in Smith chart, while the corresponding θ values are labeled at the outermost circle. Sometimes the electric length is labeled instead of θ values.

By combing the circles and lines in Figure 2.22, we can get a full Smith chart. As shown in Figure 2.22, to make the chart clear and simple, ρ circles and θ lines are usually not shown.

Smith chart is often used to find the impedance of a point at a transmission from the impedance of another point at the transmission line. For a lossless transmission line, the module of the reflection coefficient does not change, and only the phase angle changes. So the normalized impedance points at different positions on the transmission line are on a ρ circle. When we know the impedance at one point, the impedance at another point can be obtained by rotating the point along

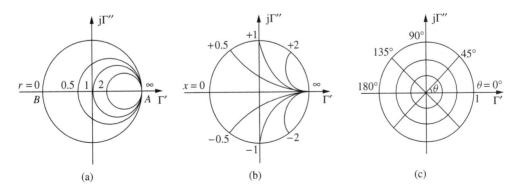

Figure 2.21 Typical circles and lines at the Γ plane. (a) r circles, (b) x circles, (c) ρ circles and θ lines

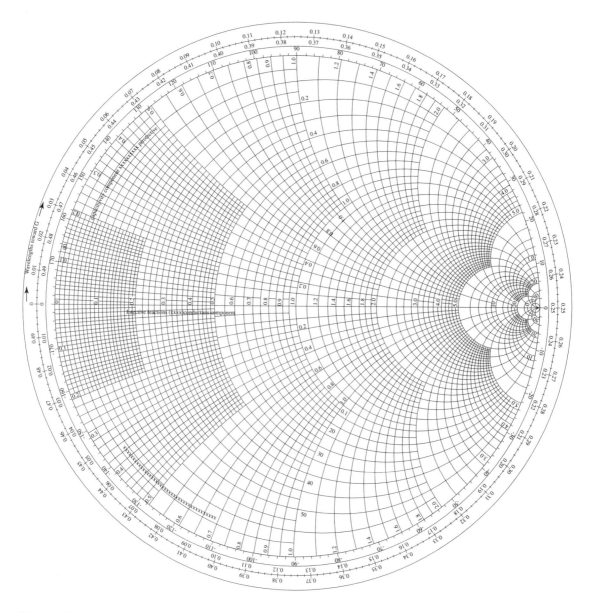

Figure 2.22   A Smith chart. Source: Pozar, D. M. (1998). *Microwave Engineering*, 2nd ed., John Wiley & Sons, Inc., New York

the ρ circle. In the rotation, it should be noted that the θ value is calculated starting from the positive Γ' axis. The θ value increases along the counter clockwise direction, while it decreases along the clockwise direction. As shown in Figure 2.15, in a transmission line, the zero position is chosen at the end of the transmission line, and the positive

direction of the z-axis is from the load to the signal generator.

2.2.2.2 Admittance Smith chart

In some cases, it is more convenient to use admittance. For a complex admittance $y = g + jb$, its normalized conductance is $g = GZ_c$, and its normalized susceptance is $b = BZ_c$. According to Table 2.1, the relationship between admittance and reflection coefficient is

$$y = \frac{1 - \Gamma}{1 + \Gamma} = \frac{1 + (-\Gamma)}{1 - (-\Gamma)} \qquad (2.101)$$

Equations (2.94) and (2.101) indicate that the relationship between y and $(-\Gamma)$ are the same as those between Z and Γ. So the g lines and b lines are also two groups of orthogonal tangent circles, and they have the same shapes as r lines and x lines. As every point on the impedance Smith chart can be converted into its admittance counterpart by taking a 180° rotation around the origin of the Γ

complex plane, an admittance Smith chart can be obtained by rotating the whole impedance Smith chart by 180°, as shown in Figure 2.23. So a Smith chart can be used as an impedance Smith chart or an admittance Smith chart.

However, it is necessary to give special attention to some special points and lines on the two Smith charts. The main differences of the two Smith charts are listed in Table 2.2.

Smith charts are powerful tools widely used in microwave engineering. With known impedance or admittance, we can calculate the reflection coefficient and standing-wave coefficient. We can also calculate the impedance or admittance from the standing-wave coefficient and position of voltage

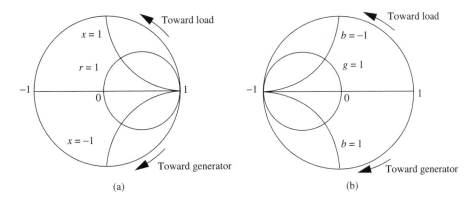

(a)   (b)

Figure 2.23   Two types of Smith charts. (a) Impedance Smith chart and (b) admittance Smith chart

Table 2.2   Comparison between the impedance Smith chart shown in Figure 2.23(a) and the admittance Smith chart shown in Figure 2.23(b)

	Impedance Smith chart	Admittance Smith chart
Point "1"	Open point, $\Gamma = 1$, $r = \infty$, $x = \infty$	Open point, $\Gamma = 1$, $g = 0$, $b = 0$
Point "-1"	Short point $\Gamma = -1$, $r = 0$, $x = 0$	Short point $\Gamma = -1$, $g = \infty$, $b = \infty$
Point "0"	Matching point $\Gamma = 0$, $r = 1$, $x = 0$	Matching point $\Gamma = 0$, $g = 1$, $b = 0$
Line "-1" to "0"	Voltage nodes $r < 1$, $x = 0$	Voltage nodes $g > 1$, $b = 0$
Line "0" to "1"	Voltage antinodes $r > 1$, $x = 0$	Voltage antinodes $g < 1$, $x = 0$
Upper half circle	Inductive impedance $x > 0$	Inductive admittance $b < 0$
Lower half circle	Capacitive impedance $X < 0$	Capacitive admittance $b > 0$

node. Using Smith chart, we can realize impedance transformation at two points on a transmission line, and we can also design impedance matching. Besides its applications in transmission lines, as will be discussed in Section 2.4.7, Smith charts can also be used in analyzing resonant structures.

2.2.3 Guided transmission lines

Here, we discuss several typical kinds of transmission lines often used in materials property characterization, including coaxial lines, planar transmission lines, and hollow metallic waveguides. Coaxial lines and planar transmission lines can support TE mode, TM mode, and TEM mode or quasi-TEM mode, while hollow metallic waveguides cannot support TEM mode, but can support TE or TM modes.

In materials property characterization, both the equivalent lumped parameters and the field distributions of transmission lines are important. In the following discussions, line approach and the field approach are used in combination.

2.2.3.1 Coaxial line

As shown in Figure 2.24, a coaxial line mainly consists of a central conductor with diameter a and an outer conductor with inner diameter b. For coaxial cables used in microwave circuits, the space between the central conductor and outer conductor is filled with a dielectric material, such as Teflon. If the dielectric material between the

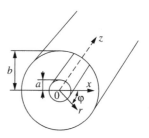

Figure 2.24   The structure of coaxial line

central conductor and outer conductor is air, the coaxial line is usually called coaxial air line. In materials property characterization, coaxial air lines are often used, and the toroidal samples under test are inserted in the space between the central conductor and outer conductor.

Coaxial lines can support TEM, TE, and TM modes, and TEM mode is its fundamental mode. As in most of the microwave applications, single mode is required, most of the coaxial lines work at the TEM mode. In the following discussion, we focus on TEM mode.

According to Eq. (2.28), for a coaxial line shown in Figure 2.25, its potential function Φ satisfies the two-dimensional Laplace's equation:

$$\nabla_T^2 \Phi = \frac{1}{r}\frac{\partial}{\partial r}\left(r\frac{\partial \Phi}{\partial r}\right) + \frac{1}{r^2}\frac{\partial^2 \Phi}{\partial \varphi^2} = 0 \quad (2.102)$$

As the potential function Φ does not change with $\varphi(\partial \Phi/\partial \phi = 0)$, Eq. (2.102) becomes

$$\frac{1}{r}\frac{d}{dr}\left(r\frac{d\Phi}{dr}\right) = 0 \quad (2.103)$$

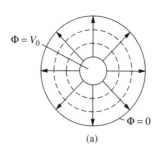

(a)

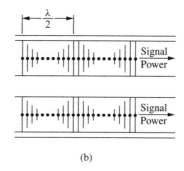

(b)

Figure 2.25   Field distributions of TEM mode in a coaxial line. (a) Field distribution at a transverse cross section and (b) the field distribution along the *z*-axis. Modified from Ishii, T. K. (1995). *Handbook of Microwave Technology*, vol 1, Academic Press, San Diago, CA, 1995

The general solutions for Eq. (2.103) is

$$\Phi(r) = C_1 \ln r + C_2 \qquad (2.104)$$

According to the boundary conditions: $\Phi(a) = V_0$ and $\Phi(b) = 0$, we can get

$$C_1 = \frac{V_0}{\ln(a/b)} \qquad (2.105)$$

$$C_2 = -C_1 \ln b \qquad (2.106)$$

So the electric and magnetic fields of a TEM wave propagating in $+z$ direction are

$$\boldsymbol{E}_{\mathbf{T}} = -\hat{\boldsymbol{r}}\frac{\partial \Phi}{\partial r}e^{-jkz} = \hat{\boldsymbol{r}}\frac{1}{r}\frac{V_0}{\ln(b/a)}e^{-jkz} \qquad (2.107)$$

$$\boldsymbol{H}_{\mathbf{T}} = \frac{1}{\eta}\hat{\boldsymbol{z}} \times \boldsymbol{E}_r e^{-jkz} = \hat{\boldsymbol{\varphi}}\frac{1}{r}\sqrt{\frac{\varepsilon}{\mu}}\frac{V_0}{\ln(b/a)}e^{-jkz} \qquad (2.108)$$

The field distributions are shown in Figure 2.25.

The characteristic impedance Z_c is defined by

$$Z_c = \frac{V_0^2}{2P} = \frac{V_0}{I} = \frac{2P}{I^2}, \qquad (2.109)$$

where I is the current flowing in the coaxial line and P is the power transmitted by the line. For a coaxial line, its characteristic impedance is

$$Z_c = \frac{\eta}{2\pi}\ln\left(\frac{b}{a}\right) = \frac{1}{2\pi}\sqrt{\frac{\mu}{\varepsilon}}\ln\left(\frac{b}{a}\right) \qquad (2.110)$$

As the filling medium in a coaxial line is usually dielectric and the wave impedance of free space is 377 Ω, we have

$$Z_c = \frac{60}{\sqrt{\varepsilon_r}}\ln\left(\frac{b}{a}\right) \quad (\Omega) \qquad (2.111)$$

The attenuation of a coaxial line consists of conductor attenuation α_c and dielectric attenuation α_d:

$$\alpha = \alpha_c + \alpha_d \qquad (2.112)$$

with

$$\alpha_c = \frac{4.34R_s}{2b\eta}\frac{1 + (b/a)}{\ln(b/a)} \quad \text{(dB/unit length)} \qquad (2.113)$$

$$\alpha_d = 27.3\sqrt{\varepsilon_r}\frac{\tan\delta}{\lambda_0} \quad \text{(dB/unit length)}, \qquad (2.114)$$

where R_s is the surface resistance of the conductor and λ_0 is the free-space wavelength. If the conductor used is copper at $20\,^\circ$C, the conductor attenuation α_c can be calculated using the following equation (Chang 1989):

$$\alpha_c = \frac{9.5 \times 10^{-5}\sqrt{f}(a + b)\sqrt{\varepsilon_r}}{ab\ln(b/a)} \quad \text{(dB/unit length)}, \qquad (2.115)$$

where f is the operating frequency.

2.2.3.2 Planar transmission line

As the characteristics of a planar transmission line can be controlled by the dimensions in a single plane, the circuit fabrication can be conveniently carried out by photolithography and photoetching techniques. The application of these techniques at microwave frequencies has led to the development of microwave integrated circuits. As will be discussed in Chapter 7, planar transmission lines are also used in materials property characterization.

As shown in Figure 2.26, three types of planar transmission lines are often used in microwave electronics and materials characterization: stripline, microstrip, and coplanar waveguide. The stripline shown in Figure 2.26(a) has an advantage that the radiation losses are negligible. The propagation in a stripline is in pure TEM mode, and stripline circuits are usually quite compact. The problem with stripline is the difficulty of construction. Usually, two substrates are required to be sandwiched together, and the air gaps between the substrates may cause perturbation to the impedance. Microstrip line, shown in Figure 2.26(b), is the most widely used planar transmission structures. Usually the propagation mode on a microstrip circuit is quasi-TEM. For the development high-density microstrip circuits, thin substrates are often

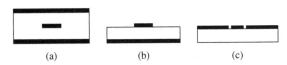

(a)   (b)   (c)

Figure 2.26 Cross-sectional views of three types of transmission planar lines. (a) Stripline, (b) microstrip, and (c) coplanar waveguide

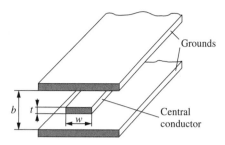

Figure 2.27   Structure of a stripline

used to maintain reasonable impedance and to reduce the coupling between different parts of the circuit. As shown in Figure 2.26(c), the circuit line and the grounding of a coplanar waveguide are on the same plane, and the wave propagation mode is also quasi-TEM.

Stripline

As shown in Figure 2.27, a stripline consists of upper and down grounding plates, and the central conductor. Between the grounding plates and the central conductor is air or dielectric materials. This structure can be taken as a derivation from a coaxial line, by cutting the outer conductor into two pieces and flattening them. Usually, the filling medium is dielectric ($\mu_r' = 1$), and the dimensions of the transverse cross section (b and w defined in Figure 2.27) of a stripline are much less than the wavelength.

The fundamental propagation mode for a stripline is TEM. For the TEM wave propagating in a stripline, the phase velocity is

$$v_p = \frac{1}{\sqrt{L_1 C_1}} = \frac{c}{\sqrt{\varepsilon_r}}, \qquad (2.116)$$

where ε_r is the dielectric constant of the filling medium, C_1 and L_1 are distributed capacitance and inductance, respectively, and c is the speed of light.

The characteristic impedance is given by

$$Z_c = \sqrt{\frac{L_1}{C_1}} = \frac{1}{v_p C_1} \qquad (2.117)$$

As the calculations of phase velocity and distributed capacitance are quite complicated, it is difficult to give a general equation for the calculation

of Z_c. In the following, we discuss two special cases.

Thin central conductor ($t/b \ll 1$)

For a stripline with a thin central conductor, if the width of the control conductor w is much larger than the distance between the ground plate and the central conductor ($b/2$), the field between the central conductor and the ground plate is uniform except the fields at the edges. Using conformal mapping techniques, we can get an appropriate equation for the stripline with zero-thickness central conductor ($t = 0$):

$$Z_c = \frac{120\pi^2}{8\sqrt{\varepsilon_r}\cosh^{-1}e^{(\pi w)/(2b)}} \quad (\Omega) \qquad (2.118)$$

As the central conductor of an actual stripline has certain thickness, Eq. (2.118) can be modified into

$$Z_c = \frac{120\pi^2(1 - t/b)}{8\sqrt{\varepsilon_r}\cosh^{-1}e^{(\pi w)/(2b)}} \qquad (2.119)$$

Thick central conductor

If the condition ($t/b \ll 1$) cannot be satisfied, the calculation of the distributed capacitance becomes complicated. We consider two conditions: wide central conductor and narrow central conductor. If the width of the central conductor satisfies the condition $w/(b - t) \geq 0.35$, we can assume that the field at right side and left side do not interfere. As shown in Figure 2.28, the distributed capacitance C_1 mainly consists of two parallel-plate capacitors and four edge capacitors:

$$C_1 = 2C_p + 4C_f' \qquad (2.120)$$

with

$$C_p = \frac{0.0885\varepsilon_r w}{(b - t)/2} \quad (\text{pF/cm}) \qquad (2.121)$$

$$C_f' = \frac{0.0885\varepsilon_r}{\pi}\left\{\frac{2}{1 - t/b}\ln\left(\frac{1}{1 - t/b} + 1\right)\right.$$

$$\left. -\left(\frac{1}{1 - t/b} - 1\right)\ln\left[\frac{1}{(1 - t/b)^2} - 1\right]\right\}$$

$$(\text{pF/cm}). \qquad (2.122)$$

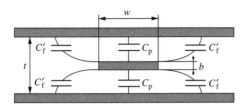

Figure 2.28 The distribution capacitance of stripline with thick central conductor

Therefore the characteristic impedance can be calculated using

$$Z_c = \frac{94.15}{\sqrt{\varepsilon_r}\left(\dfrac{w/b}{1-t/b} + \dfrac{C_f'}{0.0885\varepsilon_r}\right)} \quad (\Omega)$$

(2.123)

If the central conductor is narrow, the interference between the fields at the two edges cannot be neglected. We may take the central conductor as a cylinder by introducing equivalent diameter

$$d = \frac{w}{2}\left\{1 + \frac{t}{w}\left[1 + \ln\frac{4\pi w}{t} + 0.51\pi\left(\frac{t}{w}\right)^2\right]\right\}$$

(2.124)

and the characteristic impedance can be calculated using the following equation:

$$Z_c = \frac{60}{\sqrt{\varepsilon_r}}\ln\left(\frac{4b}{\pi d}\right) \quad (\Omega)$$

(2.125)

Similar to coaxial line, the attenuation also consists of conductor attenuation and dielectric attenuation: $\alpha = \alpha_c + \alpha_d$. An approximate expression for attenuation resulting from conductor surface resistance is (Ramo *et al.* 1994)

$$\alpha_c = \frac{R_s}{\eta b}\left(\frac{\pi w/b + \ln(4b/\pi t)}{\ln 2 + \pi w/2b}\right)$$

(nepers/unit length) (2.126)

Equation (2.126) is valid for $w > 2b$ and $t < b/10$. Approximations for other dimensions can be found in (Hoffmann 1987). The attenuation caused by dielectric loss is (Collin 1991)

$$\alpha_d = \frac{\pi \varepsilon_r''}{\lambda_0\sqrt{\varepsilon_r'}} = \frac{\pi\sqrt{\varepsilon_r'}\tan\delta}{\lambda_0} \quad \text{(nepers/unit length)},$$

(2.127)

where λ_0 is the free-space wavelength.

Here we make some explanations on two ratio units: "dB" and "neper". The definitions for "dB" and "neper" are

$$\text{dB} = 10 \cdot \log_{10}\,(\text{power ratio})$$

$$= 20 \cdot \log_{10}\,(\text{voltage ratio}) \quad (2.128)$$

$$\text{neper} = \ln\,(\text{voltage ratio}). \quad (2.129)$$

The conversion relations between dB and neper are

$$\text{neper} = \text{dB} \times 0.115129255, \quad (2.130)$$

$$\text{dB} = \text{neper} \times 8.685889638. \quad (2.131)$$

Microstrip

As shown in Figure 2.29, a microstrip line consists of a strip conductor and a ground plane separated by a dielectric substrate. It can be taken as a transformation of coaxial line by cutting the outer conductor and flattening it. As the dielectric constant of the substrate is usually high, the field is concentrated near the substrate.

In a strict meaning, the wave propagating on a microstrip line is not a pure TEM wave, nor a simple TE wave or TM wave. The wave propagating on a microstrip is in a quasi-TEM mode. Accurate determination of the wave propagation on a microstrip line requires intense numerical simulations. But in engineering design, we may take the wave in a microstrip line as TEM wave, and use the quasi-static method to calculate the distributed capacitance, and then calculate its propagation constant, wavelength, and characteristic impedance.

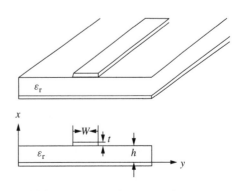

Figure 2.29 Geometry of a microstrip

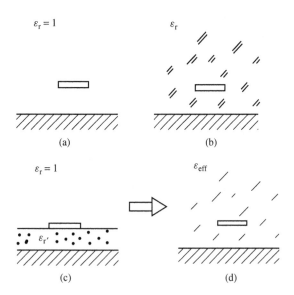

Figure 2.30   The concept of effective dielectric constant. (a) Microstrip fully filled with air, (b) microstrip fully filled with dielectric with permittivity ε_r, (c) microstrip partially filled with dielectric with permittivity ε_r, and (d) microstrip fully filled with dielectric with permittivity ε_{eff}

In the analysis of microstrip lines using quasi-static method, we introduce the concept of the effective dielectric constant, as shown in Figure 2.30. If the filling medium is air ($\varepsilon_r = 1$), as shown in Figure 2.30(a), the microstrip line can support the TEM wave, and its phase velocity equals the speed of light c. If the transmission system is fully filled with a dielectric material with $\varepsilon_r > 1$, as shown in Figure 2.30(b), the microstrip can support TEM wave, and its phase velocity:

$$v_p = c/\sqrt{\varepsilon_r}.   (2.132)$$

If a microstrip line is partially filled with a dielectric material with dielectric constant ε_r, as shown in Figure 2.30(c), we introduce the concept of effective dielectric permittivity ε_{eff} to calculate the transmission parameters of the transmission line: wavelength λ_g, phase velocity v_p, and characteristic impedance Z_c:

$$\lambda_g = \frac{\lambda_0}{\sqrt{\varepsilon_{eff}}}   (2.133)$$

$$v_p = \frac{c}{\sqrt{\varepsilon_{eff}}}   (2.134)$$

$$Z_c = \frac{Z_c^0}{\sqrt{\varepsilon_{eff}}} = \frac{1}{v_p C_1},   (2.135)$$

where c is the speed of light, C_1 is the distributed capacitance of the microstrip, and Z_c^0 is the characteristic impedance of the microstrip when the filling medium is air.

In most cases, the thickness of the strip is negligible ($t/h \leq 0.005$). The characteristic impedance and effective permittivity can be calculated using appropriate equations. If we define the relative strip width

$$u = \frac{w}{h},   (2.136)$$

the effective dielectric constant and characteristic impedance are given by (Ishii 1995)

$$\varepsilon_{eff} = \frac{\varepsilon_r + 1}{2} + \frac{\varepsilon_r - 1}{2} \left[\frac{1}{\sqrt{1 + 12/u}} \right.$$
$$\left. + 0.041(1 - u)^2 \right] \quad \text{(for } u \leq 1\text{)}   (2.137)$$

$$\varepsilon_{eff} = \frac{\varepsilon_r + 1}{2} + \frac{\varepsilon_r - 1}{2} \frac{1}{\sqrt{1 + 12/u}}$$
$$\text{(for } u > 1\text{)}   (2.138)$$

$$Z_0 = \frac{60}{\sqrt{\varepsilon_{eff}}} \ln \left(\frac{8}{u} + 0.25u \right) \quad (\Omega)$$
$$\text{(for } u \leq 1\text{)}   (2.139)$$

$$Z_c = \frac{120\pi}{\sqrt{\varepsilon_{eff}}} \frac{1}{1.393 + u + \ln(u + 1.4444)} \quad (\Omega)$$
$$\text{(for } u > 1\text{)}   (2.140)$$

Actually, the thickness of strip conductor affects the transmission properties of the microstrip line. We assume $t < h$ and $t < w/2$. If the thickness of the strip t is not negligible, the effective dielectric constant should be modified (Ishii 1995):

$$\varepsilon_{eff}(t) = \varepsilon_{eff} - \delta\varepsilon_{eff}   (2.141)$$

with

$$\delta\varepsilon_{eff} = (\varepsilon_r - 1)\frac{t}{4.6h\sqrt{u}}   (2.142)$$

We should also introduce a concept of effective relative strip width u_{eff}:

$$u_{\text{eff}} = u + \frac{1.25t}{\pi h}\left(1 + \ln\left(\frac{4\pi w}{t}\right)\right)$$
$$\text{(for } u \leq 1/(2\pi)) \tag{2.143}$$

$$u_{\text{eff}} = u + \frac{1.25t}{\pi h}\left(1 + \ln\left(\frac{2h}{t}\right)\right)$$
$$\text{(for } u > 1/(2\pi)) \tag{2.144}$$

The attenuation factor of microstrip consists of dielectric loss factor and conductor loss factor: $\alpha = \alpha_c + \alpha_d$ (Ishii 1995). The dielectric loss factor is given by

$$\alpha_d = 27.3 \frac{\varepsilon_r}{\varepsilon_r - 1} \frac{\varepsilon_{\text{eff}} - 1}{\sqrt{\varepsilon_{\text{eff}}}} \frac{\tan\delta}{\lambda_0}, \quad \text{(dB/unit length)} \tag{2.145}$$

and the conductor loss factor can be calculated by

$$\alpha_c = 1.38A \frac{R_s}{hZ_c} \frac{32 - u_{\text{eff}}^2}{32 + u_{\text{eff}}^2}$$
$$\text{(dB/unit length)} \quad \text{(for } u \leq 1) \tag{2.146}$$

$$\alpha_c = 6.1 \times 10^{-5} A \frac{R_s Z_c \varepsilon_{\text{eff}}}{h}\left(u_{\text{eff}} + \frac{0.667 u_{\text{eff}}}{1.444 + u_{\text{eff}}}\right)$$
$$\text{(dB/unit length)} \quad \text{(for } u > 1) \tag{2.147}$$

with

$$A = 1 + \frac{1}{u_{\text{eff}}}\left[1 + \frac{1}{\pi}\ln\left(\frac{2B}{t}\right)\right], \tag{2.148}$$

$$B = \begin{cases} h & \text{for } u \geq (1/2\pi) \\ 2\pi w & \text{for } u \leq (1/2\pi) \end{cases}, \tag{2.149}$$

where R_s is the surface resistance of the conductor.

Coplanar waveguide

As shown in Figure 2.31, in a coplanar waveguide, all the conductors are on the top surface of a dielectric substrate. Similar to the microstrip, the fundamental mode of propagation in the coplanar waveguide is a quasi-TEM mode. As shown in Figure 2.31(b), the pattern of the electric field in the space above the substrate is the same as in the substrate if the thickness of the strip is negligible and the substrate is thick enough. Therefore, we can get the effective dielectric constant:

$$\varepsilon_{\text{eff}} = \frac{\varepsilon_r + 1}{2}, \tag{2.150}$$

and the phase velocity is given by

$$v_p = \frac{1}{\sqrt{\mu\varepsilon_0\varepsilon_{\text{eff}}}} \tag{2.151}$$

By assuming that the thickness of the conductors is zero, ground conductors are infinitely wide, and the substrate has infinite thickness, we have the following approximate formulas (Ramo *et al.* 1994):

$$Z_c = \frac{\eta_0}{\pi\sqrt{\varepsilon_{\text{eff}}}}\ln\left(2\sqrt{\frac{a}{w}}\right) \quad (\Omega)$$
$$\text{for } 0 < w/a < 0.173 \tag{2.152}$$

$$Z_c = \frac{\pi\eta_0}{4\sqrt{\varepsilon_{\text{eff}}}}\left(\ln\left(2\frac{1 + \sqrt{w/a}}{1 - \sqrt{w/a}}\right)\right)^{-1} \quad (\Omega)$$
$$\text{for } 0.173 < w/a < 1 \tag{2.153}$$

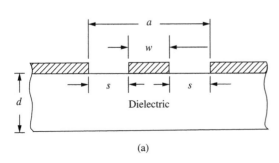

(a)

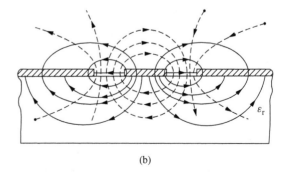

(b)

Figure 2.31 Coplanar line. (a) Structural dimensions and (b) field distributions. The solid lines represent electric field and the dashed lines represent magnetic field

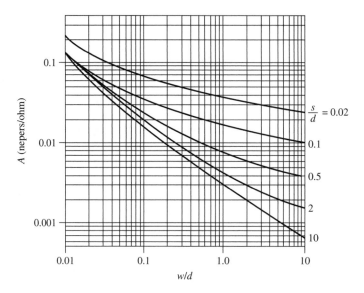

Figure 2.32 Factor for calculation of conductor loss in coplanar waveguide (Hoffmann 1987). Source: Müller, E.: Wellenwiderstand und Mittlere Dielektrizitätskonstante von koplanaren Zwei-und Dreidrahtleitungen auf einem dielektrischen Träger und deren Beeinflussung durch Metallwände. Dissertation, Techn. Universität Stuttgart (1977)

The relations for coplanar waveguides with thick conductor and thin substrate are very complicated (Hoffmann 1987; Gupta *et al.* 1979).

The attenuation factor of a coplanar waveguide consists of dielectric loss factor and conductor loss factor: $\alpha = \alpha_c + \alpha_d$. The attenuation from conductor loss factor in the coplanar waveguide can be calculated from the following equation (Ramo *et al.* 1994):

$$\alpha_c = A \frac{R_s}{d} \sqrt{\varepsilon_{\text{eff}}} \quad \text{(nippers/unit length)} \quad (2.154)$$

The value of A can be found from Figure 2.32 on the condition that all the conductors have the same R_s and have thickness satisfying $t > 3\delta$, where δ is the penetration depth of the conductor. If $d > a$, the dielectric loss factor can be calculated using (Hoffmann 1987)

$$\alpha_d = \frac{\pi f \sqrt{\varepsilon_{\text{eff}}}}{c} \left(\frac{1 - 1/\varepsilon_{\text{eff}}}{1 - 1/\varepsilon_r} \right) \tan \delta \quad \text{(nepers/m)} \quad (2.155)$$

Finally, it should be noted that in the above discussions on microstrip and coplanar waveguide, we do not consider the dispersion of the effective dielectric constant. In a strict meaning, the operating frequency also affects the value of effective

dielectric constant. More discussions on this topic can be found in (Ramo *et al.* 1994).

2.2.3.3 Hollow metallic waveguides

Hollow metallic waveguides are widely used in microwave engineering and materials property characterization. We first introduce the parameters describing the propagation properties of hollow metallic waveguides, and then discuss two types of hollow metallic waveguides: rectangular waveguide and circular waveguide. Finally, we introduce a transition between a circular waveguide and a rectangular waveguide.

Propagation parameters

As shown in Figure 2.33, a hollow metallic waveguide refers to a straight metal tube which has infinite length and whose cross section does change along the z-axis. The wave propagation in a general hollow metallic waveguide can be described by

$$\boldsymbol{E}(u_1, u_2, z, t) = C_1 \boldsymbol{E}(u_1, u_2) e^{j\omega t - \gamma z} \quad (2.156)$$

$$\boldsymbol{H}(u_1, u_2, z, t) = C_2 \boldsymbol{H}(u_1, u_2) e^{j\omega t - \gamma z} \quad (2.157)$$

$$\gamma^2 = -(k^2 - k_c^2), \quad (2.158)$$

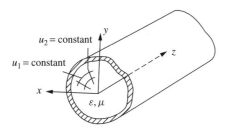

Figure 2.33   A general hollow metallic waveguide

where E and H are the electric and magnetic fields propagating in the waveguide along the Z-axis, and u_1 and u_2 are two orthogonal coordinates perpendicular to Z-axis. The propagation parameter γ is related to wave frequency, medium properties, and field distributions.

If the frequency is high enough ($k > k_c$), then γ is an imaginary number:

$$\gamma = j\beta \qquad (2.159)$$

with

$$\beta = k\sqrt{1 - (k_c/k)^2} \qquad (2.160)$$

So Eqs. (2.156) and (2.157) can be rewritten as

$$E(u_1, u_2, z, t) = C_1 E(u_1, u_2)e^{j(\omega t - \beta z)} \qquad (2.161)$$

$$H(u_1, u_2, z, t) = C_2 H(u_1, u_2)e^{j(\omega t - \beta z)} \qquad (2.162)$$

Equations (2.161) and (2.162) show that β represents the phase change of a unit length along z-axis, and is usually called the *phase constant*.

If the frequency is low ($k < k_c$), γ becomes a real number:

$$\gamma = \alpha = k_c\sqrt{1 - (k/k_c)^2} \qquad (2.163)$$

So Eqs. (2.156) and (2.157) become

$$E(u_1, u_2, z, t) = C_1 E(u_1, u_2)e^{-\alpha z}e^{j\omega t} \qquad (2.164)$$

$$H(u_1, u_2, z, t) = C_2 H(u_1, u_2)e^{-\alpha z}e^{j\omega t} \qquad (2.165)$$

Equations (2.164) and (2.165) indicate that the phase of E and H does not change with z-axis, and the fields decrease along the z-axis. So the wave is in a cutoff state.

There is a critical state between the transmission state and cutoff state: $k = k_c$ and so $\gamma = 0$. The frequency corresponding to the critical state is

called the *cutoff frequency* f_c, and its corresponding wavelength is called the *cutoff wavelength* λ_c. The relationship between k_c and λ_c is

$$k_c = \frac{2\pi}{\lambda_c} \qquad (2.166)$$

Both k_c and λ_c are related to the transverse field distribution in the waveguide. The transmission requirement can be described as $k > k_c$, $\lambda < \lambda_c$, and Eq. (2.160) can be rewritten as

$$\beta = \frac{2\pi}{\lambda}\sqrt{1 - \left(\frac{\lambda}{\lambda_c}\right)^2} \qquad (2.167)$$

For a TEM wave, as $\beta = k$, the cutoff wavelength is infinity, so TEM waves with any frequency satisfy the propagation requirement.

In the following, we discuss parameters often used in describing the propagation properties of hollow metallic waveguides, including phase velocity, group velocity, and the wave impedances for TE and TM waves.

According to the definition, phase velocity is the velocity of the movement of phase planes. From Eqs. (2.161) and (2.162), for a certain phase plane moving along the z-axis, following requirement is satisfied:

$$\omega t - \beta z = \text{constant} \qquad (2.168)$$

From Eq. (2.168), we have

$$\frac{d}{dt}(\omega t - \beta z) = \omega - \beta\frac{dz}{dt} = 0 \qquad (2.169)$$

So the phase velocity is

$$v_p = \frac{dz}{dt} = \frac{\omega}{\beta} \qquad (2.170)$$

From (2.167), we have

$$v_p = \frac{\omega}{\beta} = \frac{v}{\sqrt{1 - (\lambda/\lambda_c)^2}} = \frac{v}{\sqrt{1 - (\lambda/\lambda_c)^2}}, \qquad (2.171)$$

where v is the velocity of electromagnetic wave:

$$v = \frac{\omega}{2\pi}\lambda = f\lambda \qquad (2.172)$$

Equation (2.171) indicates that the phase velocity of TE or TM wave is frequency-dependent, while for TEM wave, the phase velocity does not change with frequency:

$$v_p = \omega/k = 1/\sqrt{\varepsilon\mu}. \qquad (2.173)$$

The waveguide wavelength refers to the wavelength of a wave propagating along a waveguide. For TEM wave, the waveguide wavelength

$$\lambda_g = v/f \qquad (2.174)$$

So it is the same as the wavelength when the TEM wave propagates in a free space filled with the same medium. However, for TE and TM wave, the waveguide wavelength λ_g is given by

$$\lambda_g = \frac{v_p}{f} = \frac{\lambda}{\sqrt{1 - (\lambda/\lambda_c)^2}}, \qquad (2.175)$$

where λ is the wavelength of wave propagation in free space filled with the same medium. Equation (2.175) indicates that λ_g is related to the shape and size of the waveguide, and the propagation mode in the waveguide.

Equation (2.171) indicates that the phase velocity may be greater than the speed of light. Actually, phase velocity is defined to a single frequency and endless signal ($-\infty < t < +\infty$), and such a signal does not transmit any information. Information is transmitted through modulation, and the speed of the transmitting information is the speed of transmitting the information component in a modulated wave. A modulated wave is not a single frequency wave but a group of waves with different frequencies, so its transmission velocity is called the *group velocity*.

Here we discuss a simple example. We assume that the wave group consists of two signals with the same magnitude and very close frequencies and phase constants:

$$E_1 = E_0 e^{j((\omega_0+\Delta\omega)t-(\beta_0+\Delta\beta)z)} \qquad (2.176)$$

$$E_2 = E_0 e^{j((\omega_0-\Delta\omega)t-(\beta_0-\Delta\beta)z)} \qquad (2.177)$$

So the modulated wave is

$$E = E_1 + E_2 = 2E_0 \cos(\Delta\omega \cdot t - \Delta\beta z)e^{j(\omega_0 t-\beta_0 z)} \qquad (2.178)$$

Equation (2.178) can also be written as

$$E = 2E_0 \cos(\Delta\omega \cdot t - \Delta\beta z) \cos(\omega_0 t - \beta_0 z) \qquad (2.179)$$

It is an amplitude-modulated wave with two cosine factors. The factor $\cos(\omega_0 t - \beta_0 z)$ corresponds to the transmission of the wave group along z-axis, and the factor $\cos(\Delta\omega t - \Delta\beta z)$ corresponds to the change of magnitude along the z-axis. The information transmitted is the change of magnitude (envelope) along the z-axis. So the velocity of information transmission is the velocity of the envelope transmission. For a plane in the envelope,

$$\Delta\omega \cdot t - \Delta\beta z = \text{constant} \qquad (2.180)$$

Differentiating Eq. (2.180) with t gives

$$\frac{dz}{dt} = \frac{\Delta\omega}{\Delta\beta} \qquad (2.181)$$

By taking $\Delta\omega \to 0$, we can get the group velocity

$$v_g = \frac{d\omega}{d\beta} \qquad (2.182)$$

For TEM mode, as $\beta = k = \omega(\varepsilon\mu)^{1/2}$, we have

$$v_g = \frac{d\omega}{d\beta} = \frac{1}{\sqrt{\varepsilon\mu}} = v_p \qquad (2.183)$$

However, for TE and TM modes, as $\beta^2 = -\gamma^2 = k^2 - k_c^2 = \omega^2\varepsilon\mu - k_c^2$, we have

$$v_g = \frac{d\omega}{d\beta} = \frac{\beta}{k\sqrt{\varepsilon\mu}} = v\sqrt{1 - \left(\frac{\lambda}{\lambda_c}\right)^2} \qquad (2.184)$$

It is clear that the group velocity is less than the speed of light. From Eqs. (2.171) and (2.184), we find that

$$v_p \cdot v_g = v^2 \qquad (2.185)$$

Figure 2.34 shows the relationship between group velocity, phase velocity, and frequency of electromagnetic waves propagating along a hollow metallic waveguide.

Now, we discuss the wave impedances of TE and TM waves. As discussed earlier, the wave impedance is defined as the ratio between the

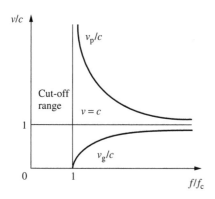

Figure 2.34 Relationship between phase velocity, group velocity, and frequency

transverse electric field and magnetic field. For TE mode at the transmission state, we have

$$Z_{TE} = \frac{j\omega\mu}{\gamma} = \frac{\omega\mu}{\beta} = \frac{\sqrt{\mu/\varepsilon}}{\sqrt{1 - (\lambda/\lambda_c)^2}}$$

$$= \frac{Z_{TEM}}{\sqrt{1 - (\lambda/\lambda_c)^2}} \qquad (2.186)$$

So, the impedance is pure resistance at transmission state. But in the cutoff state, $\gamma = \alpha$, the wave impedance is an inductive reactance:

$$Z_{TE} = \frac{j\omega\mu}{\alpha} \qquad (2.187)$$

Similarly, the wave impedance for TM modes at the transmission state is

$$Z_{TM} = \frac{\gamma}{j\omega\varepsilon} = \sqrt{\frac{\mu}{\varepsilon}}\sqrt{1 - \left(\frac{\lambda}{\lambda_c}\right)^2}$$

$$= Z_{TEM}\sqrt{1 - \left(\frac{\lambda}{\lambda_c}\right)^2} \qquad (2.188)$$

It is pure resistance. But for the cutoff state, the wave impedance is a capacitive reactance:

$$Z_{TM} = \frac{\alpha}{j\omega\varepsilon} \qquad (2.189)$$

In the following, we discuss two typical types of waveguides: rectangular waveguide and circular waveguide, which are widely used in materials property characterization.

Rectangular waveguide

Figure 2.35 shows a rectangular waveguide with width a and height b. Rectangular waveguides can transmit TE and TM modes. Usually, two subscripts m and n are used to specify TE or TM modes, so the propagation mode is often denoted as TE_{mn} or TM_{mn}. The subscript "m" indicates the number of changing cycles along the width a, while the subscript "n" indicates the number of changing cycles along the height b.

The field components of a TE_{mn} wave are

$$H_x = A\frac{\gamma_{mn}}{k_c^2}\frac{m\pi}{a}\sin\left(\frac{m\pi}{a}x\right)\cos\left(\frac{n\pi}{b}\right) \qquad (2.190)$$

$$H_y = A\frac{\gamma_{mn}}{k_c^2}\frac{n\pi}{b}\cos\left(\frac{m\pi}{a}x\right)\sin\left(\frac{n\pi}{b}\right) \qquad (2.191)$$

$$H_z = A\cos\left(\frac{m\pi}{a}x\right)\cos\left(\frac{n\pi}{b}\right) \qquad (2.192)$$

$$E_x = Z_{TE}H_y \qquad (2.193)$$

$$E_y = -Z_{TE}H_x \qquad (2.194)$$

$$E_z = 0 \qquad (2.195)$$

The constant A is related to the power of the wave. The parameters γ_{mn}, k_c, and other parameters are listed in Table 2.3.

The field components of a TM_{mn} wave are:

$$E_x = B\frac{\gamma_{mn}}{k_c^2}\frac{m\pi}{a}\cos\left(\frac{m\pi}{a}x\right)\sin\left(\frac{n\pi}{b}\right) \qquad (2.196)$$

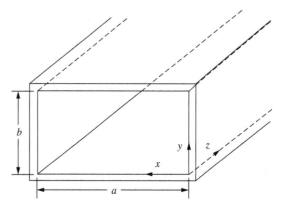

Figure 2.35 Rectangular waveguide. Source: Ramo, S. Whinnery, J. R. and Van Duzer, T. (1994). *Fields and Waves in Communication Electronics*, 3rd ed., John Wiley & Sons, Inc., New York

Table 2.3　Properties of empty rectangular waveguide

	TE$_{mn}$ mode	TM$_{mn}$ mode
Cutoff wave number, k_c	$\sqrt{\left(\dfrac{m\pi}{a}\right)^2 + \left(\dfrac{n\pi}{b}\right)^2}$	$\sqrt{\left(\dfrac{m\pi}{a}\right)^2 + \left(\dfrac{n\pi}{b}\right)^2}$
Propagation constant, γ_{mn}	$\sqrt{k_c^2 - k_0^2}$	$\sqrt{k_c^2 - k_0^2}$
Guided wavelength, λ_g	$\dfrac{\lambda_0}{\sqrt{1 - (k_c/k_0)^2}}$	$\dfrac{\lambda_0}{\sqrt{1 - (k_c/k_0)^2}}$
Group velocity, v_g	$c\dfrac{\lambda_0}{\lambda_g}$	$c\dfrac{\lambda_0}{\lambda_g}$
Phase velocity, v_p	$c\dfrac{\lambda_g}{\lambda_0}$	$c\dfrac{\lambda_g}{\lambda_0}$
Wave impedance, Z	$\dfrac{jk_0\eta_0}{\gamma_{mn}}$	$-\dfrac{j\gamma_{mn}\eta_0}{k_0}$
Attenuation for TE$_{mn}$ modes (Ramo *et al.* 1994)	$\dfrac{2R_s}{b\eta\sqrt{1 - (f_c/f)^2}}\left\{\left(1 + \dfrac{b}{a}\right)\left(\dfrac{f_c}{f}\right)^2 \right.$ $\left. + \left[1 - \left(\dfrac{f_c}{f}\right)^2\right]\left[\dfrac{(b/a)((b/a)m^2 + n^2)}{(b^2m^2/a^2 + n^2)}\right]\right\}$ (nepers/unit length)　　($n \neq 0$) $\dfrac{R_s}{b\eta\sqrt{1 - (f_c/f)^2}}\left[1 + \dfrac{2b}{a}\left(\dfrac{f_c}{f}\right)^2\right]$ (nepers/unit length)　　($n = 0$)	
Attenuation for TM$_{mn}$ modes (Ramo *et al.* 1994)	$\dfrac{2R_s}{b\eta\sqrt{1 - (f_c/f)^2}}\dfrac{m^2(b/a)^3 + n^2}{m^2(b/a)^2 + n^2}$ (nepers/unit length)	

$$E_y = B\frac{\gamma_{mn}}{k_c^2}\frac{n\pi}{b}\sin\left(\frac{m\pi}{a}x\right)\cos\left(\frac{n\pi}{b}\right) \quad (2.197)$$

$$E_z = B\sin\left(\frac{m\pi}{a}x\right)\sin\left(\frac{n\pi}{b}\right) \quad (2.198)$$

$$H_x = \frac{-1}{Z_{TM}}H_y \quad (2.199)$$

$$H_y = \frac{1}{Z_{TM}}H_x \quad (2.200)$$

$$H_z = 0 \quad (2.201)$$

The constant B is related to the power of the wave. The parameters γ_{mn}, k_c, and other parameters are listed in Table 2.3.

The constants A and B in Eqs. (2.190)–(2.201) affect the amplitude of the fields, but do not affect the field distribution. The field distributions of several typical TE and TM modes are shown in Figure 2.36.

Figure 2.37 shows the sequence in which various modes come into existence as the operation frequency increases for aspect ratio b/a equals 1 and 0.5. In the cutoff range, no mode can propagate in the waveguide. TE$_{10}$ is the most often used mode, and in most cases we should ensure waveguides work in single mode state. In microwave engineering, waveguides with $b/a = 0.5$ is more widely used.

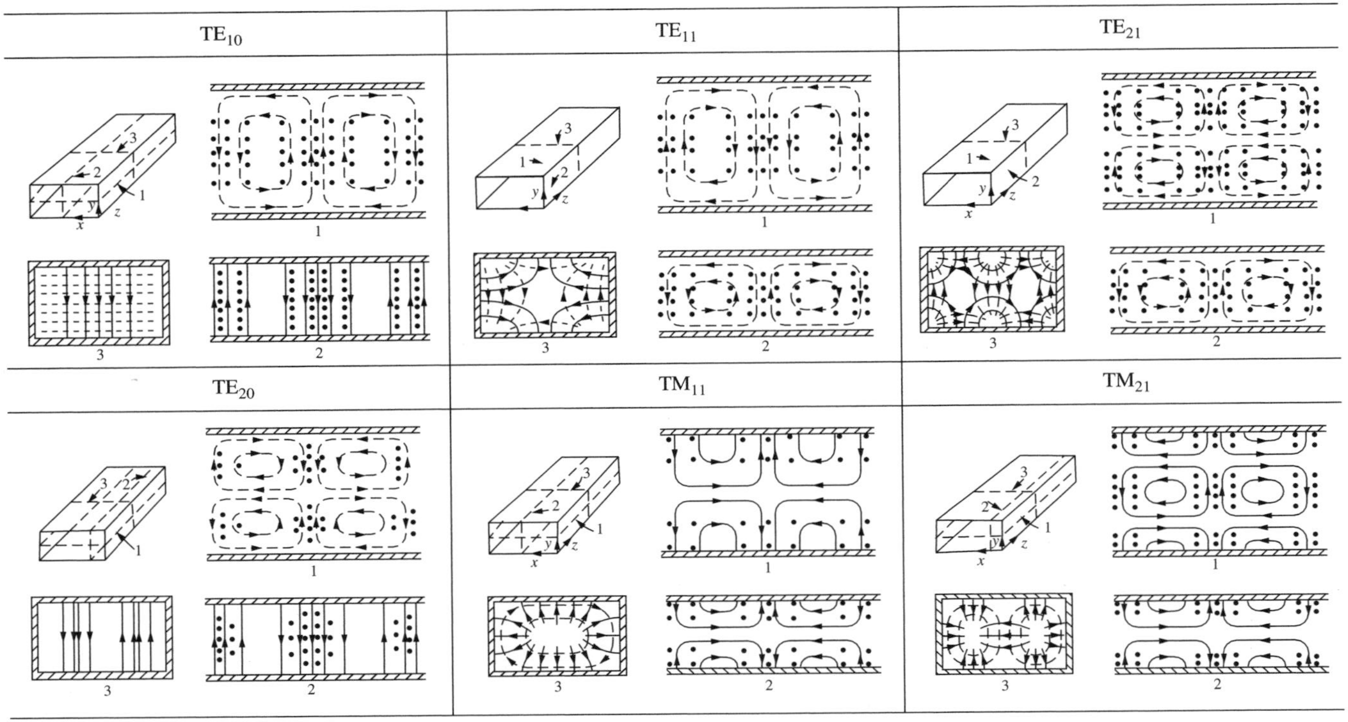

Figure 2.36 Summary of wave types of rectangular waveguides. Electric field lines are shown solid and magnetic field lines are dashed (Ramo *et al.* 1994). Source: Ramo, S. Whinnery, J. R. and Van Duzer, T. (1994). *Fields and Waves in Communication Electronics*, 3rd ed., John Wiley & Sons, Inc., New York

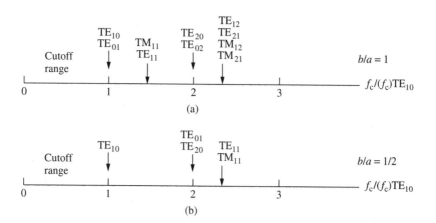

Figure 2.37   Relative cutoff frequencies of rectangular guides (Ramo *et al.* 1994). Modified from Ishii, T. K. (1995). *Handbook of Microwave Technology*, vol 1, Academic Press, San Diago, CA, 1995; Ramo, S. Whinnery, J. R. and Van Duzer, T. (1965). *Fields and Waves in Communication Electronics*, John Wiley & Sons, Inc., New York

Circular waveguide

As shown in Figure 2.38, in the analysis of a circular waveguide, it is more convenient to use cylindrical coordinate (r, φ, z). The dimension of a circular waveguide is its radius a. Circular waveguide can transmit TE and TM modes. Usually, the propagation mode in a circular waveguide is denoted as TE_{ni} or TM_{ni}. The subscript "n" indicates the number of changing periods in φ-direction, while the subscript "i" indicates the number of the changing periods in r-direction.

The field components of a TE_{ni} wave are

$$H_r = -A\gamma_{ni}\left(\frac{a}{\mu_{ni}}\right) J_n'\left(\frac{\mu_{ni}}{a}r\right)\cos n\varphi \quad (2.202)$$

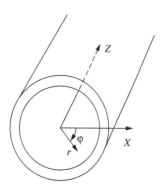

Figure 2.38   Circular waveguide

$$H_\varphi = -An\gamma_{ni}\left(\frac{a}{\mu_{ni}}\right)^2 J_n\left(\frac{\mu_{ni}}{a}r\right)\sin n\varphi \quad (2.203)$$

$$H_z = -AJ_n\left(\frac{\mu_{ni}}{a}r\right)\cos n\varphi \quad (2.204)$$

$$E_r = Z_{TE}H_\varphi \quad (2.205)$$

$$E_\varphi = -Z_{TE}H_r \quad (2.206)$$

$$E_z = 0 \quad (2.207)$$

The constant A is related to the microwave power transmitted in the waveguide, J_n is nth-order Bessel function, and μ_{ni} is the ith root of J_n'. Some characteristic parameters are listed in Table 2.4.

The field components of a TM_{ni} wave are

$$E_r = -B\gamma_{ni}\left(\frac{a}{v_{ni}}\right) J_n'\left(\frac{v_{ni}}{a}r\right)\cos n\varphi \quad (2.208)$$

$$E_\varphi = -B\frac{n}{r}\gamma_{ni}\left(\frac{a}{v_{ni}}\right)^2 J_n\left(\frac{v_{ni}}{a}r\right)\sin n\varphi \quad (2.209)$$

$$E_z = BJ_n\left(\frac{v_{ni}}{a}r\right)\cos n\varphi \quad (2.210)$$

$$H_r = -\frac{1}{Z_{TM}}E_\varphi \quad (2.211)$$

$$H_\varphi = \frac{1}{Z_{TM}}E_r \quad (2.212)$$

$$H_z = 0, \quad (2.213)$$

Table 2.4   Properties of empty rectangular waveguide

	TE$_{ni}$ mode	TM$_{ni}$ mode
Cutoff wave number, k_c	$\dfrac{\mu_{ni}}{a}$ $(i = 0, 1, 2, 3, \ldots)$	$\dfrac{v_{ni}}{a}$ $(i = 1, 2, 3, \ldots)$
Propagation constant, γ_{ni}	$\sqrt{k_c^2 - k_0^2}$	$\sqrt{k_c^2 - k_0^2}$
Guided wavelength, λ_g	$\dfrac{\lambda_0}{\sqrt{1 - (k_c/k_0)^2}}$	$\dfrac{\lambda_0}{\sqrt{1 - (k_c/k_0)^2}}$
Group velocity, v_g	$c\,\dfrac{\lambda_0}{\lambda_g}$	$c\,\dfrac{\lambda_0}{\lambda_g}$
Phase velocity, v_p	$c\,\dfrac{\lambda_g}{\lambda_0}$	$c\,\dfrac{\lambda_g}{\lambda_0}$
Wave impedance, Z	$\dfrac{\mathrm{j}k_0\eta_0}{\gamma_{ni}}$	$-\dfrac{\mathrm{j}\gamma_{ni}\eta_0}{k_0}$
Attenuation for TE$_{ni}$ modes (Ishii 1995)	$\dfrac{R_s}{a\eta_0}\dfrac{1}{\sqrt{1 - (k_c/k_0)^2}}\left[\left(\dfrac{k_c}{k_0}\right)^2 + \dfrac{n^2}{(k_c a)^2 - i^2}\right]$ (nepers/unit length)	
Attenuation for TM$_{ni}$ modes (Ishii 1995)	$\dfrac{R_s}{a\eta_0}\dfrac{k_0^2}{\sqrt{k_0^2 - k_c^2}}$ (nepers/unit length)	

where v_{ni} is the is the ith root of J_n, and B is related to the microwave power transmitted in the waveguide. Some characteristic parameters are listed in Table 2.4.

Figure 2.39 shows the field distributions of several typical circular waveguide modes. We can see that along the φ direction, the field changes in a sinuous way, and the number n indicates the period number in the range of 0 to 2π. In the radius direction, the field changes according to Bessel function or differentiated Bessel functions, and i indicates the number of zeros along the radius $(0 < r < a)$.

In circular waveguides, there exist degeneration phenomena. There are two kinds of degenerations: polar degeneration and E-H degeneration. For a TE$_{ni}$ or a TM$_{ni}$ $(n \neq 0)$, mode there are two kinds of field distributions that have the same shape, but their polarization planes are perpendicular to each other. Such degeneration is called *polar degeneration*. Meanwhile, because

$$J_0'(x) = -J_1(x), \qquad (2.214)$$

the roots for $J_0'(x)$ and $J_1(x)$ are the same: $\mu_{0i} = v_{1i}$. Therefore, TE$_{0i}$ and TM$_{1i}$ have the same wavelength, and this is called *E-H degeneration*.

Figure 2.40 shows the cutoff frequencies for different modes of circular waveguides. Similar to rectangular waveguides, there is a cutoff range where no mode can propagate, and the fundamental mode is TE$_{11}$ mode. In the design and selection of circular waveguides, we should ensure single mode requirement.

Transition from rectangular waveguide to circular waveguide

In microwave engineering, we often use the prime modes of rectangular and circular waveguides. The prime mode of rectangular waveguide is TE$_{10}$, and that of circular waveguide is TE$_{11}$. Owing to the E-H degeneration of circular waveguides, rectangular waveguides are more widely used while circular waveguides are often used in antennas, polarization attenuators, ferrite isolators, and circulators.

Wave type	TM_{01}	TM_{02}	TM_{11}	TE_{01}	TE_{12}
Field distributions in cross-sectional plane, at plane of maximum transverse fields	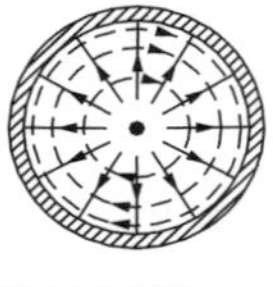		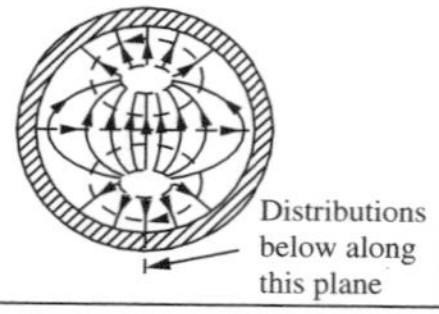 Distributions below along this plane	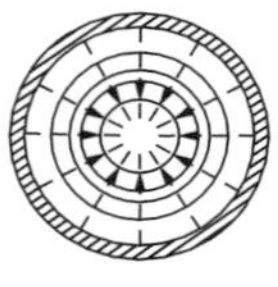	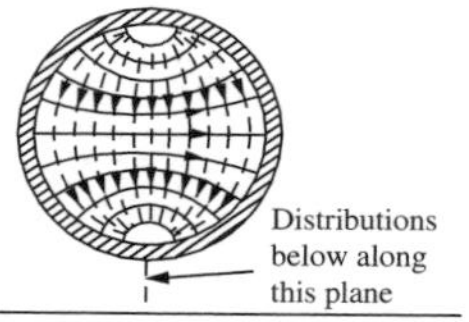 Distributions below along this plane
Field distributions along guide	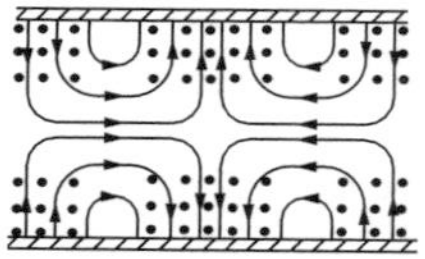	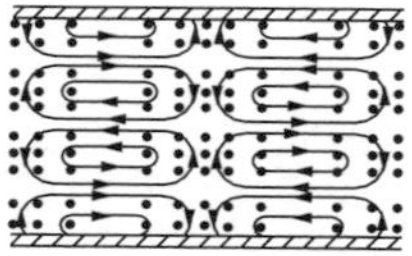	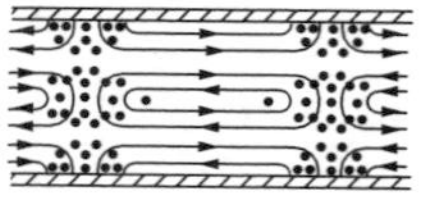	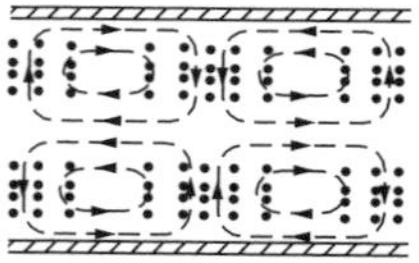	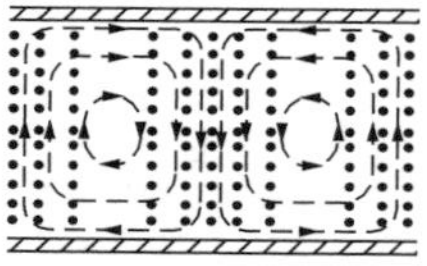
Field components present	$E_{s1}E_{r1}H_\phi$	$E_{s1}E_{r1}H_\phi$	$E_{s1}E_{r1}E_{\phi1}H_{r1}H_\phi$	$H_{s1}H_{r1}E_\phi$	$H_{s1}H_{r1}H_{\phi1}E_{r1}E_\phi$
p_al or p'_al	2.405	5.52	3.83	3.83	1.84
$(k_c)_al$	$\dfrac{2.405}{a}$	$\dfrac{5.52}{a}$	$\dfrac{3.83}{a}$	$\dfrac{3.83}{a}$	$\dfrac{1.84}{a}$
$(\lambda_c)_al$	$2.61a$	$1.14a$	$1.04a$	$1.64a$	$3.41a$
$(f_c)_al$	$\dfrac{0.383}{a\sqrt{\mu\varepsilon}}$	$\dfrac{0.877}{a\sqrt{\mu\varepsilon}}$	$\dfrac{0.609}{a\sqrt{\mu\varepsilon}}$	$\dfrac{0.609}{a\sqrt{\mu\varepsilon}}$	$\dfrac{0.293}{a\sqrt{\mu\varepsilon}}$
Attenuation due to imperfect-conductors	$\dfrac{R_s}{a\eta}\dfrac{1}{\sqrt{1-(f_c/f)^2}}$	$\dfrac{R_s}{a\eta}\dfrac{1}{\sqrt{1-(f_c/f)^2}}$	$\dfrac{R_s}{a\eta}\dfrac{1}{\sqrt{1-(f_c/f)^2}}$	$\dfrac{R_s}{a\eta}\dfrac{(f_c/f)^2}{\sqrt{1-(f_c/f)^2}}$	$\dfrac{R_s}{a\eta}\dfrac{1}{\sqrt{1-(f_c/f)^2}}\left[(f_c/f)^2+0.420\right]$

Figure 2.39 Summary of wave types of circular waveguides. Electric field lines are shown solid and magnetic field lines are dashed (Ramo *et al.* 1994). Source: Ramo, S. Whinnery, J. R. and Van Duzer, T. (1994). *Fields and Waves in Communication Electronics*, 3rd ed., John Wiley & Sons, Inc., New York

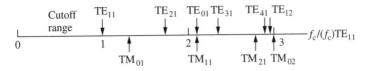

Figure 2.40 Relative cutoff frequencies of waves in a circular guide (Ramo *et al.* 1994). Source: Ramo, S. Whinnery, J. R. and Van Duzer, T. (1994). *Fields and Waves in Communication Electronics*, 3rd ed., John Wiley & Sons, Inc., New York

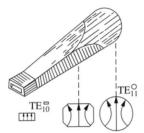

Figure 2.41 Transition between a rectangular waveguide and a circular waveguide

As the field distributions of rectangular TE_{10} and circular TE_{11} are similar, it is easy to realize the transition between them. Figure 2.41 shows an example for the transition between a rectangular waveguide and a circular waveguide.

2.2.3.4 Transitions between different types of transmission lines

In building microwave measurement circuits, it is necessary to make transitions between different types of transmission structures, for example, transitions between waveguide and coaxial line, and transitions between waveguide and microstrip line. The function of a microwave transition is to couple the electromagnetic wave in one type of transmission structure into another. Meanwhile, a transition between two different transmission structures transforms the electromagnetic field distributions in one transmission structure to conform to the boundary conditions of another transmission structure. Here, we discuss the basic requirements for the design of transitions and then give several transition examples. Detailed analysis and more examples can be found in (Izadian and Izadian 1988).

Field matching and impedance matching

The objective of transition design is to make the transformation between different types of transmission structures as efficient as possible. To obtain good transformation, two requirements should be satisfied: impedance matching and field matching.

A practical approach to realize the efficient field transition from one transmission structure to another is to smoothly and gradually change the physical boundary conditions. A transition example is the transition between rectangular and circular waveguides, as shown in Figure 2.41. Here, we discuss the transitions between different types of transmission structures.

Waveguide, coaxial line, and microstrip are three types of transmission structures often used in microwave engineering and materials characterization, and their electric field and magnetic field distributions are shown in Figure 2.42. To ensure field matching, it is necessary to transform the field geometry by reshaping the transmission structures. Figure 2.43 illustrates the evolution procedure of modifying a coaxial transmission line into a microstrip line by cutting the coaxial line along the longitudinal direction and unfolding it. Meanwhile, the transition between two transmission structures must provide impedance matching

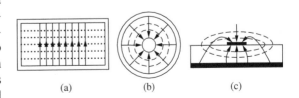

Figure 2.42 Field distributions of three types of transmission structures. (a) Rectangular waveguide, (b) coaxial line, and (c) microstrip

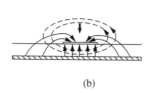

(a)   (b)

Figure 2.43 The evolution from a coaxial line to a microstrip line. (a) Coaxial line and (b) microstrip line. Reprinted with permission from *Microwave Transition Design*, by Izadian, J. S. and Izadian, S. M. (1988). Artech House Inc., Norwood, MA, USA. www.artechhouse.com

between the two transmission structures to reduce the reflection at the transition and improve the transition efficiency. To achieve both field matching and impedance matching, most of the transitions use step transition or continuous taper transition approaches.

In the following, we discuss two typical transitions: transition between rectangular waveguide and coaxial line, and transition between coaxial line and microstrip line.

Transition between rectangular waveguide and coaxial line

Rectangular waveguide is a non-TEM wave transmission structure, which only supports TE or TM modes, so a rectangular waveguide may be taken as a high-pass or band-pass structure, while a coaxial line supports a TEM mode and can be used from dc. The upper frequency limit of coaxial lines is the increase of loss of the transmission line and the higher order modes.

Two classic approaches have been used in the design of transition between a waveguide and a coaxial line: the electric probe and the magnetic probe. The coaxial line is usually 50 Ohms, but the impedance of the TE_{10} mode in a rectangular waveguide is usually several hundred Ohms. So, the design procedure is to get an optimum impedance matching by changing the location, height, and diameter of the electric or magnetic probe.

Figure 2.44 shows a transition in magnetic probe approach. The inner conductor of the coaxial line is connected to the top side of the waveguide wall,

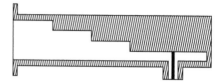

Figure 2.44 Magnetic-dipole approach for the transition between a rectangular waveguide and coaxial line (Izadian and Izadian 1988). Reprinted with permission from *Microwave Transition Design*, by Izadian, J. S. and Izadian, S. M. (1988). Artech House Inc., Norwood, MA, USA. www.artechhouse.com

and the shield of the coaxial line is connected to the bottom side of the waveguide wall. At the connection region, the height of the waveguide is decreased to achieve an impedance value close to that of the coaxial line (50 Ω), and so the first critical translation is made between coaxial line and waveguide. Quarter-wavelength waveguide sections between the normal waveguide and the connection region are often used to make a transition between the normal waveguide and 50 Ω waveguide.

Figure 2.45 shows an electric probe approach for making a transition between a rectangular waveguide and a coaxial line. In this approach, the central conductor of the coaxial line extends into the waveguide, but is not connected to the opposite waveguide wall. The central conductor acts as a small monopole antenna exciting the propagation mode in the rectangular waveguide.

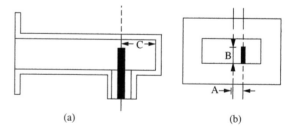

(a)   (b)

Figure 2.45 Electric-probe approach for a transition between a rectangular waveguide and a coaxial line. (a) Side view and (b) front view (Izadian and Izadian 1988). Reprinted with permission from *Microwave Transition Design*, by Izadian, J. S. and Izadian, S. M. (1988). Artech House Inc., Norwood, MA, USA. www.artechhouse.com

The impedance matching can be achieved by varying the dimensional parameters, including off-center position of the probe A, probe length B, and probe position in the waveguide C.

Transition between coaxial line and microstrip line

The transitions between coaxial and microstrip line are the most frequently used transitions in microwave electronics. In materials property characterization, these types of transitions are often required in developing planar circuit method, which will be discussed in Chapter 7. The transitions between coaxial line and microstrip lines are often required to have broad working frequency range, high return loss, and low insertion loss.

The basic principle for a transition between coaxial line and microstrip line has been shown in Figure 2.43. As microstrip circuits are usually hosted in a casing, the transition between microstrip line and coaxial line is often designed in the package casing. In the development of such transitions, commercially available standard connectors, including the central pins and shields, are often used. In a transition structure, the central pin is connected to the microstrip circuit, and the shield is connected to the wall of the casing, as shown in Figure 2.46(a). It is preferable to choose a connector with a dielectric insulator whose thickness is close to the thickness of the microstrip substrate, and the dielectric insulator of the connector and the microstrip substrate have close dielectric constant.

Meanwhile, to obtain better transition efficiency, the width of the microstrip must be close to the diameter of the coaxial central pin.

Figure 2.46(b) shows a right-angle transition between a coaxial line and a microstrip line. Such a transition can be fabricated by drilling a hole in the microstrip substrate. The pin of the connector is inserted through the hole and connected to the microstrip circuit and the shield of the connector is connected to the ground of the microstrip line. This kind of transition is often used in antennas, but it requires high fabrication techniques.

Figure 2.47 shows another example of transition between a coaxial line and a microstrip line. In such a transition, the central conductor of the coaxial line goes gradually from the axis position to the strip of the microstrip, and the grounding plate of the microstrip is connected to the outer conductor of the coaxial line. The field distributions at different cross sections of the transition structure indicate that the fields change gradually, and so high transition efficiency can be achieved using this transition structure.

2.2.4 Surface-wave transmission lines

Besides the guided transmission lines discussed in Section 2.2.3, there exists a class of open-boundary structures which can also be used in guiding electromagnetic waves. Such structures are capable of supporting a mode that is closely bound to the surfaces of the structures. The field distributions of the electromagnetic waves on such structures are characterized by an exponential decay away

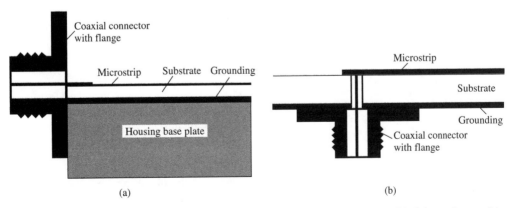

Figure 2.46 Transitions from coaxial line to microstrip. (a) Straight transition and (b) right angle transition

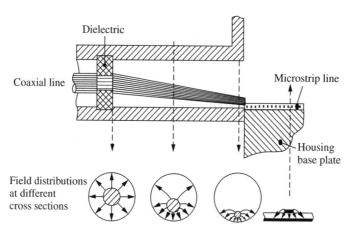

Figure 2.47   A transition between a coaxial line and a microstrip line (Modified from Hoffmann, R. K. (1987). *Handbook of Microwave Integrated Circuits*, Artech House, Norwood, MA, 1987. © 2003 IEEE

from the surface and having the usual propagation function $\exp(\pm j\beta z)$ along the axis of the structure. Such an electromagnetic wave is called a *surface wave*, and the structure that guides this wave is often called a *surface waveguide*. One of the most characteristic properties of a surface wave is that it does not have low-frequency limit.

Sometimes, surface waveguides are also called *dielectric waveguides*, as in most cases, the key components consisting a surface waveguide are dielectrics. In a surface waveguide, the wave travels because of the total internal reflections at the boundary between two different dielectric materials. Figure 2.48 shows a cross section of a generalized surface waveguide. The conductor loss in a surface waveguide is usually very low, while the loss due to the curvature, junction, and discontinuities, and so on, may be quite large. The loss of a dielectric waveguide can be decreased using

high permittivity and extremely low loss dielectric materials. But the use of high permittivity materials may result in very small size of the surface waveguide and severe fabrication tolerance requirements.

In the following, we mainly discuss the surface waves at dielectric interfaces, dielectric slabs, rectangular dielectric waveguides, cylindrical dielectric waveguides, and coaxial surface-wave transmission structures.

2.2.4.1 Dielectric interface

The simplest surface waveguide structure is a dielectric interface between two dielectric materials with different dielectric permittivities as shown in Figure 2.49. For an electromagnetic wave incident on the interface, we have Snell's laws of reflection and refraction:

$$\theta_i = \theta_r \tag{2.215}$$

$$k_1 \sin \theta_i = k_2 \sin \theta_t, \tag{2.216}$$

where k_1 and k_2 are the wave-numbers in the two dielectric media, given by

$$k_i = \omega \sqrt{\mu_0 \varepsilon_i} \quad (i = 1, 2), \tag{2.217}$$

where ω is the operating frequency. Other parameters are defined in Figure 2.49. In the following discussion, we assume $\varepsilon_1 > \varepsilon_2$.

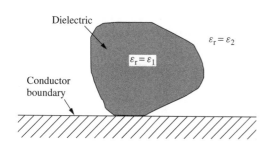

Figure 2.48   Cross section of a generalized dielectric waveguide

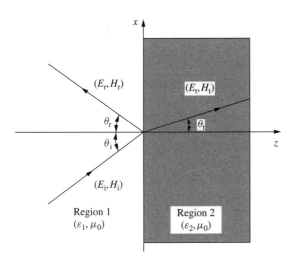

Figure 2.49 Geometry for a plane wave obliquely incident at the interface between two dielectric regions

From Eq. (2.216), we have

$$\sin\theta_t = \sqrt{\varepsilon_1/\varepsilon_2}\,\sin\theta_i \qquad (2.218)$$

Equation (2.218) indicates that when the incident angle θ_i increases from $0°$ to $90°$, the refraction angle θ_t will increase, in a faster rate, from $0°$ to $90°$. At a critical incident angle θ_c defined by

$$\sin\theta_c = \sqrt{\varepsilon_2/\varepsilon_1}, \qquad (2.219)$$

$\theta_t = 90°$. When the incident angle is equal to or larger than the critical angle, the transmitted wave does not propagate into region 2.

When $\theta_i > \theta_c$, the angle θ_t loses its physical meaning defined in Figure 2.49. We write the incident fields as

$$E_i = E_0(\hat{x}\cos\theta_i - \hat{z}\sin\theta_i)$$
$$\exp[-jk_1(x\sin\theta_i + z\cos\theta_i)] \qquad (2.220)$$
$$H_i = \frac{E_0}{\eta_1}\hat{y}\exp[-jk_1(x\sin\theta_i + z\cos\theta_i)] \qquad (2.221)$$

When $\theta_i > \theta_c$, the transmitted fields are usually expressed as

$$E_t = E_0 T\left(\frac{j\alpha}{k_2}\hat{x} - \frac{\beta}{k_2}\hat{z}\right)\exp(-j\beta x)\exp(-\alpha z) \qquad (2.222)$$
$$H_t = \frac{E_0 T}{\eta_2}\hat{y}\exp(-j\beta x)\exp(-\alpha z), \qquad (2.223)$$

where T is the transmission coefficient, β is the propagation constant, and η_i is the wave impedance given by

$$\eta_i = \sqrt{\mu_0/\varepsilon_i} \quad (i=1,2) \qquad (2.224)$$

From the boundary condition, we can get

$$\beta = k_1\sin\theta_i = k_1\sin\theta_r \qquad (2.225)$$
$$\alpha = \sqrt{\beta^2 - k_2^2} = \sqrt{k_1^2\sin^2\theta_i - k_2^2}. \qquad (2.226)$$

The reflection and transmission coefficients can then be obtained (Pozar 1998):

$$\Gamma = \frac{(j\alpha/k_2)\eta_2 - \eta_1\cos\theta_i}{(j\alpha/k_2)\eta_2 + \eta_1\cos\theta_i} \qquad (2.227)$$
$$T = \frac{2\eta_2\cos\theta_i}{(j\alpha/k_2)\eta_2 + \eta_1\cos\theta_i} \qquad (2.228)$$

The magnitude of Γ is unity as it is of the form $(a-jb)/(a+jb)$, so all the incident power is reflected.

Equations (2.222) and (2.223) indicate that the transmitted wave propagates in the x-direction along the interface, while it decays in the z-direction. As the field is tightly bound to the interface, the transmitted wave is called *surface wave*. From Eqs. (2.222) and (2.223), we can calculate the complex Poynting vector (Pozar 1998):

$$S_t = E_t \times H_t^* = \frac{|E_0|^2|T|^2}{\eta_2}\left(\hat{z}\frac{j\alpha}{k_2} + \hat{x}\frac{\beta}{k_2}\right)\exp(-2\alpha z) \qquad (2.229)$$

Equation (2.229) indicates that no real power flow occurs in the z-direction. The real power flow in the x-direction is that of the surface wave field, which decays exponentially with distance into region 2. So, even though no real power is transmitted into region 2, a nonzero field does exist there in order to satisfy the boundary conditions at the interface.

2.2.4.2 Dielectric slab

Surface waves can propagate on dielectric slabs, including ungrounded and grounded dielectric slabs.

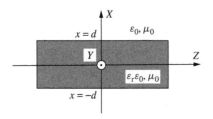

Figure 2.50   Cross section of an ungrounded dielectric slab

Ungrounded dielectric slab

An ungrounded dielectric slab is also called *symmetrical dielectric slab* due to its structural symmetry. Figure 2.50 shows an ungrounded dielectric slab with a thickness $2d$, and at the regions $x > d$ and $x < -d$, the medium is air. We assume that the dielectric loss of the slab is negligible and the dielectric constant of the slab is ε_r. For a plane wave propagating from the slab to the interface between the dielectric and air, if the incident angle satisfies

$$\theta_i > \sin^{-1}(1/\sqrt{\varepsilon_r}), \qquad (2.230)$$

the wave energy will be totally reflected, resulting in surface wave propagation.

We assume that the dielectric slab is infinitely wide, the electromagnetic field does not change along the y-direction, and the propagation factor along the z-direction is $\exp(-j\beta z)$. According to Maxwell's equations and the boundary conditions, it can be verified that there are two types of surface waves: TM modes with components H_y, E_x, and E_z, and TE modes with components E_y, H_x, and H_z. Detailed discussions on TM and TE modes can be found in (Collin 1991).

Owing to the symmetrical structure of the dielectric slab, the surface waves also fall into symmetrical modes and antisymmetrical modes. For a symmetrical TM mode, as the distribution of H_y along x-direction is symmetrical for the plane $x = 0$, we have

$$\frac{\partial H_y}{\partial x}\bigg|_{x=0} = 0. \qquad (2.231)$$

Equation (2.231) indicates that the tangent electric field component along the $x = 0$ plane equals zero, so we can put an electric wall at the $x = 0$ plane.

For an antisymmetrical TM mode, at $x = 0$ plane, we have

$$H_y = 0, \qquad (2.232)$$

so we can put a magnetic wall at the $x = 0$ plane.

For TE modes, we have opposite conclusions. For a symmetrical TE mode, we can put a magnetic wall at the $x = 0$ plane; and for an antisymmetrical TE mode, we can put an electrical wall at the $x = 0$ plane.

The cutoff wavelength for both TM_n and TE_n modes are given by

$$\frac{2d}{\lambda_c} = \frac{n}{2(\varepsilon_r - 1)^{1/2}} \quad (n = 0, 1, 2, 3, \ldots) \ (2.233)$$

Even values of $n(0, 2, 4, \ldots)$ correspond to even TM or TE modes, while odd values of $n(1, 3, 5, \ldots)$ correspond to odd TM or TE modes. Equation (2.233) indicates that for an ungrounded dielectric slab, the first even mode ($n = 0$) has no low-frequency cutoff.

Grounded dielectric slabs

Figure 2.51 shows a dielectric slab grounded by a metal plate. A grounded dielectric slab with thickness d can be taken as a special case of ungrounded dielectric slab with thickness $2d$ as shown in Figure 2.50, with an electric wall placed at the plane $x = 0$. Detailed discussion on grounded dielectric slab can be found in (Pozar 1998).

The surface waves propagating on a grounded dielectric slab can also classified into TM and TE modes. The cutoff wavelength for TM_n mode is

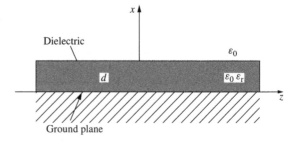

Figure 2.51   Geometry of a grounded dielectric slab

given by

$$\frac{2d}{\lambda_c} = \frac{n}{(\varepsilon_r - 1)^{1/2}} \quad (n = 0, 1, 2, \ldots), \quad (2.234)$$

while the cutoff wavelength for TE$_n$ mode is given by:

$$\frac{2d}{\lambda_c} = \frac{2n - 1}{2(\varepsilon_r - 1)^{1/2}} \quad (n = 1, 2, 3, \ldots) \quad (2.235)$$

Equations (2.234) and (2.235) indicate that the order of propagation for the TM$_n$ and TE$_n$ modes is TM$_0$, TE$_1$, TM$_1$, TE$_2$, TM$_2$, ….

2.2.4.3 Rectangular dielectric waveguide

A rectangular dielectric waveguide can be taken as a modification from a dielectric slab, by limiting the width of the slab. Corresponding to ungrounded and grounded dielectric slabs, we have isolated dielectric waveguides and image guides. The determination of propagation properties of surface waves on dielectric waveguides usually requires numerical techniques, among which the mode-matching method is often used. Detailed discussion on rectangular dielectric waveguide can be found in (Ishii 1995; Goal 1969). In the following, we discuss the propagation constants of isolated rectangular dielectric waveguides and image guides.

Isolated rectangular waveguide

Figure 2.52 shows the geometrical structure of an isolated rectangular waveguide and its field distributions along the x-direction and y-direction. The axis of the dielectric waveguide is along the z-direction, and dimensions along the x-direction and y-direction are $2a$ and $2b$, respectively.

The propagation constant for the surface wave along the rectangular waveguide is given by (Ishii 1995)

$$k_z = (\varepsilon_r k_0^2 - k_x^2 - k_y^2)^{1/2} \quad (2.236)$$

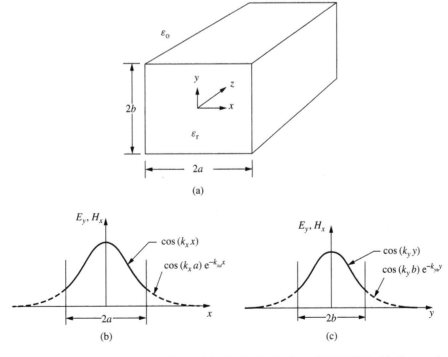

Figure 2.52 Rectangular dielectric waveguide and its field distributions (Ishii 1995). (a) Geometrical structure, (b) field distribution along x-direction, and (c) field distribution along y-direction. Source: Ishii, T. K. (1995). *Handbook of Microwave Technology*, Vol 1, Academic Press, San Diago, CA, 1995

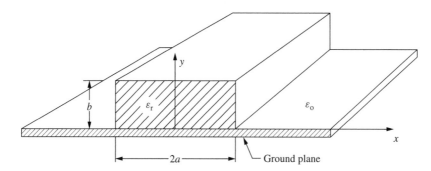

Figure 2.53   Configuration of an image guide (Ishii 1995). Source: Ishii, T. K. (1995). *Handbook of Microwave Technology*, Vol 1, Academic Press, San Diago, CA, 1995

with

$$k_x = \frac{m\pi}{2a}\left\{1 + \frac{1}{a[(\varepsilon_r - 1)k_0^2 - k_y^2]}\right\}^{-1} \tag{2.237}$$

$$k_y = \frac{n\pi}{2b}\left\{1 + \frac{1}{\varepsilon_r b[(\varepsilon_r - 1)k_0]^{1/2}}\right\}^{-1} \tag{2.238}$$

$$k_{x0} = (\varepsilon_r - 1)k_0^2 - k_x^2 - k_y^2 \tag{2.239}$$

$$k_{y0} = (\varepsilon_r - 1)k_0^2 - k_y^2, \tag{2.240}$$

where k_x and k_y, k_{x0} and k_{y0} are the transverse propagation constants inside and outside the dielectric waveguide respectively, and k_0 is the free-space propagation constant.

Rectangular image guide

A rectangular image guide can be taken as a modification from the grounded dielectric slab by limiting the width of the slab. Figure 2.53 shows the configuration of an image guide whose axis is along the z-direction. The width of the dielectric is $2a$ while the height of the dielectric is b.

The propagation constant of a surface wave on an image guide is also given by Eq. (2.236), where the value of k_x is the solution of the following set of equations (Ishii 1995):

$$\tan(k_x a) = k_{x0}/k_x \tag{2.241}$$

$$k_x^2 = \varepsilon_{re}(y)k_0^2 - k_z^2 \tag{2.242}$$

$$k_{x0}^2 = k_z^2 - k_0^2$$
$$= [\varepsilon_{re}(y) - 1]k_0^2 - k_x^2 \tag{2.243}$$

$$\varepsilon_{re}(y) = \varepsilon_r - (k_y/k_0)^2 \tag{2.244}$$

and the value of k_y is the solution of the following set of equations (Ishii 1995):

$$\tan(k_y b) = \varepsilon_{re}(x)k_{y0}/k_y \tag{2.245}$$

$$k_y^2 = \varepsilon_{re}(x)k_0^2 - k_z^2 \tag{2.246}$$

$$k_{y0}^2 = [\varepsilon_{re}(x) - 1]k_0^2 - k_y^2 \tag{2.247}$$

$$\varepsilon_{re}(x) = \varepsilon_r - (k_x/k_0)^2 \tag{2.248}$$

Dielectric microstrip

Figure 2.54 shows the geometry of a dielectric microstrip, which is a modified image guide with a dielectric slab interposed between a dielectric ridge and the grounding plane. In this structure, the dielectric constant of the ridge (ε_{r2}) is usually greater than that of the substrate (ε_{r1}). The fields are thus mostly confined to the area around the dielectric ridge, resulting in low attenuation. On the basis of this basic geometry, many variations can be made for different purposes.

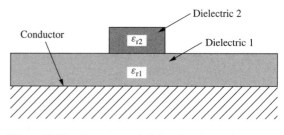

Figure 2.54   Geometry of dielectric microstrip

2.2.4.4 Cylindrical dielectric waveguide

Figure 2.55(a) shows a cylindrical dielectric waveguide whose cross section is a circle with radius a. The dielectric constant of the cylinder is ε_{r1}, and that of the environment is ε_{r2}. In some cases, the dielectric cylinder is covered with a layer of another dielectric material, and such a structure is often used in optical communications, and is usually called optical cable, as shown in Figure 2.55(b). For optical cables, usually the refraction index $n = (\varepsilon_r)^{1/2}$ is used. Usually, the refraction index of the core n_1 is larger than that of the cover n_2. Both dielectric cylinders and optical cables can support surface waves. As electromagnetic fields decay quickly in the cover layer along the r-direction, if the cover layer is thick enough, the fields outside the cover can be neglected and we can assume that the cover layer has infinite thickness. Therefore, for the propagation of surface waves, optical cables shown in Figure 2.55(b) can be taken as a dielectric cylinder shown in Figure 2.55(a). In the following discussion, we concentrate on the surface waves propagating along a dielectric cylinder.

As shown in Figure 2.55(a), we assume that the axis of the dielectric cylinder is along the z-axis, and the propagation factor of electromagnetic wave along the z-direction is $\exp(-j\beta z)$. The longitudinal field components $E_z(r, \varphi)$ and $H_z(r, \varphi)$ satisfy the following equation:

$$\frac{\partial^2}{\partial r^2}(E_z H_z) + \frac{1}{r}\frac{\partial}{\partial r}(E_z H_z) + \frac{1}{r^2}\frac{\partial^2}{\partial \varphi^2}(E_z H_z)$$
$$+ k_c^2(E_z H_z) = 0 \qquad (2.249)$$

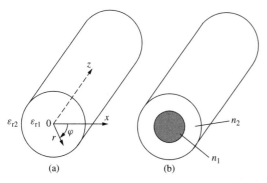

ε_{r2} ε_{r1} 0 x

(a)   (b)

n_2 n_1

Figure 2.55 Cylindrical surface waveguides. (a) Dielectric cylinder and (b) optical cable

with

$$k_c^2 = n_1^2 k_0^2 - \beta^2 = h^2 \qquad (r < a) \qquad (2.250)$$
$$k_c^2 = n_2^2 k_0^2 - \beta^2 = -p^2 \qquad (r > a) \qquad (2.251)$$
$$n_i = \sqrt{\varepsilon_{ri}} \qquad (i = 1, 2) \qquad (2.252)$$

By assuming

$$(E_z H_z) = (AB) R(r)\Phi(\varphi), \qquad (2.253)$$

from Eq. (2.249), we can get

$$\frac{d^2\Phi}{d\varphi^2} + n^2\Phi = 0 \qquad (2.254)$$

$$r^2\frac{d^2 R}{dr^2} + r\frac{dR}{dr} + (h^2 r^2 - n^2)R = 0 \quad (r < a) \qquad (2.255)$$

$$r^2\frac{d^2 R}{dr^2} + r\frac{dR}{dr} - (p^2 r^2 + n^2)R = 0 \quad (r > a) \qquad (2.256)$$

Equations (2.254)–(2.256) indicate that the longitudinal field components are in the following forms:

$$E_z = A_n J_n(hr)\exp(jn\varphi)\exp(-j\beta z) \quad (r < a) \qquad (2.257)$$

$$H_z = B_n J_n(hr)\exp(jn\varphi)\exp(-j\beta z) \quad (r < a) \qquad (2.258)$$

$$E_z = C_n K_n(pr)\exp(jn\varphi)\exp(-j\beta z) \quad (r > a) \qquad (2.259)$$

$$H_z = D_n K_n(pr)\exp(jn\varphi)\exp(-j\beta z) \quad (r > a), \qquad (2.260)$$

where A_n, B_n, C_n, and D_n are amplitude constants, $J_n(hr)$ is the first type Bessel function, and $K_n(pr)$ is the second type modified Bessel function. According to wave propagation equations, we can get the transverse field components (E_r, E_φ, H_r, and H_φ) from the longitudinal field components (E_z and H_z).

According to the boundary conditions at $r = a$, we can determine the relative amplitudes of the field components and get the eigenvalue equation:

$$\left[\frac{k_1^2 J_n'(u_1)}{u_1 J_n(u_1)} + \frac{k_2^2 K_n'(u_2)}{u_2 K_n(u_2)}\right]\left[\frac{J_n'(u_1)}{u_1 J_n(u_1)} + \frac{K_n'(u_2)}{u_2 K_n(u_2)}\right]$$
$$= n^2\beta^2\left(\frac{1}{u_1^2} + \frac{1}{u_2^2}\right)^2, \qquad (2.261)$$

where

$$k_i^2 = \omega \varepsilon_{ri} \varepsilon_0 \mu_0 \quad (i = 1, 2) \qquad (2.262)$$

and the two parameters ($u_1 = ha$ and $u_2 = pa$) satisfy the following equation:

$$u_1^2 + u_2^2 = (n_1^2 - n_2^2)(k_0 a)^2 \qquad (2.263)$$

From Eqs. (2.261) and (2.263), we can calculate the values of u_1 and u_2 from which we can further get the values of h, p, and β. The results obtained are related to the value of n.

For Eq. (2.261), when $n = 0$, the right-hand side vanishes, and one of the two factors should be equal to zero. Actually, the two factors are the eigenvalue equations for the axially symmetric TM and TE modes, respectively:

$$\frac{k_1^2 J_n'(u_1)}{u_1 J_n(u_1)} + \frac{k_2^2 K_n'(u_2)}{u_2 K_n(u_2)} = 0 \quad \text{(TM modes)} \qquad (2.264)$$

$$\frac{J_n'(u_1)}{u_1 J_n(u_1)} + \frac{K_n'(u_2)}{u_2 K_n(u_2)} = 0 \quad \text{(TE modes)} \qquad (2.265)$$

TM_{0i} and TE_{0i} modes are degenerate, and their cutoff wavelength is given by

$$\lambda_{c,0i} = \frac{2\pi a \sqrt{n_1^2 - n_2^2}}{v_{0i}}, \qquad (2.266)$$

where v_{0i} ($i = 1, 2, 3, \ldots$) is the root of zero order Bessel function. TM_{01} and TE_{01} modes have the longest cutoff wavelength:

$$\lambda_{c,01} = \frac{2\pi a \sqrt{n_1^2 - n_2^2}}{2.405} \qquad (2.267)$$

It should be indicated that pure TM or TE modes are possible only if the field is independent of the angular coordinate ($n = 0$). As the radius of the rod increases, the number of TM and TE modes also increases. When the field depends on the angular coordinate ($n \neq 0$), pure TM or TE modes no longer exist. All modes with angular dependence are a combination of a TM and a TE mode, and are classified as hybrid EH or HE modes, depending on whether the TM or TE mode predominates, respectively. For hybrid EH_{ni} and HE_{ni} modes, the solutions for Eq. (2.261) are quite complicated, and usually numerical methods are needed.

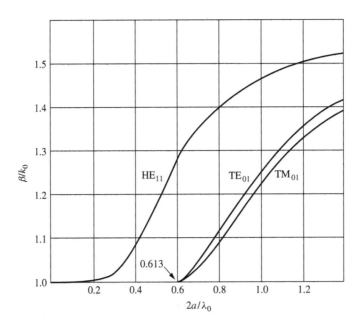

Figure 2.56   Ratio of β to k_0 for the first three surface-wave modes on a polystyrene rod with $\varepsilon_r = 2.56$ (Collin 1991, p722). Source: Collin, R. E. (1991). *Field Theory of Guided Waves*, 2nd ed., IEEE Press, Piscataway, NJ, 1991. © 2003 IEEE

All the hybrid modes ($n \neq 0$), with the exception of the HE_{11} mode, exhibit cutoff phenomena similar to those of the axially symmetric modes. For $n = 1$, the cutoff condition for HE_{1i} mode is $J_1(u_1) = 0$, and the cutoff wavelength for HE_{11} is infinity. Since the HE_{11} mode has no low-frequency cutoff, it is the dominant mode. Figure 2.56 shows the relationship between β and λ_0 of the three lowest surface wave modes (HE_{11}, TM_{01}, TE_{01}) of a polystyrene rod in air. It is clear that if the diameter of the rod is less than $0.613\lambda_0$, only the HE_{11} mode can propagate.

Figure 2.57 shows the field distribution of HE_{11} mode. The field distribution of HE_{11} mode is quite similar to that of TE_{11} mode in a circular waveguide. So the HE_{11} mode of a dielectric rod can be excited using a circular waveguide, as shown in Figure 2.58.

As shown in Figure 2.59, if we place an infinitely large ideal conducting plane at the center of the dielectric cylinder, as the electric field is perpendicular to the conducting plane, the field distribution is not affected. So we can move away the half below the conducting plane, and such a structure is usually called *cylindrical image guide*. As the electromagnetic energy is concentrated on the space close to the dielectric material, such structure does not require very wide conducting plane. In practical applications, the conducting plane is also used as a support to the image guide.

2.2.4.5 Coaxial surface-wave transmission structure

As shown in Figure 2.60, a conducting cylinder covered with a dielectric layer can also support surface waves. Among the possible propagation modes, the TM_{01} one has no low-frequency cutoff. Coaxial surface-wave structures have the advantage that Maxwell's equations for the structures can be solved rigorously. Usually, a perfectly conducting cylinder and a lossless dielectric coating are assumed.

In a cylindrical coordinate system, if we assume longitudinal field components in the form $G(r, \varphi) = R(r) \exp(jv\varphi)$, we have the following differential equation (Marincic *et al.* 1986):

$$\frac{d^2 R}{dr^2} + \frac{1}{r} \cdot \frac{dR}{dr} + \left(\varepsilon_r k_0^2 - \beta^2 - \frac{v^2}{r^2} \right) R = 0 \tag{2.268}$$

Equation (2.268) has to be solved in two regions: the dielectric region and the outer space. In the dielectric region, ε_r is the relative permittivity of the dielectric, while in the outer region $\varepsilon_r = 1$.

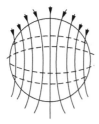

Figure 2.57 Field distribution of HE_{11} mode

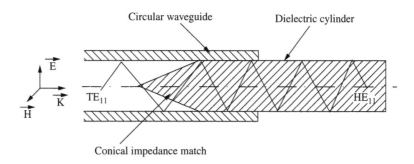

Figure 2.58 Excitation of HE_{11} mode surface wave on a dielectric cylinder by a circular waveguide in TE_{11} mode (Musil and Zacek 1986). Reprinted from Musil, J. and Zacek, F. (1986). *Microwave Measurements of Complex Permittivity by Free Space Methods and Their Applications* with permission from Elsevier, Amsterdam

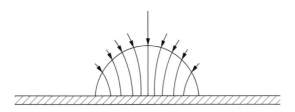

Figure 2.59   Cylindrical image guide

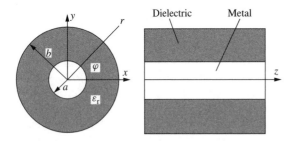

Figure 2.60   Cross section of a coaxial surface-wave guide

The solutions for Eq. (2.268) are either Bessel functions of the first and second kind, $J_v(x)$ and $Y_v(x)$, or modified Bessel functions of the first and second kind, $I_v(x)$ and $K_v(x)$. The selection of solutions depends on the sign of $(k_0^2 \varepsilon_r - \beta^2)$. If this term is positive, the solutions are Bessel functions of the first and second kind. In the opposite case, the solutions are the modified Bessel functions.

It can be shown that the phase coefficient β must lie between the limits (Marincic *et al.* 1986):

$$1 \le \beta / k_0 \le \sqrt{\varepsilon_r} \qquad (2.269)$$

Here, we introduce following two parameters u and w:

$$u^2 = k_0^2 \varepsilon_r - \beta^2 \qquad (2.270)$$

$$w^2 = \beta^2 - k_0^2 \qquad (2.271)$$

If β satisfies Eq. (2.269), the parameters u and w are real.

It can be shown that (Collin 1991) the characteristic equation for TM modes is

$$\frac{u}{\varepsilon_r} \frac{J_0(ua)Y_0(ub) - Y_0(ua)J_0(ub)}{Y_0(ua)J_1(ub) - J_0(ua)Y_1(ub)} = \frac{w K_0(wb)}{K_1(wb)}, \qquad (2.272)$$

and the characteristic equation for TE modes is:

$$-u \frac{J_1(ua)Y_0(ub) - Y_1(ua)J_0(ub)}{J_1(ua)Y_1(ub) - Y_1(ua)J_1(ub)} = \frac{w K_0(wb)}{K_1(wb)} \qquad (2.273)$$

Equation (2.272) gives a solution for the wave that has no low-frequency cutoff, and in fact it represents the eigenvalue equation for the TM_{0m} modes. The lowest-order mode is TM_{01}, which closely resembles coaxial line TEM mode in the dielectric region. This type of wave is known as the Sommerfeld–Goubau wave (Goubau 1950). Equation (2.273) is the eigenvalue equation for the TE_{0m} modes. If $v \ne 0$, the boundary conditions for TE or TM modes cannot be satisfied, while those for hybrid HE and EH modes can be satisfied. The TE_{0m} and all hybrid modes have low-frequency cutoff.

Usually, TM_{01} mode is the prime mode. The field components of TM_{01} mode in the dielectric and free-space region are (Marincic *et al.* 1986)

$$E_\varphi(r) = 0, \; H_r(r) = 0 \qquad (2.274)$$

In the dielectric region ($a \le r \le b$)

$$E_z(r) = A J_0(ur) + B Y_0(ur) \qquad (2.275)$$

$$E_r(r) = j \frac{\beta}{u} [A J_1(ur) + B Y_1(ur)] \qquad (2.276)$$

$$H_\varphi(r) = j \frac{\omega \varepsilon_r \varepsilon_0}{u} [A J_1(ur) + B Y_1(ur)] \qquad (2.277)$$

In the free space region ($r \ge b$)

$$E_z(r) = C K_0(wr) \qquad (2.278)$$

$$E_r(r) = -j \frac{\beta}{w} C K_1(wr) \qquad (2.279)$$

$$H_\varphi(r) = -j \frac{\omega \varepsilon_0}{w} C K_1(wr) \qquad (2.280)$$

The constants A, B, and C satisfy the following relations:

$$B = -A [J_0(ua)/Y_0(ua)] \qquad (2.281)$$

$$C = A \frac{J_0(ub)Y_0(ua) - J_0(ua)Y_0(ub)}{K_0(wb)Y_0(ua)} \qquad (2.282)$$

The surface waves on a coaxial surface waveguide are usually launched and received using horns. Figure 2.61 shows an example for launching

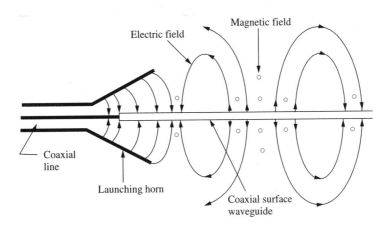

Figure 2.61 Launching of surface waves on a coaxial surface waveguide. Modified from Friedman, M. and Fernsler, R. F. (2001). "Low-loss RF transport over long distance", *IEEE Transactions on Microwave Theory and Techniques*, **49** (2), 341–348. © 2003 IEEE

surface waves on a coaxial surface waveguide (Friedman and Fernsler 2001).

In the design of a coaxial surface waveguide, it is important to know the radius that determines the contour through which a certain specified amount of power is transmitted. Another important factor is the cross section through which a certain specified amount of power is transmitted. Discussions on these two issues can be found in (Marincic *et al.* 1986).

2.2.5 Free space

Free space is an important wave transmission scheme in communication and materials research. In radar and satellite communication, electromagnetic waves propagate through free-space. In materials property characterization, free-space provides much flexibility in studying electromagnetic materials under different conditions. In this section, we first introduce antennas as transitions from guided lines to free-space, and then we discuss two types of electromagnetic waves in free-space: parallel electromagnetic beams and focused electromagnetic beams.

2.2.5.1 Antenna as transition

Antennas are designed to efficiently radiate electromagnetic energy into free-space, while transmission lines are designed to efficiently transport the

energy from one point to another without significant loss or dispersion. As shown in Figure 2.62, antennas can be taken as a transition from a transmission line to free space. In actual applications, a piece of transmission line is used to transport electromagnetic energy from the signal source to the antenna, or from the antenna to the receiver. So antennas can be classified into transmitting antennas and receiving antennas. Sometimes, an antenna is used as both transmitting and receiving antenna.

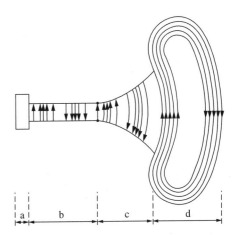

Figure 2.62 Antenna as a transition device. (a) Source, (b) transmission line, (c) antenna, and (d) radiated free-space wave (Balanis 1997). Source: Balanis, C. A. (1997), *Antenna Theory: Analysis and Design*, John Wiley, New York

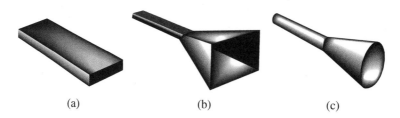

(a)   (b)   (c)

Figure 2.63 Configurations of typical aperture antennas. (a) Rectangular waveguide, (b) pyramidal horn, and (c) conical horn (Balanis 1997). Source: Balanis, C. A. (1997), *Antenna Theory: Analysis and Design*, John Wiley, New York

Figure 2.63 shows three types of aperture antennas often used in microwave electronics and the characterization of electromagnetic materials. An open-end hollow pipe, such as rectangular waveguide, can be taken as an antenna, and such antennas can be used in materials property measurement using reflection method. In materials characterization, pyramidal horn and conical horn are often used in reflection and transmission measurements in free-space. Besides, antennas sometimes are covered with a dielectric material to protect them from hazardous conditions of the environment. In addition, as will be discussed later, sometimes dielectric lens is used to control the beam shape.

For the characterization of electromagnetic properties of materials, important electromagnetic parameters describing antennas mainly include polarization, radiation pattern, power gain, bandwidth, and reciprocity. In the following, we make a brief discussion on these parameters, and detailed discussions on them can be found in (Balanis 1997; Connor 1989; Wait 1986).

Generally speaking, the electromagnetic wave radiated from an antenna is in elliptical polarization. Linear and circular polarizations are two special cases of the general form of elliptical polarization. In experiments, linear polarizations are further classified into vertical or horizontal polarizations. Circular polarization can be taken as a combination of vertical and horizontal polarizations. In materials property characterization, electromagnetic waves are usually linearly polarized though other types of polarization may be used for specific purposes on occasions. For example, in the measurement of chirality, which will be discussed

in Chapter 10, circularly polarized electromagnetic waves are often used.

Radiation pattern is one of the most important properties of an antenna. For a transmitting antenna, the pattern is a graphical plot of the power or field strength radiated by the antenna in different angular directions. Based on the principle of reciprocity, the transmitting and receiving radiation patterns of a given antenna are the same. Omnidirectional and pencil-beam patterns are two typical polar patterns. In an omnidirectional pattern, the antenna radiates or receives energy equally in all directions, while in a pencil-beam pattern, the antenna radiates or receives energy mainly in one direction. A typical pencil-beam pattern consists of a main lobe and a number of side lobes. The level of the side lobes in a pencil-beam pattern must be kept to a minimum as the energy in the side lobes is wasted when the antenna is transmitting radiations and the side lobes can pick up noise and interference when the antenna is receiving radiations.

Two gain parameters are related to the radiation pattern: power gain and directive gain. Power gain G is normally defined in the direction of maximum radiation per unit area as

$$G = \frac{P}{P_0}, \qquad (2.283)$$

where P is the power radiated by the antenna under study, while P_0 is the power radiated by a reference antenna. A reference antenna usually is assumed to be lossless and radiates equally in all directions. The power gain G is expressed as a pure number or in dB. Equation (2.283) assumes that the input power to the given antenna is the same as the input power to the reference antenna.

If the antenna is lossless, its directive gain D is defined by

$$D = \frac{P_{max}}{P_{ave}}, \qquad (2.284)$$

where P_{max} is the maximum power radiated per unit solid angle, while P_{ave} is the average power radiated per unit solid angle. From their definitions, G is slightly less than D.

Working bandwidth is also an important parameter describing an antenna. Three expressions for the working frequency band are often used. For an antenna with the lowest working frequency f_1 and the highest working frequency f_2, the absolute bandwidth of the antenna is

$$\Delta f = f_2 - f_1 \qquad (2.285)$$

The relative bandwidth is

$$\frac{\Delta f}{f} = 2 \frac{f_2 - f_1}{f_2 + f_1} \qquad (2.286)$$

The bandwidth ratio is expressed as: $f_2 : f_1$.

Reciprocity is another important consideration in selecting antennas for communication and materials characterization. The properties of a transmitting antenna are very similar to those of a receiving antenna because of the theorem of reciprocity. It is often assumed that the antenna parameters, such as polar power gain and bandwidth, are the same whenever the antenna is used for transmitting or receiving electromagnetic waves.

2.2.5.2 Parallel microwave beam

The control of the beam shape is important for microwave communication and materials property characterization (Musil and Zacek 1986). In free-space measurements, the sample under test is often put between the transmitting and receiving antennas. In the development of the algorithms for materials property characterization, we assume that all the energy of the plane electromagnetic wave interacts with the sample, and the sample is infinite in the direction perpendicular to the direction of wave propagation so that the diffractions by the edges of the sample are negligible. However, in actual experiments, the dimensions of samples are limited. In some cases, we need to measure samples with small transverse dimensions, for example, in the order of a few centimeters. Therefore, to achieve accurate measurement results, we should control the transverse cross section of the beam. To ensure that the algorithm is applicable for practical measurements, following requirements should be satisfied:

$$D \gg h \gg \lambda_0, \qquad (2.287)$$

where D stands for the transverse dimensions of the plate or specimen, h denotes the transverse dimensions of the probing beam, λ_0 is the wavelength of the electromagnetic wave.

To satisfy the requirements of Eq. (2.287), two kinds of microwave beams are often used: parallel microwave beams and focused microwave beams. As shown in Figure 2.64, a parallel microwave beam can be achieved by using specially designed lenses or reflectors. The function of the lens and the reflector is to transform the spherical wave from a point-source feed to a plane wave.

Though a lens and a reflector may have the same functions in producing parallel beams, the positions of the point-source feeds for a lens and a reflector are different. As shown in Figure 2.64, the point-source feed for a lens is located on one side of the lens and the generated plane wave emerges from the other side of the lens, while for a reflector system, the point-source feed and the generated plane wave lie on the same side of a reflector. As the point-source feed is placed at the region for the outgoing wave, the output parallel beam is disturbed by the source feed. So in materials property characterization, we often use plane waves generated by dielectric lenses. However, in some special cases such as the measurement of radar cross section (RCS), as the object to be measured is large, the required parallel beam should also be large. As it is impractical to build a very large lens, usually reflectors are used to produce parallel beams, and the perturbation of the source feed can be minimized by proper arrangement of the source feed and the reflector.

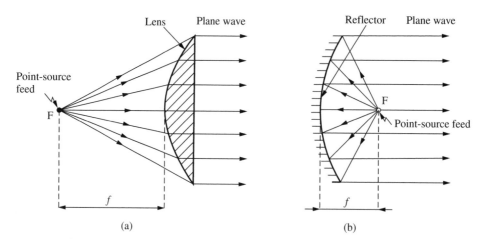

Figure 2.64   The generation of a parallel microwave beam by means of (a) a lens and (b) a reflector. Source: Musil, J. and Zacek, F. (1986). *Microwave Measurements of Complex Permittivity by Free Space Methods and Their Applications*, Elsevier, Amsterdam

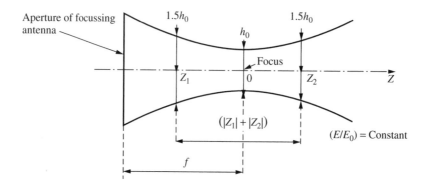

Figure 2.65   Parameters describing a focused microwave beam. Source: Musil, J. and Zacek, F. (1986). *Microwave Measurements of Complex Permittivity by Free Space Methods and Their Applications*, Elsevier, Amsterdam

It should be noted that the focus length f of a lens or a reflector may be different for different frequencies. If we want to generate parallel beams using the configurations shown in Figure 2.64 at frequencies far separated, it may be necessary to adjust the position of the source feed for different frequencies.

2.2.5.3 Focused microwave beam

For the characterization of small samples, focused microwave beams are often required. A focused microwave beam can be produced by focusing microwave energy transmitted from an antenna into a focus point, and its geometry is determined by the spatial distribution of the electric field strength in the beam (Musil and Zacek 1986). As shown in Figure 2.65, a focused microwave beam is usually described by three parameters: focus distance f that determines the position of the focal plane $(z = 0)$, beam width h_0 at the focal plane, and beam depth that is the distance between the two planes $(z = z_1)$ and $(z = z_2)$, where the beam width increases by 50 % compared to the beam width at the focal plane $(z = 0)$. It should be noted that the magnitudes used in defining these parameters are determined at a certain level of field intensity with respect to the field intensity on the axis of the focused beam. The beam width increases with the decrease of the field strength level at which we determine the beam width.

Beam width determines minimum transverse dimensions a specimen should have. If the samples have large transverse dimensions, the beam width determines the spatial resolution for the measurement of local inhomogeneities. Generally speaking, a system with smaller wavelength λ_0, smaller focus distance l, and larger radius of the lens aperture has smaller beam width. The smallest beam width that can be obtained in practice at a level of $-10\,\text{dB}$ is approximately $1.5\lambda_0$ (Muzil 1986).

The phase distribution in a focused microwave beam is also important for materials property characterization using focused microwave beams. In the place near the focal plane $z = 0$, the electromagnetic field can be taken as a plane wave. So we use focused microwave beams in nonresonant methods; we should ensure that the sample under test is placed close to the focal plane ($z = 0$). Therefore beams with larger depths are favorable.

In summary, for materials characterization, we hope to have minimum beam-width and meanwhile a maximum beam depth. Practical configurations using focused microwave beams in materials property characterization will be given in later chapters.

2.3 MICROWAVE RESONANCE

2.3.1 Introduction

The resonant methods for materials property characterization are based on microwave resonance. Generally speaking, there are two kinds of resonant structures: transmission type, which is made from transmission structures, and non-transmission type, such as ring resonators and sphere resonators. In the following discussions, we focus on transmission type resonators, such as rectangular resonator, cylindrical resonator, coaxial resonator, and microstrip resonator.

Similar to microwave transmission, microwave resonance can be studied in both field approach

and line approach. After introducing the basic parameters describing a resonator, we use equivalent circuits to analyze the general properties of microwave resonance, then we discuss the field distributions of several types of resonators often used in materials characterization, including coaxial resonators, planar-circuit resonators, waveguide resonators, dielectric resonators, and open resonators.

2.3.1.1 Resonant frequency and quality factor

Resonance is related to energy exchange, and electromagnetic resonance can be taken as a phenomenon when electric energy and magnetic energy can periodically change totally from one to the other. If the resonance is lossless, the sum of electric energy and magnetic energy does not change with time:

$$W_e(t) + W_m(t) = W_0 \qquad (2.288)$$

Resonant frequency is the frequency when the electric energy can be totally changed to magnetic frequency, and vice versa. Resonant frequency f_0 is the most important parameter for a resonator. The electric energy and magnetic energy can be calculated from the field distributions in the resonator. The electromagnetic field distribution and the resonant frequency can be found by solving the wave functions with certain boundary conditions:

$$\nabla^2 \mathbf{E} + k^2 \mathbf{E} = 0 \qquad (2.289)$$

$$\nabla^2 \mathbf{H} + k^2 \mathbf{H} = 0, \qquad (2.290)$$

where $k^2 = \omega^2 \varepsilon \mu$, and ε and μ are the permittivity and permeability of the medium in the resonator. For an ideal lossless resonator, k is a series of discrete real numbers $k_1, k_2, \ldots$, and they are called the eigenvalues of Eqs. (2.289) and (2.290). The resonant frequencies can be calculated from the eigenvalues:

$$f_i = \frac{c}{2\pi} k_i \quad (i = 1, 2, \ldots), \qquad (2.291)$$

where c is the speed of light in free space.

Quality factor is defined as

$$Q_0 = 2\pi \frac{W}{P_L T_0} = \omega_0 \frac{W}{P_L}, \qquad (2.292)$$

where W is the total energy storage in the cavity, P_L is the average energy dissipation within the cavity, and T_0 is the resonant period. At resonance, the total energy storage equals the maximum electric energy or maximum magnetic energy:

$$W = W_{e,max} = \frac{\varepsilon}{2} \iiint_V |E|^2 dV$$

$$= \frac{\mu}{2} \iiint_V |H|^2 dV = W_{m,max} \qquad (2.293)$$

The above integration is made over the whole resonator.

Here we consider a hollow metallic cavity filled with a dielectric medium. If the medium within the cavity is lossless the energy dissipation power is caused by the cavity wall:

$$P_L = \frac{R_s}{2} x \oiint_S |H_t|^2 dS \qquad (2.294)$$

The above integration is made over the whole cavity wall. The surface resistance of the cavity wall R_s in Eq. (2.294) is given by:

$$R_s = \frac{\omega \mu_1 \delta}{2}, \qquad (2.295)$$

where μ_1 is the permeability of the conductor, δ is the penetration depth of the conductor. According to Eqs. (2.292)–(2.295), we have

$$Q_0 = \frac{2}{\delta} \cdot \frac{\displaystyle\iiint_V |H|^2 dV}{\displaystyle\oiint_S |H_t|^2 dS} \qquad (2.296)$$

If we also consider the dielectric loss of the medium in the cavity, the value of Q_0 will be decreased:

$$Q_0 = \frac{\omega_0 W}{P_L} = \frac{\omega_0 W}{P_c + P_d}$$

$$= \frac{1}{[P_c/(\omega_0 W) + P_d/(\omega_0 W)]}$$

$$= \frac{1}{1/Q_{0c} + 1/Q_{0d}}, \qquad (2.297)$$

where P_c is the energy dissipation due to conductor loss, P_d is the energy dissipation due to the dielectric loss, Q_{0c} and Q_{0d} are the quality factors of the resonator if we only consider the conductor loss and dielectric loss, respectively. If the conductivity of the dielectric medium within the cavity is σ, we have

$$P_d = \frac{1}{2}\sigma \iiint_V |E|^2 dV \qquad (2.298)$$

So the value of Q_{0d} can be calculated from

$$Q_{0d} = \frac{\omega_0 W}{P_d} = \frac{\omega_0 \cdot \frac{1}{2}\varepsilon'_r \displaystyle\iiint_V |E|^2 dV}{\frac{1}{2}\sigma \displaystyle\iiint_V |E|^2 dV}$$

$$= \frac{\omega_0 \varepsilon'_r}{\sigma} = \frac{1}{\tan \delta}, \qquad (2.299)$$

where ε'_r is the dielectric constant of the medium in the cavity, $\sigma = \omega_0 \varepsilon''$, and $\tan \delta = \varepsilon''_r / \varepsilon'_r$.

The relationship between the energy storage and energy dissipation can also be described by the attenuation parameter α, which is related to the attenuation rate of a resonator after the source is removed. By defining $E = E_0 e^{-\alpha t}$, we have

$$W = W_0 e^{-2\alpha t}, \qquad (2.300)$$

where W_0 is the energy storage when $t = 0$. So the energy dissipation is given by

$$P_L = -\frac{dW}{dt} = 2\alpha W \qquad (2.301)$$

From Eqs. (2.292) and (2.301), we can get

$$\alpha = \frac{P_L}{2W} = \frac{\omega_0}{2Q_0} \qquad (2.302)$$

From Eqs. (2.300) and (2.302), we have

$$W = W_0 e^{-\frac{\omega_0}{Q_0} t} \qquad (2.303)$$

Equation (2.303) indicates that the higher the quality factor, the slower the resonance attenuation.

2.3.1.2 Equivalent circuits of resonant structure

As shown in Figure 2.66, depending on the selection of reference plane, a resonator can be represented by a series equivalent circuit or a parallel equivalent circuit. If the reference plane

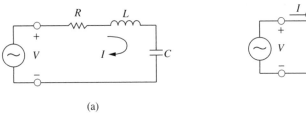

(a)   (b)

Figure 2.66 Equivalent RLC circuits for a resonator. (a) Series circuit and (b) parallel circuit

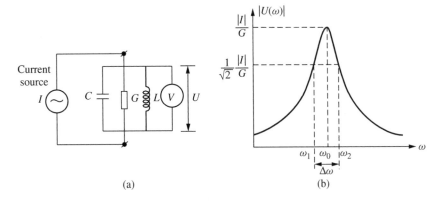

(a)   (b)

Figure 2.67 Frequency response of a resonator. (a) Parallel equivalent circuit of a resonator. The relationship between conductance G and resistance R is: $G = 1/R$. (b) Characteristic frequencies on a resonant curve

is chosen at a place where the electric field integration is zero (correspondingly the magnetic field integration there is the maximum), the resonator can be represented by a series RLC circuit, and if the reference plane is chosen at a plane where the electric field integration is the maximum (correspondingly the magnetic field integration there is zero), the resonator can be represented by a parallel RLC equivalent circuit. The following discussions focus on parallel equivalent circuits.

As shown in Figure 2.67, when a current source is connected to the resonator, a voltage U is built across the resonator. The voltage U will change with the change of frequency:

$$U = \frac{I}{G + j\omega C + 1/(j\omega L)} \qquad (2.304)$$

For a parallel circuit, the voltage reaches its highest value at resonance. From Eq. (2.304), we have

$$j\omega_0 C + \frac{1}{j\omega_0 L} = 0 \qquad (2.305)$$

So we can get the resonant frequency:

$$\omega_0 = \frac{1}{\sqrt{LC}} \qquad (2.306)$$

From Eqs. (2.304)–(2.306), we can get

$$U = \frac{1}{G + jC[\omega - 1/(\omega LC)]} = \frac{1}{G + jC\Delta\omega} \qquad (2.307)$$

with

$$\Delta\omega = 2(\omega - \omega_0) \qquad (2.308)$$

If $\Delta\omega = \pm G/C$, the voltage amplitude decreases to $1/\sqrt{2}$ of the maximum value:

$$|U(\omega)| = \frac{|I|}{\sqrt{2}G} = \frac{|U(\omega_0)|}{\sqrt{2}} \qquad (2.309)$$

From Eq. (2.309), we can get

$$U(\omega)U^*(\omega) = \frac{1}{2}U(\omega_0)U^*(\omega_0) \qquad (2.310)$$

As the energy dissipation at G is given by

$$P_d = \frac{1}{2}UU^*G, \qquad (2.311)$$

it also decreases to half of the maximum value. Consider the average energy storage at the capacitor and inductor:

$$W_e = \frac{1}{4}UU^*C \qquad (2.312)$$

$$W_m = \frac{1}{4}I_L I_L^* L = \frac{1}{4}L\left(\frac{U}{j\omega L}\right)\left(\frac{U}{j\omega L}\right)^*$$

$$= \frac{1}{4}\frac{1}{\omega^2 L}UU^* = \frac{1}{4}CUU^*\frac{\omega_0^2}{\omega^2} \qquad (2.313)$$

Equation (2.312) indicates that the energy storage at the capacitor decreases to half of its maximum value. As the frequency (ω) is close to the resonant frequency (ω_0), Eq. (2.313) indicates that the energy storage at the inductor also decreases to half the value of its maximum value.

The difference between two half-power frequencies ($\omega_1 = \omega_0 - G/(2C)$ and $\omega_1 = \omega_0 + G/(2C)$) is called the *half-power bandwidth*:

$$\Delta\omega = \omega_2 - \omega_1 = \frac{G}{C} \qquad (2.314)$$

Sometimes, the half power bandwidth is described as

$$\Delta f = \frac{\Delta\omega}{2\pi} = \frac{1}{2\pi}\frac{G}{C} \qquad (2.315)$$

It is clear that the narrower the half-power bandwidth, the better the frequency selectivity of the resonator.

The quality of the resonator can also be defined as

$$Q = \frac{\omega_0}{\Delta\omega} = \omega_0\frac{C}{G} = \omega_0\frac{(1/2)UU^*C}{(1/2)UU^*G} = \omega_0\frac{W}{P_d} \qquad (2.316)$$

Comparing Eqs. (2.292) and (2.316) indicates that the two definitions are consistent, and in experiments, Eq. (2.316) is more often used.

Equations (2.312) and (2.313) indicate that at resonant frequency, the electric field energy in the

resonator equals the magnetic field energy in the resonator.

$$W_e = W_m = \frac{1}{4}CUU^* \qquad (2.317)$$

Actually, Eq. (2.317) can be used as a criteria to determine the resonant frequency of a resonator. For a parallel equivalent circuit, if the working frequency is higher than the resonant frequency ($\omega > \omega_0$), the electric field energy is larger than the magnetic field energy; if the working frequency is lower than the resonant frequency ($\omega < \omega_0$), the magnetic field energy is higher than the electric field energy. For a series equivalent circuit, contrary conclusions can be obtained.

2.3.1.3 Coupling to external circuit

An actual resonator is always coupled to external circuits. Through coupling, the source provides energy to the resonator, and such a procedure is usually called *excitation*. A resonator can also provide energy to an external load through coupling, and such a procedure is usually called *loading*.

The coupling mechanisms often used generally fall into three categories: electrical coupling, magnetic coupling, and mixed coupling. In electrical coupling, an electrical dipole, usually a coaxial needle made from the central conductor of a coaxial line, is inserted into the place with maximum electric field. In magnetic coupling, a magnetic dipole, usually a coaxial loop made by connecting the central conductor to the outer conductor is often placed at the place with maximum magnetic field. A typical example of mixed coupling is the coupling iris between a hollow cavity and a metallic waveguide. In the equivalent circuits, a coupling structure is often represented by a transformer.

Cavity coupled to one transmission line

Figure 2.68 shows the equivalent circuits of a resonator coupled to a transmission line. As shown in Figure 2.68(b), if the position of the reference plane T_s is suitably chosen, the resonator is represented by a parallel RLC circuit, and the coupler is represented by an ideal transformer with transforming ratio $(1 : n)$. As shown in Figure 2.68(c), the parallel RLC circuit can be transformed to T_s plane through the transformer: $C' = n^2 C$, $R'_P = R_P/n^2$, and $L' = L/n^2$.

After the transformation, the resonant frequency and quality factor of the equivalent circuit do not change:

$$\omega'_0 = \frac{1}{\sqrt{L'C'}} = \frac{1}{\sqrt{(L/n^2)(n^2 C)}} = \frac{1}{\sqrt{LC}} = \omega_0 \tag{2.318}$$

$$Q'_0 = \omega'_0 C' R'_P = \omega_0 (n^2 C) \left(\frac{R_P}{n^2}\right) = \omega_0 C R_P = Q_0 \tag{2.319}$$

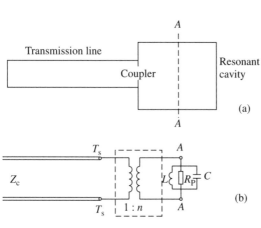

(a)

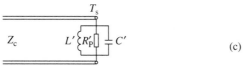

(c)

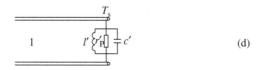

(d)

Figure 2.68 A resonant cavity coupled to a transmission line and its equivalent circuits. (a) A cavity coupled to a transmission line, (b) an equivalent circuit of the cavity with coupling structure and transmission line, (c) the equivalent circuit transformed to the T_s plane, (d) normalized equivalent circuit

The coupling coefficient β describes the relationship between the energy dissipation in the cavity P_d and the energy dissipation of the external circuit P_e, and indicates the extent to which the external circuit is coupled to the resonator:

$$\beta = \frac{P_e}{P_d} = \frac{\dfrac{1}{2}\dfrac{U^2}{Z_c}}{\dfrac{1}{2}\dfrac{U^2}{R_P/n^2}} = \frac{R_P}{n^2 Z_c} = \frac{R'_P}{Z_c} = r'_P, \tag{2.320}$$

where U is voltage applied to the resonator. Equation (2.320) indicates that the coupling coefficient equals the normalized resonance resistance of the cavity transformed to the plane T_s. At a critical coupling state ($\beta = 1$), the energy dissipation of the external circuit equals the energy dissipation in the cavity ($P_e = P_d$). At an over-coupling state ($\beta > 1$), the energy dissipation of the external load is larger than the energy dissipation in the cavity ($P_e > P_d$). At an under-coupling state ($\beta < 1$), the energy dissipation of the external load is less than the energy dissipation in the cavity ($P_e < P_d$).

The external quality factor of a resonator, coupled to an external circuit, describes the energy dissipation in the external circuit:

$$Q_e = \omega_0 \frac{W}{P_e}, \tag{2.321}$$

where W is the energy storage in the resonator and P_e is the energy dissipation in the external circuit coupled to the resonator. From Eqs. (2.297), (2.320), and (2.321), we can get

$$\beta = \frac{Q_0}{Q_e} \tag{2.322}$$

So the coupling coefficient equals the ratio between the intrinsic quality factor and external quality factor of the resonator.

In experiments, loaded quality factor Q_L is often used:

$$Q_L = \omega_0 \frac{W}{P_0 + P_e} \tag{2.323}$$

The relationships between the loaded quality factor Q_L, intrinsic quality factor Q_0, external quality

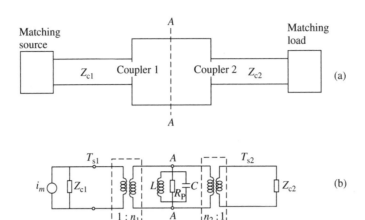

Figure 2.69   (a) A resonator coupled to two transmission lines and (b) its equivalent circuit

factor Q_e, and coupling coefficient β are

$$Q_L = Q_0 \frac{1}{1+\beta} = Q_e \frac{\beta}{1+\beta} \quad (2.324)$$

$$\frac{1}{Q_L} = \frac{1}{Q_0} + \frac{1}{Q_e} \quad (2.325)$$

Cavity coupled to two transmission lines

Figure 2.69(a) shows a resonator coupled to two transmission lines. One transmission line is connected to a matching source and the other transmission line is connected to a matching load. The equivalent circuit of the resonator is shown in Figure 2.69(b), and the two transformers represent the two couplers.

Using a similar method used in analyzing the cavity coupled to one transmission line, we can get the parameters describing the resonant and coupling properties of the resonator. The intrinsic quality factor Q_0 of the resonator is given by

$$Q_0 = \omega_0 C R_P \quad (2.326)$$

The external quality factors for the two transmission lines are

$$Q_{e1} = \omega_0 C n_1^2 Z_{c1} \quad (2.327)$$

$$Q_{e2} = \omega_0 C n_2^2 Z_{c2} \quad (2.328)$$

The coupling coefficients of the two couplers are

$$\beta_1 = \frac{Q_0}{Q_{e1}} = \frac{R_P}{n_1^2 Z_{c1}} \quad (2.329)$$

$$\beta_2 = \frac{Q_0}{Q_{e2}} = \frac{R_P}{n_2^2 Z_{c2}} \quad (2.330)$$

The loaded quality factor

$$Q_L = \omega_0 C \frac{1}{1/R_P + 1/(n_1^2 Z_{c1}) + 1/(n_2^2 Z_{c2})}$$

$$= Q_0 \frac{1}{1 + \beta_1 + \beta_2} \quad (2.331)$$

$$\frac{1}{Q_L} = \frac{1}{Q_0} + \frac{1}{Q_{e1}} + \frac{1}{Q_{e2}} \quad (2.332)$$

Equations (2.331) and (2.332) are the extensions of Eqs. (2.324) and (2.325), respectively. We can analyze a resonator coupled to more transmission lines in a similar way, and similar conclusions can be drawn.

In the following, we discuss several typical types of resonators, including coaxial resonators, planar circuit resonators, waveguide resonators, dielectric resonators, surface wave resonators, and open resonators. As will be discussed in Chapters 5 and 6, these resonators are often used in the characterization of materials properties

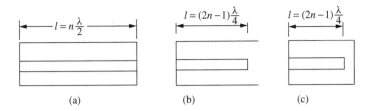

Figure 2.70 Three typical types of coaxial resonators. (a) Half-wavelength resonator, (b) quarter wavelength resonator, and (c) capacitor-loaded resonator

using resonator methods and resonant-perturbation methods.

2.3.2 Coaxial resonators

Coaxial resonators are made from coaxial transmission lines. As shown in Figure 2.70, there are three typical types of coaxial resonators: half-wavelength resonator, quarter wavelength resonator, and capacitor-loaded resonator.

2.3.2.1 Half-wavelength resonator

As shown in Figure 2.70(a), a half-wavelength coaxial resonator is a segment of coaxial line with length l terminated at two ends. Usually, a coaxial resonator works at TEM mode. To avoid the possible resonance along φ direction, following requirements should be satisfied:

$$\pi(a + b) < \lambda_{\min}, \tag{2.333}$$

where a and b are the radius of the inner and outer conductors respectively, and $\lambda_{\min}$ is the shortest wavelength corresponding to the highest working frequency.

The field distributions can be calculated according to the boundary conditions:

$$E_r = -j\frac{2E_0}{r}\sin\beta z \tag{2.334}$$

$$H_\varphi = Y_0\frac{2E_0}{r}\cos\beta z, \tag{2.335}$$

where $E_0 = V_0/\ln(b/a)$, $Y_0 = (\varepsilon/\mu)^{1/2}$, and $\beta = n\pi/l$ ($n = 1, 2, 3, \ldots$). As $\beta = 2\pi/\lambda_0$, we can get

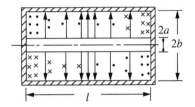

Figure 2.71 Field distribution of a half-wavelength resonator ($n = 1$). The solid lines stand for electric fields. The dots stand for the magnetic fields coming out from the paper, while the crosses stand for magnetic field going into the paper

the relationship between the resonant wavelength λ_0 and the length of the resonator l:

$$l = n\frac{\lambda_0}{2} \tag{2.336}$$

Equation (2.336) indicates that the length of the resonator is integral times of the half wavelength, so such a resonator is called *half-wavelength resonator*, and its field distribution is shown in Figure 2.71.

In microwave engineering, the waveform factor is often used, which is defined as: $Q_0\delta/\lambda_0$, where δ is the penetration depth of the conductor. Waveform factor is only related to the geometry, size, and working mode of the resonator. The waveform factor of a half-wavelength resonator ($n = 1$) is given by

$$Q_0\frac{\delta}{\lambda_0} = \frac{1}{4 + \dfrac{l}{b} \cdot \dfrac{1 + (b/a)}{\ln(b/a)}} \tag{2.337}$$

It can be proven that when $(b/a) = 3.6$, the waveform factor has the highest value. The quality factor can be easily calculated from the waveform factor.

2.3.2.2 Quarter-wavelength resonator

As shown in Figure 2.70(b), in a quarter-wavelength resonator, one end is shorted, and the other end is open. As an open load also causes total reflection, pure standing wave is also built. According to the boundary conditions, we have: $\beta l = (2n+1)\pi/2$ $(n = 0, 1, 2, \ldots)$, so the relationship between the length of the resonator l and resonant wavelength λ_0 is given by

$$l = (2n + 1)\frac{\lambda_0}{4} \quad (n = 0, 1, 2, \ldots) \quad (2.338)$$

Equation (2.338) indicates that the length of the resonator is odd number times of the quarter wavelength. The field distribution of a quarter-wavelength resonator is shown in Figure 2.72.

In an actual structure, an open end has some radiation loss. To minimize the radiation loss, usually the outer conductor is extended, forming a segment of cutting-off TM_{01} circular waveguide. However, owing to the capacitance introduced by extending the outer conductor, the inner conductor is usually made a bit shorter than the quarter wavelength.

The quality factor of a quarter-wavelength resonator can be calculated from the waveform factor, which is given by

$$Q_0\frac{\delta}{\lambda_0} = \frac{1}{4} \cdot \frac{1}{4 + \dfrac{l}{2b}\dfrac{1 + (b/a)}{\ln(b/a)}} \quad (2.339)$$

Similar to half-wavelength resonator, the waveform factor has the highest value at $(b/a) = 3.6$.

2.3.2.3 Capacitor-loaded resonator

In a capacitor-loaded resonator, shown in Figure 2.70(c), the two ends of the coaxial line are shorted, but at one end of the resonator, there is a small gap between the inner conductor and the short plate. As shown in Figure 2.73(a), the gap between the inner conductor and short plate is equivalent to a lumped capacitor, so such a resonator can be regarded as a hybrid resonator with both lumped and distributed elements. The field distribution of the basic mode is shown in Figure 2.73(b). At the gap between the inner conductor and the short plate, electric field dominates, and the magnetic field is very weak.

The susceptance B_1 at plane AA' looking into the left is $B_1 = -\cot(\beta l)/Z_c$, with $\beta = \omega/c$, where Z_c is the characteristic impedance of the coaxial line and c is the speed of light. The susceptance B_2 of the lumped capacitor C_0 is $B_2 = \omega C_0$. At resonance, $B_1 + B_2 = 0$: therefore we have

$$Z_c\omega C_0 = \cot(\beta l) \quad (2.340)$$

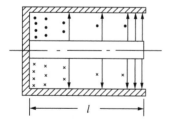

Figure 2.72 The field distribution of quarter wavelength resonator $(n = 0)$. The solid lines stand for electric fields. The dotes stand for the magnetic fields coming out from the paper, while the crosses stand for magnetic field going into the paper

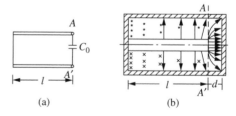

Figure 2.73 Capacitor-loaded resonator. (a) Equivalent circuit and (b) field distribution

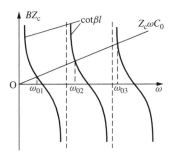

Figure 2.74 Graphical method for solving Eq. (2.340)

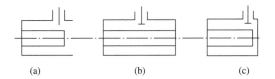

Figure 2.76 Electric couplings. (a) Excitation of quarter-wavelength resonator using a needle dipole, (b) excitation of half-wavelength resonator, and (c) excitation of capacitor-loaded resonator. The small metal plates on the tips of the needles in (b) and (c) are used to increase the couplings

Equation (2.340) can be solved by numerical method and graphical method. The graphical method is shown in Figure 2.74. We draw the graph $BZ_c = \cot(\beta l)$ and $BZ_c = Z_c \omega C_0$. Their crossing points are the solutions of Eq. (2.340), and each point corresponds to a resonant mode.

2.3.2.4 Coupling to external circuit

Coupling method is an important consideration in the design of resonators. Here we introduce magnetic coupling and electric coupling often used for coaxial resonators.

Magnetic coupling

Figure 2.75 shows a magnetic coupling by a co-axial loop often used in coaxial resonators. The loop is placed at the place where the magnetic field dominates, and usually the plane of the loop

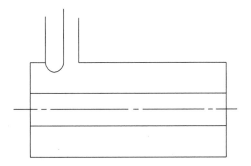

Figure 2.75 Excitation of a half-wavelength resonator using a magnetic coupling

is perpendicular to the magnetic field distribution of the mode. The loop is equivalent to a magnetic dipole, and the coupling can be adjusted by adjusting the orientation and position of the loop.

Electric coupling

As shown in Figure 2.76, electric coupling is often achieved by a needle made from the inner conductor of a coaxial line. The needle should be put at a place where electric field dominates, and the orientation of the needle should be parallel to the electric field. As shown in Figures 2.76(b) and (c), adding a small metal plate on the tip of a coupling needle may increase the coupling.

2.3.3 Planar-circuit resonators

The planar circuits used in microwave electronics mainly include stripline, microstrip, and coplanar. As microstrip is most widely used in microwave integrated circuits, we concentrate our discussion on microstrip circuits, and similar conclusions can be obtained for other types of planar circuits. The application of planar resonators in materials characterization will be discussed in Chapter 7.

In the following, we will make a brief discussion on the three types of microstrip resonators shown in Figure 2.77: straight ribbon resonator, ring resonator, and circular resonator. However, it should be noted that the equations given below are approximate and can only be used for estimation. To make accurate analysis, numerical methods are often required.

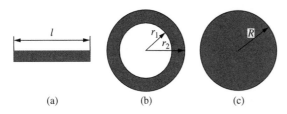

Figure 2.77 Three types of microstrip resonators: (a) straight ribbon resonator, (b) ring resonator, and (c) circular resonator

2.3.3.1 Straight ribbon resonator

A straight ribbon resonator shown in Figure 2.77(a) is a segment of microstrip line with two open ends, and the relationship between the length of the microstrip line l and the resonant wavelength λ_g is given by

$$l = n\frac{\lambda_g}{2} \qquad (n = 1, 2, 3, \ldots) \qquad (2.341)$$

In actual structures, the fields extend slightly beyond the ends of the microstrip line on both ends. The field effects can be represented by a grounding capacitor C_{end} at each end, and C_{end} can be transformed to a length of transmission line Δl. So the resonant condition becomes

$$l + 2\Delta l = n\frac{\lambda_g}{2} \qquad (n = 1, 2, 3, \ldots). \qquad (2.342)$$

The values of C_{end} and Δl can be calculated using some empirical equations, and can also be measured experimentally.

Straight ribbon resonators are a type of transmission type resonators made from their corresponding transmission line, microstrip. However, owing to the radiations at the two open ends, the quality factors of these types of resonators are usually not very high.

2.3.3.2 Ring resonator

A ring resonator shown in Figure 2.77(b) does not have open ends, so the radiation loss is greatly decreased. Usually, a ring resonator has higher quality factor than a straight ribbon resonator. Mainly due to its high quality factor, in materials property characterization, the ring resonator

method has higher accuracy and sensitivity than the straight ribbon resonator method.

Most of the ring resonators work at TM_{mn0}, and the prime mode is TM_{110}. The resonant condition for TM_{m10} is

$$\pi(r_1 + r_2) = m\lambda_g, \qquad (2.343)$$

where r_1 and r_2 are the inner and outer radius of the ring respectively. As $\lambda_g = \lambda_0/(\varepsilon_{eff})^{1/2}$, where ε_{eff} is the effective dielectric constant of a microstrip line, we have

$$\lambda_0 = \frac{\pi(r_1 + r_2)}{m}\sqrt{\varepsilon_{eff}} \qquad (2.344)$$

To avoid higher order modes, usually the dimensions of the ring resonator should satisfy the following requirements:

$$\frac{r_2 - r_1}{r_2 + r_1} < 0.05 \qquad (2.345)$$

2.3.3.3 Circular resonator

As shown in Figure 2.77(c), a circular resonator can be taken as a special case of a ring resonator when $r_2 = 0$. Its resonant modes are also TM_{mno}, and the prime mode is TM_{110} mode.

As the structure of the circular resonator is quite simple, and has high quality factor, it is widely used as resonant element in microwave electronic circuits, and is also often used in materials property characterization.

2.3.3.4 Coupling to external circuit

Here, we discuss the coupling methods for microstrip half-wavelength resonator, while the methods can be extended for other kinds of planar resonators. More discussions on the coupling of planar circuits can be found in (Chang 1989).

Figure 2.78 schematically shows three types of coupling methods for the half wavelength resonators: capacitive coupling, parallel-line coupling, and tap coupling. In the following, we focus on the coupling mechanisms of these methods. However, it should be indicated that to get accurate knowledge about the coupling properties, numerical simulations are often needed.

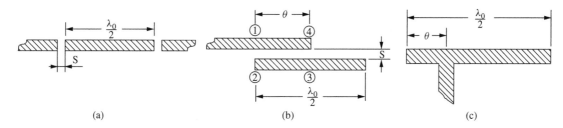

Figure 2.78 Three types of coupling methods. (a) Capacitive coupling, (b) parallel-line coupling, and (c) tap coupling (Chang 1989). Source: Chang, K. (1989). *Handbook of Microwave and Optical Components*, Vol. 1, John Wiley, New York

As shown in Figure 2.78(a), capacitive coupling is achieved by the capacitance resulted from the fringing fields at the open ends of the resonator and the coupling port. The impedance of the coupling ports is usually chosen as the characteristic impedance of the external transmission line. The coupling coefficient is closely related to the width of the gap (S) between the open ends of the resonator and the coupling port. Generally speaking, the coupling coefficient increases with the decrease of the gap. However, when the gap is very small, a small variation in the gap width may result in a large change in the coupling coefficient, so this method requires high fabrication accuracy.

As shown in Figure 2.78(b), a parallel-line coupler consists of a coupled line of an electrical length θ with two open ports (2 and 4). The port 3 is connected to an open-circuited line of an electrical length $(\pi - \theta)$, forming the resonator, and the input port (port 1) is connected to a transmission line with characteristic impedance Z_c. The coupling coefficient is mainly determined by the electrical length θ and the spacing between the parallel lines S. Compared to capacitive coupling, coupled line coupling has more flexibility in adjusting the coupling coefficient.

As shown in Figure 2.78(c), a tap coupling is achieved by directly connecting the coupling port to the resonator at an appropriate position. The coupling coefficient is mainly related to the electrical length θ between the open end of the resonator and taping point. Theoretically, if the taping point is at the center of the resonator, the coefficient is zero because the center point is at the electric field node. The smaller the θ value, the stronger the coupling. Theoretical analysis shows that the coupling

coefficient is proportional to $\cos^2 \theta$. In practical structures, to minimize the effects of the T-junction at the taping point to the resonant properties of the resonator, the input line is often tapered, or a quarter-wavelength transformer is used (Chang 1989).

2.3.4 Waveguide resonators

Waveguide resonators are widely used in materials property characterization, especially in cavity perturbation methods. A waveguide resonator is usually made by shorting the two ends of a segment of waveguide. According to the types of waveguides from which the resonators are made, two kinds of waveguide resonators are often used: rectangular cavity resonators and circular cavity resonators. Corresponding to the wave propagation modes in waveguides, resonant modes include TE and TM modes.

2.3.4.1 Rectangular cavity resonator

Figure 2.79 shows a rectangular cavity resonator made from a rectangular waveguide by shorting the two ends. Its structural dimensions include width a, height b, and length c. For a rectangular cavity resonator, there are two groups of resonant modes (TE_{mnp} and TM_{mnp}), which correspond to the TE_{mn} and TM_{mn} propagation modes in a rectangular waveguide respectively. The first two subscripts m and n come from the wave propagation modes TE_{mn} and TM_{mn}, and they represent the changing cycles along the x and y directions. The last

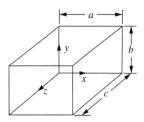

Figure 2.79   Structure of a rectangular cavity resonator

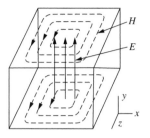

Figure 2.80   Field distribution of TE_{101} rectangular cavity

subscript p represents the changing cycles along the z direction.

TE resonant modes

On the basis of our discussions on TE propagation modes of rectangular waveguides and the boundary conditions, we can get the field distributions of the TE_{mnp} mode:

$$H_x = j\frac{2A}{k_c^2} \cdot \frac{p\pi}{c} \cdot \frac{m\pi}{a} \sin\left(\frac{m\pi}{a}x\right)$$
$$\times \cos\left(\frac{n\pi}{b}y\right)\cos\left(\frac{p\pi}{c}z\right) \quad (2.346)$$

$$H_y = j\frac{2A}{k_c^2} \cdot \frac{p\pi}{c} \cdot \frac{m\pi}{b} \cos\left(\frac{m\pi}{a}x\right)$$
$$\times \sin\left(\frac{n\pi}{b}y\right)\cos\left(\frac{p\pi}{c}z\right) \quad (2.347)$$

$$H_z = -j2A\cos\left(\frac{m\pi}{a}x\right)$$
$$\times \cos\left(\frac{n\pi}{b}y\right)\sin\left(\frac{p\pi}{c}z\right) \quad (2.348)$$

$$E_x = 2A\frac{\omega\mu}{k_c^2} \cdot \frac{n\pi}{b}\cos\left(\frac{m\pi}{a}x\right)$$
$$\times \sin\left(\frac{n\pi}{b}y\right)\sin\left(\frac{p\pi}{c}z\right) \quad (2.349)$$

$$E_y = -2A\frac{\omega\mu}{k_c^2} \cdot \frac{m\pi}{a}\sin\left(\frac{m\pi}{a}x\right)$$
$$\times \cos\left(\frac{n\pi}{b}y\right)\sin\left(\frac{p\pi}{c}z\right) \quad (2.350)$$

$$E_z = 0 \quad (2.351)$$

with

$$k_c^2 = \left(\frac{m\pi}{a}\right)^2 + \left(\frac{n\pi}{b}\right)^2 \quad (2.352)$$

A special resonant mode, TE_{101} mode is widely used in the characterization of electromagnetic materials. From Eqs. (2.346)–(2.352), we can get the field components of TE_{101} mode:

$$E_y = -2\frac{A\omega\mu a}{\pi}\sin\left(\frac{\pi}{a}x\right)\sin\left(\frac{\pi}{c}z\right) \quad (2.353)$$

$$H_x = j\frac{Aa}{c}\sin\left(\frac{\pi}{a}x\right)\cos\left(\frac{\pi}{c}z\right) \quad (2.354)$$

$$H_z = -j2A\cos\left(\frac{\pi}{a}x\right)\sin\left(\frac{\pi}{c}z\right) \quad (2.355)$$

$$E_x = E_z = H_y = 0 \quad (2.356)$$

The field distribution is shown in Figure 2.80.

The resonant wavelength for TE_{mnp} mode resonator can be calculated:

$$\lambda_0 = \frac{2}{\sqrt{\left(\frac{m}{a}\right)^2 + \left(\frac{n}{b}\right)^2 + \left(\frac{p}{c}\right)^2}} \quad (2.357)$$

So the resonant wavelength for TE_{101} mode is given by

$$\lambda_0 = \frac{2ac}{\sqrt{a^2 + c^2}} \quad (2.358)$$

Equation (2.358) indicates that, if b is the shortest among the three sides (a, b, and c), TE_{101} mode has the largest resonant wavelength, so it is the lowest mode.

It should be indicated that, for a rectangular cavity resonator with $a = b$, TE_{101} and TE_{011} are degenerate modes. They have the same resonant frequency and the field distribution pattern, but their fields are perpendicular to each other. As will be discussed in Chapter 11, these two degenerate modes can be used in the measurement of microwave Hall effects.

The quality factor of the resonance can be estimated by (Ishii 1995)

$$Q\frac{\delta}{\lambda_0} = \frac{abc}{4}$$

$$\times \frac{\left[\left(\frac{m}{a}\right)^2 + \left(\frac{n}{b}\right)^2\right]\left[\left(\frac{m}{a}\right)^2 + \left(\frac{n}{b}\right)^2 + \left(\frac{p}{c}\right)^2\right]^{\frac{3}{2}}}{ac\left\{\left(\frac{m}{a}\right)^2\left(\frac{p}{c}\right)^2 + \left[\left(\frac{m}{a}\right)^2 + \left(\frac{n}{b}\right)^2\right]^2\right\}}$$

$$+ bc\left\{\left(\frac{n}{b}\right)^2\left(\frac{p}{c}\right)^2 + \left[\left(\frac{m}{a}\right)^2 + \left(\frac{n}{b}\right)^2\right]^2\right\}$$

$$+ ab\left(\frac{p}{c}\right)^2\left[\left(\frac{m}{a}\right)^2 + \left(\frac{n}{b}\right)^2\right]$$

(2.359)

So the waveform factor for TE_{101} mode is given by

$$Q_0\frac{\delta}{\lambda_0} = \frac{b}{2} \cdot \frac{(a^2 + c^2)^{\frac{3}{2}}}{2b(a^3 + c^3) + ac(a^2 + c^2)} \quad (2.360)$$

TM resonant modes

According to the properties of TM wave propagation modes and the boundary conditions, we can get the field components of the TM_{mnp} mode:

$$E_x = \frac{-2B}{k_c^2} \cdot \frac{p\pi}{c} \cdot \frac{m\pi}{a} \cos\left(\frac{m\pi}{a}x\right)$$

$$\times \sin\left(\frac{n\pi}{b}y\right)\sin\left(\frac{p\pi}{c}\right) \quad (2.361)$$

$$E_y = \frac{-2B}{k_c^2} \cdot \frac{p\pi}{c} \cdot \frac{n\pi}{b} \sin\left(\frac{m\pi}{a}x\right)$$

$$\times \cos\left(\frac{n\pi}{b}y\right)\sin\left(\frac{p\pi}{c}z\right) \quad (2.362)$$

$$E_z = 2B\sin\left(\frac{m\pi}{a}x\right)$$

$$\times \sin\left(\frac{n\pi}{b}y\right)\cos\left(\frac{p\pi}{c}z\right) \quad (2.363)$$

$$H_x = j2B\frac{\omega\varepsilon}{k_c^2} \cdot \frac{n\pi}{b} \sin\left(\frac{m\pi}{a}x\right)$$

$$\times \cos\left(\frac{n\pi}{b}y\right)\cos\left(\frac{p\pi}{c}z\right) \quad (2.364)$$

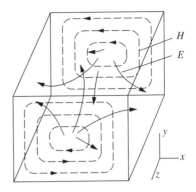

Figure 2.81 The field distribution of TM_{111} mode

$$H_y = -j2B\frac{\omega\varepsilon}{k_c^2} \cdot \frac{m\pi}{a} \cos\left(\frac{m\pi}{a}x\right)$$

$$\times \sin\left(\frac{n\pi}{b}y\right)\cos\left(\frac{p\pi}{c}z\right) \quad (2.365)$$

$$H_z = 0, \quad (2.366)$$

where k_c can be calculated from Eq. (2.352). As an example, the field distribution of TM_{111} is shown Figure 2.81.

For TM_{mnp} mode, the resonant wavelength can be calculated from Eq. (2.357), and the waveform factor can be calculated according to following equation (Ishii 1995):

$$Q_0\frac{\delta}{\lambda_0} = \frac{abc}{4} \cdot \frac{\left[\left(\frac{m}{a}\right)^2 + \left(\frac{n}{b}\right)^2\right]^{\frac{3}{2}}}{b(a + c)\left(\frac{m}{a}\right)^2 + a(b + 2c)\left(\frac{n}{b}\right)^2}$$

(2.367)

TE_{101}/TM_{110} mode

As discussed earlier, a rectangular cavity working at TE_{101} mode can be made by shorting both ends of a segment of rectangular waveguide working at the TE_{10} mode with length $\lambda_g/2$. As shown in Figure 2.82, if we change the coordinate system from xyz to XYZ, the TE_{101} mode in the xyz coordinate system becomes TM_{110} mode in the XYZ coordinate system. For a rectangular cavity resonator, its TE or TM modes with zero subscripts can always be taken as TM or TE modes, respectively, in another coordinate system.

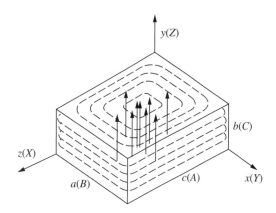

Figure 2.82   A TE_{101} mode rectangular cavity in the *xyz* coordinate system can be taken as a TM_{110} mode rectangular cavity in the *XYZ* coordinate system

2.3.4.2 Cylindrical cavity resonator

Figure 2.83 shows a cylindrical cavity resonator made from a segment of circular waveguide by shorting the two ends. Cylindrical cavity resonators usually have higher quality factors than corresponding rectangular resonators. The structural parameters of a cylindrical cavity resonator include the radius a and the length l. The resonant modes of a cylindrical cavity resonator include TE_{nip} and TM_{nip}, corresponding to TE_{ni} and TM_{ni} propagation modes in a cylindrical waveguide. The first two subscripts n and i come from the wave propagation modes TE_{ni} and TM_{ni}, representing the changing cycles in φ-direction and r-direction respectively, and the subscript p represents the changing cycles in z-direction.

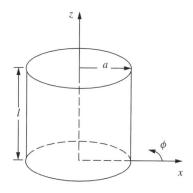

Figure 2.83   Structure of a cylindrical cavity resonator

TE resonant modes

On the basis the boundary conditions, we can get the field components of the TE_{nip} resonant modes in a cylindrical cavity with radius a and length l:

$$H_r = -j\frac{2A}{k_c} \cdot \frac{p\pi}{l} J_n'(k_c r)\cos n\varphi \cos\left(\frac{p\pi}{l}z\right)$$
(2.368)

$$H_\varphi = j\frac{2An}{k_c^2 r} \cdot \frac{p\pi}{l} J_n(k_c r)\sin n\varphi \cos\left(\frac{p\pi}{l}z\right)$$
(2.369)

$$H_z = -j2AJ_n(k_c r)\cos n\varphi \sin\left(\frac{p\pi}{l}z\right) \quad (2.370)$$

$$E_r = \frac{2A\omega\mu n}{k_c^2 r} J_n(k_c r)\sin n\varphi \sin\left(\frac{p\pi}{l}z\right) \quad (2.371)$$

$$E_\varphi = \frac{2A\omega\mu n}{k_c^2 r} J_n'(k_c r)\cos n\varphi \sin\left(\frac{p\pi}{l}z\right) \quad (2.372)$$

$$E_z = 0 \qquad (2.373)$$

with

$$k_c = \frac{\mu_{ni}}{a}, \qquad (n = 0, 1, 2, \ldots; i = 1, 2, 3, \ldots)$$
(2.374)

It should be noted that p should be a positive integer. If p equals zero, all the field components in Eqs. (2.368)–(2.373) become zero.

The resonant wavelength for a TE_{nip} mode is given by

$$\lambda_0 = \frac{1}{\sqrt{\left(\frac{\mu_{ni}}{2\pi a}\right)^2 + \left(\frac{p}{2l}\right)^2}} \qquad (2.375)$$

The waveform factor for a TE_{nip} mode resonator can be calculated (Ishii 1995):

$$Q_0\frac{\delta}{\lambda_0}$$

$$= \frac{\left[1 - \left(\frac{l}{\mu_{ni}}\right)^2\right]\left[\mu_{ni} + \left(\frac{p\pi a}{l}\right)^2\right]^{3/2}}{2\pi\left[\mu_{ni}^2 + \left(\frac{p\pi}{2}\right)^2\left(\frac{2a}{l}\right)^3 + (1-r)\left(\frac{p\pi a}{\mu_{ni}}\right)^2\right]}$$
(2.376)

Two TE modes are widely used in microwave engineering and materials property characterization: TE_{011} and TE_{111}. If $l > 2.1a$, TE_{111} mode is the basic mode. From Eqs. (2.368)– (2.373), we can get the field components for TE_{111} mode by letting $n = 1$, $j = 1$, $p = 1$, and $\mu_{11} = 1.841$. The field distribution of TE_{111} mode is schematically shown in Figure 2.84. As TE_{111} mode has degenerate modes with electric fields perpendicular to each other, this mode is often used in the measurement of microwave Hall effect. Detailed discussions on the measurement of microwave Hall effect can be found in Chapter 11.

From Eqs. (2.368)–(2.373), we can get the field components of TE_{011} mode:

$$E_\varphi = -2A\frac{\omega\mu a}{3.832} J_1\left(\frac{3.832}{a}r\right) \sin\left(\frac{\pi}{l}z\right) \quad (2.377)$$

$$H_r = j2A\frac{a}{3.832} \cdot \frac{\pi}{l} J_1\left(\frac{3.832}{a}r\right) \cos\left(\frac{\pi}{l}z\right) \quad (2.378)$$

$$H_z = -j2A J_0\left(\frac{3.832}{a}r\right) \sin\left(\frac{\pi}{l}z\right) \quad (2.379)$$

$$H_\varphi = E_r = E_z = 0 \quad (2.380)$$

As shown in Figure 2.85, in a TE_{011} mode cavity resonator, all the electric fields are in φ-direction. Near the sidewall, magnetic field is in z-direction, and near the end wall, magnetic field is in r-direction. So all the electric currents are in φ-direction, and there is no current flowing between the sidewall and end wall. Therefore this mode does not require good electric contact between the sidewall and end wall. This mode usually has very

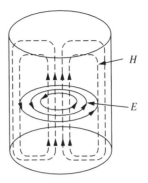

Figure 2.85 Field distribution of TE_{011} mode

high quality factor, and is often used in frequency meters.

TM resonant modes

According to the boundary conditions, we can get the field components of a TM_{nip} mode:

$$E_r = -\frac{2B}{k_c} \cdot \frac{p\pi}{l} J_n'(k_c r) \cos n\varphi \sin\left(\frac{p\pi}{l}z\right) \quad (2.381)$$

$$E_\varphi = \frac{2Bn}{k_c^2 r} \cdot \frac{p\pi}{l} J_n(k_c r) \sin n\varphi \sin\left(\frac{p\pi}{l}z\right) \quad (2.382)$$

$$E_z = 2B J_n(k_c r) \cos n\varphi \cos\left(\frac{p\pi}{l}z\right) \quad (2.383)$$

$$H_r = -j\frac{2B\omega\varepsilon n}{k_c^2 r} J_n(k_c r) \sin n\varphi \cos\left(\frac{p\pi}{l}z\right) \quad (2.384)$$

$$H_\varphi = -j\frac{2B\omega\mu}{k_c^2} J_n'(k_c r) \cos n\varphi \cos\left(\frac{p\pi}{l}z\right) \quad (2.385)$$

with

$$k_c = \frac{v_{ni}}{a} \quad (n = 0, 1, 2, 3, \ldots; i = 1, 2, 3, \ldots) \quad (2.386)$$

The resonant wavelength of a cylindrical cavity in TM_{nip} mode is given by

$$\lambda_0 = \frac{1}{\sqrt{\left(\frac{v_{ni}}{2\pi a}\right)^2 + \left(\frac{p}{2l}\right)^2}} \quad (2.387)$$

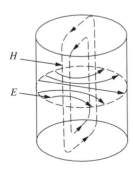

Figure 2.84 Field distribution of TE_{111} mode

The waveform factor a TM_{nip} mode can be calculated from (Ishii 1995)

$$Q_0 \frac{\delta}{\lambda_0} = \frac{\sqrt{v_{ni}^2 + \left(\frac{p\pi a}{l}\right)^2}}{2\pi \left(1 + \frac{2a}{l}\right)} \qquad (2.388)$$

The TM_{010} mode is often used in materials property characterization. By letting $n = 0$, $I = 1$, and $p = 0$, we can get the field distributions of TM_{010} mode from Eqs. (2.381)–(2.385):

$$E_z = 2B J_0 \left(\frac{v_{01}}{a} r\right) \qquad (2.389)$$

$$H_\varphi = -j \frac{2B\omega\varepsilon a}{v_{01}} J_0' \left(\frac{v_{01}}{a} r\right) = j \frac{2B\omega\varepsilon a}{v_{01}} J_1 \left(\frac{v_{01}}{a} r\right) \qquad (2.390)$$

$$E_r = E_\varphi = H_r = H_z = 0 \qquad (2.391)$$

The field distribution of TM_{010} mode is shown in Figure 2.86. As the electric field at the central part of the cavity is quite uniform, in the resonant-perturbation method for permittivity measurement, the dielectric sample under test is often placed at the central part of the cavity.

Mode chart

In the design of cavity resonators, usually we should ensure cavities working in a single mode state. The mode chart is often used to check how many modes may exist in a cylindrical cavity.

Mode chart is drawn on the basis of Eqs. (2.375) and (2.387). If we use u_{ni} to represent μ_{ni} and v_{ni},

Eq. (2.375) and (2.387) can be rewritten as

$$\lambda_0 = \frac{1}{\sqrt{\left(\frac{u_{ni}}{2\pi a}\right)^2 + \left(\frac{p}{2l}\right)^2}} \qquad (2.392)$$

Equation (2.392) can be further modified as

$$(2af_0)^2 = \left(\frac{c u_{ni}}{\pi}\right)^2 + \left(\frac{cp}{2}\right)^2 \left(\frac{2a}{l}\right)^2, \qquad (2.393)$$

where f_0 is the resonant frequency, c is the speed of light. So there are linear relationships between $(2af_0)^2$ and $(2a/l)^2$ with the tangent $(cp/2)^2$ and the cross value $(c u_{ni}/\pi)^2$. These lines form the mode chart, as shown in Figure 2.87.

The mode chart can be used in the following three ways. First, if we know the diameter and resonant mode of the cavity, we can get the relationship between the resonant frequency and the length of the cavity. Second, if we know the diameter and the length of the cavity, we can obtain the resonant frequencies for different working modes. Third, we can build a working rectangular from which we can check whether the

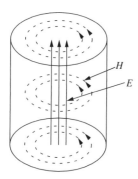

Figure 2.86   Field distribution of TM_{010} mode

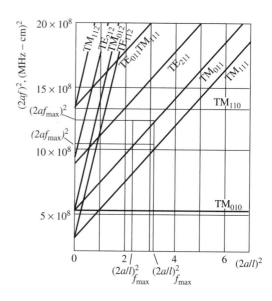

Figure 2.87   Resonant mode chart for a cylindrical cavity. Modified from Pozar, D. M. (1998). *Microwave Engineering*, 2nd ed., John Wiley & Sons, Inc., New York

Table 2.5 Changes in resonant frequency due to materials perturbation and cavity shape perturbation. In the table, ω_0 is the resonant frequency before perturbation and ω is the resonant frequency after the perturbation

	Materials perturbation		Cavity shape perturbation	
	Metal	Dielectric or magnetic materials	Inward	Outward
Electric field dominates	$\omega < \omega_0$	$\omega < \omega_0$ $(\varepsilon_r > 1)$	$\omega < \omega_0$	$\omega > \omega_0$
Magnetic field dominates	$\omega > \omega_0$	$\omega < \omega_0$ $(\mu_r > 1)$	$\omega > \omega_0$	$\omega < \omega_0$

cavity is in single-mode state. For a resonator with a given working mode, diameter $(2a)$, and working frequency range (f_{min} and f_{max}), there are two cross points between the resonant line and two horizontal lines $(2af_{min})^2$ and $(2af_{max})^2$: $(2a/l)^2_{fmin}$ and $(2a/l)^2_{fmax}$. The two horizontal lines $(2af_{min})^2$ and $(2af_{max})^2$ and two vertical lines $(2a/l)^2_{fmin}$ and $(2a/l)^2_{fmax}$ form a working rectangular. If there is no other resonant line in the working rectangular, the cavity works at a single mode state.

If different resonant modes have the same resonant line, for example TE_{011} and TM_{111}, these modes are degenerate. In the design of resonators, special methods are needed to eliminate degenerate modes, which will be discussed in later chapters.

2.3.4.3 Resonant perturbation

In some cases, the boundary conditions of a cavity may change slightly, or the properties of the medium in the cavity may change slightly. In materials property characterization, we may introduce a small sample into a resonant cavity. Such changes may result in small changes in the resonant frequency and quality factor of the cavity.

The changes in resonant frequency and quality factor can be approximately obtained on the basis of the resonant-perturbation theory. Resonant-perturbation theory focuses on the change of the energy due to the perturbation, not the change of the electromagnetic field distribution. Some useful conclusions about the change of the resonant frequency due to different types of resonant perturbations are listed in Table 2.5. Further discussions on resonant perturbation can be found in Chapter 6.

2.3.4.4 Coupling to external circuit

Aperture coupling is often used for coupling between a waveguide and a cavity. Usually, the waveguide works at TE_{10} mode, whose magnetic field and electric current distribution is shown in Figure 2.88. The coupling is mainly based on the continuity of the magnetic field or electric current.

Figure 2.89 shows two examples of the excitation of resonant cavities using a rectangular waveguide. In Figure 2.89(a), TM_{010} resonator is excited by the magnetic field, while in Figure 2.89(b), TE_{011} resonator is excited by the current along the inside wall of the waveguide.

2.3.5 Dielectric resonators

Figure 2.90 shows several typical dielectric resonators, including spherical dielectric resonator, cylindrical dielectric resonator, ring dielectric resonator, and rectangular dielectric resonator.

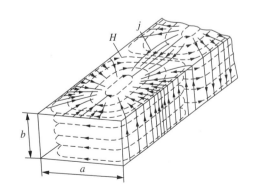

Figure 2.88 Distributions of magnetic field and electric current of a rectangular waveguide at TE_{10} mode

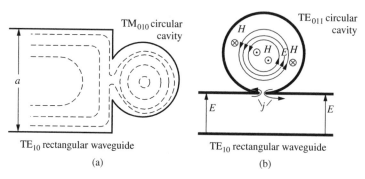

Figure 2.89 Aperture coupling using rectangular waveguide. (a) Excitation of TM_{010} mode and (b) excitation of TE_{011} mode

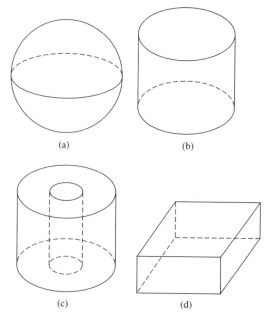

Figure 2.90 Typical configurations of dielectric resonators. (a) Sphere, (b) cylinder, (c) ring, and (d) rectangular

Dielectric resonators also fall into transmission type, such as cylinder resonator and rectangular resonator, and non-transmission type, such as sphere resonator. In the following discussion, we concentrate on transmission-type dielectric resonators, especially the cylindrical dielectric resonators. A transmission-type dielectric resonator is made from its corresponding dielectric waveguide, and a cylindrical dielectric resonator is actually a segment of cylindrical dielectric waveguide with two open ends.

In a closed metallic cavity, all the electromagnetic fields are confined within the space enclosed by the metallic cavity; however, there is field fringing or leakage from the boundaries of a dielectric resonator. The resonators shown in Figure 2.90 are isolated dielectric resonators without any support. Isolated dielectric resonators are convenient for theoretical analysis. However, in actual cases, dielectric resonators should be supported or shielded. Owing to field fringing and leakage, the support or shield affects the resonant properties of the dielectric resonator. Usually, a dielectric resonator with support or shield is called a *shielded dielectric resonator.*

The dielectric permittivity ε_r of the dielectric material is an important parameter in determining the resonant properties of a dielectric resonator. For a given resonant frequency, the higher the dielectric constant, the smaller the dielectric resonator, and the microwave energy is more concentrated within the dielectric resonator and the effects of the external circuit to the dielectric resonator become less. For microwave dielectric resonators, the dielectric constant is usually in the range of 10–100. If the dielectric constant is very large, the dielectric resonator becomes very small, and the fabrication of dielectric resonators becomes very difficult. There are a type of resonators working at whispering-gallery modes, which have relatively large sizes. Such resonators are often used in the characterization of high-dielectric-constant materials.

In the following, we start with the resonant properties of isolated dielectric resonators, followed by the resonant properties of shielded

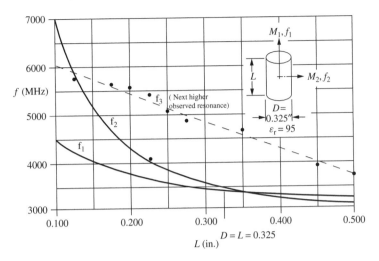

Figure 2.91 First three resonant frequencies versus lengths of a dielectric cylinder of constant diameter (Cohn 1968). Source: Cohn, S. B. (1968). "Microwave bandpass filters containing high-Q dielectric resonators", *IEEE Transactions on Microwave Theory and Techniques*, **16** (4), 218–227. © 2003 IEEE

dielectric resonators. We then introduce the techniques for realizing the couplings between dielectric resonators and external circuits, and finally the resonant properties of whispering-gallery dielectric resonators will be discussed.

2.3.5.1 Isolated dielectric resonators

Cylindrical dielectric resonators are widely used in microwave electronics and materials property characterizations. Exact analyses of cylindrical dielectric resonators are quite complicated, and approximation techniques and experimental methods are often used. Figure 2.91 shows some experimental results for the resonant frequencies of cylindrical dielectric resonators with same diameter but with different lengths. The curve labelled f_1 represents the TE mode with zero axial electric field, and the curve labelled f_2 represents the TM mode with zero axial magnetic field.

In most cases, the length L of the dielectric resonator is chosen to be less than the diameter ($D = 2a$) of the dielectric resonator. In microwave electronics and materials characterization, the prime resonance of interest is usually the lowest order mode. From Figure 2.91, the lowest mode for cylinder dielectric resonator with $L < D$ is $TE_{01\delta}$ mode, where the first two subscript integers denote

the waveguide mode and δ is the non-integer ratio $2L/\lambda_g < 1$. As shown in Figure 2.92, for a distant observer, this mode appears as a magnetic dipole, so this mode is also called *magnetic dipole mode*.

For a given dielectric resonator, its accurate resonant frequency of $TE_{01\delta}$ mode can be calculated only by complicated numerical procedures. For an appropriate estimation of the resonant frequency of the isolated dielectric resonator, the following simple formula can be used (Kajfez and Guillon 1986):

$$f = \frac{34}{a\sqrt{\varepsilon_r'}}\left(\frac{a}{L} + 3.45\right) \quad \text{(GHz)}, \qquad (2.394)$$

where a and L are the radius and length of the dielectric resonator, ε_r' is the dielectric constant of the material. In Eq. (2.394), the unit for frequency is GHz, and the unit for a and L is mm. The accuracy for Eq. (2.394) is about 2 % in the range of $0.5 < (a/L) < 2$ and $30 < \varepsilon_r' < 50$.

For the case with L greater than D, the prime mode has an equivalent magnetic dipole moment transverse to the axis, as shown in Figure 2.93 (Cohn 1968). Resonant properties for cylindrical resonators with $L = D$ can be found in (Tsuji *et al.* 1984).

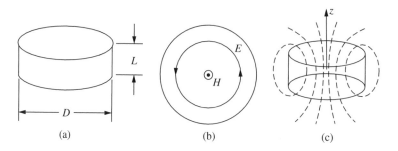

Figure 2.92   Prime mode ($\text{TE}_{01\delta}$) of a cylindrical dielectric resonator with $L < D$. (a) Dimensions of the resonator, (b) top view of the field distribution, and (c) three-dimensional view of magnetic field

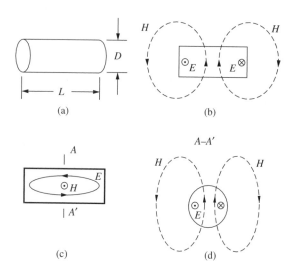

Figure 2.93   Prime mode of a cylindrical dielectric resonator with $L > D$. (a) Dimensions of the resonator, (b) side view of the field distributions, (c) top view of the field distributions, and (d) A–A' cross view of the field distributions. Source: Cohn, S. B. (1968). "Microwave bandpass filters containing high-Q dielectric resonators", *IEEE Transactions on Microwave Theory and Techniques*, **16** (4), 218–227 © 2003 IEEE

2.3.5.2 Shielded dielectric resonators

In actual applications, dielectric resonators are often supported or enclosed by metal shields, and different ways of shielding may result in different resonant properties. In this part, we discuss several typical shielding methods, including parallel plates, asymmetrical parallel plates, cut-off waveguide, and closed shields.

Parallel-plate dielectric resonator

As shown in Figure 2.94, a parallel-plate dielectric resonator consists of an isotropic cylindrical dielectric between two parallel conducting plates perpendicular to the axis of the dielectric cylinder. It is a symmetrical structure, and in theoretical analysis of its $\text{TE}_{01\delta}$ mode, the following assumptions are made. Firstly, the dielectric cylinder can be treated as a lossless cylindrical dielectric waveguide with length L excited in TE_{01} mode. Secondly, the z-dependence of the z-component magnetic field (H_z) for $|z| \geq L/2$ can be described by $\sinh \alpha(L_1 + (L/2) \pm z)$. Thirdly, the cross-sectional field distribution of H_z for $|z| \geq (L/2)$ is the same as for the TE_{01} mode in cylindrical dielectric waveguide at the "cutoff" frequency. We follow the method proposed by Pospieszalski (1979).

We define the normalized resonant frequency F_0:

$$F_0 = \frac{\pi D}{\lambda_0} \sqrt{\varepsilon_{\text{r}}}, \qquad (2.395)$$

where λ_0 is the free-space wavelength corresponding to the resonant frequency f_0, D is the diameter, and ε_{r} is the relative dielectric constant. From the first assumption, F_0 satisfies the following equations:

$$F_0^2 = (u^2 + w^2) \frac{\varepsilon_{\text{r}} - 1}{\varepsilon_{\text{r}}} \qquad (2.396)$$

$$\frac{J_1(u)}{u J_0(u)} = -\frac{K_1(w)}{w K_0(w)}, \qquad (2.397)$$

where J_n is the Bessel function of the first kind of nth order and K_n is the modified Hankel function of nth order. The parameters w and u are defined by

$$\left(\frac{w}{a}\right)^2 = \beta^2 - \left(\frac{F_0}{a}\right)^2 \frac{1}{\varepsilon_r} \tag{2.398}$$

$$\left(\frac{u}{a}\right)^2 = \left(\frac{F_0}{a}\right)^2 - \beta^2, \tag{2.399}$$

where β is the propagation constant of the mode in cylindrical dielectric waveguide and a is the radius of dielectric cylinder.

The second assumption indicates that α and β satisfy the following transcendental equation:

$$\cot \beta \frac{L}{2} = \frac{\beta}{\alpha} \tanh \alpha L_1. \tag{2.400}$$

According to the third assumption, α is given by

$$\alpha^2 = \left(\frac{\rho_{01}}{a}\right)^2 - \frac{1}{\varepsilon_r} \left(\frac{F_0}{a}\right)^2 \tag{2.401}$$

where ρ_{01} is the first root of J_0. From Eqs. (2.398)–(2.401), we can get

$$\tan\left(\frac{L}{D}\sqrt{F_0^2 - u^2}\right)$$

$$= \frac{\sqrt{\rho_{01}^2 - \dfrac{1}{\varepsilon_r}F_0^2}}{\sqrt{F_0^2 - u^2}\,\tanh\left(2\dfrac{L_1}{D}\cdot\sqrt{\rho_{01}^2 - \dfrac{1}{\varepsilon_r}F_0^2}\right)} \tag{2.402}$$

The normalized resonant frequency F_0 can be computed from the set of Eqs. (2.396), (2.397) and (2.402) as a function of (D/L), (L_1/L), and ε_r. Figure 2.95 shows the $TE_{01\delta}$ mode chart, which is often used in the design of cylindrical dielectric resonators.

As will be discussed later, in materials property characterizations, usually the conducting plates directly shorts the dielectric resonator ($L_1 = 0$). So the resonator is in TE_{011} mode, and it corresponds to the line ($S_1 = 0$) in Figure 2.95.

Asymmetrical parallel-plate dielectric resonator

Figure 2.96 shows an asymmetrical dielectric resonator structure with parallel conducting plates.

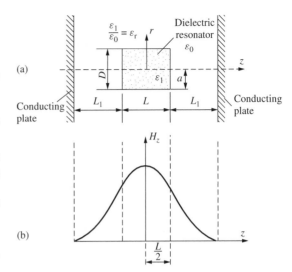

Figure 2.94 Parallel-plate dielectric resonator at $TE_{01\delta}$ mode. (a) Configuration of the resonator and (b) distribution of magnetic field H_z along z-axis (Pospieszalski 1979). Source: Pospieszalski, M. W. (1979). "Cylindrical dielectric resonators and their applications in TEM line microwave circuits", *IEEE Transactions on Microwave Theory and Techniques*, **27** (3), 233–238 © 2003 IEEE

The present structure can be regarded as modified from the one shown in Figure 2.94 by inserting a dielectric layer in one side of the cylindrical dielectric resonator, and the thickness values at the two sides are not equal. This structure is often used in microwave integrated circuits, and is also called microstrip dielectric resonator.

On the basis of similar assumptions and with the similar analysis method, we can get (Pospieszalski 1979)

$$2\frac{L}{D}\sqrt{F_0^2 - u^2}$$

$$= \tan^{-1}\frac{\sqrt{\rho_{01}^2 - \dfrac{1}{\varepsilon_r}F_0^2}}{\sqrt{F_0^2 - u^2}\,\tanh\left(2\dfrac{L_1}{D}\cdot\sqrt{\rho_{01}^2 - \dfrac{1}{\varepsilon_r}F_0^2}\right)}$$

$$+ \frac{\sqrt{\rho_{01}^2 - \dfrac{1}{\varepsilon_r}F_0^2}}{\sqrt{F_0^2 - u^2}\,\tanh\left(2\dfrac{L_2}{D}\cdot\sqrt{\rho_{01}^2 - \dfrac{\varepsilon_p}{\varepsilon_r}F_0^2}\right)}, \tag{2.403}$$

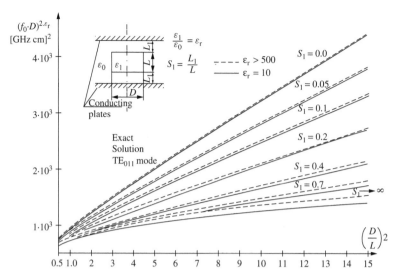

Figure 2.95 The $TE_{01\delta}$ mode chart for cylindrical dielectric resonator between parallel plates (Pospieszalski 1979). Source: Pospieszalski, M. W. (1979). "Cylindrical dielectric resonators and their applications in TEM line microwave circuits", *IEEE Transactions on Microwave Theory and Techniques*, **27** (3), 233–238. © 2003 IEEE

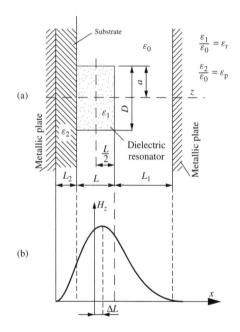

Figure 2.96 An asymmetrical parallel-plate dielectric resonator. (a) Configuration of the dielectric resonator and (b) distribution of magnetic field H_z along z-axis for the $TE_{01\delta}$ mode (Pospieszalski 1979). Source: Pospieszalski, M. W. (1979). "Cylindrical dielectric resonators and their applications in TEM line microwave circuits", *IEEE Transactions on Microwave Theory and Techniques*, **27** (3), 233–238. © 2003 IEEE

where ε_p is the relative dielectric constant of the dielectric layer (substrate). From the set of Eqs. (2.396), (2.397), and (2.403), F_0 can be found as a function of (D/L), (L_1/L), (L_2/L), ε_r, and ε_p.

There are approximate and straightforward equations for the design of this kind of dielectric resonators (Ishii 1995). To design a cylindrical dielectric resonator with resonant frequency f_0, we should determine its diameter D and length L. The first step is to select D satisfying

$$\frac{5.4}{k_0\sqrt{\varepsilon_r}} \leq D \leq \frac{5.4}{k_0\sqrt{\varepsilon_p}}, \tag{2.404}$$

where $k_0 = (\varepsilon_0\mu_0)^{1/2}$ is the free-space propagation constant. The second step is to determine the length L according to Eq. (2.405):

$$L = \frac{1}{\beta}\left[\tan^{-1}\left(\frac{\alpha_1}{\beta}\coth\alpha_1 L_1\right)\right.$$
$$\left. + \tan^{-1}\left(\frac{\alpha_2}{\beta}\coth\alpha_2 L_2\right)\right] \tag{2.405}$$

with

$$\alpha_1 = \sqrt{h^2 - k_0^2} \tag{2.406}$$

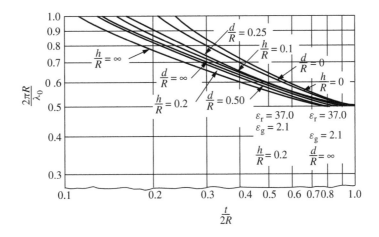

Figure 2.97 Generalized normalized design curves for microstrip dielectric resonators (Chang 1989). Source: Chang, K. (1989). *Handbook of Microwave and Optical Components*, Vol. 1, John Wiley, New York

$$\alpha_2 = \sqrt{h^2 - k_0^2 \varepsilon_p} \qquad (2.407)$$

$$\beta = \sqrt{k_0^2 \varepsilon_r - h^2} \qquad (2.408)$$

$$h = \frac{2}{D} \cdot \left(2.405 + \frac{y_0}{2.405(1 + 2.43/y_0 + 0.291 y_0)}\right) \qquad (2.409)$$

$$y_0 = \sqrt{\left(k_0 \frac{D}{2}\right)^2 (\varepsilon_r - 1) - 2.405^2} \qquad (2.410)$$

Figure 2.97 shows a generalized normalized design curve for asymmetrical parallel-plate dielectric resonators. In the figure, λ_0 is the wavelength in free space, t is the height of the dielectric resonator, R is the radius of the dielectric resonator, d is the distance of air gap between the dielectric resonator and metal plate, h is the thickness of the substrate, ε_r is the dielectric constant of the resonator, and ε_g is the dielectric constant of the substrate. Figure 2.97 shows the case when $\varepsilon_r = 37.0$ and $\varepsilon_g = 2.1$. There are two sets of curves. One set of curves are for the condition of ($h/R = 0.2$), and they are indicated by $d/R = 0$, 0.25, 0.50, and ∞ respectively. Another set of curves are for the condition of ($d/R = \infty$), and they are indicated by $h/R = 0$, 0.1, 0.2, and ∞, respectively.

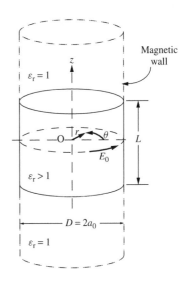

Figure 2.98 Dielectric cylinder in magnetic-wall waveguide boundary. Source: Cohn, S. B. (1968). "Microwave bandpass filters containing high-Q dielectric resonators", *IEEE Transactions on Microwave Theory and Techniques*, **16** (4), 218–227. © 2003 IEEE

Dielectric resonator in cutoff magnetic-wall waveguide

Figure 2.98 shows a dielectric cylinder contained in a contiguous magnetic-wall waveguide. In the dielectric region, the waveguide is above its cutoff frequency, while in the air regions, the waveguide is below the cut-off frequency. At resonance, a standing wave exists in the dielectric region, while exponentially attenuating waves exist in air regions. The equivalent circuit is a piece of transmission line with length L, terminated at both ends by reactance equal to the pure imaginary characteristic impedance of the cut-off air-filled waveguide.

According to Cohn's mode (Cohn 1968), we can get the field distributions for $TE_{01\delta}$ mode:

$$E_{\phi} = 2E_0 \frac{J_1(k_c r)}{J_1(p_{01})} \cos \beta_d z \left(|z| \leq \frac{L}{2} \right) \quad (2.411)$$

$$E_{\phi} = 2E_0 \frac{J_1(k_c r)}{J_1(p_{01})} \cos \frac{\beta_d L}{2}$$
$$\times e^{-\alpha_a(|z|-L/2)} \left(|z| \geq \frac{L}{2} \right) \quad (2.412)$$

$$H_r = j \frac{\beta_d}{\omega \mu} 2E_0 \frac{J_1(k_c r)}{J_1(p_{01})} \sin \beta_d z \left(|z| \leq \frac{L}{2} \right) \quad (2.413)$$

$$H_r = \pm j \frac{\alpha_a}{\omega \mu} 2E_0 \frac{J_1(k_c r)}{J_1(p_{01})} \cos \beta_d z$$
$$\times e^{-\alpha_a(|z|-L/2)} \left(|z| \geq \frac{L}{2} \right) \quad (2.414)$$

where

$$k_c = \frac{p_{01}}{a} = \frac{2.405}{a} \quad (2.415)$$

$$\beta_d = \sqrt{\varepsilon_r k_0^2 - k_c^2} = \sqrt{\frac{(2\pi f)^2 \varepsilon_r}{c^2} - \left(\frac{p_{01}}{a} \right)^2} \quad (2.416)$$

$$\alpha_a = \sqrt{k_c^2 - k_0^2} = \sqrt{\left(\frac{p_{01}}{a} \right)^2 - \left(\frac{2\pi f}{c} \right)^2} \quad (2.417)$$

The upper sign in Eq. (2.414) applies at $z = L/2$ and the lower at $z = -L/2$. The matching tangent fields at $z = \pm L/2$ lead to

$$\tan \frac{\beta_d L}{2} = \frac{\alpha_a}{\beta_d} \quad (2.418)$$

This equation can be solved numerically for the resonant frequency. As α_a and β_d should be real numbers, the possible frequency range is from f_1 to f_2, where

$$f_1 = \frac{p_{01}}{a} \cdot \frac{c}{2\pi \sqrt{\varepsilon_r}} \quad (2.419)$$

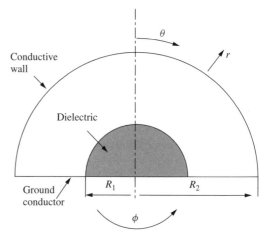

Figure 2.99 Semispherical dielectric resonator in a semispherical metal cavity

$$f_2 = \frac{p_{01}}{a} \cdot \frac{c}{2\pi} \quad (2.420)$$

Because of the magnetic-wall waveguide, the resonant frequency of this configuration is about 10 % lower than those of an isolated dielectric resonator and a parallel-plate dielectric resonator with same dimensions and dielectric constant.

Dielectric resonator in closed metal shields

Dielectric resonators can be enclosed in closed metal shields, and the study of the resonant properties of such resonant structures is important for the design of dielectric resonators and materials property characterization.

A spherical dielectric resonator enclosed in a metal sphere is often used in studying the effects of a metal shield to the resonant properties of a resonator, because the space between the dielectric resonator and the shield can be described by one parameter: radius. Imai and Yamamoto studied the resonant properties a semi-spherical dielectric resonator in a semi-spherical conductive shield, as shown in Figure 2.99 (Imai and Yamamoto 1984). The properties of a semi-spherical resonator at TE_{011} mode are similar with those for TE_{021} spherical resonator. One advantage of using a semi-spherical dielectric resonator is that the ground conductor is also a support to the dielectric resonator, and no other support is needed.

In Figure 2.99, the radius of the dielectric resonator is R_1, and that of the metal sphere is R_2. When R_2 gradually increases while R_1 keeps constant, the change of the resonant frequency of the resonant structure is shown in Figure 2.100. In region a, R_2 is slightly larger than R_1, and the resonant frequency changes rapidly as a function of R_2. When R_2 is approaching R_1, the resonant frequency approaches that of a cavity fully filled with a dielectric material. In region b, the resonant frequency is almost independent of R_2. In region c, the resonant frequency again changes as a function of R_2. When R_2 is much larger than R_1, the resonant structure can be taken as a semi-spherical dielectric resonator mounted on a ground plane, which is discussed in (Collin 1992).

According to the incremental frequency rule, the quality factor Q_c related to conductor losses of the shield is large when the slope df/dR_2 (frequency vs. the size of the cavity) is small (Kajfez and Guillon 1986). Therefore, in the region b, Q_c is large and so the effect of the conductor loss to the overall quality factor Q of the resonant structure can be neglected. Therefore in this region Q is mainly determined by the losses in the dielectric.

In materials property characterization, usually cylindrical dielectric resonators are used, and the cylindrical dielectric resonators are often enclosed in metallic cylinders. The conclusions for spherical resonators discussed above can be extended to cylindrical dielectric resonators. Figure 2.101 shows two typical configurations. In

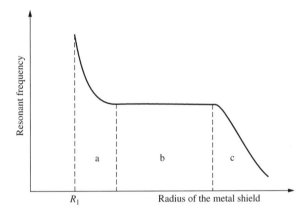

Figure 2.100 Resonant frequency of the semispherical dielectric resonator. (a) Transient mode, (b) dielectric resonator mode, and (c) hollow cavity mode

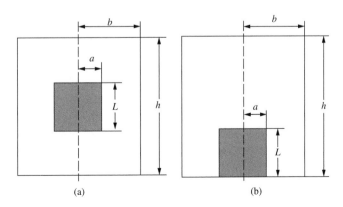

Figure 2.101 Cylindrical resonators in cylindrical cavities. (a) The dielectric resonator is at the center of the cavity. (b) The dielectric resonator contacts the end wall of the cavity

Figure 2.101(a), the dielectric resonator is at the center of the cavity, and the contribution of the conductor losses of the metal shield to the overall quality factor of the resonant structure is small. This configuration is often used in characterizing the dielectric permittivity of dielectric cylinders. In Figure 2.101(b), one end of the dielectric cylinder directly contacts the end-wall of the resonant cavity. In this configuration, the surface resistance of the end-wall contacting the dielectric cylinder affects the overall quality factor of the resonant structure. This configuration is often used in characterizing the surface resistance of conductors.

Kobayashi *et al.* made discussions on the quality factors of the circularly symmetric TE_0 modes for dielectric cylinders placed between two parallel conductor plates and in a closed conductor shield (Kobayashi *et al.* 1985). The techniques allowing separate estimation of the quality factors due to radiation, conductor, and dielectric losses are proposed. The conclusions obtained are helpful for realizing high-quality dielectric resonators and for the measurement of dielectric properties of low-loss materials.

2.3.5.3 Coupling to external circuit

Generally speaking, all kinds of coupling methods can be used for dielectric resonators. The coupling probes often used for dielectric resonators include electric dipole, magnetic dipole, microstrip line, and waveguide. It should be noted that, for a coupling probe, different coupling positions and different orientations result in different resonant modes. As shown in Figure 2.102, in materials property characterization, electric dipole probe and magnetic dipole probe are often used in couplings to dielectric resonators. Using these kinds of probes, the coupling coefficients can be easily adjusted. Detailed discussions on the coupling methods for dielectric resonators can be found in (Kajfez and Guillon 1986).

2.3.5.4 Whispering-gallery dielectric resonator

Whispering-gallery mode (WGM) dielectric resonators form a class of rotationally invariant

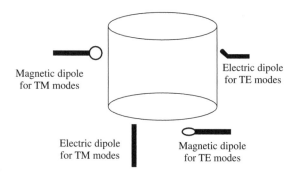

Figure 2.102   Electric and magnetic dipole probes for couplings of dielectric TE and TM resonant modes

Figure 2.103   Typical WGM resonance in a dielectric resonator. The dark regions indicate areas of high energy density

resonant structures, typically characterized by high values of the azimuthal index n. The name is applied due to the similarity with Lord Rayleigh's observations of whispers that would travel around the inside wall of St. Catherine's cathedral in England in early 20th century. As shown in Figure 2.103, WGM resonators exhibit a multiplicity of sharp resonances and they are natural candidates for the realization of an ultra wideband resonant structure. Since WGM dielectric resonators are running essentially in the azimuthal direction, they offer the possibility of new microwave devices, such as directional filters and power combiners with very low reflection coefficient. In materials property characterization, WGM dielectric resonators are often used in characterizing extremely low-loss dielectrics, high-dielectric-constant materials, anisotropic dielectrics, and ferrites.

In a WGM dielectric resonator, due to the total internal reflections, the electromagnetic wave bounces around inside the dielectric/air interface.

For given boundary conditions, resonance occurs only at certain wavelengths (λ) given by

$$n\lambda = \pi d \sqrt{\varepsilon_r \mu_r}, \qquad (2.421)$$

where the integer n is the azimuthal index, d is the diameter of the crystal, ε_r is the relative permittivity, and μ_r is the relative permeability. However, as will be indicated later, this relation is only true at high frequency where n is large. At lower frequencies, the electrical path length is extended by reflections from an internal caustic surface (Cros and Guillon 1990).

WGM in a dielectric rod can be described as comprising waves running against the concave side of the cylindrical boundary of the rod (Cros and Guillon 1990). The waves move essentially in the plane of the circular cross section. As shown in Figure 2.104(a), most of the energy is confined between the cylindrical boundary ($r = a$) and an inner modal caustic ($r = a_c$). Near this region, the electromagnetic fields are evanescent. The modal energy confinement can be explained from a ray optics point of view. As shown in Figure 2.104(b), a ray is totally reflected at the dielectric–air interface; it is then tangent to an inner circle called the caustic. Thus the ray moves merely in a small region within the rod near the rod boundaries.

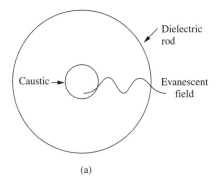
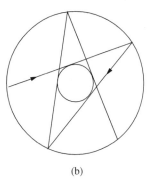

(a)   (b)

Figure 2.104 WGM in a dielectric rod. (a) Electric field variation as a function of the radius, (b) representation by ray optic techniques of WGM propagation in cylindrical rod (Cros and Guillon 1990). Source: Cros, D. and Guillon, P. (1990). "Whispering gallery dielectric resonator modes for W-band devices", *IEEE Transactions on Microwave Theory and Techniques*, **38** (11), 1667–1674. © 2003 IEEE

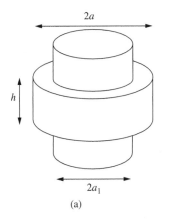
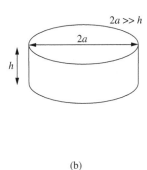

(a)   (b)

Figure 2.105 Two types of WGM dielectric resonators. (a) Cylindrical dielectric resonator and (b) planar dielectric resonator. Source: Cros, D. and Guillon, P. (1990). "Whispering gallery dielectric resonator modes for W-band devices", *IEEE Transactions on Microwave Theory and Techniques*, **38** (11), 1667–1674. © 2003 IEEE

As shown in Figure 2.105, WGM can be excited in cylindrical dielectric resonators and planar dielectric resonators. In the cylindrical dielectric resonator shown in Figure 2.105(a), its WGM is characterized by a second energy confinement (axial) ensured by the enlargement of the resonator radius in the central zone. In this area, the modes propagate axially with a small propagation constant and they decay exponentially away from the discontinuity of the resonator radius. In Figure 2.105(b), WGM is excited in a thin dielectric disk with the diameter $2a$ much larger than its thickness h. Such resonators have been extensively investigated because they are compatible with millimeter-wavelength integrated circuits.

WGM dielectric resonators are very interesting for a number of reasons (Cros and Guillon 1990). First, their dimensions are relatively large, even in the millimeter-wavelength band. Second, the quality factors are very high: the unloaded quality factor of a WGM dielectric resonator is limited only by the value of the loss tangent of the material used to realize the dielectric resonator, and the radiation losses are negligible. This is an important feature of WGM compared with the conventional TE or TM modes, whose unloaded quality factor depends not only on the material loss tangent but also on the metallic shields in which they are enclosed. Third, WGM resonators have good suppression of spurious modes because

the propagation constant along the Z-axis is very small and the unwanted modes leak out axially and can be absorbed without perturbation. Last, they offer a high level of integration. They can be easily integrated into planar circuits.

As shown in Figure 2.106, the WGM of dielectric resonators can be classified as WGE_{nml} and WGH_{nml}. In a WGE_{nml} mode, the electric field is essentially transversal, while in a WGH_{nml} mode, the electric field is essentially axial. The integer n denotes the azimuthal variations, m the radial variations, and l the axial ones. As an example, the field distributions of WGE_{800} and WGE_{810} modes are shown in Figure 2.107.

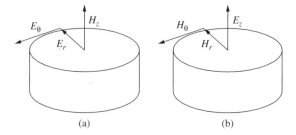

Figure 2.106 Electromagnetic fields of WGE and WGH modes in planar dielectric resonator. (a) WGE mode and (b) WGH mode. Source: Cros, D. and Guillon, P. (1990). "Whispering gallery dielectric resonator modes for W-band devices" *IEEE Transactions on Microwave Theory and Techniques*, **38** (11), 1667–1674. © 2003 IEEE

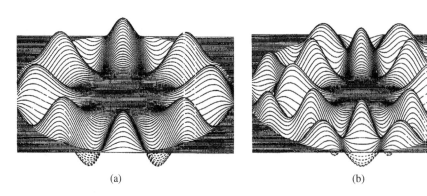

(a)   (b)

Figure 2.107 Field distributions of WGM. (a) Electric field amplitude for WGE_{800} mode and (b) electric field distribution of WGE_{810} mode (Niman 1992). Source: Niman, M. J. (1992). "Comments on 'Whispering gallery dielectric resonator modes for W-band devices' ", *IEEE Transactions on Microwave Theory and Techniques*, **40** (5), 1035–1036. © 2003 IEEE

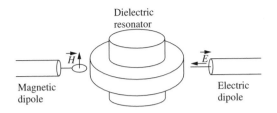

Figure 2.108 Whispering-gallery modes excited by electric and magnetic dipole. Source: Cros, D. and Guillon, P. (1990). "Whispering gallery dielectric resonator modes for W-band devices", *IEEE Transactions on Microwave Theory and Techniques*, **38** (11), 1667–1674. © 2003 IEEE

WGM dielectric resonators can be excited in various ways. As shown in Figure 2.108, in the low microwave frequency range we use an electric dipole or a magnetic dipole. Using this type of excitation we can obtain stationary WGM. In the high microwave frequency range we can use dielectric image waveguides or microstrip transmission lines as shown in Figure 2.109. Using this method, we can build travelling WGM.

Finally, it should be indicated that WGM could also be built in other types of resonant structures. Figure 2.110 shows a travelling microstrip resonator (Harvey 1963). If the length of the microstrip ring is a multiple of the wavelength, absorption can be observed from the transmission line, indicating the occurrence of a resonance. Correspondingly, we may also take the resonance in a microstrip ring resonator discussed in Section 2.3.3.2 as a stationary microstrip WGM.

2.3.6 Open resonators

2.3.6.1 Concept of open resonator

As discussed earlier, free space can also be taken as a type of transmission line. We can also make resonators from free-space transmission line by terminating it with two parallel metal plates, forming a Fabry–Perot resonator, as shown in Figure 2.111. In millimeter-wave and sub millimeter-wave frequency range, the conductor loss is a main part of energy dissipation of a resonant cavity. Compared to closed resonators, open resonators may have higher quality factors because the conductor loss is decreased as some of the conductor walls are moved away.

We assume the parallel plates in Figure 2.111 to be infinite in extent, and a TEM mode standing wave is built between the plates. From the boundary conditions, we can get the resonant frequency f_0:

$$f_0 = \frac{cn}{2d}, \qquad (2.422)$$

where d is the distance between the two parallel plates, n is the mode number ($n = 1, 2, 3, \ldots$), c is the speed of light. Equation (2.422) indicates that the resonant frequency can be adjusted by changing the distance between the two plates. In an actual Fabry–Perot resonator, usually one of the reflectors is movable so that the resonant frequency can be changed continuously. This is an attractive advantage for materials property research.

The quality factor Q_0 of the resonator can be estimated by

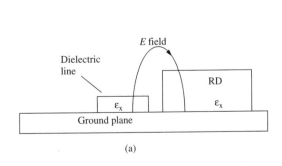

(a)

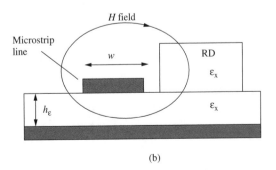

(b)

Figure 2.109 Whispering gallery modes excited by (a) dielectric image waveguide and (b) microstrip line. Source: Cros, D. and Guillon, P. (1990). "Whispering gallery dielectric resonator modes for W-band devices", *IEEE Transactions on Microwave Theory and Techniques*, **38** (11), 1667–1674. © 2003 IEEE

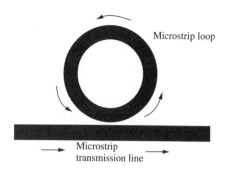

Microstrip loop

Microstrip
transmission line

Figure 2.110   A traveling-wave microstrip resonator

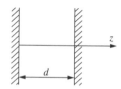

Figure 2.111   An ideal Fabry–Perot resonator

$$Q_0 = \frac{\pi n \eta_0}{4 R_s}, \qquad (2.423)$$

where n is the mode number, η_0 is the intrinsic impedance of the free space, and R_s is the surface resistance of the reflector. Equation (2.423) indicates that the quality factor increases with the increase of the mode number. As n is often several thousand or more, very high quality factors could be achieved.

2.3.6.2 Stability requirements

Equation (2.423) gives the quality factor of an ideal case: the two plates are infinite and they are strictly parallel. However, in an actual case as shown in Figure 2.112, the sizes of the plates are limited, and the two plates may be not strictly parallel, microwave energy will be radiated. Therefore, the actual quality factor is less than what Eq. (2.423) gives.

As shown in Figure 2.113, significant improvements can be made if one or both of the reflecting plane plates is replaced by a concave reflector (Musil and Zacek 1986). In these cases, the field is focused in a smaller volume and the requirements for arranging the two reflectors are not very high. Furthermore, if one or both reflectors have a concave spherical surface, the field

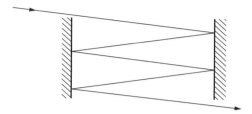

Figure 2.112   Leakage of Fabry–Perot resonator due to limited size of the plates. Reprinted from Musil, J. and Zacek, F. (1986). Microwave Measurements of Complex Permittivity by Free Space Methods and Their Applications, Permission from Elsevier, Amsterdam

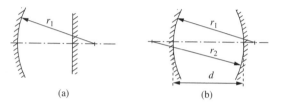

(a)   (b)

Figure 2.113   Resonator geometries with planar and spherical reflectors (Musil and Zacek 1986). Reprinted from Musil, J. and Zacek, F. (1986). Microwave Measurements of Complex Permittivity by Free Space Methods and Their Applications, Permission from Elsevier, Amsterdam

inside the resonator has a well-known Gaussian distribution. This fact can be used to simulate free-space field conditions, quite similar with the case of focused beams described in Section 2.2.5.

An open resonator consisting of two spherical reflectors with radii of r_1 and r_2 can support a stable mode if the following condition is met (Pozar 1998):

$$0 \le \left(1 - \frac{d}{r_1}\right)\left(1 - \frac{d}{r_2}\right) \le 1 \qquad (2.424)$$

On the basis of Eq. (2.424), a stability diagram can be drawn, as shown in Figure 2.114. In the stability diagram, the stable regions are shadowed. There are three special conditions, representing parallel-plane resonator, concentric resonator, and symmetrical confocal resonator respectively.

Parallel plate can be taken as a spherical reflector with infinite radius. The point corresponding to a parallel plate resonator ($d/r_1 = d/r_2 = 0$) is at the boundary between the stable region and unstable region. If there are any irregularities, the system becomes unstable.

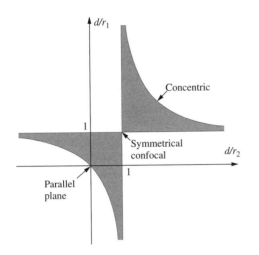

Figure 2.114   Stability diagram for open resonators

For a confocal resonator ($r_1 = r_2 = d$), the corresponding point is at the boundary between the stable region and unstable region. For a concentric resonator ($r_1 = r_2 = (d/2)$), the corresponding point is at the stability boundary. So these two kinds of resonators are also not very stable. To increase the stability, confocal and concentric resonators are often modified into near-confocal and near-concentric geometries, as shown in Figure 2.115.

When open resonators are used in materials property characterization, we should consider the beam-width in the resonator, as the beam-width determines the minimum dimensions of samples that can be measured. Similar to focused microwave beam, in the determination of beam width, we should define a certain level of field intensity with respect to the field intensity on the axis of the resonator.

2.3.6.3 Coupling to external circuit

In a microwave Fabry–Perot resonator, coupling may be achieved by quasi-optical methods or typical microwave methods. Figure 2.116 diagrammatically shows two common geometries for coupling in a quasi-optical manner: through one or both mirrors or normal to the resonator axis (Clarke and Rosenberg 1982). In the through-mirror method, mirrors of the open resonator are perforated metal plates, the RF signal is injected by means of a collimating lens fed by a small horn placed at its focus, and a similar lens-and-horn arrangement is used to collect the small signals passing through the system. The method shown in Figure 2.116(b) utilizes a dielectric beam splitter set at 45° to the resonator axis, and usually the beam splitter is made of polyethylene. In this system, the quality-factor of the resonator can be continuously adjusted by rotating the incident polarization. The quality factor reaches its maximum value when the electric vector lies in the plane of incidence and reaches its minimum value when the electric vector is perpendicular to the plane of incidence. Meanwhile, the quality factor of the resonator is also affected by the thickness of the beam splitter.

Fabry–Perot resonators can also be coupled using typical microwave methods, such as the coupling apertures in the mirror reflectors. Figure 2.117 shows an open resonator coupled to waveguides through apertures. The details of the coupling aperture are shown in Figure 2.118. It comprises a

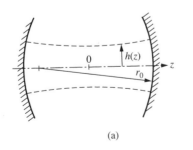

(a)

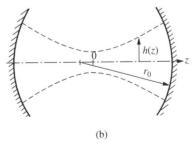

(b)

Figure 2.115   Modification of confocal and concentric resonators. (a) Near-confocal resonator and (b) near-concentric resonator (Musil and Zacek 1986). Reprinted from Musil, J. and Zacek, F. (1986). Microwave Measurements of Complex Permittivity by Free Space Methods and Their Applications, Permission from Elsevier, Amsterdam

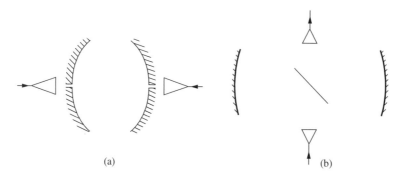

(a)   (b)

Figure 2.116   Two types of quasi-optical coupling methods. (a) Through-mirror method and (b) normal-to-axis method. Modified from Clarke, R. N. and Rosenberg, C. B. (1982). "Fabry–Perot and open resonators at microwave and millimeter wave frequencies, 2–300 GHz", *Journal of Physics E: Scientific Instruments*, **15** (1), 9–24

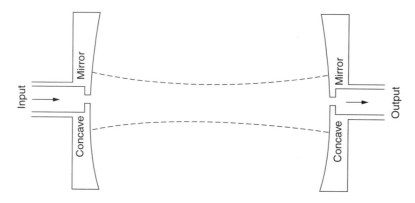

Figure 2.117   An open resonator coupled to waveguides through apertures

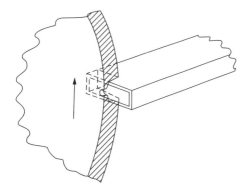

Figure 2.118   Structure of a coupling aperture (Clarke and Rosenberg 1982). Source: Clarke, R. N. and Rosenberg, C. B. (1982). "Fabry–Perot and open resonators at microwave and millimeter wave frequencies, 2–300 GHz", *Journal of Physics E: Scientific Instruments*, **15** (1), 9–24, by permission of The Institute of Physics

standard TE_{10} mode feed-waveguide butted into a rectangular recess in the back surface of one reflector. Loops of magnetic field emanate from the aperture which thus acts as a magnetic dipole radiating into a half-space (Clarke and Rosenberg 1982). The coupling coefficient is mainly determined by the diameter of the coupling hole and the thickness of the wall.

Measurement of dielectric properties is a natural application for open resonators since easily measured parameters of an open resonator, including the resonant frequency f, quality factor Q, and resonator length d, are simply related to the dielectric constant and loss tangent of the material included in the resonator. Actually, materials property characterization was one of the first applications for microwave open resonators, and open-resonator methods are presently among the most sensitive for low-loss dielectric measurements at millimeter-wave frequencies (Cullen 1983). More discussions on the application of open resonators in materials property characterization can be found in Chapter 5.

2.4 MICROWAVE NETWORK

As mentioned earlier, field method and line method are two important methods in microwave theory and engineering. In the field method, the distributions of electric field and magnetic field are analyzed. In the line method, the microwave properties of the transmission lines or resonant structures are represented by their equivalent lumped elements. Network approach is developed from the line method. In the network approach, we do not care the distributions of electromagnetic fields within a microwave structure, and we are only interested in how the microwave structure responds to external microwave signals.

In this section, we first introduce the concept of microwave network and the parameters describing microwave networks. We then introduce network analyzer, followed by a discussion on the methods for the measurements of reflection, transmission, and resonant properties of microwave networks.

2.4.1 Concept of microwave network

The concept of microwave network is developed from the transmission line theory, and is a powerful tool in microwave engineering. Microwave network method studies the responses of a microwave structure to external signals, and it is a complement to the microwave field theory that analyzes the field distribution inside the microwave structure.

Two sets of physical parameters are often used in network analysis. As shown in Figure 2.119, one set of parameters are voltage V (or normalized voltage v) and current I (or normalized current i). The other set of parameters are the input wave a (the wave going into the network) and the output wave b (the wave coming out of the network). Different network parameters are

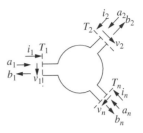

Figure 2.119   Two definitions of a network

used for different sets of physical parameters. For example, impedance and admittance matrixes are used to describe the relationship between voltage and current, while scattering parameters are used to describe the relationships between the input waves and output waves.

In the following discussion, we focus on two-port networks, and the conclusions obtained for two-port networks can be extended to multiport networks. We will discuss impedance matrix, admittance matrix, and scattering matrix, and we will also discuss the conversions between these parameters.

2.4.2 Impedance matrix and admittance matrix

2.4.2.1 Unnormalized impedance and admittance matrixes

There are two types of impedance matrixes and admittance matrixes: unnormalized ones and normalized ones. Unnormalized impedance and admittance describe the relationships between unnormalized voltage and unnormalized current. For a two-port network as shown in Figure 2.120, we have

$$[V] = [Z][I] \tag{2.425}$$

$$[I] = [Y][V], \tag{2.426}$$

where $[V]$ is the unnormalized voltage: $[V] = [V_1, V_2]^T$, $[I]$ is the unnormalized current: $[I] = [I_1, I_2]^T$, and the impedance matrix is

$$[Z] = \begin{bmatrix} Z_{11} & Z_{12} \\ Z_{21} & Z_{22} \end{bmatrix} \tag{2.427}$$

Figure 2.120   A two-port network with voltages and currents defined

From Eq. (2.425), we know that if $I_i = 0$ $(i \neq j)$

$$Z_{jj} = \frac{V_j}{I_j}, Z_{ij} = \frac{V_i}{I_j} \quad (i = 1, 2; j = 1, 2).$$

(2.428)

Equation (2.428) indicates that Z_{jj} is the input impedance at port j when the other port is open, and Z_{ij} $(i \neq j)$ is the transition impedance from j port to i port when port i is open.

The admittance matrix $[Y]$ in Eq. (2.426) is in the form of

$$[Y] = \begin{bmatrix} Y_{11} & Y_{12} \\ Y_{21} & Y_{22} \end{bmatrix}$$

(2.429)

Similarly, Y_{jj} is the input admittance at port j when the other port is shorted, and $Y_{ij}(i \neq j)$ is the transition admittance from j port to i port when port i shorted. From Eqs. (2.425) and (2.426), we can get the relationship between $[Y]$ and $[Z]$

$$[Z][Y] = [1]$$

(2.430)

2.4.2.2 *Normalized impedance and admittance matrixes*

Normalized impedance matrix $[z]$ and normalized admittance matrix $[y]$ defines the relationships between the normalized currents and normalized voltages:

$$[v] = [z][i]$$

(2.431)

$$[i] = [y][v]$$

(2.432)

From Eqs. (2.425), (2.426), (2.431), and (2.432), we can get

$$[z] = [\sqrt{Y_c}][Z][\sqrt{Y_c}]$$

(2.433)

$$[y] = [\sqrt{Z_c}][Y][\sqrt{Z_c}]$$

(2.434)

with:

$$[\sqrt{Z_c}] = \begin{bmatrix} \sqrt{Z_{c1}} & 0 \\ 0 & \sqrt{Z_{c2}} \end{bmatrix}$$

(2.435)

$$[\sqrt{Y_c}] = \begin{bmatrix} \sqrt{Y_{c1}} & 0 \\ 0 & \sqrt{Y_{c2}} \end{bmatrix},$$

(2.436)

where Z_{c1} and Z_{c2} are the characteristic impedances of the transmission lines connected to port 1 and port 2 respectively; Y_{c1} and Y_{c2} are the characteristic admittances of the transmission lines connected to port 1 and port 2 respectively. It is clear that

$$\left[\sqrt{Z_c}\right]\left[\sqrt{Y_c}\right] = [1]$$

(2.437)

For single components in Eqs. (2.433) and (2.434), we have

$$z_{ij} = Z_{ij}\sqrt{Y_{ci}Y_{cj}} = \frac{Z_{ij}}{\sqrt{Z_{ci}Z_{cj}}}$$

(2.438)

$$y_{ij} = Y_{ij}\sqrt{Z_{ci}Z_{cj}} = \frac{Y_{ij}}{\sqrt{Y_{ci}Y_{cj}}}$$

(2.439)

When $i = j$,

$$z_{ii} = \frac{Z_{ii}}{Z_{ci}}$$

(2.440)

$$y_{ii} = \frac{Y_{ii}}{Y_{ci}}$$

(2.441)

Equations (2.440) and (2.441) agree with the definitions in transmission line theory.

2.4.3 Scattering parameters

As shown in Figure 2.121, the responses of a network to external circuits can also be described by the input and output microwave waves. The input waves at port 1 and port 2 are denoted as a_1 and a_2 respectively, and the output waves from port 1 and port 2 are denoted as b_1 and b_2 respectively. These parameters (a_1, a_2, b_1, and b_2) may be voltage or current, and in most cases, we do not distinguish whether they are voltage or current.

The relationships between the input wave $[a]$ and output wave $[b]$ are often described by scattering parameters $[S]$:

$$[b] = [S][a],$$

(2.442)

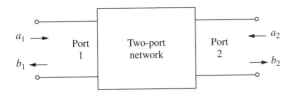

Figure 2.121 A two-port network with "*a*"s and "*b*"s defined

where $[a] = [a_1, a_2]^T$, $[b] = [b_1, b_2]^T$, and the scattering matrix $[S]$ is in the form of

$$[S] = \begin{bmatrix} S_{11} & S_{12} \\ S_{21} & S_{22} \end{bmatrix} \qquad (2.443)$$

For a scattering parameter S_{ij}, if $a_i = 0$ ($i \neq j$), from Eq. (2.442), we have

$$S_{jj} = \frac{b_j}{a_j} \quad (j = 1, 2) \qquad (2.444)$$

$$S_{ij} = \frac{b_i}{a_j} \quad (i \neq j; i = 1, 2; j = 1, 2) \qquad (2.445)$$

Equation (2.444) shows that when port j is connected to a source and the other port is connected to a matching load, the reflection coefficient at port j is equal to S_{jj}:

$$\Gamma_j = S_{jj} = \frac{b_j}{a_j} \qquad (2.446)$$

Equation (2.445) shows that when port j is connected to a source, and port i is connected to a matching load, the transmission coefficient from port j to port i is equal to S_{ij}:

$$T_{j \to i} = S_{ij} = \frac{b_i}{a_j} \qquad (2.447)$$

2.4.4 Conversions between different network parameters

From the relationships between different physical parameters, we can get conversions between different network parameters (Frickey 1994). From the definitions of normalized voltage and current, input and output waves, we have

$$v = a + b \qquad (2.448)$$

$$i = a - b \qquad (2.449)$$

and

$$a = \tfrac{1}{2}(v + i) \qquad (2.450)$$

$$b = \tfrac{1}{2}(v - i). \qquad (2.451)$$

From Eqs. (2.448)–(2.451), we can get the relationships between $[S]$, $[z]$, and $[y]$ as listed in Table 2.6. If we know any one of $[S]$, $[z]$, and $[y]$, from Table 2.6, we can find the other two. As will be discussed later, in experiments, usually the scattering parameters are measured, and other parameters including impedance parameters and admittance parameters can be calculated from scattering parameters.

2.4.5 Basics of network analyzer

Network analyzer is one of the most important tools for analyzing analog circuits. By measuring the amplitudes and phases of transmission and reflection coefficients of an analog circuit, a network analyzer reveals all the network characteristics of the circuit. In microwave engineering, network analyzers are used to analyze a wide variety of materials, components, circuits, and systems.

In most of the methods, which will be discussed in later chapters, for materials property characterization, measurements are conducted by network analyzers. In this part, we discuss the basic principle of network analyzers and the error correction techniques. More detailed information about network analyzers can be found in (Ballo 1998).

2.4.5.1 Principle of network analyzers

Network analyzers are widely used to measure the four elements in a scattering matrix: S_{11}, S_{12}, S_{21}, and S_{22}. As shown in Figure 2.122, a network analyzer mainly consists of a source, signal separation devices, and detectors. Basically, a network analyzer can measure the four waves independently: two forward traveling waves a_1 and a_2, and two reverse traveling waves b_1 and b_2. The scattering parameters can then be obtained by the combinations of these four waves according to Eqs. (2.444) and (2.445). The four detectors, labeled by a_1, a_2, b_1, and b_2 are used to measure the four corresponding waves respectively, and the

Table 2.6   Conversions between [S], [z], and [y]

	Expression in [S]	Expression in [z]	Expression in [y]
[S]	$[S] = \begin{bmatrix} S_{11} & S_{12} \\ S_{21} & S_{22} \end{bmatrix}$	$s_{11} = \dfrac{\|z\| + z_{11} - z_{22} - 1}{\|z\| + z_{11} + z_{22} + 1}$	$s_{11} = \dfrac{1 - y_{11} + y_{22} - \|y\|}{1 + y_{11} + y_{22} + \|y\|}$
		$s_{12} = \dfrac{2z_{12}}{\|z\| + z_{11} + z_{22} + 1}$	$s_{12} = \dfrac{-2y_{12}}{1 + y_{11} + y_{22} + \|y\|}$
		$s_{21} = \dfrac{2z_{21}}{\|z\| + z_{11} + z_{22} + 1}$	$s_{21} = \dfrac{-2y_{21}}{1 + y_{11} + y_{22} + \|y\|}$
		$s_{22} = \dfrac{\|z\| - z_{11} + z_{22} - 1}{\|z\| + z_{11} + z_{22} + 1}$	$s_{22} = \dfrac{1 + y_{11} - y_{22} - \|y\|}{1 + y_{11} + y_{22} + \|y\|}$
[z]	$z_{11} = \dfrac{1 + S_{11} - S_{22} - \|S\|}{1 - S_{11} - S_{22} + \|S\|}$	$[z] = \begin{bmatrix} z_{11} & z_{12} \\ z_{21} & z_{22} \end{bmatrix}$	$[z] = \dfrac{1}{\|y\|} \begin{bmatrix} y_{22} & -y_{12} \\ -y_{21} & y_{22} \end{bmatrix}$
	$z_{12} = \dfrac{-2S_{12}}{1 - S_{11} - S_{22} + \|S\|}$		
	$z_{21} = \dfrac{-2S_{21}}{1 - S_{11} - S_{22} + \|S\|}$		
	$z_{22} = \dfrac{1 - S_{11} + S_{22} - \|S\|}{1 - S_{11} - S_{22} + \|S\|}$		
[y]	$y_{11} = \dfrac{1 - S_{11} + S_{22} - \|S\|}{1 + S_{11} + S_{22} + \|S\|}$	$[y] = \dfrac{1}{\|z\|} \begin{bmatrix} z_{22} & -z_{12} \\ -z_{21} & z_{22} \end{bmatrix}$	$[y] = \begin{bmatrix} y_{11} & y_{12} \\ y_{21} & y_{22} \end{bmatrix}$
	$y_{12} = \dfrac{1 - S_{11} + S_{22} - \|S\|}{1 + S_{11} + S_{22} + \|S\|}$		
	$y_{21} = \dfrac{1 - S_{11} + S_{22} - \|S\|}{1 + S_{11} + S_{22} + \|S\|}$		
	$y_{22} = \dfrac{1 - S_{11} + S_{22} - \|S\|}{1 + S_{11} + S_{22} + \|S\|}$		

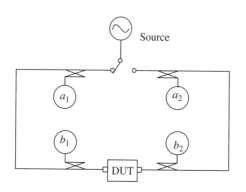

Figure 2.122   The block diagram for a network analyzer

signal separation devices ensure the four waves be measured independently.

2.4.5.2 Types of measurement errors

The measurement errors of a network analyzer generally fall into three categories: systematic errors, random errors, and drift errors.

Systematic errors

Systematic errors of a network analyzer mainly include match, directivity, cross talk, and frequency response. Match errors arise from multiple reflections of the device under test (DUT) that are not sensed at the incident wave detector. Directivity errors are due to leakage signals that are sensed at the reflected wave detector but are not reflected from the DUT. Cross-talk errors are due to leakage signals that are sensed at the transmitted wave detector without passing through the DUT. The frequency response errors arise from path loss, phase delay, and detector response. These errors are caused by imperfections in the measurement systems. Most of these errors do not vary with time, so they can be characterized through calibration and mathematically removed during the measurement process.

The dynamic ranges of a measurement system are limited by systematic errors. The dynamic range for reflection measurements is mainly limited by directivity, while the dynamic range for transmission measurement is mainly limited by noise floor or cross talk.

Random errors

Random errors are usually unpredictable and cannot be removed by calibration. The possible sources of random errors include instrument noise, switch repeatability, and connector repeatability. Random errors can be minimized by making measurements several times and taking the average values.

Drift errors

Drift errors are mainly caused by the change of working conditions of the measurement system after a calibration has been done. Temperature variation is one of the main sources for drift errors. In experiments, we should try to keep the working conditions as close to the calibration conditions as possible. To remove drift errors, further calibrations are needed.

2.4.5.3 Corrections of systematic errors

The systematic errors of a network analyzer can be corrected by calibration. This process computes the systematic errors from measurements on known reference standards. When subsequent measurements are made, the effects of the systematic errors are mathematically removed from the measurement results. The two main types of error corrections that can be done are response corrections and vector corrections. Response calibration is simple to perform, but only corrects for a few of the possible systematic error terms. Response calibration is essentially a normalized measurement where a reference trace is stored in memory, and subsequent measurement data is divided by this memory trace. A more advanced form of response calibration is open/short averaging for reflection measurements using broadband diode detectors. In this case, two traces are averaged together to derive the reference trace. Vector-error correction requires the network analyzer can measure both magnitude and phase data. Vector-error correction can account for all major sources of systematic error, and requires more calibration standards. It should be noted that response calibration can be performed on a vector network analyzer, in which case we store a vector reference trace in memory, so that we can display normalized magnitude or phase data. This is not the same as vector-error correction, because we cannot remove the individual systematic errors, all of which are vector quantities.

Vector-error correction is a process of characterizing systematic vector error-terms by measuring known calibration standards, and then removing the effects of these errors from subsequent measurements. In microwave electronics and materials property characterization, one-port and two-port measurements are often used. In the following, we discuss one-port and two port calibrations. More discussions on vector-error corrections can be found in (Ballo 1998).

One-port calibration

In reflection methods for materials property characterization, one-port reflection measurements are required. One-port calibration can measure and remove three systematic error terms in one-port measurements: directivity, source match, and reflection tracking.

Figure 2.123 shows the equivalent circuits of an ideal case and an error adapter. The relationship

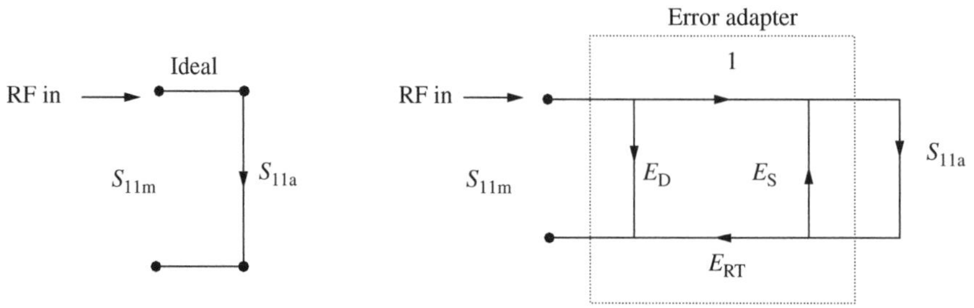

Figure 2.123 Model for one-port calibration. In the figure, S_{11m} is the measured S_{11} value, S_{11a} the actual S_{11} value, E_D the directivity, E_{RT} the reflection tracking, and E_S the source match. Source: Ballo, D. (1998). *Network Analyzer Basics*, Hewlett-Packard Company. Santa Rosa, CA

between the actual S-parameter S_{11a} and the measurement result S_{11m} is given by

$$S_{11m} = E_D + \frac{E_{RT}S_{11a}}{1 - E_S S_{11a}} \qquad (2.452)$$

In order to get the three systematic error terms so that the actual reflection S-parameters can be derived from our measurements, it is necessary to create three equations with three unknowns and solve them simultaneously. So three known standards, for example, a short, an open, and a Z_0 load should be measured. Solving these equations will yield the systematic error terms.

Two-port calibration

Two-port error correction accounts for all the major sources of systematic errors. There are two types of two-port calibrations often used: full two-port calibration and through-reflect-line (TRL) calibration. These two calibration methods are based on similar models, and we focus on full two-port calibration.

Full two-port calibration

All the systematic error terms are measured and removed by full two-port calibration. After full two-port calibration, the network analyzer can be used for both reflection and transmission measurements. The error model for a two-port device is shown in Figure 2.124. There are six types of systematic errors: directivity and cross-talk

errors relating to signal leakage, source, and load; impedance mismatches relating to reflections; and frequency response errors caused by reflection and transmission tracking within the test receivers. The full two-port error model includes all six of these terms for the forward direction and the same six terms in the reverse direction, for a total of 12 error terms. So full two-port calibration is often referred to as 12-term error correction.

The relationships between the actual device S-parameters and the measured S-parameters are given by (Ballo 1998)

$$S_{11a} = \frac{\left(\dfrac{S_{11m} - E_D}{E_{RT}}\right)\left(1 + \dfrac{S_{22m} - E'_D}{E'_{RT}}E'_S\right) - E_L\left(\dfrac{S_{21m} - E_X}{E_{TT}}\right)\left(\dfrac{S_{12m} - E'_X}{E'_{TT}}\right)}{\left(1 + \dfrac{S_{11m} - E_D}{E_{RT}}E_S\right)\left(1 + \dfrac{S_{22m} - E'_D}{E'_{RT}}E'_S\right) - E'_L E_L\left(\dfrac{S_{21m} - E_X}{E_{TT}}\right)\left(\dfrac{S_{12m} - E'_X}{E'_{TT}}\right)}$$

$$(2.453)$$

$$S_{21a} = \frac{\left(\dfrac{S_{21m} - E_X}{E_{TT}}\right) \times \left(1 + \dfrac{S_{22m} - E'_D}{E'_{RT}}(E'_S - E_L)\right)}{\left(1 + \dfrac{S_{11m} - E_D}{E_{RT}}E_S\right)\left(1 + \dfrac{S_{22m} - E'_D}{E'_{RT}}E'_S\right) - E'_L E_L\left(\dfrac{S_{21m} - E_X}{E_{TT}}\right)\left(\dfrac{S_{12m} - E'_X}{E'_{TT}}\right)}$$

$$(2.454)$$

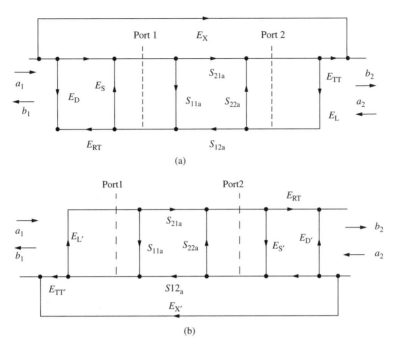

Figure 2.124 Systematic errors of network analyzer. (a) Forward model. In the figure, E_D is the forward directivity; E_S, forward source match; E_{RT}, forward reflection tracking; E_L, forward load match; E_{TT}, forward transmission tracking; and E_X, forward isolation and (b) reverse model. In the figure, E_D' is the reverse directivity; E_S', reverse source match; E_{RT}', reverse reflection tracking; E_L', reverse load match; E_{TT}', reverse transmission tracking; and E_X', reverse isolation. Source: Ballo, D. (1998). *Network Analyzer Basics*, Hewlett-Packard Company. Santa Rosa, CA

To obtain the 12 terms of systematic error terms, 12 independent measurements on known standards are required. Usually, four known standards, short-open-load-thru (SOLT) are used, and thus this calibration is also called SOLT calibration. Some standards are measured multiple times, for example, the thru standard is usually measured four times. Once the systematic error terms have been characterized, the actual device S-parameters can be derived from the measured S-parameters using Eqs. (2.453)–(2.456). It should be noted that each actual S-parameter is a function of all four measured S-parameters, so the network analyzer must make a forward and reverse sweep to update any one of the four S-parameters.

$$S_{12a} = \frac{\left(\dfrac{S_{12m} - E_X'}{E_{TT}'}\right) \times \left(1 + \dfrac{S_{22m} - E_D}{E_{RT}}(E_S - E_L')\right)}{\left(1 + \dfrac{S_{11m} - E_D}{E_{RT}}E_S\right)\left(1 + \dfrac{S_{22m} - E_D'}{E_{RT}'}E_S'\right) - E_L'E_L\left(\dfrac{S_{21m} - E_X}{E_{TT}}\right)\left(\dfrac{S_{12m} - E_X'}{E_{TT}'}\right)}$$

(2.455)

$$S_{22a} = \frac{\left(\dfrac{S_{22m} - E_D'}{E_{RT}}\right)\left(1 + \dfrac{S_{11m} - E_D}{E_{RT}}E_S\right) - E_L'\left(\dfrac{S_{21m} - E_X}{E_{TT}}\right)\left(\dfrac{S_{12m} - E_X'}{E_{TT}'}\right)}{\left(1 + \dfrac{S_{11m} - E_D}{E_{RT}}E_S\right)\left(1 + \dfrac{S_{22m} - E_D'}{E_{RT}'}E_S'\right) - E_L'E_L\left(\dfrac{S_{21m} - E_X}{E_{TT}}\right)\left(\dfrac{S_{12m} - E_X'}{E_{TT}'}\right)}$$

(2.456)

TRL calibration

To perform a two-port calibration, there are several choices based on the type of calibration standards to be used. Besides the full two-port calibration discussed above, TRL calibration is also widely used. The name comes from the three standards used: thru-reflect-line (TRL). TRL calibration is a two-port calibration technique primarily used in noncoaxial systems, such as waveguide, fixtures, and wafer probing. It solves for the same 12 error terms as the full two-port calibration, using a slightly different error model. One advantage of TRL calibration is that the required standards can be easily designed, fabricated, and characterized.

For a four-receiver network analyzer as shown in Figure 2.122, true TRL calibration can be performed, while for a three-receiver analyzer, the version of TRL calibration is usually called TRL*. There are several variations of TRL calibration, such as Line-Reflect-Line (LRL), Line-Reflect-Match (LRM), Thru-Reflect-Match (TRM), and all these calibration methods share error models similar to the ones for TRL calibration. More discussions on TRL calibration can be found in (Focus Microwaves Inc. 1994) and (Metzger 1995).

2.4.5.4 Improvement of the accuracy for materials property characterization

Network analyzers are widely used in materials property characterization. As discussed earlier, there are two types of materials property characterization methods: resonant methods and nonresonant methods. Resonant methods usually have higher accuracy and sensitivity, but require higher frequency stability. To improve the accuracy and sensitivity of resonant methods, the primary choice is using synthesized source, and the secondary is error correction. Nonresonant methods can characterize materials over a frequency range. For nonresonant methods, the primary choice is error correction, and the secondary is using synthesized source.

2.4.6 Measurement of reflection and transmission properties

In nonresonant methods for materials property characterization, the sample under test is inserted in a segment of transmission line, and the reflection from the sample and/or the transmission through the sample are measured for the calculation of materials properties. In the following, we discuss one-port method and two-port method for reflection and transmission measurements. In one-port and two-port measurements, the measurement results may be affected by some unwanted signals, such as the reflections from the mismatch between the connectors in coaxial-line measurement and the reflections from ground in free-space measurement. So we will also discuss the time-domain techniques, which are often used in reflection and transmission measurement to eliminate unwanted signals.

2.4.6.1 One-port method

One-port method is mainly for the reflection measurement. In this method, the single port transmits signals, and meanwhile receives the signals reflected from the sample under study. The S-parameter measured is S_{11} or S_{22}.

The dynamic range of reflection measurements is mainly limited by the directivity of the measurement port. To improve the measurement accuracy and sensitivity, it is usually required to conduct one-port calibration, which requires three known standards, as discussed in Section 2.4.5.3. However, in some cases, when we do not require high accuracy and sensitivity, one known standard, such as "short" or "air", may be enough to conduct a simple calibration.

2.4.6.2 Two-port method

Two-port method is suitable for the transmission/reflection method for materials property characterization. In a two-port method, one port transmits signal and the other port receives the signal from the sample under study. As shown in Figure 2.125, two-port method can be used for transmission/reflection measurement and bi-static reflection or scattering measurement.

The dynamic range of a two-port method is mainly limited by the noise floor of the signal and the cross talk between the two ports. Transmission measurement usually requires two-port calibration discussed in Section 2.4.5.3.

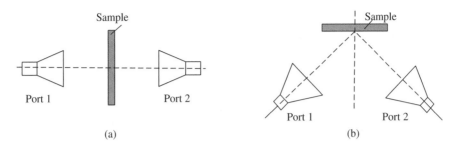

Figure 2.125 Two-port measurements. (a) Transmission/reflection measurement and (b) bistatic reflection or scattering measurement

If we do not need accurate measurement, in some cases a "through" calibration may be enough. For bistatic reflection or scattering measurement, special calibrations and measurement techniques are needed (Knott 1993; Knott *et al.* 1993).

2.4.6.3 Time-domain techniques

The Fourier transform techniques provide a method for transforming data between frequency domain and time domain. In network analyzers, usually a type of fast Fourier transform (FFT) known as the Chirp Z Transform is used, which permits users to "zoom in" on a specific time (distance) range of interest. Here, we discuss the main properties of the transform process and how the various processing options can be used to obtain optimum results. More discussions on the applications of time-domain techniques in network analyzers can be found in (Anritsu Company 1998; Rohde & Schwarz 1998).

By means of the inverse Fourier transform, the measurement results in the frequency domain can be transformed to the time domain. The time-domain results give clear representations of the characteristics of the device under test (DUT). For instance, the faults in cables can be directly localized. Furthermore, special time domain filters, called gates, can be used to suppress unwanted signal components such as multireflections. The measured data "gated" in the time domain can be transformed back to the frequency domain and an *S*-parameter representation without the unwanted signal components can be obtained as a function of frequency.

Fundamentals

The responses of a linear and invariant network to electromagnetic waves can be represented in the time domain by its impulse response $h(t)$ or in the frequency domain by its transfer function $H(f)$. Fundamentally speaking, $h(t)$ and $H(f)$ give the same information, and they can be transformed from one to the other through the Fourier transform:

$$H(f) = \int_{-\infty}^{\infty} h(t) \exp(-\mathrm{j}2\pi f t)\, \mathrm{d}t \qquad (2.457)$$

It is clear that the data measured in the frequency domain by a network analyzer can be transformed to the time domain using inverse Fourier transform:

$$h(t) = \int_{-\infty}^{\infty} H(f) \exp(-\mathrm{j}2\pi f t)\, \mathrm{d}f \qquad (2.458)$$

Impulse response and step response

In time domain, besides the impulse response, signals can also be represented by the step response, which can be obtained by integration of the impulse response $h(t)$. Figure 2.126 shows a stepped coaxial line and its time-domain responses in impulse and step forms. In general, the relationship between the measured reflection coefficient S_{11} and the impedance Z is given by

$$S_{11} = \frac{Z - Z_0}{Z + Z_0}, \qquad (2.459)$$

where Z_0 is the characteristic impedance. In Figure 2.126, the characteristic impedance is 50 Ω.

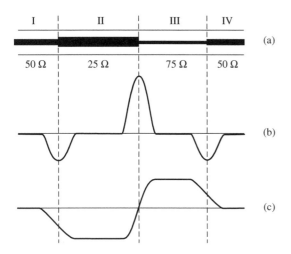

Figure 2.126   Time-domain responses (S_{11} in low-pass mode) of a stepped coaxial line. (a) Impedance of a stepped coaxial line, (b) impulse response, and (c) step response

From time-domain responses, we can locate the impedance discontinuities along a transmission line. Furthermore, the step representation clearly shows the variation of the impedance along the coaxial line. There is no reflection in the regions (I) and (IV), so the zero value in Figure 2.126(c) represents the reference impedance Z_0. Positive values stand for higher impedances than the reference impedance ($Z > Z_0$) and negative values for lower impedances ($Z < Z_0$).

Mathematically, these two forms of representations are equivalent. They can be converted into each other by differentiation or integration. It is recommended to use the step response if the impedance characteristics of the DUT are of interest, while, in most of the other cases, the impulse response should be used, especially for the determination of discontinuities.

Now, we consider the time domain responses of reactive DUTs. As the signal components of the step response occurring later in time correspond to lower frequency components down to DC, only the low-frequency behavior of the DUT has an effect on the later part of the step response. A capacitor now reacts like an interruption whereas an inductor for low frequencies is similar to a through connection.

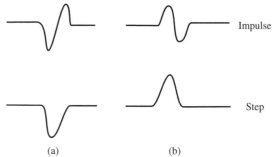

Figure 2.127   Ideal low-pass responses associated with (a) shunt capacitance and (b) series inductance

Figure 2.127 shows the low-pass responses associated with a series inductance and a shunt capacitance. The meaning of "low-pass" will be discussed later. Figure 2.127(a) shows the response of a shunt capacitance. Before the capacitance, the impedance of the line is equal to the reference impedance (Z_0). The capacitor first acts as a short and is responsible for the negative edge of the step response. The capacitor charges up gradually. When the capacitor is fully charged, it acts as an open. The parallel connection consisting of capacitor and resistor is then equivalent to a single resistor. Since the value of the resistor in the example is equal to the reference impedance (Z_0), no reflection will occur due to the matching. Thus the step response again attains zero. The characteristic of the impulse response can be imagined by a differentiation of the step response. In a similar way, we can understand the response of an inductance in series, as shown in Figure 2.127(b).

Finite pulse width

Mathematically, in time domain, an infinitely wide frequency range is assumed. By Fourier transform, infinitely narrow Dirac pulses can be obtained. However, in actual Fourier transform, as the frequency span is limited, the frequency domain data are multiplied by a rectangular weighting function that takes the value 1 for the actual frequency range of the network analyzer and which is otherwise zero. This multiplication in the frequency domain corresponds to a convolution of ideal Dirac pulses with a si function in the time domain:

$$\text{si}(x) = \frac{\sin x}{x}, \qquad (2.460)$$

and the width ΔT of the si impulses is inversely proportional to the frequency span ΔF of the frequency range:

$$\Delta T = \frac{2}{\Delta F} \qquad (2.461)$$

The principal property of time domain techniques for most microwave applications is resolution, which reflects the ability to locate a specific signal in the presence of other signals. The above discussion indicates that the basic limitation of resolution is inversely related to data collection bandwidth in the frequency domain. A rule of thumb: resolution is on the order of 150 mm/[frequency span (GHz)]. For example, a 10-GHz frequency span will provide resolution of about 15 mm. Besides, it should be noted that resolution is also influenced by the processing method and window selection which will be discussed later.

Figure 2.128(a) shows another characteristic of si pulses: the occurrence of side lobes to the left and right of the main pulse. In experiments, side lobes are perceived as interference. According to the si function, the highest (negative) side amplitudes to the left and right of the main pulse are given by

$$\frac{\sin(3\pi/2)}{3\pi/2} = -0.212 \qquad (2.462)$$

This corresponds to a side-lobe suppression of -13.46 dB. Similar to the width of the impulse, the side lobes can also be reduced by suitable windows in the frequency domain, as will be discussed later.

Alias

The actual measurement results in the frequency domain are not obtained continuously with frequency but only at a finite number of discrete frequency points. The frequency discrete measurements can be regarded as a modified continuous spectrum by multiplying the continuous spectrum with a comb function in the frequency domain. In the time domain, this corresponds to a convolution of the time response with a periodic Dirac impulse sequence. This results in the alias effect of a frequent repetition of the original time response, as shown in Figure 2.129. The time interval Δt between the repetitions in the time domain response is called the alias-free range, which is related to the frequency step width Δf in the frequency domain:

$$\Delta t = 1/\Delta f. \qquad (2.463)$$

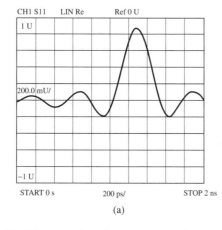
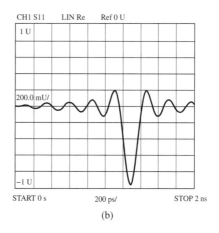

(a)   (b)

Figure 2.128 The time-domain responses of a shorted line. (a) Widened si impulse due to finite span ($\Delta F = 4$ GHz). The pulse width Δt is about 500 ps and (b) the width of si pulse is halved in low-pass mode. The pulse width Δt is about 250 ps. Source: Rohde & Schwarz (1998). *Time domain measurements using vector network analyzer ZVR*, Application Note 1EZ44_0E, Olaf Ostwald, Munich

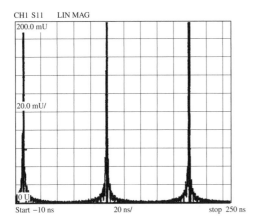

CH1 S11   LIN MAG
200.0 mU

20.0 mU/

0 U

Start −10 ns   20 ns/   stop 250 ns

Figure 2.129 Example of alias. The alias-free range is $\Delta t = 100$ ns. Source: Rohde & Schwarz (1998). *Time domain measurements using vector network analyzer ZVR*, Application Note 1EZ44_0E, Olaf Ostwald, Munich

In circuit measurements, alias is usually not a serious problem. But in fault location, alias should be taken into serious consideration. For example, with a 10-GHz, 201-point frequency collection, the alias free range is $1/(50\,\text{MHz}) = 20$ ns. This is large for most circuits, but if one wants to locate a fault in a 100-meter long cable, the range is inadequate. For such applications, the step size must be reduced either by decreasing the frequency span or increasing the number of points.

Low-pass and band-pass transformations

There are many methods for time-domain conversion, and the methods often used are mainly lowpass and bandpass. To obtain accurate time-domain results, suitable windows are often applied to the data in the frequency domain before the conversion, and after time-domain data are obtained, suitable gating are often used to eliminate the unwanted signals. In this part, we discuss lowpass and bandpass modes, and windowing and gating will be discussed in later parts.

The bandpass mode allows an arbitrary number of points and an arbitrary frequency range. It is suitable for displaying the magnitude of the impulse response, and is generally recommended for scalar applications. However, it does not

provide any information about zero frequency (dc), and the spectrum is limited to positive frequencies only. In some situations such as waveguide or band-limited devices, bandpass processing is often used. In this situation, vector information is lost as there is no phase reference; but useful magnitude information is still available. This type of processing is often used in fault location of transmission lines.

Actually, the impulse and step responses are complex and their phases depend upon the distance between the DUT and the reference plane. It is therefore generally not recommended to use the bandpass mode when the sign of the measured reflection coefficient is of interest. In these cases, the low-pass mode is often used.

Low-pass mode requires a special frequency plan: a harmonically related set of frequencies with starting frequency as low as possible. A dc term is extrapolated which provides a phase reference, so that the true nature of a discontinuity can be obtained. In the frequency grid for low-pass mode, the step width Δf between the frequency points is equal to the start frequency, so that an exact extrapolation to zero frequency is possible:

$$f_{\text{start}} = \Delta f \qquad (2.464)$$

A frequency grid meeting this requirement is called a harmonic grid since the frequency value at each frequency point is an integer multiple of the start frequency. Sometimes, a more general definition for a harmonic grid can be used:

$$f_{\text{start}} = n\Delta f \qquad (2.465)$$

In this case, the start frequency is an integer multiple n of the step width.

After generation of the required harmonic grid, the network analyzer is able to add an additional frequency point at zero frequency to the frequency grid and mirror the data measured at positive frequencies around zero frequency to the negative frequencies in a conjugate complex way, as shown in Figure 2.130. So, in the low-pass mode, the width of the frequency domain is doubled, and in contrast to the bandpass mode, the resolution in the time domain is improved by the factor of 2, as shown in Figure 2.128.

Figure 2.130 Harmonic frequency grid for lowpass mode

Figure 2.128 also indicates that the pulse in low-pass mode has a negative amplitude in contrast to the pulse in bandpass mode. Since the DUT is in a shorted line ($S_{11} = -1$), the measured negative amplitude of Figure 2.128(b) better complies with the expectations. As the response of the DUT in bandpass mode is complex, the bandpass time domain cannot give correct sign as shown in Figure 2.128(a), and usually people are only interested in the amplitude of bandpass time domain response, neglecting its sign. On the other hand, the low-pass mode provides a real time response (imaginary part $= 0$), and so it always gives the correct sign and amplitude of the reflection coefficient.

Windowing

As discussed earlier, due to the limited frequency range, the pulses in the time domain are widened and side lobes occur. The side lobes are very disadvantageous for time domain measurements since fraudulent echoes may be originated, and widening of the pulse and the occurrence of side lobes also decrease the resolution as well as the accuracy of measurements.

The occurrence of side lobes can be remedied by suitably windowing the measured frequency domain data. Windowing is essentially an attenuation of spectral components in the vicinity of the start and stop frequencies. It is a curve derived from a mathematical function that tapers off from unity gain at the center of the frequency domain data to a low value at the ends. Figures 2.131(a) and (b) show the effect of applying the widely used two-term Hamming window to the data obtained by measuring a short circuit. Figures 2.131(c) and (d) indicate that windowing obviously decreases the levels of the side lobes. But, at the same time, the window has the effect of widening the main lobe, thus decreasing the effective resolution.

In actual experiments, it is good to have a range of different window types available so that a compromise between resolution and side-lobe level can be reached. Table 2.7 shows several types of windows often used in microwave measurements, and detailed discussion on various windows can be found in (Harris 1978; Anritsu Company 1998; Rohde & Schwarz 1998).

Gating

Gate is essentially a filter in the time domain. It can be used to select a special discontinuity, and it can also be used to gate out a specific discontinuity thus observing the performance of a microwave circuit with an imperfection removed. Therefore, when a suitable gate is applied, we can observe the selected range without the influence of unwanted elements, such as connectors, and in the case of transmission measurements, we can eliminate the effects of multipath signals.

The time gate allows filtering out the components, impedance discontinuities or line faults that are spatially separated in the time domain. Owing to the different distances to the reference plane of the network analyzer, the associated reflections arrive at the test port at different times can thus be measured separately from each other in the time domain. For transmission measurements, direct transmissions can be distinguished from indirect transmissions (such as multireflected transmissions) in the time domain, and the signal components with different propagation speeds can be identified. One of the real benefits of the gate function is the transformation of the impulse response of the DUT filtered in the time domain back to the frequency domain. The interesting part of the DUT can thus be displayed without unwanted discontinuities. Figure 2.132 shows the application of a gate and the frequency response with gating on the initial discontinuity of a transmission line.

It should be indicated that for different gating requirements, different shapes of gate should be used. Table 2.7 shows several filtering shapes often used in gating. Detailed discussions on these gate shapes can be found in (Harris 1978; Anritsu Company 1998; Rohde & Schwarz 1998).

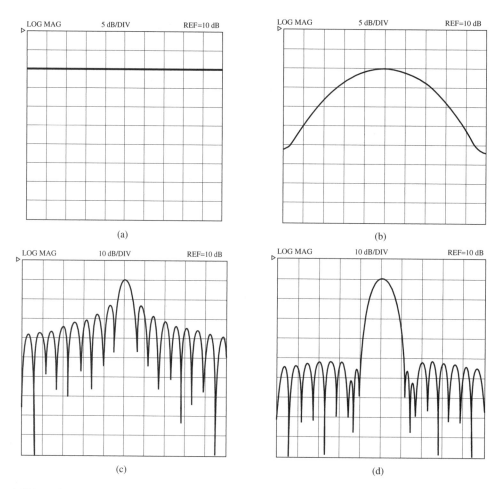

Figure 2.131   Effect of Hamming window. (a) Frequency domain data before application of Hamming window, (b) frequency domain data after application of Hamming window, (c) time-domain data before application of Hamming window, and (d) time-domain data after application of Hamming window. Source: Anritsu (1998). *Time domain for vector network analyzers*, Application Note, Anritsu Company, Morgan Hill, CA

Table 2.7   Filter functions and corresponding windows and gates

Filter function	Window in frequency domain	Date in time domain
Rectangular	No profiling	Steepest edges
Hamming	Low first side lobe	Steep edges
Hann	Normal profile	Normal gate
Bohman	Steep falloff	Maximum flatness
Dolph-Chebyshev	Arbitrary side lobes	Arbitrary gate shape

Application examples

Time-domain techniques are widely used in microwave measurements. In the characterization of materials properties, time domain techniques are usually applied in the following procedure. After raw data in frequency domain are obtained, apply suitable windowing functions on the frequency domain data, and then convert the frequency domain data to the time domain. Once the data is converted to the time domain, a suitable gating function may be applied to select the data of interest. The processed data may then be displayed in the time domain with start and stop times or in

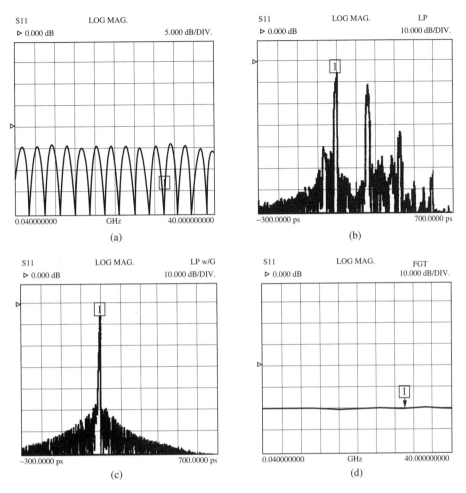

Figure 2.132 Effects of gating. (a) Frequency response, (b) lowpass time-domain response, (c) time-domain response with gate applied, and (d) gated frequency domain display. Source: Anritsu (1998). *Time domain for vector network analyzers*, Application Note, Anritsu Company, Morgan Hill, CA

the distance domain with start and stop distances. The data may also be converted back to the frequency domain with a time gate, so the frequency response we are interested in can be obtained. In the following, we give two application examples of time-domain techniques.

Transmission/reflection method for materials characterization

Coaxial air lines are widely used in the characterization of permittivity and permeability of low-conductivity materials using transmission/reflection method, and the parameters used in calculating

permittivity and permeability are reflection (S_{11}) and transmission (S_{21}). However, as shown in Figure 2.133, due to the inevitable mismatch at the interfaces between the connectors and coaxial air-line, there are reflections (Γ_1 and Γ_2) at the two interfaces. At the frequencies when the length of the coaxial air-line equals integer times of the half-wavelength, the reflections at the interfaces will cause obvious errors to the measurement results.

The effects of reflections at the interfaces can be minimized by using time-domain technique. As shown in Figure 2.133, if we place the sample at the center of the coaxial line, the reflections due to the interfaces (Γ_1 and Γ_2) and the reflection

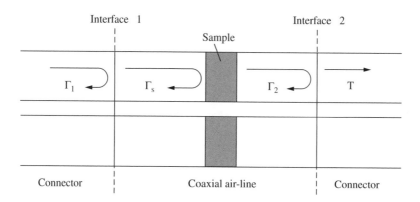

Figure 2.133   Coaxial air-line for transmission/reflection measurement

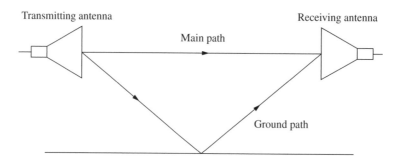

Figure 2.134   Free-space measurement set-up with a transmitting antenna and receiving antenna

due to the sample (Γ_s) occur at different positions, so they can be identified in time domain. The reflections due to the interfaces (Γ_1 and Γ_2) can be gated out by time-domain gating. Using the S_{11} data in the frequency domain with time gating, accurate permittivity and permeability results can be obtained.

Free-space measurement

Time-domain techniques play especially impor-tant roles in free-space measurements. As shown in Figure 2.134, most of the free-space measure-ment systems are subject to multipath reflections. Time domain provides an effective method of eliminating the effects of multipath reflections, and displays the information in time-gated fre-quency domain. Therefore the distortion of mul-tipath reflections to the DUT performance can be eliminated.

2.4.7 Measurement of resonant properties

2.4.7.1 Introduction

In resonant methods, including resonator methods and resonant-perturbation methods, we measure the resonant frequency and quality factor of the measurement fixture. As the resonant frequency can be easily determined from *S*-parameters in the frequency domain, in the following we focus on the measurement of quality factor.

As discussed in Section 2.3.1, the quality factor can be calculated according to

$$Q_L = \frac{f_0}{\Delta f}, \qquad (2.466)$$

where f_0 is the resonant frequency, Δf is the half-power bandwidth. As shown in Figure 2.135, Δf can be determined in different ways (Sucher and Fox 1963).

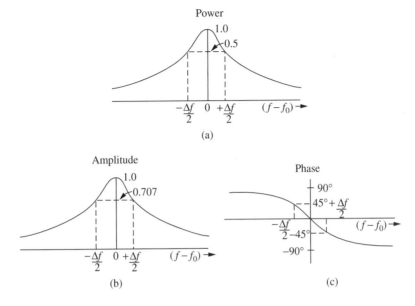

Figure 2.135 Determination of Δf for the measurement of loaded quality factor. (a) Power, (b) amplitude, and (c) phase (Sucher and Fox 1963). Modified from Sucher, M. and Fox, J. (1963), *Handbook of microwave measurements*, 3rd edition, vol. 2, by permission of Polytechnic Press of the Polytechnic Institute of Brooklyn

In actual measurements, the resonator under test is coupled to the external measurement circuit. The coupling between the resonator and the measurement circuit is often described by the coupling coefficient β. The resonant quality factor obtained from Eq. (2.466) is the loaded quality factor. For a resonator coupled to n pieces of transmission lines, with coupling coefficients β_i ($i = 1, 2, \ldots, n$) respectively, the relationship between the loaded quality factor $Q_{\rm L}$ and the unloaded quality factor Q_0 is given by

$$Q_0 = \left(1 + \sum_{i=1}^{n} \beta_i\right) Q_{\rm L} \qquad (2.467)$$

The measurement of the unloaded quality factor is crucial in resonant methods for the determination of loss tangent of materials. In the following, we discuss two methods often used for quality factor measurement: reflection method and transmission method. At the end of this part, we discuss the nonlinear phenomena in quality-factor measurements.

2.4.7.2 Reflection method

This is a one-port method, and the parameter we directly measure is the scattering parameter S_{11}. As shown in Figure 2.136, the S_{11} value for determining the half-power width is

$$S_{11,\Delta f} = 10 \cdot \log_{10} \left(\frac{10^{S_{11,b}/10} + 10^{S_{11,f_0}/10}}{2} \right) \ \ (\text{dB}), \qquad (2.468)$$

where $S_{11,b}$ is the S_{11} value of the base line of the resonance, and S_{11,f_0} is the S_{11} value at the resonant frequency.

As shown in Figure 2.137, the coupling coefficient β can be determined from the reflection Smith chart (Kajfez 1995):

$$\beta = \frac{1}{(d_2/d) - 1}, \qquad (2.469)$$

where d is the diameter of the "resonant circle", and d_2 is the diameter of the loss circle. After obtaining $Q_{\rm L}$ and β, the unloaded quality factor Q_0 can be calculated from

$$Q_0 = (1 + \beta)Q_{\rm L}. \qquad (2.470)$$

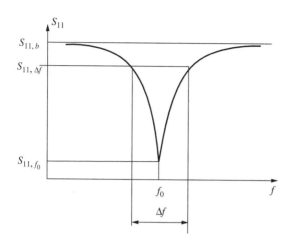

Figure 2.136 Measurement of quality factor from S_{11}

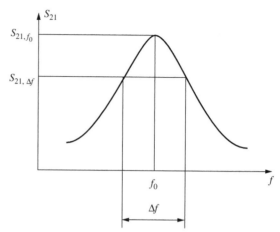

Figure 2.138 Measurement of quality factor from S_{21}

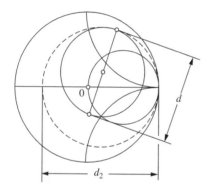

Figure 2.137 Measurement of coupling coefficient (Kajfez 1995). Modified from Kajfez, D. (1995). Q-factor measurement with a scalar analyser, *IEE Proceedings-Microwaves, Antennas and Propagations*, **142**, 369–372, by permission of IEE

2.4.7.3 Transmission method

This is a two-port method, and the quality factor is obtained from the scattering parameter S_{21}. As shown in Figure 2.138, the S_{21} value for determining the half-power width is

$$S_{21,\Delta f} = S_{21,f_0} - 3 \quad \text{(dB)}, \quad (2.471)$$

where S_{21,f_0} is the S_{21} value at the resonant frequency.

The unloaded quality factor can be calculated from

$$Q_0 = (1 + \beta_1 + \beta_2) \cdot Q_L, \quad (2.472)$$

where β_1 and β_2 are the coupling coefficients of the cavity to the two ports respectively. If the magnitudes of all the four S-parameters are measured at the resonant frequency f_0, we have

$$\beta_1 = \frac{1 - S_{11,f_0}}{S_{11,f_0} + S_{22,f_0}} \quad (2.473)$$

$$\beta_2 = \frac{1 - S_{22,f_0}}{S_{11,f_0} + S_{22,f_0}} \quad (2.474)$$

$$S_{21,f_0} = S_{12,f_0} = \frac{2\sqrt{\beta_1\beta_2}}{1 + \beta_1 + \beta_2} \quad (2.475)$$

By assuming that the two couplings between the cavity and the two transmission lines are weak and equal, we can calculate the unloaded quality factor Q_0 from the loaded quality factor Q_L:

$$Q_0 = \frac{Q_L}{1 - |S_{21,f_0}|} = \frac{Q_L}{1 - 10^{S_{21,f_0}(\text{dB})/20}} \quad (2.476)$$

Further discussions on the transmission method for quality factor measurement can be found in (Kajfez *et al.* 1999) and the references given therein.

2.4.7.4 Nonlinear phenomena in quality-factor measurement

In the study of nonlinear materials, for example, high-temperature superconducting thin films, after the microwave power reaches a certain level, the resonance curve shows nonlinear phenomena, changing from Lorentzian shape to non-Lorentzian shape. In this case, the traditional 3-dB bandwidth measurement may result in large errors. Rao *et al.* proposed a method to extract Q_L with improved accuracy for the case when nonlinear responses are involved (Rao *et al.* 2000).

In the case involving only linear responses, the frequency dependence of the transmission loss $T(f) = |S_{21}(f)|^2$ for a transmission mode resonator is given by (Ginzton 1957)

$$T(f) = \frac{|T(f_0)|}{1 + [2Q_L(f - f_0)/f_0]^2}, \quad (2.477)$$

where $T(f_0) = 4\beta_1\beta_2/(1 + \beta_1 + \beta_2)^2$, β_1 and β_2 are the coupling coefficients. The resonance curve of $T(f)$ is in Lorentzian shape and the loaded Q-factor, Q_L, of the resonator can be obtained by measuring the 3-dB (half-power) bandwidth, $\Delta f_{3\,dB}$, of the transmission curve and the resonant frequency f_0 using Eq. (2.466).

When the resonator undergoes a nonlinear response, the resonance curve becomes asymmetric. Figure 2.139 shows the resonance curves of a microstrip resonator made from double-sided superconducting YBa$_2$Cu$_3$O$_{7-d}$ (YBCO) thin film, and measurements are made when the input power

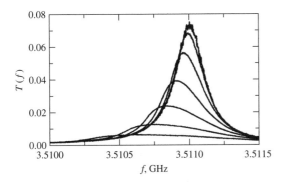

Figure 2.139 The transmission loss of a superconducting microstrip resonator against frequency for input power levels ranging from -18 to 12 dBm in 5 dB increments. Reproduced from Rao, X. S. Ong, C. K. and Feng, Y. P. (2000). "Q-factor measurement of nonlinear superconducting resonators", *Electronics Letter*, **36**, pp. 271–273, by permission of IEE

is increased in 5-dB steps from -18 to 12 dBm. When the input power is low, the resonance curves are symmetric about the resonant frequency and can be fitted well with the Lorentzian functions though fluctuations due to small signal-to-noise ratio are noticeable. As the input power increases, the resonance curve broadens gradually and becomes asymmetric and non-Lorentzian, with the peak resonant frequency shifting to lower frequency and the insertion loss increasing. Strictly speaking, when the resonance curve is clearly non-Lorentzian, the traditional 3-dB bandwidth measurement of Q_L is no longer applicable.

The above non-Lorentzian resonance mainly originates from following two reasons. Firstly, due to the $T(f)$ response of the resonator, the microwave power (P) coupled into the microstrip are different at different frequencies around the resonance. Secondly, when microwave power is high enough, the surface impedance of the superconducting microstrip shows a power dependent behavior: $Z_S = Z_S(P)$. Therefore in a nonlinear response, Z_S is dependent on frequency. The variations of R_S and X_S make the overall resistance R, capacitance C, and inductance L of the equivalent circuit of the resonator also change with frequency. So the different points on the resonance curve correspond to different LCR resonant circuits and thus have different effective values of Q_L and f_0.

By rewriting Eq. (2.477), Q_L can be generally calculated from (Rao *et al.* 2000)

$$Q_L = \sqrt{1/\tau - 1}\,\frac{f_0}{f_R - f_L} \quad (2.478)$$

with the relative power transmission ratio

$$\tau = \frac{T(f_R)}{T(f_0)} = \frac{T(f_L)}{T(f_0)} \quad (2.479)$$

In Eq. (2.478), $(f_R - f_L)$ is the bandwidth measured at power transmission ratio τ. If we choose the value of τ to be 0.5, $f_R - f_L = \Delta f_{3\,dB}$ and Eq. (2.478) becomes Eq. (2.466). If we choose a larger value of τ, smaller difference between $P(f_R)$ and $P(f_L)$ is expected and Eq. (2.478) gives a better approximation for the actual $Q_L(f_0)$ value.

Figure 2.140 shows a typical non-Lorentzian transmission curve where P indicates the resonant

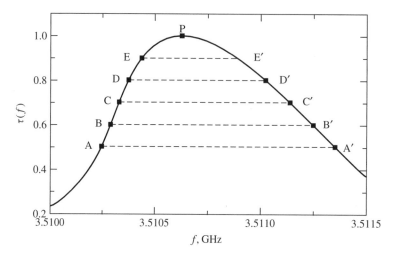

Figure 2.140 Typical non-Lorentzian resonance curve. Reproduced from Rao, X. S. Ong, C. K. and Feng, Y. P. (2000). "Q-factor measurement of nonlinear superconducting resonators", *Electronics Letter*, **36**, pp. 271–273, by permission of IEE

peak; A and A′ are the 3-dB points used in the traditional Q_L measurement; BB′, CC′, and so on, indicate other pairs of reference points with different τ values. For each pair of points, an approximate $Q_L(f_0)$ value can be obtained. The accuracy of the obtained $Q_L(f_0)$ increases when the τ value becomes larger. Theoretically, when τ is approaching unity, Eq. (2.478) gives an accurate $Q_L(f_0)$ value. However, following the error analysis similar to that in (Kajfez *et al.* 1999), we can get the relative uncertainty in measuring Q_L, given by

$$\left|\frac{\Delta Q_L}{Q_L}\right| = \frac{1}{2\tau\,(1-\tau)}\Delta\tau \qquad (2.480)$$

The error in τ is mainly caused by the inaccuracy of the amplitude reading of the instrument. As τ approaches unity, the error in τ may cause a very large error in Q_L. This problem can be circumvented using an extrapolation method. First, we measure a set of Q_L values as a function of τ and then extrapolate the results to a common intercept at $\tau = 1$. The resulting value is a reasonable approximation for $Q_L(f_0)$.

Figure 2.141 shows the calculated Q_L versus the τ values of the reference points for the curve in Figure 2.140. When the τ value becomes large, the resulting Q_L value gets smaller. This is mainly

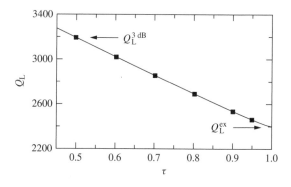

Figure 2.141 The Q_L values measured as a function of τ of reference points for the resonance curve shown in Figure 2.140. Reproduced from Rao, X. S. Ong, C. K. and Feng, Y. P. (2000). "Q-factor measurement of nonlinear superconducting resonators", *Electronics Letter*, **36**, pp. 271–273, by permission of IEE

due to the fact that the YBCO thin film has a larger surface resistance at the frequency point with larger τ value. The difference between the Q_L value from traditional 3-dB bandwidth measurement ($Q_L^{3\,dB}$) and that obtained from the method presented above (Q_L^{ex}) is prominent.

The value of $(Q_L^{3\,dB} - Q_L^{ex})/Q_L^{ex}$ for the resonance curves in Figure 2.139 are plotted as a function of the input power in Figure 2.142 and the absolute values of $Q_L^{3\,dB}$ and Q_L^{ex} are shown in

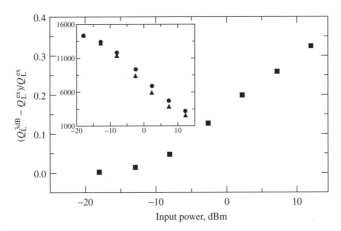

Figure 2.142 $(Q_L^{3\,dB} - Q_L^{ex})/Q_L^{ex}$ versus input power obtained from the resonance curves in Figure 2.139. The insert shows the corresponding $Q_L^{3\,dB}$ (circle) and Q_L^{ex} (triangle). Reproduced from Rao, X. S. Ong, C. K. and Feng, Y. P. (2000). "Q-factor measurement of nonlinear superconducting resonators", *Electronics Letter*, **36**, pp. 271–273, by permission of IEE

the insert. When the input power is not large, the $Q_L^{3\,dB}$ is close to Q_L^{ex}, and both of them give good approximations for $Q_L(f_0)$. When the input power increases, the difference between $Q_L^{3\,dB}$ and Q_L^{ex} becomes large. In this case, the application of the 3-dB method should be avoided.

REFERENCES

Afsar, M. U. Birch, J. R. Clarke, R. N. and Chantry, G. W. (1986). "The measurement of the properties of materials", *Proceedings of the IEEE*, **74**, 183–199.

Anritsu Company (1998). *Time Domain for Vector Network Analyzers*, Application Note, Anritsu Company, Morgan Hill, CA.

Baker-Jarvis, J. Janezic, M. D. Grosvenor, J. H. Jr. and Geyer, R. G. (1993). *Transmission/Reflection and Short-circuit Line Methods for Measuring Permittivity and Permeability*, NIST Technical Note 1355 (revised), U. S. Department of Commerce.

Balanis, C. A. (1997). *Antenna Theory: Analysis and Design*, John Wiley, New York.

Ballo, D. (1998). *Network Analyzer Basics*, Hewlett-Packard Company, Santa Rosa, CA.

Booth, J. C. Wu, D. H. and Anlage, S. M. (1994). "A broadband method for the measurement of the surface resistance of thin films at microwave frequencies", *Review of Scientific Instruments*, **65**, 2082–2090.

Chang, K. (1989). *Handbook of Microwave and Optical Components*, Vol. 1, John Wiley, New York.

Clarke, R. N. and Rosenberg, C. B. (1982). "Fabry-Perot and open resonators at microwave and millimeter wave frequencies, 2-300 GHz", *Journal of Physics E: Scientific Instruments*, **15** (1), 9–24.

Cohn, S. B. (1968). "Microwave bandpass filters containing high-Q dielectric resonators", *IEEE Transactions on Microwave Theory and Techniques*, **16** (4), 218–227.

Cohn, S. B. and Kelly, K. C. (1966). "Microwave measurement of high-dielectric constant materials", *IEEE Transactions on Microwave Theory and Techniques*, **14**, 406–410.

Collin, R. E. (1991). *Field Theory of Guided Waves*, 2nd edition, IEEE Press, Piscataway, NJ.

Collin, R. E. (1992). *Foundations for Microwave Engineering*, 2nd edition, McGraw-Hill, New York.

Connor, F. R. (1989). *Antennas*, 2nd edition, Edward Arnold, London.

Cros, D. and Guillon, P. (1990). "Whispering gallery dielectric resonator modes for W-band devices", *IEEE Transactions on Microwave Theory and Techniques*, **38** (11), 1667–1674.

Cullen, A. L. (1983). "Chapter 4 Millimeter-wave open-resonator techniques", in *Infrared and Millimeter Waves*, Vol. 10, Part II, K. J. Button, Ed., Academic Press, Orlando, 233–281.

Dew-Hughes, D. (1997). "Microwave properties and applications of high temperature superconductors", in *Microwave Physics and Techniques*, H. Groll and I. Nedkov, Eds., Kluwer Academic Publishers, Dordrecht, 83–114.

Fannis, P. C. Relihan, T. and Charles, S. W. (1995). "Investigation of ferromagnetic resonance in magnetic fluids by means of the short-circuited coaxial line techniques", *Journal of Physics D: Applied Physics*, **28**, 2002–2006.

Focus Microwaves Inc. (1994). "Coaxial TRL calibration kits for network analyzers up to 40 GHz", *Microwave Journal*, **37** (10), 160–162.

Frickey, D. A. (1994). "Conversions between S, Z, Y, h, ABCD, and T parameters which are valid for complex source and load impedances", *IEEE Transactions on Microwave Theory and Techniques*, **42**, 205–211.

Friedman, M. and Fernsler, R. F. (2001). "Low-loss RF transport over long distance", *IEEE Transactions on Microwave Theory and Techniques*, **49** (2), 341–348.

Geyer, R. G. and Krupka, J. (1995). "Microwave dielectric properties of anisotropic materials at cryogenic properties", *IEEE Transactions on Microwave Theory and Techniques*, **44**, 329–331.

Ginzton, E. L. (1957). *Microwave measurements*, McGraw-Hill, New York.

Goal, J. E. (1969). "A circular harmonic computer analysis of rectangular dielectric waveguides", *Bell System Technical Journal*, **48**, 2133–2160.

Goubau, G. (1950). "Surface waves and their applications to transmission lines", *Journal of Applied Physics*, **21**, 1119–1128.

Guillon, P. (1995). "Microwave techniques for measuring complex permittivity and permeability of materials", in *Materials and Processes for Wireless Communications*, T. Negas and H. Lings, Eds., The American Ceramic Society, Westerville, 65–71.

Gupta, K. C. Garg, R. and Bahl, I. J. (1979). *Microstrip Lines and Slotlines*, Artech House, Dedham, MA.

Harris, F. J. (1978). "On the use of windows for harmonic analysis with discrete Fourier transform", *Proceedings of the IEEE*, **66**, 51–83.

Harvey, A. F. (1963). *Microwave Engineering*, Academic Press, London.

Hoffmann, R. K. (1987). *Handbook of Microwave Integrated Circuits*, Artech House, Norwood, MA.

Imai, N. and Yamamoto, K. (1984). "A design of high-Q dielectric resonators for MIC applications", *Electronics and Communications in Japan*, **67B**, 59–67.

Ishii, T. K. (1995). *Handbook of Microwave Technology*, Vol. 1, Academic Press, San Diago, CA.

Izadian, J. S. and Izadian, S. M. (1988). *Microwave Transition Design*, Artech House, Norwood, MA.

Kajfez, D. (1995). "Q-factor measurement with a scalar analyser", *IEE Proceedings-Microwaves, Antennas and Propagations*, **142**, 369–372.

Kajfez, D. Chebolu, S. Abdul-Gaffoor, M. R. and Kishk, A. A. (1999). "Uncertainty analysis of the transmission-type measurement of Q-factor", *IEEE Transactions on Microwave Theory and Techniques*, **47** (3), 367–371.

Kajfez, D. and Guillon, P. (1986). *Dielectric Resonators*, Artech House, Dedham, MA.

Knott, E. F. (1993). *Radar Cross Section Measurements*, Van Nostrand Reinhold, New York.

Knott, E. F. Shaeffer, J. F. and Tuley, M. T. (1993). *Radar Cross Section*, 2nd edition, Artech House, Boston.

Kobayashi, Y. and Tanaka, S. (1980). "Resonant modes of a dielectric rod resonator short-circuited at both ends by parallel conducting plates", *IEEE Transactions on Microwave Theory and Techniques*, **28**, 1077–1085.

Kobayashi, Y. Aoki, T. and Kabe, Y. (1985). "Influence of conductor shields on the Q-factors of a TE_0 dielectric resonator", *IEEE Transactions on Microwave Theory and Techniques*, **33** (12), 1361–1366.

Krupka, J. Geyer, R. G. Kuhn, M. and Hinden, J. H. (1994). "Dielectric properties of Al_2O_3, $LaAlO_3$, $SrTiO_3$, and MgO at cryogenic temperature", *IEEE Transactions on Microwave Theory and Techniques*, **42**, 1886–1890.

Krupka, J. and Weil, C. (1998). "Recent advances in metrology for the electromagnetic characterization of materials at microwave frequencies", *12th International Conference on Microwaves and Radar*, **4**, 243–253.

Li, C. C. and Chen, K. M. (1995). "Determination of electromagnetic properties of materials using flanged open-ended coaxial probe – full-wave analysis," *IEEE Transactions on Instrumentation and Measurement*, **44**, 19–27.

Marincic, A. Benson, F. A. and Tealby, J. M. (1986). "Measurements of permittivity by the use of surface waves in open and closed structures", *IEE Proceedings-H*, **133**, 441–449.

Metzger, D. (1995). "Improving TRL* calibrations of vector network analyzers", *Microwave Journal*, **38** (5), 56–68.

Musil, J. and Zacek, F. (1986). *Microwave Measurements of Complex Permittivity by Free Space Methods and Their Applications*, Elsevier, Amsterdam.

Nicolson, A. M. and Rose, G. F. (1970). "Measurement of the intrinsic properties of materials by time domain techniques,", *IEEE Transactions on Instrumentation and Measurement*, **19**, 377–382.

Niman, M. J. (1992). "Comments on 'Whispering gallery dielectric resonator modes for W-band devices'", *IEEE Transactions on Microwave Theory and Techniques*, **40** (5), 1035–1036.

Nyfors, E. and Vainikainen, P. (1989). *Industrial Microwave Sensors*, Artech House, Norwood.

Pospieszalski, M. W. (1979). "Cylindrical dielectric resonators and their applications in TEM line microwave circuits", *IEEE Transactions on Microwave Theory and Techniques*, **27** (3), 233–238.

Pozar, D. M. (1998). *Microwave Engineering*, 2nd edition, John Wiley & Sons, New York.

Ramo, S. Whinnery, J. R. and Van Duzer, T. (1994). *Fields and Waves in Communication Electronics*, 3rd edition, John Wiley & Sons, New York.

Rao, X. S. Ong, C. K. and Feng, Y. P. (2000). "Q-factor measurement of nonlinear superconducting resonators", *Electronics Letter*, **36**, 271–273.

Rohde & Schwarz (1998). *Time Domain Measurements Using Vector Network Analyzer ZVR*, Application Note 1EZ44_0E, Olaf Ostwald, Munich.

Stuchly, M. A. and Stuchy, S. S. (1980). "Coaxial line reflection methods for measuring dielectric properties of biological substances at radio and microwave frequencies-a review", *IEEE Transactions on Instrumentation and Measurement"*, **29**, 176–183.

Sucher, M. and Fox, J. (1963), *Handbook of microwave measurements*, Vol. 2, 3rd edition, Polytechnic Press of the Polytechnic Institute of Brooklyn, New York.

Tsuji, M. Shigesawa, H. and Takiyama, K. (1984). "Analytical and experimental investigations on several resonant modes in open dielectric resonators", *IEEE Transactions on Microwave Theory and Techniques*, **32** (6), 628–633.

Wait, J. R. (1986). *Introduction to Antennas & Propagation*, Peter Peregrinus, London.

Weil, C. W. (1995). "The NIST metrology program on electromagnetic characterisation of materials", in *Materials and Processes for Wireless Communications*, T. Negas and H. Lings, Eds., The American Ceramic Society, Westerville.

Weir, W. B. (1974). "Automatic measurement of complex dielectric constant and permeability at microwave frequencies", *Proceedings of the IEEE*, **62**, 33–36.

Wu, R. X. and Qian, M. (1997). "A simplified power transmission method used for measuring the complex conductivity of superconducting thin films", *Review of Scientific Instruments*, **68**, 155–158.

Zaki, K. A. and Wang, C. (1995). "Accurate measurements of electrical properties of dielectric resonators and substrate materials", in *Materials and Processes for Wireless Communications*, T. Negas and H. Lings, Eds., The American Ceramic Society, Westerville, 49–63.

Zoughi, R. (2000). *Microwave Non-Destructive Testing and Evaluation*, Kluwer Academic Publishers, Boston.

3

Reflection Methods

In a reflection method, the properties of a sample are obtained from the reflection due to the impedance discontinuity caused by the presence of the sample in a transmission structure. We first discuss the reflection methods developed from coaxial line and free space. Subsequently, we discuss the measurement of both permittivity and permeability using modified reflection methods. Following that, we discuss the reflection method for the measurement of surface resistance. In the last section, we discuss a near-field scanning probe based on the reflection principle.

3.1 INTRODUCTION

As indicated in Chapter 2, reflection method is a type of nonresonant method. From the view of transmission line, in a reflection method, the sample under test is introduced into a certain position of a transmission line, and so the impedance loading to the transmission line is changed. The properties of the sample are derived from the reflection due to the impedance discontinuity caused by the sample loading. The two types of reflections in transmission line theory, short-circuited reflection and open-circuited reflection, can be used in reflection methods. As will be indicated later, different ways of loading samples require different calculation algorithms for the derivation of materials properties.

In a reflection method, the measurement fixture made from a transmission line is usually called measurement probe or sensor. In order to increase the measurement accuracy and sensitivity, or

to satisfy special measurement requirements, the measurement probes are often specially designed. Though they share the same principle, probes with different designs require different calculation algorithms. Many technical literatures have been published for different probe designs.

In the following, we make a general discussion on open-circuited reflection and short-circuited reflection. In the discussion, we assume that the transmission line used is coaxial line, and the conclusions obtained can be extended to other types of transmission lines, such as waveguide and free space.

3.1.1 Open-circuited reflection

Figure 3.1 shows an open-circuited flanged coaxial dielectric probe, and the sample under test directly contacts the open end of the coaxial line. The impedances at the two sides of the interface are different, so there is a reflection when electromagnetic wave propagates through the interface, and the reflectivity is determined by the impedances of the media at the two sides of the interface. As the impedance of the side occupied by the sample is related to the electromagnetic properties of the sample, from the reflectivity at the interface, the properties of the sample can be obtained.

As the impedance of the space filled with the sample is related to the permittivity and permeability of the sample, in principle, both the permittivity and permeability can be obtained provided sufficient independent reflection measurements are

Microwave Electronics: Measurement and Materials Characterization   L. F. Chen, C. K. Ong, C. P. Neo, V. V. Varadan and V. K. Varadan
© 2004 John Wiley & Sons, Ltd   ISBN: 0-470-84492-2

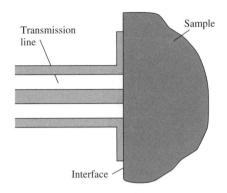

Figure 3.1 Open-circuited reflection

made. However, in most cases, only one independent measurement is made, so only one materials property parameter, either permittivity or permeability, can be obtained. In the following discussion, we focus on the measurement of permittivity. The measurement of both permittivity and permeability using reflection methods will be discussed in Section 3.4.

Figure 3.2 shows the equivalent circuits for a general reflection method. In the general equivalent circuit shown in Figure 3.2(a), the impedance $Z(\varepsilon_r)$ of the coaxial aperture is a function of the relative permittivity ε_r of the sample under test. The impedance $Z(\varepsilon_r)$ can be obtained from the reflection measurement, and the relative permittivity ε_r of the sample can be deduced from impedance $Z(\varepsilon_r)$. As terminating the open-ended coaxial probe with a dielectric sample is equivalent to introducing the sample into an equivalent capacitor, the equivalent circuit shown in Figure 3.2(a) can be simplified to Figure 3.2(b), by replacing the impedance $Z(\varepsilon_r)$ with capacitance $C(\varepsilon_r)$. From the

change of capacitance due to the insertion of the sample, the relative dielectric permittivity of the sample can be obtained.

Iterative schemes are often used to find out the relative permittivity ε_r of the sample. Usually, an objective function is defined as

$$F(\varepsilon_r) = Y_L(\varepsilon_r) - Y_m \qquad (3.1)$$

where $Y_L(\varepsilon_r)$ is the admittance calculated by using an aperture model and Y_m is the admittance of the aperture obtained by measurement. The permittivity of the sample can be calculated by finding the zero of the function. It is clear that the model for the calculation of $Y_L(\varepsilon_r)$ is crucial for a reflection method, and many models have been developed for various cases, some of which will be discussed later. As an example, here we discuss the simple model shown in Figure 3.2(b). In this model, we assume $C(\varepsilon_r) = \varepsilon_r C_0$, where C_0 is the capacitance of the capacitor when it is filled with air. So the expression for the aperture admittance can be represented in terms of the dielectric constant and loss factor of the sample (Burdette *et al.* 1980):

$$Y_L(\varepsilon_r) = j\omega\varepsilon_0\varepsilon_r C_0 = j\omega\varepsilon_0\varepsilon_r' \cdot (1 - j\tan\delta)C_0 \qquad (3.2)$$

where ε_r' and $\tan\delta$ are the dielectric constant and loss factor of the sample respectively. The value of C_0 is determined by the structure of the probe. For a given probe with known C_0, the dielectric properties of the sample can be calculated from the measured admittance Y_m, according to Eqs. (3.1) and (3.2).

3.1.2 Short-circuited reflection

Figure 3.3 shows a general case of short-circuited reflection. A piece of sample is inserted in a segment of shorted transmission line, with the end face of the sample located at a distance Δl from the short. We assume that only the prime mode exists in the transmission line, the sample is homogeneous and isotropic, and there exists only transverse electric field in the transmission line. The electric fields in the three regions are

$$E_I = \exp(-\gamma_0 z) + C_1 \exp(\gamma_0 z) \qquad (3.3)$$

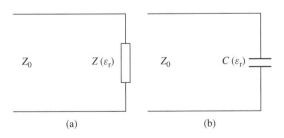

Figure 3.2 Equivalent circuits for reflections methods. (a) General equivalent circuit and (b) simplified equivalent circuit

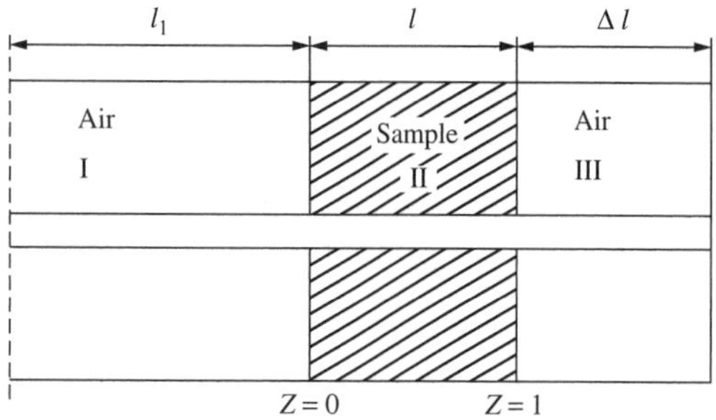

Figure 3.3 A sample is inserted in a short-circuited transmission line

$$E_{II} = C_2 \exp(-\gamma z) + C_3 \exp(\gamma z) \qquad (3.4)$$

$$E_{III} = C_4 \exp(-\gamma_0(z - l)) + C_5 \exp(\gamma_0(z - l)) \qquad (3.5)$$

where γ_0 and γ are the propagation constants in air and the sample respectively. The five constants (C_1 to C_5) can be obtained from the boundary conditions.

If we assume that the sample plane coincides with the calibration plane, the matching boundary conditions of the field equations at the interfaces yield an equation for the reflection coefficient (Baker-Jarvis *et al.* 1993)

$$S_{11} = C_1 = \frac{-2\beta\delta + [(\delta + 1) + (\delta - 1)\beta^2] \tanh \gamma l}{2\beta + [(\delta + 1) - (\delta - 1)\beta^2] \tanh \gamma l} \qquad (3.6)$$

with

$$\beta = \frac{\gamma \mu_0}{\gamma_0 \mu} \qquad (3.7)$$

$$\delta = \exp(-2\gamma_0 \Delta l) \qquad (3.8)$$

Eq. (3.6) can be modified in terms of hyperbolic functions as

$$S_{11} = \frac{\begin{array}{l} \tanh \gamma l + \beta \tanh \gamma_0 \Delta l \\ - \beta(1 + \beta \tanh \gamma l \tanh \gamma_0 \Delta l) \end{array}}{\begin{array}{l} \tanh \gamma l + \beta \tanh \gamma_0 \Delta l \\ + \beta(1 + \beta \tanh \gamma l \tanh \gamma_0 \Delta l) \end{array}} \qquad (3.9)$$

We can get the general expressions for the reflection coefficient if the calibration plane is not at the sample plane:

$$S_{11(trans)} = \exp(-2\gamma_0 l_1) S_{11} \qquad (3.10)$$

where $S_{11(trans)}$ is the reflection coefficient transformed at the reference plane of calibration, l_1 is

the distance from the calibration reference plane to the sample front face, and S_{11} can be calculated from Eq. (3.6) or (3.9).

The effect of distance l_1 can be eliminated. We measure the reflection of the empty sample holder:

$$S_{11(empty)} = -\exp(-2\gamma_0(l_1 + \Delta l + l)) \qquad (3.11)$$

From Eqs. (3.10) and (3.11), we can get

$$\frac{S_{11(trans)}}{S_{11(empty)}} = -\exp(2\gamma_0(\Delta l + l)) S_{11} \qquad (3.12)$$

So the reflection coefficient S_{11} at the front face of the sample can be measured experimentally. From Eq. (3.9), we can obtain the propagation constant in the sample, from which the electromagnetic properties of the sample can be deduced. Usually, reflection methods can only measure one complex parameter, either permittivity or permeability.

For the configuration shown in Figure 3.3, standing waves are built in the region between the sample and the short circuit and between the calibration plane and the sample front face. Depending on the sample length and the other lengths, at certain frequencies permittivity can be measured accurately, and at other frequencies permeability can be measured accurately. Meanwhile, according to the parameters to be measured, we put the samples at suitable positions. The position of the short termination is a low electric field and high magnetic field region, and the position $\lambda/4$ away from the short termination is a high electric field and low magnetic field region. For permittivity measurements the sample should be moved away from the short termination, while for permeability measurements the sample should be moved near to the short termination. Detailed discussion can be found in (Baker-Jarvis 1990; Baker-Jarvis *et al.* 1993).

3.2 COAXIAL-LINE REFLECTION METHOD

Among various types of transmission lines, coaxial line is most widely used in the reflection method for materials property characterization. A coaxial probe made from a coaxial line has several advantages over the ones made from other types of transmission lines, and the most obvious one is its wide working frequency range.

There are many excellent papers on this method published by several groups of researchers, and there are several in-depth review papers on this topic (Stuchly and Stuchly 1980; Pournaropoulos and Misra 1997).

In the past decades, several models for determining the aperture admittance of open-ended coaxial probe have been developed by assuming that the radial electric field distribution at the aperture region is inversely proportional to the radius. These formulae have been further improved by considering the effects of the higher-order modes excited at the aperture when the electrical dimensions of the probe are large, so that these formulae can be extended to higher frequencies. Along with the improvements in computational methods and speed, numerical methods have also been used to rigorously analyze the field distributions at the aperture, so more exact solutions to the aperture admittance become possible. Furthermore, in recent years, many efforts have been made in investigating the radiation of open-ended probe into layered media, and based on the results in this topic, many variations in the conventional open-end reflection method have been made for various purposes.

In this section, we discuss the coaxial probes mainly with respect to six aspects. Firstly, we discuss different admittance/capacitance models used in analyzing open-ended coaxial probe terminated by a semi-infinite material. Secondly, we discuss the radiations of end-ended apertures into multilayer medium. Thirdly, we take the aperture as a short monopole antenna, and analyze several typical monopole antennas used in materials property characterization. Fourthly, we discuss the case when a truncated coaxial line is connected to a circular waveguide, which may be in a propagating or cutoff state. Fifthly, we discuss the shielded coaxial lines. Finally, we discuss dielectric-filled cavities adapted at the end of a coaxial line.

It should be noted that the working principles of coaxial reflection methods are applicable for other types of transmission lines. Most of the coaxial reflection methods discussed in the following text have their waveguide counterparts.

3.2.1 Open-ended apertures

In this part, we discuss the radiation of a coaxial open-ended aperture into a homogeneous medium. The theoretical background of this topic can be traced to the work of Levine and Papas, who analyzed an air-filled coaxial line with an infinite conducting flange, radiating into free space (Levine and Papas 1951).

In practical applications, coaxial dielectric probes sometimes are connected to grounding flanges. In theoretical analysis, some models are developed for open-ended probe with infinitely large grounding flange, while some models are developed for open-ended probe without grounding flange. In the discussion of various models, we do not distinguish whether the probe is connected to a grounding flange. After discussing the frequently used models, we analyze the effects of the grounding flange. Finally, we discuss some variations in the structure of the conventional coaxial open-ended apertures.

3.2.1.1 Modeling of the open-ended coaxial probe

Many models have been built to analyze open-ended coaxial probes terminated by semi-infinite homogeneous materials. There are four typical models (Pournaropoulos and Misra 1997; Berube *et al.* 1996): capacitive model (Stuchly *et al.* 1982; Athey *et al.* 1982), antenna model (Brady *et al.* 1981), virtual line model (Ghannouchi and Bosisio 1989), and rational function model (Stuchly *et al.* 1994; Anderson *et al.* 1994). In recent years, full-wave simulation method has also been used in analyzing open-ended coaxial lines.

Capacitance model

A typical measurement configuration of coaxial open-ended reflection is shown in Figure 3.4(a), and the equivalent circuit for the capacitive model is shown in shown in Figure 3.4(b). The difference between Figure 3.2(b) and Figure 3.4(b) is that, in the present model, the capacitance consists of two parts: $C(\varepsilon_r)$ which is related to the dielectric properties of the sample, and C_f which is independent of the dielectric properties of the

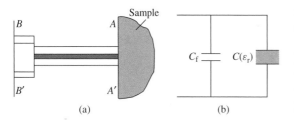

Figure 3.4 Open-ended reflection method. (a) Coaxial probe terminated by a semi-infinite sample and (b) capacitive equivalent circuit at plane $A - A'$

sample. When the coaxial probe is connected to a dielectric sample with complex relative permittivity ε_r, the equivalent capacitor will be changed, and the reflection coefficient at the tip of the open-ended probe can be obtained from the capacitances:

$$\Gamma^* = \Gamma e^{j\Phi} = \frac{1 - j\omega Z_0 \cdot [C(\varepsilon_r) + C_f]}{1 + j\omega Z_0 \cdot [C(\varepsilon_r) + C_f]} \quad (3.13)$$

where $C(\varepsilon_r) = \varepsilon_r C_0$, and C_0 is the capacitance of the capacitor filled with air, C_f is the capacitance independent of the material, ω is the measurement angular frequency, and Z_0 is the characteristic impedance of the coaxial line connected to the open-ended probe.

From Eq. (3.13), we get

$$\varepsilon_r = \frac{1 - \Gamma^*}{j\omega Z_0 C_0 (1 + \Gamma^*)} - \frac{C_f}{C_0} \quad (3.14)$$

In order to calculate the complex relative permittivity ε_r from the complex reflection coefficient Γ^*, we should know the values of C_f and C_0. These two parameters are usually obtained by calibrating the open-ended probe with a standard sample with known dielectric permittivity, for example, deionized water (Berube *et al.* 1996):

$$C_0 = \frac{(1 - |\Gamma^*_{\text{diel}}|^2)}{\omega Z_0 (1 + 2|\Gamma^*_{\text{diel}}| \cos(\Phi_{\text{diel}}) + |\Gamma^*_{\text{diel}}|^2) \varepsilon''_{\text{diel}}} \quad (3.15)$$

$$C_f = \frac{-2|\Gamma^*_{\text{diel}}| \sin(\Phi_{\text{diel}})}{\omega Z_0 (1 + 2|\Gamma^*_{\text{diel}}| \cos(\Phi_{\text{diel}}) + |\Gamma^*_{\text{diel}}|^2)} - \varepsilon'_{\text{diel}} C_0 \quad (3.16)$$

where $\varepsilon'_{\text{diel}}$ and $\varepsilon''_{\text{diel}}$ are the real and imaginary parts of the complex permittivity of the standard sample

respectively, and $|\Gamma^*_{\text{diel}}|$ and Φ_{diel} are the magnitude and phase of the complex reflection coefficient Γ^*_{diel} respectively. After obtaining C_0 and C_f from Eqs. (3.15) and (3.16), the dielectric properties of the sample under test can be calculated from the complex reflection coefficient according to Eq. (3.14).

However, it should be noted that the reflection coefficient should be measured at the plane $A - A'$. If the reference plane has previously been defined at the entrance of the probe ($B - B'$ plane), then we should find the phase difference between the $B - B'$ and $A - A'$ planes:

$$\Gamma^*_{A-A'} = \Gamma^*_{B-B'} e^{j2\theta} \quad (3.17)$$

with

$$2\theta = \Phi_{A-A'} - \Phi_{B-B'} \quad (3.18)$$

where $\Phi_{A-A'}$ and $\Phi_{B-B'}$ are the reflection phase angles of planes $A - A'$ and $B - B'$ respectively. The value of $\Phi_{B-B'}$ can be measured using vector network analyzer directly. The determination of the round-trip phase factor 2θ can be made by measuring the complex reflection coefficient in air, $\Phi_{B-B'(\text{air})}$ (Berube *et al.* 1996).

Radiation model

The coaxial probe can be considered as a radiation source, and the equivalent circuit of radiation model is shown in Figure 3.5. The capacitance C_1 is mainly determined by the structure of the coaxial probe, and is independent of the material under test. The sample under study can be modeled as a capacitance $\varepsilon_r C_2$ and a resistance R ($R = 1/G$) connected in parallel to the capacitances (Brady

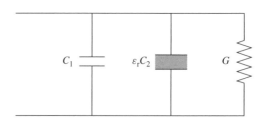

Figure 3.5 Equivalent circuit for antenna model

et al. 1981), which is mainly related to the radiation from the coaxial aperture. So, the normalized admittance is given by

$$\frac{Y}{Y_0} = j\omega C_1 Z_0 + j\omega \varepsilon_r C_2 + Z_0 G(\omega, \varepsilon_r) \quad (3.19)$$

where Z_0 is the characteristic impedance of the coaxial line, Y_0 is the characteristic admittance of the coaxial line ($Y_0 = 1/Z_0$), ω is the angular frequency, and ε_r is the complex permittivity of the material under test.

It has been shown that for an infinitesimal antenna, the radiation conductance can be expressed as (Burdette *et al.* 1980; Deschamps 1972)

$$G(\omega, \varepsilon_r) = \varepsilon_r^{\frac{5}{2}} G(\omega, \varepsilon_0) \quad (3.20)$$

From Eqs. (3.19) and (3.20), we can get

$$\frac{Y}{Y_0} = j\omega C_1 Z_0 + j\omega \varepsilon_r C_2 Z_0 + \varepsilon_r^{\frac{5}{2}} G(\omega, \varepsilon_0) Z_0 \quad (3.21)$$

Eq. (3.21) can be further modified into

$$\frac{Y}{Y_0} = K_1 + K_2 \varepsilon_r + K_3 \varepsilon_r^{\frac{5}{2}} \quad (3.22)$$

The factors K_1, K_2, and K_3 are generally complex. To determine these three factors, one must use three media with known permittivity values to do calibration. Similar to the capacitive mode, if the measurement is made at $B - B'$ plane, the complex admittance should be transferred to $A - A'$ plane.

It should be noted that there are several admittance expressions similar to Eq. (3.22), obtained through different approaches and with different approximations. If we consider the frequency dependence of C_1, we can get a more accurate model (Gajda 1983):

$$\frac{Y}{Y_0} = K_1 + K_2 \varepsilon_r + K_3 \varepsilon_r^2 + K_4 \varepsilon_r^{\frac{5}{2}} \quad (3.23)$$

From quasi-static analysis, we can get (Staebell and Misra 1990)

$$\frac{Y}{Y_0} = K_1 \varepsilon_r + K_2 \varepsilon_r^2 + K_3 \varepsilon_r^{\frac{5}{2}} \quad (3.24)$$

At a very low frequency, Eq. (3.24) can be further approximated (Staebell and Misra 1990):

$$\frac{Y}{Y_0} = K_1 \varepsilon_r + K_2 \varepsilon_r^2 \quad (3.25)$$

The applications of Eqs. (3.22)–(3.25) could be found in their corresponding references. In actual measurements, for each model, calibrations are needed to determine their corresponding configuration parameters K_i.

Virtual line model

When an open-ended coaxial probe is terminated by a dielectric sample, the fringing field at the extremity of the probe can be modeled as a segment of equivalent transmission line (Ghannouchi and Bosisio 1989). As shown in Figure 3.6, in this model, the transmission line consists of a segment of physical line with length D and a segment of virtual line with length L modeling the dielectric medium.

The complex admittance at $A - A'$ plane is given by

$$Y_L = Y_d \frac{Y_E + jY_d \tan(\beta_d L)}{Y_d + jY_E \tan(\beta_d L)} \quad (3.26)$$

where Y_L is the admittance of the virtual transmission line, Y_d is the characteristic admittance of the virtual transmission line, Y_E is the terminating admittance of the virtual transmission line; β_d is the propagation constant in the test medium. As in this model, the virtual line is terminated by an open circuit ($Y_E = 0$), and Eq. (3.26) becomes

$$Y_L = jY_d \tan(\beta_d L) \quad (3.27)$$

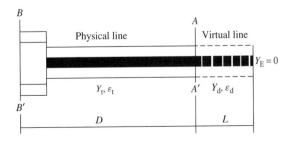

Figure 3.6   Virtual line model

The characteristic admittance Y_d of the virtual line can be expressed as a function of the physical parameters of the effective transmission line as

$$Y_d = \frac{\sqrt{\varepsilon_d}}{60 \ln(b/a)} \qquad (3.28)$$

where a and b are the inner and outer diameters of the coaxial probe respectively.

The admittance Y_L can be related to the characteristic admittance of the probe Y_t and the measured reflection coefficient Γ_m at plane $B - B'$:

$$Y_L = \left(\frac{1 - \Gamma_m e^{2j\beta_t D}}{1 + \Gamma_m e^{2j\beta_t D}} \right) Y_t \qquad (3.29)$$

with

$$Y_t = \frac{\sqrt{\varepsilon_t}}{60 \ln(b/a)} \qquad (3.30)$$

where β_t is the propagation constant in coaxial probe, Γ_m is the complex reflection coefficient measured at plane $B - B'$, and ε_t is the permittivity of the dielectric material inside the coaxial line. Usually the dielectric materials inside coaxial lines are Teflon.

From Eqs. (3.27), (3.28), and (3.29), we can get the complex permittivity of the material under test:

$$\varepsilon_d = \frac{-jc\sqrt{\varepsilon_t}}{2\pi f L} \cdot \frac{1 - \Gamma_m e^{2j\beta_t D}}{1 + \Gamma_m e^{2j\beta_t D}} \cdot \cot\left(\frac{2\pi f L \sqrt{\varepsilon_d}}{c} \right) \qquad (3.31)$$

where c is speed of light and f is the measurement frequency. According to Eq. (3.31), the complex permittivity of the sample can be calculated from the measured reflection coefficient.

The calculation of complex permittivity requires the values of D and L. The values of D and L can be determined by calibration using two dielectric media with known dielectric permittivity. From Eqs. (3.27) and (3.29), we can get

$$\Gamma_m e^{2j\beta_t D} = \frac{\rho + e^{-2j\beta_d L}}{1 + \rho e^{-2j\beta_d L}} \qquad (3.32)$$

with

$$\rho = \frac{\sqrt{\varepsilon_t} - \sqrt{\varepsilon_d}}{\sqrt{\varepsilon_t} + \sqrt{\varepsilon_d}} \qquad (3.33)$$

The two parameters D and L can be obtained from Eqs. (3.29) and (3.32) with an iterative computation procedure. The two standard media often used are air and deionized water.

Rational function model

In the geometry shown in Figure 3.7, the coaxial probe is immersed in the dielectric medium under test. In this model, the complex admittance of the coaxial probe is computed with the moment method. The radiation effects, the energy storage in the near-field region, and the evanescent mode of the guide are taken into consideration (Stuchly et al. 1994; Anderson et al. 1994). This model is usually applied for samples whose relative complex permittivity is in the range of ($1 \leq \varepsilon_r' \leq 80$) and ($1 \leq \varepsilon_r'' \leq 80$), and the measurement frequency is in the range of ($1 \leq f \leq 20\,\text{GHz}$).

The admittance of the probe can be expressed as

$$\frac{Y}{Y_0} = \frac{\sum_{n=1}^{4} \sum_{p=1}^{8} \alpha_{np} (\sqrt{\varepsilon_r})^p (j\omega a)^n}{1 + \sum_{m=1}^{4} \sum_{q=0}^{8} \beta_{mq} (\sqrt{\varepsilon_r})^q (j\omega a)^m} \qquad (3.34)$$

where Y is the admittance at the end of the coaxial probe, Y_0 is the characteristic admittance of the coaxial probe, α_{np} and β_{mq} are the coefficients of the model, ε_r is the complex relative permittivity of the material under test, and a is the inner diameter of the coaxial line.

Equation (3.34) gives the value of the complex admittance at the end of the probe as a function of the complex permittivity of the material under test and the dimensions of the probe. The inverse problem for calculating the complex permittivity of the material under test from the measured complex admittance can be solved in the following way (Anderson et al. 1994):

$$\sum_{i=0}^{8} (b_i - Y c_i) \sqrt{\varepsilon_r}^{-i} = 0 \qquad (3.35)$$

Dielectric sample under test ε_r

$2a$ $2b$

Figure 3.7   Rational functional model

with

$$b_p = \sum_{m=1}^{4} \alpha_{mp}(\mathrm{j}\omega a)^m \quad (p = 1, 2, \ldots, 8) \quad (3.36)$$

$$b_0 = 0 \quad (3.37)$$

$$c_q = \sum_{m=1}^{8} \beta_{mp}(\mathrm{j}\omega a)^m \quad (q = 1, 2, \ldots, 8) \quad (3.38)$$

$$c_0 = 1 + \sum_{m=1}^{8} \beta_{m0}(\mathrm{j}\omega a)^m \quad (3.39)$$

It should be noted that it is necessary to use the complex admittance values referred at the end of the probe.

In this model, it is not necessary to determine any calibration parameters, and so we do not need to use any standard dielectric samples. Actually, the parameters α_{np} and β_{mq} have been established and optimized by using 56 dielectric media in the range ($1 \leq \varepsilon_r' \leq 80$) (Berube *et al.* 1996).

Full-wave simulation method

Recently, with easy access to fast computers, full-wave simulation method has been used in analyzing aperture admittance (Pournaropoulos and Misra 1994; Stuchly *et al.* 1994; Panariello *et al.* 2001). With this method, more accurate value of admittance can be obtained, and rigorous analysis on the sensitivity and uncertainty of the probe can be made. However, this method usually requires much calculation time, and many efforts are being made in improving this method.

3.2.1.2 Effect of flange

Among the various models discussed above, some are developed for a probe with an infinite grounding flange, and some are developed for a probe without a grounding flange. Though, with certain approximations, all these models could be applied to coaxial probes with and without grounding flanges, it is still necessary to discuss the effects of flanges to the accuracies of models.

Figure 3.8 shows open-ended coaxial probes with and without grounding flange. Zheng and

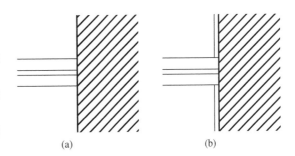

(a)   (b)

Figure 3.8 Two types of open-ended coaxial probes. (a) Probe without flange and (b) probe with flange. Modified from Zheng, H. M. and Smith, C. E. (1991). "Permittivity measurements using a short open-ended coaxial line probe", *IEEE Microwave and Guided Wave Letters*, **1** (11), 337–339. © 2003 IEEE

Smith compared the two models described by Eqs. (3.24) and (3.25) (Zheng and Smith 1991). Equation (3.24) is a three-term admittance formula, while Eq. (3.25) is a two-term admittance formula. It is found that, for a coaxial probe without grounding flange, the three-term formula does indeed give more accurate results than the two-term formula. For a flanged probe, the results from the two-term formula and the three-term formula have quite close accuracy. It can be further concluded that the use of flanged probes results in more accurate measurements than the use of probes without flange. Other researchers also recommend that in the design of coaxial probes, it is better to use grounding flange (Langhe *et al.* 1994).

3.2.1.3 Modifications of open-ended probe

The conventional coaxial dielectric probes have been widely used in the study of the dielectric properties of composite materials and biological tissues. Meanwhile, open-ended probes have also been modified to satisfy special measurement requirements or to improve the measurement accuracy and sensitivity.

Large open-ended coaxial probe

In the measurement of the effective complex permittivity of composite materials with large-grain heterogeneities, such as rocks and concretes,

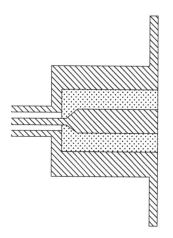

Figure 3.9 Large open-ended probe. Modified from Otto, G. P. and Chew, W. C. (1991). "Improved calibration of a large open-ended coaxial probe for dielectric measurement", *IEEE Transactions on Instrumentation and Measurement*, **40** (4), 742–746. © 2003 IEEE

large sample sizes are needed. As the diameter of commercially available dielectric probe is quite small, for example 3.5 mm, large coaxial probes should be specially designed. Figure 3.9 shows a large open-ended coaxial probe. A transition is needed between the enlarged coaxial line and the normal coaxial line. As it is difficult to ensure that the impedance at the transition region is 50 Ω,

a calibration procedure is often required for a large probe.

Otto *et al.* developed a large probe with outer diameter 32.5 mm, inner diameter 10.0 mm, and flange diameter 100 mm, and the probe they developed can work up to about 2 GHz (Otto and Chew 1991). They also proposed a calibration method utilizing a short, an open into air, and a short-cavity termination. With the calibration method they proposed, reliable and accurate measurement results can be obtained.

Coaxial probe with an elliptic aperture

Theoretical analysis indicates that an open-ended elliptic coaxial probe has higher sensitivity than a conventional coaxial probe (Xu *et al.* 1992). However, the fabrication of an elliptic line is quite complicated. As shown in Figure 3.10, a simple method to obtain an elliptic configuration is to bevel a circular coaxial line. The resulted aperture may be considered as a transition from circular coaxial line to elliptic coaxial line, and subsequently the radiation aperture also becomes elliptic.

Theoretical and experimental studies indicate that the measurement sensitivity increases with respect to the bevel angle φ for a circular coaxial line with given dimension. A slant-cut aperture

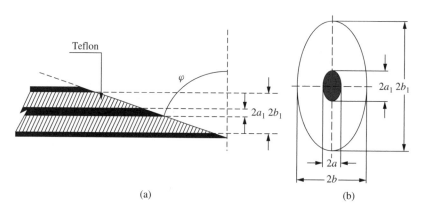

(a)   (b)

Figure 3.10 Fabrication of coaxial probe with elliptic aperture. (a) Coaxial open end with a bevel angle and (b) bevel section view of a coaxial line. Modified from Xu, Y. S. Ghannouchi, F. M. and Bosisio, R. G. (1992). "Theoretical and experimental study of measurement of microwave permittivity using open-ended elliptical coaxial probes", *IEEE Transactions on Microwave Theory and Techniques*, **40** (1), 143–150. © 2003 IEEE

from a circular coaxial line (the beveled probe) is a practical approach for the realization of probe tips with elliptical aperture. Actually, a beveled probe is more valuable than a right angle–cut aperture made from an elliptical coaxial line. The fabrication of the beveled probe is much simpler than that of the elliptical probe. Furthermore, a beveled probe is very useful in biological as well as biochemical applications, as *in vivo* measurements of biological tissues on a living body are often essential in these fields.

3.2.2 Coaxial probes terminated into layered materials

In the above discussion, the coaxial probe is terminated by a homogeneous semi-infinite medium. In some situations, the semi-infinite requirement cannot be satisfied, and sometimes the samples under study may be of multilayer structure. So the knowledge about the admittance of coaxial probe terminated into layered dielectric materials is important for materials property characterization.

3.2.2.1 Admittance of coaxial probe terminated into layered materials

The admittance of a coaxial probe terminated into layered materials has been studied by lumped parameter approach and quasi-static approximation approach (Fan *et al.* 1990; Anderson *et al.* 1986).

Here, we introduce a general formulation for an open-ended coaxial transmission line terminated by a multilayered dielectric (Bakhtiari *et al.* 1994).

As shown in Figure 3.11, we assume that the open end of a coaxial transmission line is connected to a perfectly conducting infinite flange, and the inner and outer diameters of the coaxial line are $2a$ and $2b$ respectively. As shown in Figure 3.12, the multilayer composite may be terminated into an infinite half-space or backed by a conductor.

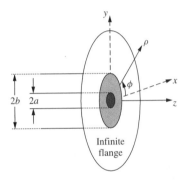

Figure 3.11 Coaxial line with inner diameter $2a$ and outer diameter $2b$ opening onto a perfectly conducting infinite flange. Source: Bakhtiari, S. Ganchev, S. I. and Zoughi, R. (1994). "Analysis of radiation from an open-ended coaxial line into stratified dielectrics", *IEEE Transactions on Microwave Theory and Techniques*, **42** (7), 1261–1267. © 2003 IEEE

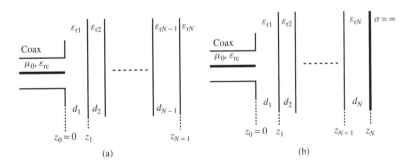

Figure 3.12 Cross sections of coaxial line radiating into layered media. (a) The multilayer material is terminated into an infinite half-space and (b) cross section of a coaxial line radiating into a layered media terminated into a perfectly conducting sheet. (Bakhtiari *et al.* 1994). Modified from Bakhtiari, S. Ganchev, S. I. and Zoughi, R. (1994). "Analysis of radiation from an open-ended coaxial line into stratified dielectrics", *IEEE Transactions on Microwave Theory and Techniques*, **42** (7), 1261–1267. © 2003 IEEE

If only the fundamental transverse electromagnetic (TEM) mode propagates inside the coaxial line, the terminating admittance of the line can be obtained using the continuity of the power flow across the aperture (Bakhtiari *et al.* 1994):

$$
y_s = g_s + jb_s = \frac{\varepsilon_{r1}}{\sqrt{\varepsilon_{rc}}\ln(b/a)} \int_0^\infty
$$

$$
\frac{[J_0(k_0\xi b) - J_0(k_0\xi a)]^2}{\xi} F(\xi)\,d\xi \qquad (3.40)
$$

where g_s and b_s are the normalized aperture conductance and susceptance, ε_{rc} is the relative permittivity of the dielectric filling inside the coaxial line, J_0 is the zero-order first kind of Bessel function, and the function $F(\xi)$ is given by

$$
F(\xi) = \frac{1}{\sqrt{\varepsilon_{r1} - \xi^2}}\left(\frac{1 + \rho_1}{1 - \rho_1}\right) \qquad (3.41)
$$

For an N-layer medium, the value of ρ_1 may be calculated from the following recurrence relations:

$$
\rho_i = \frac{1 - \kappa_i\beta_{i+1}}{1 + \kappa_i\beta_{i+1}} e^{-j2k_0z_i\sqrt{\varepsilon_{ri}-\xi^2}} \qquad (3.42)
$$

where

$$
\kappa_i = \frac{\varepsilon_{r(i+1)}}{\varepsilon_{ri}} \cdot \frac{\sqrt{\varepsilon_r - \xi^2}}{\sqrt{\varepsilon_{r(i+1)} - \xi^2}} \qquad (3.43)
$$

$$
\beta_{i+1} = \frac{1 - \rho_{i+1}e^{j2k_0z_i\sqrt{\varepsilon_{r(i+1)}-\xi^2}}}{1 + \rho_{i+1}e^{j2k_0z_i\sqrt{\varepsilon_{r(i+1)}-\xi^2}}} \qquad (3.44)
$$

$$
z_n = \sum_{i=1}^n d_i \qquad (3.45)
$$

In Eq. (3.45), if the Nth layer is infinite in $+z$-direction, $1 \le n \le (N-1)$; if the Nth layer is backed by conducting sheet, $1 \le n \le N$.

If the Nth layer is infinite in $+z$-direction,

$$
\rho_N = 0 \qquad (3.46)
$$

If the Nth layer is terminated into a conducting sheet,

$$
\rho_N = e^{-j2k_0z_N\sqrt{\varepsilon_{rN}-\xi^2}} \qquad (3.47)
$$

The above calculation must start from $i = N - 1$ and carried out backward to $i = 1$. The value of

ρ_N is chosen from Eq. (3.46) or (3.47) depending on whether the Nth medium is an infinite half-space or is of finite thickness backed by a conducting sheet.

3.2.2.2 Applications in materials property characterization

The conclusions for the aperture admittance of coaxial probe terminated into stratified dielectrics can be used to modify the conventional reflection method using coaxial dielectric probe. In the following, we discuss several examples of modifications often used in materials property characterization.

Dielectric samples with finite thickness backed by metal plate

For a laminar sample, its thickness is usually not thick enough for semi-infinite requirement. As shown in Figure 3.13, the dielectric properties of a laminar sample backed by a conducting plane can be measured by using an open-ended coaxial probe. Actually, it is a special case of Figure 3.12(b) with $N = 1$. The terminating admittance of the probe can be calculated according to Eq. (3.40), and the function $F(\xi)$ can be calculated according to Eq. (3.41) with ρ_1 given by

$$
\rho_1 = e^{-j2k_0d_1\sqrt{\varepsilon_r-\xi^2}} \qquad (3.48)
$$

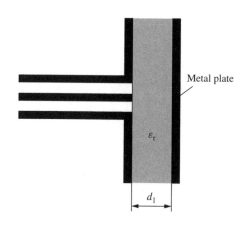

Figure 3.13 Geometry for the measurement of a laminar dielectric sample backed by a conducting plane

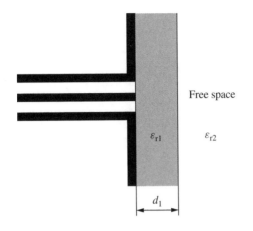

Figure 3.14 Geometry for the measurement of a laminar dielectric sample backed by free space

Dielectric samples with finite thickness backed by free space

As shown in Figure 3.14, in the measurement of a laminar sample, the sample can also be backed by free space. This is a special case of Figure 3.12(a) when $N = 2$. The second layer is infinite in $+z$-direction. The explicit form of $F(\xi)$ is given by

$$F(\xi) = \frac{1}{\sqrt{\varepsilon_{r1} - \xi^2}} \cdot \frac{\kappa_1 + j\tan(k_0 d_1 \sqrt{\varepsilon_{r1} - \xi^2})}{1 + j\kappa_1 \tan(k_0 d_1 \sqrt{\varepsilon_{r1} - \xi^2})} \tag{3.49}$$

with

$$\kappa_1 = \frac{\varepsilon_{r2}}{\varepsilon_{r1}} \cdot \frac{\sqrt{\varepsilon_{r1} - \xi^2}}{\sqrt{\varepsilon_{r2} - \xi^2}} \tag{3.50}$$

where d_1 is the thickness of the laminar sample under test, ε_{r1} is the relative permittivity of the sample, and ε_{r2} is the relative permittivity of the half-infinite sample backing the sample. For a sample backed by free space, $\varepsilon_{r2} = 1$.

Two-layered media terminated by metal plate

In the measurement of dielectric properties using coaxial probes, good contact between the coaxial aperture and the sample surface is required. If there is an air gap between the sample and the probe, the discontinuity of the electric field causes a large error in the calculation of permittivity. In actual measurements, the air gap between the coaxial

aperture and the sample surface is inevitable. To minimize the effect of the air gap, Baker-Jarvis *et al.* developed a model for the electromagnetic response of a coaxial probe with liftoff (Baker-Jarvis *et al.* 1994).

Figure 3.15 shows a coaxial probe terminated by a two-layer dielectric sample backed by a conducting plane, and the dielectric permittivity values of the two layers are ε_{r1} and ε_{r2} respectively. It is a special case of Figure 3.12(b) when $N = 2$. The explicit form of $F(\xi)$ is given by

$$F(\xi) = \frac{j}{\sqrt{\varepsilon_{r1} - \xi^2}} \cdot \frac{\tan(k_0 d_1 \sqrt{\varepsilon_{r1} - \xi^2})}{\kappa_1 \tan(k_0 d_1 \sqrt{\varepsilon_{r1} - \xi^2}) + \tan(k_0 d_2 \sqrt{\varepsilon_{r2} - \xi^2})} \tag{3.51}$$

For the case of coaxial probe with liftoff, the first layer with thickness d_1 is air ($\varepsilon_{r1} = 1$), and the second layer with thickness d_2 is the sample under test. Probes with liftoff are quite useful for nondestructive material property testing and material thickness testing. However, in order to increase the interaction of the field with the material under test, either larger diameter probes or higher frequencies need to be used.

The configuration shown in Figure 3.15 can also be used in the measurement of high-dielectric-constant materials. When a dielectric probe is used to measure materials with high dielectric constant, for example, higher than 50, because of the serious

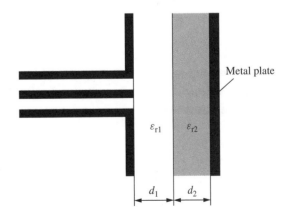

Figure 3.15 Geometry for the measurement of two-layer dielectric sample backed by a conducting plate

impedance mismatch between the sample and the coaxial line, only a small part of the microwave signal is radiated into the sample, and so the measurement accuracy and sensitivity are low. To increase the measurement accuracy and sensitivity, we may introduce a matching layer with known dielectric properties between the probe aperture and the sample, so that more microwave signal can be radiated into the sample under test.

Two-layered media terminated by free space

Figure 3.16 shows a coaxial probe terminated by a two-layer sample backed by free space. The free space can be taken as a third layer with half-infinite thickness; so the medium terminating the probe can be taken as a structure consisting of three layers with relative permittivity $\varepsilon_{r1}, \varepsilon_{r2}$, and ε_{r3} respectively and thickness d_1, d_2, and d_3 respectively. The explicit expression for $F(\xi)$ when the third layer is an infinite half-space extending in the z-direction with a dielectric constant ε_{r3} is given by (Ganchev *et al.* 1995)

$$F(\xi) = \frac{1}{\sqrt{\varepsilon_{r1} - \xi^2}} \cdot \frac{1 + \rho_1}{1 - \rho_1} \tag{3.52}$$

where

$$\rho_1 = \frac{\kappa_1 - \beta_2}{\kappa_1 + \beta_2} e^{-j2k_0 d_1 \sqrt{\varepsilon_{r1} - \varsigma^2}} \tag{3.53}$$

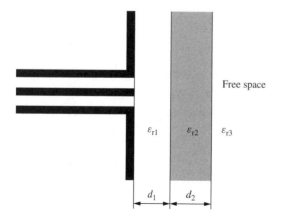

Figure 3.16 Geometry for the measurement of two-layer dielectric sample backed by free space

$$\kappa_1 = \frac{\varepsilon_{r2}}{\varepsilon_{r1}} \cdot \frac{\sqrt{\varepsilon_{r1} - \varsigma^2}}{\sqrt{\varepsilon_{r2} - \varsigma^2}} \tag{3.54}$$

$$\beta_2 = \frac{1 - \rho_2 e^{j2k_0 d_1 \sqrt{\varepsilon_{r2} - \varsigma^2}}}{1 + \rho_2 e^{j2k_0 d_1 \sqrt{\varepsilon_{r2} - \varsigma^2}}} \tag{3.55}$$

$$\rho_2 = \frac{\kappa_2 - 1}{\kappa_2 + 1} \cdot e^{-j2k_0(d_1 + d_2)\sqrt{\varepsilon_{r2} - \varsigma^2}} \tag{3.56}$$

$$\kappa_2 = \frac{\varepsilon_{r3}}{\varepsilon_{r2}} \cdot \frac{\sqrt{\varepsilon_{r2} - \varsigma^2}}{\sqrt{\varepsilon_{r3} - \varsigma^2}} \tag{3.57}$$

Similar to the case of two-layer sample backed by a conducting plane, the above model can be used in improving the conventional coaxial dielectric probe in two ways. One is to minimize the uncertainties caused by the inevitable air gap between the dielectric probe, and the other is to measure samples with high dielectric constants.

3.2.3 Coaxial-line-excited monopole probes

Short monopole probes are widely used in the measurement of dielectric properties of materials, especially biological samples. The theoretical basis of monopole probe stems from the application of an antenna-modeling theorem to the characterization of unknown dielectric media. Monopole probes have obvious advantages in materials property characterization. The manufacturing of monopole is easy, and the equations relating the permittivity to the measured parameters are simple (Stuchly and Stuchly 1980).

3.2.3.1 Theoretical basis

In a nonmagnetic medium with $\mu = \mu_0$, the antenna-modeling theorem can be expressed as (Burdette *et al.* 1980)

$$\frac{Z(\omega, \varepsilon_0 \varepsilon_r)}{\eta} = \frac{Z(n\omega, \varepsilon_0)}{\eta_0} \tag{3.58}$$

where $Z(\omega, \varepsilon)$ is the terminal impedance of the antenna. The complex intrinsic impedance of the dielectric medium is given by

$$\eta = \sqrt{\mu_0 / (\varepsilon_0 \varepsilon_r)} \tag{3.59}$$

and the intrinsic impedance of free space is given by

$$\eta_0 = \sqrt{\mu_0/\varepsilon_0} \qquad (3.60)$$

The complex index of refraction of the dielectric medium is given by

$$n = \sqrt{\varepsilon_r}. \qquad (3.61)$$

Equation (3.58) is based on the assumption that the medium surrounding the antenna is infinite in extent, or conversely, the theorem is valid as long as the probe's radiation field is contained completely within the medium. This theorem is applicable for any probe provided that an analytical expression for the terminal impedance of the antenna is known both in free space and in dielectric medium.

When the length of the probe antenna is approximately one-tenth wavelength or larger, a radiation field exists. In cases where the penetration depth in the medium under study is greater than the sample thickness, errors are introduced in the measurement of the complex impedance of the medium because the field is not totally within the sample. Usually, short monopole antennas with length less than one-tenth wavelength are used in materials characterization, as shown in Figure 3.17.

The terminal impedance of a short antenna in free space can be expressed as

$$Z(\omega, \varepsilon_0) = A\omega^2 + \frac{1}{jC\omega} \qquad (3.62)$$

where A and C are constants determined by the physical dimensions of the antenna. The antenna constants A and C can be determined analytically

and experimentally. From the knowledge of the antenna constants and the complex impedance $Z(\omega, \varepsilon_0\varepsilon_r)$ of the antenna in a lossy medium, the complex permittivity of the medium ε_r can be obtained from Eq. (3.58).

From Eqs. (3.58) and (3.62), we can get the antenna impedance in the medium:

$$Z(\omega, \varepsilon_0\varepsilon_r) = A\omega^2\sqrt{\varepsilon_r} + \frac{1}{jC\omega\varepsilon_r} \qquad (3.63)$$

In terms of dielectric constant and loss tangent, Eq. (3.63) can be rewritten as

$$Z(\omega, \varepsilon_0\varepsilon_r) = A\omega^2\sqrt{\varepsilon_r'(1 - j\tan\delta)}$$
$$+ \frac{1}{jC\omega\varepsilon_r'(1 - j\tan\delta)} \qquad (3.64)$$

The antenna impedance $Z(\omega, \varepsilon_0\varepsilon_r)$ can be determined through the measurements of the input reflection coefficient S_{11}. The complex Eq. (3.64) can be written in the form $Z = R + jX$, which results in two real equations

$$R = \frac{\sin 2\delta}{2\varepsilon_r'\omega C} + A\sqrt{\varepsilon_r'}\omega^2\sqrt{\frac{\sec\delta + 1}{2}} \qquad (3.65)$$

$$X = \frac{\cos^2\delta}{\varepsilon_r'\omega C} + A\sqrt{\varepsilon_r'}\omega^2\sqrt{\frac{\sec\delta - 1}{2}} \qquad (3.66)$$

In Eqs. (3.65) and (3.66), the parameters R and X are the real and imaginary parts of the measured impedance, A and C are the physical constants of the probe, and all the other parameters are known except ε_r' and δ. Because the inverse pair of equations corresponding to Eqs. (3.65) and (3.66) cannot be easily obtained, the solutions for ε_r' and δ are often obtained using iterative method.

When the probe length decreases, the configuration approaches to the one shown in Figure 3.17(b), which is quite similar to an open-ended coaxial probe, and the method for analyzing the open-ended coaxial line can also be used.

3.2.3.2 Typical monopole antennas used in materials property characterization

To achieve higher accuracy and sensitivity and to satisfy various measurement requirements, many

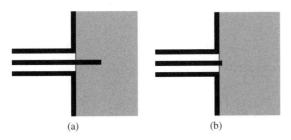

(a)   (b)

Figure 3.17 Short antennas used for materials property characterization. (a) Short coaxial monopole antenna and (b) very short coaxial monopole antenna

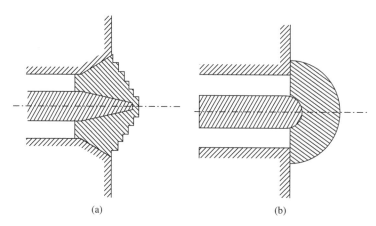

(a)   (b)

Figure 3.18   Two coaxial antennas that can be used for materials characterization. (a) Coaxial antenna with step transition and (b) hemispherical antenna. Modified from Wang, Y. and Fan, D. (1994). "Accurate global solutions of EM boundary-value problems for coaxial radiators", *IEEE Transactions Antennas and Propagation*, **42** (5), 767–770. © 2003 IEEE

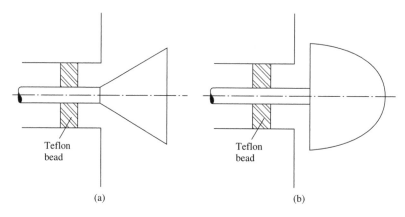

(a)   (b)

Figure 3.19   Two types of coaxial antennas used in materials characterization. (a) Conical monopole antenna. Modified from Smith, G. S. and Nordgard, J. D. (1985). "Measurement of the electrical constitutive parameters of materials using antennas", *IEEE Transactions Antennas and Propagation*, **33** (7), 783–792. © 2003 IEEE. (b) Spheroidal antenna. Modified from Stuchly, M. A. and Stuchly, S. S. (1980). "Coaxial line reflection methods for measuring dielectric properties of biological substances at radio and microwave frequencies – A review", *IEEE Transactions on Instrumentation and Measurement*, **29** (3), 176–183. © 2003 IEEE

types of antennas have been used in materials property characterization. Figure 3.18 shows two examples of coaxial antennas that can be used in materials property characterization (Wang and Fan 1994). A coaxial antenna mainly consists of three regions: feedline, transition region, and radiation region. The main purpose of the transition is to improve the radiation properties of the

antenna so that higher accuracy and sensitivity can be achieved.

Figure 3.19 shows two other types of coaxial antennas: conical monopole antenna and spheroidal antenna (Smith and Nordgard 1985; Bucci and Franceschetti 1974). Such antennas have the advantage that they can get more accurate values of effective permittivity of composite materials.

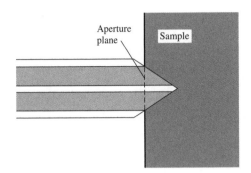

Figure 3.20 Cross-sectional view of a conical-tip coaxial-line probe

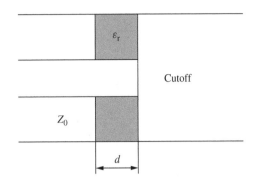

Figure 3.21 Coaxial line open into a cutoff circular waveguide

However, the insertion of the probe to the material, unless it is a liquid, creates serious difficulty, especially in assuring that the gap is filled with the material under test.

There are many other types of monopole antennas used in materials property characterization. Figure 3.20 shows a conical-tip coaxial-line probe (Keam and Holdem 1997). Compared to conventional monopole antennas, a conical-tip coaxial-line probe has obvious advantages. Since the conical-tip probe has a cone formed out of the coaxial-line dielectric, it tends to push aside material in the insertion procedure, so the influence of moisture due to cell damage is minimized. It does not require a flat sample surface, and it is suitable for measurements on deformable materials. Furthermore, the length of the probe may be optimized for a specific permittivity and frequency range.

3.2.4 Coaxial lines open into circular waveguides

In materials property characterization, the measurement fixture can be a coaxial line open into a circular waveguide (Stuchly and Stuchly 1980). As will be discussed below, in such a fixture, the sample may be placed at the coaxial line part or the circular waveguide part.

3.2.4.1 Measurement of small-thickness sample

Figure 3.21 shows a coaxial line open into a circular waveguide below cutoff. The sample under test in a ring shape with length d is placed at the open end of the coaxial line. The equivalent circuit of

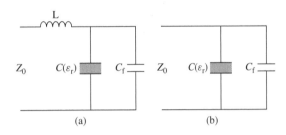

Figure 3.22 Equivalent circuits for the structure shown in Figure 3.21. (a) Accurate equivalent circuit and (b) simplified equivalent circuit

this configuration is shown in Figure 3.22(a). It has an inductance component L and two capacitance components $C(\varepsilon_r)$ and C_f. The component C_f is mainly caused by the fringing field, and is independent of the sample. The component $C(\varepsilon_r)$ is related to the permittivity of the sample. Owing to the dielectric losses, $C(\varepsilon_r)$ also includes a resistive component.

The dielectric properties of the sample under measurement can be calculated from the reflection coefficient S_{11} and its phase angle ϕ (Stuchly and Stuchly 1980):

$$\varepsilon_r' = \frac{\lambda}{2\pi d}$$

$$\cdot \left[\frac{(1 + S_{11}^2)(2\pi d/\lambda) + 2S_{11}\sin(-\phi) - 2S_{11}\cos\phi(2\pi d/\lambda)}{1 + S_{11}^2 + 4S_{11}\sin(-\phi)(2\pi d/\lambda) + 2S_{11}\cos\phi} - \frac{\omega C_f Z_0}{(2\pi d/\lambda)} \right]$$

$$(3.67)$$

$$\varepsilon_r'' = \frac{\lambda}{2\pi d} \cdot \frac{1 - S_{11}^2}{1 + S_{11}^2 + 4S_{11}\sin(-\phi)(2\pi d/\lambda) + 2S_{11}\cos\phi}$$

$$(3.68)$$

The fringe capacitance C_f is equal to $0.16\,\mathrm{pF}$ and $0.08\,\mathrm{pF}$ for 14-mm and 7-mm coaxial lines respectively.

For a thin sample with high dielectric constant ($d/\lambda \leq 0.01$ and $\varepsilon_r' > 2$), the inductance component in Figure 3.22(a) is negligible in comparison with the capacitance. So the equivalent circuit can be simplified into Figure 3.22(b), and the permittivity can be calculated from

$$\varepsilon_r' = \frac{2S_{11}\sin(-\phi)}{(2\pi d/\lambda)(1 + 2S_{11}\cos\phi + S_{11}^2)} - \frac{\omega C_f Z_0}{(2\pi d/\lambda)} \qquad (3.69)$$

$$\varepsilon_r'' = \frac{1 - S_{11}^2}{(2\pi d/\lambda)(1 + 2S_{11}\cos\phi + S_{11}^2)} \qquad (3.70)$$

The highest accuracy can be achieved when the sample thickness satisfies the following requirement (Stuchly and Stuchly 1980):

$$d = \frac{\lambda}{2\pi\sqrt{\varepsilon_r'^2 + \varepsilon_r''^2}} \qquad (3.71)$$

Figure 3.23 shows another configuration of coaxial line open into a circular waveguide. In this configuration, the circular sample with thickness d is placed in the circular waveguide. The advantage of this configuration is that the sample can be easily fabricated; however, the theoretical analysis is rather complicated as the ratio of the fringe capacitance to the capacitance resulting from the presence of the dielectric sample cannot be easily determined. A general solution for this configuration is difficult, as two cases should be considered; the wave propagation and lack of wave propagation in the dielectric-filled section of the circular waveguide (Stuchly and Stuchly 1980).

3.2.4.2 Semi-infinite sample

If the sample thickness increases, the configuration shown in Figure 3.23 changes to those of the cases shown in Figure 3.24. As shown in Figure 3.24, the circular waveguide filled with dielectric material may be in propagation state and cutoff state. These configurations have been used to measure the dielectric properties of biological materials. Such configurations may be analyzed in general terms (Razaz and Davies 1979), but the analyzing method is not convenient for permittivity determination due to its complexity. In actual measurements, calibration method is often used, and the uncertainty of measurements is usually large.

3.2.5 Shielded coaxial lines

In this part, we discuss coaxial lines shielded by a short circuit. In this configuration, the inner conductor may directly contact the short circuit, or there may be a gap between the end of the central conductor and the short circuit.

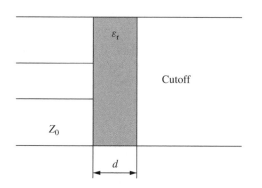

Figure 3.23 Another configuration of coaxial line open into a circular waveguide for materials property characterization

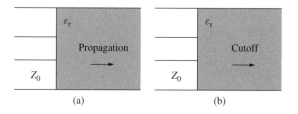

Figure 3.24 Coaxial lines opened into (a) propagation waveguide, and (b) cutoff waveguide for measurement of semi-infinite samples

3.2.5.1 Short-circuited coaxial-line method

The configuration shown in Figure 3.25 is a special case of the configuration shown in Figure 3.3 with $\Delta l = 0$. This configuration was first used by von Hippel *et al.* to measure the dielectric properties of materials (Von Hippel 1995). The input reflection coefficient of the sample holder S_{11}^* at plane $A - A'$ is given by

$$S_{11}^* = S_{11}e^{j\phi} = \frac{Z_i - Z_0}{Z_i + Z_0} \tag{3.72}$$

where Z_i is the input impedance at the plane $A - A'$ and Z_0 is the characteristic impedance of the coaxial line. The input impedance is a function of the relative permittivity:

$$Z_i = Z_0 \frac{\tanh(2\pi d\sqrt{\varepsilon_r}/\lambda)}{\sqrt{\varepsilon_r}} \tag{3.73}$$

An obvious advantage of this method is the simplicity of the sample holder, and this method has been successfully applied to measure the permittivity of biological materials. However, the method has its disadvantages (Stuchly and Stuchly 1980). First, it is necessary to solve a transcendental equation. Second, proper corrections are needed to achieve permittivity results with reasonable accuracy. Third, the sample thickness has to be selected properly. Measurements of samples with various thickness values may be needed before an optimum thickness is selected. The highest accuracy is obtained when the sample thickness is equal to an odd number of quarter wavelengths in the dielectric for low-loss dielectrics and one-quarter

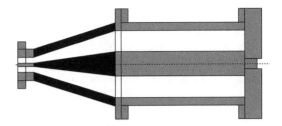

Figure 3.26 Enlarged short-circuited coaxial-line reflection method. Modified from Huang, Y., "Design, calibration and data interpretation for a one-port large coaxial dielectric measurement cell", *Measurement Science and Technology*, **12** (1), 111–115 (2001), by permission of IOP Publishing Ltd

wavelength for high-loss dielectrics. Besides, for liquid samples, a window has to be placed at the interface $A - A'$. Unless the window is thin and made of a material with low dielectric constant and low loss factor, it may cause errors to the measurement results.

To measure the dielectric properties of composites with large inclusions, one-port large coaxial measurement cell can be used, as shown in Figure 3.26 (Huang 2001). This configuration can be taken as the counterpart of the configuration shown in Figure 3.9. Similar to the configuration shown in Figure 3.9, the enlarged short-circuited coaxial line requires a transition. In the system developed by Huang, measurements can be made up to 1 GHz, and a three-position short-circuit calibration is used.

3.2.5.2 Lumped capacitance method

In the short-circuited coaxial lines shown in Figure 3.27, there is a small gap between the

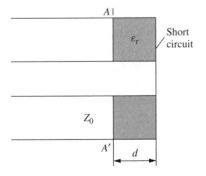

Figure 3.25 Dielectric-filled section of a TEM line

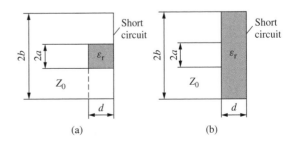

Figure 3.27 Lumped capacitance methods. (a) Sample with diameter $2a$ and (b) sample with diameter $2b$

end of the central conductor and the short circuit, forming a lumped capacitor. The sample under test is inserted into the gap, resulting in the change of the lumped capacitance between the inner conductor and the short circuit. The configurations shown in Figure 3.27 can also be analyzed using the equivalent circuit shown in Figure 3.22(b).

If there is no sample in the gap between the inner conductor and the short plate, the total capacitance is given by (Marcuvitz 1965)

$$C_T = C_0 + C_f = \frac{\pi a^2}{4d} \varepsilon_0 + 2a\varepsilon_0 \ln\left(\frac{b-a}{2d}\right) \tag{3.74}$$

where C_0 represents the capacitance of the parallel-plate capacitor formed by the inner conductor and the terminating short circuit, C_f is the fringe capacitance, "d" is the distance between the end of the central conductor and the short plate, "a" and "b" are the radii of the inner and outer conductor of the coaxial line respectively, and ε_0 is the permittivity of free space. Equation (3.74) is based on the assumptions that $\lambda \gg (b-a)$ and $d \ll (b-a)$, where λ is the wavelength.

As shown in Figure 3.27(a), when a dielectric sample is filled in the capacitor, we assume that introducing the dielectric sample into the parallel-plate capacitor causes a negligibly small variation to the fringe capacitance C_f; so the reflection coefficient is given by

$$S_{11}^* = S_{11}e^{j\phi} = \frac{1 - j\omega Z_0[C(\varepsilon_r) + C_f]}{1 + j\omega Z_0[C(\varepsilon_r) + C_f]} \tag{3.75}$$

where $C(\varepsilon_r^*) = \varepsilon_r^* C_0$. In experiments, C_0 and C_f are usually determined by calibration.

According to Eq. (3.75), we can get the relative dielectric constant and loss factor (Stuchly and Stuchly 1980):

$$\varepsilon_r' = \frac{2S_{11}\sin(-\phi)}{\omega Z_0 C_0(1 + 2S_{11}\cos\phi + S_{11}^2)} - \frac{C_f}{C_0} \tag{3.76}$$

$$\varepsilon_r'' = \frac{1 - S_{11}^2}{\omega Z_0 C_0(1 + 2S_{11}\cos\phi + S_{11}^2)} \tag{3.77}$$

The uncertainty in ε_r can be minimized by a proper selection of C_0 (Rzepecka and Stuchly 1975), and

the optimum capacitance value is given by

$$C_0 = \frac{1}{2\pi f Z_0} \cdot \frac{1}{\sqrt{\varepsilon_r'^2 + \varepsilon_r''^2}} \tag{3.78}$$

The configuration shown in Figure 3.27(b) can be analyzed in a similar way, and similar conclusions can also be obtained. But the analysis procedure is quite complicated.

3.2.6 Dielectric-filled cavity adapted to the end of a coaxial line

Saed *et al.* proposed a technique utilizing a cylindrical cavity completely filled with a sample of the material under test (Saed *et al.* 1990). As shown in Figure 3.28, the cavity is adapted to the end of a coaxial line. In the measurement, the cavity works at off-resonance state, and the dielectric properties of the sample are derived from the measured reflection coefficient.

The configuration shown in Figure 3.28 mainly consists of two regions: the transmission line region and the cavity region. There is no need to cut or shape the sample to fit fixed structures. The cavity walls can be molded or deposited on a cylindrically shaped sample, so the presence of any possible air gaps is eliminated. To establish the relationship between the measured reflection coefficient and the complex permittivity of the dielectric sample, a full-wave analysis is needed. The full-wave simulation mainly consists of two problems: the forward problem of calculating the

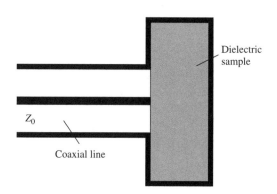

Figure 3.28   Dielectric-filled cylindrical cavity adapted to the end of a coaxial line

reflection coefficient for given dielectric permittivity, and the inverse problem of determining the dielectric permittivity for a given reflection coefficient. The detailed discussions on these two problems can be found in (Saed *et al.* 1990).

This method offers advantages over the conventional methods in which the sample is placed in a resonant cavity or inserted in a transmission line. First, this technique is usable over a wide band of frequencies that can extend into the millimeter wave region depending on the sample thickness, the sample permittivity, and the measuring equipment. Second, air gaps that might lead to significant errors are eliminated. Third, there is a well-defined reference plane necessary for the calibration of the measurement system. Fourth, higher-order modes can be taken into consideration by precise field analysis. However, for the low-loss dielectric materials, this method cannot give very precise value of dielectric loss. Besides, in the analysis of the structure, the losses in the conductor walls of the cavity should be taken into consideration since these losses are often on the same order as the losses in the dielectric sample.

This technique is suitable for the sample preparation and thus characterization of thick-film dielectric materials (Saed *et al.* 1989). As shown in Figure 3.29, as thick-film materials cannot support themselves, they must be printed or layered on a substrate. To construct the terminating cavity structure, the thick-film dielectric layer along with the thick-film conductor walls of the cavity are printed on a ceramic substrate. The substrate

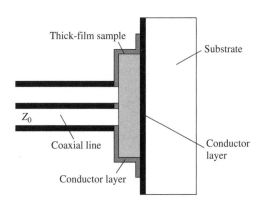

Figure 3.29 Terminating cavity configuration for thick-film materials

is isolated from the dielectric layer by the back-conductor wall of the cavity, so it merely acts as a support and does not contribute to the electrical parameters of the cavity.

3.3 FREE-SPACE REFLECTION METHOD

In the measurement of the effective permittivity of composite materials, it is required that the sample dimensions should be much larger than the sizes of the inclusions. For composites with inclusions whose sizes are comparable with the wavelength of the microwave signal, for example, fiber composites, the conventional coaxial line, and waveguide methods cannot be used. In these cases, free-space methods are often used. In the following, we discuss the general requirements for free-space measurements, and we then discuss three examples of free-space methods.

3.3.1 Requirements for free-space measurements

To achieve accurate measurement results using free-space method, several requirements should be satisfied, which mainly include far-field requirement, sample size, and measurement environment.

3.3.1.1 Far-field requirement

In free-space measurement, to ensure that the wave incident to the sample from the antenna can be taken as a plane wave, the distance d between the antenna and the sample should satisfy the following far-field requirement:

$$d > \frac{2D^2}{\lambda} \qquad (3.79)$$

where λ is the wavelength of the operating electromagnetic wave and D is the largest dimension of the antenna aperture. For an antenna with circular aperture, D is the diameter of the aperture, and for an antenna with rectangular aperture, D is the diagonal length of the rectangular aperture. When the far-field requirement is fulfilled, free space can be taken as a uniform transmission line, and most of the measurement schemes discussed above for coaxial reflection measurements can be realized by free space.

3.3.1.2 Sample size

In the measurement of permittivity and permeability of planar samples using free-space method, if the sample size is much smaller than the wavelength, the responses of the sample to electromagnetic waves are similar to those of a particle object. To achieve convincing results, the size of the sample should be larger than the wavelength of the electromagnetic wave. To further minimize the effects of the scatterings from the sample boundary, the sample size should be twice larger than the wavelength.

3.3.1.3 Measurement environment

In a free-space transmission structure, as the electromagnetic wave is not limited by a sharp boundary, the measurement results may be affected by the environments. At lower frequencies, the effects of environments are more serious. To minimize the effects of the environments, it is recommended to conduct free-space measurements in an anechoic chamber. Meanwhile, we can also use time-domain gating to eliminate the unwanted signal caused by environment reflections and multireflections, as discussed in Section 2.4.6.

3.3.2 Short-circuited reflection method

Figure 3.30 shows a typical setup for short-circuited free-space reflection measurement. The sample backed by a metal plate is placed in front of an antenna with a distance satisfying the far-field requirement.

The complex reflectivity S_{11} at the interface between the free space and the sample is given by

$$S_{11} = \frac{jz\tan(\beta d) - 1}{jz\tan(\beta d) + 1}, \qquad (3.80)$$

where z is the wave impedance of material under test normalized to the wave impedance of free space and β is the phase constant in the

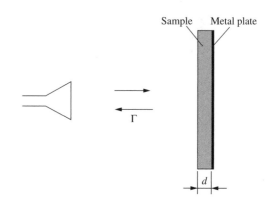

Figure 3.30   Free-space reflection method

material under test. For nonmagnetic materials, the normalized wave impedance is given by

$$z = \frac{1}{\sqrt{\varepsilon_r}}, \qquad (3.81)$$

and the phase constant β is given by

$$\beta = \frac{2\pi}{\lambda}\sqrt{\varepsilon_r} \qquad (3.82)$$

where λ is the free-space wavelength and d is the thickness of the sample. Therefore, the dielectric permittivity of the sample can be obtained from the complex reflectivity.

However, Eqs. (3.80)–(3.82) indicate that the permittivity of the sample cannot be expressed explicitly in terms of S_{11} and d, and numerical methods are often used in the calculation of dielectric permittivity.

3.3.3 Movable metal-backing method

Kalachev *et al.* proposed a movable metal-backing method, whose measurement configuration is shown in Figure 3.31 (Kalachev *et al.* 1991). The sample under test is placed directly against the aperture of the horn, so the sample size does not need to be very large, and the microwave anechoic chamber is not required.

The measurement structure consisting of the sample under test and the metal backing is quite similar to an open resonator. As the metal backing is movable, the resonant frequency can be adjusted. The dielectric sample is semitransparent

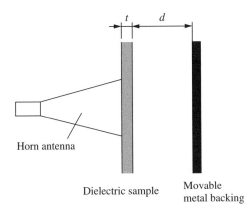

Figure 3.31 Measurement configuration of dielectric permittivity for semitransparent samples

to electromagnetic wave. When an electromagnetic wave is incident to the dielectric sample, some of the energy is reflected back and some of the energy passes through the sample. In the resonant structure, the dielectric sample serves as a reflection mirror, and meanwhile it also provides coupling to the antenna.

This method is based on the reflectivity measurement near the resonant frequency of the open resonator, and so it is a method between a nonresonant method and a resonant method. It should be noted that, in this method, as the sample directly contacts the antenna aperture, the far-field requirement is not satisfied, and the residual reflections of the horn should be taken into consideration. The residual reflections can be eliminated by time-domain techniques, and they can also be corrected by an additional measurement when the distance between the sample and the horn is changed by a quarter wavelength (Kalachev *et al.* 1991).

To determine the permittivity of the dielectric sample, we measure the reflection coefficient of the system consisting of the sample and the metal backing. The distance d between the sample and the metal backing can be varied, and the distance dependence of the reflection coefficient $\Gamma(d)$ exhibits a resonance form. For an optically thin sample satisfying

$$\frac{2\pi t \sqrt{|\varepsilon_r|}}{\lambda} \ll 1 \qquad (3.83)$$

where t is the thickness of the sample, we can get (Kalachev *et al.* 1991)

$$\cot \frac{2\pi d_0}{\lambda} = \frac{2\pi t \varepsilon_r'}{\lambda} \qquad (3.84)$$

$$\frac{2\pi t \varepsilon_r''}{\lambda} = \frac{1 \pm \Gamma_0}{1 \mp \Gamma_0} \qquad (3.85)$$

where d_0 is the distance satisfying the resonance conditions and Γ_0 is the reflection coefficient at resonance.

So the dielectric properties of the sample can be calculated from Eqs. (3.84) and (3.85). However, there are two solutions for Eq. (3.85). This is a common disadvantage of using reflection method to measure resonant structures. One convenient method to resolve this ambiguity is to introduce an additional source of low dielectric loss into the structure and find out whether the additional loss results in an increase or decrease in the reflection coefficient.

To avoid the limitation on the optical thickness of the sample, and to reduce the random experimental errors, we can measure the reflection coefficient $\Gamma(d)$ at several values of d near the resonance. The reflection coefficient of the resonant structure can be calculated using the multilayer interferometer model (Brekhovskikh 1980). The value of complex permittivity is selected so that the agreement between the measured reflection constants $\Gamma_m(d)$ and the calculated ones $\Gamma_c(d)$ is the best. The coincidence between the calculated and the measured values of reflection coefficients can be described by

$$F = \sum_{i=1}^{n} [\Gamma_m(d_i) - \Gamma_c(d_i)]^2 \qquad (3.86)$$

The dielectric permittivity of the sample corresponding to the minimum value of F can be found by two-dimensional simplex minimization routine with the use of Eqs. (3.84) and (3.85) as a starting point.

The applicability of this method depends on the value of Γ_0 of the sample. When the dielectric losses are absent ($\varepsilon_r'' = 0$) or very high ($\varepsilon_r'' \to \infty$), the value of Γ_0 is close to unity and there are secondary wave reflections between the sample and the horn, which increase the errors of reflection measurement as well as the errors of the determination of the dielectric permittivity values. Supposing that accurate measurement of reflection coefficient is possible when $\Gamma < -5\,\text{dB}$, from Eq. (3.85) we can find the range of ε_r'' suitable for this method (Kalachev 1991):

$$0.5 < \frac{2\pi t \varepsilon_r''}{\lambda} < 4 \qquad (3.87)$$

If the above requirements cannot be satisfied, this method should be modified. For a low-loss sample with

$$\frac{2\pi t \varepsilon_r''}{\lambda} \ll 1, \qquad (3.88)$$

to obtain accurate results, it is recommended to characterize the sample together with an additional dielectric-sheet material with known properties. The additional sample should provide the dielectric loss that is sufficient for obtaining low reflectivity.

For an extremely high-loss material with

$$\frac{2\pi t \varepsilon_r''}{\lambda} \gg 1, \qquad (3.89)$$

the sample is practically nontransparent and the reflection coefficient does not depend on the presence and position of the metal backing behind the sample. To measure such nontransparent samples, Kalachev *et al.* modified the measurement technique discussed above (Kalachev *et al.* 1991). As the loss factor of a material is proportional to the conductivity of the material, we may use surface impedance to describe the electromagnetic properties of an extremely high-loss material. The reflection methods used for the measurement of surface impedance will be discussed in Section 3.5.

3.3.4 Bistatic reflection method

In most of the reflection methods, only one probe is used, and this probe transmits signal and meanwhile receives signal. In such a configuration, it is difficult to change the incident angle of the transmitted signal. Figure 3.32 shows a bistatic system for free-space reflection measurement. In this configuration, two antennas are used for transmitting and receiving signals respectively,

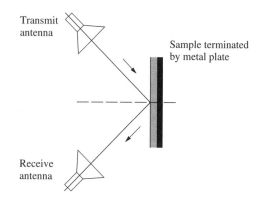

Figure 3.32   Bistatic reflection measurement

and the reflections at different incident angles can be measured. Using this configuration, the properties of materials at different directions can be characterized.

It should be noted that in bistatic reflection measurements, the reflection is dependent on the polarization of the incident wave. Incident waves with parallel and perpendicular polarization usually result in different reflection coefficients. Besides, special calibration is needed for free-space bistatic reflection measurements (Umari *et al.* 1991).

3.4 MEASUREMENT OF BOTH PERMITTIVITY AND PERMEABILITY USING REFLECTION METHODS

Usually, in a reflection method, only one complex reflection coefficient is measured, from which only one complex materials property parameter, either permittivity or permeability, can be derived. To obtain both the complex permittivity and complex permeability, at least two independent reflection measurements are required. Here, we discuss six examples for the measurement of both permittivity and permeability using reflection method.

3.4.1 Two-thickness method

As shown in Figure 3.33, in a two-thickness method, the two independent reflection measurements are made by measuring the reflections from two samples made of the same material under test but with different thickness values. We assume that

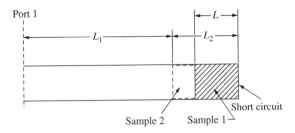

Port 1

L_1

L

L_2

Short circuit

Sample 2   Sample 1

Figure 3.33 Two samples in a short-circuited transmission line. Source: Baker-Jarvis, J. Domich, M. D. and Geyer, R. G. (1993), *Transmission/reflection and short-circuit line methods for measuring permittivity and permeability*, NIST Technical Note 1355 (revised), National Institute of Standards and Technology, U.S. Department of Commerce

the thickness values of the two samples are L and αL respectively, and their corresponding reflections measurements are $S_{11(1)}$ and $S_{11(2)}$ respectively.

According to the transmission theory, the two independent reflections are given by

$$S_{11(1)} = \frac{\Gamma - Z^2}{1 - \Gamma Z^2} \qquad (3.90)$$

$$S_{11(2)} = \frac{\Gamma - Z^{2\alpha}}{1 - \Gamma Z^{2\alpha}} \qquad (3.91)$$

with the reflection coefficient Γ given by

$$\Gamma = \frac{(\mu/\gamma) - (\mu_0/\gamma_0)}{(\mu/\gamma) + (\mu_0/\gamma_0)} \qquad (3.92)$$

and the transmission coefficient Z given by

$$Z = \exp(-\gamma L) \qquad (3.93)$$

From Eq. (3.90), we have

$$Z^2 = \frac{S_{11(1)} - \Gamma}{S_{11(1)}\Gamma - 1} \qquad (3.94)$$

Substituting Eq. (3.94) into Eq. (3.61) we get

$$S_{11(2)} = \frac{\Gamma - \left(\dfrac{S_{11(1)} - \Gamma}{S_{11(1)}\Gamma - 1}\right)^{\alpha}}{1 - \Gamma\left(\dfrac{S_{11(1)} - \Gamma}{S_{11(1)}\Gamma - 1}\right)^{\alpha}} \qquad (3.95)$$

Eq. (3.95) can be solved iteratively for Γ.

With Γ given by Eq. (3.95) and Z given by Eq. (3.93), we can calculate the properties of the material using the following equations (Baker-Jarvis *et al.* 1993):

$$\varepsilon_r = \frac{\lambda_0^2}{\mu_r}\left(\frac{1}{\lambda_c^2} - \frac{1}{\Lambda^2}\right) \qquad (3.96)$$

$$\mu_r = \frac{1 + \Gamma}{(1 - \Gamma)\Lambda\sqrt{1/\lambda_0^2 - 1/\lambda_c^2}} \qquad (3.97)$$

with

$$\frac{1}{\Lambda^2} = -\left[\frac{1}{2\pi L}\ln\left(\frac{1}{Z}\right)\right]^2 \qquad (3.98)$$

where λ_0 is the free-space wavelength and λ_c is the cutoff wavelength of the transmission structure. It should be noted that Eq. (3.98) has an infinite number of roots because the logarithm of a complex number is multivalued. In order to choose the correct root, it is necessary to compare the measured group delay to the calculated group delay.

3.4.2 Different-position method

The two independent reflections for the determination of permittivity and permeability can be obtained when the sample is placed at different positions of a transmission line (Baker-Jarvis *et al.* 1993). As shown in Figure 3.34, if measurements are made when the sample is placed at two different positions at a short-circuited transmission line, explicit solution to Eq. (3.6) can be obtained by solving the Eq. (3.6) at a given short-circuit position (position 1) for $\tanh \gamma L$ and then substituting this expression into Eq. (3.6) at another short-circuit position (position 2).

The reflection coefficients corresponding to the two different positions shown in Figure 3.34 are given by

$$\Gamma_1 = \frac{2\beta\delta_1 - [(\delta_1 + 1) + (\delta_1 - 1)\beta^2]\tanh \gamma L}{-2\beta + [(\delta_1 - 1)\beta^2 - (\delta_1 + 1)]\tanh \gamma L} \qquad (3.99)$$

$$\Gamma_2 = \frac{2\beta\delta_2 - [(\delta_2 + 1) + (\delta_2 - 1)\beta^2]\tanh \gamma L}{-2\beta + [(\delta_2 - 1)\beta^2 - (\delta_2 + 1)]\tanh \gamma L} \qquad (3.100)$$

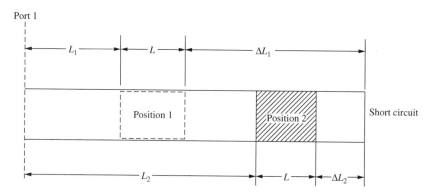

Figure 3.34 Two-position measurements in a short-circuited transmission line. Source: Baker-Jarvis, J. Janezic, M. D. Grosvenor, J. H. Jr. and Geyer, R. G. (1993), *Transmission/reflection and short-circuit line methods for measuring permittivity and permeability*, NIST Technical Note 1355 (revised), National Institute of Standards and Technology, U.S. Department of Commerce

where δ_1 and δ_2 denote the phases calculated from Eq. (3.8) for ΔL_1 and ΔL_2 respectively. From Eqs. (3.99) and (3.100), we can get

$$\tanh \gamma L$$
$$= \frac{2\beta(\delta_1 + \Gamma_1)}{\beta^2(\Gamma_1 + 1)(\delta_1 - 1) + (1 - \Gamma_1)(\delta_1 + 1)}$$

$$(3.101)$$

$$\gamma = \frac{1}{L}$$
$$\left\{ \tanh^{-1} \left[\frac{2\beta(\delta_1 + \Gamma_1)}{\beta^2(\Gamma_1 + 1)(\Gamma_1 - 1) + (1 - \Gamma_1)(\delta_1 + 1)} \right] \right.$$
$$\left. + 2n\pi j \right\}$$

$$(3.102)$$

where n is an integer, whose correct value can be determined from the group delay.

Meanwhile, we have

$$\beta^2 = \frac{\substack{\delta_1[\delta_2(\Gamma_1 - \Gamma_2) + \Gamma_1\Gamma_2 + 1 - \Gamma_2] \\ -\{\delta_2[\Gamma_1(\Gamma_2 - 2) + 1] + \Gamma_2 - \Gamma_1\}}}{\substack{\delta_1[\delta_2(\Gamma_1 - \Gamma_2) + \Gamma_1\Gamma_2 + 1 + 2\Gamma_2] \\ -\{\delta_2[\Gamma_1(\Gamma_2 + 2) + 1] + \Gamma_2 - \Gamma_1\}}}$$

$$(3.103)$$

Once the value of β is obtained, the permittivity and permeability of the material can be calculated according to Eqs. (3.102) and (3.103).

3.4.3 Combination method

The two independent complex parameters can be obtained by the method shown in Figure 3.35. First, measure the complex reflection Γ_1 by directly terminating the probe with the sample shorted by a metal plate. Second, measure the complex reflection Γ_2 by inserting a material with known electromagnetic properties, such as Teflon, between the probe and the sample under test. From the two complex reflection coefficients Γ_1 and Γ_2, the materials complex parameters ε_r and μ_r can be obtained.

In the calculation of the permittivity and permeability of the sample, we need to know the

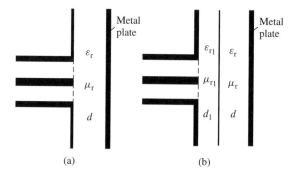

Figure 3.35 Two independent reflection measurements. (a) Sample directly contacts the probe and (b) a layer of material with known permittivity and permeability is inserted between the probe and the sample

admittance values at the two conditions shown in Figure 3.35. The admittance of the probe connected with the layered media can be analyzed using the method discussed in Section 3.2.2.

3.4.4 Different backing method

As shown in Figure 3.36, the two independent reflection measurements can be made by backing the sample with free space and metal plate respectively. We directly terminate probe with the sample under test and measure the reflection coefficients when the sample is backed by free space and metal plate respectively. From the two complex reflection coefficients, the two complex material property parameters ε_r and μ_r can be obtained.

Similar to the combination method discussed above, in the calculation of materials properties, we need to know the admittance values when the sample is backed by free space and metal plate respectively. The admittance of the probe at these two conditions can be done using the method discussed in Section 3.2.2.

3.4.5 Frequency-variation method

The frequency-variation method was proposed by Wang *et al.* for the measurement of both permittivity and permeability using a reflection probe (Wang *et al.* 1998). This method takes frequency as an independent variable, and only needs one frequency-sweeping measurement. This method has some similarity with the two-thickness method discussed in Section 3.4.1. For two measurements made at two different frequencies, though the physical thickness of the sample does not change, the sample has different values of electric thickness at different frequencies. If we assume that the sample has the same permittivity and permeability at different frequencies, the two reflections measured at two frequencies, corresponding to two values of electric thickness of the sample, can be used to determine the permittivity and permeability of the sample.

In this method, the open-ended probe may be made from a coaxial line, rectangular or circular waveguide. The material under test may be semi-infinite, or it can be a single layer or multiple-layer sample backed by free space or metal plate. As an example, Figure 3.37 shows a flanged open-ended coaxial probe terminated by a single-layer sample backed by a metal plate. In this configuration, the reflection coefficient $\Gamma(a, b, d, \varepsilon_r, \mu_r, f)$ is a function of the coaxial dimensions a and b, sample thickness d, electromagnetic property parameters ε_r and μ_r, and measurement frequency f. From the reflection coefficients Γ_1 and Γ_2 measured at frequency f_1 and f_2 respectively, the permittivity and permeability of the sample can be calculated. The obtained permittivity and permeability can be taken as the averages of the values at frequencies f_1 and f_2.

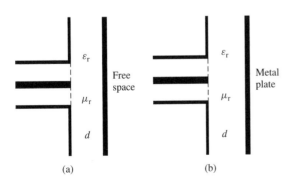

Figure 3.36 Two independent reflection measurements. (a) Sample backed by free space and (b) sample backed by a metal plate

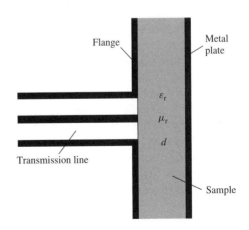

Figure 3.37 Configuration for a typical reflection measurement using a coaxial probe

In some cases, we need to consider the frequency dependence of the electromagnetic properties of materials. In these cases, the permittivity $\varepsilon_r(f)$ and permeability $\mu_r(f)$ are functions of frequency, and so interpolation techniques are needed for extracting permittivity and permeability from the frequency-sweeping reflection data. Linear interpolation is the simplest interpolation:

$$\varepsilon_r(f) = af + b \qquad (3.104)$$

$$\mu_r(f) = cf + d \qquad (3.105)$$

By using the linear interpolation in Eqs. (3.104) and (3.105), the determination of permittivity and permeability becomes the determination of four complex parameters a, b, c, and d, and so the reflection coefficients at four frequency points are needed. If we use the parabolic interpolation, reflection coefficients at six frequency points are needed to determine the six complex parameters involved. As at microwave frequencies, most of the polarizations are Debye type, Debye equations are often used in interpolation.

In the application of the frequency-variation method, after the reflection coefficients are measured over a wide frequency range, the complex permittivity and permeability are reconstructed from the reflection coefficients at frequency points in a certain frequency band according to the adopted interpolation technique. This process continues throughout the entire measurement frequency range.

The choice of the interpolation and frequency interval Δf depends on many factors such as the natural characteristics of the electromagnetic properties of the materials and the requirements on measurement speed and accuracy. In general, for highly dispersive media and for critical accuracy requirements, higher-order interpolation should be used. At the same time, the inverse problem will become more difficult and time consuming. A trade-off between accuracy and speed must be made in practical real-time and on-site measurements. The frequency interval should guarantee that the reflection coefficient is changed enough to be distinguished by the network analyzer. Meanwhile, to improve the measurement resolution, the frequency interval selected should be as small

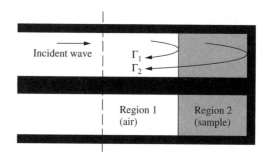

Figure 3.38   A short-ended coaxial line with a sample loaded at the end. The time histories of the incident wave and the first and second reflections (Γ_1 and Γ_2) are measured at a location in front of the interface between the air and the sample

as possible, especially for the highly dispersive materials whose permittivity and permeability vary rapidly with frequency.

3.4.6 Time-domain method

Besides the five methods discussed below, Courtney and Motil proposed a time-domain measurement in a short-circuited coaxial geometry and data-reduction procedure to experimentally determine the complex permittivity and permeability of materials (Courtney and Motil 1999). As shown in Figure 3.38, using time-domain techniques, this method measures the first and second reflection waves (partial components of the complete reflection) in the short-circuited coaxial transmission line that holds the material under test. In this method, the two independent parameters for the determination of permittivity and permeability are the first and second reflections (Γ_1 and Γ_2). Detailed discussions on this method can be found in (Courtney and Motil 1999; Courtney 1998).

3.5 SURFACE IMPEDANCE MEASUREMENT

Figure 3.39 shows the measurement of surface impedance of a conducting film using reflection method. To ensure the electrical contact between the end of coaxial line and the film sample, gold contacts with a thickness of a few thousand angstroms are evaporated on the film. The portion of the film exposed between the contacts forms

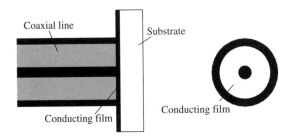

Figure 3.39 Measurement of surface resistance using reflection method. (a) Measurement configuration using coaxial line and (b) shape of a Corbino disc

the Corbino disc, as shown in Figure 3.39(b). In this method, the surface impedance Z_s is extracted from the reflection coefficient S_{11}, which is related to the load impedance Z_L of the sample:

$$S_{11} = \frac{Z_L - Z_0}{Z_L + Z_0} \tag{3.106}$$

where Z_0 is the characteristic impedance of the coaxial transmission line. The reflection coefficient is a complex, dimensionless quantity that measures the impedance mismatch between the transmission line and the load, and is bounded in magnitude between 0 and 1.

The load impedance Z_L in Eq. (3.106) is the ratio of the total voltage across the Corbino disc to the total current flowing through the disc. The load impedance Z_L depends on the surface impedance of the conducting film and the dimensions of the Corbino disc. If only the TEM mode exists in the coaxial line, Z_L can be expressed in terms of the surface impedance Z_s of the film (Booth *et al.* 1994):

$$Z_L = \alpha Z_s \tag{3.107}$$

The scale factor α in this expression simply relates the "ohmic" impedance Z_{IV} and the field impedance Z_{field} for the TEM mode in a coaxial system:

$$Z_{IV} = \frac{V}{I} \tag{3.108}$$

$$Z_{field} = \frac{E_r}{H_\phi} \tag{3.109}$$

where V and I represent the voltage and current across the coaxial cable respectively, and E_r and H_ϕ represent the electric and magnetic fields within the coaxial cable, respectively. The surface impedance Z_s of the film is the field impedance Z_{field} evaluated at the surface of the film. For a coaxial line with inner radius a and outer radius b, the scale factor is given by

$$\alpha = 2\pi \ln\left(\frac{b}{a}\right) \tag{3.110}$$

Therefore, once the reflection coefficient S_{11} of the sample has been measured, the surface impedance of the conducting film can be obtained from Eqs. (3.106) and (3.107).

The technique takes advantage of a special geometry in which the self-fields from currents flowing in the film are parallel to the film surface everywhere, making it an ideal configuration for the study of vortex dynamics in superconductors. Furthermore, its broadband nature makes it suitable for investigating superconductivity in a broad-frequency region. However, it should be noted that this method requires very good contact between the superconducting film and the inner and outer conductors of the coaxial line. Figure 3.40 shows a superconducting thin-film/connector interface structure employing a spring-loaded inner conductor pin. Direct contact is made between the outer

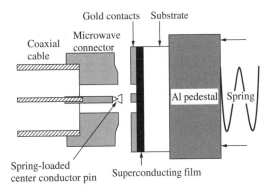

Figure 3.40 The thin-film/connector interface, just prior to contact. Pressure is exerted on the aluminum pedestal by a spring in order to maintain contact between the film and connector throughout the temperature range (Booth *et al.* 1994). Source: Booth, J. C. Wu, D. H. and Anlage, S. M. (1994). "A broadband method for the measurement of the surface impedance of thin films at microwave frequencies", Review of Scientific Instruments, **65** (6), 2082–2090

conductor of the connector and the outer gold contact of the film, while the contact between the inner conductor of the connector and the inner gold contact of the film is ensured by using a small pin inserted in the center conductor of the connector. The contacts between the connector and the sample are maintained by using an aluminum pedestal and spring assembly that applies pressure to the backside of the substrate. The contacts can reliably hold from room temperature to very low temperature (several Kelvin), and the changes in contact resistance with temperature are negligibly small (Booth *et al.* 1994).

3.6 NEAR-FIELD SCANNING PROBE

Scanning techniques for local characterization of conducting and insulating films are attracting much interest. Many efforts have been made on developing microwave near-field scanning techniques, and various types of near-field microwave microscopes have been developed for different purposes (Rosner and van der Weide 2002). The near-field microwave microscopes generally fall into the resonant type and the nonresonant type. In a resonant near-field microwave microscope, the probe is a resonator, and it works on the basis of resonant-perturbation theory. While a nonresonant near-field microwave microscope is usually based on reflection method, and the properties of a sample are obtained from the reflectivity due to the presence of the sample. Here we focus on nonresonant near-field microwave microscopes, and the discussions on resonant near-field microwave microscopes can be found in Chapter 6.

In principle, any type of transmission lines can be used to develop near-field microwave microscopes. In a near-field microwave microscope developed from rectangular waveguide, the most important part is an aperture in the form of a narrow rectangular slit. As shown in Figure 3.41, a thin metal diaphragm is mounted across a rectangular waveguide, and a rectangular slit is cut in the diaphragm parallel to the wide side. For transverse electric (TE) mode propagation, this slit is transparent at a certain wavelength λ,

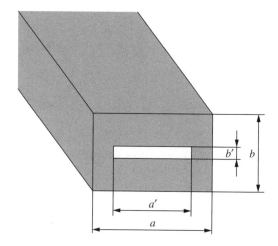

Figure 3.41   A rectangular slit on a metal diaphragm mounted across a rectangular waveguide

which can be found by solving the following equation (Golosovsky *et al.* 1996)

$$\frac{a}{b} \cdot \sqrt{1 - \left(\frac{\lambda}{2a}\right)^2} = \frac{a'}{b'} \cdot \sqrt{1 - \left(\frac{\lambda}{2a'}\right)^2} \quad (3.111)$$

where λ is the free-space wavelength, a and b are the wide and narrow sides of the waveguide, and a' and b' are the wide and narrow sides of the slit. Equation (3.111) has solution even for arbitrarily narrow slit. When b' approaches zero, the solution for λ approaches $2a'$. We can also take a rectangular waveguide terminated by a rectangular slit window as a junction of two rectangular waveguides with different dimensions. The reflection from such a junction is determined by the waveguide impedance (Golosovsky *et al.* 1996)

$$Z = \frac{Z_0 \pi b'}{2a' \sqrt{1 - [\lambda/(2a')]^2}} \quad (3.112)$$

where Z_0 is the free-space impedance ($Z_0 = 377\,\Omega$). By taking the slit as a section of a rectangular waveguide, we can find that for given a' and arbitrarily small b', there is always a certain wavelength $\lambda \approx 2a'$ at which $Z \approx Z_0$. Therefore, even a very narrow slit may be matched to free space at a narrowband.

When a conducting surface is in the near-field zone of the slit, the microwave is reflected mostly from the region under the slit. Since reflection from a conducting surface is determined by the resistivity, by measuring the amplitude and phase of the reflected wave while raster scanning the surface, it is possible to map the microwave resistivity of the surface. For conductive layers with thicknesses much larger than the skin depth, we can get surface impedance, while for thin layers, we can get sheet resistance. In the determination of microwave resistivity, it is necessary to measure layer thickness independently.

In measurement, the sample is placed close to the slit. If the distance between the slit and the sample is smaller than the narrow side of the slit, the spatial resolution in the direction perpendicular to the slit is determined by the narrow side of the slit b' and maybe as small as $\lambda/100$. The spatial resolution in the direction parallel to the slit is determined by the field pattern in the slit and by the wide side of the slit a', and for a flat slit it is about $\lambda/2$. However, the resolution in this direction may be considerably improved (down to $\lambda/60$) if the slit is fabricated in a curved surface. This improvement originates from the strong decay of the field upon increasing distance from the slit, so that the central part of the slit, which is most close to the tested surface, plays a dominant role. Generally speaking, the spatial resolutions in the directions of the wide and narrow sides of the slit are different. However, by doing scans in several directions and using deconvolution techniques and image processing, it is possible to achieve two-dimensional resistivity maps with equal resolution in all directions.

Figure 3.42 shows a dual-frequency electromagnetic scanning probe for quantitative mapping of sheet resistance of conducting films (Lann *et al.* 1998). The high-frequency (82 GHz) mode is used for image acquisition, while the low-frequency (5 MHz) mode is used for distance control. The key component is a thin-slit aperture in a convex end plate of a rectangular waveguide. This aperture operates as a transmitting/receiving antenna at 82 GHz. A fundamental TE wave is excited in the waveguide and the reflected wave is analyzed. The electric field lines in the TE mode stretch from one side of the slit to the

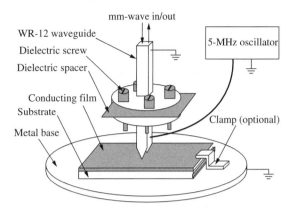

Figure 3.42 A mm-wave near-field probe with a capacitive distance control (Lann *et al.* 1998). Source: Lann, A. F. Golosovsky, M. Davidov, D. and Frenkel, A. (1998). "Combined millimeter-wave near-field microscope and capacitance distance troll for the quantitive mapping of sheet resistance of conducting layers", *Applied Physics Letters*, **73** (19), 2832–2834

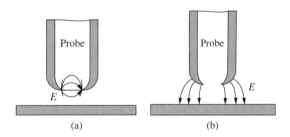

Figure 3.43 Electric fields near the slit. (a) Electric field of mm-wave operation (inductive mode) and (b) electric field of rf operation (capacitive mode). Source: Lann, A. F. Golosovsky, M. Davidov, D. and Frenkel, A. (1998). "Combined millimeter-wave near-field microscope and capacitance distance troll for the quantitive mapping of sheet resistance of conducting layers", *Applied Physics Letters*, **73** (19), 2832–2834

other, as shown in Figure 3.43(a), and they are mostly tangential to the sample surface and induce currents in the sample. The magnitude of the induced currents is determined by the sample resistivity and thickness. In this high-frequency operation mode, the probe behaves inductively and provides information on the resistance of the sample.

In order to enable a simultaneous low-frequency operation mode, a short waveguide section containing the probe is electrically isolated from the rest of the mm-wave circuitry by a thin mylar sheet, as shown in Figure 3.42. A 5-MHz oscillator is connected to the waveguide section with the probe, while the ground of the oscillator is connected to the rest of the mm-wave circuitry and to the conducting sample. As the slit size for the rf-operation mode is much smaller than the rf wavelength, the surface of the probe is equipotential. So, as shown in Figure 3.43(b), the electric field lines stretch from the probe to the sample surface, and they are almost normal to the sample and induce charges rather than currents. Thus the probe behaves capacitively, and is almost insensitive to the sample resistance. The resonant frequency of the oscillator strongly depends upon the probe-sample capacitance that is determined by the probe-sample distance. Therefore by measuring the resonant frequency of the oscillator, the probe-sample distance can be precisely controlled.

REFERENCES

Anderson, L. S. Gajda, G. B. and Stuchly, S. S. (1986). "Analysis of an open-ended coaxial line sensor in layered dielectrics", *IEEE Transactions on Instrumentation and Measurement*, **35** (1), 13–18.

Anderson, L. S. Gajda, G. B. and Stuchly, S. S. (1994). "Dielectric measurements using a rational functional model", *IEEE Transactions on Microwave Theory and Technique*, **42**, 199–204.

Athey, T. W. Stuchly, M. A. and Stuchly S. S. (1982). "Measurement of radio-frequency permittivity of biological tissues with an open-ended coaxial line-Part I", *IEEE Transactions on Microwave Theory and Techniques*, **30** (1), 82–86.

Baker-Jarvis, J. (1990). *Transmission/reflection and short-circuit line permittivity measurements*, NIST Technical Note 1341, National Institute of Standards and Technology, U.S. Department of Commerce.

Baker-Jarvis, J. Janezic, M. D. Grosvenor, J. H. Jr. and Geyer, R. G.
(1993). *Transmission/Reflection and Short-Circuit Line Methods for Measuring Permittivity and Permeability*, NIST Technical Note 1355 (revised), National Institute of Standards and Technology, U.S. Department of Commerce.

Baker-Jarvis, J. Domich, M. D. and Geyer, R. G. (1994). "Analysis of an open-ended coaxial probe with lift-off for nondestructive testing", *IEEE Transactions on Instrumentation and Measurement*, **43** (5), 711–718.

Bakhtiari, S. Ganchev, S. I. and Zoughi, R. (1994). "Analysis of radiation from an open-ended coaxial line into stratified dielectrics", *IEEE Transactions on Microwave Theory and Techniques*, **42** (7), 1261–1267.

Berube, D. Ghannouchi, F. M. and Savard, P. (1996). "A comparative study of four open-ended coaxial probe models for permittivity measurements of lossy dielectric biological materials at microwave frequencies", *IEEE Transactions on Microwave Theory and Techniques*, **44** (10), 1928–1934.

Booth, J. C. Wu, D. H. and Anlage, S. M. (1994). "A broadband method for the measurement of the surface impedance of thin films at microwave frequencies", *Review of Scientific Instruments*, **65** (6), 2082–2090.

Brady, M. M. Symons, S. A. and Stuchly, S. S. (1981). "Dielectric behavior of selected animal tissues in vitro at frequencies from 2 to 4 GHz", *IEEE Transactions on Biomedical Engineering*, **28** (3), 305–307.

Brekhovskikh, L. M. (1980). *Waves in Layered Media*, 2nd edition, Academic Press, New York.

Bucci, O. M. and Franceschetti, G. (1974). "Input admittance and transient response of spheroidal antennas in dispersive media", *IEEE Transactions Antennas and Propagation*, **22** (4), 526–536.

Burdette, E. C. Clain, F. L. and Seals, J. (1980). "In vivo probe measurement technique for determining dielectric properties at VHF through microwave frequencies", *IEEE Transactions on Microwave Theory and Techniques*, **28**, 414–427.

Courtney, C. C. (1998). "Time-domain measurement of the electromagnetic properties of materials", *IEEE Transactions on Microwave Theory and Techniques*, **46** (5), 517–522.

Courtney, C. C. and Motil, W. (1999). "One-port time-domain measurement of the approximate permittivity and permeability of materials", *IEEE Transactions on Microwave Theory and Techniques*, **47** (5), 551–555.

Deschamps, A. (1972). "Impedance of an antenna in a conducting medium", *IEEE Transactions on Antennas and Propagation*, **10**, 648–650.

Fan, S. Staebell, K. F. and Misra, D. (1990). "Static analysis of an open-ended coaxial line sensor in layered dielectrics", *IEEE Transactions on Instrumentation and Measurement*, **39**, 435–437.

Gajda, G. and Stuchly, S. S. (1983). "An equivalent circuit of an open-ended coaxial line", *IEEE Transactions on Instrumentation and Measurement*, **32** (4), 506–508.

Ganchev, S. I. Qaddoumi, N. Bakhtiari, S. and Zoughi, R. (1995). "Calibration and measurement of dielectric properties of finite thickness composite sheets with open-ended coaxial sensors", *IEEE Transactions on Instrumentation and Measurement*, **44** (6), 1023–1029.

Ghannouchi, F. M. and Bosisio, R. G. (1989). "Measurement of microwave permittivity using six-port reflector with an open-ended coaxial line", *IEEE Transactions on Instrumentation and Measurement*, **38** (2), 505–508.

Golosovsky, M. Galkin, A. and Davidov, D. (1996). "High-spatial resolution resistivity mapping of large-area YBCO films by a near-field millimeter-wave microscope", *IEEE Transactions on Microwave Theory and Techniques*, **44** (7), 1390–1392.

Huang, Y. (2001). "Design, calibration and data interpretation for a one-port large coaxial dielectric measurement cell", *Measurement Science and Technology*, **12** (1), 111–115.

Kalachev, A. A. Kukolev, I. V. Matytsin, S. M. Novogrudsiy, L. N. Rozanov, K. N. and Sarychev, A. K. (1991). "The methods of investigation of complex dielectric permittivity of layer polymers containing conductive inclusions", in *Optical and Electrical Properties of Polymers, Materials Research Society Symposia Proceedings*, Vol. 214, J. A. Emerson and J. M. Torkelson, Ed., Materials Research Society, Pittsburgh, pp. 119–124.

Keam, R. B. and Holdem, J. R. (1997). "Permittivity measurements using a coaxial-line conical-tip probe", *Electronics Letters*, **33** (5), 353–355.

Langhe, P. D. Martens, L. and Zutter, D. D. (1994). "Design rules for an experimental setup using an open-ended coaxial probe based on theoretical modeling", *IEEE Transactions on Instrumentation and Measurement*, **43** (6), 810–817.

Lann, A. F. Golosovsky, M. Davidov, D. and Frenkel, A. (1998). "Combined millimeter-wave near-field microscope and capacitance distance troll for the quantitive mapping of sheet resistance of conducting layers", *Applied Physics Letters*, **73** (19), 2832–2834.

Levine, H. and Papas, C. H. (1951). "Theory of the circular diffraction antenna", *Journal of Applied Physics*, **22**, 29–43.

Marcuvitz, N. (1965). *Waveguide Handbook*, Dover Publications, New York.

Otto, G. P. and Chew, W. C. (1991). "Improved calibration of a large open-ended coaxial probe for dielectric measurement", *IEEE Transactions on Instrumentation and Measurement*, **40** (4), 742–746.

Panariello, G. Verolino, L. and Vitolo, G. (2001). "Efficient and accurate full-wave analysis of the open-ended coaxial cable", *IEEE Transactions on Microwave Theory and Techniques*, **49** (7), 1304–1309.

Pournaropoulos, C. L. and Misra, D. K. (1994). "A study on the coaxial aperture electromagnetic sensor and its application in material characterization", *IEEE Transactions on Instrumentation and Measurement*, **43** (2), 111–115.

Pournaropoulos, C. L. and Misra, D. K. (1997). "The co-axial aperture electromagnetic sensor and its application in material characterization", *Measurement Science and Technology*, **8** (11), 1191–1202.

Razaz, M. and Davies, J. B. (1979). "Capacitance of the abrupt transition from coaxial-to-circular waveguide", *IEEE Transaction on Microwave Theory and Techniques*, **27** (6), 564–569.

Rosner, B. T. and van der Weide, D. W. (2002). "High-frequency near field microscopy", *Review of Scientific Instruments*, **73** (7), 2505–2525.

Rzepecka, M. A. and Stuchly, S. S. (1975). "A lumped capacitance method for the measurement of the permittivity in the frequency and time domain – a further analysis", *IEEE Transactions on Instrumentation and Measurement*, **24** (1), 27–32.

Saed, M. A. Riad, S. M. and Elshabini-Riad, A. (1989). "Wide-band measurement of the complex permittivity of dielectric materials using a wide-band cavity", *IEEE Transactions on Instrumentation and Measurement*, **38** (2), 488–495.

Saed, M. A. Riad, S. M. and Davis, W. A. (1990). "Wide-band characterization using a dielectric filled cavity adapted to the end of a transmission line", *IEEE Transactions on Instrumentation and Measurement*, **39** (3), 485–489.

Smith, G. S. and Nordgard, J. D. (1985). "Measurement of the electrical constitutive parameters of materials using antennas", *IEEE Transactions Antennas and Propagation*, **33** (7), 783–792.

Staebell, K. F. and Misra, D. (1990). "An experimental technique for in vivo permittivity measurement of materials at microwave frequencies", *IEEE Transactions on Microwave Theory and Techniques*, **38** (3), 337–339.

Stuchly, M. A. Athey, T. W. Samaras, C. M. and Taylor, G. E. (1982). "Measurement of radio frequency permittivity of biological tissues with an open-ended coaxial line: Part II-experimental results", *IEEE Transactions on Microwave Theory and Techniques*, **30** (1), 87–91.

Stuchly, S. S. Sibbald, C. L. and Anderson, J. M. (1994). "A new admittance model for open-ended waveguides", *IEEE Transactions on Microwave Theory and Techniques*, **42** (2), 192–198.

Stuchly, M. A. and Stuchly, S. S. (1980). "Coaxial line reflection methods for measuring dielectric properties of biological substances at radio and microwave frequencies – a review", *IEEE Transactions on Instrumentation and Measurement*, **29** (3), 176–183.

Umari, M. H. Ghodgaonkar, D. K. Varadan, V. V. and Varadan, V. K. (1991). "A free-space bistatic calibration technique for the measurement of parallel and perpendicular reflection coefficients of planar samples", *IEEE Transactions on Instrumentation and Measurement*, **40** (1), 19–24.

Von Hippel, A. R. Ed. (1995). *Dielectric Materials and Applications*, Artech House, Boston.

Wang, Y. and Fan, D. (1994). "Accurate global solutions of EM boundary-value problems for coaxial radiators", *IEEE Transactions Antennas and Propagation*, **42** (5), 767–770.

Wang, S. J. Niu, M. D. and Xu, D. M. (1998). "A frequency-varying method for simultaneous measurement of complex permittivity and permeability with an open-ended coaxial probe", *IEEE Transac-* *tions on Microwave Theory and Techniques*, **46** (12), 2145–2147.

Xu, Y. S. Ghannouchi, F. M. and Bosisio, R. G. (1992). "Theoretical and experimental study of measurement of microwave permittivity using open-ended elliptical coaxial probes", *IEEE Transactions on Microwave Theory and Techniques*, **40** (1), 143–150.

Zheng, H. M. and Smith, C. E. (1991). "Permittivity measurements using a short open-ended coaxial line probe", *IEEE Microwave and Guided Wave Letters*, **1** (11), 337–339.

4

Transmission/Reflection Methods

In a transmission/reflection method, the sample under test is inserted into a segment of transmission line, and the permittivity and permeability of the sample are derived from the reflection and transmission of the sample-loaded unit. After analyzing the working principle and calculation algorithms, we discuss four types of transmission/reflection methods, including coaxial airline method, hollow metallic waveguide method, surface-wave method and free-space method. We then make a brief review of the modifications of the conventional transmission/reflection methods. At the end of this chapter, we discuss the measurement of complex conductivity of superconductors using transmission/reflection methods.

4.1 THEORY FOR TRANSMISSION/REFLECTION METHODS

As nonresonant methods can cover certain frequency bands, they can be used for the measurements of electromagnetic properties of materials for a multitude of applications, such as the study of polarization mechanisms of materials and the development of broadband functional materials and structures. Owing to their relative simplicity, the transmission/reflection methods, as a category of nonresonant methods, are widely used in various fields of materials research and engineering.

In a transmission/reflection method, the sample under test is inserted into a segment of transmission line, such as waveguide or coaxial line. From the relevant scattering equations relating the scattering parameters of the segment of transmission

line filled with the sample under study to the permittivity and permeability of the sample, we can get the electromagnetic properties of the sample. In a transmission/reflection method, all the four scattering parameters can be measured, so we have more data at our disposal than in reflection measurements. For a transmission/reflection measurement, the relevant scattering equations contain variables including the complex permittivity and permeability of the sample, the positions of the two reference planes, and the sample length. These relevant scattering equations are generally overdetermined and therefore can be solved in various ways.

4.1.1 Working principle for transmission/reflection methods

The working principles for transmission/reflection methods have been systematically analyzed in literatures (Baker-Jarvis *et al*. 1993, 1990; Baker-Jarvis 1990). Here, we only discuss the basic principle for transmission/reflection measurements.

Figure 4.1 shows a typical measurement configuration for a transmission/reflection method. The sample under study is inserted into a segment of transmission line, whose axis is in x-direction. Scattering equations are often used to analyze the electric fields at the sample interfaces. We assume the electric fields at the three sections of the transmission line as E_{I}, E_{II}, and E_{III}. For a normalized incident wave, we have (Baker-Jarvis 1990)

$$E_{\mathrm{I}} = \exp(-\gamma_0 x) + C_1 \exp(\gamma_0 x) \qquad (4.1)$$

Microwave Electronics: Measurement and Materials Characterization L. F. Chen, C. K. Ong, C. P. Neo, V. V. Varadan and V. K. Varadan
© 2004 John Wiley & Sons, Ltd ISBN: 0-470-84492-2

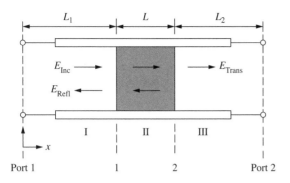

Figure 4.1 Electromagnetic waves transmitting through and reflected from a sample in a transmission line

$$E_{II} = C_2 \exp(-\gamma x) + C_3 \exp(\gamma x) \quad (4.2)$$

$$E_{III} = C_4 \exp(-\gamma_0 x) \quad (4.3)$$

with

$$\gamma = j\sqrt{\frac{\omega^2 \mu_r \varepsilon_r}{c^2} - \left(\frac{2\pi}{\lambda_c}\right)^2} \quad (4.4)$$

$$\gamma_0 = j\sqrt{\left(\frac{\omega}{c}\right)^2 - \left(\frac{2\pi}{\lambda_c}\right)^2} \quad (4.5)$$

where ω is the angular frequency, c is the speed of light in vacuum, γ_0 and γ are the propagation constants in the transmission lines filled with free space and the sample respectively. λ_0 is the cutoff wavelength of the transmission line, and for a transmission line in TEM mode, for example, coaxial line, $\lambda_c = \infty$.

The constants $C_i (i = 1, 2, 3, 4)$ in Eqs. (4.1)–(4.3) can be determined from the boundary conditions on the electric field and the magnetic field. The boundary condition on the electric field is the continuity of the tangential component at the interfaces:

$$E_I|_{x=L_1} = E_{II}|_{x=L_1} \quad (4.6)$$

$$E_{II}|_{x=L_1+L} = E_{III}|_{x=L_1+L} \quad (4.7)$$

where L_1 and L_2 are the distances from the respective ports to the sample faces and L is the sample length. The total length of the transmission line is denoted as $L_{air} = L_1 + L_2 + L$. The boundary condition on the magnetic field requires the additional

assumption that no surface currents are generated, so the tangent component of magnetic field is continuous across the interface:

$$\frac{1}{\mu_0} \cdot \frac{\partial E_I}{\partial x}\bigg|_{x=L_1} = \frac{1}{\mu_0 \mu_r} \cdot \frac{\partial E_{II}}{\partial x}\bigg|_{x=L_1} \quad (4.8)$$

$$\frac{1}{\mu_0 \mu_r} \cdot \frac{\partial E_{II}}{\partial x}\bigg|_{x=(L_1+L)} = \frac{1}{\mu_0} \cdot \frac{\partial E_{III}}{\partial x}\bigg|_{x=(L_1+L)} \quad (4.9)$$

The scattering parameters of the two-port network shown in Figure 4.1 can be obtained by solving Eqs. (4.1)–(4.3) subject to the boundary conditions (Eqs. (4.6)–(4.9)). As the scattering matrix is symmetric ($S_{12} = S_{21}$), we have (Baker-Jarvis 1990)

$$S_{11} = R_1^2 \cdot \frac{\Gamma(1 - T^2)}{1 - \Gamma^2 T^2} \quad (4.10)$$

$$S_{22} = R_2^2 \cdot \frac{\Gamma(1 - T^2)}{1 - \Gamma^2 T^2} \quad (4.11)$$

$$S_{21} = R_1 R_2 \cdot \frac{\Gamma(1 - T^2)}{1 - \Gamma^2 T^2} \quad (4.12)$$

where R_1 and R_2 are the reference plane transformations at two ports:

$$R_i = \exp(-\gamma_0 L_i) \quad (i = 1, 2) \quad (4.13)$$

The transmission coefficient T is given by

$$T = \exp(-\gamma L) \quad (4.14)$$

The reflection coefficient Γ is given by

$$\Gamma = \frac{(\gamma_0/\mu_0) - (\gamma/\mu)}{(\gamma_0/\mu_0) + (\gamma/\mu)} \quad (4.15)$$

For coaxial line, the cutoff wavelength is infinity, so Eq. (4.15) can be rewritten as

$$\Gamma = \frac{\sqrt{\mu_r/\varepsilon_r} - 1}{\sqrt{\mu_r/\varepsilon_r} + 1} \quad (4.16)$$

Additionally, S_{21} for the empty sample holder is

$$S_{21}^0 = R_1 R_2 \exp(-\gamma_0 L) \quad (4.17)$$

For nonmagnetic materials, Eqs. (4.10), (4.11), and (4.12) contain ε_r', ε_r'', L, and the reference plane transformations R_1 and R_2 as unknown quantities. We have four complex Eqs. (4.10), (4.11), (4.12), and (4.17), plus the equation for the length of the air line, so we have equivalently nine real equations for the five unknowns. In many applications we know the sample length. For magnetic materials we have seven unknowns. Thus, the system of equations is overdetermined and it is possible to solve the equations in various combinations. Therefore, the complex relative permittivity (ε_r) and complex relative permeability (μ_r) of the sample can be determined using different ways. In the following, we discuss the algorithms often used for the calculation of ε_r and μ_r.

4.1.2 Nicolson–Ross–Weir (NRW) algorithm

Nicolson and Ross (1970) and Weir (1974) combined the Eqs. (4.10) and (4.11) for S_{11} and S_{21}, and derived explicit formulas for the calculation of permittivity and permeability. The algorithm is usually called Nicolson–Ross–Weir (NRW) algorithm.

In the NRW algorithm, the reflection and transmission are expressed by the scattering parameters S_{11} and S_{21}. The reflection coefficient Γ is given by

$$\Gamma = K \pm \sqrt{K^2 - 1} \qquad (4.18)$$

with

$$K = \frac{(S_{11}^2 - S_{21}^2) + 1}{2S_{11}} \qquad (4.19)$$

The correct choice of positive or negative sign in Eq. (4.18) is made by requiring $|\Gamma| \leq 1$. The transmission coefficient T is given by

$$T = \frac{(S_{11} + S_{21}) - \Gamma}{1 - (S_{11} + S_{21})\Gamma} \qquad (4.20)$$

So the permittivity and permeability are calculated from

$$\mu_r = \frac{1 + \Gamma}{(1 - \Gamma)\Lambda\sqrt{(1/\lambda_0^2) - (1/\lambda_c^2)}} \qquad (4.21)$$

$$\varepsilon_r = \frac{\lambda_0^2}{\mu_r[(1/\lambda_c^2) - (1/\Lambda^2)]} \qquad (4.22)$$

with

$$\frac{1}{\Lambda^2} = -\left[\frac{1}{2\pi D} \ln\left(\frac{1}{T}\right)\right]^2 \qquad (4.23)$$

where λ_0 is the free-space wavelength; λ_c is the cutoff wavelength of the transmission line section, and for a coaxial line $\lambda_c = \infty$.

It should be noted that Eq. (4.23) has an infinite number of roots since the imaginary part of a complex quantity (T) is equal to the angle of the complex value plus $2\pi n$, where n is an integer. Equation (4.23) is ambiguous because the phase of the transmission coefficient T does not change when the length of the material is increased by a multiple of wavelength.

As the group delay through the material is strictly a function of the total length of the sample, the phase ambiguity can be resolved by finding a solution for ε_r and μ_r from which a value of group delay is computed that corresponds to the value determined from measurement data at two or more frequencies. For this method to work, the discrete frequency steps at which measurements are made must be small enough so that the phase of the transmission coefficient (T) changes less than 2π from one measurement frequency to the next. The group delay at each frequency may be computed for each solution of ε_r and μ_r, assuming that the changes in ε_r and μ_r are negligible over a small increment of frequency:

$$\tau_{g,n} = L\frac{d}{df}\left(\frac{\varepsilon_r\mu_r}{\lambda_0^2} - \frac{1}{\lambda_c^2}\right)_n^{\frac{1}{2}} \qquad (4.24)$$

where f is the frequency and $\tau_{g,n}$ is the group delay for the nth solution of Eqs. (4.21) and (4.22). The measured group delay is determined from the slope of the phase of the transmission coefficient versus frequency:

$$\tau_g = -\frac{1}{2\pi} \cdot \frac{d\phi}{df} \qquad (4.25)$$

where ϕ is the phase in radians of T. The correct root ($n = k$) should satisfy

$$\tau_{g,k} - \tau_g = 0 \qquad (4.26)$$

Besides the group-delay method, the problem of phase ambiguity can also be solved using phase-unwrapping method (Hock 2002). It is possible to

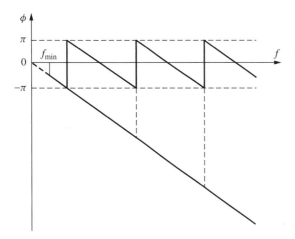

Figure 4.2 Wrapping and unwrapping of the phase of the transmission coefficient

divide this problem into two parts: determination of the initial phase, and phase unwrapping. Consider a sample with constant permittivity over a wide frequency range, such as Teflon. Figure 4.2 shows the phase of T calculated from Eq. (4.20), which gives an answer between $\pm\pi$. It is clear from the figure that phase ambiguity arises from the phase-wrapping effect. Obtaining the correct additive constant of $2n\pi$ for the logarithm in Eq. (4.22) is then equivalent to phase unwrapping. The unwrapped phase is also shown in Figure 4.2.

As the dielectric permittivity of an actual sample may be frequency dependent, it is not easy to determine the initial value correctly. One way to resolve this is to choose the starting working frequency (f_{min}) as low as possible, so that the calculated phase can be seen to extrapolate to the origin. In a low-loss sample, assuming that ε_r and μ_r are fairly constant at low frequencies, the maximum values of ε_r and μ_r that can be determined unambiguously with a single sample is

$$\varepsilon_r\mu_r = \left(\frac{c}{2lf_{min}}\right)^2 \qquad (4.27)$$

where f_{min} is the lowest frequency measurable and l is the length of the sample. Equation (4.27) is based on the assumption that the frequency of the first wrapping of phase is equal to the starting working frequency (f_{min}).

The phase-unwrapping can be done in a simple way. The phase can be unwrapped by detecting a jump in phase value of, say, more than π from one measurement frequency to the next, and then shifting all the subsequent phases by 2π in the opposite direction. This is a commonly used phase-unwrapping method. As long as the noise is less than π, such an unwrapping is reliable.

Besides the phase ambiguities discussed above, as will be discussed later, spurious peaks for permittivity and permeability can be observed when the sample length is a multiple of half wavelength of the microwave within the sample. In most of the transmission/reflection measurements, the determinations of permittivity and permeability are based on the NRW algorithm. To get the correct permittivity and permeability results, the phase ambiguities and spurious peaks should be eliminated. In the following, we discuss three methods to solve these problems.

4.1.3 Precision model for permittivity determination

The NRW algorithm does not work well at frequencies where the sample length is a multiple of half wavelength in the material. The solid line in Figure 4.3 displays the typical results calculated from the Nicolson–Ross–Weir equations. At frequencies corresponding to integer multiples of half wavelength in low-loss materials, the scattering parameter $|S_{11}|$ becomes very small. The equations are algebraically unstable as $S_{11} \to 0$; meanwhile, for small $|S_{11}|$, the uncertainty in phase measurement is large. To bypass this problem, many researchers resort to using short samples. However, use of short samples lowers the measurement sensitivity. In fact, to minimize the uncertainty in low-loss materials a relatively long sample is preferred (Baker-Jarvis 1990; Baker-Jarvis *et al.* 1990).

Several ways can be used in solving the scattering equations depending on the information available. If the sample length and the positions of the reference planes are exactly known, by taking various linear combinations of the scattering equations and solving the equations in an iterative fashion, we can get stable solutions on samples of

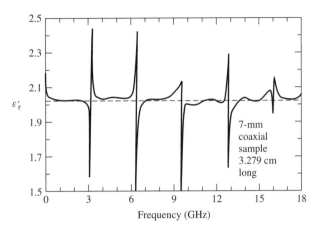

Figure 4.3 The determination of the permittivity of a PTFE sample as a function of frequency using Nicolson–Ross–Weir equations (solid line) and the iteration procedure (dashed line) (Baker-Jarvis 1990). Source: Baker-Jarvis, J. (1990), *Transmission/Reflection and Short-circuit Line Permittivity Measurements*, NIST Technical Note, National Institute of Standards and Technology Boulder, CO

arbitrary length. A useful combination is (Baker-Jarvis 1990)

$$\frac{1}{2}[(S_{12} + S_{21}) + \beta(S_{11} + S_{22})]$$

$$= \frac{T(1 - \Gamma^2) + \beta\Gamma(1 - T^2)}{1 - P^2\Gamma^2} \quad (4.28)$$

In Eq. (4.28), the S-parameters to be used need to be transformed from the calibration planes to the sample faces. Here β is a constant that varies as a function of the sample length, uncertainty in scattering parameters, and loss characteristics of the material. For low-loss materials, the S_{21} signal is strong and so we can set β equal to zero, whereas for high-loss materials, S_{11} dominates and a large value of β is appropriate. A general relation for β is given by the ratio of the uncertainty in S_{21} divided by the uncertainty in S_{11}. In Figure 4.3, the iterative solution (in the dashed line) is compared to the Nicolson–Ross–Weir procedure for a sample of polytetrafluoroethylene (PTFE) in 7-mm coaxial line. The contrast between the two solutions is striking.

The measurement uncertainty for this method has been analyzed (Baker-Jarvis 1990). Generally speaking, for low-loss materials, the uncertainty decreases as a function of increasing sample length; for high-loss materials the uncertainty increases as the sample length increases.

4.1.4 Effective parameter method

Boughriet *et al.* proposed a way to suppress the inaccuracy peaks at multiple half-wavelength frequencies, by introducing effective electromagnetic parameters ε_{eff} and μ_{eff} that satisfy the following equations (Boughriet *et al.* 1997):

$$\Gamma = \frac{\sqrt{\mu_{eff}/\varepsilon_{eff}} - 1}{\sqrt{\mu_{eff}/\varepsilon_{eff}} + 1} \quad (4.29)$$

$$T = \exp\left(-j\frac{2\pi}{\lambda_{0g}}\sqrt{\mu_{eff}\varepsilon_{rff}}L\right) \quad (4.30)$$

$$\gamma = \gamma_0\sqrt{\varepsilon_{eff}\mu_{eff}} \quad (4.31)$$

$$Z = Z_0\sqrt{\frac{\mu_{eff}}{\varepsilon_{eff}}} \quad (4.32)$$

with

$$\lambda_{0g} = \frac{1}{\sqrt{(1/\lambda_0^2) - (1/\lambda_c^2)}} \quad (4.33)$$

Though ε_{eff} and μ_{eff} presuppose a TEM propagation mode in the measurement cell, the following conclusions obtained can be extended to other measurement cells, such as microstrip or coplanar lines, or rectangular waveguide with TM_{01} propagation mode (Boughriet *et al.* 1997).

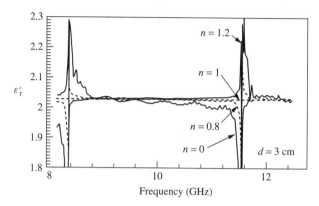

Frequency (GHz)

Figure 4.4   Influence of the n parameter on the calculated real permittivity for PTFE sample (observed for $n = 0$ and, respectively, $n = 1 + x$, $n = 1 - x$ with $x = 0.2$) (Boughriet *et al.* 1997). Source: Boughriet, A. H. Legrand, C. Chapoton, A. (1997), "Noniterative stable transmission/reflection method for low-loss material complex permittivity determination", *IEEE T Microw Theory*, **45** (1) 52–57. © 2003 IEEE

From Eqs. (4.29) and (4.30), we can get

$$\mu_{\text{eff}} = \frac{\lambda_{0g}}{\Lambda} \cdot \frac{1 + \Gamma}{1 - \Gamma} \tag{4.34}$$

$$\varepsilon_{\text{eff}} = \frac{\lambda_{0g}}{\Lambda} \cdot \frac{1 - \Gamma}{1 + \Gamma} \tag{4.35}$$

From Eqs. (4.31) and (4.32), we can get the relationship between the effective parameters and sample properties:

$$\mu_r = \mu_{\text{eff}} \tag{4.36}$$

$$\varepsilon_r = \left(1 - \frac{\lambda_0^2}{\lambda_c^2}\right)\varepsilon_{\text{eff}} + \frac{\lambda_0^2}{\lambda_c^2} \cdot \frac{1}{\mu_{\text{eff}}} \tag{4.37}$$

On the basis of the above preparation, Boughriet *et al.* established a new expression for the determination of permittivity of dielectric materials. For a dielectric material ($\mu_r = \mu_{\text{eff}} = 1$), from Eqs. (4.34) and (4.35), we can get

$$\varepsilon_{\text{eff}} = \varepsilon_{\text{eff}}\mu_{\text{eff}} = \frac{\lambda_{0g}^2}{\Lambda^2} \tag{4.38}$$

In Eq. (4.38), as the term $[(1 - \Gamma)/(1 + \Gamma)]$ in Eqs. (4.34) and (4.35), which is the source of source of the inaccuracy peaks, is eliminated, so the inaccuracy peaks are suppressed (Boughriet *et al.* 1997).

For dielectric material, a more general equation can be obtained by combining Eqs. (4.34) and (4.35):

$$\varepsilon_{\text{eff}} = \varepsilon_{\text{eff}}(\mu_{\text{eff}})^n = \left(\frac{1 - \Gamma}{1 + \Gamma}\right)^{n-1} \cdot \left(\frac{\lambda_{0g}}{\Lambda}\right)^{n+1} \tag{4.39}$$

The exponent n is a real number and can be positive or negative. This general equation includes the Stuchly method (with $n = -1$) (Stuchly and Matuszewski 1978), the NRW method (with $n = 0$), and Eq. (4.38) (with $n = 1$). Figure 4.4 shows the calculated permittivity of PTFE versus the frequency at various values of the n parameters. We find that inaccuracy peak amplitudes decrease as n approaches the value of one, and completely disappear when $n = 1$. Figure 4.4 also shows a symmetry in the inaccuracy peaks for the cases $n = 1 + x$ and $n = 1 - x$. Detailed uncertainty analysis can be found in Boughriet *et al.* (1997).

4.1.5 Nonlinear least-squares solution

Usually, the determination of permittivity and permeability by the reduction of scattering data is on a frequency-by-frequency or point-by-point basis, that is, by the explicit or implicit solution of a system of nonlinear scattering equations at each particular frequency. The point-by-point reduction techniques for magnetic materials contain large random uncertainties because of the propagation of uncertainties through the equations.

Baker-Jarvis *et al.* proposed a viable alternative solution using nonlinear processes, which minimize the square error (Baker-Jarvis *et al.* 1992).

This optimization-based data reduction has an advantage over point-by-point schemes in that correlations are allowed between frequency measurements. In nonlinear regression, if deemed appropriate, it is not necessary to even include S_{11} in the set of constraint equations. Another advantage of regression is that constraints such as causality and positivity can be incorporated into the solution.

This method is suitable for obtaining complex permittivity and permeability spectra of isotropic, homogeneous materials, from scattering parameters, using a nonlinear regression model. The scattering equations are solved in a nonlinear least-squares sense with a regression algorithm over the entire frequency measurement range. The complex permittivity and permeability are obtained by determining the coefficients of a pole-zero model for these parameters consistent with linearity and causality constraints. The method can be extended to the analysis of multimode problems and the determination of systematic uncertainty of measurements (Baker-Jarvis *et al.* 1992).

In this method, an explicit frequency-dependent form for ε_r and μ_r is used for the determination of ε_r and μ_r. Usually, the general form for ε_r and μ_r should satisfy a Kramers–Kronig relation. The zeros and poles of a complex function determine the function.

The Laplace transform of the real, time-dependent permittivity satisfies

$$\varepsilon(r, s) = \int_0^\infty \varepsilon(r, t)e^{-st}dt \qquad (4.40)$$

For stability, there can be no poles in the right of the *s*-plane. Since $\varepsilon(t)$ is real, the poles and zeros are confined to the negative real *s*-axis of the *s*-plane, and the poles which are off the real *s*-axis must occur in complex conjugate pairs (Baker-Jarvis *et al.* 1992).

According to the Debye model, the permittivity of a material can be expressed in the form

$$\varepsilon_r = \frac{1}{1 + jB\omega} + C \qquad (4.41)$$

where B and C are constants and ω is the angular frequency. On the basis of the Debye model, we can assume a series of poles of the first and second order for ε_r and μ_r:

$$\mu_r(\omega) = A_0 + \sum_i \frac{A_{1i}}{1 + jB_{1i}\omega}$$

$$+ \sum_i \frac{A_{2i}}{(1 + jB_{2i}\omega)^2} \qquad (4.42)$$

$$\varepsilon_r(\omega) = D_0 + \sum_i \frac{A_{3i}}{1 + jB_{3i}\omega}$$

$$+ \sum_i \frac{A_{4i}}{(1 + jB_{4i}\omega)^2} \qquad (4.43)$$

The above expansions can yield excellent results where B_i are real numbers. The poles should all reside in the left half plane. In the algorithm, a couple of poles per frequency decade is assumed.

The approach for determining the parameters A_j and B_j is to minimize the sum of the squares of the differences between the predicted and observed *S*-parameters (Baker-Jarvis *et al.* 1992):

$$\min \left\| \sum_{ij} (S_{ij} - P_{ij}) \right\| \qquad (4.44)$$

where the measured *S*-parameter vector is denoted by

$$S_{ij} = (S_{ij}(\omega_1), S_{ij}(\omega_2), \ldots, S_{ij}(\omega_n)) \qquad (4.45)$$

and the predicted *S*-parameter vector is expressed as

$$P_{ij} = (P_{ij}(\omega_1), P_{ij}(\omega_2), \ldots, P_{ij}(\omega_n)) \qquad (4.46)$$

So the problem becomes one of finding the minimum normal solution to these equations. A comparison of the optimized solution to a point-by-point solution is shown in Figure 4.5. For more complicated polarization phenomena, other relations for permittivity can be used.

This technique has been successful for many isotropic magnetic and relatively high-dielectric-constant materials. The reflection data (S_{11}) are usually of lesser quality than the transmission data (S_{21}) for low-loss, low-permittivity materials. Therefore, it is not necessary to include S_{11} in the solution for low-loss materials. The use of analytic

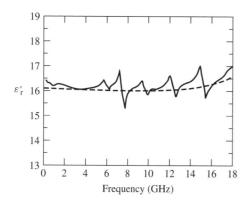

Figure 4.5 Permittivity for a leaded glass over 0.045–18 GHz for the optimized solution (dashed line) and point-by-point techniques (solid line) (Baker-Jarvis *et al.* 1992). Source: Baker-Jarvis, J. Geyer, R. G. and Domich, P. D. (1992). "A nonlinear least-squares solution with causality constraints applied to transmission line permittivity and permeability determination", *IEEE Transactions on Instrumentation and Measurement*, **41** (5), 646–652. © 2003 IEEE

functions for the expansion functions allows a correlation between the real and imaginary parts of the permittivity and permeability.

This method can be used to reduce the scattering data of high-dielectric-constant materials. In some cases, the optimized procedure yields solutions when the point-by-point technique fails completely. The optimized technique can also be used to treat problems where sample lengths, sample holder lengths, and sample positions are not known to be highly accurate. More discussions on this method can be found in Baker-Jarvis *et al.* (1992).

4.2 COAXIAL AIR-LINE METHOD

Coaxial air line is the most widely used transmission line in the characterization of permittivity and permeability of materials. In a coaxial air-line method, the toroidal sample is inserted between the inner and outer conductors of the coaxial line. Coaxial air-line method has obvious advantages in that it, theoretically, can work down to zero frequency and can cover a wide frequency range. In the choice of a coaxial air line, we should consider the characteristic impedance and the working frequency range.

4.2.1 Coaxial air lines with different diameters

As discussed in Chapter 2, the characteristic impedance Z_c of a coaxial line is given by

$$Z_c = \frac{60}{\sqrt{\varepsilon_r}} \ln\left(\frac{b}{a}\right) \qquad (4.47)$$

where b and a are the outer and inner radii of the coaxial line and ε_r is the relative dielectric permittivity of the insulator filled between the inner and outer conductors. As in most measurement systems, the characteristic impedance of the measurement circuit is 50 Ω and the impedance of coaxial line is often chosen as 50 Ω. To decrease the insertion loss of a coaxial line, low-loss dielectric, usually Teflon, is used as the insulator between the inner and outer conductors.

For a coaxial air line often used in materials property characterization, the insulation dielectric between the inner and outer conductors is air. From Eq. (4.47), we can get the relationship between the inner and outer diameters of a 50 Ω coaxial line:

$$b = 2.3 \cdot a \qquad (4.48)$$

Usually, a coaxial line is named according to its outer diameter, for example, 3.5-mm coaxial line, 7-mm coaxial line, and 14-mm coaxial line.

Besides the fundamental TEM mode, TE and TM modes can also propagate in a coaxial cable. To ensure the measurement accuracy and sensitivity, the coaxial line for materials property characterization should work in a pure TEM mode. The working frequency ranges for coaxial air lines with different dimensions are listed in Table 4.1. The coaxial air line with smaller outer diameter has wider working frequency range, but it has more strict requirements on sample fabrication than the coaxial air line with larger outer diameter.

Table 4.1 The working frequency ranges of 50 Ω coaxial air lines with different outer diameters

Outer diameter (mm)	Working frequency range (GHz)
3.5	0–34.5
7.0	0–18.2
14.0	0– 8.6

For composite samples with large size inclusions, coaxial air lines with larger outer diameters are more suitable.

4.2.2 Measurement uncertainties

As transmission/reflection methods have closed-form solutions for the calculation of complex permittivity and permeability, its uncertainty analysis can be conducted systematically. The uncertainty sources of transmission/reflection method mainly include algorithm uncertainty, air gap, uncertainty of sample position, and uncertainties of S-parameter measurement.

4.2.2.1 Differential analysis of algorithm

Differential analysis on the algorithms for transmission/reflection method has been conducted by many researchers, including Baker-Jarvis (1990) Baker-Jarvis *et al.* (1993), Smith (1995) and Youngs (1996). As an example, we discuss Nicolson–Ross–Weir algorithm. In Eqs. (4.18)–(4.23), five parameters are used to deduce the complex permittivity and complex permeability: amplitude and phase of complex reflection (S_{11} and ϕ_{11}), amplitude and phase of complex transmission (S_{21} and ϕ_{21}), and the thickness of sample D. In the uncertainty analysis, we should consider the contributions from each of the five parameters. The uncertainties of real and imaginary parts of permittivity and permeability can be generally expressed as

$$U_{\varepsilon_r'} = \sqrt{\sum_{n=1}^{n} \mathrm{Re}\left(\frac{\partial \varepsilon_r}{\partial e_i} \cdot U_{e_i}\right)^2} \qquad (4.49)$$

$$U_{\varepsilon_r''} = \sqrt{\sum_{n=1}^{n} \mathrm{Im}\left(\frac{\partial \varepsilon_r}{\partial e_i} \cdot U_{e_i}\right)^2} \qquad (4.50)$$

$$U_{\mu_r'} = \sqrt{\sum_{n=1}^{n} \mathrm{Re}\left(\frac{\partial \mu_r}{\partial e_i} \cdot U_{e_i}\right)^2} \qquad (4.51)$$

$$U_{\mu_r''} = \sqrt{\sum_{n=1}^{n} \mathrm{Im}\left(\frac{\partial \mu_r}{\partial e_i} \cdot U_{e_i}\right)^2} \qquad (4.52)$$

where e_i are S_{11}, ϕ_{11}, S_{21}, ϕ_{21} and D; U_a indicates the uncertainty of parameter a; and parameter

a refers to the real and imaginary parts of permittivity and permeability, or parameters e_i.

4.2.2.2 Effect of air gaps

For a coaxial air-line method, the uncertainties caused by the air gaps are serious. As shown in Figure 4.6, in a coaxial measurement cell, there may exist air gaps between the sample and the inner conductor, and between the sample and the outer conductor.

The effects of air gaps, shown in Figure 4.6, to the measurement results of permittivity and permeability can be analyzed using the layered capacitor model (Baker-Jarvis *et al.* 1993). In this model, the air gaps are assumed to be uniform, with circular symmetry, and the segment of coaxial air-line filled with sample can be taken as capacitors in series. The relationships between the measurement results (ε_m and μ_m) and the corrected results (ε_c and μ_c) are given by (Youngs 1996)

$$\varepsilon_c' = \varepsilon_m' \cdot \left(\frac{L_2}{L_3 - \varepsilon_m' L_1}\right) \qquad (4.53)$$

$$\varepsilon_c'' = \left(\varepsilon_c' \frac{\varepsilon_m''}{\varepsilon_m'}\right) \cdot \frac{L_3}{L_3 - L_1 \varepsilon_m'[1 + (\varepsilon_m''/\varepsilon_m')^2]} \qquad (4.54)$$

$$\mu_c' = \mu_m' \cdot \frac{L_3 - L_1}{L_2} \qquad (4.55)$$

$$\mu_c'' = \mu_m'' \cdot \frac{L_3}{L_2} \qquad (4.56)$$

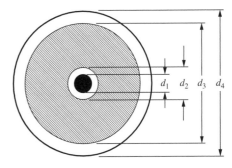

Figure 4.6 Air gaps in a coaxial-line sample holder. Source: Baker-Jarvis, J. Janezic, M. D. Grosvenor, J. H. Jr. and Geyer, R. G. (1993). *Transmission/Reflection and Short-circuit Line Methods for Measuring Permittivity and Permeability*, NIST Technical Note 1355 (revised), National Institute of Standards and Technology, Boulder, CO

where

$$L_1 = \ln\left(\frac{d_2}{d_1}\right) + \ln\left(\frac{d_4}{d_3}\right) \qquad (4.57)$$

$$L_2 = \ln\left(\frac{d_4}{d_1}\right) \qquad (4.58)$$

$$L_3 = \ln\left(\frac{d_3}{d_2}\right) \qquad (4.59)$$

The correction factor is taken as the ratio of the corrected component to the measured uncorrected component. Figure 4.7 shows the relationship between the correction component of dielectric constant and the air gap between the inner conductor and the sample. It shows that when the gap between the inner conductor and the sample increases, the correction component increases; and when the dielectric constant value increases, the correction factor increases. Similar conclusions can also be obtained for the correction components of other components of materials' intrinsic properties.

It should be noted that the air gap between the inner conductor and the sample could cause larger uncertainties than the one between the sample and the outer conductor. As the electric and magnetic field are concentrated near the inner conductor, even a small air gap between the inner

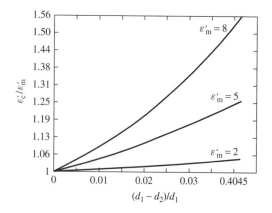

Figure 4.7   The gap correction calculated for various values of dielectric constants, where d_1 and d_2 are defined in Figure 4.6 (Baker-Jarvis *et al.* 1993). Source: Baker-Jarvis, J. Domich, M. D. and Geyer, R. G. (1993). *Transmission/Reflection and Short-circuit Line Methods for Measuring Permittivity and Permeability*, NIST Technical Note 1355 (revised), National Institute of Standards and Technology, Boulder, CO

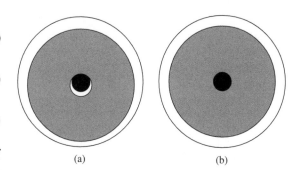

(a)   (b)

Figure 4.8   Air gaps in a coaxial air line filled with a sample. (a) A sample with an air gap between the inner conductor and the sample, and an air gap between the outer conductor and the sample and (b) a sample without an air gap between the inner conductor and the sample, but with an air gap between the outer conductor and the sample

conductor and the sample may result in large errors. Furthermore, as shown in Figure 4.8(a), because of the existence of the air gap between the inner conductor and the sample, the sample may be not symmetrical in the coaxial air line, so the above corrections could not be applied. Therefore, we should try to eliminate the air gap between the inner conductor and the sample. Besides, in experiments, the outer diameter of the sample is often made a little smaller than the inner diameter of the outer conductor of the coaxial air-line, as shown in Figure 4.8(b), so that the sample could be easily inserted into the measurement fixture, and the effect of the air gap between the sample and outer conductor can be corrected using Eqs. (4.53)–(4.59).

4.2.2.3 Effect of sample placement

In actual measurements, there may be some distances between the sample ends and the calibration planes. A phase correction is required when any length of transmission line is added beyond the calibration plane. For the general case shown in Figure 4.9, the phase corrections for S_{11} and S_{21} are given by

$$\Delta\phi_{11} = 2a \cdot \frac{2\pi f}{c} \qquad (4.60)$$

$$\Delta\phi_{21} = (a + b) \cdot \frac{2\pi f}{c} \qquad (4.61)$$

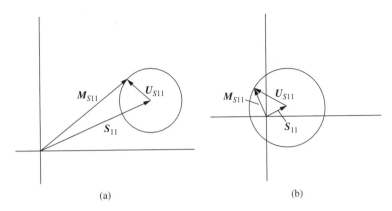

(a)   (b)

Figure 4.10   Effects of measurement uncertainties of S_{11}. (a) $|S_{11}| > |U_{S11}|$ and (b) $|S_{11}| < |U_{S11}|$

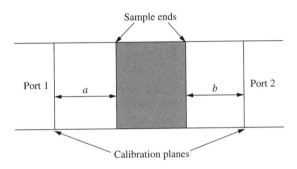

Figure 4.9   Sample placement in the sample holder

where f is the measurement frequency and c is speed of light in vacuum.

In most of the algorithms for transmission/reflection methods, S_{11} and S_{21} are used in the calculation of the properties of materials. So the uncertainties of the sample position in the transmission line may cause uncertainties to the measurement results. Meanwhile, it should be noted that only the phases of S_{11} and S_{21} varies with the position of the sample within the transmission line, provided that the overall length of the transmission line is accurately known.

4.2.2.4 S-parameter measurement uncertainties

The measurement uncertainties of S-parameters directly cause uncertainties of permittivity and permeability values. The uncertainties of S-parameters can be minimized by calibration, but cannot be eliminated because of the limited dynamic range in an actual measurement instrument.

Figure 4.10 schematically shows the effects of measurement uncertainties of S_{11}. In the figure, the actual S_{11} value is denoted as S_{11}, the measurement uncertainty of S_{11} is denoted as U_{S11}, and the measured value of S_{11} is denoted as M_{S11}. The vector M_{S11} is the vector sum of S_{11} and U_{S11}. The amplitude and phase difference between the measured S_{11} and actual S_{11} is due to the measurement uncertainty of S_{11}. When $|S_{11}| > |U_{S11}|$ as shown in Figure 4.10(a), the amplitude and phase differences between the measured S_{11} and the actual S_{11} are not large. Whereas, if $|S_{11}| < |U_{S11}|$ as shown in Figure 4.10(b), the amplitude and phase differences between the measured S_{11} and actual S_{11} are large, and such differences cause great uncertainties in the calculation of permittivity and permeability. In transmission/reflection method, when the sample length is an integral times of the half wavelength of the microwave in the sample, the value of S_{11} is very small, so the uncertainties of S_{11} cause great uncertainties in the results of permittivity and permeability.

4.2.3 Enlarged coaxial line

For samples with coarse grains, such as soil, concrete, and rock, large samples are needed to average out the fluctuations in the dielectric properties of the heterogeneous materials. So, the coaxial lines need to be enlarged to host large samples. Two kinds of enlarged coaxial lines are often used: enlarged circular coaxial line and enlarged square coaxial lines.

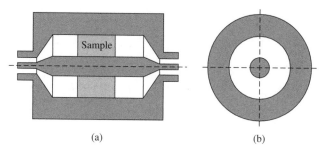

(a)   (b)

Figure 4.11   Schematic drawings of an enlarged circular coaxial line. (a) Longitudinal view and (b) cross-section view

4.2.3.1 Enlarged circular coaxial line

Chew *et al.* proposed an enlarged circular coaxial-line structure (Chew *et al.* 1991). As shown in Figure 4.11, the measurement cell mainly consists of three parts: transition from normal coaxial line to large coaxial line, large coaxial sample holder, and transition from large coaxial line to normal coaxial line.

In the design of an enlarged measurement cell, the outer conductor radius b is made as large as possible, while the inner conductor radius a is adjusted for a characteristic impedance of $50\,\Omega$. The restriction for this is that the value of $(b - a)$ should be no larger than the half-sample wavelength at the highest operating frequency to prevent the existence of higher-order propagating modes. Meanwhile, the desired sample length is as long as possible for low-frequency-phase accuracy, but short enough to avoid resonance and to ensure adequate transmission at high frequencies. For the measurement cell developed by Chew *et al.*, the diameter of the inner conductor is 1.535 cm, and the diameter of the outer conductor is 4.992 cm.

Similar to a conventional coaxial measurement cell, an enlarged coaxial measurement cell requires calibration. As calibration standards for large coaxial lines are not commercially available, special calibration method should be developed in the design of large coaxial measurement cell. As a coaxial line can cover a wide frequency range, in calibration and measurement procedures, we can divide the whole frequency range into several frequency subranges. Chew *et al.* proposed three methods for three frequency subranges respectively: low frequency (1–30 MHz),

medium frequency (30–800 MHz), and high frequency (800 MHz–3 GHz).

Standard "short" is often used in different kinds of calibration methods. A standard "short" should provide good electrical contact to both inner and outer conductors, as there are electrical currents flowing between the inner and outer conductors through the "short", and a small gap between the standard "short" and the inner or outer conductor may result in large errors. In the design of the structure of a large coaxial line, the calibration procedures and techniques should be taken into consideration.

4.2.3.2 Enlarged square coaxial line

Enlarged square coaxial line has also been developed for the characterization of electromagnetic properties of materials. As shown in Figure 4.12, the structure of an enlarged square coaxial line is similar to that of an enlarged circular coaxial line. The measurement cell also consists mainly of three parts: transition from normal coaxial line to large square coaxial line, large square coaxial sample holder, and transition from large square coaxial line to normal coaxial line. In the design of a square coaxial line, the side length b of the outer conductor and the side length a of the inner conductor should be chosen to ensure suitable characteristic impedance and to avoid higher-order propagation modes. The impedance of square coaxial line is usually chosen as $50\,\Omega$ or $60\,\Omega$. Similar to the enlarged circular coaxial line method, in the design of a square coaxial line, we should consider the calibration techniques used in the enlarged square coaxial line method.

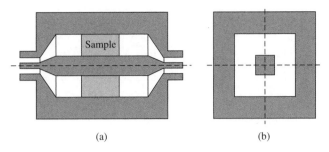

Figure 4.12   Schematic drawings of an enlarged rectangular coaxial line. (a) Longitudinal view and (b) cross view

The samples for square coaxial lines may consist of several pieces of rectangle-shaped materials. The fabrication of samples for square coaxial lines is usually easier than the fabrication of toroid-shaped samples for circular coaxial lines. Square coaxial lines are also ideal for the characterization of periodic structures such as pyramidal absorbers, honeycombs, circuit-analog (CA) sheets, and frequency selective surfaces (FSS). For these applications, square coaxial lines with 60-Ω characteristic impedance have obvious advantage. For a 60-Ω square coaxial line, the side length b of the outer conductor is about three times of the side length a of the inner conductor. To simulate a sample with infinite array, we can arrange eight pieces of square samples with side length b in the space between the outer and inner conductors as shown in Figure 4.12(b).

4.3 HOLLOW METALLIC WAVEGUIDE METHOD

Two types of hollow metallic waveguides are often used in microwave electronics: rectangular waveguide and circular waveguide. Owing to the possible degenerations in circular waveguides, rectangular waveguides are more widely used, while circular waveguides have advantages in the characterization of chiral materials, which will be discussed in Chapter 10. Here we focus on rectangular waveguides.

Compared to the samples for coaxial-line method, the samples for rectangular waveguide method can be fabricated more easily, while a coaxial-line method can cover much wider frequency range than a waveguide method.

4.3.1 Waveguides with different working bands

In most cases, the width a and height b of a rectangular waveguide satisfies $b/a = 1/2$, and the waveguide usually works at TE$_{10}$ mode. As discussed in Chapter 2, to ensure the single-mode requirement in materials property characterization, the wavelength should be larger than a and should be less than $2a$, so that for a given waveguide, there are limits for minimum frequency and maximum frequency. To ensure good propagations, about 10 % of the frequency range next to the minimum and maximum frequency limits is not used. Table 4.2 lists several bands of waveguides often used in microwave electronics and materials property characterization.

4.3.2 Uncertainty analysis

The uncertainties of waveguide transmission/reflection methods can be analyzed in a similar way as that

Table 4.2   Properties of rectangular waveguides often used in materials property characterization

Band	a (mm)	b (mm)	Cutoff frequency (GHz)	Maximum frequency (GHz)	Operating frequency range (GHz)
L	165.10	82.55	0.908	1.816	1.12–1.70
W	109.22	54.61	1.372	2.744	1.70–2.60
S	72.14	34.04	2.078	4.156	2.60–3.95
G	47.55	22.15	3.152	6.304	3.95–5.85
J	34.85	15.80	4.302	8.604	5.75–8.20
X	22.86	10.16	6.557	13.114	8.20–12.4
P	15.80	7.90	9.487	18.974	12.4–18.0
K	10.67	4.32	14.048	28.096	18.0–26.5
R	7.112	3.556	21.082	42.164	26.5–40.0

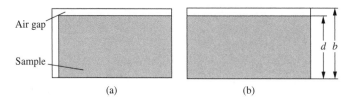

Figure 4.13   Air gaps in a waveguide sample holder. (a) Waveguide with air gaps along the width and height and (b) waveguide with air along the width only

conducted for coaxial-line transmission/reflection method. In the following, we discuss the uncertainties due to air gap and sample groove.

4.3.2.1 Air gap

As shown in Figure 4.13, for a waveguide filled with sample, there may exist air gaps along the width and height of the waveguide. As the effect of air gap along the height is not as serious as the one along the width, in the following discussion, we consider the case with air gap along the width only, as shown in Figure 4.13(b).

Many researchers have studied the effects of air gap by representing the sample with air gap as a layered capacitor (Champlin and Glover 1966). In this approach, the gaps between the transmission line and the sample are effectively modeled by a set of capacitors in series. It should be noted that this model is frequency independent and thus is strictly valid only at low frequencies, and it breaks down at high frequencies because the wavelength decreases with increasing frequency to a point where multiple scattering dominates.

In order to account for multiple scattering, Baker-Jarvis *et al.* studied the effect of air gap in a frequency-dependent approach (Baker-Jarvis *et al.* 1993). From the transverse resonance condition for the structure shown in Figure 4.13(b), we can get (Marcuvitz 1951)

$$\tan(k_{1c}d) + X \tan[k_{2c}(b - d)] = 0 \qquad (4.62)$$

with

$$k_{1c} = \frac{\omega}{c}\sqrt{\varepsilon_{rs} - \varepsilon_{r0}} \qquad (4.63)$$

$$k_{2c} = \frac{\omega}{c}\sqrt{\varepsilon_{rg} - \varepsilon_{r0}} \qquad (4.64)$$

$$X = \frac{\varepsilon_{rs}}{\varepsilon_{rg}} \cdot \frac{\sqrt{\varepsilon_{rg} - \varepsilon_{r0}}}{\sqrt{\varepsilon_{rs} - \varepsilon_{r0}}} \qquad (4.65)$$

where ε_{rs} is the dielectric constant of the sample, ε_{rg} is the dielectric constant of the gap, and ε_{r0} is the measured value of dielectric constant. By solving Eq. (4.62) using iteration method, we can get the dielectric constant ε_{rs} of the sample from the measured dielectric constant ε_{r0}. The relationship between ε_{r0} and ε_{rs} can also be approximately expressed by (Musil and Zacek 1986)

$$\varepsilon'_{rs} = \varepsilon'_{r0} \cdot \frac{d}{b - (b - d)\varepsilon'_{r0}} \qquad (4.66)$$

$$\tan \delta_s = \tan \delta \cdot \frac{b}{b - (b - d)\varepsilon'_{r0}} \qquad (4.67)$$

It should be noted that, if the air gaps in waveguide sample holders are very small, the errors due to the air gaps could essentially be eliminated by using a conducting paste to fill the gaps (Wilson 1988).

4.3.2.2 Sample groove

When a thin sample, much thinner than the transverse dimensions of the waveguide, is held in a waveguide measurement fixture, we may encounter an error source that is difficult to be removed (Luebbers 1993). This situation occurs when measuring the complex permittivity and permeability for very thin material samples at low frequencies. Since very large waveguides must be used, the sample thickness is only a small fraction of the waveguide transverse dimensions, and it is very difficult to reliably hold the samples perpendicular to the waveguide axis.

One approach for holding thin samples perpendicular to the measurement waveguide axis is to cut a small groove in the waveguide walls with the same thickness as the sample and just deep enough to hold the sample in place, as shown

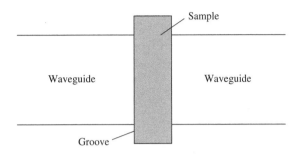

Figure 4.14 Waveguide with material sample in the groove

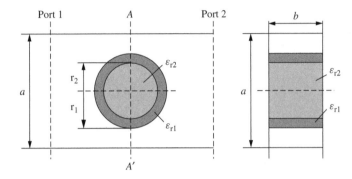

Figure 4.15 Dielectric rod within a cylindrical holder placed inside a rectangular waveguide. (a) Upper view and (b) cross view $A - A'$

in Figure 4.14. The inclusion of the waveguide groove for holding the sample introduces errors to the measurement results. These errors are mainly due to the excitation of higher-order modes in the sample and in the waveguides. These higher-order modes mainly have unwanted effects, as the determination of the complex constitutive parameters of the material assumes single-mode propagation through the waveguide.

In actual cases, it is usually difficult to ensure that the groove strictly fits the thin sample under study, for example, the thickness of the groove may be larger than the thickness of the sample. In these cases, the effects of the groove become more serious. Detailed discussion on the effects of the groove can be found in (Luebbers 1993).

4.3.3 Cylindrical rod in rectangular waveguide

Introducing a dielectric rod into a rectangular waveguide is a kind of variation of the conventional waveguide transmission/reflection method

(Abdulnour *et al.* 1995; Esteban *et al.* 2000). As shown in Figure 4.15, a rod dielectric sample is placed along the *y*-axis inside a rectangular waveguide. The dielectric properties of the rod sample can be calculated from the scattering parameters measured at port 1 and port 2. For liquid samples or granular samples, a sample holder is needed.

Using this method, the air gaps existing in conventional waveguide method are eliminated, and this method has wider working frequency ranges as the possible higher-order modes can be taken into consideration in the algorithms for the calculation of properties of materials. Abdulnour *et al.* (1995) proposed a genetic method based on a combination of the boundary integral equation technique with a modal expansion approach, while Esteban *et al.* (2000) proposed a hybrid iterative method. In this section, we discuss Esteban's method, and Abdulnour's method is discussed in Chapter 7 when this method is used in microstrip lines.

We consider the case when the dielectric pie inside the waveguide is filled with the material under test, as shown in Figure 4.15. In the simulation of the behavior of the multilayer dielectric rod placed inside a rectangular waveguide, a three-step analysis procedure is performed. First, the plates of the rectangular guide are split into smaller strips that are characterized individually by using a numerical method. Next, the multilayer dielectric rod is characterized using the analytic spectral method (Kolawole 1992). The scattering behavior of each element, including the strips and the multilayer rod, is provided by an "individual scattering matrix", which only relates incident and scattered spectral waves to that object. Finally, the electromagnetic coupling among all the elements previously characterized is solved by an iterative method, initially proposed by Esteban *et al.* (1997) for open-space problems, and which has been revisited by Esteban *et al.* (2000) for the accurate analysis of guided problems. Following this method, each scattering object is finally characterized by a "combined scattering matrix", which relates incident and scattered spectra to that object but taking into account the presence of the other scattering objects.

These combined scattering matrices give a full-wave solution of the structure in an open-space spectral domain. However, the scattering parameters of the guided structure are needed to compare with the measurement results. To obtain the desired scattering parameters, a short circuit is placed in the output reference plane (port 2 in Figure 4.15). The "combined scattering matrices" of the objects are used to compute the electric and magnetic fields in both the reference planes, taking the fundamental mode in the input reference plane as the excitation of the structure. Finally, the admittance parameters Y_{11} and Y_{21} can be expressed as

$$Y_{11} = \frac{\int_0^a \boldsymbol{H}_1 \cdot \boldsymbol{h}_i'' \mathrm{d}x}{\int_0^a \boldsymbol{E}_1 \cdot \boldsymbol{e}_i'' \mathrm{d}x}\Bigg|_{E_2=0} \qquad (4.68)$$

$$Y_{21} = \frac{\int_0^a \boldsymbol{H}_2 \cdot \boldsymbol{h}_i'' \mathrm{d}x}{\int_0^a \boldsymbol{E}_1 \cdot \boldsymbol{e}_i'' \mathrm{d}x}\Bigg|_{E_2=0} \qquad (4.69)$$

where $\boldsymbol{E}_1$, $\boldsymbol{H}_1$ and $\boldsymbol{H}_2$ are the electric and magnetic fields in the input and output reference planes, $\boldsymbol{e}_i''$ and $\boldsymbol{h}_i''$ are the TE_{10} vector mode functions (Marcuvitz 1951), x is a coordinate in the transversal direction, and a is the width of the waveguide.

Once the admittance matrix has been obtained, the scattering matrix is directly computed from the admittance matrix. More detailed information about the algorithm can be found in Esteban *et al.* (2000).

4.4 SURFACE WAVEGUIDE METHOD

Regarding geometrical structure and field distribution, surface waveguide is a type of transmission line between free space and closed transmission line such as coaxial line and hollow metallic waveguide. Surface waveguides have been used in the characterization of the electromagnetic properties of materials. In the following, we discuss the applications of circular and rectangular dielectric waveguides in the characterizations of materials properties.

4.4.1 Circular dielectric waveguide

Figure 4.16 shows a configuration for the measurement of complex permittivity using dielectric waveguides. The sample under test is placed between the ends of two dielectric waveguides, and it is in direct contact with the two waveguides. This configuration is suitable for the measurement of samples in the form of plates. It is usually required that the sample must be plane parallel. The sample may be of an arbitrary shape, but its transverse dimensions must be greater than, or in the limiting case equal to the transverse dimensions of the dielectric waveguide.

This system is based on two basic properties of dielectric waveguides (Musil and Zacek 1986). The first property is that the energy transferred in a dielectric waveguide is practically entirely concentrated inside the dielectric waveguide. The ability of dielectric waveguides to transfer practically the whole power inside themselves permits us to measure specimens with the transverse cross section practically equal to the cross section

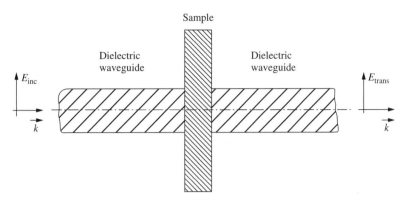

Figure 4.16 Principle of the nondestructive measurement of the complex permittivity with dielectric waveguides (Musil and Zacek 1986, p69). Reprinted from Musil, J. and Zacek, F. (1986). *Microwave Measurements of Complex Permittivity by Free Space Methods and Their Applications*, with permission from Elsevier, Amsterdam

of the dielectric waveguide. So, with the help of dielectric waveguides, we can measure plates of small transverse dimensions, and we can also make measurements on local inhomogeneities of large specimens.

The other property is that the phase velocity v_ϕ of the wave propagation in a dielectric waveguide is equal to the phase velocity of a plane wave in an unbounded dielectric medium:

$$v_\phi = c/\sqrt{\varepsilon_{\rm rd}} \qquad (4.70)$$

where $\varepsilon_{\rm rd}$ is the dielectric constant of the dielectric waveguide. So, we can use the algorithms discussed in the first section of this chapter for

evaluating the complex permittivity of the sample being measured.

However, it should be noted that these formulae may be used only for evaluating the complex permittivity of specimens whose dielectric constant is greater than, or at least equal to the dielectric constant of the material of which the dielectric waveguides are made. Only in this case, the phase velocity of the wave passing through the sample is equal to the phase velocity in the infinite dielectric medium whose dielectric constant equals the dielectric constant of the sample under test (Musil and Zacek 1986). This can be seen from Figure 4.17. Consider dielectric waveguides

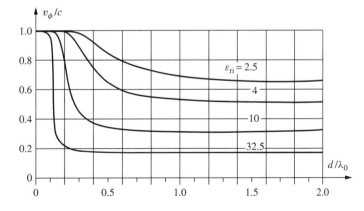

Figure 4.17 Dependence of the normalized phase velocity v_ϕ/c of the hybrid HE_{11} mode on the diameter d/λ_0 of a dielectric rod for different values of dielectric constant of the rod material (Musil and Zacek 1986, p67). Reprinted from Musil, J. and Zacek, F. (1986). *Microwave Measurements of Complex Permittivity by Free Space Methods and Their Applications*, with permission from Elsevier, Amsterdam

made of the material with dielectric constant $\varepsilon_{rd} > 10$ and diameter $d = 0.8\lambda_0$. For samples with $\varepsilon_{rs} \geq 10$, the phase velocity of waves passing through the specimen is given by Eq. (4.70). For samples with $\varepsilon_{rs} < 10$, the phase velocity of waves passing through the specimen is larger than what is given by Eq. (4.70). As the phase velocity of waves passing through the specimens with $\varepsilon_{rs} < \varepsilon_{rd}$ does not equal the phase velocity in an unbounded dielectric medium having the same dielectric constant ε_{rs}, the complex permittivity of such specimens cannot be accurately measured in this way.

In practice, dielectric waveguides of the circular shape are used most frequently. The design of a cylindrical dielectric waveguide can easily be performed by means of the dependences of v_ϕ/c on d/λ_0, as shown in Figure 4.17. The graphs of Figure 4.19 may also be used for approximate design of dielectric waveguides of noncircular cross sections by defining the d/λ_0 on the abscissa axis as

$$d/\lambda_0 = 1.13\sqrt{S}/\lambda_0 \qquad (4.71)$$

where S is the area of the transverse cross section of the waveguide (Schlesinger and King 1958).

4.4.2 Rectangular dielectric waveguide

A rectangular dielectric waveguide technique has been proposed to determine the dielectric constant of materials of various thickness and cross sections at the Q and W bands (Abbas *et al.* 1998a, b, 2001). In a similar manner to the cylindrical dielectric waveguide technique described above, a parallelepiped-shaped sample is placed in direct mechanical contact between two rectangular dielectric waveguides, as shown in Figure 4.18. Using this technique for the measurement of dielectric materials, the dielectric constant of the sample is usually determined iteratively from the effective refractive index measurements by using the solution of the wave equation. However, due to the open discontinuity problem between the rectangular dielectric waveguide and sample, this technique cannot give accurate value of loss tangent ($\tan\delta$), especially for low-loss samples.

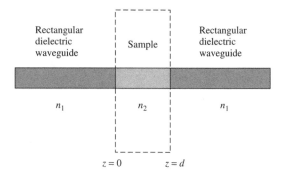

Figure 4.18   Effective index model of rectangular dielectric waveguide and sample

We follow Abbas's approach for measuring the dielectric constant and loss tangent of a sample using the rectangular dielectric waveguide technique (Abbas *et al.* 2001). In the following, we discuss the properties of rectangular dielectric waveguide, the calculation of the dielectric constant and loss tangent, and the measurement setup.

4.4.2.1 Rectangular dielectric waveguide

The propagation characteristics in rectangular dielectric waveguide and its derivatives have been studied for quite a long time. The solution of each mode lies between two extreme formulations given by Marcatili's method and the conventional effective index method (Abbas *et al.* 2001). Consider the E_{pq}^y mode propagating along the z-direction with the propagation constant β_z. The electric field is polarized along the y-direction with subscripts p and q indicating the number of extrema of the electric field in the y and x directions, respectively. In the conventional effective index method, the propagation constant β_z of the E_{pq}^y mode can be found by solving the following equations:

$$\beta_x a = p\pi - 2\tan^{-1}\left(\frac{\beta_x}{\beta_{x0}}\right) \qquad (4.72)$$

$$\beta_y a = q\pi - 2\tan^{-1}\left(\frac{\beta_y}{\beta_{y0}}\right) \qquad (4.73)$$

$$\beta_z^2 = k_0^2\varepsilon_r - \beta_x^2 - \beta_y^2 \qquad (4.74)$$

with

$$\beta_{x0} = [(\varepsilon_{eff} - 1)^2 k_0^2 - \beta_x^2]^{1/2} \qquad (4.75)$$

$$\beta_{y0} = [(\varepsilon_r - 1)^2 k_0^2 - \beta_y^2]^{1/2} \qquad (4.76)$$

$$k_0 = \frac{2\pi}{\lambda} \tag{4.77}$$

$$\varepsilon_{\text{eff}} = \varepsilon_{\text{r}} - \left(\frac{\beta_y}{k_0}\right)^2 \tag{4.78}$$

The effective index method reduces to Marcatili's method if Eq. (4.78) is replaced by

$$\varepsilon_{\text{eff}} = \varepsilon_{\text{r}}'. \tag{4.79}$$

Figure 4.19 shows the dispersion relation for E_{11}^y and E_{21}^y modes for a rectangular dielectric waveguide made of Teflon ($\varepsilon_{\text{r}} = 2.0 - j0.0002$) with dimensions equal to the Ka-band waveguide. It can be seen that according to the Marcatili's method, the E_{21}^y modes do not appear in the waveguide over the Ka-band frequency range. However, when using the effective index method, the cutoff frequency for the E_{21}^y mode is approximately 37 GHz. The latter is used as the condition for single-mode E_{11}^y propagation in the waveguide.

4.4.2.2 Calculation of dielectric constant and loss tangent

Using the Nicolson–Ross–Weir algorithm, the dielectric constant can be calculated from effective refractive index measurements from S-parameter data

$$n_{\text{eff}}^* = -\left(\frac{c}{\omega \text{d}} \ln \frac{1}{T}\right)^2 \tag{4.80}$$

where T is the complex transmission coefficient obtained from S-parameter measurement data. It is assumed that the sample has a homogeneous material composition and is nonmagnetic, linear, and isotropic. We also assume that only the single mode E_{11}^y propagates in the waveguide and the sample. The effective complex refractive index of the sample is defined as

$$n_{\text{eff}}^* = n_{\text{eff}} - jk_{\text{eff}} \tag{4.81}$$

where

$$n_{\text{eff}} = \left[\frac{1}{2}\left(\sqrt{\varepsilon_{\text{effR}}^2 + \varepsilon_{\text{effI}}^2} + \varepsilon_{\text{effR}}\right)\right]^{1/2} \tag{4.82}$$

$$k_{\text{eff}} = \left[\frac{1}{2}\left(\sqrt{\varepsilon_{\text{effR}}^2 + \varepsilon_{\text{effI}}^2} - \varepsilon_{\text{effR}}\right)\right]^{1/2} \tag{4.83}$$

where $\varepsilon_{\text{effR}}$ and $\varepsilon_{\text{effI}}$ are the real and imaginary parts of the effective complex permittivity respectively. The true dielectric constant ε_{r}' can be recovered iteratively from the effective refractive index n_{eff} by using the effective index method. Conversely, the effective index method can be used to calculate n_{eff} for given values of the cross section and the dielectric constant of a sample at a specified frequency.

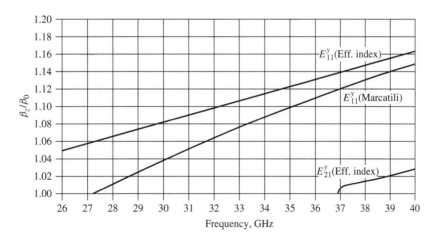

Figure 4.19 Variation in β_z/β_0 with frequency for E_{11}^y and E_{21}^y modes for the rectangular dielectric waveguide (Abbas *et al.* 2001). Source: Abbas, Z. Pollard, R. D. Kelsall, R. W. (2001). "Complex permittivity measurements at Ka-band using rectangular dielectric waveguide", *IEEE Transactions on Instrumentation and Measurement*, **50** (5), 1334–1342. © 2003 IEEE

A more accurate way to determine the true dielectric constant is by means of an optimization procedure, from which the loss tangent and accurate sample thickness can be determined by using a suitable objective function. Using the effective index model, we can express the reflection and transmission coefficients in simpler forms compared to other solutions to the discontinuity problem in an open dielectric waveguide:

$$\Gamma = \frac{1 - n_{\text{eff}}^*}{1 + n_{\text{eff}}^*} \tag{4.84}$$

$$T = \exp\left(-j\frac{\omega d}{c} n_{\text{eff}}^*\right) \tag{4.85}$$

The following objective function F is the most efficient to determine ε_r', ε_r'' and d (Abbas *et al.* 2001):

$$F = \sum_{i=1}^{a} \left\{\left[\ln\left(\frac{1}{|T_{\text{m}}|}\right) - \ln\left(\frac{1}{|T_{\text{c}}|}\right)\right]^2\right.$$
$$\left. + [\angle\phi_{\text{m}} - \angle\phi_{\text{c}}]^2\right\} \tag{4.86}$$

where $|T|$ and $\angle\phi$ are the magnitude and principal value of the phase angle of the transmission coefficient, a is the number of measurement points, and the subscripts m and c denote the measured and calculated values, respectively. The first square difference component in Eq. (4.74) is related to the attenuation, while the second component is associated with the phase angle of the probing wave. It is appropriate to set $|T_{\text{m}}| = 1$ in the calculation of

objective function F to justify the lossless assumption when using Eqs. (4.72)–(4.78). The problem of multiple solutions for ε_r' and ε_r'' can be reduced by applying the constraints on these parameters to stay within the desired tolerance level.

4.4.2.3 Measurement setup

Figure 4.20 illustrates the setup for transmission/reflection method using the rectangular dielectric waveguide, and usually TRL calibration is used. The rectangular dielectric waveguide and its one-quarter wavelength spacer (standard "Line" for TRL calibration) have cross-sectional dimensions equal to the standard hollow metallic waveguide. The dielectric waveguides are usually made of PTFE because of its ease of fabrication, very low loss and low dielectric constant. The low dielectric constant of PTFE provides a wide coverage of single-mode propagation in the waveguide. On the other hand, the low-loss factor is an important criterion for direct application of the effective index method that assumes lossless material.

The length of the dielectric waveguide beyond the horn aperture is approximately $5\lambda_0$. The dielectric waveguide is tapered only at the feed section to reduce the reflection coefficient between the dielectric waveguide and the standard hollow metallic waveguide. Only the *H*-plane of the dielectric waveguide is tapered to allow a natural transition from center-loaded, partially dielectric–filled to completely dielectric–filled metallic waveguide. Usually, an extra length of PTFE is further allocated within the metallic waveguide to form a tight

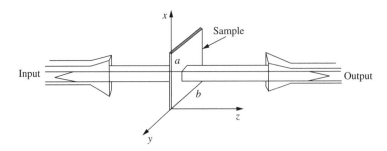

Figure 4.20   The rectangular dielectric waveguide measurement setup. Source: Abbas, Z. Pollard, R. D. Kelsall, R. W. (2001). "Complex permittivity measurements at Ka-band using rectangular dielectric waveguide", *IEEE Transactions on Instrumentation and Measurement*, **50** (5), 1334–1342. © 2003 IEEE

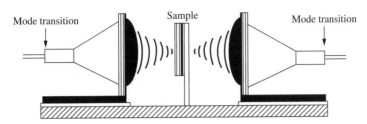

Figure 4.21 Setup for free-space measurement. Modified from Ghodgaonkar, D. K. Varadan, V. V. and Varadan, V. K. (1990). "Free-space measurement of complex permittivity and complex permeability of magnetic materials at microwave frequencies", *IEEE Trans. Instrum. Meas.* **39** (2) 387–394. © 2003 IEEE

fit to the metal walls as well as to provide support to the suspended rectangular dielectric waveguide at the waveguide opening. A metallic waveguide horn is employed to launch the E_{11}^y mode into the rectangular dielectric waveguide, as well as to serve as a mechanical support. More detailed technical specifications about the measurement setup can be found in (Abbas *et al.* 2001).

4.5 FREE-SPACE METHOD

In recent years, there has been an increasing interest in using free-space techniques for the measurement of electromagnetic properties of materials and for estimating plasma parameters of magnetoactive plasma (Varadan *et al.* 1991; Ghodgaonkar *et al.* 1990, 1989; Musil and Zacek 1986). Because of the availability of precision horn lens antennas that have far-field focusing ability, it is possible to make accurate free-space measurements at microwave frequencies. Free-space techniques for material property measurements have several advantages. First, materials such as ceramics, composites, and so on are inhomogeneous due to variations in manufacturing processes. Because of inhomogeneity, the unwanted higher-order modes can be excited at an air–dielectric interface in hollow metallic waveguides, while this problem does not exist in free-space measurement. Second, measurements using free-space techniques are nondestructive and contactless, so free-space methods can be used to measure samples under special conditions, such as high temperature. Third, in hollow metallic waveguide methods, it is necessary to machine the sample so as to fit the waveguide cross section with negligible air gaps. This

requirement limits the accuracy of measurements for materials that cannot be machined precisely; in free-space method, this problem does not exist.

Figure 4.21 shows a typical measurement setup, mainly consisting of two antennas and a sample holder. The transmit and receive antennas are spot-focusing horn lens antennas. The use of focusing antennas is to minimize the effects of sample boundaries and measurement environment. The transmit and receive antennas are mounted on a carriage and the distance between them can be adjusted. A specially fabricated sample holder is placed at the common focal plane for holding planar samples. The microwave signal incident to planar samples can be taken as plane waves, and the properties of the sample under test are obtained from the reflection from the sample and transmission through the sample.

According to the orientations of the sample in the measurement system, the microwave signal incident to the sample may be normal or oblique to the sample. In the following discussion, we concentrate on the case with normal incidence. The discussion on the cases with oblique incidence can be found in (Munoz *et al.* 1998; Friedsam and Biebl 1997).

4.5.1 Calculation algorithm

In principle, the algorithms discussed in Section 4.1 can be used in free-space methods for the determination of permittivity and permeability of the sample under test. The following discussion is concentrated on the special case for planar samples in free-space measurement.

Figure 4.22 shows a planar sample with thickness d placed in free space. It is assumed that the

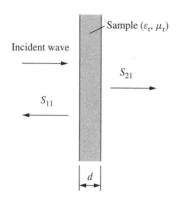

Figure 4.22 Schematic diagram of planar sample

planar sample is of infinite extent laterally so that diffraction effects at the edges can be neglected. A linearly polarized, uniform plane wave is normally incident on the sample. The reflection and transmission coefficients S_{11} and S_{21} are measured in free space for the normally incident plane wave.

By applying boundary conditions at the air-sample interface, it can be shown that the S_{11} and S_{21} parameters are related to the parameters Γ and T by the following equations (Ghodgaonkar *et al.* 1990):

$$S_{11} = \frac{\Gamma(1 - T^2)}{1 - \Gamma^2 T^2}, \qquad (4.87)$$

$$S_{21} = \frac{T(1 - \Gamma^2)}{1 - \Gamma^2 T^2}, \qquad (4.88)$$

where Γ is the reflection coefficient of the air-sample interface given by

$$\Gamma = \frac{(z - 1)}{(z + 1)}, \qquad (4.89)$$

and the T is the transmission coefficient given by

$$T = e^{-\gamma d}. \qquad (4.90)$$

In Eqs. (4.89) and (4.90), z and γ are the normalized characteristic impedance and propagation constant of the sample. They are related to permittivity and permeability by the following relationships:

$$\gamma = \gamma_0 \sqrt{\varepsilon_r \mu_r}, \qquad (4.91)$$

$$z = \sqrt{\frac{\mu_r}{\varepsilon_r}}, \qquad (4.92)$$

with

$$\gamma_0 = (j2\pi/\lambda_0) \qquad (4.93)$$

where γ_0 the propagation constant of free space, and λ_0 is the free-space wavelength.

From Eqs. (4.87) and (4.88), Γ and T can be written as

$$\Gamma = K \pm \sqrt{K^2 - 1}, \qquad (4.94)$$

$$T = \left(\frac{S_{11} + S_{21} - \Gamma}{1 - (S_{11} + S_{21})\Gamma} \right) \qquad (4.95)$$

with

$$K = \frac{S_{11}^2 - S_{21}^2 + 1}{2S_{11}}. \qquad (4.96)$$

The plus or minus sign in Eq. (4.94) is chosen such that $|\Gamma| < 1$. According to Eq. (4.90), the complex propagation constant γ can be written as

$$\gamma = \frac{\ln(1/T)}{d} \qquad (4.97)$$

From Eqs. (4.89) and (4.92), we have

$$\sqrt{\frac{\mu_r}{\varepsilon_r}} = \left(\frac{1 + \Gamma}{1 - \Gamma} \right) \qquad (4.98)$$

From Eqs. (4.91) and (4.98), we obtain

$$\varepsilon_r = \frac{\gamma}{\gamma_0} \left(\frac{1 - \Gamma}{1 + \Gamma} \right), \qquad (4.99)$$

$$\mu_r = \frac{\gamma}{\gamma_0} \left(\frac{1 + \Gamma}{1 - \Gamma} \right). \qquad (4.100)$$

As the parameter T in Eq. (4.97) is a complex number, there are multiple values of γ. If we define T as

$$T = |T|e^{j\phi}, \qquad (4.101)$$

then γ is given by

$$\gamma = \frac{\ln(1/|T|)}{d} + j\left(\frac{2\pi n - \phi}{d} \right), \qquad (4.102)$$

where $n = 0, \pm 1, \pm 2, \ldots$. The real part of γ is unique and single valued, but the imaginary part of γ has multiple values. We define the phase constant β as

$$\beta = (2\pi/\lambda_m) = \text{Im}(\gamma), \qquad (4.103)$$

where λ_m is the wavelength in the sample material. From Eqs. (4.102) and (4.103), we can get

$$d/\lambda_m = n - \frac{\phi}{2\pi}. \qquad (4.104)$$

If the sample thickness d is chosen to be less than λ_m, then we can obtain the unique value of ε_r and μ_r corresponding to $n = 0$. For $d > \lambda_m$, the ambiguity in ε_r and μ_r can be resolved by making measurements on two samples made of the same material but with different thicknesses.

For the measurement of thin and flexible materials, the sample under test is usually sandwiched between two fused-quartz plates that are half wavelength at midband, as shown in Figure 4.23. The parameters S_{11a} and S_{21a} of the quartz plate–sample–quartz plate assembly are measured in free space. Using the known complex permittivity and the thickness of the quartz plates, S_{11} and S_{21} of the sample can be calculated from S_{11a} and S_{21a} of the assembly, as will be shown in the following.

The *ABCD* matrix of the assembly can be obtained as

$$[A^a] = \begin{bmatrix} A^a & B^a \\ C^a & D^a \end{bmatrix} \qquad (4.105)$$

where

$$A^a = \frac{(1 + S_{11a})(1 - S_{22a}) + S_{12a}S_{21a}}{2S_{21a}}, \qquad (4.106)$$

$$B^a = Z_0 \frac{(1 + S_{11a})(1 + S_{22a}) - S_{12a}S_{21a}}{2S_{21a}}, \qquad (4.107)$$

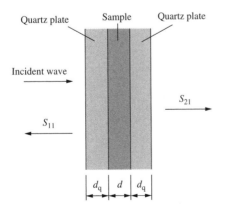

Figure 4.23 Schematic diagram of quartz plate–sample–quartz plate assembly

$$C^a = \frac{1}{Z_0} \frac{(1 - S_{11a})(1 - S_{22a}) - S_{12a}S_{21a}}{2S_{21a}}, \qquad (4.108)$$

$$D^a = \frac{(1 - S_{11a})(1 + S_{22a}) + S_{12a}S_{21a}}{2S_{21a}}, \qquad (4.109)$$

and Z_0 is the free space wave impedance.

The *ABCD* matrix for the quartz $[A^q]$ can be obtained similarly from the permittivity and thickness of the quartz. The *ABCD* matrix of the sample $[A^s]$ can be obtained as

$$[A^s] = [A^q]^{-1}[A^a][A^q]^{-1} \qquad (4.110)$$

The parameters S_{11} and S_{21} of the sample can then be obtained using the following expressions:

$$S_{11} = \frac{A^s + B^s/Z_0 - C^s Z_0 - D^s}{A^s + B^s/Z_0 + C^s Z_0 + D^s} \qquad (4.111)$$

$$S_{21} = \frac{2}{A^s + B^s/Z_0 + C^s Z_0 + D^s} \qquad (4.112)$$

After obtaining the *S*-parameters, the permittivity and permeability of the sample can be obtained according to Eqs. (4.87)– (4.100).

4.5.2 Free-space TRL calibration

To accurately measure the *S*-parameters of the sample or the sample assembly under test, it is necessary to make calibration at the measurement planes, and usually free-space TRL calibration techniques are used (Ghodgaonkar *et al.* 1989).

To implement TRL calibration in free space, it is necessary to establish TRL standards in the free-space medium. The two reference planes for port 1 and port 2 are located at the focal planes of the transmit and receive antennas, respectively. A through standard is configured by keeping the distance between the two antennas equal to twice that of the focal distance. The reflect standards for port 1 and port 2 are obtained by placing a metal plate at the focal planes of the transmit and receive antenna, respectively. The line standard is achieved by separating the focal planes of the transmit and receive antennas by a distance equal to a quarter of the free-space wavelength at the center of the working band. The specifications for TRL calibration in the network analyzer should be modified subsequently by defining TRL standards regarding wave impedance and electrical delay.

The free-space calibration can be verified for plane wave propagation by measuring S_{21} for free-space delays of different lengths. For a well-calibrated free-space system, a linear phase variation and negligible amplitude change with increasing length of delay line can be observed. So the electromagnetic fields in the neighborhood of the common focal plane are those of a plane wave.

It should be indicated that, as the TRL calibration could not fully correct the multiple reflections between the antennas and sample, time-domain gating is often needed to eliminate multiple reflections.

4.5.3 Uncertainty analysis

After free-space TRL calibration and time-domain gating of the S-parameter response, the measurement errors in S_{11} and S_{21} are mainly due to two kinds of errors. The first kind comprises the residual postcalibration errors resulting from imperfections in the calibration standards, the instrumentation, and the measurement circuit. Usually, the instrumentation errors, such as frequency instability, power variation of microwave signals, are negligible. In a TRL calibration, the error caused by the calibration standards is mainly due to the possible wave impedance of the line standard. This impedance error causes errors in the normalization of reflection and transmission coefficients to the impedance of the line standard. The residual source and load-mismatch errors due to the imperfections in the calibration standards can be minimized by time-domain gating. The errors due to the measurement circuit can be minimized by ensuring a stable configuration of the cables during the calibration and measurement procedures.

The second kind of error has to do with small changes in the reference planes (as defined by through or reflect standard) between calibration and measurement due to small changes in the positions of plates holding the sample. When the sample is translated or rotated from the reference planes by a small amount, it is possible to have errors in the S-parameter measurements. Figure 4.24 shows a schematic diagram of the translated and rotated sample. It can be seen that the path length for S_{11} is dependent on the sample position relative to the reference plane, whereas for S_{21} the path length is independent of sample position. Also, if the incident signal is at an angle ϕ, the reflected signal is at an angle of 2ϕ; so, not all reflected power can be received by the antenna. However, the transmitted signal through the sample is parallel to the incident beam with some very small lateral displacement, and so for thin samples and small angles the transmitted power loss is negligible. Therefore, if we only use S_{21} to calculate the permittivity of a dielectric material, the effect of small rotation and transition can be neglected,

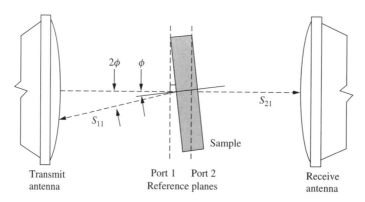

Figure 4.24 Schematic diagram of reflected and transmitted waves when the sample is translated and rotated from the reference planes defined during calibration. Source: Varadan, V. V. Hollinger, R. D. Ghodgaonkar, D. K. and Varadan, V. K. (1991). "Free-space, broadband measurements of high-temperature, complex dielectric properties at microwave frequencies", *IEEE Transactions on Instrumentation and Measurement*, **40** (5), 842–846. © 2003 IEEE

while if we use both S_{21} and S_{11} to calculate both the permittivity and the permeability of a magnetic sample, the effect of rotation and transition should be taken into consideration.

4.5.4 High-temperature measurement

As flying vehicles such as aircraft and missiles fly faster at higher Mach numbers, they experience greater aerodynamic heating, mechanical stress, and flight erosion (Dicaudo 1970; Gagnon *et al.* 1986). New materials are necessary particularly for radome applications, which are often located in areas of greatest exposure to the aforementioned elements. So the electromagnetic properties and their variation with temperature of potential radome materials must be characterized accurately (Ho 1981). A lot of efforts have been made for the development of microwave system for the measurement of permittivity and permeability at high temperatures. More discussions on the measurement of materials properties at high temperatures can be found in Chapter 12.

In a free-space measurement setup, the sample under test can be thermally insulated from the measurement equipment, so free-space techniques are suitable for the study of the electromagnetic properties of materials at high temperatures. Several free-space methods for high-temperature material characterization have been developed (Varadan *et al.* 1991; Ho 1988; Gagnon *et al.* 1986; Breeden 1969).

Figure 4.25 shows an experimental setup for high-temperature measurement using free-space method. Two spot-focusing horn lens antennas are used, and the diffraction effects at the edges of the sample are negligible if the minimum transverse dimension of the sample is greater than three times that of the E-plane 3-dB beam width of the antenna at its focus. The antennas are mounted facing each other and separated by a distance equal to twice the focal distance. For high-temperature measurements, the furnace with a high-temperature sample holder is placed at the common focal plane of the antennas. The furnace uses resistive heating panels on the top, bottom, and two sides, while the front and back of the furnace are enclosed only by thermally insulating fibrous materials that are virtually transparent to microwaves because of its low density.

To achieve accurate measurement results, the measurement system needed to be calibrated using a free-space TRL calibration, and time-domain gating techniques should be used. However, because of the presence of the furnace, it is difficult to place the sample holder exactly at the reference plane during and after calibration. Therefore, some additional steps are needed to minimize the errors due to the incorrect positioning of the sample. The first step is to perform free-space TRL calibration with the standard room-temperature sample holder and measure S_{11} and S_{21} of the sample. Next, the furnace and high-temperature sample holder are placed between the antennas and TRL calibration is performed again. Then the sample is placed in the high-temperature sample holder in the furnace. The position of the sample is adjusted such that

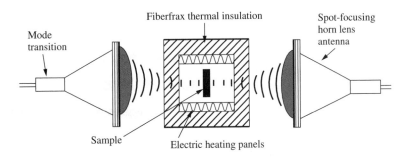

Figure 4.25 Schematic diagram of the free-space microwave measurement setup for high-temperature measurement. Modified from Varadan, V. V. Hollinger, R. D. Ghodgaonkar, D. K. and Varadan, V. K. (1991). "Free-space, broadband measurements of high-temperature, complex dielectric properties at microwave frequencies", *IEEE Transactions on Instrumentation and Measurement*, **40** (5), 842–846. © 2003 IEEE

the measured S_{11} and S_{21} are close to the values obtained with the standard sample holder. After that, we conduct measurements at high temperatures.

As it is almost impossible to place the sample exactly at the reference plane, a correction factor is often used to introduce numerically an electrical delay and magnitude offset to account for the slight mispositioning of the sample. The measured data can be corrected as follows:

$$S_{11}^c = \frac{S_{11}^{STD}}{S_{11}^{FRT}} S_{11}^{HT} \qquad (4.113)$$

$$S_{21}^c = \frac{S_{21}^{STD}}{S_{21}^{FRT}} S_{21}^{HT} \qquad (4.114)$$

where the superscripts C, STD, FRT, and HT denotes corrected, standard room-temperature sample holder, furnace room temperature, and furnace high-temperature data, respectively. From Eqs. (4.113) and (4.114), it can be seen that when the furnace room-temperature data is substituted for the high-temperature data, the corrected data is identically the standard room-temperature data.

4.6 MODIFICATIONS ON TRANSMISSION/REFLECTION METHODS

The basic principle of transmission/reflection methods is to deduce materials properties from the signals reflected from the sample and the signals transmitted through the sample. Many modifications have been made on the conventional transmission/reflection methods to fulfill various measurement requirements.

4.6.1 Coaxial discontinuity

The coaxial transmission/reflection methods can cover wide frequency ranges, but, as discussed in Section 4.2.2.2, the air gaps in the coaxial sample holders cause large uncertainties in the measurement results, especially the air gap between the sample and the inner conductor. Many configurations have been proposed to avoid the air gap between the samples and the inner conductors.

As shown in Figure 4.26, a segment of inner conductor is removed, and the sample completely

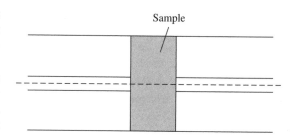

Figure 4.26   A coaxial discontinuity

fills the space within the outer conductor of the coaxial line, resulting in a coaxial discontinuity. This structure can also be considered as introducing a segment of circular waveguide between the two coaxial lines, with the sample completely filling the circular waveguide. This configuration has been investigated by two groups independently. Taherian *et al.* (1991) introduced the deduction algorithm for permittivity, and Belhadj-Tahar *et al.* (1998, 1992; 1990) introduced algorithms for both permittivity and permeability. The advantage of this method is the easiness of sample fabrication.

4.6.2 Cylindrical cavity between transmission lines

Figure 4.27 shows a dielectric-filled cylindrical cavity separating two coaxial transmission lines. The structure can be used as a building block in constructing coaxial filters. Once the scattering parameters of this building block are known,

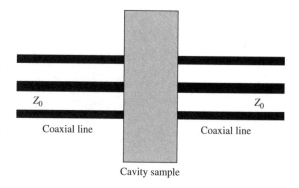

Figure 4.27   Cylindrical cavity placed between two coaxial lines

the overall characteristics of a filter consisting of several blocks can be obtained using matrix-manipulation techniques. Another application of this structure is the measurement of the complex permittivity of dielectric materials for the case where the dielectric material under test forms the dielectric-filled cavity (Saed 1991).

When this structure is used in the characterization of materials properties, it can be taken as a general case of the coaxial discontinuity method discussed in Section 4.6.1, and it can also taken as the two-port counterpart of the dielectric-filled cavity adapted at a coaxial end discussed in Section 3.2.6. The configuration shown in Figure 4.27 has an advantage in sample fabrication. A conventional coaxial transmission/reflection method requires the insertion of the sample into a coaxial air line, usually resulting in air-gap errors. The structure shown in Figure 4.27 does not need to insert the sample into a coaxial line; instead, the conductor walls of the cavity can be molded or deposited on the dielectric sample, so the possible errors due to air gaps are eliminated.

Full-wave analysis techniques based on the method of moments can be used to obtain the relationships between the scattering parameters of the cavity to its dimensions and the complex permittivity of the filling dielectric. The algorithm mainly consists of two parts. One part calculates the scattering parameters, S_{11} and S_{21}, from the permittivity of the sample and the dimensions of the cavity. The other part is an optimization program that calculates the complex permittivity of the sample from the scattering parameters and dimensions of the cavity. As higher-order modes can be taken into consideration in the development of algorithm, this method can cover a broad frequency range. To improve the accuracy of the algorithm, especially for the calculation of loss tangent, the wall loss of the dielectric-filled cavity should be taken into consideration.

The algorithm developed by Saed is for the measurement of dielectric materials (Saed 1991). As two independent *S*-parameters can be measured, in principle, this method can be extended for the determination of both permittivity and permeability of the electromagnetic materials.

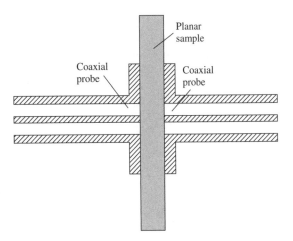

Figure 4.28 Measurement configuration for two-probe method

4.6.3 Dual-probe method

In Figure 4.28, a planar sample is placed between two coaxial open-reflection probes. This configuration is suitable for planar samples, and the most important advantage of this method is that planar samples can be characterized without any special preparation. The algorithm for permittivity deduction is reported by Scott (1992), and in principle, this configuration can measure both permittivity and permeability. Yushchenko and Chizhov (1997) proposed a similar structure for testing dielectric substrates by using circular waveguide instead of coaxial line.

4.6.4 Dual-line probe method

Figure 4.29 shows two open-circuited coaxial lines connected to a metal plate, forming a dual-line probe. Bird studied the coupling between the two coaxial lines, and showed that when a semi-infinite sample is in contact with the dual-line probe, the electromagnetic properties of the material can be deduced from the reflection back to the coaxial lines and the coupling between the coaxial open ends (Bird 1996). This configuration is an improvement of the conventional coaxial dielectric probe; however, the algorithm for extracting materials properties is quite complicated.

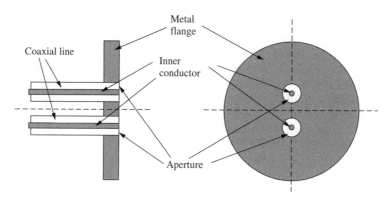

Figure 4.29   Coupling between two open-circuited coaxial lines

4.6.5 Antenna probe method

The methods for *in situ* characterization of materials properties can be generally divided into two categories: surface and drill-hole measurement techniques. Surface techniques are used whenever a drill hole is not feasible or when dielectric constant values are needed over very large areas. Drill-hole techniques, on the other hand, are usually used to get localized values of dielectric constant and to measure the depth profiles in materials.

Nassar *et al.* developed a handheld, ultra wide-band drill-hole-type probe antenna for the *in situ* measurement of the complex dielectric constant

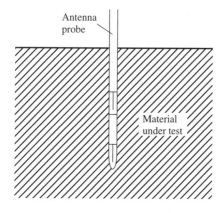

Figure 4.30   Measurement setup for electromagnetic probe (Nassar *et al.* 1999). Modified from Nassar, E. M. Lee, R. and Young, J. D. (1999). "A probe for antenna for *in situ* measurement of the complex dielectric constant of materials", *IEEE Transactions on Antennas and Propagation*, **47** (6), 1085–1093. © 2003 IEEE

(Nassar *et al.* 1999). The measurement configuration is shown in Figure 4.30, and the structure of the probe is shown in Figure 4.31. This probe uses two stub antennas mounted on a copper cylinder. To perform a measurement, the probe is inserted in the material under study. Microwave signals are transmitted by one of the antennas and received by the other. On the basis of the comparison of the received signal to the predicted signal from a numerical model of the probe antenna, the complex dielectric permittivity can be extracted.

Although the two stub antennas may have low gain, the outer surface of the copper cylinder acts as a guiding structure, so that the coupling between the two antennas can be very strong. This coupling is dependent upon the properties of the material in which the probe is immersed. If an accurate electromagnetic model can be obtained for the interaction between the probe and the material, we can accurately extract the complex permittivity of the material. A simple plane-wave approximation is assumed for the wave traveling between the two antennas. This approximation is fairly accurate because the guided wave along the surface of the copper cylinder can be considered to be quasi-TEM. There are several factors that may cause differences from this ideal model. The plane-wave model does not account for the input impedance and radiation pattern of the antennas. Furthermore, there may be reflections at the antenna/material interface. An alternative model can be developed based on the finite-difference time domain (FDTD) method, which should produce more accurate solutions, but is more computationally expensive.

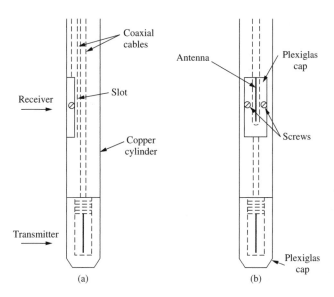

Figure 4.31 Front and side views of the electromagnetic probe showing the coaxial cables in the copper cylinder, the antennas, and plaxiglass covers (Nassar *et al.* 1999). (a) Side view and (b) front view. Modified from Nassar, E. M. Lee, R. and Young, J. D. (1999). "A probe for antenna for *in situ* measurement of the complex dielectric constant of materials", *IEEE Transactions on Antennas and Propagation*, **47** (6), 1085–1093. © 2003 IEEE

In actual measurements, calibration is often needed to check the operation of the probe by taking measurements in free space. This is done by recording three signals: the coupled signal between the two antennas (S_{21}), the reflected signal from the tip antenna (S_{11}), and the reflected signal from the middle antenna (S_{22}). According to these signals, we can check whether the antennas and the connections function properly.

In order to minimize measurement errors, it is necessary to eliminate any air gaps between

the probe and the material walls. The antennas on the probe should also be far enough from the boundaries of the sample in order to prevent the reflections at the boundaries from interfering with the direct signal between the transmitter and receiver. If these reflections are far enough in time from the direct coupled signal, they can be gated out before the signal is processed for extracting the complex dielectric permittivity.

4.7 TRANSMISSION/REFLECTION METHODS FOR COMPLEX CONDUCTIVITY MEASUREMENT

In the structure shown in Figure 4.32, a conducting thin film forms a quasi–short circuit in a waveguide transmission line. From the transmission coefficient, the conductivity of the thin film can be obtained (Dew-Hughes 1997; Wu and Qian 1997). However, this method is only suitable for extrathin films whose thickness is less than the penetration depth of the conductor, and requires a measurement system with very high dynamic range. As the thickness of most of the technologically useful conducting thin films is often larger than the penetration depth, this method is hardly used in current

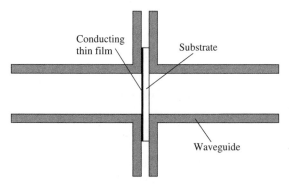

Figure 4.32 Measurement of power transmission through a conducting thin film

microwave electronics and engineering. However, this method is useful for the study of complex conductivity of high-temperature superconducting (HTS) thin films at microwave frequencies.

The principle of this method can be understood by taking a thin film deposited on a substrate as a two-layer dielectric system shown in Figure 4.33(a). We assume layer I to be a thin film with refraction index n_1 and thickness h_1, and layer II to be the substrate with refraction index n_2 and thickness h_2. For a single layer, such as layer I or layer II, as shown in Figure 4.33(b) or (c), the transmission and reflection coefficients are given by (Wu and Qian 1997)

$$t_i = \frac{-2\mathrm{j}n_i e^{\mathrm{j}k_0 h_i}}{(1 + n_i^2)\sin(k_0 n_i h_i) - 2\mathrm{j}n_i \cos(k_0 n_i h_i)}$$
(4.115)

$$r_i = \frac{(1 - n_i^2)\sin(k_0 n_i h_i)}{(1 + n_i^2)\sin(k_0 n_i h_i) - 2\mathrm{j}n_i \cos(k_0 n_i h_i)}$$
(4.116)

with $i = 1, 2$.

The relationship between the total transmission coefficient T of the two-layer system and (t_i, r_i) $(i = 1, 2)$ is given by

$$\frac{1}{t_1} \cdot \frac{1}{t_2} = \frac{1}{T} + \frac{r_1}{t_1} \cdot \frac{r_2}{t_2}$$
(4.117)

Since the total transmission coefficient T and the coefficients (t_2, r_2) of a substrate can be measured directly in experiments, we retain T, r_2, and t_2 in Eq. (4.117) and replace r_1 and t_1 by Eqs. (4.115) and (4.116). For convenience, from now on we use (t, r) to represent (t_2, r_2) and (n_s, d) to represent (n_1, h_1). From Eq. (4.117), we have the basic equation of simplified power transmission measurement (Wu and Qian 1997):

$$\frac{1}{T} = \frac{e^{-\mathrm{j}k_0 d}}{2t} \left\{ 2\cos(k_0 n_s d) + \mathrm{j}\sin(k_0 n_s d) \right.$$

$$\left. \times \left[\left(\frac{1}{n_s} + n_s \right) - r \left(\frac{1}{n_s} - n_s \right) \right] \right\}$$
(4.118)

For a metal with conductivity σ, the wave number should be

$$k_s^2 = (k_0 n_s)^2 = -\mathrm{j}\omega\mu\sigma$$
(4.119)

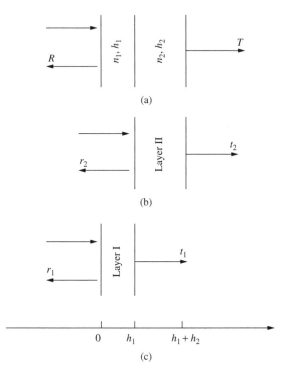

Figure 4.33   Dielectric layers in free space. (a) Two-layer structure, (b) layer II separated from (a), and (c) layer I separated from (a). Source: Wu, R. X. and Qian, M. (1997). "A simplified power transmission method used for measuring the complex conductivity of superconducting thin films", *Review of Scientific Instruments*, **68** (1), 155–158

On the basis of the two-fluid model, the conductivity of a superconductor is a complex parameter

$$\sigma = \sigma_1 - \mathrm{j}\sigma_2.$$
(4.120)

The equivalent reflection index n_s of a superconductor is given by

$$n_s = \mathrm{j}\frac{\lambda_0}{2\pi\lambda_L} \cdot \sqrt{1 + \mathrm{j}\frac{\sigma_1}{\sigma_2}}$$
(4.121)

where λ_0 is the wavelength in free space, and the penetration depth is given by

$$\lambda_L = 1/\sqrt{\omega\mu\sigma_2}.$$
(4.122)

Using Eqs. (4.118) and (4.121), the refraction index of the thin film and its conductivity can be obtained by the following procedures: first, measuring the r and t of a bare substrate; next, depositing the superconducting film on the substrate; finally, measuring again the total transmission coefficient of the superconducting sample.

In practical situations, the measurement is done in a waveguide. Therefore, it is necessary to make some changes in the basic equations. For rectangular waveguides, the propagation term is expressed in the form of $\exp(-jk_z z)$, where the k_z is the propagation constant. If the refraction index of filling dielectric in a waveguide is n, the propagation constant can be written as

$$k_z^2 = k^2 - k_c^2 = k_0^2 \left(n^2 - \frac{k_c^2}{k_0^2} \right) \qquad (4.123)$$

where k_c is the cutoff wave number. If we define an equivalent refraction index $\tilde{n}$ in the waveguide

$$\tilde{n} = \sqrt{n^2 - \frac{k_c^2}{k_0^2}} \qquad (4.124)$$

then the propagation constant k_z can be written as

$$k_z = \tilde{n} k_0 \qquad (4.125)$$

Therefore the propagation term in the waveguide is the same as that in free space with refraction index $\tilde{n}$. The basic equations are still correct if the n_s and n_d in the equation are replaced by the normalized equivalent refraction index $\tilde{n}'_s = \tilde{n}_s/\tilde{n}_0$, $\tilde{n}'_d = \tilde{n}_d/\tilde{n}_0$, and k_0 is replaced by $\tilde{n}_0 k_0$. Here $\tilde{n}_0$, $\tilde{n}_s$, and $\tilde{n}_d$ denote the equivalent refraction index of vacuum, superconducting film, and substrate in the waveguide, respectively.

REFERENCES

Abbas, Z. Pollard, R. D. and Kelsall, R. W. (1998a). "Determination of the dielectric constant of materials from effective refractive index measurements", *IEEE Transactions on Instrumentation and Measurement*, **47** (1), 148–152.

Abbas, Z. Pollard, R. D. and Kelsall, R. W. (1998b). "A rectangular dielectric waveguide technique for determination of permittivity of materials at W-band", *IEEE Transactions on Microwave Theory*, **46** (12), 2011–2015.

Abbas, Z. Pollard, R. D. and Kelsall, R. W. (2001). "Complex permittivity measurements at Ka-band using rectangular dielectric waveguide", *IEEE Transactions on Instrumentation and Measurement*, **50** (5), 1334–1342.

Abdulnour, J. Akyel, C. and Wu, K. (1995). "A generic approach for permittivity measurement of dielectric materials using discontinuity in a rectangular waveguide or a microstrip line", *IEEE Transactions on Microwave Theory and Techniques*, **43** (5), 1060–1066.

Baker-Jarvis, J. (1990). *Transmission/Reflection and Short-Circuit Line Permittivity Measurements*, NIST Technical Note, National Institute of Standards and Technology, Boulder, CO.

Baker-Jarvis, J. Geyer, R. G. and Domich, P. D. (1992). "A nonlinear least-squares solution with causality constraints applied to transmission line permittivity and permeability determination", *IEEE Transactions on Instrumentation and Measurement*, **41** (5), 646–652.

Baker-Jarvis, J. Vanzura, E. J. and Kissick, W. A. (1990). "Improved technique for determining complex permittivity with the transmission/reflection method", *IEEE Transactions on Microwave Theory and Techniques*, **38** (8), 1096–1103.

Baker-Jarvis, J. Janezic, M. D. Grosvenor, J. H. Jr. and Geyer, R. G. (1993). *Transmission/Reflection and Short-Circuit Line Methods for Measuring Permittivity and Permeability*, NIST Technical Note 1355 (revised), National Institute of Standards and Technology, Boulder, CO.

Belhadj-Tahar, N. E. Dubrunfaut, O. and Fourrier-Lamer, A. (1998). "Broad-band microwave characterization of a tri-layer structure using a coaxial discontinuity with applications for magnetic liquids and films", *IEEE Transactions on Microwave Theory*, **46** (12), 2109–2116.

Belhadj-Tahar, N. E. and Fourrier-Lamer, A. (1992). "Broadband simultaneous measurement of complex permittivity and permeability for uniaxial or isotropic materials using a coaxial discontinuity", *Journal of Electromagnetic Waves and Applications*, **6** (9), 1225–1245.

Belhadj-Tahar, N. Fourrier-Lamer, A. and de Chanterac, H. (1990). "Broadband simultaneous measurement of complex permittivity and permeability using a coaxial discontinuity", *IEEE Transactions on Microwave Theory and Techniques*, **38** (1), 1–7.

Bird, T. S. (1996). "Cross-coupling between open-ended coaxial radiators", *IEE Proceedings – Microwave and Antenna Propagations*, **143** (4), 265–271.

Boughriet, A. H. Legrand, C. and Chapoton, A. (1997). "Noniterative stable transmission/reflection method for low-loss material complex permittivity determination", *IEEE Transactions on Microwave Theory*, **45** (1), 52–57.

Breeden, K. H. (1969). "Error analysis for waveguide-bridge dielectric-cons'tant measurements at millimeter wavelengths", *IEEE Transactions on Instrumentation and Measurement*, **18**, 203–208.

Champlin, K. S. and Glover, G. H. (1966). "Influence of waveguide contact on measured complex permittivity of semiconductors", *Journal of Applied Physics*, **37**, 2355–2360.

Chew, W. C. Olp, K. J. and Otto, G. P. (1991). "Design and calibration of a large broadband dielectric measurement cell", *IEEE Transactions on Geoscience and Remote Sensing*, **29** (1), 42–47.

Dew-Hughes, D. (1997). "Microwave properties and applications of high temperature superconductors", in *Microwave Physics and Techniques*, H. Groll and I. Nedkov, Eds., Kluwer Academic Publishers, Dordrecht, 83–114.

Dicaudo, V. J. (1970). "Randomes", in *Radar Handbook*, M. I. Skolnix, Ed., McGraw-Hill, New York, 14-1–14-32.

Esteban, H. Boria, V. E. Baquero, M. and Ferrando, M. (1997). "Generalized iterative method for solving 2d multiscattering problems using spectral techniques", *IEE Proceedings on Microwave, Antennas and Propagation*, **144** (2), 73–80.

Esteban, H. Catala-Civera, J. M. Cogollos, S. and Boria, V. E. (2000). "Characterization of complex permittivity properties of materials in rectangular waveguides using a hybrid iterative method", *IEEE Microwave and Guided Wave Letters*, **10** (5), 186–188.

Friedsam, G. L. and Biebl, E. M. (1997). "A broadband free-space dielectric properties measurement system at millimeter wavelengths", *IEEE Transactions on Instrumentation and Measurement*, **46** (2), 515–518.

Gagnon, D. R. White, D. J. Everett, G. E. and Banks, D. J. (1986). *Technique for Microwave Dielectric Measurements*, Final Rep. NWC TP 6643, Naval Weapons Center, China Lake, CA, 39–50.

Ghodgaonkar, D. K. Varadan, V. V. and Varadan, V. K. (1989). "A free-space method for measurement of dielectric constants and loss tangents at microwave frequencies", *IEEE Transactions on Instrumentation and Measurement*, **38**, 789–793.

Ghodgaonkar, D. K. Varadan, V. V. and Varadan, V. K. (1990). "Free-space measurement of complex permittivity and complex permeability of magnetic materials at microwave frequencies", *IEEE Transactions on Instrumentation and Measurement*, **39**, 387–394.

Ho, W. W. (1981). *High Temperature Millimeter Wave Characterization of the Dielectric Properties of Advanced Window Materials*, Tech. Int. Rep. AMMRC Contract No. DAAG46-79-C-0077, SC5235.5IRD.

Ho, W. W. (1988). "High-temperature dielectric properties of polycrystalline ceramics", in *Microwave Processing of Materials, Materials Research Society Symposia Proceedings, vol. 124*, W. H. Sutton, M. H. Brooks and I. J. Chabinsky, Eds., pp. 137–148, Materials Research Society, Pittsburgh.

Hock, K. M. (2002). Private communication.

Kolawole, M. O. (1992). "Scattering from dielectric cylinders having radially layered permittivity", *Journal of Electromagnetic Waves and Applications*, **6** (2), 235–239.

Luebbers, R. (1993). "Effects of waveguide wall grooves used to hold samples for measurement of permittivity and permeability", *IEEE Transactions on Microwave Theory and Techniques*, **41** (11), 1959–1964.

Marcuvitz, N. (1951). *Waveguide Handbook*, Dover Publications, New York.

Munoz, J. Rojo, M. Parreno, A. and Margineda, J. (1998). "Automatic measurement of permittivity and permeability at microwave frequencies using normal and oblique free-wave incidence with focused beam", *IEEE Transactions on Instrumentation and Measurement*, **47** (4), 886–892.

Musil, J. and Zacek, F. (1986). *Microwave Measurements of Complex Permittivity by Free Space Methods and their Applications*, Elsevier, Amsterdam.

Nassar, E. M. Lee, R. and Young, J. D. (1999). "A probe for antenna for *in situ* measurement of the complex dielectric constant of materials", *IEEE Transactions on Antennas and Propagation*, **47** (6), 1085–1093.

Nicolson, A. M. and Ross, G. F. (1970). "Measurement of the intrinsic properties of materials by time domain techniques", *IEEE Transactions on Instrumentation and Measurement*, **19** (4), 377–382.

Saed, M. A. (1991). "A method of moments solution of a cylindrical cavity placed between two coaxial transmission lines", *IEEE Transactions on Microwave Theory and Techniques*, **39** (10), 1712–1717.

Schlesinger, S. P. and King, D. D. (1958). "Dielectric image lines", *IRE Transactions on Microwave Theory and Techniques*, **6**, 291–299.

Scott Jr., W. R. (1992). "A new technique for measuring the constitutive parameters of planar materials", *IEEE Transactions on Instrumentation and Measurement*, **41** (5), 639–645.

Smith, F. C. (1995). "Uncertainty in the reflection/transmission method for measuring permeability and permittivity and its effects on radar absorber design", *Proceedings of 7th British Electromagnetic Measurement Conference*, **54**, 1–4.

Stuchly, S. and Matuszewski, M. (1978). "A combined total reflection transmission method in application to dielectric spectroscopy", *IEEE Transactions on Instrumentation and Measurement*, **27**, 285–288.

Taherian, M. R. Yuen, D. J. Habashy, T. M. and Kong, J. A. (1991). "A coaxial-circular waveguide for dielectric measurement", *IEEE Transactions on Geoscience and Remote Sensing*, **29** (2), 321–330.

Varadan, V. V. Hollinger, R. D. Ghodgaonkar, D. K. and Varadan, V. K. (1991). "Free-space, broadband measurements of high-temperature, complex dielectric properties at microwave frequencies", *IEEE Transactions on Instrumentation and Measurement*, **40** (5), 842–846.

Weir, W. B. (1974). "Automatic measurement of complex dielectric constant and permeability at microwave frequencies", *Proceedings of the IEEE*, **62** (1), 33–36.

Wilson, S. B. (1988). "Model analysis of the gap effect in waveguide dielectric measurements", *IEEE Transactions on Microwave Theory and Techniques*, **36** (4), 752–756.

Wu, R. X. and Qian, M. (1997). "A simplified power transmission method used for measuring the complex conductivity of superconducting thin films", *Review of Scientific Instruments*, **68** (1), 155–158.

Youngs, I. J. (1996). "NAMAS-accredited microwave permittivity/permeability measurement system", *IEE Proceedings: Science, Measurement and Technology*, **143** (4), 247–253.

Yushchenko, A. G. and Chizhov, V. V. (1997). "Precision microwave testing of dielectric substrates", *IEEE Transactions on Instrumentation and Measurement*, **46** (2), 507–510.

5

Resonator Methods

In a resonator method, the sample serves as a resonator or a key part of a resonator in the measurement circuit, and the properties of the sample are derived from the resonant properties of the resonator. Three major types of resonators are discussed for the characterization of dielectric materials, including conventional dielectric resonator, coaxial surface-wave resonator, and split resonator. The measurements of surface impedance of high-conductivity materials using dielectric resonator methods are discussed in the last section.

5.1 INTRODUCTION

Though resonant methods can only measure materials properties at single or several discrete frequencies, resonant methods are widely used because of their high accuracy and sensitivity. Resonant methods include resonator methods and resonant-perturbation methods. In a resonator method, the sample under test serves as a resonator or a key part of a resonator, and the properties of the sample are determined from the resonant properties of the resonator. In a resonant-perturbation method, the sample is introduced into a resonant structure, and the properties of the sample are calculated from the change of resonant properties of the resonant structure caused by the introduction of the sample. In this chapter, we focus on resonator methods, and the resonant-perturbation methods are discussed in Chapter 6. The planar resonators used in materials property measurements are discussed in Chapter 7.

The main requirement of a resonator method is that the sample forms a resonator or a key part of resonator, and the algorithm for materials property calculation is related to the field distributions in the resonator and the sample. In principle, any kind of resonator can be used in materials property characterization. In a resonator method, if the electric energy is more concentrated on the sample, this method usually has higher sensitivity and accuracy. In this chapter, we discuss three types of resonators often used in materials property characterization: dielectric resonators, coaxial surface-wave resonators, and split resonators.

In actual measurement fixtures for materials property characterization, resonators are often shielded by conductors. In the measurement of dielectric materials, the surface impedance of the shielding conductors, usually metals, is assumed be known. If the properties of the dielectric material are known, from the resonant properties of the resonator, the surface impedance of the shielding conductors can be obtained. In the final part of this chapter, we discuss the measurement of surface impedance of high-conductivity materials using dielectric resonator methods.

5.2 DIELECTRIC RESONATOR METHODS

Dielectric resonator methods are widely used in the characterization of low-loss dielectric materials (Kent 1988a). The basic properties of dielectric resonators have been discussed in Chapter 2. In the design of dielectric resonator methods for the characterization of the dielectric properties of materials, the structures of both the dielectric material to be studied, and the metal shield should be designed. Different dielectric resonator methods have different structures of metal shield and dielectric sample, and thus have different algorithms for the calculation of materials properties.

5.2.1 Courtney resonators

This method was first described by Hakki and Coleman (1960), and later fully analyzed and developed by Courtney (1970). Now this method is simply known as "Courtney method."

5.2.1.1 Configuration

As shown in Figure 5.1, the structure is essentially a shorted dielectric waveguide. A finite length of a cylindrical dielectric rod waveguide is shorted by placing conducting plates at each end of the rod, turning the transmission line into a resonator. This structure can be taken as a special case of the parallel-plate dielectric resonator discussed in Section 2.3.5.2, with $L_1 = 0$.

The theoretical model for this configuration assumes that the two conducting plates are infinitely large. The characteristic equation for the normal modes is (Courtney 1970)

$$\left[\frac{\varepsilon_r J_m'(\alpha)}{\alpha J_m(\alpha)} + \frac{K_m'(\beta)}{\beta K_m(\beta)} \right] \left[\frac{J_m'(\alpha)}{\alpha J_m(\alpha)} + \frac{K_m'(\beta)}{K_m(\beta)} \right]$$
$$= m^2 \left[\frac{\varepsilon_r}{\alpha^2} + \frac{1}{\beta^2} \right] \left[\frac{1}{\alpha^2} + \frac{1}{\beta^2} \right] \qquad (5.1)$$

with

$$\alpha = \frac{\pi D}{\lambda} \left[\varepsilon_r - \left(\frac{l\lambda}{2L} \right)^2 \right]^{\frac{1}{2}} \qquad (5.2)$$

$$\beta = \frac{\pi D}{\lambda} \left[\left(\frac{l\lambda}{2L} \right)^2 - 1 \right]^{\frac{1}{2}} \qquad (5.3)$$

where $J_m(\alpha)$ and $K_m(\beta)$ are Bessel functions of the first and second kind respectively, λ is the free-space wavelength, D and L are the diameter and the length of the dielectric specimen respectively, ε_r is the dielectric constant of the dielectric specimen, and $l = 1, 2, 3, \ldots$, and so on, corresponds to the multiple half wavelengths in the resonator along the axial direction. Figure 5.2 shows the mode chart for a dielectric resonator with dielectric constant 15 (Courtney 1970). More detailed discussions on the resonant modes of this kind of resonators can be found in (Kobayashi and Tanaka 1980).

The TE$_{011}$ mode is widely used in materials property characterization because in this mode there is no current crossing the dielectric and the conducting plates, so the possible air gaps between the dielectric and the conducting plates do not have much effects on the resonant properties of this mode. The mode chart shown in Figure 5.2 indicates that TE$_{011}$ is the second low-frequency mode. For the TE$_{011}$ mode, the relationship between α and β is given by

$$\varepsilon_r = 1.0 + \left(\frac{c}{\pi D f_1} \right)^2 (\alpha_1^2 + \beta_1^2) \qquad (5.4)$$

where α_1 and β_1 are the first roots of the characteristic equation with $m = 0, l = 1$, and f_1 is the resonant frequency. For a specimen with known dimensions, by measuring the frequency of the TE$_{011}$ mode, the real part of the dielectric permittivity can be found.

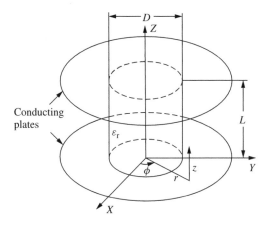

Figure 5.1 Configuration of a Courtney resonator. Modified from Kobayashi, Y. Tanaka, S. (1980). "Resonant modes of a dielectric rod resonator short-circuited at both ends by parallel conducting plates", *IEEE Trans. Microwave Theory Tech.*, **28**, 1077–1086. © 2003 IEEE

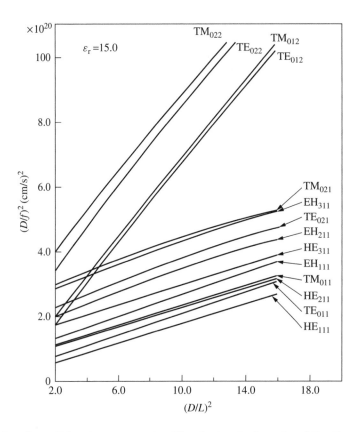

Figure 5.2 Mode chart for a dielectric postresonator. The abscissa is the ratio of the diameter to the length of the dielectric post. The ordinate is the diameter multiplied by the frequency (Courtney 1970). Source: Courtney, W. E. (1970). "Analysis and evaluation of a method of measuring the complex permittivity and permeability of microwave insulators", *IEEE Transactions on Microwave Theory and Techniques*, **18** (8), 476–485. © 2003 IEEE

The loss tangent of the specimen can also be obtained (Hakki and Coleman 1960):

$$\tan \delta = \frac{A}{Q_0} - BR_s \qquad (5.5)$$

where

$$A = 1 + \frac{W}{\varepsilon_r} \qquad (5.6)$$

$$B = \left(\frac{l\lambda}{2L}\right)^3 \frac{1 + W}{30\pi^2 \varepsilon_r l} \qquad (5.7)$$

$$W = \frac{J_1^2(\alpha_1)}{K_1^2(\beta_1)} \cdot \frac{K_0(\beta_1)K_2(\beta_1) - K_1^2(\beta_1)}{J_1^2(\alpha_1) - J_0(\alpha_1)J_2(\alpha_1)} \qquad (5.8)$$

$$R_s = \sqrt{\frac{\pi f_1 \mu}{\sigma}} \qquad (5.9)$$

where σ is the conductivity of the shorting plate, Q_0 is the unloaded quality factor of the dielectric resonator, and the function W is the ratio of electric field energy stored outside to inside the rod. Using Eqs. (5.5)–(5.9), the loss tangent can be evaluated from the resonant frequency, unloaded quality factor, and sample dimensions.

For low-loss materials ($\tan \delta \ll 1$), the terms A/Q_0 and BR_s in Eq. (5.5) may be of the same orders of magnitude when the TE_{011} mode is used. Therefore, the accurate effective value of R_s is required for the measurement of $\tan \delta$ with high accuracy.

Detailed discussion on the measurement of surface resistance can be found in Section 5.5 and Chapter 6. Here we introduce a technique for the measurement of R_s using the same instrument

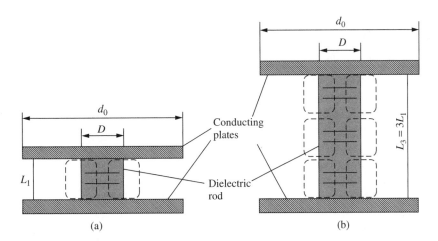

Figure 5.3 Dielectric rod resonators placed between two parallel conducting plates. (a) TE$_{011}$ mode and (b) TE$_{013}$ mode. Modified from Kobayashi, Y. Imai, T. and Kayano, H. (1990). "Microwave measurement of surface impedance of high-Tc superconductor", *IEEE MTT-S Digest*, 281–284. © 2003 IEEE.

as used for the dielectric measurement discussed above. As shown in Figure 5.3, we use two rod samples cut from a dielectric rod, which have the same diameters but different lengths. The rod for a TE$_{01l}$ resonator is l times as long as the other for a TE$_{011}$ resonator, where $l \geq 2$. We denote the quantities for both resonators by subscripts l and 1, respectively. If $f_{0l} = f_{01}$, because of the different effects of conductor loss on the values of unloaded quality factor, we have $Q_{0l} > Q_{01}$. Owing to the fact that both rods have the same $\tan \delta$ values, from Eqs. (5.5)–(5.9), we know that R_s can be determined from the measured values of Q_{01} and Q_{0l} (Kobayashi and Katoh 1985; Kobayashi *et al.* 1985; Kobayashi *et al.* 1990):

$$R_s = 30\pi^2 \left(\frac{2L}{l\lambda}\right)^3 \cdot \frac{\varepsilon_r + W}{1 + W} \cdot \frac{l}{l-1} \cdot \left(\frac{1}{Q_{01}} - \frac{1}{Q_{0l}}\right) \tag{5.10}$$

It is clear that the difference between the Q_{01} and Q_{0l} values increases with ε_r and l. The substitution of Eq. (5.10) into Eq. (5.5) yields

$$\tan \delta = \frac{A}{l-1} \left(\frac{l}{Q_{0l}} - \frac{1}{Q_{01}}\right) \tag{5.11}$$

which is independent of R_s and therefore allows precise measurement of $\tan \delta$.

As it is difficult to precisely machine samples to the designed lengths, we often use the following expressions instead of Eqs. (5.10) and (5.11) (Kobayashi and Katoh 1985):

$$R_s(f_0) = \frac{1}{C_1 - C_l} \cdot \left(\frac{A_1}{Q_{01}} - \frac{A_l}{Q_{0l}}\right) \tag{5.12}$$

$$\tan \delta = \frac{C_1}{C_1 - C_l} \cdot \left(\frac{A_l}{Q_{0l}} - \frac{C_l}{C_1} \cdot \frac{A_1}{Q_{01}}\right) \tag{5.13}$$

which are derived from Eq. (5.5) by using relations

$$R_{sl}(f_{0l}) = \sqrt{\frac{f_{0l}}{f_0}} R_s(f_0) \tag{5.14}$$

$$C_l = B_l \sqrt{\frac{f_{0l}}{f_0}} \tag{5.15}$$

where f_0 is chosen arbitrarily near f_{01} and f_{0l} and the parameters A_l and B_l are defined by Eqs. (5.6) and (5.7) for a TE$_{01l}$ mode resonator.

Figure 5.4 shows a measurement configuration of Courtney resonator method. The upper and bottom plates are the two conducting plates shorting the dielectric rod transmission line into a resonant structure. The upper plate can be raised or lowered to accommodate various lengths of cylindrical samples. Couplings to the sample are achieved by the right-angle E-field probes, both of which can be moved in and out along the radius direction, with respect to the sample, to vary the

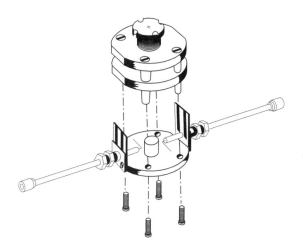

Figure 5.4 Explored view of the dielectric postresonator showing the sample and probes in position (Courtney 1970). Source: Courtney, W. E. (1970). "Analysis and evaluation of a method of measuring the complex permittivity and permeability of microwave insulators", *IEEE Transactions on Microwave Theory and Techniques*, **18** (8), 476–485. © 2003 IEEE

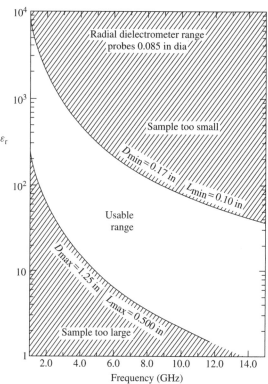

Figure 5.5 Usable range of dielectric resonator with 0.085-in.-diameter probes and 2.5-in.-diameter conducting plates. The dielectric constant, ε_r, as a function of the resonant frequency of the TE_{011} mode for maximum and minimum values of dielectric samples (Courtney 1970). Source: Courtney, W. E. (1970). "Analysis and evaluation of a method of measuring the complex permittivity and permeability of microwave insulators", *IEEE Transactions on Microwave Theory and Techniques*, **18** (8), 476–485. © 2003 IEEE

coupling coefficients. To increase the accuracy and sensitivity in the measurement of loss tangent, both the upper and bottom plates are gold plated.

In this configuration, placing the specimen exactly in the center is not critical. The specimen merely needs to be placed approximately symmetrical with the two probes. Since the unloaded quality factor of the TE_{011} mode must be measured, the coupling of each probe should be reduced until no change is detectable in the resonant frequency or 3-dB line width of the resonance. If the sample is isotropic, the TE_{011} mode can be identified by finding the second low-frequency mode, as indicated by the mode chart in Figure 5.2. Further proof of a TE_{0nl} mode can be made by raising and lowering the upper plate. As the plate is raised, the TM modes move rapidly to higher frequencies, while the TE_{011} mode remains stationary. In most situations, the dielectric constant is known approximately so that the frequency of the TE_{011} resonance can be estimated. A combination of the above three conditions enables the TE_{0l1} mode to be identified unambiguously.

A dielectric resonator fixture for the measurement of dielectric permittivity has a usable range. Figure 5.5 shows the usable range of the particular

dielectric resonator fixture. In the figure, the dielectric constant is plotted as a function of the resonant frequency of the TE_{011} mode for the maximum and minimum values of specimen dimensions. The maximum dimensions are determined by the diameter of the shorting plates and the minimum dimensions by the diameter of the coupling probes.

5.2.1.2 Effects of the size of conducting plates

In the design of dielectric resonator measurement fixtures, we should make an estimation of the dielectric constant of the sample to be tested and determine suitable dimensions for sample rod. Meanwhile, we should also consider the sizes of conducting plates, surface resistance of conducting plates, and higher resonant modes. In the following, we discuss the effects of the size of conducting plates.

In the above analysis, the conducting plates are assumed to be infinitely large ($d = \infty$). In actual structures, since the field outside the dielectric rod decays rapidly in radial direction, a suitable finite size of the plates is permitted, provided the influence on the measurement is negligible.

Let ε_r and $\tan \delta$ be the values obtained from given values of f_0 and Q_0 for a structure in the case of $d = \infty$. We define

$$\Delta \varepsilon_r(S) = \varepsilon_r - \varepsilon_{r1} \qquad (5.16)$$

where ε_{r1} is obtained from the same f_0 value for a structure short circuited with a conducting ring of diameter d, as shown in Figure 5.6(a). We also define

$$\Delta \tan \delta(S) = \tan \delta - \tan \delta_1, \qquad (5.17)$$

where $\tan \delta_1$ is obtained from the same Q_0 value for a structure enclosed with an ideal wave absorber, in which the energy is perfectly dissipated without reflection, as shown in Figure 5.6(b). It is clear that, if $d = \infty$, $\Delta \varepsilon_r(S) = 0$ and $\Delta \tan \delta(S) = 0$.

To make the influence of the finite size of plates negligible, the required size ratio $S = d/D$ can be determined analytically from the view point of the possible magnitudes of $\Delta \varepsilon_r(S)$ and

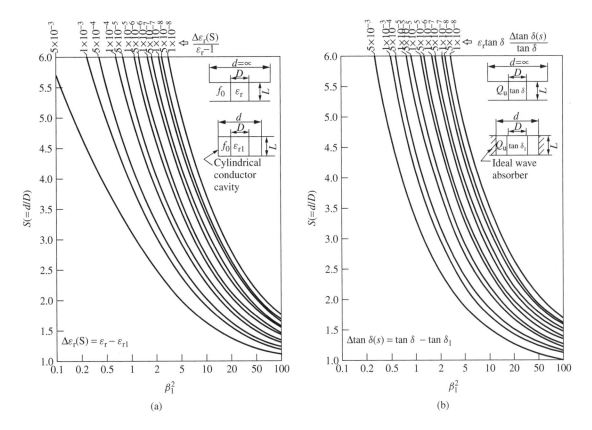

Figure 5.6 Size ratio S required for permissible error. (a) $\Delta \varepsilon_r(S)$ and (b) $\Delta \tan \delta(S)$ (Kobayashi and Katoh 1985). Source: Kobayashi Y. and Katoh, M. (1985). "Microwave measurement of dielectric properties of low-loss materials by the dielectric rod resonator method", *IEEE Transactions on Microwave Theory and Techniques*, **33** (7), 586–592. © 2003 IEEE

$\Delta \tan \delta(S)$ (Kobayashi and Katoh 1985a):

$$\frac{\Delta \varepsilon_{\mathrm{r}}(S)}{\varepsilon_{\mathrm{r}} - 1} = \frac{g(\beta_1)}{f'(\alpha_1)} \cdot \frac{2\alpha_1}{\beta_1^2 + \alpha_1^2}$$

$$\cdot \left[\frac{I_0(\beta_1)}{K_0(\beta_1)} + \frac{I_1(\beta_1)}{K_1(\beta_1)} \right] \cdot \frac{K_1(S\beta_1)}{I_1(S\beta_1)} \tag{5.18}$$

$$\varepsilon_{\mathrm{r}} \Delta \tan \delta(S) = SW \frac{K_0(S\beta_1)K_2(S\beta_1) - K_1^2(S\beta_1)}{K_0(\beta_1)K_2(\beta_1) - K_1^2(\beta_1)} \tag{5.19}$$

with

$$g(\beta_1) = -\alpha_1 \frac{K_0(\beta_1)}{K_1(\beta_1)} \tag{5.20}$$

$$f'(\alpha_1) = -\alpha_1 \frac{J_1^2(\alpha_1) - J_0(\alpha_1)J_2(\alpha_1)}{J_1^2(\alpha_1)} \tag{5.21}$$

where $I_n(\beta_1)$ is the modified Bessel function of the first kind and W is given by Eq. (5.8). The results computed from Eqs. (5.18) and (5.19) are shown in Figure 5.5. For a resonator with smaller β_1, more energy is stored outside the rod, and then the larger value of S is needed. In experiments, $\Delta \varepsilon_{\mathrm{r}}/\varepsilon_{\mathrm{r}}$ of 0.1 % can be obtained easily. In high-accuracy measurements, it is often required that the S value should be sufficiently large to satisfy $\Delta \varepsilon_{\mathrm{r}}(S)/\varepsilon_{\mathrm{r}} < 0.01$ % at least. For low-loss materials, the influence of S on $\tan \delta$ is more severe than that on ε_{r}.

5.2.1.3 Higher resonant modes

When we measure samples using Courtney resonator method, usually each sample is measured at only one resonant frequency, corresponding to the TE_{011} mode. In experiments, we can observe the existence of other resonant frequencies. If we can identify other resonant modes, it is possible to measure the value of ε_{r} at the frequencies corresponding to these resonant modes. Information on other modes can be found in Figure 5.2. Usually, it is not difficult to identify the resonant modes including TE_{011}, TE_{02l}, TE_{03l}, and TE_{04l}. Besides, quasi-TE modes are also well suited for the measurement of materials properties, in particular, the modes HEM_{12l}, HEM_{22l}, and HEM_{14l} (Wheless Jr. and Kajfez 1985).

The major challenge in using higher resonant modes is the identification of different resonant modes. Wheless Jr. and Kajfez proposed a modified Courtney resonator structure than can be used for mode identification, and detailed information about the modified Courtney resonator can be found in (Wheless Jr. and Kajfez 1985; Kajfez and Guillon 1986).

5.2.2 Cohn resonators

The basic properties of Cohn resonators have been discussed in Section 2.3.5.2, and here we concentrate on its application in the characterization of dielectric permittivity. In a Cohn resonator method, a sample in the shape of a right circular cylinder resonates within a close-fitting circular waveguide that is below cutoff in its air-filled regions, and the dielectric properties of the sample are obtained from the resonant properties of the resonator.

For a Cohn resonator, the operating frequency is above the cutoff frequency of the region filled with the dielectric cylinder, but is below the cutoff frequency of the air region. As shown in Figure 5.7, because of the exponential decay of the fields in the air regions, the total stored energy of the resonance is mainly within the dielectric sample. Resonance occurs when the following condition is satisfied (Cohn and Kelly 1966):

$$B_{\mathrm{a}} + B_{\mathrm{d}} = 0 \tag{5.22}$$

where B_{a} and B_{d} are the susceptances at one of the transverse surfaces of the sample looking toward the air region and dielectric region respectively. The lowest-order resonant mode for which the normal component of electric field E_n is zero over the surface of the dielectric, as required to make the air-gap effects negligible, is the TE_{011}. The characteristic admittance Y_0 of the waveguide is real in the propagating dielectric region and imaginary in the cutoff air region. For the TE_{01} mode, the susceptance B_{a} is equal to Y_0/j of the air-filled waveguide:

$$B_{\mathrm{a}} = \frac{Y_{0\mathrm{a}}}{\mathrm{j}} = -\frac{1}{\eta} \left[\left(\frac{\lambda}{0.820D} \right)^2 - 1 \right]^{\frac{1}{2}} \tag{5.23}$$

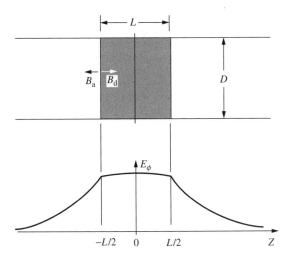

Figure 5.7 Resonant electric field distribution in a Cohn resonator at TE_{011} mode. Source: Cohn, S. B. and Kelly, K. C. (1966). "Microwave measurement of high-dielectric constant materials", *IEEE Transactions on Microwave Theory and Techniques*, **14** (9), 406–410. © 2003 IEEE

where $\eta = 376.7\,\Omega$ is the characteristic impedance of free space, λ is free-space wavelength, and D is the diameter of the waveguide and sample. Equation (5.23) is obtained under the assumption that the cutoff waveguide extends to plus and minus infinity. In practice, when the air-filled waveguide is terminated at points at which the evanescing wave is attenuated by at least 30 dB, errors caused by the termination are not significant.

As shown in Figure 5.7, in the TE_{011} mode, at position $z = 0$, the transverse field E_ϕ reaches its maximum value, while $H_t = 0$. So a magnetic wall, or open-circuiting plane, may be assumed to exist at $z = 0$. The susceptance B_d is therefore given by

$$B_d = Y_{0d}\tan\frac{\beta L}{2} = \frac{\lambda}{\eta\lambda_{gd}}\tan\frac{\pi L}{\lambda_{gd}} \qquad (5.24)$$

where

$$\lambda_{gd} = \frac{\lambda}{\{\varepsilon_r - [\lambda/(0.820D)]^2\}^{\frac{1}{2}}} \qquad (5.25)$$

By substituting Eqs. (5.23)–(5.25) into Eq. (5.22), we can get (Cohn and Kelly 1966)

$$\left(\varepsilon_r - \left(\frac{\lambda}{0.820D}\right)^2\right)^{\frac{1}{2}}\tan\left[\frac{\pi L}{\lambda}\left(\varepsilon_r - \left(\frac{\lambda}{0.820D}\right)^2\right)^{\frac{1}{2}}\right]$$

$$= \left(\left(\frac{\lambda}{0.820D}\right)^2 - 1\right)^{\frac{1}{2}} \qquad (5.26)$$

Equation (5.26) is a transcendental equation that may be solved for ε_r as a function of directly measurable dimensions and the resonant frequency $f = c/\lambda$, where c is the speed of light in free space.

Employment of this technique is straightforward, but the desired TE_{011} resonance mode must be identified among the various resonance modes that may also appear, such as TE_{111}, TM_{011}, and TM_{111}. In experiments, this identification can be easily accomplished. All the modes other than circular-electric modes produce longitudinal current on the conducting walls of the circular waveguide. Therefore, when the two bored blocks shown in Figure 5.8 are separated slightly, the TE_{011} resonance mode is hardly affected, while the other resonance modes are severely detuned. It should be noted that, to make a clear mode identification, the sample under test is placed off-center with respect to the gap. Otherwise, the gap would coincide with the null of longitudinal current at the midplane of the sample in some other resonance modes, such as TE_{111}, TM_{012}, and so on, making mode identification difficult. Though higher TE_{0mn} modes are also unaffected by the gap, as their resonant frequencies are separated widely from the TE_{011} resonance, they generally do not present any ambiguity to TE_{011} resonance mode.

As shown in Figure 5.9, electric dipole and magnetic dipole are two methods often used for excitation of Cohn resonator and coupling to external circuits. It is usually required that the coupling coefficient should be adjustable for the measurement of different samples.

This technique may be extended to measure loss tangent in addition to dielectric constant. For high-dielectric-constant samples, it is more accurate and convenient to measure the unloaded quality factor, Q_u, of the sample as a band-stop

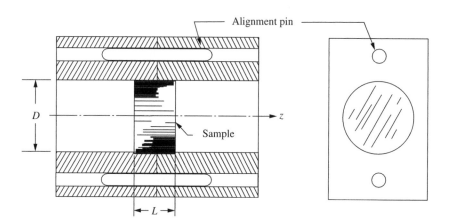

Figure 5.8   Split-block structure of a dielectric resonator shielded by a cutoff circular waveguide (Cohn and Kelly 1966). Source: Cohn, S. B. and Kelly, K. C. (1966). "Microwave measurement of high-dielectric constant materials", *IEEE Transactions on Microwave Theory and Techniques*, **14** (9), 406–410. © 2003 IEEE

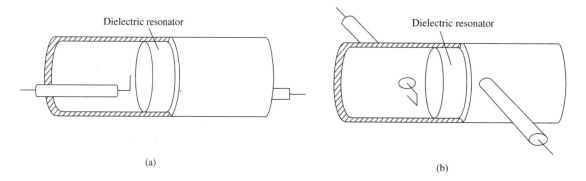

(a)   (b)

Figure 5.9   Coupling methods to a Cohn resonator. (a) Electric dipole coupling and (b) magnetic dipole coupling. Reproduced from Kent, G. (1988a). "Dielectric resonances for measuring dielectric properties", *Microwave Journal*, **31** (10), 99–114, by permission of Horizon House Publications, Inc.

resonator at the center of a large propagating waveguide. If the electric energy were completely stored inside the sample, and the dissipation loss outside the sample were zero, the loss tangent would be exactly $1/Q_u$. These conditions are not met perfectly in practice, but when ε_r is greater than 50, virtually all of the electric energy is stored in the resonant sample, and dissipation loss on the relatively distant waveguide walls is very small. So the value of $(1/Q_u)$ can be approximately taken as the loss tangent of the sample. The above analysis is based on the assumption that the magnetic loss is zero.

5.2.3 Circular-radial resonators

Figure 5.10 shows the $\rho - z$ plane of a circular-radial resonator that is rotationally symmetric along the z-axis. It mainly consists of two circular metallic plates placed opposing each other, where each plate has a center hole to accommodate a cylindrical rod of the material to be measured. The fixture has circular and radial waveguide regions with dimensions denoted by subscripts "c" and "r" respectively. The circular waveguide contains the dielectric rod and the radial waveguide contains the surrounding medium, generally air. At the junction

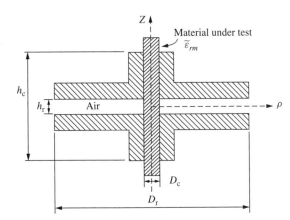

Figure 5.10 Structure of a circular-radial resonator. Modified from Humbert, W. R. and Scott, Jr. W. R. (1996). "A new technique for measuring the permittivity and loss tangent of cylindrical dielectric rods", *IEEE Microwave and Guided Wave Letters*, **6** (7), 262–264. © 2003 IEEE

of these waveguide regions is the core of the resonator. With properly chosen dimensions of h_r and D_c, the resonant mode will be confined in the region near the core of the resonator and produce exponentially decaying fields in the circular and radial waveguide regions. So the height of the fixture h_c and the diameter of the fixture D_r can be chosen to be of finite sizes. The relative permittivity and loss tangent of the rod sample can be derived from the measured resonant frequency and quality factor of the resonator (Humbert and Scott 1996).

We assume that the dielectric rod under test is linear, isotropic, homogeneous, and nonmagnetic. The fields and the resonant frequency for the lossless case can be calculated using a numerical method. Because the fixture is rotationally symmetric, it is an axis-symmetric problem; therefore, the structure can be reduced to a two-dimensional problem with a known ϕ dependence.

The quality factor is determined using the electromagnetic fields approximated by the perturbational analysis. Because the stored energies in the E- and H-fields are equal at resonance, we can write the loaded quality factor as

$$Q = \frac{2\omega_0 W_e}{P_l} \qquad (5.27)$$

where ω_0 is the angular resonant frequency, W_e is the time-average energy stored in the electric field, and P_l is the time-average power loss in the system. Both the losses in the dielectric rod and conduction losses in the metallic walls of the measurement fixture are considered. To achieve accurate results, it is necessary to ensure that the coupling of the probes to the resonator is weak and the dimensions of h_c and D_r are sufficiently large so that no appreciable fields leak out of the ends of the waveguide regions. So it can be assumed that the power extracted from the probes and the energy radiated out are negligible. Therefore, we have

$$P_l = P_d + P_c, \qquad (5.28)$$

where P_d is the power dissipated in the dielectric sample given by

$$P_d = \frac{\omega_0}{2} \int_V \varepsilon'_r \varepsilon_0 \tan\delta |E|^2 dv \qquad (5.29)$$

and P_c is the power loss due to surface currents J_s on the walls of the fixture

$$P_c = \frac{R_s}{2} \int_S |J_s|^2 ds \qquad (5.30)$$

where R_s is the surface resistance of the walls of the fixture. From Eq. (5.27), we can get

$$\frac{1}{Q} = \frac{1}{Q_d} + \frac{1}{Q_c} \qquad (5.31)$$

where Q_d and Q_c are the quality factors corresponding to the dielectric loss of the sample and the conducting loss of the fixture walls, respectively.

The permittivity of dielectric rod can be determined from the resonant frequency by using a root-finding technique (Humbert and Scott 1996), and the electromagnetic fields should be calculated. The calculated fields can also be used to determine the loss tangent of the dielectric rod from the quality factor, using Eqs. (5.27)–(5.31). The lowest-measurable dielectric loss tangent is limited by the conduction loss. The loss tangent can be accurately determined when the dielectric loss is comparable to or greater than the conduction loss. The highest-measurable dielectric loss tangent is limited by the lowest-measurable quality factor of experimental instrument.

The small air gaps between the dielectric rod and the measurement fixture may produce significant measurement errors for some resonance modes. Having only H_z, H_ρ, and E_ϕ field components, the TE_{011} mode does not have an E-field component normal to the air gaps; therefore, measurement results using this mode will not be strongly affected by the air gaps. In the following, we concentrate our discussion on the TE_{011} mode using the method by Humbert and Scott (1996).

The TE_{011} mode is ϕ-independent and has one variation in both the ρ- and z-directions. In and near the resonator core, the field distributions are complicated because of the complex geometry, and thus no closed-form expression for the field exists. However, sufficiently far from the resonator core, the field distributions are essentially zero due to a single waveguide mode in the circular and radial waveguide regions.

In the upper circular-waveguide region, the field is essentially the TE_{01} circular-waveguide mode. The mode is ϕ-independent, has one variation in the ρ-direction and has the z-dependence $\exp(-k_z z)$. As the structure is in a resonant state, the fields must decay, therefore $k_z z$ must be positive. In the radial waveguide region, the field

is essentially the TE_{01} radial waveguide mode. It is ϕ-independent, has one variation in the z-direction and has the ρ dependence $K_0(k_\rho \rho)$. Similarly, the fields must decay, and so $k_\rho \rho$ must be positive.

Figure 5.11 shows all the possible operating points for the TE_{011} mode. The operating boundaries are set by the cutoff conditions in the circular and radial waveguide regions, $k_z D_c = 0$, and $k_\rho D_c = 0$, respectively. For example, for $\varepsilon_r' = 3$, the TE_{011} mode will only resonate when h_r/D_c is between 0.23 and 0.97. The TE_{011} mode will not resonate for any point outside the $k_z D_c = 0$ and $k_\rho D_c = 0$ curves. The curves corresponding to $k_z D_c$ and $k_\rho D_c = 1$, 2, and 4, respectively, are also included. This allows the rates of decay in each of the waveguide regions to be determined at particular operating points. For example, with $\varepsilon_r' = 10$ and $h_r/D_c = 0.5$, the mode will resonate with decay rates $2 < k_z D_c < 4$ and $k_\rho D_c > 4$.

Figure 5.12 shows the normalized resonant frequencies as a function of h_r/D_c with ε_r' as a parameter. In this graph, the left endpoints of these curves correspond to $k_z D_c = 0$ and right endpoints are where $k_\rho D_c = 0$. Continuing the previous examples, for $\varepsilon_r' = 3$ the normalized resonant frequency ranges from 3.15 to 4.39. For $\varepsilon_r' = 10$ and $h_r/D_c = 0.5$, the normalized resonant frequency is 2.2.

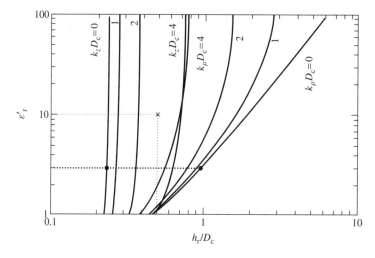

Figure 5.11   Diagram depicting valid operating regions of the TE_{011} mode (Humbert and Scott 1996). Source: Humbert, W. R. and Scott, Jr. W. R. (1996). "A new technique for measuring the permittivity and loss tangent of cylindrical dielectric rods", *IEEE Microwave and Guided Wave Letters*, **6** (7), 262–264. © 2003 IEEE

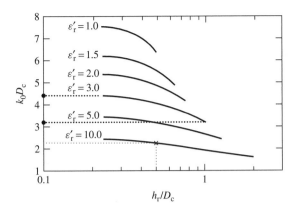

Figure 5.12 Normalized resonant frequencies of the TE$_{011}$ mode (Humbert and Scott 1996). Source: Humbert, W. R. and Scott, Jr. W. R. (1996). "A new technique for measuring the permittivity and loss tangent of cylindrical dielectric rods", *IEEE Microwave and Guided Wave Letters*, **6** (7), 262–264. © 2003 IEEE

For accurate determination of the loss tangent, an accurate value for the surface resistance of the metal walls is needed. The surface resistance of the metal walls can be measured using a known low-loss dielectric material; in this case, the majority of the loss is the conduction loss in the metal walls. In most cases, the dielectric loss of air is negligible, and the measurement fixture will resonate when it is filled with air. Therefore, by measuring the resonant frequency and quality factor of a fixture filled with air, we can determine the permittivity of air and the surface resistance of the metal walls of the fixture.

5.2.4 Sheet resonators

Dielectric sheet materials are widely used as substrates in microwave electronic circuits, and so the characterization of the dielectric properties of sheet materials is important. If a resonance is excited in a sheet sample, the dielectric properties of the sheet can be calculated from the resonant frequency and the quality factor of the sheet resonator. Figure 5.13 schematically shows a method of exciting a sheet resonator. The sheet sample is placed between the flat flanges of the two metal circular waveguides. Usually, the waveguides work in the TE$_{01}$ mode and the dielectric sheet resonates at a TE mode. Therefore, intimate contact between the sample and the flat flanges is not required. The thickness of the sheet can be thinner than the gap between the two flat flanges, and the measurement results are not sensitive to the position of the sheet in the gap.

In the structure shown in Figure 5.13, the transmission through the sheet sample of TE$_{01}$ mode reaches its maximum at frequency below the cutoff frequency of the waveguides and above the cutoff frequency of the cylindrical portion of the sample. If the sheet sample is large enough, the frequency at which the maximum transmission occurs is mainly determined by the waveguide radius, the thickness, and the dielectric constant of the sheet sample. The resonant frequency of the sheet resonator is close to that of the TE$_{01\delta}$ mode of a dielectric resonator of diameter $2a$ and height $2d$.

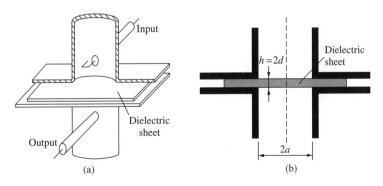

Figure 5.13 A sheet resonator for the characterization of dielectric sheets. (a) Three-dimensional view and (b) cross-section view. Reproduced from Kent, G. (1988a). "Dielectric resonances for measuring dielectric properties", *Microwave Journal*, **31** (10), 99–114, by permission of Horizon House Publications, Inc.

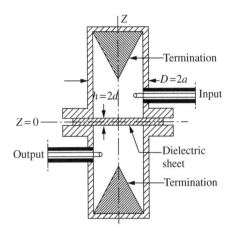

Figure 5.14 Absorbers are used to eliminate the possible resonance of the metal shield (Kent 1988b). Source: Kent, G. (1988b). "An evanescent-mode tester for ceramic dielectric substrates", *IEEE Transactions on Microwave Theory and Techniques*, **36** (10), 1451–1454. © 2003 IEEE

Figure 5.13(a) shows an excitation and coupling method for sheet resonators. Usually there is a coupling port at the cylinder waveguide at each side of sheet resonator. In the measurement system, the waveguides are not in resonant states. As shown in Figure 5.14, when shorted waveguides are used, absorbers are often used at the shorting plates of the waveguides to avoid the possible resonance of the metal shield.

We follow the analysis method presented in (Kent 1988b). We are interested in the condition when dielectric sheet is in a resonant state while the waveguides are in a cutoff state. It is not needed to consider the modes that propagate or are evanescent both in the empty portions of the waveguide and in the dielectric sheet sample. It is also not necessary to consider any narrowband resonance phenomena of the cylinder waveguides. Furthermore, we assume that a mode that is evanescent in the waveguide but propagates in the dielectric is evanescent in the radial waveguide in the gap beyond the radius a. So we can use the conducting wall boundary condition at $r = a$ in the dielectric, and the error that follows from this approximation can be estimated by a perturbation calculation.

The boundary conditions at the dielectric surfaces lead to the eigenvalue equations (Kent 1988b). For TE modes

$$(\theta \sin \theta - \gamma \cos \theta)(\theta \cos \theta + \gamma \sin \theta) = 0 \quad (5.32)$$

and for TM modes

$$(\theta \sin \theta - \varepsilon_r' \gamma \cos \theta)(\theta \cos \theta + \varepsilon_r' \gamma \sin \theta) = 0 \quad (5.33)$$

The notations for above two equations are as follows:

$$\theta^2 = \varepsilon_r' \theta_0^2 - \theta_c^2 \quad (5.34)$$

$$\varepsilon_r' = (\varepsilon'/\varepsilon_0) \quad (5.35)$$

$$\gamma^2 = \theta_c^2 - \theta_0^2 \quad (5.36)$$

$$\theta_c = (x_{lm} d/a) \quad (5.37)$$

$$\theta_0 = (\omega d/c) \quad (5.38)$$

For a TE_{lm} mode, x_{lm} is the mth zero of the derivative of J_l, and for a TM_{lm} mode, it is the mth zero of J_l. A resonance exists when Eq. (5.36) is positive, and Eq. (5.32) or (5.33) has a solution for θ in terms of γ. Then, the dielectric constant ε_r' can be calculated from Eq. (5.34).

In Eqs. (5.32) and (5.33), the first term is zero for even modes and the second term vanishes for odd modes. For the reasonable dielectric constant constraints of $\varepsilon_r' < 100$ and $(d/a) < 0.05$, there are no odd-mode solutions. The effect of the substrate on TM modes is relatively small, and the even-mode TM resonant frequencies are close to their cutoff frequencies. For the TE_{01} and TM_{11} modes, which have the same cutoff frequency, the substrate produces a wide separation between the two even resonant modes.

The modes one expects to observe in order of increasing frequency are TE_{11}, TE_{21}, TE_{01}, TE_{31}, ... With small coupling loops in the transverse plane, coupling to the TM modes is very weak. The resonant modes of the first four TE modes are well separated and can be easily identified. All these frequencies can be used to calculate the dielectric constant, but only the TE_{01} mode gives reliable results. Figure 5.15 shows the relationship between the dielectric constant and the resonant frequency of a TE_{01} mode resonator.

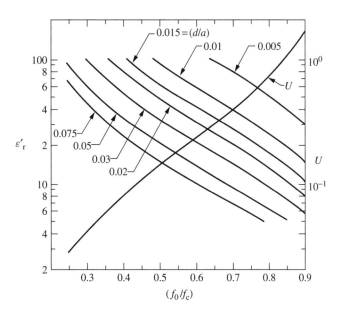

Figure 5.15 Curves of ε_r' versus (f_0/f_c) of the TE_{01} mode with (d/a) as parameter. As shown is the energy ratio U for $(d/a) = 0.03$. (Kent 1988b). Source: Kent, G. (1988b). "An evanescent-mode tester for ceramic dielectric substrates", *IEEE Transactions on Microwave Theory and Techniques*, **36** (10), 1451–1454. © 2003 IEEE

For TE_{01} mode, the reciprocal of the unloaded quality factor is given by

$$\frac{1}{Q_0} = \frac{(\varepsilon_r''/\varepsilon_r') + (\delta/a)(\theta_c/\theta_0)^2 U}{1 + U} \qquad (5.39)$$

where δ is the skin depth of the waveguide wall, $(\varepsilon_r' - j\varepsilon_r'')$ is the relative complex permittivity of the substrate. We also have (Kent 1988b)

$$\varepsilon_r' \cdot U = \frac{\cos^3 \theta}{(\theta + \sin \theta \cos \theta) \sin \theta} \qquad (5.40)$$

where the quantity U is the ratio of the electric energy in the waveguide to that in the substrate. The value of U increases when the resonant frequency f_0 approaches the cutoff frequency f_c and the dielectric constant ε_r' decreases. The value of U is not sensitive to (d/a) over its practical range. The contribution to $(1/Q_0)$ from conductor losses, represented by the second term in the numerator of Eq. (5.39), cannot be made negligibly small, and it cannot be measured at the correct frequency in a simple way. The calculations of that term, using the bulk conductivity of brass, indicate

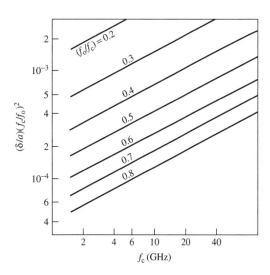

Figure 5.16 Plot of $(\delta/a)(f_c/f_0)^2$ versus f_c for brass waveguide (Kent 1988b). Source: Kent, G. (1988b). "An evanescent-mode tester for ceramic dielectric substrates", *IEEE Transactions on Microwave Theory and Techniques* **36** (10), 1451–1454. © 2003 IEEE

that it increases more or less as the root of the cutoff frequency f_c of the waveguide increases, as shown in Figure 5.16.

Three sources of error in the determination of dielectric constant are uncertainties in the measurement of a, d, and f. They can be estimated by the following expressions (Kent 1988b):

$$\left(\frac{\Delta \varepsilon_r'}{\varepsilon_r'}\right)_a = -2\left(\frac{\theta_c}{\theta_0}\right)^2\left(U + \frac{1}{\varepsilon_r'}\right)\left(\frac{\Delta a}{a}\right) \quad (5.41)$$

$$\left(\frac{\Delta \varepsilon_r'}{\varepsilon_r'}\right)_d = -\frac{2}{\varepsilon_r'}\left(\frac{\theta_c}{\theta_0}\right)^2 \frac{1}{1 + (\sin 2\theta)/(2\theta)}\left(\frac{\Delta d}{d}\right) \quad (5.42)$$

$$\left(\frac{\Delta \varepsilon_r'}{\varepsilon_r'}\right)_{\theta_0} = -2(1 + U)\left(\frac{\Delta \theta_0}{\theta_0}\right) \quad (5.43)$$

A fourth source of error is the assumption that the electric field in the substrate is zero at $r = a$. This can be estimated as a frequency perturbation produced by removing the stored energy of the evanescent radial modes in the space $r > a$. The approximate result is (Kent 1988b)

$$\frac{\Delta \varepsilon_r'}{\varepsilon_r'} = \frac{4}{\pi \varepsilon_r'} \cdot \left(\frac{\theta_c}{\theta_0}\right)^2 \left[1 - \varepsilon_r' \cdot \left(\frac{2\theta_0}{\pi}\right)^2\right]^{-\frac{1}{2}}\left(\frac{d}{a}\right) \quad (5.44)$$

The approximation is made by truncating the Fourier expansion in evanescent modes and equating the remaining coefficient to its asymptotic form. The validity depends on how small the value of (d/a) is. For practical parameter values, $(\varepsilon_r' \cdot d/a) \sim 0.25$, and an overestimate of Eq. (5.44) is about $10(d/a)^2$. Errors from this source are usually below 1 %.

With different radii of sheet resonators, sheet sample can be measured over a frequency range. The maximum radius must be somewhat less than the width of the substrate, and the minimum radius is determined by the available instrumentation and the maximum possible value of (d/a). Figure (5.15) suggests that it should be possible to span the frequency range up to one octave.

It should be noted that, if we want to make an exact analysis on this measurement configuration, full-wave simulation method is needed (Humbert and Scott 1997). Using full-wave simulation method, there is no limit on the sample thickness and dielectric constant.

5.2.5 Dielectric resonators in closed metal shields

In the configurations of the dielectric resonators discussed above, the metal shields are physically open, though cut off for the working microwave signals. Here, we discuss dielectric resonators shielded by closed metal shields. We mainly discuss two types of dielectric resonators in closed metal shields: TM_{0n0} resonators and dielectric-loaded cavities.

5.2.5.1 TM_{0n0} resonators

Most of the dielectric resonators for the characterization of dielectric properties of materials work at TE modes. Generally speaking, neither the quasi-TM modes nor the TM modes are suitable for the measurement of materials properties. The main reason is that the presence of an air gap between the dielectric sample and the metal shield may considerably alter the resonant frequency, f, thereby ruining the accuracy of measurement. Another reason is that these resonant modes are leaky, in the sense that a part of electromagnetic energy may propagate radially outward. The leaky TM_{0np} modes have no radial cutoff frequency, and consequently their quality factors are relatively low.

However, as shown in Figure 5.17, cylindrical TM_{0n0} cavities can be used in the characterization of dielectric materials (Gershon *et al.* 2000; Damaskos 1995). In this configuration, the sample is put at the center of the cavity, and the sample has good contacts with the top and bottom end plates of the metal shield. Usually for this configuration, the diameter of the cavity is larger than

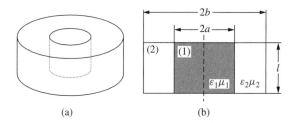

(a)   (b)

Figure 5.17 Cylindrical TM_{0n0} cavity loaded with cylindrical samples. (a) Three-dimensional view and (b) cross view

its height ($2b > l$). The coupling to the resonator can be realized by coupling loops at the sidewalls or electric dipoles at the two end plates. A small air gap between the end plates of the cavity and the sample may cause great uncertainties to the measurement results. If the cavity has a movable wall, samples with different thickness can be measured (Gershon *et al.* 2000).

The resonant frequency of the resonator shown in Figure 5.17 is related to the radius of the dielectric rod a and that of the metal shield b. Conclusions to semispherical dielectric resonator shown in Figure 2.99 can be extended to this configuration. If $a \ll b$, the resonant frequency of the resonator approaches that of the resonator fully filled with materials with dielectric constant ε_2. If $a \approx b$, the resonant frequency of the resonator approaches that of the resonator fully filled with materials with dielectric constant ε_1. In a certain range, the resonant frequency of the resonator changes little with the radius of the dielectric rod. To increase the measurement accuracy and sensitivity of loss tangent, we should use this region.

In the following, we analyze this measurement configuration in a general case, so that we can compute the complex permittivity from the resonant frequency and quality factor of the resonator loaded with the sample. As shown in Figure 5.17, the sample, which is centered in the cavity and labeled as region 1, has material properties of ε_1 and μ_1. Region 2, which is air, has material properties of ε_2 and μ_2. The resonant cavity has an inside diameter of $2b$ and the sample has a diameter of $2a$. The height of the cavity and the sample is l.

The electric and magnetic fields within the cavity interact with the material under test. To relate the shift and broadening of the resonant modes to the intrinsic properties of the material, we match the electromagnetic fields at the sample–air interface. The general expressions for the electric field in axial direction in region i ($i = 1, 2$) are (Gershon *et al.* 2000)

$$
\begin{aligned}
E_{zi} = &[A_i J_m(\alpha_i \rho) + B_i N_m(\alpha_i \rho)] \\
&\times (C_i e^{im\phi} + D_i e^{-im\phi}) \\
&\times [E_i \cos(kz) + F_i \sin(kz)]
\end{aligned} \tag{5.45}
$$

with α_i defined by

$$
\alpha_i^2 = \mu_i \varepsilon_i \omega^2 - k^2 \tag{5.46}
$$

where J_m and N_m are the Bessel and Newman functions of order m. A, B, C, D, E, and F are constants. ρ is the radius and ϕ is the angle around the axis of symmetry, α is the transverse wave number, ω is the angular frequency, and i indicates the region in the cavity ($i = 1, 2$).

As the diameter of the cavity is larger than its height, the lowest-order modes in a cylindrical cavity are TM modes ($H_z = 0$). Boundary conditions at the top and bottom ends of the cavity require the electric field parallel to a conducting surface to be zero. So we have

$$
kl = p\pi \qquad (p = 0, 1, 2, \ldots) \tag{5.47}
$$

Applying this boundary condition at the cavity walls and matching the tangential electric and magnetic fields at the sample interface provides the following relationship (Gershon *et al.* 2000):

$$
\begin{aligned}
&\frac{\varepsilon_1}{\alpha_1} \cdot \frac{J_m'(\alpha_1 a)}{J_m(\alpha_1 a)} \\
&= \frac{\varepsilon_2}{\alpha_2} \cdot \left[\frac{J_m(\alpha_2 b) N_m'(\alpha_2 a) - J_m'(\alpha_2 a) N_m(\alpha_2 b)}{J_m(\alpha_2 b) N_m(\alpha_2 a) - J_m(\alpha_2 a) N_m(\alpha_2 b)} \right]
\end{aligned} \tag{5.48}
$$

The complex permittivity of the sample can then be calculated from the measured complex angular frequency of the cavity:

$$
\omega = 2\pi \left(f_0 + \frac{f_0}{2Q_s} \right) \tag{5.49}
$$

and numerically solving Eq. (5.48) for complex permittivity. Here, Q_s is the quality factor of the sample:

$$
\frac{1}{Q_s} = \frac{1}{Q_l} - \frac{1}{Q_e} \tag{5.50}
$$

where Q_l is the Q-factor of a loaded cavity and Q_e is the Q-factor of the empty cavity. In the determination of the complex permittivity of the material, an initial estimation is needed, and a two-variable minimization routine is required. In actual measurements, the accuracy and reproducibility in the measurement of f_0 and Q_e are mainly

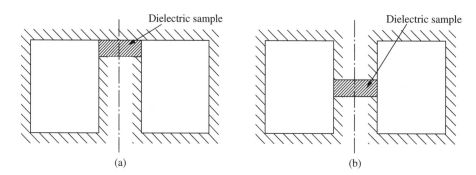

Figure 5.18 Capacitively loaded TM_{010} mode cavity for permittivity measurement. (a) Single-post cavity and (b) two-post cavity. Modified from Hakki, B. W. Coleman, P. D. (1960). "A dielectric resonator method of measuring inductive capacities in the millimeter range", *IRE Trans. Microwave Theory Tech*, **8**, 402–410. © 2003 IEEE

determined by whether the electrical contacts between the sample and the end plates of the metal shield are good and consistent.

Sometimes, cylindrical TM_{0n0} cavities are modified into reentrant cavity by capacitive loading. Figure 5.18 shows two configurations of TM_{010} mode cavity with capacitive loading. In these structures, the electric field is concentrated on the area between the postends or between the postend and the end plate of the cavity. The sample under test is put at the place with the strongest electric field, so the measurement sensitivity and accuracy can be improved. Xi *et al.* (1994, 1992) analyzed the cases when the gap is partially filled with a dielectric material.

Figure 5.19 shows a modified cylindrical cavity in the TM_{010} mode with two apertures at the two end plates for sample loading. In this configuration, a large part of the electromagnetic energy is concentrated in the vicinity of the center. The electric field lies along the axis, and the two apertures hardly affect the current lines. As the moisture of a material is related to its conductivity, which is further related to the imaginary part of permittivity, the moisture of the material could be derived from the quality factor of the resonator. This technique has been applied for the measurement of the moisture content of products on a fabrication line, for example, tobacco in cigarettes, and this technique can obtain the moisture-content information in real time (Bourdel *et al.* 2000).

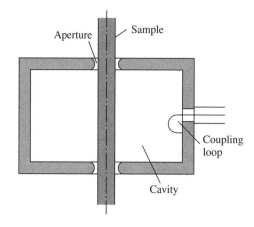

Figure 5.19 The structure of a modified TM_{010} mode cavity. Modified from Bourdel, E. Pasquet, D. Denorme, P. and Roussel, A. (2000). "Measurement of the moisture content with a cylindrical resonating cavity in TM_{010} mode", *IEEE Transactions on Instrumentation and Measurement*, **49** (5), 1023–1028. © 2003 IEEE

5.2.5.2 Dielectric-loaded cavities

Here, we discuss a general case of a closed cavity loaded with a dielectric sample. The cavity and the dielectric rod can be in any shape, but usually the cavity and the dielectric rod are in simple shapes, such as cylinder, sphere, and rectangle. Figure 5.20 shows a cylinder cavity loaded with a dielectric rod, which is supported by a low dielectric constant, and a low-loss dielectric material.

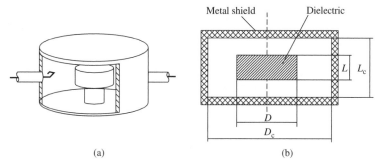

(a)   (b)

Figure 5.20 A closed cavity loaded with dielectric. (a) Three-dimensional view. Reproduced from Kent, G. (1988a). "Dielectric resonances for measuring dielectric properties", *Microwave Journal*, **31** (10), 99–114, by permission of Horizon House Publications, Inc. (b) Cross-section view. Source: Krupka, J. Derzakowski, K. Abramowicz, A. Tobar, M. E. and Baker-Jarvis, J. (1999). "Use of whispering-gallery modes for complex permittivity determinations of ultra-low-loss dielectric materials", *IEEE Transactions on Microwave Theory and Techniques*, **47** (6), 752–759. © 2003 IEEE

Measurement of dielectric constant

In a resonator method, the dielectric constant is derived from the resonant frequency of the dielectric resonator. Generally speaking, measurement accuracies for dielectric constant are limited by three factors (Krupka *et al.* 1999). First, in the resonant structure that requires contact between the sample and the metal shield, the presence of air gaps between the sample and the metal shield may cause uncertainties. Second, computational inaccuracies limit the accuracy of measurement results. Third, the uncertainties in physical dimensions of the sample under test and resonant structure are transferred to the final results of dielectric constant.

The presence of air gaps limits the measurement accuracy of high-permittivity solid materials when the distribution of electromagnetic field in the measurement system has an electric field component normal to the sample surface. The depolarizing effects of air gaps can be mitigated by metallizing the sample surfaces. The metallization of sample surfaces improves the measurement accuracy of dielectric constant, but can substantially degrade measurement accuracy of dielectric loss factor.

When the applied electric field is continuous across a sample boundary, such as with cylindrical samples in TE_{0np} or quasi-TE_{0np}-mode resonant fixtures, high measurement accuracies can be achieved, as in these cases, air gaps do not play a significant role. In the structure shown in Figure 5.20, the sample under test is far from the metal shield. The problem of air gap does not exist, and furthermore, the effect of conductor losses on the complex permittivity determination is negligible.

Numerical methods are often required for the analysis of the resonant structure, and computational inaccuracies can be an important source of the overall measurement uncertainty. Exact relations between permittivity, sample dimensions, and measured resonant frequency exist when resonant structures permit theoretical analysis by separation of variables. Practically, this is only possible when the resonant structure has simple cylindrical, spherical, or rectangular geometry. Numerical methods, for example finite element, Rayleigh–Ritz, and mode-matching methods, are often required for analyzing more complicated resonant structures. All these techniques allow an improvement of accuracy by taking into account more terms in the field expansions or by refining the mesh. In principle, if the convergence of a numerical method used to analyze a particular structure is sufficiently fast, the computations may have higher accuracy than the measurement accuracy of physical dimensions (Krupka *et al.* 1999). However, depending on the numerical method, substantial computational effort is usually required.

If the effects of air gaps are not severe, and the numerical method used for the analysis is sufficiently accurate, the accuracy in the measurement of dielectric constant depends essentially on the measurement accuracy of the physical dimensions of the sample and measurement system. It can be expected that the most accurate permittivity-measurement techniques would be dominated by the measurement accuracy of the physical dimensions.

Measurement of dielectric loss tangent

For a resonant structure containing an isotropic dielectric, the unloaded quality factor Q_u can be expressed in the form of (Krupka *et al.* 1999)

$$\frac{1}{Q_u} = p_{es} \tan \delta_s + p_{ed} \tan \delta_d + \frac{R_s}{G} + \frac{1}{Q_r} \quad (5.51)$$

where R_s is the surface resistance of the metal shield and Q_r is the radiation quality factor of the resonant structure. The parameter p_{es} is the sample partial electric-energy filling factor defined by

$$p_{es} = \frac{W_{es}}{W_{et}} = \frac{\iiint_{V_s} \varepsilon_s E \cdot E * \mathrm{d}v}{\iiint_V \varepsilon(v)E \cdot E * \mathrm{d}v} \quad (5.52)$$

where W_{es} is the electric energy stored in the sample, W_{et} is the total electric energy stored in the resonant-measurement fixture, ε_s is the dielectric constant of the sample, $\varepsilon(v)$ is the spatially dependent dielectric constant in the resonant structure, and * denotes complex conjugate. The parameter p_{ed} is the electric-energy filling factors of dielectric supports with dielectric constant ε_d and loss tangent $\tan \delta_d$, inside the resonant-measurement system:

$$p_{ed} = \frac{W_{ed}}{W_{et}} = \frac{\iiint_{V_d} \varepsilon_d E \cdot E * \mathrm{d}v}{\iiint_V \varepsilon(v)E \cdot E * \mathrm{d}v} \quad (5.53)$$

where W_{ed} is the electric energy stored in the dielectric support. The parameter G is the geometrical factor of the resonant system defined by

$$G = \frac{\omega \iiint_V \mu_0 H \cdot H * \mathrm{d}v}{x \oiint_S H_t \cdot H_t * \mathrm{d}v} \quad (5.54)$$

It is clear that, to obtain high accuracy and sensitivity, the first term on the right-hand side of Eq. (5.51) should be larger than other terms. The estimation of the last three terms often requires the rigorous numerical computation methods. Generally speaking, there are three methods to increase the measurement accuracy. The first is to optimize the structure of the metal shield, as will be discussed below in details. The second is to minimize the conductor loss of the metal shield by using low-surface-resistance conductors. We may use superconducting thin films, or coat the shield with a layer of gold. The third is to use resonant mode with low radiation loss, such as whispering-gallery modes, which is further discussed in Section 8.3.3.4.

For most of the actual resonant-measurement fixtures, the measurement uncertainties of the dielectric loss tangent are limited by conductor losses. Equation (5.51) indicates that conductor losses decrease as the surface resistance becomes small and as the geometric factor increases. One common procedure by which conductor losses are minimized is to situate the dielectric sample under test in a position away from the conducting walls, as shown in Figure 5.20.

In most cases, quasi-TE$_{011}$ mode (TE$_{01\delta}$ mode) is used. Figure 5.21 shows the geometric factor for TE$_{01\delta}$-mode dielectric resonators as a function of permittivity and the diameter of the metal shield. There is a maximum value for the geometric factors as a function of D_c/D with the position and value depending on the sample permittivity. So, the optimal shield dimensions for dielectric loss-tangent measurements depend on the permittivity of the sample and its diameter. Besides, when D_c/D becomes larger than the optimum value, the electric-energy filling factor of the sample decreases rapidly and the field distribution converges to that for an empty cavity, and thus the geometric factor converges to that of an empty cavity.

As shown in Figure 5.21, for an empty cavity having an aspect ratio equal to two ($D_c/D \gg 1$), the geometric factor is approximately 667. The other limiting case is that the dimensions of the shield are equal to those of the sample under test ($D_c/D = 1$). For this case, the geometrical factors are the same as that of the empty cavity normalized

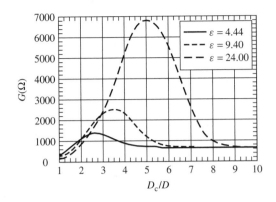

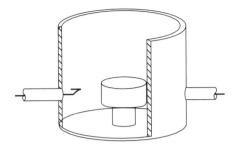

Figure 5.21 Geometrical factors versus relative dimensions of metal shield for $TE_{01\delta}$-mode dielectric resonators with an aspect ratio $D_c/L_c = D/L = 2$ (Krupka *et al.* 1999). Source: Krupka, J. Derzakowski, Abramowicz, A. Tobar, M. E. and Baker-Jarvis, J. (1999). "Use of whispering-gallery modes for complex permittivity determinations of ultra-low-loss dielectric materials", *IEEE Transactions on Microwave Theory and Techniques*, **47** (6), 752–759. © 2003 IEEE

Figure 5.22 Open cavity for dielectric property characterization. Reproduced from Kent, G. (1988a). "Dielectric resonances for measuring dielectric properties", *Microwave Journal*, **31** (10), 99–114, by permission of Horizon House Publications, Inc.

loss-tangent measurements as a result of very large radiation losses.

In some cases, the closed metal shield is not convenient for sample loading and unloading. Figure 5.22 shows a resonator structure with one of the end plates of the metal shield replaced by a cutoff circular waveguide. For a carefully designed structure, the radiation loss can be neglected.

5.3 COAXIAL SURFACE-WAVE RESONATOR METHODS

Coaxial surface-wave resonators are developed from coaxial surface-wave transmission structures. As shown in Figure 5.23, two types of coaxial surface-wave transmission lines are often used: open coaxial surface-wave transmission line and closed coaxial surface-wave transmission

by the square root of the sample permittivity, so the geometrical factor becomes small, especially for high-permittivity samples. Therefore, the sensitivity for the measurement of dielectric loss-tangent is decreased. In a typical fixture for the measurement of dielectric loss tangent, shield dimensions are usually three times larger than that of the dielectric resonator under test. It should be noted that $TE_{01\delta}$-mode dielectric resonators without any shield ($D_c/D \to \infty$) are not suitable for dielectric

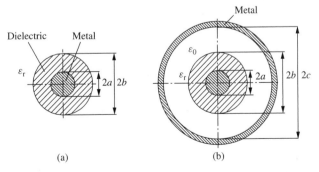

Figure 5.23 Cross sections of coaxial surface waveguides. (a) Open coaxial surface waveguides and (b) closed coaxial surface waveguide. Reproduced from Marincic, A. Benson, F. A. and Tealby, J. M. (1986). Measurements of permittivity by the use of surface waves in open and closed structures, *IEE Proceedings-H*, **133**, 441–449, by permission of IEE

line (Marincic *et al.* 1986). For an open coaxial surface-wave transmission line, there are two-dimensional parameters: the radius of the metal cylinder *a* and the outer diameter of the dielectric cylinder *b*, while for a closed coaxial surface-wave transmission line, the radius of the outer screen *c* has also to be included. The disadvantage of having one more parameter in the latter case is usually well compensated for by the fact that the measuring system is more compact and less subject to the effects of the surrounding medium and finite short-circuit plates. The discussions on the properties of open surface-wave transmission lines can be found in Section 2.2.4.5. In the following, we focus on the properties of coaxial surface-wave resonators and their applications in the characterization of dielectric properties.

5.3.1 Coaxial surface-wave resonators

A coaxial surface-wave resonator is formed by terminating a segment of surface-wave transmission lines with short circuits at both ends. By introducing an input signal probe at one end of the resonator, and another probe at some other place in the resonator, a transmission-type resonator is produced. In building a coaxial surface-wave resonator, it is important to minimize the direct coupling between the probes. If the couplings are weak, a high quality factor of the resonator can be obtained, and accurate determination on the resonant frequency is possible. At resonance, the length of the resonator is a multiple of half-waveguide wavelength, and the phase coefficient of the surface wave is given by

$$\beta = n\frac{\pi}{L} \qquad (5.55)$$

where *n* is the number of half waves along the length *L* of the resonator. When resonant frequency *f* and phase coefficient β are known, the relative permittivity of the dielectric ε_r can be obtained. The relationships between β, f, a, b, c, and ε_r are usually quite complex, and so have to be obtained numerically. However, in some special cases, by using a simple quasi-static approach, it is

possible to obtain less accurate results for closed structures at lower frequencies.

The basic requirement for the measurement of dielectric properties using a coaxial conductor-dielectric structure is that this structure should always support a TM_{01} surface wave at any operating frequencies. The coaxial surface-wave resonator may have one or two ports. For a one-port resonator, it is possible to detect resonance by observing the reflected wave magnitude. However, if a weak-probe coupling is preferable, the detection of the resonance can be very difficult because of the relatively small change of in the magnitude of the reflected wave. In the following discussion, we concentrate on the two-port transmission-type resonators, which are more convenient and easier to use.

5.3.2 Open coaxial surface-wave resonator

As shown in Figure 5.24, an open surface-wave resonator is a segment of open coaxial surface waveguide terminated by two metal plates, and usually there are several dielectric spacers fixing the two metal plates. The plate size has to be sufficiently large to act as an efficient reflector of the surface waves. Its efficiency can be determined by the decay rate of the field in the radial direction. The determination of the plate size is based on

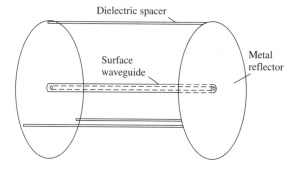

Figure 5.24   Open surface-wave resonator (Marincic *et al.* 1986). Reproduced from Marincic, A. Benson, F. A. and Tealby, J. M. (1986). Measurements of permittivity by the use of surface waves in open and closed structures, *IEE Proceedings-H*, **133**, 441–449, by permission of IEE

the power transfer through the space around the surface waveguide.

The calculation of relative permittivity ε_r of a dielectric in the open surface-wave resonator is based on the solution of characteristic equation Eq. (2.272). The input parameters in this case are a, b, f, and L. Parameters a, b, and L are obtained from the geometry of the open surface-wave resonator, and the resonant frequencies f of the resonator can be obtained from actual microwave measurements.

The experimental β/k_0 values, which are the basis for any calculation of ε_r at a number of resonant frequencies f_n, are obtained from (Marincic *et al.* 1986)

$$\frac{\beta}{k_0} = \frac{150n}{Lf_n}, \quad (n = 1, 2, 3, \ldots) \qquad (5.56)$$

where L is the resonator length in meters, n is the number of the resonance, and f_n is in megahertz. Calculation of ε_r, for known a, b, and β/k_0 determined from Eq. (5.56), requires solution of Eqs. (2.270), (2.271), and (2.272) in terms of ε_r. In the calculation, initial values of permittivity are taken:

$$\varepsilon_1 \leq \varepsilon(x) \leq \varepsilon_2 \qquad (5.57)$$

where the lower limit is ε_1 and the upper limit is ε_2. A search through the numerical solution of the characteristic equation for the input parameters can be carried out using the bisection technique. The calculated values of β/k_0 for several assumed $\varepsilon(x)$ values are compared with β/k_0 values from Eq. (5.56). The logical procedure of searching narrower limits for $\varepsilon(x)$ is repeated until the value β/k_0 calculated from $\varepsilon(x)$ and the value β/k_0 obtained from Eq. (5.56) differ by a specified small amount.

Figure 5.25 shows the structure of an open surface-wave resonator with two ports. Two metal plates serve as the flat reflectors. The plates are separated by dielectric spacers with low loss and low dielectric constant, such as wooden rods, at the edges. One loop probe is attached to the reflector, and the other probe is fixed on one of the dielectric spacers.

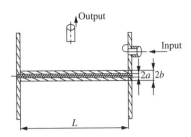

Figure 5.25 Structure of open surface-wave resonator (Marincic *et al.* 1986). Reproduced from Marincic, A. Benson, F. A. and Tealby, J. M. (1986). Measurements of permittivity by the use of surface waves in open and closed structures, *IEE Proceedings-H*, **133**, 441–449, by permission of IEE

5.3.3 Closed coaxial surface-wave resonator

In practical experiments, as the reflecting plates in an open coaxial surface-wave resonator have limited size, it is difficult to realize the total reflection at the two ends of the resonator. This problem can be solved by using a closed surface-wave structure, and a closed surface-wave resonator is usually more compact than an open one. The same idea for measurement of permittivity as with the open resonator is used with closed structures.

The geometry of the waveguide considered is shown in Figure 5.23(b). Similar to the open surface waveguide case discussed above, the wave equation of Eq. (2.268) should be solved in the dielectric and the surrounding region. In this case, as there is another conducting surface at $r = c$, the air region is limited. The solutions for the fields $E_z(r, \varphi)$ and $H_z(r, \varphi)$ are also given in terms of Bessel functions, but in the air region the modified Bessel function of the first kind, $I_v(wr)$, has to be retained. The characteristic equation for the TM_{0m} modes is (Marincic *et al.* 1986)

$$\frac{u}{\varepsilon_r} \frac{J_0(ua)Y_0(ub) - Y_0(ua)J_0(ub)}{Y_0(ua)J_1(ub) - J_0(ua)Y_1(ub)}$$
$$= w \frac{K_0(wb)I_0(wc) - K_0(wc)I_0(wb)}{K_1(wb)I_0(wc) + K_0(wc)I_1(wb)} \qquad (5.58)$$

It should be noted that the left side of Eq. (5.58) is identical to that of Eq. (2.272) for the open surface-wave structure. The right side would be identical if $K_0(wc) = 0$, corresponding to $c \rightarrow \infty$.

The characteristic equation Eq. (5.58) is also a transcendental equation that can only be solved in a numerical way.

For the case with $c \gg b$, in the low-frequency range, a simple quasi-static model can be used to relate β/k_0 with a, b, c, and ε_r (Marincic *et al.* 1986). The model is based on equality of capacitances of the coaxial structures, as shown in Figure 5.26. As shown in Figure 5.26(b), if the coaxial structure is filled with dielectric of effective relative permittivity ε_{eff}, the capacitance per unit length is given by

$$C_{\text{eff}} = \frac{2\pi \varepsilon_{\text{eff}} \varepsilon_0}{\ln(c/a)} \tag{5.59}$$

The capacitance of the structure in Figure 5.26(a) is equivalent to the series connection of the capacitances C_d and C_0 given by

$$C_d = \frac{2\pi \varepsilon_r \varepsilon_0}{\ln(b/a)} \tag{5.60}$$

$$C_0 = \frac{2\pi \varepsilon_0}{\ln(c/b)} \tag{5.61}$$

By equalizing the capacitance of the two structures, we have

$$\frac{\ln(b/a)}{\varepsilon_r} = \frac{\ln(c/a)}{\varepsilon_{\text{eff}}} - \ln(c/b) \tag{5.62}$$

The structure shown in Figure 5.26(b) is actually a coaxial line. When a wave in TEM mode propagates along the line, the phase constant is

$$\beta_0 = \omega\sqrt{\mu_0\varepsilon_0\varepsilon_{\text{eff}}} = k_0\sqrt{\varepsilon_{\text{eff}}} \tag{5.63}$$

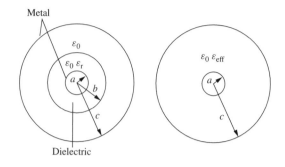

Figure 5.26 Low-frequency approximate mode of closed surface-wave guiding structure. (a) Actual structure and (b) equivalent quasi-static structures. (Marincic *et al.* 1986). Reproduced from Marincic, A. Benson, F. A. and Tealby, J. M. (1986). Measurements of permittivity by the use of surface waves in open and closed structures, *IEE Proceedings-H*, **133**, 441–449, by permission of IEE

By substituting ε_{eff} from Eq. (5.63) into Eq. (5.62), we get

$$\varepsilon_r = \frac{\ln(b/a)}{(k_0/\beta)^2 \ln(c/a) - \ln(c/b)} \tag{5.64}$$

So ε_r can be calculated from the values of (β/k_0), (b/a), and (c/a). It should be noted that, only for relatively low frequencies, Eq. (5.64) gives the same answer as the complex characteristic equation Eq. (5.58). The result of ε_r from Eq. (5.64) is always higher than the actual value. But, in any case, Eq. (5.64) can be used to find an approximate solution for ε_r prior to making the calculation on the basis of Eq. (5.58).

Figure 5.27 shows the structure of a closed surface-wave resonator for the measurement of dielectric constant. Usually, the input port is fixed

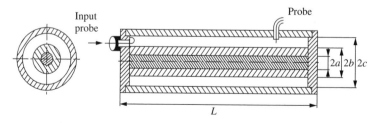

Figure 5.27   Geometry of closed surface-wave resonator (Marincic *et al.* 1986). Reproduced from Marincic, A. Benson, F. A. and Tealby, J. M. (1986). Measurements of permittivity by the use of surface waves in open and closed structures, *IEE Proceedings-H*, **133**, 441–449, by permission of IEE

to an end plate, and the other probe is fixed on the outer conductor near the end of the resonator opposite to where the input port is located.

A closed coaxial surface-wave resonator has several advantages over an open one (Marincic *et al.* 1986). The closed structure can be used from the lowest resonance when the resonator is half-wavelength long. Although the design and fabrication of an open structure are simpler and easier, the size of the end reflectors may be prohibitively large because of the loosely bound waves at low microwave frequencies, for example, less than 5 GHz. Furthermore, for a closed structure, a simple and accurate calculation of relative permittivity is possible at low microwave frequencies. As discussed above, the quasi-static model is applicable for a closed coaxial surface-wave resonator at low-microwave-frequency range, so the dielectric constant of the material can be obtained using Eq. (5.64), and it is not necessary to perform a complex numerical calculation of the characteristic equation. An open coaxial surface-wave structure can be used at high-microwave-frequency range, but the characteristic equation has to be solved using numerical techniques. However, it should be noted that, if the low-frequency approximation for ε_r, obtained from Eq. (5.64) cannot give sufficient accuracy, a numerical technique can be used in the calculation of ε_r from Eq. (5.58). The calculation procedure is practically identical to that for open surface-wave resonator method, except that a different characteristic equation is involved.

5.4 SPLIT-RESONATOR METHOD

Split-resonator methods are suitable for the characterization of dielectric sheet samples, including dielectric substrates for planar circuits. In a split-resonator method, the measurement fixture usually has a cylindrical structure working at a TE mode, and the resonator is split into two parts at the electric current node along a plane perpendicular to the cylinder axis. The sample under test is placed in the gap between the two parts of the resonator, and usually the sample is at the place of maximum electric field. The loading of a dielectric sheet sample changes the resonant properties of a split resonator, and the dielectric properties

of the sample can be derived from the resonant properties of the resonator loaded with sample and the dimensions of the resonator and the sample. The most distinguished advantage of this method is that sample preparation is not required, and dielectric sheet samples with enough area can be measured directly.

Calibration technique is often used in a split-resonator method, and the dielectric properties of the sheet sample can be derived from the differences of the resonant properties of the split resonator with and without the sample. There are two major differences between the split-resonator method using calibration technique and the resonant-perturbation method, which is discussed in Chapter 6. One is that in a split-resonator method, the change of the resonant frequency due to the introduction of the sample may be large, while in a resonant-perturbation method, the frequency change due to sample insertion should be small. The other is that in a split-resonator method, the field distributions outside the sample may be changed obviously, while the resonant-perturbation method assumes that the field distributions outside the sample remain unchanged.

In this section, we discuss several typical split-resonator methods, including split-cylinder-cavity method, split-coaxial-resonator method, split-dielectric-resonator method, and open-resonator method. Besides, planar split resonators, for example, microstrip split resonators, discussed in Section 9.4.3.2, have also been developed for the characterization of the properties of dielectric thin films.

5.4.1 Split-cylinder-cavity method

As shown in Figure 5.28(a), Kent proposed the split-cylinder-cavity method using a cylindrical empty cavity, which is separated into two shorted cylindrical-waveguide sections, and the sheet sample is placed in the gap between the two halves (Kent 1996). The parameters measured in this method are the resonant frequency and quality factor. These data together with the specimen thickness, the cavity dimensions, and the cavity losses are used for the calculation of the dielectric

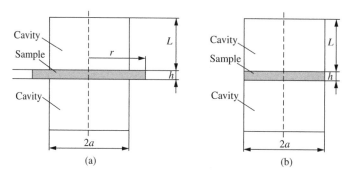

Figure 5.28   Models for split-cylinder-cavity method. (a) Split cylindrical cavity loaded with a dielectric sample. The axis of rotation is the z-axis, and the $x-y$ plane is at the center of the sample and (b) approximation structure of the split cylindrical resonator

constant and loss tangent of the sheet sample. The geometry requirement on samples is only that the sample must be flat and must extend beyond the diameter of the two cavity sections. Therefore, this method requires little or no sample preparation.

This method requires a comprehensive model for the split cylinder cavity. In a conventional resonator method, the boundary condition is relatively straightforward since the electric and magnetic fields are confined to the sample or within a metallic cavity. However, for a split cylinder cavity, a gap exists between the two cylindrical-waveguide sections, and the electric and magnetic fields extend into the sample outside of the cylindrical cavity regions. In this sense, the model of this method is similar to that of the circular-radial resonator method discussed in Section 5.2.3. Janezic and Baker-Jarvis made an exact full-wave analysis on this configuration (Janezic and Baker-Jarvis 1999). In the following, we follow the approximate model proposed by Kent (1996). But it should be noted that this approximate model is only suitable for optically thin samples whose thickness h should satisfy

$$\frac{2\pi}{\lambda} \cdot h\sqrt{\varepsilon_r'} \ll 1, \tag{5.65}$$

while the full-wave analysis method does not have a strict requirement on sample thickness.

In principle, any TE_{0mn} mode can be used for this method. But, if higher-order modes are used, mode identification may be difficult and there might be some overlap between the working mode

and its neighboring modes. So the TE_{011} mode is often used. In the excitation of TE_{011} mode, TM modes that are affected very little by the dielectric constant of the sample should be avoided, and this can be achieved by coupling only to the axial magnetic field.

In Kent's approximate model, the dielectric properties of the sheet sample are obtained in two steps. In the first step, we approximate the split resonator with a closed cylinder cavity, and do not consider the effect of the air gap. In the second step, correction is made to consider the effect of the air gap, so that accurate dielectric constant value of the sheet sample can be obtained.

In the first step, we assume that the air gap between the two parts of the cylinder cavity be shorted, and it is equivalent to that the sample be enclosed by a conducting ring at the cavity diameter. So the structure can be approximated as a closed cylinder cavity with diameter $2a$ and length $(2L + h)$, as shown in Figure 5.28(b). The continuity of electric and magnetic fields at the surface of the dielectric sample requires an eigenvalue θ satisfying (Kent 1996)

$$\theta \tan \theta = \psi \cot \psi \cdot \left(\frac{h}{2L}\right) \tag{5.66}$$

with

$$\theta = \beta h/2 \tag{5.67}$$

$$\psi = \beta_0 L \tag{5.68}$$

$$\beta_0^2 = \left(\frac{2\pi}{\lambda_0}\right)^2 - \left(\frac{2\pi}{\lambda_c}\right)^2 \tag{5.69}$$

$$\beta^2 = \varepsilon'_r \left(\frac{2\pi}{\lambda_0}\right)^2 - \left(\frac{2\pi}{\lambda_c}\right)^2 \qquad (5.70)$$

where L is the length of the half cavity, h is the thickness of the sample, λ_0 is the wavelength of the resonant frequency, λ_c is the cutoff wavelength of the TE_{01} mode. The right side of Eq. (5.66) can be determined by experimental and structural data, so the value of θ can be determined from Eq. (5.66). When the value of θ is known, the dielectric constant can be calculated from the defining relation

$$\varepsilon'_r = \frac{1 + (\theta/\theta_c)^2}{\eta^2} \qquad (5.71)$$

with

$$\eta = \lambda_c/\lambda_0 \qquad (5.72)$$

However, in the actual case shown in Figure 5.28(a), the gap is not shorted and the specimen extends some distance beyond the cavity radius, and the resonant frequency becomes lower. So the results from Eqs. (5.66) and (5.71) are erroneously large.

The magnetic field H_z in the gap region is a sum of waves with the approximate exponential decrease:

$$H_z \approx \frac{2r}{h} \exp\{-[(\pi/2)^2 - (\theta^2 + \theta_c^2)]^{\frac{1}{2}}\} \qquad (5.73)$$

We assume the dimensions of the specimen satisfy the following two requirements:

$$h/(2a) \le 0.15 \qquad (5.74)$$

$$r \ge 4a/3 \qquad (5.75)$$

where r is the radius of the sample. Equation (5.73) indicates that the reflections from the outer edges of the sample can be ignored. Under these assumptions, the gap correction can be obtained by calculating the frequency difference between the two structures shown in Figure 5.28. The frequency difference can be calculated in terms of the energies stored at the gap in the outside of the radius of the cavity. With the corrected frequency, the right side of Eq. (5.66) has a different value, and its solution in Eq. (5.71) yields the correct value for the dielectric constant. Detailed discussions on

gap correction and typical measurement results can be found in (Kent and Bell 1996).

For the determination of loss tangent, the unloaded quality factor of the split cylinder cavity is needed. To separate the dielectric loss of the sample from the losses of the cavity itself, we introduce two partition functions. The ratio of the electric energy in the cavity outside the dielectric sample to the electric energy in the dielectric sample is given by

$$U = \frac{1}{\varepsilon'_r} \cdot \frac{\cos^3 \theta}{(\theta + \sin \theta \cos \theta) \sin \theta}, \qquad (5.76)$$

and the ratio of the end-wall loss to sidewall loss is given by

$$W = j_{11} \cdot \frac{(\psi/\psi_c)^3}{\psi - \sin \psi \cos \psi} \qquad (5.77)$$

with

$$\psi_c = \frac{\pi h}{\lambda_c} = j_{11} \cdot \frac{h}{2a} \qquad (5.78)$$

where $j_{11} = 3.83171$ is the first zero of the Bessel function derivative $J'_0(x)$. With these partition functions, the loss tangent can be calculated using (Kent 1996)

$$\tan \delta = \frac{(1 + U)}{Q_0} - SU(1 + W) \qquad (5.79)$$

where S is a measure of the loss of the cavity itself, which depends on the frequency and conductivity σ of the cavity:

$$S = 0.722/(a\sigma \varsigma_0 \eta^3)^{1/2} \qquad (5.80)$$

where ς_0 is the free-space wave impedance, and η is given by $\eta = \lambda_c/\lambda_0$. The term $SU(1 + W)$ in Eq. (5.79) is the inverse quality factor due to the loss of the cavity itself. Because the calculation ignores losses in the gap region and the radiation from the outer edge of the sample, the value of $SU(1 + W)$ is an underestimate of the inverse quality factor due to fixture loss. Discussions on the measurement uncertainties about this method can be found in (Kent 1996; Kent and Bell 1996).

5.4.2 Split-coaxial-resonator method

Matytsin *et al.* proposed a split coaxial resonator for the characterization of dielectric sheet

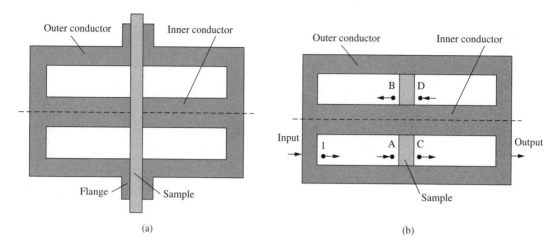

Figure 5.29 (a) Scheme of the split coaxial resonator and (b) approximation structure of split resonator. Letters A–D denote amplitudes of partial waves propagating in the resonator

samples (Matytsin *et al.* 1996). As shown in Figure 5.29(a), in this method, the coaxial resonator is split into two halves. The slot is situated at a peak of electrical field and it almost does not interrupt electrical currents in the frame of the resonator. The sheet sample under test is inserted at the slot between the two parts of the resonator, and the dielectric properties of the sample can be derived from the resonant properties of the split resonator loaded with the sample under test.

The working principle of this method is similar to that of the split-cylinder-cavity method discussed in Section 5.4.1, and the most obvious difference is that the split coaxial resonator is made from coaxial line, while split cylinder cavity is made from circular waveguide. Full-wave analysis of this configuration can also be conducted in a similar way discussed in (Janezic and Baker-Jarvis 1999), but such an analysis is complicated and not convenient since it involves a great deal of computation. Here, we introduce the approximate model proposed by Matytsin *et al.* (1996). This model can be used for determining dielectric constant up to several hundred, and this range covers most of the practical applications. But it should be noted that, similar to the mode proposed by Kent (1996), this model is applicable only to optically thin samples:

$$\frac{2\pi}{\lambda_0} \cdot d\sqrt{\varepsilon_r'} \ll 1 \qquad (5.81)$$

where λ_0 is the free-space wavelength, d is the thickness of the sample, and ε_r' is the dielectric constant of the sample.

Calibration measurement is often used in this method. We first measure the resonance frequency f_c of the empty resonator. When a sample is inserted into the resonator, the resonance frequency shifts to the other value f_m. The complex permittivity of the sample can be extracted from the frequency shift due to the presence of the sample. The development of the calculation algorithm mainly consists of two steps. In the first step, we consider the case of a coaxial resonator loaded with a ring sample, as shown in Figure 5.29(b). In the second step, we modify the conclusions obtained from the first step by considering the effects of the slot.

The propagation of an electromagnetic wave through a symmetrical resonator can be analyzed using the method of partial waves based on a section of transmission line. In Figure 5.29(b), characters A, B, C, and D correspond to the amplitudes of partial waves near the sample boundaries, and we assume the amplitude of the electromagnetic wave entering the resonator to be equal to unity. The sample has reflectivity R and transmittance T, and the complex transmittance coefficient of one half of the resonator is Y. We assume that the sample is optically thin and the quality factor of the resonator is high ($Q \gg 1$). The quantities A, B, C, and D are related by equations

$$A = Y + BY^2 \qquad (5.82)$$

$$B = AR + DT \qquad (5.83)$$

$$C = AT + DR \qquad (5.84)$$

$$D = CY^2 \qquad (5.85)$$

Transmission coefficient of the resonator is proportional to C, which can be found from Eqs. (5.82–5.85):

$$C = \frac{YT}{[1 - (R + T)Y^2][1 + (T - R)Y^2]} \qquad (5.86)$$

The two terms in the denominator of Eq. (5.86) correspond to two different resonant modes. The first term is related to an electric mode of the resonator, while the second term corresponds to a magnetic mode. Near an electrical resonance, frequency dependence of the transmittance of the resonator is mostly related to the first term of the denominator in Eq. (5.86). Thus, we can neglect the frequency dependence of other terms, and the transmission coefficient of the resonator $W(f)$ can be expressed by

$$W(f) \propto \frac{1}{1 - (R + T)Y^2} \qquad (5.87)$$

Resonance is achieved when the denominator of Eq. (5.87) equals zero. So the shift of complex resonance frequency caused by inserting the sample into the resonator can be found from the equation (Matytsin *et al.* 1996)

$$\exp(\text{j}2\pi f_\text{m}L/c) \cdot (R + T)$$
$$= \exp(\text{j}2\pi f_\text{c}L/c) \cdot (R_0 + T_0) \qquad (5.88)$$

where $(R_0 + T_0)$ is the sum of the reflection and transmission coefficients of the air gap of thickness d:

$$R_0 + T_0 = \exp(-\text{j}2\pi d/\lambda_0) \qquad (5.89)$$

where L is the length of the resonator, c is the velocity of light, and $\exp(\text{j}\omega t)$, the time dependence of the fields, is assumed. Values R and T can also be calculated with the partial waves

approach. For an optically thin sample inserted into a coaxial line, we have

$$R = -\frac{\text{j}(\varepsilon_\text{r} - 1) \cdot \pi d/\lambda_0}{1 + \text{j}(\varepsilon_\text{r} + 1) \cdot \pi d/\lambda_0} \qquad (5.90)$$

$$T = \frac{1}{1 + \text{j}(\varepsilon_\text{r} + 1) \cdot \pi d/\lambda_0}. \qquad (5.91)$$

By substituting Eqs. (5.90) and (5.91) into Eq. (5.88), after some manipulations, we can get (Matytsin *et al.* 1996)

$$\varepsilon_\text{r} = \frac{\tan[(d + L\delta) \cdot \pi f_\text{c}/c]}{\pi d f_\text{m}/c} \qquad (5.92)$$

with

$$\delta = (f_\text{c} - f_\text{m})/f_\text{c} \qquad (5.93)$$

In the case of small permittivity, the frequency shift is also small ($|\delta| \ll 1$). So from Eq. (5.92) we can get

$$\varepsilon_\text{r} \approx 1 + \frac{L}{d} \cdot \frac{f_\text{c} - f_\text{m}}{f_\text{c}} \qquad (5.94)$$

The effect of the slot can be analyzed by comparing two cases. In the first case, a sample is fitted into unslotted coaxial line, and in the second case, the sample is inserted through a slot in the line. The first case has been considered above. In the second case, we assume that the slot width is equal to the sample thickness d and hence is small in comparison with λ_0 and the transversal sizes of the line. We can assume an electromagnetic field corresponding to the second case to be a small perturbation of the field in the unslotted line. Thus, Eq. (5.94) can be modified as (Matytsin *et al.* 1996)

$$\varepsilon_\text{r} = \frac{\tan[d(1 + L\delta/d - \alpha d) \cdot \pi f_\text{c}/c] \cdot (1 + \alpha d)}{\pi d f_\text{m}/c} \qquad (5.95)$$

with

$$\alpha = \left(\frac{2}{\pi}\right)^3 \cdot \frac{1}{\ln(b/a)} \cdot \left(\frac{1}{a} + \frac{1}{b}\right) \qquad (5.96)$$

where a and b are the inner and outer radii of the coaxial resonator respectively. Equation (5.95) relates the permittivity of the sample and the shift of the resonance frequency in the case of the slotted resonator. By comparing Eqs. (5.92) and (5.95), we can find the error if the slot effect is neglected. Denoting the permittivity values obtained from Eqs. (5.95) and (5.92) as ε_1 and ε_2 respectively, we have

$$\frac{\varepsilon_2 - \varepsilon_1}{\varepsilon_2} \approx \alpha d \cdot \frac{\varepsilon_1 - 1}{\varepsilon_1}. \tag{5.97}$$

It should be noted that in the above discussion the permittivity and frequency are all complex parameters, so using this method both the resonant frequency and quality factor should be measured and both the real and the imaginary part of the permittivity can be obtained (Matytsin *et al.* 1996).

5.4.3 Split-dielectric-resonator method

Krupka *et al.* proposed a split-dielectric-resonator method for the measurements of the complex permittivity of dielectric sheet samples (Krupka *et al.* 1996). The proposed geometry of a split-dielectric-resonator fixture is shown in Figure 5.30. The resonator mainly consists of two dielectric disks in a metal enclosure. The dielectric disks are thin and the height of metal enclosure is relatively small, so the evanescent electromagnetic field character is strong not only in the air-gap region outside the cavity but also in the cavity region for radii

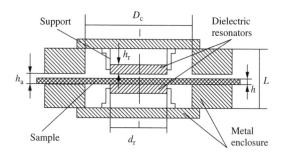

Figure 5.30 Schematic diagram of a split dielectric resonator fixture (Krupka *et al.* 1996). Reproduced from Krupka, J. Geyer, R. G. Baker-Jarvis, J. and Ceremuga, J. (1996). "Measurements of the complex permittivity of microwave circuit board substrates using split dielectric resonator and reentrant cavity techniques", *Seventh International Conference on Dielectric Materials Measurements & Applications*, IEE, London, 21–24, by permission of IEE

greater than the radius of the dielectric resonator. Therefore, the electromagnetic fields are also attenuated in the cavity so it is usually not necessary to take them into account in the air gap. This simplifies the numerical analysis and reduces possible radiation.

Such a split dielectric resonator usually operates with the $TE_{01\delta}$ mode, which only has the azimuthal electric field component, so the electric field remains continuous on the dielectric interfaces. Figure 5.31 shows the distributions of the electric field component, E_φ, in the split dielectric resonator operating at the $TE_{01\delta}$ mode, with and without a sheet sample. Figure 5.31 shows that the field distributions are affected by the introduction of the sample, and the changes of the field distributions are related to changes of the resonant frequency and the unloaded Q-factor of the split resonator. The dielectric properties of the sample are derived from the changes of resonant frequency and unloaded quality factor due to the insertion of sample.

For low-loss materials, the influence of losses on the resonant frequencies is negligible, so the real part of permittivity of the sample under test is related to the resonant frequencies and physical dimensions of the cavity and sample only. In this method, calibration technique is used, and we compare the difference of resonant frequency of the split dielectric resonator before and after the sample is inserted. The dielectric constant of the sample is an iterative solution to the following equation (Krupka *et al.* 1996):

$$\varepsilon_r' = 1 + \frac{f_0 - f_s}{h f_0 K_\varepsilon(\varepsilon_r', h)} \tag{5.98}$$

where h is the thickness of the sample under test, f_0 is the resonant frequency of empty resonant fixture, f_s is resonant frequency of the resonant fixture with dielectric sample, K_ε is a function of ε_r' and h.

For a given resonant fixture and known ε_r' and h, from Eq. (5.98), we have

$$K_\varepsilon(\varepsilon_r', h) = \frac{f_0 - f_s}{(\varepsilon_r' - 1) \cdot h f_0}. \tag{5.99}$$

In the development of the calculation algorithm, exact resonant frequencies and then the values

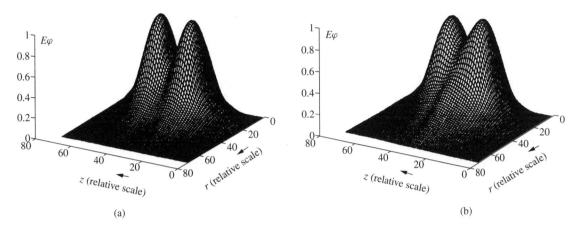

(a)   (b)

Figure 5.31 Distributions of the electric field in a cross section of a split dielectric resonator with the following parameters: $\varepsilon_d = 28.9$, $L = 13$ mm, $h_a = 1.6$ mm, $D_c = 30.2$ mm, $d_r = 13.85$ mm, and $h_r = 1.98$ mm. (a) Without sample and (b) with a sample of the following parameters: $\varepsilon_r = 10$ and $h = 1.2$ mm. (Krupka *et al.* 1996). Reproduced from Krupka, J. Geyer, R. G. Baker-Jarvis, J. and Ceremuga, J. (1996). "Measurements of the complex permittivity of microwave circuit board substrates using split dielectric resonator and reentrant cavity techniques", *Seventh International Conference on Dielectric Materials Measurements & Applications*, IEE, London, 21–24, by permission of IEE

of K_ε are computed for a number of ε_r' and h, and interpolation is used to compute K_ε for any other values of ε_r' and h. In the calculation of dielectric constant using Eq. (5.98), the initial value of K_ε is usually taken to be the same as its corresponding value for a given h and $\varepsilon_r' = 1$. Subsequent values of K_ε are found for the subsequent dielectric constant values obtained in the iterative procedure. As K_ε is a slowly varying function of ε_r' and h, the iterations using Eq. (5.98) converge rapidly (Krupka *et al.* 1996).

The dielectric loss tangent of the sample can be determined by (Krupka *et al.* 1996)

$$\tan\delta = \frac{1}{p_{es}}\left(\frac{1}{Q} - \frac{1}{Q_{DR}} - \frac{1}{Q_c}\right) \qquad (5.100)$$

with

$$p_{es} = h\varepsilon_r' K_1(\varepsilon_r', h) \qquad (5.101)$$

$$Q_c = Q_{c0} K_2(\varepsilon_r', h) \qquad (5.102)$$

$$Q_{DR} = Q_{DR0} \cdot \frac{f_0}{f_s} \cdot \frac{p_{eDR0}}{p_{eDR}} \qquad (5.103)$$

where p_{es} and p_{eDR} are the electric-energy filling factors for the sample and for the dielectric split resonator respectively; p_{eDR0} is the electric-energy filling factor of the dielectric split resonator

for empty resonant fixture; Q_{c0} is the quality factor depending on metal enclosure losses for empty resonant fixture; Q_{DR0} is the quality factor depending on dielectric losses in dielectric resonators for empty resonant fixture; and Q is the unloaded quality factor of the resonant fixture containing the dielectric sample.

The values of p_{eDR}, p_{es}, and Q_c for a given resonant structure can be calculated using numerical techniques. In the development of calculation algorithm, these parameters usually are computed for a number of h and ε_r', and then the functions K_1 and K_2 are evaluated and tabulated. Because K_1 and K_2 vary slowly with h and ε_r', interpolation can be used to evaluate these parameters for any values of h and ε_r' with acceptable accuracy.

The position of the sample in z-direction in the gap is not sensitive to the measurement results. Furthermore, the measurement results for stacked dielectric films are independent of the number of films so the split dielectric resonator method is not sensitive to the presence of air gaps between stacked films. Detailed discussion on measurement accuracy and sensitivity can be found in (Krupka *et al.* 1996).

5.4.4 Open resonator method

The use of cavity resonators may run into difficulties at high microwave frequencies, at which wavelengths lie in the millimeter range (Cullen 1983). First, as the size of cavity resonator is directly proportional to the wavelength, cavity resonators are inconveniently small for the characterization of materials properties at millimeter-wave frequencies. In particular, for solid dielectric specimens, it becomes increasingly difficult to ensure that the specimen fits the cavity with a negligible air gap. Second, the quality factor of a cavity resonator varies with the wavelength as $\lambda^{3/2}$, and so decreases rather rapidly as the wavelength decreases. For a cavity with low quality factor, the additional loss due to the specimen may be swamped by the inherent loss of the cavity. Because the loss tangent of a dielectric specimen is determined by measuring the change in the quality factor produced when the specimen is inserted, the accuracy of measurement must decrease as the wavelength decreases. These difficulties can be solved by using open resonators (Afsar *et al.* 1990). The essential feature of an open resonator is that the number of modes is proportional to the length rather than the volume of the resonator. Thus, the number of modes is proportional to L/λ rather than $(L/\lambda)^3$, where L is the characteristic length of the resonator. Even for a fairly large value of L/λ, the mode separation is sufficient for single-mode operation to be used.

The basic properties of open resonators have been discussed in Chapter 2. When an open resonator is used for the characterization of materials properties, the sample under study is placed between the two reflecting mirrors, and in this meaning, an open resonator can be taken as a special type of split resonator. This method usually works in the frequency range of 10 to 200 GHz, and generally speaking, it is easier to load and unload samples to an open resonator than to a cavity resonator. Using the open resonator method, flat disc-shaped solid samples can be measured, and liquids or powders can also be measured provided they are restrained within a solid annulus, which is outside the microwave beam area and

covered by plates with low dielectric constant and low loss tangent.

In the characterization of materials properties using open resonator method, the wavefront of electromagnetic beam in the open resonator is an important factor. If the face of a plane dielectric specimen is situated at a point where the wavefront is curved, the parameters of the beam, such as its radius of curvature r, transform through the surface in a way that is inaccessible to simple theories. So the sample is usually placed at the beam waist, where r is large, but corrections are still needed to obtain accurate results.

Many resonator geometries have been used in materials property characterization. Each of them has its advantages and disadvantages. For example, a hemispherical resonator can be used to measure small-diameter specimens (not less than $5 w_0$ for negligible diffraction), but, if the specimen rests on the plane mirror, which is at an electric field minimum, such a resonator is not suitable for samples with a thickness much less than $\lambda/2$. Lynch made a comparison about the performances of "short" ($D < r_0$), confocal, semiconfocal and hemispherical geometries for permittivity measurements (Lynch 1979).

In the following, we discuss the applications of biconcave resonator and plano-concave resonator in characterization of materials properties. These types of resonators usually have high accuracy and sensitivity. Such structures exhibit TEM resonances, and the electromagnetic fields in the resonators are in the form of a standing-wave Gaussian beam. The resonant fields have maximum amplitude on the axis and decrease when the position moves away from the axis. When the sample is placed at the "waist" of the resonator, sample diameters down to six wavelengths are quite adequate for measurements.

It should be indicated that in an open resonator method for materials property characterization, usually the refractive index n is used. For a dielectric material, the relationship between refractive index and dielectric permittivity is given by

$$n = \sqrt{\varepsilon'_r} \qquad (5.104)$$

5.4.4.1 Biconcave open resonator

Figure 5.32 shows the measurement configuration using a biconcave open resonator. A parallel-plane-sided slab of dielectric material with thickness $2t$ is placed midway between two identical spherical mirrors with radius R_0. The total length of the resonator is D and the distance between the sample plane and the spherical mirror is d. There are two techniques for the measurement of materials properties. In one technique, the length of the cavity is fixed, and the resonant frequency is measured with and without sample. In the other technique, the length of the cavity is adjusted to establish resonance with and without the sample at the same frequency. Here, we discuss the basic algorithms for these two techniques, and more detailed discussions can be found in (Cullen 1983; Yu and Cullen 1982).

Before we discuss the calculation algorithms, consider the open resonator shown in Figure 5.32, but with no sample present and with two spherical mirrors of equal radii of curvature R_0, located at $z = \pm D/2$. The radius w_z of the beam at position z is given by

$$w_z^2 = w_0^2 \cdot (1 + z^2/z_0^2) \tag{5.105}$$

with

$$w_0^2 = \frac{\lambda}{2\pi}\sqrt{D(2R_0 - D)} \tag{5.106}$$

$$z_0^2 = \frac{D}{2} \cdot \left(R_0 - \frac{D}{2}\right) \tag{5.107}$$

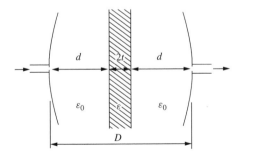

Figure 5.32 An open resonator loaded with dielectric sheet sample. Reproduced from Cullen, A. L. (1983). "Chapter 4 Millimeter-wave open-resonator techniques", in *Infrared and Millimeter Waves*, Volume 10, Part II, edited by Button, K. J. (Academic Press, Orlando), 233–281, by permission of Elsevier

where w_0 is the radius of the "waist" of the resonator. For a symmetrical biconcave open resonator, the "waist" is at the position of $z = 0$. As will be discussed later, when a sample is inserted into the open resonator, the radius of the beam should be modified.

Frequency-variation technique

The first step of this technique is to accurately determine the total length D of the open resonator by measuring the resonant frequency f_e of the empty open resonator. The value of D can be derived according to following equation (Yu and Cullen 1982):

$$f_e = \frac{c}{2D}\left[q + 1 + \frac{1}{\pi}\arccos\left(1 - \frac{D}{R_0}\right) - \frac{1}{2\pi k_e R_0}\right] \tag{5.108}$$

where q is the axial mode number, R_0 is the radius of the curvature at $z = t + d$, $k_e = 2\pi f_e/c$, and c is the speed of light in air. It is assumed that R_0 is sufficiently accurately known from mechanical measurement, and, actually, as indicated by Eq. (5.108), f_e does not depend strongly on R_0. It is also assumed that the length D is sufficiently well known in advance to enable the mode number q to be determined unambiguously.

In the next step, we place the dielectric sample at the "waist" of the resonator as shown in Figure 5.32 and measure the new resonant frequency. The mode number is now uncertain unless an approximate value of the refractive index is available, and such ambiguity can be resolved by using samples of different thickness values. The measured frequency should be corrected by applying the interface correction and mirrored in reverse to obtain an effective frequency from which the wave number k is calculated. The measured frequency should be decreased by δf given by (Yu and Cullen 1982)

$$\frac{\delta f}{f_0} = \frac{t(n - \Delta)}{n^2 k^2 w_t^2 \cdot (t\Delta + d)} + \frac{3}{4k^2 \cdot (t\Delta + d)R_0} \tag{5.109}$$

with

$$\Delta = \frac{n^2}{n^2 \sin^2(nkt - \Phi_T) + \cos^2(nkt - \Phi_T)} \tag{5.110}$$

for symmetric modes and

$$\Delta = \frac{n^2}{n^2 \cos^2(nkt - \Phi_\mathrm{T}) + \sin^2(nkt - \Phi_\mathrm{T})} \quad (5.111)$$

for antisymmetric modes. Also, the radius of the beam should also be modified:

$$w_\mathrm{t}^2 = w_0^2 \cdot (1 + t^2/z_0^2) \quad (5.112)$$

with

$$z_0 = \sqrt{d' \cdot (R_0 - d')} \quad (5.113)$$

$$d' = d + t/n \quad (5.114)$$

The value of k derived from the effective frequency is then substituted in Eq. (5.115) or Eq. (5.116), depending on whether the mode is symmetric or antisymmetric. And the value of n can be subsequently obtained by solving the transcendental equation, and the value of dielectric constant $\varepsilon_\mathrm{r}' = n^2$ can then be obtained. The transcendental equation for symmetrical modes is

$$(1/n)\cot(nkt - \Phi_\mathrm{T}) = \tan(kd - \Phi_\mathrm{D}) \quad (5.115)$$

and the one for antisymmetrical modes is

$$(1/n)\tan(nkt - \Phi_\mathrm{T}) = \tan(kd - \Phi_\mathrm{D}) \quad (5.116)$$

with

$$\Phi_\mathrm{T} = \tan^{-1}(t/nz_0) - \tan^{-1}[1/nkR_1(t)] \quad (5.117)$$

$$\Phi_\mathrm{D} = \tan^{-1}(d'/z_0) - \tan^{-1}(1/kR_0)$$
$$\quad - \tan^{-1}(t/nz_0) - \tan^{-1}[1/nkR_2(t)] \quad (5.118)$$

$$R_1(t) = t + n^2 z_0^2/t \quad (5.119)$$

$$R_2(t) = t/n + nz_0^2/t \quad (5.120)$$

An approximate value of n must be assumed at the outset in evaluating the correction terms given by Eq. (5.109), and, if necessary, these corrections can be recalculated using the value of n deduced from the transcendental equation. This equation may then be solved again to obtain an improved value of n if necessary.

Length-variation technique

In this technique, we keep the frequency constant and change the length of the resonator to reestablish resonance after the dielectric specimen is inserted. The starting point is to establish the resonant length of the empty resonator, adjusting the length at fixed frequency until the desired mode is resonant. We denote the half-length by d_0. After inserting the specimen, the new resonant length is found. We denote this by $(d_0 - p)$. The modified half-length, $(d + t)$, is then calculated using the previously derived interface and mirror corrections in a suitable form as follows (Yu and Cullen 1982):

$$d + t = d_0 - p + \frac{t(n - \Delta)}{n^2 k^2 w_\mathrm{t}^2} + \frac{3}{4k^2 R_0} \quad (5.121)$$

The rest of the calculation follows along the same lines as the frequency-variation method, solving the transcendental equation for n and iterating if necessary.

Measurement of dielectric loss

As discussed earlier, each contribution to the losses of a resonant mode can be described by an individual quality factors, Q_i. By combining these individual quality factors, we can get the overall resonator quality factor Q_0 of the resonant mode:

$$\frac{1}{Q_0} = \sum_i \frac{1}{Q_i} \quad (5.122)$$

The quality factor associated with the sample loss is inversely related to the loss tangent of the sample:

$$Q_\mathrm{d} = F/\tan\delta \quad (5.123)$$

where F is a filling factor. Except for perturbation methods, it is desirable that the quality factor associated with the sample loss Q_d is comparable with the overall quality factor Q_0. The quality factors for plane-parallel resonators are often an order of magnitude less than those for confocal resonators of similar dimensions and at similar frequencies. For this reason, planar resonators are more suitable for medium-loss samples, while confocal resonators are more suitable for low-loss samples.

The formulae relating the loss tangent $\tan\delta$ to the quality factor of the dielectric-loaded resonator for both symmetric and antisymmetric modes can be derived from Eqs. (5.115) and (5.116) by working first in terms of a complex refractive index and a complex phase coefficient $k(=\omega/c)$. The imaginary part of the refractive index is simply related to $\tan\delta$, and the imaginary part of k or ω is simply related to the quality factor. If the losses are small, with first-order approximations, we have (Yu and Cullen 1982)

$$\tan\delta = \frac{2nk(d + t\Delta_{\mathrm{s}})}{Q[2nkt\,\Delta_{\mathrm{s}} + \Delta_{\mathrm{s}}\sin 2(nkt - \Phi_{\mathrm{T}})]} \tag{5.124}$$

for symmetric modes, where Δ_{s} is given by Eq. (5.110), while for antisymmetric modes, we have

$$\tan\delta = \frac{2nk(d + t\Delta_{\mathrm{a}})}{Q[2nkt\,\Delta_{\mathrm{a}} - \Delta_{\mathrm{a}}\sin 2(nkt - \Phi_{\mathrm{T}})]} \tag{5.125}$$

where Δ_{a} is given by Eq. (2.111).

5.4.4.2 Plano-concave resonator

Figure 5.33 shows a plano-convex resonator, and the sample under test is placed on the plane mirror. The axis of the resonator is usually vertical, so liquid samples can be measured. Usually it works in a transmission mode, and small coupling apertures are situated near the pole of the curved mirror. The resonator is usually used in a configuration intermediate between semiconfocal and hemispherical. In the following, we discuss semiconfocal plano-concave resonator.

The plano-concave resonator works in a similar principle as biconcave resonator method. The measurement procedure for plano-concave resonator method also consists of two steps. First, measure the resonant frequency f_q of a fundamental TEM_{00q} mode of empty resonator with an arbitrary mode number q and calculate the resonator length D, using the following

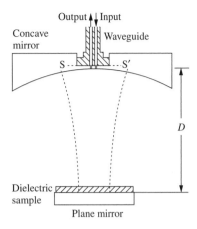

Figure 5.33 Schematic diagram of the semiconfocal open resonator used for dielectric measurement (Komiyama *et al.* 1991). Source: Komiyama, B. Kiyokawa, M. and Matsui, T. (1991). "Open resonator for precision dielectric measurements in the 100 GHz band", *IEEE Transactions on Microwave Theory and techniques*, **39** (10), 1792–1796. © 2003 IEEE

equation (Komiyama *et al.* 1991):

$$f_q = \frac{c}{2D}\left[q + 1 + \frac{1}{\pi}\tan^{-1}\left(\frac{D}{R_0 - D}\right)^{1/2}\right] \tag{5.126}$$

The next step is to place a sample with thickness t on the plane mirror and measure the new resonant frequency f_{s}, of the TEM_{00s} mode with the sample present. The refractive index n can be determined by solving the following equation (Komiyama *et al.* 1991):

$$(1/n)\tan(nkt - \Phi_{\mathrm{t}}) = -\tan(kd - \Phi_{\mathrm{d}}) \tag{5.127}$$

with

$$k = 2\pi f_{\mathrm{s}}/c \tag{5.128}$$

$$d = D - t \tag{5.129}$$

$$\Phi_{\mathrm{t}} = \tan^{-1}(t/nz_0) \tag{5.130}$$

$$\Phi_{\mathrm{d}} = \tan^{-1}(d'/z_0) - \tan^{-1}(t/nz_0) \tag{5.131}$$

$$d' = d + t/n \tag{5.132}$$

$$z_0 = [d'(R_0 - d')]^{1/2} \tag{5.133}$$

The loss tangent of the sample is given by (Komiyama *et al.* 1991)

$$\tan \delta = \frac{1}{Q_e} \cdot \frac{t\Delta + d}{t\Delta + [\sin 2(kd - \Phi_d)]/(2k)} \quad (5.134)$$

with

$$\Delta = \frac{\varepsilon_r}{\varepsilon_r \cos^2(nkt - \Phi_t) + \sin^2(nkt - \Phi_t)} \quad (5.135)$$

$$\frac{1}{Q_e} = \frac{1}{Q_d} - \frac{1}{Q_1} \quad (5.136)$$

where Q_d is the measured quality factor of the resonator containing the sample and Q_1 is the calculated quality factor for resonator containing a similar but loss-free sample

$$Q_1 = Q_0 \frac{2(t\Delta + d)}{D(\Delta + 1)} \quad (5.137)$$

where Q_0 is the measured quality factor of the empty resonator with the same axial mode number s. If a combination of frequency and thickness is made so that $\Delta = 1$, the measurement results of ε_r' and $\tan \delta$ are most reliable (Cullen 1983).

5.5 DIELECTRIC RESONATOR METHODS FOR SURFACE-IMPEDANCE MEASUREMENT

The basic properties of the surface impedance of normal conductors and superconductors have been discussed in Chapter 2. The measurement of the surface impedance of conductors is important for developing microwave electronic circuits, especially after the discovery of high-temperature superconductivity. Dielectric resonator methods for the measurement of the surface impedance have been studied extensively (Taber 1990; Mourachkine and Barel 1995; Shen *et al.* 1992; Mazierska 1997; Mazierska and Wilker 2001; Obara *et al.* 2001; Talanov *et al.* 1999; Kobayashi and Yoshikawa 1998; Kobayashi *et al.* 1991; Krupka *et al.* 1993; Tellmann *et al.* 1994), and dielectric resonator methods can be used as standards in the characterization of surface impedance.

The dielectric resonator methods for this purpose generally fall into two categories. In one category, only the surface resistance is measured. In the other category, both surface resistance and surface impedance are measured.

5.5.1 Measurement of surface resistance

Many dielectric resonators with different configurations have been developed for the measurement of surface resistance (Taber 1990; Mourachkine and Barel 1995; Shen *et al.* 1992; Mazierska 1997; Mazierska and Wilker 2001; Obara *et al.* 2001; Krupka *et al.* 1993; Tellmann *et al.* 1994), and they share similar working principles. The basic configuration of the measurement method is shown in Figure 5.34. The two conducting plates short the

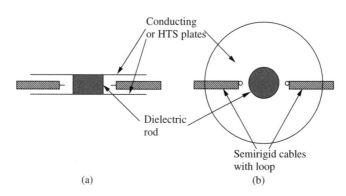

Figure 5.34 Configuration of a Courtney resonator for surface-resistance measurement. (a) Side view and (b) top view without the upper conducting plate. Modified from Mourachkine, A. P. and Barel, A. R. F. (1995). "Microwave measurement of surface resistance by the parallel-plate dielectric resonator method", *IEEE Transactions on Microwave Theory and Techniques*, **43** (3), 544–551. © 2003 IEEE

two open ends of a dielectric rod. The resonant properties of such a configuration are related to the dielectric properties of the dielectric rod and the surface impedance of the conducting plates. As discussed in Section 5.2.1, this configuration can be used to characterize the dielectric properties of the dielectric rod. If we know the dielectric properties of the dielectric rod, this configuration can be used to characterize the surface impedance of conducting plates.

In this method, the surface resistance R_s of the conducting plates can be calculated from the quality factor of the resonator using the following equation that is derived from Eq. (5.5):

$$R_s = \frac{1}{B} \cdot \left(\frac{A}{Q_0} - \tan \delta \right) \quad (5.138)$$

where the parameters A and B, defined by Eqs. (5.6)–(5.8), are related to the geometrical dimensions and the dielectric properties of the dielectric rod. Equation (5.138) indicates that the knowledge of the dielectric constant and loss tangent of the dielectric rod is needed. If the dielectric constant and loss tangent of the dielectric rod is unknown, two measurements on two sets of conducting plates with known surface resistance values are needed to calibrate the values of A and B.

Usually, the dielectric constant of the dielectric rod is accurately known, while the loss tangent of a low-loss dielectric has large uncertainty. As discussed in Section 5.2.1, we can use a two-resonator method to avoid the uncertainty due to the uncertainty in loss-tangent value, and calculate the surface resistance using Eq. (5.12). More discussion on this method can be found in (Obara *et al.* 2001).

In an actual measurement system, the resonant structure shown in Figure 5.34 is often housed in a metal shield. Besides, the conducting plates have limited sizes. With these considerations, the unloaded quality factor Q_0 of the resonant structure can be expressed as

$$\frac{1}{Q_0} = \frac{1}{Q_s} + \frac{1}{Q_m} + \frac{1}{Q_d} + \frac{1}{Q_{rad}} \quad (5.139)$$

where Q_s, Q_m, Q_d, and Q_{rad} are the quality factors related to the sample, metallic, dielectric, and

radiation losses respectively. The sample and metallic losses can be directly determined by the respective surface resistance (R_s, R_m) and the dielectric losses can be determined by the loss tangent of the dielectric rod:

$$\frac{1}{Q_s} = \frac{R_s}{A_s} \quad (5.140)$$

$$\frac{1}{Q_m} = \frac{R_m}{A_m} \quad (5.141)$$

$$\frac{1}{Q_d} = p_d \tan \delta \quad (5.142)$$

where A_s and A_m are geometrical factors of the sample and metallic part respectively and p_d is the electric-energy filling factor of the dielectric part. The geometrical factors depend on the type, structure, and dimensions of the resonator. Further discussion on these aspects can be found in (Mazierska 1997; Mazierska and Wilker 2001; Krupka *et al.* 1993).

5.5.2 Measurement of surface impedance

In order to understand the conduction mechanism and the electronic application of normal conductors and superconductors, it is necessary to measure the surface impedance, including the surface reactance and surface resistance, of conducting plates or superconducting thin films. Though there are many methods for the measurement of surface resistance, there are only a few methods for the measurement of surface reactance. Talanov *et al.* proposed a modified parallel-plate transmission-line resonator with a smoothly variable thickness of dielectric spacer for the measurement of the absolute penetration depth and surface resistance of superconductors (Talanov *et al.* 1999). In the following, we discuss two methods proposed by Kobayashi *et al*: two-resonator method (Kobayashi *et al.* 1991) and dual-mode resonator method (Kobayashi and Yoshikawa 1998).

5.5.2.1 Two-resonator method

Figure 5.35 (a) shows a TE$_{011}$ mode dielectric rod resonator placed between a perfect-conductor

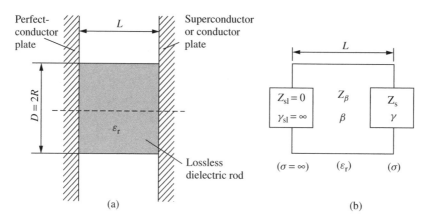

Figure 5.35   Analytical mode for surface resistance measurement. (a) TE_{011} mode dielectric rod resonator and (b) equivalent circuit

plate (at the left side) with surface impedance $Z_{sl} = R_{sl} + jX_{sl} = 0$, where the subscript l denotes the left side and a superconductor or conductor plate (at the right side) with a finite value of Z_s, such as a superconductor or a metal plate. The dielectric rod with relative permittivity ε_r, diameter $D = 2R$, and length L is assumed to be lossless. Figure 5.35(b) shows an equivalent circuit for the resonator, whose resonance condition is given by

$$Z_s + jZ_\beta \tan \beta L = 0 \qquad (5.143)$$

with

$$Z_\beta = \frac{\omega \mu_0}{\beta} \qquad (5.144)$$

where $\mu_0 = 4\pi \times 10^{-7}$ (H/m), and Z_β and β are the characteristic impedance and the phase constant in a dielectric waveguide respectively.

We introduce a perturbational quantity of complex angular frequency $\Delta \varpi / \omega$:

$$\frac{\Delta \varpi}{\omega} = \frac{\Delta f}{f} + j\frac{1}{2Q_s} \qquad (5.145)$$

with

$$\Delta f = f_0 - f \qquad (5.146)$$

where f is the resonant frequency when $Z_s = 0$ and f_0 and Q_s are the resonant frequency and quality factor due to the loss in the upper conductor when $Z_s \neq 0$. By taking the first-order approximation of Eq. (5.143), we can derive a perturbation formula for Z_s (Kobayashi *et al.* 1991):

$$Z_s = 960\pi^2 \left(\frac{L}{\lambda_0}\right)^3 \cdot \frac{\varepsilon_r + W}{1 + W} \cdot \left(-j\frac{\Delta \varpi}{\omega}\right) \qquad (5.147)$$

where $\lambda_0 = c/f_0$, c is the speed of light, and W is given by Eq. (5.8).

If the upper conductor plate is in the normal state $(R_s = X_s)$, the resonant frequency for $Z_s = 0$ can be determined from the measured values of f_0 and quality factor Q_s by

$$f = \frac{2Q_s}{2Q_s - 1}f_0 \qquad (5.148)$$

In an actual resonator, R_{sl} and the loss tangent of dielectric $\tan \delta$ are not zero. The value of Q_s can be obtained by removing the effects caused by these losses from a measured Q_u value (Kobayashi *et al.* 1991):

$$\frac{1}{Q_s} = \frac{1}{Q_u} - \frac{R_{sl}}{480\pi^2}\left(\frac{\lambda_0}{L}\right)^3\frac{1 + W}{\varepsilon_r + W} - \frac{\tan \delta}{1 + W/\varepsilon_r} \qquad (5.149)$$

Equation (5.149) shows that in order to calculate the surface impedance, the values of R_{sl}, ε_r, and $\tan \delta$ should be determined first. As shown in Figure 5.3, by measuring the resonant properties of TE_{011} and TE_{0ll} $(l \geq 2)$ mode dielectric resonators placed between two metal plates, the values of R_{sl}, ε_r, and $\tan \delta$ can be determined. As two dielectric resonators are used, this method for the measurement of the surface impedance is usually called *two-resonator method*.

5.5.2.2 Dual-mode resonator method

Kobayashi and Yoshikawa proposed the dual-mode resonator method for the measurement of surface impedance (Kobayashi and Yoshikawa 1998). In this method, two resonant modes (TE$_{021}$ and TE$_{012}$) in a dielectric rod resonator are used, and this method is also called the *one-resonator method*.

Properties of dielectric rod resonator

Consider a dielectric rod resonator placed between two parallel conducting plates. The dielectric rod has relative permittivity ε_r, loss tangent $\tan \delta$, diameter D, and length L, and the conducting plates have surface impedance Z_s and diameter d. In this method, the TE$_{012}$ and TE$_{021}$ modes are used, whose field distributions are shown in Figure 5.36.

Figure 5.37 shows a mode chart for a dielectric rod with dielectric constant $\varepsilon_r = 24$ (Kobayashi and Yoshikawa 1998). The values of D and L of the dielectric rod are chosen so that the resonant frequency f_1 for the TE$_{021}$ mode is in proximity to f_2 for the TE$_{012}$ mode, as indicated by the dashed lines in Figure 5.37.

Figure 5.38 shows an analytical model for TE$_{0ml}$ mode dielectric resonator. We assume that $d = \infty$, $\tan \delta = 0$ and $R_s = X_s = 0$. The values of ε_{rp}, with $p = 1$ representing TE$_{021}$ mode and $p = 2$ representing the TE$_{012}$ mode, can be obtained from the measured values f_p by (Kobayashi and Yoshikawa 1998)

$$\varepsilon_{rp} = \left(\frac{c}{\pi D f_p}\right)^2 (u_p^2 + v_p^2) + 1 \qquad (5.150)$$

with

$$v_p^2 = \left(\frac{\pi D f_p}{c}\right)^2 \left[\left(\frac{cp}{2Lf_p}\right)^2 - 1\right] \qquad (5.151)$$

$$u_p \frac{J_0(u_p)}{J_1(u_p)} = -v_p \frac{K_0(v_p)}{K_1(v_p)} \qquad (5.152)$$

where c is the speed of light, J_n is the Bessel function of the first kind, and K_n is the modified Bessel function of the second kind. It should be noted that u_1 for the TE$_{021}$ mode is the second solution of Eq. (5.152) and u_2 for the TE$_{012}$ mode is the first solution of Eq. (5.152).

The loss tangent of the dielectric rod can be calculated from the unloaded quality factor Q_{up}

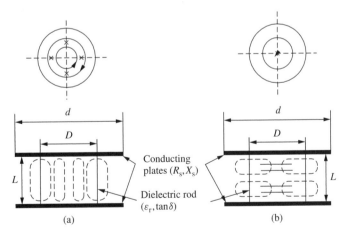

Figure 5.36 Field distributions of (a) TE$_{021}$ and (b) TE$_{012}$ modes in a dielectric rod resonator placed between two parallel conducting plates. The solid lines represent the electric force, while the dashed lines represent the magnetic force. The resonant frequency and unloaded quality factor for TE$_{021}$ mode are f_1 and Q_{u1} respectively, and the resonant frequency and unloaded quality factor for TE$_{012}$ mode are f_2 and Q_{u2} respectively (Kobayashi 1998). Source: Kobayashi, Y. Yoshikawa, H. (1998). "Microwave measurements of surface impedance of high-T-c superconductors using two modes in a dielectric rod resonator", *IEEE Transactions on Microwave Theory and Techniques*, **46** (12), 2524–2530. © 2003 IEEE

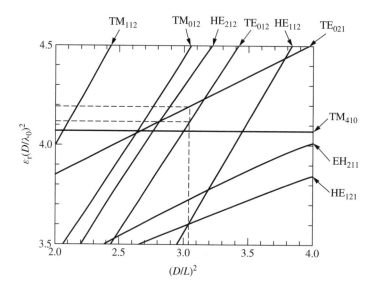

Figure 5.37 A mode chart calculated for $\varepsilon_r = 24$ (Kobayashi and Yoshikawa 1998). Source: Kobayashi, Y. Yoshikawa, H. (1998). "Microwave measurements of surface impedance of high-T-c superconductors using two modes in a dielectric rod resonator", *IEEE Transactions on Microwave Theory and Techniques*, **46** (12), 2524–2530. © 2003 IEEE

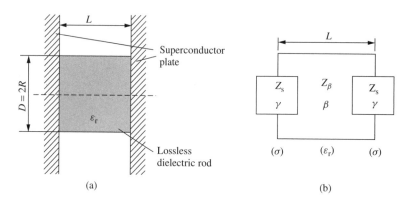

Figure 5.38 Analytical model for TE_{0ml} dielectric resonator. (a) Resonator configuration and (b) equivalent circuit

measured at f_p using the following equation similar to Eq. (5.5):

$$\tan \delta_p = \frac{A_p}{Q_{up}} - B_p R_{sp} \quad (5.153)$$

with

$$A_p = 1 + \frac{W_p}{\varepsilon_{rp}} \quad (5.154)$$

$$B_p = \left(\frac{cp}{2Lf_p}\right)^3 \frac{1 + W_p}{30\pi^2 p\varepsilon_{rp}} \quad (5.155)$$

$$W_p = \frac{J_1^2(u_p)[K_0(v_p)K_2(v_p) - K_1^2(v_p)]}{K_1^2(v_p)[J_1^2(u_p) - J_0(u_p)K_2(u_p)]} \quad (5.156)$$

where R_{sp} is the surface resistance of the conducting plates.

Measurement of surface resistance

In the following discussion, we use ε_{r0}, $\tan \delta_0$, R_{s0}, and X_{s0} to represent the values of ε_r, $\tan \delta$, R_s, and X_s at frequency f_0 respectively, taking their frequency dependences into account. Here, f_0 is an arbitrarily given frequency near f_1 and f_2. For ion-crystallized material such as polycrystalline ceramics and sapphire, in the frequency region including f_1, f_2, and f_0, we assume

$$\varepsilon_{rp} = \varepsilon_{r0} \tag{5.157}$$

$$\tan\delta_p = \frac{f_p}{f_0}\tan\delta_0 \tag{5.158}$$

For conducting plates, we assume

$$R_{sp} = \left(\frac{f_p}{f_0}\right)^n R_{s0} \tag{5.159}$$

$$X_{sp} = \left(\frac{f_p}{f_0}\right)^m X_{s0} \tag{5.160}$$

where $n = m = 1/2$ for a normal conductor and $n = 2$ and $m = 1$ for a superconductor, according to the two-fluid model. By substituting Eqs. (5.158) and (5.159) into Eq. (5.153), we get

$$\frac{f_p}{f_0}\tan\delta_0 = \frac{A_p}{Q_{up}} - B_p\left(\frac{f_p}{f_0}\right)^n R_{sp} \tag{5.161}$$

By eliminating $\tan\delta_0$ from two equations derived from Eq. (5.161) for $p = 1$ and 2, we obtain

$$R_{s0} = \frac{f_0^n f_2 \cdot (A_1/Q_{u1}) - f_0^n f_1 \cdot (A_2/Q_{u2})}{B_1 f_1^n f_2 - B_2 f_2^2 f_1} \tag{5.162}$$

Measurement of surface reactance

As indicated earlier, Eq. (5.150) is based on the assumptions that $d = \infty$, $\tan\delta = 0$ and $R_s = X_s = 0$. The values ε_{r1} and ε_{r2} calculated using Eq. (5.150) values are different from each other because of different effects of X_s depending on the modes. Therefore, an intrinsic relative permittivity of the dielectric rod ε_i can be calculated by

$$\varepsilon_i = \varepsilon_{r1} - \Delta\varepsilon_{r1} = \varepsilon_{r2} - \Delta\varepsilon_{r2} \tag{5.163}$$

where $\Delta\varepsilon_{r1}$ and $\Delta\varepsilon_{r2}$ are correction terms due to X_{s1} for the TE_{021} mode and due to X_{s2} for the TE_{012} mode respectively given by

$$\Delta\varepsilon_{rp} = 4(1 + W_p)\left(\frac{\beta_p}{k_p}\right)^2 \frac{X_{sp}}{\omega_p \mu L} \tag{5.164}$$

with

$$k_p = \frac{\omega_p}{c} \tag{5.165}$$

$$\beta_p = \frac{\pi p}{L} \tag{5.166}$$

From Eqs. (5.160), (5.163), and (5.164), we obtain

$$X_{s0} = \frac{\varepsilon_{r2} - \varepsilon_{r1}}{C_2 - C_1} \tag{5.167}$$

with

$$C_p = 4(1 + W_p)\left(\frac{\beta_p}{k_p}\right)^2 \frac{1}{\omega_p \mu L}\left(\frac{f_p}{f_0}\right)^m \tag{5.168}$$

Equation (5.167) indicates that the value of X_{s0} can be obtained from the measured values of ε_{r1} and ε_{r2}.

REFERENCES

Afsar, M. N. Li, X. H. and Chi, H. (1990). "An automated 60 GHz open resonator system for precision dielectric measurement", *IEEE Transactions on Microwave Theory and Techniques*, **38** (12), 1845–1853.

Bourdel, E. Pasquet, D. Denorme, P. and Roussel, A. (2000). "Measurement of the moisture content with a cylindrical resonating cavity in TM_{010} mode", *IEEE Transactions on Instrumentation and Measurement*, **49** (5), 1023–1028.

Cohn, S. B. and Kelly, K. C. (1966). "Microwave measurement of high-dielectric constant materials", *IEEE Transactions on Microwave Theory and Techniques*, **14** (9), 406–410.

Courtney, W. E. (1970). "Analysis and evaluation of a method of measuring the complex permittivity and permeability of microwave insulators", *IEEE Transactions on Microwave Theory and Techniques*, **18** (8), 476–485.

Cullen, A. L. (1983). "Chapter 4 Millimeter-wave open-resonator techniques", in *Infrared and Millimeter Waves*, Vol. 10, Part II, K. J. Button, Ed., Academic Press, Orlando, 233–281.

Damaskos, N. (1995). "Measuring dielectric constants of low loss materials using a broadband cavity technique", *Microwave Journal*, **38** (9), 140–149.

Gershon, D. Calame, J. P. Carmel, Y. and Antonsen Jr., T. M. (2000). "Adjustable resonant cavity for measuring the complex permittivity of dielectric materials", *Review of Scientific Instruments*, **71** (8), 3207–3209.

Hakki, B. W. and Coleman, P. D. (1960). "A dielectric resonator method of measuring inductive capacities in the millimeter range", *IRE Transactions on Microwave Theory and Techniques*, **8**, 402–410.

Humbert, W. R. and Scott Jr., W. R. (1996). "A new technique for measuring the permittivity and loss tangent of cylindrical dielectric rods", *IEEE Microwave and Guided Wave Letters*, **6** (7), 262–264.

Humbert, W. R. and Scott Jr., W. R. (1997). "Measurement of the permittivity and loss tangent of dielectric sheets", *Microwave and Optical Technology Letters*, **15** (6), 355–358.

Janezic, M. D. and Baker-Jarvis, J. (1999). "Full-wave analysis of a split-cylinder resonator for nondestructive permittivity measurements", *IEEE Transactions on Microwave Theory and Techniques*, **47** (10), 2014–2020.

Kajfez, D. and Guillon, P. (1986). *Dielectric Resonators*, Artech House, Dedham, MA.

Kent, G. (1996). "Nondestructive permittivity measurement of substrates", *IEEE Transactions on Instrumentation and Measurement*, **45** (1), 102–106.

Kent, G. (1988a). "Dielectric resonances for measuring dielectric properties", *Microwave Journal*, **31** (10), 99–114.

Kent, G. (1988b). "An evanescent-mode tester for ceramic dielectric substrates", *IEEE Transactions on Microwave Theory and Techniques*, **36** (10), 1451–1454.

Kent, G. and Bell, S. M. (1996). "The gap correction for the resonant-mode dielectrometer", *IEEE Transactions on Instrumentation and Measurement*, **45** (1), 98–101.

Kobayashi, Y. Aoki, T. and Kabe, Y. (1985). "Influence of conductor shields on the Q-factors of a TE_0 dielectric resonator", *IEEE Transactions on Microwave Theory and Techniques*, **33** (12), 1361–1366.

Kobayashi, Y. Imai, T. and Kayano, H. (1990). "Microwave measurement of surface impedance of high-T_c superconductor", *IEEE MTT-S Digest*, 281–284.

Kobayashi, Y. Imai, T. and Kayano, H. (1991). "Microwave measurement of temperature and current dependences of surface impedance for high-T_c superconductors", *IEEE Transactions on Microwave Theory and Techniques*, **39** (9), 1530–1538.

Kobayashi, Y. and Katoh, M. (1985). "Microwave measurement of dielectric properties of low-loss materials by the dielectric rod resonator method", *IEEE Transactions on Microwave Theory and Techniques*, **33** (7), 586–592.

Kobayashi, Y. and Tanaka, S. (1980). "Resonant modes of a dielectric rod resonator short-circuited at both ends by parallel conducting plates", *IEEE Transactions on Microwave Theory and Techniques*, **28**, 1077–1086.

Kobayashi, Y. and Yoshikawa, H. (1998). "Microwave measurements of surface impedance of high-T-c superconductors using two modes in a dielectric rod resonator", *IEEE Transactions on Microwave Theory and Techniques*, **46** (12), 2524–2530.

Komiyama, B. Kiyokawa, M. and Matsui, T. (1991). "Open resonator for precision dielectric measurements in the 100 GHz band", *IEEE Transactions on Microwave Theory and techniques*, **39** (10), 1792–1796.

Krupka, J. Derzakowski K., Abramowicz, A. Tobar, M. E. and Baker-Jarvis, J. (1999). "Use of whispering-gallery modes for complex permittivity determinations of ultra-low-loss dielectric materials", *IEEE Transactions on Microwave Theory and Techniques*, **47** (6), 752–759.

Krupka, J. Geyer, R. G. Baker-Jarvis, J. and Ceremuga, J. (1996). "Measurements of the complex permittivity of microwave circuit board substrates using split dielectric resonator and reentrant cavity techniques", *Seventh International Conference on Dielectric Materials Measurements & Applications*, IEE, London, 21–24.

Krupka, J. Klinger, M. Kuhn, M. Baranyak, A. Stiller, M. Hinken, J. and Modelski, J. (1993). "Surface resistance measurements of HTS films by means of sapphire dielectric resonators", *IEEE Transactions on Applied Superconductivity*, **3** (3), 3043–3048.

Lynch, A. C. (1979). "Measurement of dielectric properties in an open resonator", *Proc. IEE Conf. On Dielectr. Mater., Meas. And Appl., 1979, IEE Conf. Publ.*, Vol. **177**, Institute of Electrical Engineers, London, 373–376.

Marincic, A. Benson, F. A. and Tealby, J. M. (1986). Measurements of permittivity by the use of surface waves in open and closed structures, *IEE Proceedings-H*, **133**, 441–449.

Matytsin, S. M. Rozanov, K. N. and Simonov, N. A. (1996). "Permittivity measurement using slotted coaxial resonator", *IEEE Instrumentation and Measurement Technology Conference*, June 4–6, Brussels, Belgium, 987–990.

Mazierska, J. and Wilker, C. (2001). "Accuracy issues in surface resistance measurements of high temperature superconductors using dielectric resonators", *IEEE Transactions on Applied Superconductivity*, **11** (4), 4140–4147.

Mazierska, J. (1997). "Dielectric resonator as a possible standard for characterization of high temperature superconducting films for microwave applications", *Journal of Superconductivity*, **10** (2), 73–84.

Mourachkine, A. P. and Barel, A. R. F. (1995). "Microwave measurement of surface resistance by the parallel-plate dielectric resonator method", *IEEE Transactions on Microwave Theory and Techniques*, **43** (3), 544–551.

Obara, H. Kosaka, S. Sawa, A. Yamasaki, H. Kobayashi, Y. Hashimoto, T. Ohshima, S. Kusunoki, M. and Inadomaru, M. (2001). "Precise surface resistance measurements of $YBa_2Cu_3O_y$ films with the dielectric resonator method", *Physica C*, **357–360**, 1511–1515.

Shen, Z. Y. Wilker, C. Pang, P. Holstein, W. L. Face, D. and Kountz, D. J. (1992). "High T_c

superconductor-sapphire microwave resonator with extremely high Q-values up to 90 K", *IEEE Transactions on Microwave Theory and Techniques*, **40** (12), 2424–2432.

Taber, R. C. (1990). "A parallel plate resonator technique for microwave loss measurement on superconductors", *Review of Scientific Instruments*, **60** (8), 2200–2206.

Talanov, V. V. Mercaldo, L. and Anlage, S. M. (1999). "Measurement of the absolute penetration depth and surface resistance of superconductors using the variable parallel plate resonator", *IEEE Transactions on Applied Superconductivity*, **9** (2), 2179–2182.

Tellmann, N. Klein, N. Dahne, U. Scholen, A. Schulz, H. and Chaloupka, H. (1994). "High-Q LaAlO3 dielectric resonator shielded by YBCO-films", *IEEE Transactions on Applied Superconductivity*, **4** (3), 143–148.

Wheless Jr., P. and Kajfez, D. (1985). "The use of higher resonant modes in measuring the dielectric constant of dielectric resonators", *1985 IEEE MTT-S Digest*, 473–476.

Xi, W. G. Tian, B. Q. and Tinga, W. R. (1994). "Numerical analysis of a movable dielectric gap in coaxial resonators for dielectric measurements", *IEEE Transactions on Instrumentation and Measurement*, **43** (3), 486–487.

Xi, W. G. Tinga, W. R. Geoffrey Voss, W. A. and Tian, B. Q. (1992). "New results for coaxial reentrant cavity with partially dielectric filled gap", *IEEE Transactions on Microwave Theory and Techniques*, **40** (4), 747–753.

Yu, P. K. and Cullen, A. L. (1982). "Measurement of permittivity by means of an open resonator. I. Theoretical", *Proceedings of the Royal Society of London, Series A, Mathematical and Physical Sciences*, **380**, 49–71.

6

Resonant-perturbation Methods

In a resonant perturbation method, the sample is inserted into a resonator, and the properties of the sample are calculated from the changes of the resonant frequency and the quality factor of the resonator caused by the sample. After analyzing the resonant perturbation theory, we discuss the cavity-perturbation method and the dielectric resonator perturbation method for permittivity and permeability measurement. We then discuss the measurement of surface impedance using the resonant-perturbation method. At the final section, we discuss the resonant near-field microwave microscopes, which can be used for mapping various properties of materials, including permittivity, permeability, and sheet resistance.

6.1 RESONANT PERTURBATION

6.1.1 Basic theory

As discussed in Section 2.1, resonant methods, including dielectric resonator methods and resonant perturbation methods, usually have higher accuracy and sensitivity than nonresonant methods. In a dielectric resonator method, the sample under test resonates in the measurement circuit, and the dielectric properties of the sample are deduced from its resonant frequency and quality factor. In a resonant perturbation method, the sample under test is introduced to a resonator, and the electromagnetic properties of the sample are deduced from the change of the resonant frequency and quality factor of the resonator. Owing to its high accuracy and sensitivity, and its flexibility

in sample preparation, resonant perturbation methods are widely used for low-loss bulk samples, powders, small-size samples, and samples of irregular shapes.

Hollow metallic cavities and dielectric resonators are two types of resonators often used in resonant perturbation methods, and the corresponding resonant perturbation methods are called the cavity-perturbation method and the dielectric resonator perturbation method, respectively. In this section, we focus on the perturbation to hollow metallic cavities, and the conclusions can be extended to the perturbations to dielectric resonators.

Generally speaking, there are three types of cavity perturbations: cavity-shape perturbation, wall-impedance perturbation, and material perturbation. Cavity-shape perturbation can be achieved by pulling out or pushing in part of the cavity wall. Usually, cavity-shape perturbations change the resonant frequency of the cavity and the stored energy in the cavity, but do not change the energy dissipation in the cavity. The cavity-shape perturbation is often used to retune the resonant frequency of a resonant cavity. Wall-impedance perturbation can be achieved by replacing part of the cavity wall, keeping the shape of the cavity unchanged. The wall-impedance perturbation is often used for measuring the surface impedance of conductors. In material perturbation, a material is introduced into a cavity, and the resonant frequency and quality factor of the cavity are thus changed. The complex permittivity or complex permeability of the material can be determined from the changes of the resonant frequency and the quality factor of the cavity due to the introduction of the sample.

Microwave Electronics: Measurement and Materials Characterization   L. F. Chen, C. K. Ong, C. P. Neo, V. V. Varadan and V. K. Varadan
© 2004 John Wiley & Sons, Ltd   ISBN: 0-470-84492-2

In the following discussion, "cavity-perturbation method" refers to "material-perturbation method."

Perturbation method is an approximate method for finding the eigenvalues of the perturbed resonant system, which does not have much difference from the original system with known eigenvalues. Perturbation method assumes that, if the eigenvalues of system S are known and system S' is closely similar to system S, the difference of the eigenvalues of system S' from the corresponding eigenvalues of system S is small, and the eigenvalues of system S' can be estimated from those of system S.

An electromagnetic system can be defined by a number of parameters, such as dimensions, conductivity of wall, dielectric permittivity, and magnetic permeability of the medium filling the cavity space. The dielectric permittivity and magnetic permeability may be real or complex numbers, and they can also be tensors. When we say that the system S is close to the system S', it means that most of the parameters of these two systems are the same, and only one or a few of the parameters of system S' are slightly different from the corresponding parameters of system S. For example, a superconductor may be replaced by a metal with large but finite conductivity; a vacuum region may be filled with gas of dielectric permittivity close to unity; a lossless dielectric with real dielectric permittivity may be replaced by a lossy dielectric with a complex permittivity; and the geometrical parameters of the two cavities may be slightly different.

Consider a resonant cavity made from a perfectly conducting material, enclosed by a surface S, with volume V. Before perturbation, the electric field is E_1, and the magnetic field is H_1. According to Maxwell's equations, we have

$$\nabla \times E_1 = -j\omega_1\mu_1 H_1 \qquad (6.1)$$

$$\nabla \times H_1 = j\omega_1\varepsilon_1 E_1 \qquad (6.2)$$

where ω_1 is the angular resonant frequency of the cavity before perturbation, and ε_1 and μ_1 are the permittivity and permeability of the medium in the cavity before perturbation. After a small perturbation, the electric field becomes E_2, and the magnetic field becomes H_2. The small perturbation may be a very small change in properties of a large volume of material, such as when the cavity is filled with gas whose permittivity is to be measured, or a large change in properties of a material with very small volume, such as when a small solid object is introduced into a cavity. Similarly, we have

$$\nabla \times E_2 = -j\omega_2\mu_2 H_2 \qquad (6.3)$$

$$\nabla \times H_2 = j\omega_2\varepsilon_2 E_2 \qquad (6.4)$$

where ω_2 is the resonant angular frequency of the cavity after perturbation, and ε_2 and μ_2 are the permittivity and permeability of the medium in the cavity after perturbation.

From Eqs. (6.1)–(6.4), we have

$$H_2 \cdot \nabla \times E_1^* = j\omega_1\mu_1 H_2 \cdot H_1^* \qquad (6.5)$$

$$-E_2 \cdot \nabla \times H_1^* = j\omega_1\mu_1 E_2 \cdot E_1^* \qquad (6.6)$$

$$H_1^* \cdot \nabla \times E_2 = j\omega_2\mu_2 H_2 \cdot H_1^* \qquad (6.7)$$

$$-E_1^* \cdot \nabla \times H_2 = j\omega_2\mu_2 E_2 \cdot E_1^* \qquad (6.8)$$

By adding Eqs. (6.5)–(6.8), we obtain

$$\begin{aligned} H_2 \cdot \nabla \times E_1^* &- E_2 \cdot \nabla \times H_1^* \\ &+ H_1^* \cdot \nabla E_2 - E_1^* \cdot \nabla \times H_2 \\ &= j[(\omega_2\varepsilon_2 - \omega_1\varepsilon_1)E_2 \cdot E_1^* \\ &\quad + (\omega_2\mu_2 - \omega_1\mu_1)H_2 \cdot H_1^*] \end{aligned} \qquad (6.9)$$

According to the vector identity:

$$B \cdot \nabla \times A - A \cdot \nabla \times B = \nabla \cdot (A \times B) \qquad (6.10)$$

Eq. (6.9) can be rewritten as

$$\begin{aligned} \nabla \cdot (H_2 \times E_1^* &+ H_1^* \times E_2) \\ &= j[(\omega_2\varepsilon_2 - \omega_1\varepsilon_1)E_2 \cdot E_1^* \\ &\quad + (\omega_2\mu_2 - \omega_1\mu_1)H_2 \cdot H_1^*] \end{aligned} \qquad (6.11)$$

Integrating both sides of Eq. (6.11) over the volume of the cavity V, we can get

$$\begin{aligned} \int_V \nabla \cdot (H_2 \times E_1^* &+ H_1^* \times E_2)\, dV \\ &= j\left[(\omega_2 - \omega_1) \int_V (\varepsilon_1 E_2 \cdot E_1^* + \mu_1 H_2 \cdot H_1^*)\, dV \right. \\ &\quad \left. + \omega_2 \int_V (\Delta\varepsilon E_2 \cdot E_1^* + \Delta\mu H_2 \cdot H_1^*)\, dV \right] \end{aligned} \qquad (6.12)$$

with

$$\Delta\varepsilon = \varepsilon_2 - \varepsilon_1 \tag{6.13}$$

$$\Delta\mu = \mu_2 - \mu_1 \tag{6.14}$$

Till now, no approximation has been made, and Eq. (6.12) is exact if the cavity is made from a perfectly conducting material. Equation (6.12) is the fundamental equation in cavity-perturbation theory. In the following, we discuss cavity-shape perturbation, material perturbation, and wall-impedance perturbation.

6.1.2 Cavity-shape perturbation

As shown in Figure 6.1, before the perturbation, the volume and surface of the cavity are V_1 and S_1. After the perturbation, the new volume and the new surface become $V_2 = V_1 - \Delta V$ and $S_2 = S_1 - \Delta S$ respectively. The permittivity ε and permeability μ of the medium in the cavity do not change. So by assuming $\Delta\varepsilon = \Delta\mu = 0$, Eq. (6.12) becomes

$$\int_{V_2} \nabla \cdot (\mathbf{H}_2 \times \mathbf{E}_1^* + \mathbf{H}_1^* \times \mathbf{E}_2) \, dV$$

$$= j \left[(\omega_2 - \omega_1) \int_{V_2} (\varepsilon \mathbf{E}_2 \cdot \mathbf{E}_1^* + \mu \mathbf{H}_2 \cdot \mathbf{H}_1^*) \, dV \right] \tag{6.15}$$

As at the surface S_2, $\mathbf{n} \times \mathbf{E}_2 = 0$, and at the surface S_1, $\mathbf{n} \times \mathbf{E}_1^* = 0$, according to the divergence theorem, we have

$$\int_{V_2} \nabla \cdot (\mathbf{H}_2 \times \mathbf{E}_1^* + \mathbf{H}_1^* \times \mathbf{E}_2) \, dV$$

$$= -\oint_{\Delta S} \mathbf{H}_2 \times \mathbf{E}_1^* \cdot dS \tag{6.16}$$

From Eqs. (6.15) and (6.16), we get the perturbation formula for cavity-shape perturbation:

$$\Delta\omega = \omega_2 - \omega_1 = \frac{j \oint_{\Delta S} \mathbf{H}_2 \times \mathbf{E}_1^* \cdot dS}{\int_{V_2} (\varepsilon \mathbf{E}_2 \cdot \mathbf{E}_1^* + \mu \mathbf{H}_2 \cdot \mathbf{H}_1^*) \, dV} \tag{6.17}$$

It should be noted that in cavity-shape perturbation, no energy dissipation is involved, so the angular resonant frequency is a real number: $\omega = 2\pi f$, where f is the resonant frequency.

Now we make an approximation. As the perturbation is small, by assuming: $V_1 \approx V_2 \approx V$, $\mathbf{E}_1 \approx \mathbf{E}_2$, and $\mathbf{H}_1 \approx \mathbf{H}_2$, we can get

$$\oint_{\Delta S} \mathbf{H}_2 \times \mathbf{H}_1^* \cdot dS = \oint_{\Delta S} \mathbf{H}_1 \times \mathbf{E}_1^* \cdot dS$$

$$= \int_{\Delta V} \nabla \cdot (\mathbf{H}_1 \times \mathbf{E}_1^*) \, dV$$

$$= \int_{\Delta V} (j\omega_1 \varepsilon \mathbf{E}_1 \cdot \mathbf{E}_1^* - j\omega_1 \mu \mathbf{H}_1^* \cdot \mathbf{H}_1) \, dV$$

$$= -j\omega_1 \int_{\Delta V} (\varepsilon |\mathbf{E}_1|^2 - \mu |\mathbf{H}_1|^2) \, dV \tag{6.18}$$

From Eqs. (6.17) and (6.18), we have

$$\frac{\Delta\omega}{\omega_1} = \frac{\omega_2 - \omega_1}{\omega_1} \approx \frac{\int_{\Delta V} (\mu |\mathbf{H}_1|^2 - \varepsilon |\mathbf{E}_1|^2) \, dV}{\int_V (\varepsilon |\mathbf{E}_1|^2 + \mu |\mathbf{H}_1|^2) \, dV}$$

$$= \frac{\Delta W_m - \Delta W_e}{W} \tag{6.19}$$

where W is the total stored energy in the original cavity, ΔW_e and ΔW_m are time-average electric

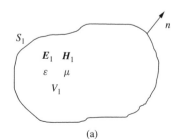
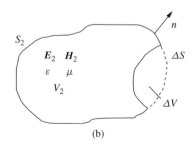

(a)   (b)

Figure 6.1   Cavity-shape perturbation. (a) Original cavity and (b) perturbed cavity

and magnetic energies originally contained in ΔV. As ΔV is very small, we assume

$$\Delta W_m = w_m \Delta V \qquad (6.20)$$

$$\Delta W_e = w_e \Delta V \qquad (6.21)$$

where w_m and w_e are the densities of the magnetic energy and electric energy at ΔV, respectively. If we assume that

$$W = wV \qquad (6.22)$$

where w is the space-average energy density in the cavity, Eq. (6.19) becomes

$$\frac{\Delta \omega}{\omega} \approx \frac{(w_m - w_e)\Delta V}{wV} = C\frac{\Delta V}{V} \qquad (6.23)$$

with

$$C = (w_m - w_e)/w \qquad (6.24)$$

The parameter C is determined by the shape of the cavity and the position of the perturbation.

Equation (6.24) indicates that an outward perturbation ($\Delta V > 0$) will increase the resonant frequency if it is made at the place where the magnetic field dominates ($C > 0$), and will increase the resonant frequency if it is made at the place where the electric field dominates ($C < 0$). Correspondingly, an inward perturbation ($\Delta V < 0$) will lower the resonant frequency if it is made at the place where the magnetic field dominates ($C > 0$) and raise the resonant frequency if it is made at a place where the electric field dominates ($C < 0$). It is evident that maximum changes in resonant frequency will occur when the perturbation is at a position of maximum E and minimum H, or maximum H and minimum E. The change of frequency due to shape perturbation is listed in Table 6.1.

In practice, inward perturbation can be achieved by pushing the cavity wall inward, while the outward perturbation can be achieved by pulling the cavity wall outward. This method is often used in adjusting cavity chains, for example, in the standing-wave accelerating tubes. However, such perturbations are difficult to reverse, and are not suitable for materials characterization. If a plunger is installed on the cavity wall, perturbations can be made by adjusting the depth of the plunger into the cavity. As will be discussed later in detail, a plunger installed on the cavity wall can be used to retune the resonant cavity for materials characterization.

Table 6.1 Change of resonant frequency due to resonant perturbation. ω_0 and ω are the angular resonant frequencies before and after perturbation respectively

	Inward perturbation	Outward perturbation
Strong magnetic field, weak electric field	$\omega > \omega_0$	$\omega < \omega_0$
Weak magnetic field, strong electric field	$\omega < \omega_0$	$\omega > \omega_0$

6.1.3 Material perturbation

Material perturbation is widely used in the characterization of materials properties. In order to analyze the validity of the conventional cavity-perturbation formulae for materials characterization, it is important to discuss the approximations made in their derivations.

At the boundary of the cavity S, the electric fields before and after perturbation satisfy $n \times E_1^* = 0$ and $n \times E_2 = 0$. According to the vector identity,

$$\int_V \nabla \cdot (A \times B)\, dV = \oint_S A \times B \cdot dS \qquad (6.25)$$

we have

$$\int_{V_c} (H_2 \times E_1^* + H_1^* \times E_2)\, dV = 0 \qquad (6.26)$$

So Eq. (6.12) can be rewritten as

$$\frac{\Delta \omega}{\omega} = \frac{\omega_2 - \omega_1}{\omega_1}$$

$$= -\frac{\int_{V_c} (\Delta \varepsilon E_2 \cdot E_1^* + \Delta \mu H_2 \cdot H_1^*)\, dV}{\int_{V_c} (\varepsilon_1 E_2 \cdot E_1^* + \mu_1 H_2 \cdot H_1^*)\, dV} \qquad (6.27)$$

where ω_1 and ω_2 are the complex angular resonant frequencies before and after the introduction of the sample, ε_1 and ε_2 are the complex permittivities of the original medium in the cavity and the sample, μ_1 and μ_2 are the complex magnetic permeabilities of the original medium in the cavity and the sample, H_1 and H_2 are the microwave magnetic fields in the cavity before and after the introduction of the sample, E_1 and E_2 are the microwave electric fields in the cavity before and after the introduction of the sample, and V_c is the region enclosed by the cavity. Equation (6.27) is the basic formula for cavity perturbation, and it stands on the assumptions that the cavity wall is perfectly conducting and the perturbation is small. The requirement of small perturbation can be satisfied in the two ways that follow.

6.1.3.1 Whole-medium perturbation

As shown in Figure 6.2, in this kind of perturbation, the whole original medium (ε_1, μ_1) is replaced by a new medium (ε_2, μ_2). As a result, the electric and magnetic fields are changed from (E_1, H_1) to (E_2, H_2). To satisfy the perturbation requirement, the differences of the dielectric permittivity and magnetic permeability between the original and new media should be small. This kind of perturbation can be used to measure the electromagnetic properties of gases, and the electromagnetic properties of the gas under test can be derived from Eq. (6.27).

6.1.3.2 Small-object perturbation

To use the cavity-perturbation method for materials characterization, as shown in Figure 6.3, a material with permittivity ε_2 and permeability μ_2 is introduced into the resonant cavity. The sample occupies a small portion of the cavity, and the electromagnetic properties of the space except the sample do not change. On the basis of three assumptions, Eq. (6.27) can be rewritten as Eq. (6.28) (Meng *et al.* 1995; Waldron 1969). First, the original medium in the cavity is lossless. Second, the sample is homogenous and is much smaller than the cavity. And third, the electromagnetic field outside the sample does not change.

$$\frac{\omega_2 - \omega_1}{\omega_1} \approx -\frac{\int_{V_s} (\Delta\varepsilon E_1^* \cdot E_2 + \Delta\mu H_1^* \cdot H_2)\, dV}{\int_{V_c} (\varepsilon_1 E_1^* \cdot E_2 + \mu H_1^* \cdot H_2)\, dV}$$

$$\approx -\frac{\int_{V_s} (\Delta\varepsilon E_1^* \cdot E_2 + \Delta\mu H_1^* \cdot H_2)\, dV}{2\int_{V_c} \varepsilon_1 E_1^* \cdot E_2\, dV}$$

$$(6.28)$$

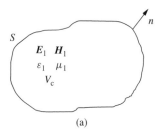
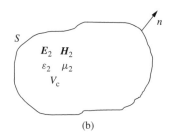

(a)

(b)

Figure 6.2   Whole-medium perturbation. (a) Original cavity with medium (ε_1, μ_1) and (b) cavity after perturbation with medium (ε_2, μ_2)

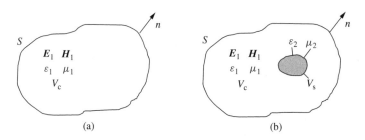

(a)

(b)

Figure 6.3   Material perturbation. (a) Original cavity and (b) perturbed cavity

where V_s is the volume of the sample. It should be noted that, in this way of perturbation, the difference in the permittivity and permeability between the original and new materials could be large, while the volume of new material V_s should be small so that the total perturbation is small.

Equation (6.28) can be further approximated by taking $E_1 \approx E_2$, $H_1 \approx H_2$, and $\omega_1 \approx \omega_2$:

$$\frac{\Delta\omega}{\omega_1} = -\frac{\displaystyle\int_{V_c} (\Delta\varepsilon|E_1|^2 + \Delta\mu|H_1|^2)\,dV}{\displaystyle\int_{V_c} (\varepsilon|E_1|^2 + \mu|H_1|^2)\,dV}$$

$$= -\frac{1}{4\,W} \int_{\Delta V_s} (\Delta\varepsilon|E_1|^2 + \Delta|H_1|^2)\,dV$$

$$(6.29)$$

Equation (6.29) indicates that any increase in ε and μ ($\Delta\varepsilon > 0$, $\Delta\mu > 0$) will decrease the resonant frequency.

6.1.4 Wall-impedance perturbation

As shown in Figure 6.4, for a resonant cavity, if the surface impedance of the boundary is changed, the resonant properties of the cavity will also be changed, and such kind of perturbation is called wall-impedance perturbation. Wall-impedance perturbation can be used to characterize the impedance of conducting plates by replacing part of the cavity wall with the sample under measurement, keeping the cavity shape unchanged. Using this method, the surface reactance of the conducting plate can be deduced from the change of resonant frequency, and the surface resistance of the conducting plate can be deduced from the change of quality factor.

In the following discussion on wall-impedance perturbation, we follow the analysis methods in (Landau and Lifshitz 1960; Ormeno *et al.* 1997; Sucher and Fox 1963). We assume that the cavity walls have finite impedance Z_s and the cavity contains conducting samples that carry a finite-free current density J. Let H_1 and E_1 be the magnetic and electric fields inside the original cavity, and H_2 and E_2 be the corresponding fields after perturbation. The integral over the cavity volume

$$\int_V (E_1 \cdot \nabla \times H_2 - H_2 \cdot \nabla \times E_1 + H_1 \cdot \nabla \\ \times E_2 - E_2 \cdot \nabla \times H_1)\,dV \qquad (6.30)$$

can be converted to a surface integral over the cavity walls using the divergence theorem. By comparing this surface integral with the volume integral using the Maxwell equations (1.3) and (1.4), we arrive at the following exact formula for the change in the resonant angular frequency if the surface impedance of the cavity wall changes by ΔZ_s:

$$\Delta\omega$$

$$= \frac{\begin{aligned} &-\omega_1 \int_V (H_1 \cdot H_2 \Delta\mu - E_1 \cdot E_2 \Delta\varepsilon)\,dV \\ &-j \int_V (E_1 \cdot J_2 - E_2 \cdot J_1)\,dV \\ &+j \int_s H_1 \cdot H_2 \Delta Z_s\,dS \end{aligned}}{\displaystyle\int_V (H_1 \cdot H_2 \mu_2 - E_1 \cdot E_2 \varepsilon_2)\,dV}$$

$$(6.31)$$

where the surface integral is taken over the cavity walls.

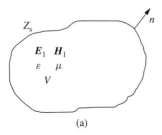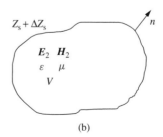

(a)   (b)

Figure 6.4 Wall-impedance perturbation. (a) Original cavity with surface impedance Z_s and (b) perturbed cavity with surface impedance $Z_s + \Delta Z_s$

In the measurement of surface impedance, we are not concerned with the changes in μ and ε. For small perturbation of a cavity with high quality factor, we may assume that over most of the cavity, $H_1 \approx H_2$, $E_1 \approx E_2$, and $J_1 \approx J_2$. If we choose the phase of H to be zero, the phase of E will be $\pm\pi/2$. We may then identify the denominator of Eq. (6.31) as $4W$, where W is the energy stored in the cavity. With these assumptions, we have

$$\Delta\omega = \frac{j \int_S H_2 \cdot H_2 \Delta Z_s \, dS}{4W}$$

$$= j \left(\frac{\int_V H_2 \cdot H_2 \, dS}{4W} \right) \Delta Z_s = j\Gamma \Delta Z_s \quad (6.32)$$

with

$$\Gamma = \left(\frac{\int_V H_2 \cdot H_2 \, dS}{4W} \right). \quad (6.33)$$

On the basis of our assumptions, Γ is a resonator constant, determined by the properties of the resonator, and is independent of the sample under study.

6.2 CAVITY-PERTURBATION METHOD

Cavity-perturbation methods are widely used in the study of the electromagnetic properties of dielectrics, semiconductors, magnetic materials, and composite materials. It works well for the measurement of low-loss and medium-loss materials. However, extremely low-loss samples often render the conventional cavity-perturbation methods less useful. If the quality factor of an empty cavity before the perturbation is not high, the power dissipation of the empty cavity may be much larger than the loss due to the introduction of the sample to be measured, and so the introduction of the sample hardly affects the quality factor of the cavity. As such, the conventional cavity-perturbation method could not correctly give the value of the imaginary part of the permittivity. The situation may be even more severe: the quality factor of the cavity may even increase after introducing an

extremely low-loss sample, so that the conventional cavity-perturbation formulae will give a negative value for the imaginary part of permittivity. The main reason for this error is that the conventional cavity-perturbation formulae for permittivity measurements incorporate many approximations and assumptions as discussed above. This problem is also discussed in (Waldron 1969; Sucher and Fox 1963; Harrington 1961).

In this section, we discuss the measurement of permittivity and permeability using cavity-perturbation method, which works quite well for low-loss and medium-loss samples. We then analyze the change of resonant properties of resonant cavities due to the introduction of samples. Finally, we discuss the methods for modifying the conventional cavity-perturbation methods.

6.2.1 Measurement of permittivity and permeability

If we introduce a sample to the antinode of an electric field or a magnetic field in a cavity, the resonant frequency and quality factor of the cavity will be changed. If the perturbation requirements can be satisfied, the dielectric permittivity or magnetic permeability of the sample can be calculated from the changes of the resonant frequency and the quality factor.

This measurement method is based on material perturbation as discussed above. In most cases, the perturbation caused by the sample belongs to small-object perturbation, and the following discussion is based on Eq. (6.28).

6.2.1.1 Permittivity measurement

We assume that the medium inside the cavity is a vacuum: $\mu_1 = \mu_2 = \mu_0$, and $\varepsilon_1 = \varepsilon_0$. So Eq. (6.28) becomes

$$\frac{\omega_2 - \omega_1}{\omega_2} = -\left(\frac{\varepsilon_r - 1}{2} \right) \frac{\iiint_{V_s} E_1 \cdot E_2 \, dV}{\iiint_{V_c} |E_1|^2 \, dV} \quad (6.34)$$

where ε_r is the relative complex permittivity of the sample: $\varepsilon_r = \varepsilon_2/\varepsilon_0$.

The complex angular frequency ω of a resonant cavity is related to the real resonant frequency f and the quality factor Q of the cavity by (Waldron 1969; Sucher and Fox 1963)

$$\omega = \omega_r + j\omega_i \qquad (6.35)$$

$$\omega_r = 2\pi f \qquad (6.36)$$

$$Q = \frac{\omega_r}{2\omega_i} \qquad (6.37)$$

If we assume that $\omega_{r1} \approx \omega_{r2}$ and $\omega_i \gg \omega_r$, we have

$$\frac{\omega_2 - \omega_1}{\omega_2} = \frac{(\omega_{r2} - \omega_{r1}) + j(\omega_{i2} - \omega_{i1})}{\omega_{r2}\left(1 + j\dfrac{\omega_{i2}}{\omega_{r2}}\right)}$$

$$\approx \left[\left(\frac{f_2 - f_1}{f_2}\right) + j\left(\frac{1}{2Q_2} - \frac{1}{2Q_1}\right)\right]$$

$$\times \left(1 - j\frac{1}{2Q_2}\right) \qquad (6.38)$$

$$\approx \left(\frac{f_2 - f_1}{f_2}\right) + j\left(\frac{1}{2Q_2} - \frac{1}{2Q_1}\right)$$

The last assumption made in Eq. (6.38) is that $Q_2 \gg 1$. From Eqs. (6.34) and (6.38), we have

$$\left(\frac{f_2 - f_1}{f_2}\right) + j\left(\frac{1}{2Q_2} - \frac{1}{2Q_1}\right)$$

$$= -\left(\frac{\varepsilon_r - 1}{2}\right)\frac{\iiint_{V_s} \boldsymbol{E}_1^* \cdot \boldsymbol{E}_2 \, dV}{\iiint_{V_c} |\boldsymbol{E}_1|^2 \, dV} \qquad (6.39)$$

Equation (6.39) can be further rewritten as

$$2\left(\frac{f_1 - f_2}{f_2}\right) = (\varepsilon_r' - 1)C \qquad (6.40)$$

$$\frac{1}{Q_2} - \frac{1}{Q_1} = \varepsilon_r'' C \qquad (6.41)$$

with

$$C = \frac{\iiint_{V_s} \boldsymbol{E}_1^* \cdot \boldsymbol{E}_2 \, dV}{\iiint_{V_c} |\boldsymbol{E}_1|^2 \, dV} \qquad (6.42)$$

In the cavity-perturbation method for permittivity measurements, the parameter C is assumed to be a constant independent of the properties of samples. However, in Eq. (6.42), the perturbed field in the sample $\boldsymbol{E}_2$ is related to the permittivity, shape, and size of the sample under measurement, and so C changes from case to case. Therefore, in a strict sense, Eqs. (6.41) and (6.42) show that the change of quality factor due to the perturbation is not only related to the imaginary part of the permittivity of the sample but also, to some extent, dependent on the real part of the permittivity of the sample.

To make the expression clearer, we introduce two parameters A and B, instead of C, in the widely used cavity-perturbation formulae, which then read as

$$\frac{f_1 - f_2}{f_2} = A(\varepsilon_r' - 1)\frac{V_s}{V_c} \qquad (6.43)$$

$$\frac{1}{Q_2} - \frac{1}{Q_1} = B\varepsilon_r''\frac{V_s}{V_c} \qquad (6.44)$$

Similar to parameter C, parameters A and B are related to the configuration and working mode of the cavity, the shape of the sample, and the sample's location in the cavity. As it is difficult to calculate the parameters A and B analytically, A and B are usually obtained by calibration using a standard sample of known permittivity. From the changes of the resonant frequency f and the quality factor Q due to the introduction of a standard sample, A and B can be calculated using Eqs. (6.43) and (6.44). But it should be noted that the standard sample used in calibration is required to be of similar shape to the samples to be measured.

Equations (6.43) and (6.44) are widely used in permittivity measurements. However, it should be kept in mind that these equations are based on mainly three assumptions. First, the electromagnetic fields in the cavity do not change due to the introduction of the sample, and the stored energy in the empty cavity equals that in the cavity with the sample. Second, the difference between the cavity wall losses with and without the sample is negligible. Third, Q_1 and Q_2 are measured at the same frequency. The first of these three assumptions is the most fundamental. The second one is consistent with and implied by the first one. Experimentally, the first and second assumptions cannot be strictly fulfilled. The third assumption provides a way to

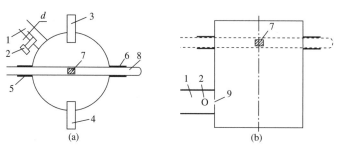

Figure 6.5 A cylinder cavity in TE$_{112}$ mode. (a) Top view and (b) side view. 1, waveguide; 2, coupling adjusting plunger; 3,4, retuning plungers; 5,6, quartz-tube holders; 7, dielectric sample; 8, quartz tube; 9, coupling iris. Source: Chen, L. F. Ong, C. K. and Tan, B. T. G. (1996). "A resonant cavity for high-accuracy measurement of microwave dielectric properties", *Measurement Science and Technology*, **7**, 1255–1259, IOP Publishing Ltd, by permission of IOP Publishing Ltd, by permission of IOP Publishing Ltd

experimentally ensure high accuracy even if the first two assumptions are not strictly fulfilled.

6.2.1.2 Permeability measurement

We assume that either the magnetic sample is placed at the position with zero electric field or the permittivity of the sample is ε_0. With the approximations made in deducing Eqs. (6.43) and (6.44), we have the following similar equations:

$$\frac{f_1 - f_2}{f_2} = A(\mu_r' - 1)\frac{V_s}{V_c} \qquad (6.45)$$

$$\frac{1}{Q_2} - \frac{1}{Q_1} = B\mu_r''\frac{V_s}{V_c} \qquad (6.46)$$

Equations (6.45) and (6.46) can be used for the characterization of magnetic materials using resonant perturbation method, and, similarly, the parameters A and B are usually determined by calibration.

6.2.2 Resonant properties of sample-loaded cavities

In order to accurately measure the electromagnetic properties of a material by inserting the material into a resonant cavity, it is necessary to analyze the resonant properties of a sample-loaded cavity, including resonant frequency, quality factor, and coupling coefficient. In the following discussion, we concentrate on the measurement of dielectric materials, while the conclusions obtained can be extended to the measurement of magnetic materials.

Figure 6.5 shows a cylinder cavity working in the TE$_{112}$ mode. The selection of the TE$_{112}$ mode was mainly based on two considerations. First, as the strength of the electric field in TE$_{112}$ mode changes along the radius directions, it can be used to study the relationship between the sample perturbation and the electric field. Second, by selecting TE$_{112}$ mode instead of TE$_{111}$ mode, the perturbation area and the coupling area are separated, as shown in Figure 6.5(b), so that the dominant field at the coupling area is almost unchanged after the perturbation is introduced.

In Figure 6.5, the cavity has diameter $D = 27.0\,\text{mm}$ and length $L = 44.2\,\text{mm}$, and its resonant frequency is near 9 GHz. The sample (labeled 7) is placed in a quartz tube (labeled 8), which is fastened by plastic screws in two quartz-tube holders (labeled 5 and 6). The sample is located at the center of the cavity where the electric field is the strongest and where the frequency shift is thus the largest. Two brass plungers (labeled 3 and 4) are used to adjust the resonant frequency of the cavity. The plungers are located at the positions where the magnetic field in the cavity is the strongest. The cavity is coupled to a waveguide (labeled 1) via an iris (labeled 9). There is a plunger (labeled 2) lying close to the coupling iris. By varying the insertion length d, the coupling coefficient β can be adjusted. In order to protect the main body of the cavity, each of the plungers is covered by a plastic screw, as shown in Figure 6.6, and the pitch of the screw is selected to be as small as possible to increase the ease with which the experiments could be repeated.

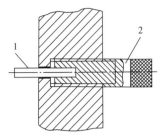

Figure 6.6 The mechanical structure of the plungers: 1, brass plunger; 2, plastic screw. Source: Chen, L. F. Ong, C. K. and Tan, B. T. G. (1996). "A resonant cavity for high-accuracy measurement of microwave dielectric properties", *Measurement Science and Technology*, **7**, 1255–1259, IOP Publishing Ltd, by permission of IOP Publishing Ltd

There are two perpendicular degenerate TE$_{112}$ modes in the cavity. Usually, only one is used and hence the other one should be eliminated. As shown in Figure 6.7, this can be achieved by inserting a slender metal needle perpendicular to the working mode. The needle hardly deforms the working mode, but disturbs the modes perpendicular to the required working mode greatly and effectively eliminates them (Maier and Slater 1952). The mechanical structure of the slender metal needle is similar to that of the plungers shown in Figure 6.6.

6.2.2.1 Resonant frequency

When a dielectric material is introduced into a resonant cavity, the change of the resonant frequency

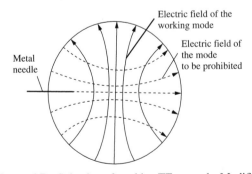

Figure 6.7 Selection of working TE$_{112}$ mode. Modified from Chen, L. F. Ong, C. K. and Tan, B. T. G. (1996). "A resonant cavity for high-accuracy measurement of microwave dielectric properties", *Measurement Science and Technology*, **7**, 1255–1259, IOP Publishing Ltd, by permission of IOP Publishing Ltd

of the cavity is related to the dielectric properties of the sample and the electric field at the position where the sample is placed. The stronger the electric field, the greater is the effect on the resonant frequency. A standard Teflon sample with dimensions $\phi = 3.0 \times 3.0$ mm and a permittivity $\varepsilon_r = 2.0008 - j0.00065$ is introduced to different positions along the radius of the cavity. Figure 6.8 shows the relationship between the change of the resonant frequency and the position. In the figure, Δf, ΔQ, and E_r are normalized to their maximum values, respectively, and the electric field E_r is calculated from Δf (Maier and Slater 1952):

$$E_r = \sqrt{\Delta f} \qquad (6.47)$$

6.2.2.2 Quality factor

Knowledge of the quality factor of a sample-loaded cavity is helpful in explaining some abnormal phenomena in the permittivity measurement of extremely low-loss samples, and it also shows the validity limits of the conventional cavity-perturbation method and justifies the amended method discussed in Section 6.2.3.

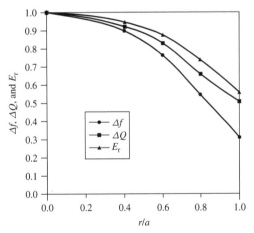

Figure 6.8 The relationships of the electric field E_r, the change of quality factor ΔQ, and the change of resonant frequency Δf. In the figure, r is the distance of the sample from the axis of the cavity. Source: Chen, L. F. Ong, C. K. and Tan, B. T. G. (1999), "Amendment of cavity perturbation method for permittivity measurement of extremely low-loss dielectrics", *IEEE Transactions on Instrumentation and Measurement*, **48** (6), 1031–1037. © 2003 IEEE

The quality factor Q of a sample-loaded cavity is directly dependent on the quality factor of the empty cavity without the sample, as well as the shape, size, dielectric properties, and location of the sample. By definition, $Q = 2\pi fW/P$, where W is the total stored energy in the cavity and P is the power dissipation of the cavity. We may regard f as a constant because the resonant perturbation requires that $(f_2 - f_1)/f_1 \ll 1$, where f_1 and f_2 are the resonant frequencies of the cavity before and after the introduction of the sample respectively. Introducing a dielectric sample to the antinode of the electric field has two effects on the quality factor of the cavity. First, the total stored energy of the whole sample-loaded cavity is increased. Second, the stored energy becomes more concentrated on the sample, and so the electromagnetic field at the cavity wall becomes weaker. Because it is impossible for the cavity walls to be perfectly conductive at microwave frequencies, the wall loss of the sample-loaded cavity may be less than that of the cavity without the sample. Therefore, the total stored energy of the cavity increases and the total power dissipation in the cavity may decrease after the introduction of an extremely low-loss sample, thus increasing the overall quality factor of the cavity.

The change of quality factor due to the insertion of a sample can also be explained using the equivalent circuit of a resonant cavity. At its resonance, a resonant cavity can be expressed by a lumped-element equivalent circuit consisting of inductance L, resistance R, and capacitance C, as shown in Figure 6.9. The equivalent capacitance C can be expressed as the sum of a major part C_1 and a minor part C_2. We then assume that the introduction of a small sample only affects the minor part C_2 while the major part C_1 remaining unperturbed. The resonant frequency f and the quality factor Q of the resonant cavity can be expressed in terms of its lumped-element circuit parameters: capacitance $C (C = C_1 + C_2)$, inductance L, and resistance R (Collin 1992):

$$f = \frac{1}{2\pi \sqrt{L(C_1 + C_2)}} \qquad (6.48)$$

$$Q = R\sqrt{\frac{C_1 + C_2}{L}} \qquad (6.49)$$

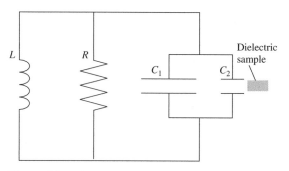

Figure 6.9   The lumped-element equivalent circuit of a resonant cavity at its resonance. Introducing a dielectric sample into the cavity is equivalent to inserting the sample into the capacitor C_2

Introducing a dielectric sample into the antinode of electric field is equivalent to inserting the sample into the minor capacitor C_2 and so increases its capacitance C_2. If the sample is a perfect dielectric $(\varepsilon_r'' = 0)$, the effect of the sample on R is negligible. So according to Eq. (6.49), the quality factor Q increases after introducing a perfect dielectric sample.

The introduction of an actual dielectric sample $(\varepsilon_r'' > 0)$ not only increases the capacitance C but also decreases the resistance R. The change of the quality factor due to the introduction of an actual dielectric sample is determined by the overall effects of the increase of capacitance C and the decrease of resistance R. For an extremely low-loss sample, if its ε_r'' is so small that the effect of the decrease of R is weaker than that of the increase of C, then, according to Eq. (6.49), the overall quality factor increases after the introduction of the sample.

The above discussions hint that the stronger the local electric field at the position where the sample is introduced, the greater is the increase of C, and thus greater is the increase of the quality factor Q. Figure 6.8 shows the measurement results of resonant frequency and quality factor when a Teflon sample (with diameter 3 mm and length 3 mm) is put at different places in a quartz tube along the radial direction of the cavity. The electric field distribution at the cross section where the Teflon sample is put is shown in Figure 6.10. Figure 6.8 indicates that the quality factor of the sample-loaded cavity is higher if the sample is at the position where the electric field is stronger.

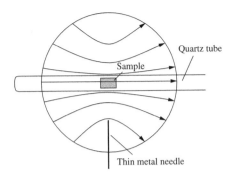

Figure 6.10 The electric field distribution in the TE$_{112}$-mode resonant cavity. Source: Chen, L. F. Ong, C. K. and Tan, B. T. G. (1999), "Amendment of cavity perturbation method for permittivity measurement of extremely low-loss dielectrics", *IEEE Transactions on Instrumentation and Measurement*, **48** (6), 1031–1037. © 2003 IEEE

6.2.2.3 Coupling coefficient

The coupling coefficient β of the cavity to external circuit is also affected by the introduction of a dielectric sample. If we choose the reference plane to be at the positions where the electric field is the strongest, the cavity can be represented by a parallel LRC circuit. If the losses of the coupling devices are negligible, the coupling devices can be represented by an ideal transformer with transforming ratio n, as shown in Figure 6.11(b). The problem can be simplified further by folding back the equivalent circuit, shown in Figure 6.11(b), to the transmission line and normalizing with respect to the characteristic impedance of the transmission line Z_0, shown in Figure 6.11(c). The value of the coupling coefficient β is equal to the normalized input resistance R'_p looking into the cavity from the transmission line (Collin 1992):

$$\beta = R'_P = \frac{R_P}{n^2 Z_0} \tag{6.50}$$

The value of R'_p decreases when an actual dielectric sample ($\varepsilon''_r > 0$) is inserted into the cavity, and so does the coupling coefficient β.

6.2.3 Modification of cavity-perturbation method

The above discussion shows that to obtain accurate results for the imaginary part of permittivity, appro-

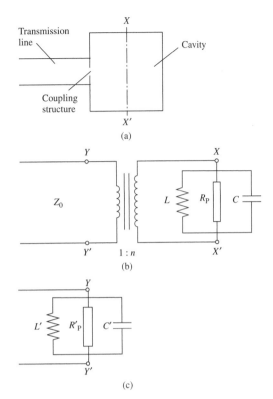

Figure 6.11 A resonant cavity coupled to a transmission line and its equivalent circuits. (a) A resonant cavity coupled to a transmission line, (b) equivalent circuit at $X-X'$ plane, where the electric field is the strongest, and (c) normalized equivalent circuit folding back to the transmission line with $Z'_0 = 1$. Modified from Chen, L. F. Ong, C. K. and Tan, B. T. G. (1996). "A resonant cavity for high-accuracy measurement of microwave dielectric properties", *Measurement Science and Technology*, **7**, 1255–1259, IOP Publishing Ltd, by permission of IOP Publishing Ltd

priate quality factor should be used in Eq. (6.44). To increase the measurement accuracy, Jow *et al.* presented a single-frequency method to measure the quality factor by retuning the cavity length to compensate for the resonant frequency shift due to the introduction of samples (Jow *et al.* 1989), and Tian and Tinga proposed a way to measure the quality factor at a single frequency by introducing a metal-tuning stub at the antinode of the magnetic field in the cavity (Tian and Tinga 1995, 1993). In the following, we discuss two modification methods: frequency-retuning method (Chen *et al.* 1996) and calibration method (Chen *et al.* 1999).

6.2.3.1 Frequency-retuning method

As the insertion of a dielectric sample at the position where electric field is dominant will increase the stored energy and decrease the resonant frequency, the values of quality factor at different frequencies before and after the introduction of the sample are not comparable. If we retune the cavity to its original frequency, the stored energy will return to its original value also. Therefore, to obtain comparable values of the quality factor that can be used in Eq. (6.44), retuning is necessary.

As shown in Figure 6.5, we use the two retuning plungers (labeled 3 and 4) to retune the resonant frequency of the sample-loaded cavity back to the frequency before the sample is introduced. In the retuning procedure, the two plungers should be inserted into the cavity to an equal extent to maintain the structural symmetry of the cavity as much as possible, thus decreasing the uncertainties due to the field deformation caused by the insertion of the plungers.

Figure 6.12 shows the Smith charts corresponding to different conditions during the retuning procedure. Before the insertion of the sample, the cavity is critically coupled to the waveguide ($\beta = 1$), and the corresponding S_{11} curve in the format of Smith chart passes through the center of the chart, as shown in Figure 6.12(a). After the introduction of the sample, along with a decrease in resonant frequency, the S_{11} curve in the format of Smith chart shrinks, as shown in Figure 6.12(b), indicating that the cavity is undercoupled ($\beta < 1$). By adjusting the retuning devices (labeled 3 and 4 in Figure 6.5), the resonant frequency can be retuned back to the value before the insertion of the sample. However, Figure 6.12(c) indicates that the cavity is still in an undercoupled condition after retuning. Further tuning by adjusting the plunger (labeled 2 in Figure 6.5) can make the cavity become critically coupled to the transmission line again, as shown in Figure 6.12(d).

The values of quality factors Q corresponding to the above different conditions shown in Figure 6.12 are listed in Table 6.2. Different values of the subscript i ($i = 1, 2, 3, 4$) refer to the different corresponding conditions shown in Figure 6.12(a) to (d) respectively. It is essential to use the appropriate value of quality factor to calculate ε_r'' from Eq. (6.44).

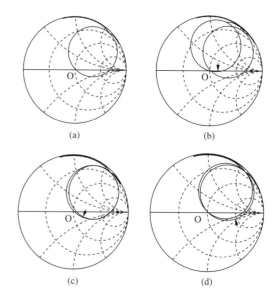

(a)   (b)

(c)   (d)

Figure 6.12 Smith charts corresponding to different conditions during the adjustment procedure. The center of each chart is labeled as O. The measurement frequency range is 9.02–9.08 GHz. (a) Before the insertion of sample ($\beta = 1$), S_{11} curve passes through O. For comparison, this chart is also shown in parts (b), (c), and (d). (b) After the insertion of sample ($\beta < 1$), the corresponding S_{11} curve shrinks and is indicated by an arrow. (c) After retuning ($\beta < 1$), the corresponding S_{11} curve is indicated by an arrow. (d) After adjusting the coupling adjustment screw ($\beta = 1$), the corresponding S_{11} curve is indicated by an arrow. Modified from Chen, L. F. Ong, C. K. and Tan, B. T. G. (1996). "A resonant cavity for high-accuracy measurement of microwave dielectric properties", *Measurement Science and Technology*, **7**, 1255–1259, IOP Publishing Ltd, by permission of IOP Publishing Ltd

Table 6.2 The measured values of f and Q corresponding to the conditions shown in Figure 6.12(a) to (d). The volume of the sample is 16.65 mm³, and the permittivity values of the sample are as follows: $\varepsilon_r' = 2.08$ and $\varepsilon_r'' = 0.0008$

	$i = 1$	$i = 2$	$i = 3$	$i = 4$
f_i(GHz)	9.0475	9.0357	9.0475	9.0475
Q_i	2469	2481	2453	2344

It should be noted that the quality factors in Table 6.2 are loaded quality factors.

Several low-loss samples have been measured, namely, quartz, polyethylene, NaCl, Rexolite, and

Table 6.3 The measurement results of ε_r'' by different techniques

	Conventional method	Retuning method	Calibration method	Reference values
Quartz	−0.007	0.0003	0.0002	0.00018 ± 0.00005^a
Polyethylene[b]	−0.004	0.0006	0.0006	0.00055 ± 0.00010^a
NaCl	−0.003	0.0006	0.0006	0.00059^c
Rexolite	0.0008	0.002	0.002	0.0016 ± 0.0002^a
Beryllia	0.003	0.004	0.004	0.0042 ± 0.0002^a

[a] The reference values are deduced from "*1991 NPL Report DES 115*" and they are for 36 GHz. Our measurements are made at 9 GHz.
[b] The sample is made from a piece of commercially available polyethylene-insulated cable.
[c] This reference value (at 10 GHz) is deduced from (Kaye and Laby 1995).

Beryllia, using the cavity described earlier. The standard sample used in calibration was a cuboid Teflon sample with volume 16.65 mm³ and permittivity $\varepsilon_r = 2.08 - j0.0008$. The samples under test had similar sizes with the standard sample. Comparisons are made between this technique and the conventional technique, and the results for ε_r'' are listed in the second and third columns of Table 6.3, respectively. Because the introduction of extremely low-loss samples does not decrease the Q-value significantly, and at times may even increase the Q-value, the conventional method cannot give the convincing value of ε_r''. Table 6.3 shows that the results ε_r'' from the retuning method are more accurate than those from the conventional cavity-perturbation method. Table 6.3 also shows that, when the dielectric losses (ε_r'') of samples become smaller and smaller, the errors of the conventional cavity-perturbation method become larger and larger.

6.2.3.2 Calibration method

As discussed above, Eq. (6.44) may lose its validity when it is used to calculate the permittivity of extremely low-loss dielectric samples, as the introduction of such extremely low-loss samples may increase the quality factor of the resonant cavity ($Q_2 > Q_1$), yielding a negative value of ε_r''.

Here we introduce a new concept, the expected quality factor Q_0, denoting the quality factor of a cavity loaded with a perfect dielectric ($\varepsilon_r'' = 0$). The introduction of a perfect dielectric sample into a resonant cavity increases the equivalent capacitance C of the cavity, and hardly affects the equivalent resistance R and equivalent inductance L of the cavity. According to Eqs. (6.48) and (6.49),

the expected quality factor Q_0 of the sample-loaded cavity is larger than the quality factor of the empty cavity, and the resonant frequency f of the sample-loaded cavity is lower than that of the empty cavity. For a given cavity, there is a certain relationship between Q_0 and f if the following requirements are satisfied. First, the perturbations caused by the samples are small. Second, all the samples are homogeneous and of the same shape, and their sizes are much smaller than that of the cavity. Third, all the samples are introduced at the same position in the cavity.

Because of the effect of ε_r'' on R, the quality factor of a cavity loaded with an actual dielectric sample ($\varepsilon_r'' > 0$) is lower than the expected quality factor Q_0 of the cavity, but it may be higher than the quality factor before the sample is introduced. Following the concept of the expected quality factor, Eq. (6.44) should be modified as

$$\frac{1}{Q_2} - \frac{1}{Q_0} = B\varepsilon_r'' \frac{V_s}{V_c} \qquad (6.51)$$

where the expected quality factor Q_0 corresponds to the resonant frequency of the cavity after the sample is introduced. So in Eq. (6.51), Q_2 and Q_0 are two parameters that correspond to the same frequency.

Usually, the relationship between Q_0 and f is quite complicated and varies from case to case. It is difficult to get a conclusive formula for the relationship between Q_0 and f. However, the relationship between Q_0 and f, for a given resonant cavity and for samples with a given shape, can be obtained by calibration, as described below.

1. Calibrate the cavity with a medium-loss standard dielectric sample with known permittivity,

obtaining constants A and B using Eqs. (6.43) and (6.44).

2. Load the cavity with a set of extremely low-loss standard dielectric samples and calculate their corresponding Q_0 using Eq. (6.51) with the B value obtained in step (1). As different standard samples correspond to different Q_0 and f, the frequency dependence curve of Q_0 is obtained.

3. To increase the accuracy of calibration, recalculate the value of B using Eq. (6.51) with the value of Q_0 obtained from the frequency dependence curve of Q_0 obtained in step (2), and recalculate the values of Q_0 at different frequencies using the more accurate B value to get a more accurate frequency dependence curve of Q_0.

Once the frequency dependence curve of Q_0 is obtained, more accurate permittivity results can be obtained by using Eqs. (6.43) and (6.51).

Some words of caution are appropriate here. Besides the shape and the location of the samples in the cavity, the real part of the permittivity ε'_r of the sample affects the relationship between Q_0 and f. So the best results would be obtained by doing a calibration using a set of extremely low-loss samples, with ε'_r close to the expected value for the sample to be measured, if a reasonably accurate initial value is available. Also, the shapes of the standard samples should be similar to those of the samples to be measured. For example, if the samples to be measured are spherical, we should calibrate the cavity with a set of spherical samples, whereas if the samples to be measured are cylindrical, the cavity should be calibrated with a set of cylindrical samples. It is important that all the standard samples used in calibration and all the samples to be measured should be placed at the same position in the cavity.

We used a floating zone growth silicon sample ($\varepsilon'_r = 12.3$, $\varepsilon''_r = 9.8$) and a set of Teflon samples ($\varepsilon'_r = 2.08$, $\varepsilon''_r = 0.00065$) to calibrate the cavity. All the samples were cylindrical. The frequency dependence curve of Q_0 is shown in Figure 6.13. We focus our interest in the portion with small perturbation, while we put the whole calibration curve in the upper part of Figure 6.13 for reference. The rightmost point ($f = 9.0508$ GHz) represents the state without perturbation, and so $Q_0 =$

$Q_1 = Q_2$. The bottom line ($Q_1 = 4812$) indicates the quality-factor value used in the conventional cavity-perturbation method. The quality factors shown in Figure 6.13 are unloaded quality factors.

The five samples measured using the frequency-retuning method are measured again using the calibration method, and the measurement results of ε''_r are shown in the fourth column of Table 6.3. It is clear that this calibration method gives results in good agreement with reliable reference values.

The difference between the calibration method and the frequency-retuning method is this: the frequency-retuning method compares the quality factors at the frequency before the introduction of the sample (f_1), while the calibration method compares the quality factors at the frequency after the introduction of the sample (f_2). Both methods take account of the increase of the total stored energy of the sample-loaded cavity due to the introduction of the sample. Experiments show that both methods are able to give accurate permittivity values of extremely low-loss dielectric samples.

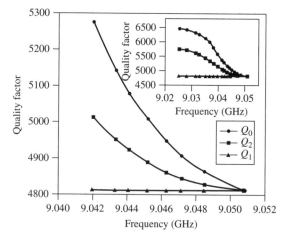

Figure 6.13   The frequency dependence curve of Q_0 for the TE_{112} cylindrical cavity. The relationships of Q_0, Q_1, and Q_2 are also shown in the figure. Modified from Chen, L. F. Ong, C. K. and Tan, B. T. G. (1999). "Amendment of cavity perturbation method for permittivity measurement of extremely low-loss dielectrics", *IEEE Transactions on Instrumentation and Measurement*, **48** (6), 1031–1037. © 2003 IEEE

6.2.4 Extracavity-perturbation method

In a cavity-perturbation method, if the sample introduced into the cavity has high dielectric constant and/or high loss, the perturbation requirements cannot be fulfilled, so the conventional cavity-perturbation method cannot be used. One way to solve this problem is to place the sample out of the cavity, but coupled to the cavity. In this case, the perturbation of the sample to the cavity can be small, and the resonant perturbation requirements can be fulfilled. This method is usually called *extracavity-perturbation method*.

The extracavity-perturbation method for the measurement of dielectric materials was proposed by Barlow (1962), and improved by Kumar (1976), Kumar and Smith (1976), Ni and Ni (1997), and Wang and Ni (1997). As shown in Figure 6.14, a TE_{01n} cylindrical resonator is coupled to an evanescent waveguide through the coupling irises at one end plane of the cavity.

When there is no sample in the waveguide, the perturbation of the waveguide to the cavity is purely reactive if we do not consider the energy dissipation at the wall of the waveguide. If there is a sample within the waveguide, part of the energy is dissipated by the sample, and the perturbation of the waveguide to the cavity has a resistive component, so the perturbation becomes complex.

Here, we follow the analyzing method proposed by Barlow (1962). The resistive and reactive components (R_e and X_e) of the complex perturbation ($Z_e = R_e + jX_e$) can be calculated from the resonant frequency and the quality factor of the cavity coupled to the evanescent waveguide loaded with the sample under test. We then derive the complex reflection coefficient ($\rho = \rho' + j\rho''$) at the interface between air and the sample from the resistive and reactive components of the perturbation. From the complex reflection coefficient,

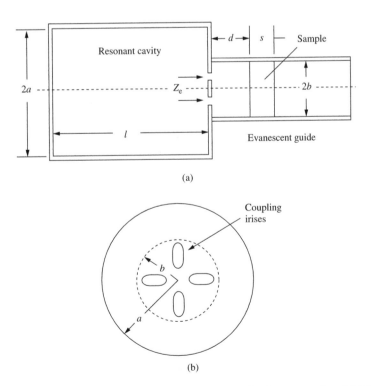

(a)

(b)

Figure 6.14 A circular-cylinder cavity coupled to an evanescent circular waveguide. (a) Dimensions of cavity and guide and (b) coupling irises (Kumar and Smith 1976). Source: Kumar, A. and Smith, D. G. (1976). "The measurement of the permittivity of sheet materials at microwave frequencies using an evanescent waveguide technique", *IEEE Transactions on Instrumentation and Measurement*, **25** (3), 190–193. © 2003 IEEE

we can obtain the complex attenuation factor $(\alpha = \alpha' + j\alpha'')$ of the segment of the waveguide filled with the sample, from which the complex dielectric permittivity $(\varepsilon = \varepsilon_r' - j\varepsilon_r'')$ of the sample can be determined.

The resistive R_e and reactive X_e components of the evanescent waveguide loaded with the sample under test are given by (Barlow 1962; Kumar 1976)

$$R_e = \frac{\lambda_g}{\lambda_0} \cdot \frac{\pi^2 a^2 f \mu_0 l}{Q_L A} - \frac{R_m \pi a [2a + (\lambda_g/\lambda_c)^2 l]}{A} \tag{6.52}$$

$$X_e = \frac{\omega \mu_0 c^2 A + 8\pi^2 \alpha_0 a^2 l^3 \mu_0 f^2 \, df}{c^2 \alpha_0 A} \tag{6.53}$$

with

$$\alpha_0 = (h^2 - \omega^2 \mu_0 \varepsilon_0)^{1/2} \tag{6.54}$$

$$h = v/b \tag{6.55}$$

where λ_g is guide wavelength, λ_c is cutoff wavelength, λ_0 is wavelength in free space, f is operation frequency, a is the radius of the cylindrical cavity, b is the radius of the evanescent waveguide, l is the length of the cylindrical cavity, A is the area of the coupling irises, R_m is the surface resistance of the endplate, ω is the angular frequency, Q_L is the quality factor of the cavity loaded with the sample, df is the shift of resonant frequency due to the presence of the sample, c is the speed of light, v is determined by the mode number in the circular waveguide, and $v = 3.831732$ for the TE$_{011}$ mode.

From the transmission line theory, we can get the relationship between the complex perturbation (Z_e) of the evanescent waveguide to the cavity and the complex reflection coefficient (ρ) at the first air–dielectric interface (Kumar and Smith 1976):

$$Z_e = R_e + jX_e = j\frac{\omega \mu_0}{\alpha_0} \cdot \frac{1 + \rho \exp(-2\alpha_0 d)}{1 - \rho \exp(-2\alpha_0 d)} \tag{6.56}$$

According to Eq. (6.56), the complex reflection coefficient $(\rho = \rho' + j\rho'')$ can be calculated from the complex perturbation $(Z_e = R_e + jX_e)$

$$\rho' = \frac{\alpha_0^2(X_e^2 + R_e^2) - \omega^2 \mu_0}{\exp(-2\alpha_0 d)[R_e^2 \alpha_0^2 + (\omega \mu_0 + \alpha_0 X_e)^2]} \tag{6.57}$$

$$\rho'' = \frac{2\alpha_0 R_e \omega \mu_0}{\exp(-2\alpha_0 d)[R_e^2 \alpha_0^2 + (\omega \mu_0 + \alpha_0 X_e)^2]} \tag{6.58}$$

The relationship between the reflection coefficient and the complex attenuation factor $(\alpha = \alpha' + j\alpha'')$ can also be obtained based on transmission line theory

$$j\frac{\omega \mu_0}{\alpha_0} \cdot \frac{(1 + \rho)}{(1 - \rho)} = j\frac{\omega \mu}{\alpha} \cdot \frac{1 - \rho \exp(-2\alpha s)}{1 + \rho \exp(2\alpha s)} \tag{6.59}$$

where s is the thickness of the dielectric sheet.

If the sample is very thick or if it has a very high dielectric constant or loss factor $(|\alpha s| \gg 1)$, according to Eq. (6.59), the complex reflection coefficient is given by

$$\rho' = \frac{\mu_r^2 \alpha_0^2 - \alpha'^2 - \alpha''^2}{(\mu_r \alpha_0 + \alpha')^2 + \alpha''^2} \tag{6.60}$$

$$\rho'' = \frac{2\alpha'' \mu_r \alpha_0}{(\mu_r \alpha_0 + \alpha')^2 + \alpha''^2} \tag{6.61}$$

From Eqs. (6.60) and (6.61), the complex attenuation factor α can be obtained from complex reflection coefficient.

If the value of $|\alpha s|$ is not much larger than unity, from Eq. (6.59), the following relationships can be obtained:

$$M\alpha' + N\alpha'' - X = 0 \tag{6.62}$$

$$M\alpha'' - N\alpha' - Y = 0 \tag{6.63}$$

with

$$M = \frac{1 - \rho' - \rho''^2}{\alpha_0[(1 - \rho')^2 + \rho''^2]} \tag{6.64}$$

$$N = \frac{2\rho''}{\alpha_0[(1 - \rho')^2 + \rho''^2]} \tag{6.65}$$

$$X = \frac{1 - q_1^2 - q_2^2}{(1 + q_1)^2 + q_2^2} \tag{6.66}$$

$$Y = \frac{2q_2}{(1 + q_1)^2 + q_2^2} \tag{6.67}$$

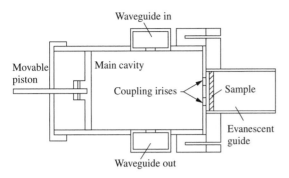

Figure 6.15 Cylindrical cavity coupled to evanescent guide (Kumar and Smith 1976). Source: Kumar, A. and Smith, D. G. (1976). "The measurement of the permittivity of sheet materials at microwave frequencies using an evanescent waveguide technique", *IEEE Transactions on Instrumentation and Measurement*, **25** (3), 190–193. © 2003 IEEE

where

$$q_1 = \exp(-2\alpha's)(\rho'\cos 2\alpha''s - \rho''\sin 2\alpha''s)$$
(6.68)

$$q_2 = \exp(-2\alpha's)(\rho'\cos 2\alpha''s + \rho''\sin 2\alpha''s)$$
(6.69)

From the values of ρ' and ρ'' obtained from Eqs. (6.57) and (6.58), the values of M, N, X, and Y can be determined, and so the values of α' and α'' can be obtained by solving Eqs. (6.62) and (6.63).

In the dielectric-filled portion of the waveguide, the relationship between attenuation factor ($\alpha = \alpha' + j\alpha''$) and permittivity ($\varepsilon = \varepsilon_r' - j\varepsilon_r''$) is given by (Kumar and Smith 1976)

$$(\alpha' + j\alpha'')^2 = h^2 - \omega^2\mu_0(\varepsilon_r' - j\varepsilon_r'')$$
(6.70)

So the dielectric permittivity can be calculated from the complex attenuation factor

$$\varepsilon_r' = \frac{h^2 + \alpha''^2 - \alpha'^2}{\omega^2\mu_0}$$
(6.71)

$$\varepsilon_r'' = \frac{2\alpha'\alpha''}{\omega^2\mu_0}$$
(6.72)

It should be noted that in the determination of X_s, besides the frequency-variation method discussed above, the length-variation method can also

be used (Ni and Ni 1997). In the length-variation method, after introducing the sample, we adjust the length of the cavity to ensure that the resonant frequency of the cavity remains the same, and the value of X_s can be obtained form the change of the cavity length. Figure 6.15 shows a fixture for extracavity-perturbation method. A movable piston is used so that the length of the cavity can be adjusted in the measurement procedure.

6.3 DIELECTRIC RESONATOR PERTURBATION METHOD

The mechanical stability is a critical parameter determining the repeatability and reliability of the cavity-perturbation method for materials measurement. Though we may employ the TM_{010} mode with nonradiative apertures so that samples may be introduced without dismantling the cavity, such approaches above are intended for a rod-shaped sample or require an elaborate procedure for sample placement. A semiopen resonant structure with good mode stability appears desirable for supporting perturbation techniques, since it would alleviate the above mentioned constraints and allow for a rapid placement of samples.

Shu and Wong proposed a method for permittivity measurement based on the perturbation of a resonant structure consisting of a dielectric resonator sandwiched between two metal plates (Shu and Wong 1995). In the following discussion, we assume the dielectric resonator is at TE_{011} mode, and the procedure can be extended to other modes.

As shown in Figure 6.16, the resonant structure consists of a dielectric resonator placed between two polished metal plates and excited by a coaxial loop. When a small sample in the form of a thin plate or disk is placed with its broad surface perpendicular to the electric field vector, the shift in resonance frequency due to the presence of the sample is given by (Shu and Wong 1995)

$$\frac{\delta\omega}{\omega_0} \approx -\frac{\varepsilon_s - 1}{\varepsilon_s} \cdot \frac{\displaystyle\int_{V_s} \boldsymbol{E}_0 \cdot \boldsymbol{D}_0 \, dV}{2\displaystyle\int_V \boldsymbol{E}_0 \cdot \boldsymbol{D}_0 \, dV}$$
(6.73)

where ε_s is the relative permittivity of the sample, $\boldsymbol{E}_0$ and $\boldsymbol{D}_0$ are the field quantities in the unperturbed system and V_s is the volume of the sample.

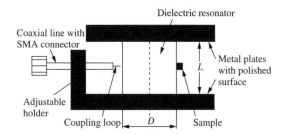

Figure 6.16 Resonant structure with excitation coaxial transmission line and test sample (Shu and Wong 1995). Reproduced from Shu, Y. and Wong, T. Y. (1995). "Perturbation of dielectric resonator for material measurement", *Electronics Letters*, **31** (9), 704–705, by permission of IEE

The integral in the denominator is made up of two parts, one over the volume of the dielectric resonator and the other over the air region. Substituting the appropriate expressions for the field quantities corresponding to the TE_{011} mode, the shift in resonant frequency is in the form of

$$
\frac{\delta\omega}{\omega_0} \approx -\frac{\varepsilon_s - 1}{\varepsilon_s}
$$

$$
\cdot \frac{2k_2^2 K_1^2(\beta)V_s}{\{\varepsilon_d K_1^2(\beta)[J_1^2(\alpha) - J_0(\alpha)J_2(\alpha)] + J_1^2(\alpha)[K_0(\beta)K_2(\beta) - K_1^2(\beta)]\}V} \quad (6.74)
$$

with

$$
k_2^2 = -\omega^2 \mu_0 \varepsilon_0 + (\pi/L)^2 \quad (6.75)
$$

$$
\alpha = \frac{\pi D}{\lambda}\sqrt{\varepsilon_d - \left(\frac{\lambda_0}{2L}\right)^2} \quad (6.76)
$$

$$
\beta = \frac{\pi D}{\lambda}\sqrt{\left(\frac{\lambda_0}{2L}\right)^2 - 1} \quad (6.77)
$$

where J_i and K_i are Bessel functions, ε_d is the relative permittivity of the dielectric resonator and λ_0 is the free-space wavelength. The dimensions of the dielectric resonator, D and L, are defined in Figure 6.16, with the volume denoted as V. From Eq. (6.74), ε_s can be calculated once the resonance frequencies have been measured.

When the change in the quality factor of the resonator is incorporated to form a complex $\delta\omega$, Eq. (6.74) enables the determination of the complex permittivity of a lossy sample. It is also a

straightforward matter to extend this method to permeability measurements (Shu and Wong 1995).

6.4 MEASUREMENT OF SURFACE IMPEDANCE

Besides the complex permittivity and complex permeability of low-conductivity materials, the complex impedance of high-conductivity materials can also be characterized using resonant perturbation methods.

6.4.1 Surface resistance and surface reactance

Surface impedance Z_s describes the interaction of a conductor with the external electromagnetic field when the penetration depth of the field is much less than the sample size. For a uniform and isotropic conductor, its surface impedance is in the form of

$$
Z_s = R_s + jX_s \quad (6.78)
$$

where R_s is the surface resistance and X_s is the surface reactance. In general, Z_s is determined by the magnetic permeability μ_r and the dielectric permittivity ε_r of a conductor:

$$
Z_s = \sqrt{\frac{\mu_0 \mu_r}{\varepsilon_0 \varepsilon_r}} \quad (6.79)
$$

where μ_0 is the magnetic permeability and ε_0 is the dielectric permittivity of vacuum. For a normal nonmagnetic conductor, the real and imaginary parts of surface impedance are equal:

$$
R_s = X_s = \sqrt{\omega\mu_0/(2\sigma)} \quad (6.80)
$$

where σ is the conductivity, which is a real number for a normal conductor. However, for a superconducting material, the conductivity is a complex number, and the surface resistance and surface reactance are not necessarily equal. The measurement of surface impedance at microwave frequency is important for the study of microwave superconductivity and the microwave applications of superconductors.

Because of their high accuracy and high sensitivity, resonant perturbation methods are widely used in the surface resistance measurement of high-temperature superconducting (HTS) thin films, though only the surface resistance at one or at several discrete frequencies can be measured. The additional advantage that makes resonant perturbation methods popular lies in the fact that these methods are nondestructive and so the films after characterization can still be used to fabricate actual devices.

Resonant perturbation methods for surface impedance measurement are based on wall-impedance perturbation. In a typical resonant method, part of the resonator wall is replaced by the sample under study. As $\omega = 2\pi f$, Eq. (6.32) can be rewritten as

$$\Delta f = \frac{j}{2\pi} \Gamma \Delta Z_s \qquad (6.81)$$

where f is the complex resonant frequency. The real part of Δf corresponds to a shift in the actual resonant frequency f_0 and is associated with the surface reactance of the sample, while the imaginary part of Δf corresponds to the change of the reciprocal of the quality factor $(1/Q)$, and is related to the surface resistance of the sample.

6.4.2 Measurement of surface resistance

For the measurement of surface resistance, from Eq. (6.81), we can get

$$R_{s2} = R_{s1} + A\left(\frac{1}{Q_2} - \frac{1}{Q_1}\right) \qquad (6.82)$$

where R_{s1} is the surface resistance of the original cavity wall, R_{s2} is the surface resistance of the sample under study; Q_1 and Q_2 are the quality factors of the resonator before and after part of the cavity wall is replaced by the sample under study respectively; and the constant A is independent of the surface resistance of the sample, and it is only related to the properties of the resonator. It is clear that a resonator with a larger A value has higher sensitivity. In experiments, the constant A is usually determined by calibrating the cavity with a sample with known surface resistance.

It should be noted that Eqs. (6.81) and (6.82) stand on the assumption that the total stored energy and the field configuration in the resonator do not change due to the wall-loss perturbation. So in Eq. (6.82), the difference between R_{s1} and R_{s2} should be small. Equation (6.82) works quite

well for the measurement of normal conducting materials because it is easy to find a calibrator whose surface resistance is close to the surface resistance of the sample under study.

In the following, we discuss two types of resonators used for surface resistance measurement: hollow metallic resonant cavity and dielectric resonator enclosed in a metal shield. After that, we discuss a mirror-image calibrator that is useful for the measurement of samples with extremely low surface resistance, such as superconducting thin films.

6.4.2.1 Cavity-perturbation method

The cavity-perturbation method for measuring the surface resistance of metals involves replacing one wall of a cylindrical cavity resonator with the sample being studied. In this method, a hollow metallic cavity is often used. As shown in Figure 6.17, the end wall of a hollow metal cavity is replaced by the sample under test. The surface resistance of the sample can be obtained from the quality factors of the cavity before and after one of the endplates is replaced by the sample, respectively (Dew-Hughes 1997; Derov *et al.* 1992). This method has acceptable accuracy and sensitivity in the measurement of normal conductors. To increase the measurement accuracy and

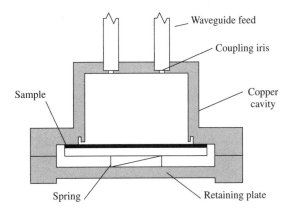

Figure 6.17 Schematic of end-wall-replacement cavity. Source: Dew-Hughes, D. (1997). "Microwave properties and applications of high temperature superconductors", in H. Groll and I Nedkov (eds). *Microwave Physics and Techniques*, Kluwer Academic Publishers, Dordrecht, 83–114

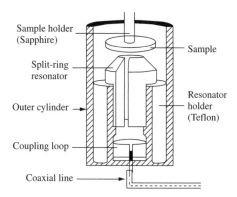

Figure 6.18 Schematic view of a split-ring resonator for measurement of surface resistance. The outer cylinder and Teflon holder have been cut away to reveal the resonator. Modified from Bonn, D. A. Morgan, D. C. and Hargy, W. H. (1991). "Split-ring for measuring microwave surface resistance of oxide superconductors", *Review of Scientific Instruments*, **62** (7), 1819–1823

sensitivity, the cavity structure can be optimized and the cavity can be made from materials with lower surface resistance, such as superconductors.

In the configuration shown in Figure 6.17, the sample under study is thermally anchored to the cavity. If we want to study the temperature dependence of the surface resistance of the sample, the temperature dependence of the loss in the cavity walls must be carefully measured in order to minimize systematic errors. The thermal connection between the sample and the resonator also makes it impossible to use high-Q superconducting resonators, putting a serious limit on the sensitivity of this technique. In the configuration shown in Figure 6.18, the split-ring resonator has microwave

magnetic field lines coming out of the central hole in the resonator, and the sample under test is positioned with its face perpendicular to these field lines (Bonn *et al.* 1991). The oscillating circular currents are induced in the surface and any resistance to these currents manifests itself as a decrease in the quality factor of the resonance. In this configuration, the sample is thermally isolated from the cavity, so we can adjust the temperature of the sample while keeping the temperature of the cavity unchanged.

6.4.2.2 Dielectric resonator method

Mainly owing to the two distinct advantages, the dielectric resonator method is much more widely used than the hollow metallic cavity method in the measurement of surface resistance. First, the dielectric resonator method has higher accuracy and sensitivity, and second, the small samples can be measured at lower frequencies, or different positions in a large sample can be measured.

As shown in Figure 6.19, usually a cylindrical dielectric pill located in a cylindrical metal shield is used, and the samples under study may be placed in contact with one end plane of the pill or with both end planes. For the configuration with samples placed in contact with both end planes of the dielectric pill, as shown in Figure 6.19(b), there is an analytic field solution, so the surface resistance can be accurately calculated from the resonant properties of the dielectric resonator (Kobayashi and Katoh 1985; Shen *et al.* 1992; Holstein *et al.* 1993). This scheme has been extensively investigated (Mazierska and Grabovickic 1998; Mazierska 1997), and it

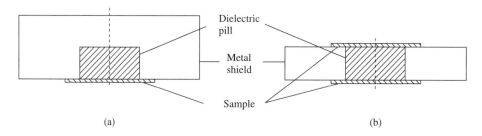

(a)   (b)

Figure 6.19 Dielectric resonators for surface resistance measurement of HTS thin films. (a) One-sample perturbation and (b) two-sample perturbation. Source: Ong, C. K. Chen, L. F. Lu, J. Xu, S. Y. Rao, X. S. and Tan, B. T. G. (1999). "Mirror-image calibrator for resonant perturbation method in surface resistance measurement of high-T$_c$ superconducting thin films", *Review of Scientific Instruments*, **70** (7), 3092–3096

is a generally accepted method for the measurement of the surface resistance of HTS thin films. However, it should be indicated that the value of surface resistance obtained directly from this scheme is the average value of the two samples.

For the configuration with a single piece of sample placed in contact with one end plane of the dielectric pill, as shown in Figure 6.19(a), the surface resistance of the sample can be measured directly, and with a higher sensitivity, though there is no analytical solution for this scheme (Edgcombe and Waldram 1994). Usually, the calibration method, based on resonant perturbation theory, is used to deduce the surface resistance of the sample. In this configuration, the sample is thermally anchored to the resonator, and this may cause difficulty in the study of the temperature dependence of the surface resistance of the sample. To avoid this problem, the configuration shown in Figure 6.20 can be used (Ormeno *et al.* 1997). In this configuration, the sample is thermally isolated from the dielectric resonator, and we can adjust the temperature of the sample while keeping the temperature of the cavity unchanged. However, the measurement sensitivity of this configuration may be lower than that of the configurations shown in Figure 6.19.

6.4.2.3 Mirror-image calibrator

The resonant perturbation methods discussed above are accurate and sensitive enough for the measurement of the surface resistance of normal conductors; however, they may experience difficulty

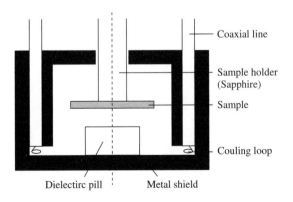

Figure 6.20 A configuration for the measurement of surface resistance

in the measurement of surface resistance of HTS thin films. As the surface resistance of the best metals, such as silver and gold, is much larger than that of HTS thin films, the surface resistance obtained from the traditional perturbation methods is not reliable. So it is not appropriate to calculate the surface resistance of HTS thin films using Eq. (6.82). Besides, if R_{s1} is much larger than R_{s2}, a small uncertainty in the resonator constant A may result in a large error in R_{s2}. The ideal condition for the measurement of HTS thin films is that the value of R_{s1} in Eq. (6.82) is zero, which requires a zero surface resistance plate. With this condition, Eq. (6.82) becomes

$$R_s = A \left(\frac{1}{Q} - \frac{1}{Q_0} \right) \tag{6.83}$$

where R_s is the surface resistance of HTS thin film under study, and Q_0 and Q are the quality factors of the dielectric resonator when the zero surface resistance plate and the HTS thin film are placed in contact with the dielectric resonator respectively.

However, no material has zero surface resistance at microwave frequencies, and in the measurement of sample with extremely low surface resistance, a mirror-image calibrator can be used (Lu *et al.* 1994, Luo *et al.* 1997; Krupka and Mazierska 1998; Ong *et al.* 1999). When the end plane of the dielectric resonator (R_s probe) is in contact with the mirror image of the probe, the quality factor of the resonator, consisting of the probe and its mirror image, is equal to the quality factor when the end plane of the probe is in contact with an ideal zero surface resistance plate. The principle of the mirror-image calibrator is explained below.

The dielectric resonators used for surface resistance measurement of HTS thin films are usually cylindrical dielectric resonators working in the TE_{011} mode. As shown in Figure 6.21(a), the TE_{011} mode is axisymmetric. It has closed loops of transverse electric field whose centers lie on the axis, and the closed loops of magnetic field lie in the planes containing the axis. There is no axial current across any possible joints between the dielectric resonator and the HTS thin film, so this mode is not sensitive to the small gaps between the dielectric resonators and HTS thin films.

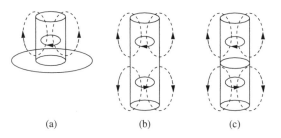

(a)   (b)   (c)

Figure 6.21 Field distributions of three dielectric resonators. (a) TE_{011} mode dielectric resonator in contact with a conducting plate, (b) TE_{012} mode dielectric resonator, and (c) TE_{012} mode dielectric resonator consisting of two TE_{011} mode dielectric resonators. Source: Ong, C. K. Chen, L. F. Lu, J. Xu, S. Y. Rao, X. S. and Tan, B. T. G. (1999). "Mirror-image calibrator for resonant perturbation method in surface resistance measurement of high-T_c superconducting thin films", *Review of Scientific Instruments*, **70** (7), 3092–3096

Figure 6.21 shows three cylindrical dielectric resonators, which are made of the same material and have the same diameter d. In Figure 6.21(a), the dielectric resonator with length l is in contact with a conducting plate, and it resonates in the TE_{011} mode. The resonator shown in Figure 6.21(b) has length $2l$, and resonates in the TE_{012} mode. The resonator shown in Figure 6.21(c) also resonates in the TE_{012} mode, and it consists of two pieces of identical dielectric cylindrical pills, each of which has length l. If the surface impedance of the conducting plate shown in Figure 6.21(a) is zero, there is no electric field in the plate, and the magnetic field near the plate is parallel to the surface of the plate. From the field distributions of TE_{011} mode and TE_{012} mode, it is clear that the field distribution in the dielectric resonator shown in Figure 6.21(a) is the same as that in the upper part of the dielectric resonator shown in Figure 6.21(b), and these two resonators have the same resonant frequency. As the plate shown in Figure 6.21(a) does not dissipate microwave energy, the dielectric resonator shown in Figure 6.21(b) has twice the stored energy and also twice the energy dissipation as the dielectric resonator shown in Figure 6.19(a). Therefore, they have the same quality factor (Kobayashi and Tanaka 1980; Kajfez and Guillon 1986). Because there is no axial current across the plane perpendicular to the axis at the middle of the cylinder, a small gap perpendicular to the axis at the middle of the cylinder hardly affects the resonant properties of the resonator. As such, the resonator shown in Figure 6.21(a) and the one shown in Figure 6.21(c) also have the same resonant frequency and quality factor. In experiments, the size of the gap between the two pieces of dielectric pills is much smaller than microwave wavelength, so the uncertainty caused by the gap is negligible.

The above discussions show that if the zero surface resistance plate shown in Figure 6.21(a) is replaced by the mirror image of the dielectric resonator, though the electromagnetic boundary condition of the dielectric resonator is changed from a short-circuit condition to an open-circuit condition, its quality factor does not change. Therefore, the quality factor of the dielectric resonator when it is connected to its mirror-image structure can be used as Q_0 in Eq. (6.83). In the following discussions, the mirror-image structure is also called the mirror-image calibrator.

Figure 6.22 shows a probe for R_s measurement in a calibration configuration: the probe (the upper part) is in contact with its mirror-image calibrator

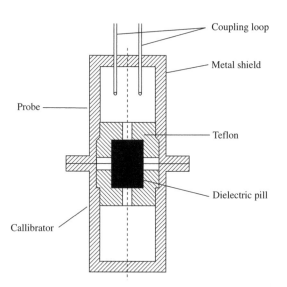

Figure 6.22 Structures of the R_s probe and its mirror-image calibrator. Source: Ong, C. K. Chen, L. F. Lu, J. Xu, S. Y. Rao, X. S. and Tan, B. T. G. (1999). "Mirror-image calibrator for resonant perturbation method in surface resistance measurement of high-T_c superconducting thin films", *Review of Scientific Instruments*, **70** (7), 3092–3096

(the lower part). The probe and its mirror-image calibrator have the same structures except that the probe has two coupling loops, while the mirror-image calibrator does not have coupling loops. The key components of the probe and the mirror-image calibrator are the dielectric pills, the Teflon holders, and the metal shields.

The probe and its mirror-image calibrator have the same dielectric pills. For the R_s probe used in the following discussions, each dielectric pill is a piece of LaAlO$_3$ single crystal with dimensions: $\Phi 6.0 \times 5.5$ mm. The two dielectric pills are made from the same piece of LaAlO$_3$ single crystal, and the symmetrical axis of each dielectric pill is the c-axis of the crystal. The two end planes of each dielectric pill are finely polished. When the two dielectric pills are connected by end plane to end plane, as shown in Figure 6.21(c) and Figure 6.22, optical interference fringes can be visually observed, indicating that the size of the gap between the two dielectric pills is of the order of optical wavelengths. Besides, experiments made at room temperature show that the two TE$_{011}$ mode dielectric resonators (formed by sandwiching each of the two dielectric pills between two silver plates) and the TE$_{012}$ mode resonator (formed by sandwiching two dielectric pills, connected by end plane to end plane, between two silver plates) have the same resonant frequency. The two TE$_{011}$ resonators have the same quality factor, while the TE$_{012}$ resonator has a higher quality factor.

The structures of the Teflon holders in the probe and the mirror-image calibrator are shown in Figure 6.23. The relative positions of the LaAlO$_3$ single crystal and the metal shield should not change when the measurement temperature changes from room temperature to liquid nitrogen temperature. Because the thermal-expansion coefficient of Teflon is much larger than those of brass and LaAlO$_3$, special attention should be paid to the design of the Teflon holders: (1) plane A and plane B are in the same plane perpendicular to the $c-c'$ axis, so the perpendicular distance between these two planes is always zero at any temperature; (2) the tolerances of the Teflon holder are carefully selected to ensure that the Teflon holder can tightly hold the LaAlO$_3$ pill at any temperature; (3) the hole in the top of the holder is used to observe whether the

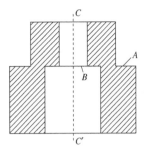

Figure 6.23 Structure of the Teflon holder. Source: Ong, C. K. Chen, L. F. Lu, J. Xu, S. Y. Rao, X. S. and Tan, B. T. G. (1999). "Mirror-image calibrator for resonant perturbation method in surface resistance measurement of high-T$_c$ superconducting thin films", *Review of Scientific Instruments*, **70** (7), 3092–3096

dielectric pill is correctly assembled; and (4) there is a groove (not shown in the figure) to release the difference in gas pressure between the upper and lower part of the probe. However, it should be noted that Teflon is a lossy material compared to single-crystal LaAlO$_3$. The Teflon holders add losses to the probe and its mirror-image calibrator, and thus lower the value of Q_0. If lower-loss materials with good mechanical and thermal properties are available, the measurement accuracy and sensitivity of the probe could be further increased.

The metal shields do not resonate at the working frequency of the dielectric resonators, preventing the microwave radiation from the dielectric resonators. In order to decrease their wall-loss contributions to the quality factors of the dielectric resonators, the brass shields are usually coated with silver or gold. In addition, the shields are sealed with indium wire and hence the samples are separated from the liquid nitrogen, and can be kept in special gas environments, such as Helium.

Equations (6.82) and (6.83) are often used for the measurement of the surface resistance of conducting plates. For the measurement of HTS thin films, Eq. (6.83) is more accurate and sensitive provided that the mirror-image calibrator corresponding to the probe is available. The calibration method is often used to determine the constant A. In order to make a comparison between the conventional perturbation technique and the technique using mirror-image calibrator, calibrations are made on the basis of Eqs. (6.82) and (6.83) respectively.

Table 6.4 Calibrations based on Eqs. (6.82) and (6.83). Two calibrations are made on the basis of Eq. (6.83), as indicated by calibration (I) and calibration (II). All the calibrations were performed at 10.65 GHz and 77 K

	Calibration based on Eq. (6.82)	Calibration (I) based on Eq. (6.83)	Calibration (II) based on Eq. (6.83)
Mirror-image calibrator	–	$Q_0 = 31,095$	$Q_0 = 31,095$
Silver	$R_{s1} = 11.5\,\text{m}\Omega$	$R_s = 11.5\,\text{m}\Omega$	–
Plate	$Q_1 = 15,126$	$Q = 15,126$	
Gold	$R_{s2} = 20.8\,\text{m}\Omega$	–	$R_s = 20.8\,\text{m}\Omega$
Plate	$Q_2 = 10,879$		$Q = 10,879$
A	3.603×10^5	3.387×10^5	3.481×10^5

In the calibration procedure, three calibrators are used: the mirror-image calibrator discussed above and two conventional calibrators. One of the two conventional calibrators is a silver plate whose surface resistance is 11.5 mΩ, and the other is a gold plate whose surface resistance is 20.8 mΩ. The silver plate and the gold plate are used in the calibration based on Eq. (6.82). Two calibrations are made on the basis of Eq. (6.83) by using the silver plate and the gold plate separately and independently to check whether the selection of the conventional calibrator is crucial if a mirror-image calibrator is used. The calibration results are shown in Table 6.4.

The uncertainties of surface resistance measurement using mirror-image calibrator mainly consist of three parts: the calibration uncertainty, the instrument measurement uncertainty, and the uncertainty due to the measurement method of unloaded quality factor. Usually, the instrument measurement uncertainty is much smaller than the calibration uncertainty. The calibration uncertainty is mainly related to two factors: the quality of the mirror-image calibrator and the repeatability of the mechanical assembling. The quality of the mirror-image calibrator is determined by the extent to which the probe and the mirror-image calibrator are identical. To minimize the uncertainty, the two dielectric pills, the two metal shields, and the two Teflon holders in the probe and the mirror-image calibrator are made to be as same as possible. The only difference between the probe and the mirror-image calibrator is that there are two coupling loops in the probe, while there is no coupling

loop in the mirror-image calibrator. However, the uncertainty caused by such a difference is negligible because the couplings are very week. So the calibration uncertainty is mainly determined by the precision of the mechanical assembly. In the measurement of the unloaded quality factor, usually the two coupling coefficients (β_1 and β_2) are assumed to be equal.

To make a comparison between the conventional perturbation method and the modified perturbation method, several YBa$_2$Cu$_3$O$_{7-\delta}$ (YBCO) thin films are characterized using these two methods. The YBCO thin films are fabricated on LaAlO$_3$ substrates by using pulsed laser ablation techniques (Low *et al.* 1997), and their thickness is around 5000 Å. The measurement results are shown in Table 6.5.

In Table 6.5, the surface resistance of each piece of YBCO thin film is calculated in three different ways according to the three calibrations listed in Table 6.4. To make a precise comparison of the calculation results according to different calibrations, we list five decimal digits for each R_s value. Table 6.5 shows that the results following two calibrations of Eq. (6.83) are obviously more accurate and credible than the ones following Eq. (6.82). The differences in the calculation results caused by the different calibrators in the calibrations (I) and (II) of Eq. (6.83) are acceptable. So, when a mirror-image calibrator is used, the selection of the conventional calibrator is not very crucial, and even if there is some error in the surface resistance value of the conventional calibrator, the effect of such an error is not very severe. For the conventional technique,

Table 6.5 Different results of surface resistance measurements following different calibrations. The samples are YBCO thin films fabricated on LaAlO$_3$ substrates. All the measurements are made at 10.65 GHz and 77 K. The unit of R_s values in the table is mΩ

Item numbers of HTS thin films	Q_2 in Eq. (6.82) or Q in Eq. (6.83)	R_s values following Eq. (6.82)	R_s values following Eq. (6.83) with Calibration (I)	R_s values following Eq. (6.83) with Calibration (II)
1	21,379	4.5331	4.9502	5.0876
2	24,835	2.1878	2.7456	2.8218
3	27,104	0.9733	1.6039	1.6484
4	28,752	0.21139	0.88762	0.91226
5	29,563	−0.13238	0.56446	0.58013
6	29,861	−0.25401	0.45013	0.46262

Eq. (6.82) could give reasonable results for the HTS thin films with large surface resistance (items 1–3). However, for HTS thin films with low surface resistance, the results following Eq. (6.82) are unreliable (items 5 and 6). Furthermore, using Eq. (6.82), a small uncertainty in the surface resistance values of the two conventional calibrators may cause larger uncertainties in the measurement results.

6.4.3 Measurement of surface reactance

As discussed earlier, for materials with normal skin effect, the surface resistance and surface reactance have the same value. However, for superconductors and semiconductors, the values for surface resistance and surface reactance are not necessarily equal. For superconductors, thin-film samples are often measured by the wall-replacement method; while for semiconductors, bulk samples are often measured by material-perturbation method.

6.4.3.1 Wall-replacement method

The measurement of the surface reactance is important for the study of microwave superconductivity. The surface reactance of a superconductor is related to the penetration depth of the superconductor λ_L:

$$X_s = \omega\mu\lambda_L. \tag{6.84}$$

In a resonant-perturbation method, the replacement of part of the cavity wall by the sample under study also results in a resonance-frequency shift that is proportional to the difference in the values of surface reactance of the sample (X_s) and the original cavity wall (X_c):

$$f_s - f_c = A \cdot (X_s - X_c) \tag{6.85}$$

where f_c and f_s are the resonant frequency of the cavity before and after part of the cavity wall is replaced by the sample respectively. The parameter A is a constant independent of the reactance of the sample, and it is determined by the properties of the cavity.

However, in actual experiments, the measurements of absolute value of X_s based on Eq. (6.85) are not feasible since the replacement of part of the cavity wall is inevitably accompanied by the uncontrolled alteration of the geometrical sizes of the cavity. The resultant spurious shift of resonance frequency may far exceed that caused by the change of X_s of the part of a cavity wall.

When it is necessary to measure X_s, such as in the investigation of metals in the superconducting state or under conditions of anomalous skin effect, the measurement results should be calibrated. We may use the relationship $R_s = X_s$ in the high-temperature range where the conditions of normal skin effect are still valid to calibrate the absolute value of X_s; subsequently, the variation of X_s with decreasing temperature can be measured.

6.4.3.2 Measurement of bulk samples

Kourov and Shcherbakov proposed a method for the measurement of surface reactance of semiconductor bulk samples with a size much less than that of the wavelength (Kourov and

Shcherbakov 1996). Consider an ideal conductor placed inside a cavity. In experiments, a metal with a sufficiently small value of dc resistivity, such as copper, silver, or gold, can be taken as an ideal conductor. If the ideal conductor is substituted by a sample under study that is of the same shape and size, and the sample is placed at the same position inside the cavity as the ideal conductor, then such a procedure is equivalent to the replacement of the part of the ideally conducting wall of the cavity by the sample being investigated. This substitution causes a shift of resonant frequency that is proportional to the surface reactance of the sample under study,

$$f - f_{id} = -X_s S/G \qquad (6.86)$$

where f and f_{id} are the resonant frequencies of the cavity with the sample and with the ideal conductor respectively, S is the surface area of the sample, and G is geometric factor given by (Kourov and Shcherbakov 1996)

$$\frac{1}{G} = \frac{1}{4\pi\mu_0 S} \cdot \frac{\oint |H|^2 \, dS}{\int |H|^2 \, dV} \qquad (6.87)$$

where H is the amplitude of the magnetic field. The integrals in the numerator and denominator are taken over the replaced area S of the cavity inner surface and the volume of the cavity respectively. But, in fact, as discussed above, such a substitution inevitably leads to a spurious resonance-frequency shift because of the difference between the actual geometrical sizes of the two samples. This spurious shift severely restricts the possibility of determining X_s.

If sample sizes are much smaller than the wavelength and if the samples are located at the antinode of the magnetic field inside a cavity, the amplitude of the field varies very slightly in distances on the order of the sample sizes. The sample appears to be placed in a region where the field distribution is nearly uniform. According to resonant-perturbation theory, placing an ideal conductor in a region of uniform electromagnetic field inside a cavity causes a resonance-frequency

shift that is proportional to the volume of the ideal conductor V_{id}:

$$f_{id} - f_0 = K \cdot V_{id} \qquad (6.88)$$

where f_0 is the resonance frequency of the empty cavity and K is the geometric factor. For a spherical sample located in a uniform ac magnetic field inside the cavity, K is given by (Kourov and Shcherbakov 1996)

$$K = \frac{3}{4} \cdot f_0 \cdot \frac{H_0^2}{\int |H|^2 \, dV} \qquad (6.89)$$

where H_0 is the amplitude of the magnetic field in the empty resonator in the region where the sample is placed; the integration in the denominator is taken over the volume of the cavity.

In the measurement of surface reactance of samples with sizes much less than the wavelength, we can use Eq. (6.88) where the volume of the sample under study is used instead of V_{id}. Then, from Eqs. (6.86) and (6.88), we have

$$X_s = \frac{G}{S}[KV - (f - f_0)] \qquad (6.90)$$

where V is the volume of the sample, and f is the resonant frequency of the cavity when the sample is loaded. Using this method, we can measure X_s without replacing part of the cavity wall and thus avoiding uncontrolled change of the cavity geometry.

Figure 6.24 shows a cylindrical cavity for the measurement of surface reactance of bulk samples. To remove degeneracy in frequency between the TE_{01p} and TM_{11p} modes, a metallic plunger is inserted along the axis of the cavity through the upper lid. The cavity is made of good conductor, such as copper, and works in the TE_{01p} modes. The sample is assumed to be a cube of side a that satisfies the condition

$$a \ll \lambda_0, \qquad (6.91)$$

and the conditions of normal skin effect are valid:

$$l \ll \delta \ll a \qquad (6.92)$$

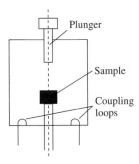

Figure 6.24 Structure of the cavity for the measurement of surface reactance

where l is the mean free path of the carriers and δ is the skin depth. The sample is usually placed on a polyfoam rod along the axis of the cavity where only H_z, the component of the magnetic field that is parallel to the cavity axis, is nonvanishing. The amplitude of H_z depends on the distance z from the bottom lid of a cylindrical cavity, as given by

$$H_z = H_0 \sin\left(\frac{p\pi z}{L}\right) \tag{6.93}$$

where L is the length of the cavity, and p is the mode number.

Usually, the sample is placed at the antinode of the magnetic field, that is, at the distance $z = L/2p$. So the useful frequency shift due to X_s of the sample under study reaches maximum value. Also, near the antinode of the magnetic field, the spatial derivatives of the field are small, resulting in maximum homogeneity of the field in the region where the sample is placed.

The determination of the geometric factor G is based on the fact that the geometric factors in the relationships for the change of bandwidth and shift of the resonant frequency are the same (Kourov and Shcherbakov 1996). As the change of bandwidth is far less sensitive to the small difference of sample sizes than the resonance-frequency shift, in the calculation of the geometric factor G, the bandwidth (Δf) of the cavity with a sample and the bandwidth of the cavity with a copper sample (Δf_{Cu}) having a size close to that of the sample under test are often used:

$$G = \frac{2S \cdot (R_s - R_s^{Cu})}{\Delta f - \Delta f_{Cu}} \tag{6.94}$$

where R_s and R_s^{Cu} are the surface resistances of the sample and copper samples respectively, which can be calculated from their dc resistivities.

The other geometric factor K can be calibrated with the copper samples having sizes close to those of the sample under test by

$$K = \frac{(f_{Cu} - f_0) + (X_s^{Cu} S_{Cu}/G)}{V_{Cu}} \tag{6.95}$$

where f_{Cu} is the resonant frequency of the cavity with the copper sample. For other values, the symbol Cu indicates that they are related to the copper sample. Note that copper can be considered as an ideal conductor, that is, within the accuracy of the experiment, one can assume that $X_s^{Cu} = 0$, because for all the modes under investigation, the shift of resonant frequency due to the surface reactance of copper ($X_s^{Cu} S_{Cu}/G$) is less than 0.1 % of the measured value of the shift ($f_{Cu} - f_0$). Thus, the neglect of the distinction between copper and the ideal conductor is responsible for the error in the determination of K, which is substantially less than what resulted from the inaccuracy of the sample volume measurement.

The geometric factors (G and K) can also be found on the basis of the known field distribution in a cylindrical cavity. The geometric factor K can be calculated according to Eq. (6.89). Besides, owing to the relationship

$$\oint |H|^2 \, dS = \frac{3}{2} H_0^2 S \tag{6.96}$$

for the integral of the magnetic field amplitude over the sample surface, where H_0 is the amplitude of the uniform ac magnetic field in the empty cavity in the region where the sample is placed, from Eq (6.87), G can be calculated from

$$\frac{1}{G} = \frac{3}{8\pi\mu_0} \cdot \frac{H_0^2}{\int |H|^2 \, dV} \tag{6.97}$$

It should be noted that Eqs. (6.89), (6.96), and (6.97) are exact only for spherical samples, but they can be used in the case of cubic samples also for the estimation of G and K within the accuracy of several percent (Kourov and Shcherbakov 1996). In the calculation, the ideally conducting cube of side

a is regarded as the sphere of the same volume, that is, the one having the radius:

$$r = a \left(\frac{3}{4\pi} \right)^{1/3}, \qquad (6.98)$$

and the cubic sample under study of side a is regarded as the sphere of the same surface area, that is, the one having the radius

$$r = a \left(\frac{3}{2\pi} \right)^{1/2}. \qquad (6.99)$$

6.5 NEAR-FIELD MICROWAVE MICROSCOPE

Many efforts have been made for the development of near-field microwave microscopes to get the spatial distribution of electromagnetic properties of materials (Rosner and van der Weide 2002). Various types of near-field microwave microscopes have been developed, and the material property parameters that can be mapped by near-field microwave microscopes mainly include conductivity, resistivity, permittivity, and permeability.

6.5.1 Basic working principle

A microwave microscope mainly consists of a sensing probe, microwave measurement system, and imaging system. The sensing probe of a near-field microwave microscope transmits microwave signal and meanwhile collects the microwave signal reflected back to the probe in the presence of the sample under study. According to the probes used, there are two general types of near-field microwave microscopes: nonresonant type and resonant type. In a nonresonant type microwave microscope, the probe is a segment of transmission in reflection state, and the properties of materials are obtained from the change of the reflectivity due to the sample. An example of nonresonant-type near-field microwave microscope is discussed in Chapter 3.

Resonant-type near-field microwave microscope is based on the resonant-perturbation theory. In a resonant near-field microwave microscope, the sensing probe is a resonator, and the presence of a sample introduces a perturbation to the resonant

probe. The properties of the sample under study are deduced from the changes of resonant properties of the probe due to the sample. Usually, a resonant near-field microwave microscope has higher sensitivity and resolution than a nonresonant one. In this part, we discuss several often-used resonant probes: tip-coaxial resonator, open-coaxial resonator, metallic waveguide resonator, and dielectric resonator. Another resonant probe often used is microstrip probe, which is discussed in Chapter 7.

Two techniques are often used in microwave measurement. In one technique, the resonant frequency and quality factor of the resonant probe are measured, and the properties of the sample are derived from the change of resonant frequency and quality factor due to the sample. In the other technique, the reflectivity from the resonant probe are measured, as shown in Figure 6.25. Usually, the cavity is critically coupled without a sample, the presence of the sample results in an increase in reflected power, and the reflected power is related to the properties of the sample. As shown in Figure 6.25, using the reflection method, the change in resonant frequency can also be obtained using reflection method, and the information on the resonant frequency can be used for accuracy and sensitivity enhancement.

In an actual measurement, a suitable operation frequency (f_x) is chosen and the change in the reflection amplitude ΔS_{11} is recorded. It should be indicated that f_x is not necessarily the resonant

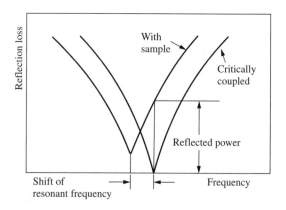

Figure 6.25   The change of resonant frequency and reflectivity due to the presence of sample

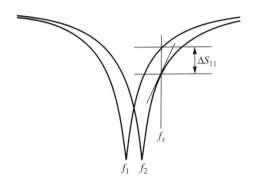

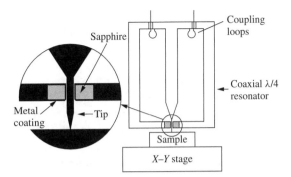

Figure 6.26 Relationship between the change of resonant frequency and change of reflection coefficient

Figure 6.27 Schematic diagram of near-field microwave microscope with tip-coaxial resonator. Modified from Gao, C. Wei, T. Duewer, F. Lu, Y. L. and Xiang, X. D. (1997). "High spatial resolution quantitative microwave impedance microscopy by a scanning tip microwave near-field microscope", *Applied Physics Letters*, **71** (13), 1872–1874

frequency of the probe when no sample is present; for different samples, the operation frequency may be different. The operation frequency f_x is often chosen to yield maximum change in the probe's reflection coefficient for a given range of property of a sample. As shown in Figure 6.26, the detected signal ΔS_{11} can be related to the slope of the resonance curve at the operation frequency f_x, which is kept fixed throughout the measurement, using the following relationship:

$$\Delta S_{11} \approx \left. \frac{\Delta S_{11}}{\Delta f} \right|_{f_x} \times \Delta f \qquad (6.100)$$

The slope $(\Delta S_{11}/\Delta f)$ at frequency f_x is the sensitivity of the probe. For a given probe, a different operation frequency can be used for different frequencies to achieve maximum sensitivity.

In a near-field microwave microscope, the scanning system and image-processing techniques play crucial roles. Besides, feedback circuits are often used in the microwave measurement systems to increase the measurement reliability. In the following discussion, we concentrate on the operation of sensing probes often used in near-field microwave microscopes.

6.5.2 Tip-coaxial resonator

As shown in Figure 6.27, in a near-field microwave microscope, the sample under test is fixed on a stage, which can scan in X- and Y-directions. The tip of the sensing probe is in the vicinity of the sample. The resonant frequency and quality

factor, which are related to the properties (e.g., dielectric constant and loss tangent) of the sample, are recorded as functions of the positions of the tip relative to the sample under test, and the distributions of the properties of the sample can then be mapped using image-processing techniques.

As shown in Figure 6.28, the sensing probe in Figure 6.27 can be taken as a modification of a coaxial cavity (Shrom 1998). The fundamental principle behind the operation of the microscope is

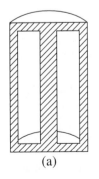

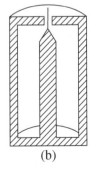

(a)   (b)

Figure 6.28 Modification of (a) a coaxial cavity into (b) a microwave microscope probe. Source: Shrom, A. J. (1998). "A scanning microwave microscope", Physics Department of Physics, Princeton University, Princeton, New Jersey

that boring a hole in the end wall and extending the center conductor past the outer surface allows materials outside the cavity to perturb the coaxial cavity, making the cavity a probe of the local dielectric constant of a sample brought near the tip. The key part of a sensing probe is the sharpened metal tip, mounted on the center conductor of a $\lambda/4$ coaxial resonator and protruding beyond an aperture formed on the end wall of the resonator. In the structure developed by Gao and Xiang (1998), a sapphire disk with a center hole of a size close to the diameter of the tip wire and a metal layer coating on the outside surface is used to shield off the far-field propagating components. This design minimizes the far-field background signal and allows submicron spatial resolution even in the quantitative analysis of complex dielectric constant. The coating layer should be as thin as the skin depth to avoid the formation of a microtransmission line with heavy loss near the aperture while still maintaining effective shielding of far-field components. To prevent vibration against the shielding hole, the tip is bonded to the sapphire disk with insulating glue.

A quasi-static model can be used to perform quantitative measurements (Gao *et al.* 1997). Since dielectric samples are placed in the near-field region of the tip, the electromagnetic interaction can be treated as quasi-static, so the wave nature can be ignored. Under this assumption, we can model the tip as a conducting sphere, and the electric field inside the dielectric sample can be solved using the image charge approach for samples with thickness much greater than the tip radius. The theory predicts that the intrinsic spatial resolution of the microwave microscope is proportional to the tip radius R_0. The ratio depends on the dielectric constant of the sample and decreases with the increase of the dielectric constant. For dielectric materials of common interest ($\varepsilon_r > 20$), the resolution can be one or two orders of magnitude smaller than the tip radius.

We consider the simplest configuration: the tip scans directly (soft contact) on the surface of a dielectric material with a thickness much larger than the tip radius. The shifts in resonant frequency (f_r) and quality factor (Q) caused by the sample can be calculated using the perturbation

theory (Gao and Xiang 1998):

$$\frac{\Delta f_r}{f_r} = A\left[\frac{\ln(1-b)}{b} + 1\right] \quad (6.101)$$

$$\Delta\left(\frac{1}{Q}\right)_d = -\frac{\Delta f_r}{f_r}\tan\delta \quad (6.102)$$

with

$$b = \frac{\varepsilon - \varepsilon_0}{\varepsilon + \varepsilon_0} \quad (6.103)$$

where A is a constant determined by the geometry of the tip-resonator assembly. For an ideal $\lambda/4$ coaxial resonator

$$A \approx \frac{16R_0\ln(R_2/R_1)}{\lambda} \quad (6.104)$$

where R_0 is the radius of the tip, R_1 and R_2 are the radii of the inner and the outer conductors of the resonator respectively, and λ is the resonant wavelength.

Besides the dielectric loss from the sample, the extra current required to support the charge redistribution induces additional resistivity loss. The corresponding change of quality factor is given by

$$\Delta\left(\frac{1}{Q}\right)_c = -B\frac{\Delta f_r}{f_r} \quad (6.105)$$

So the total shift of quality factor can be expressed by

$$\Delta\left(\frac{1}{Q}\right)_t = -(B + \tan\delta)\frac{\Delta f_r}{f_r} \quad (6.106)$$

On the basis of Eqs. (6.101) and (6.106), quantitative measurements of the local complex dielectric permittivity for samples can be made. The constants A and B can be calibrated by measuring a standard sample such as sapphire with a known dielectric constant and loss tangent.

6.5.3 Open-ended coaxial resonator

A type of sensing probe in a near-field microwave microscope is developed from open-ended coaxial resonator. The principle of such a probe is based on the perturbation of material to an open-ended coaxial resonator (Tanabe and Joines 1976; Xu *et al.* 1987). As shown in Figure 6.29, the

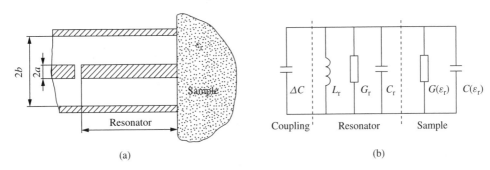

(a)   (b)

Figure 6.29 An open-ended coaxial resonator terminated by a dielectric material. (a) Configuration and (b) equivalent circuit. Modified from Xu, D. M. Liu, L. P. and Jiang, Z. Y. (1987). "Measurement of the dielectric properties of biological substances using an improved open-ended coaxial line resonator method", *IEEE Transactions on Microwave Theory and Techniques*, **35** (12), 1424–1428. © 2003 IEEE

dielectric material alters the external admittance and capacitance, so the resonant properties of the coaxial resonator are changed. From the changes of the resonant frequency and the quality factor, the dielectric constant and loss tangent of the material can be obtained. However, in a near-field microwave scope, there is a layer of air gap between the sample and the open end of the probe.

Figure 6.30 schematically shows a near-field scanning microwave microscope with an open-coaxial resonator probe. The probe is coupled to a microwave source through a capacitive coupler. Usually, the position of the probe is fixed, and the sample under measurement is put on a translation

stage. The open end of the probe is close or in soft contact with the sample. Owing to the concentration of the microwave fields at the open end of the probe, the boundary condition of the resonator, and hence, the resonant frequency f_0 and quality factor Q, are perturbed depending on the electromagnetic properties of the region of the sample immediately beneath the probe. The position of the sample is controlled using the translation stage along the X- and Y-directions. As the changes of the resonant frequency and the quality factor are related to the electromagnetic properties in the region of the sample immediately beneath the probe, images of the electromagnetic properties of the sample under study can thus be obtained from the frequency-shift image and quality-factor image.

The sensing probe plays an important role in a microwave microscope. On the basis of the structure shown in Figure 6.29, various kinds of probes have been developed. In the following, we discuss three kinds of probes for the measurements of sheet resistance, dielectric permittivity, and magnetic permeability, respectively.

6.5.3.1 Sheet resistance measurement

Figure 6.31 shows a probe for the measurement of sheet resistance R_x (Steinhauer *et al.* 1998, 1997). The sample is coupled to the probe through the capacitance formed at the gap between the probe and the sample. The sheet resistance R_x affects both the resonant frequency and the quality factor,

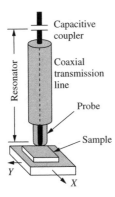

Figure 6.30 Schematic of a microwave microscope (Steinhauer *et al.* 2000). Source: Steinhauer, D. E. Vlahacos, C. P. Wellstood, F. C. *et al.* (2000). "Quantitative imaging of dielectric permittivity and tunability with a near-field scanning microwave microscope", *Review of Scientific Instruments*, **71** (7), 2751–2758

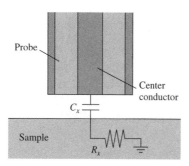

Figure 6.31   The interaction between an open probe and the sample, represented by a capacitance C_x and a resistance R_x

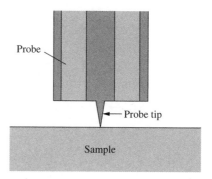

Figure 6.32   Interaction between a probe and a sample

and so can be derived from the change of resonant frequency and quality factor due to the sample. The coupling capacitance C_x is related to the thickness of the gap between the probe and the sample. When the gap becomes thinner, the coupling capacitance increases, and the change of resonant frequency due to the change of sheet resistance is greater. So the sensitivity of the probe increases as the probe-sample separation is reduced.

When the thickness of the gap is fixed, the resonant frequencies decrease with the decrease of R_x. This can be understood as follows: as R_x decreases, the boundary condition of the resonator tends from an open circuit toward a short circuit, changing the circuit from a half-wave resonator toward a quarter-wave resonator and lowering the resonant frequency.

The relationships between the unloaded quality factor (Q_0) and the sheet resistance can also be used to deduce the sheet resistance of the sample. However, R_x is a double-valued function of Q_0 (Steinhauer *et al.* 1998). This presents a problem for converting the measured Q_0 to R_x. However, R_x is a single-valued function of the frequency shift, allowing one to use the frequency-shift data to determine which branch of the $R_x(Q)$ curve should be used. In the derivation of sheet resistance, combination of frequency and quality factor results in higher accuracy.

6.5.3.2 Permittivity measurement

Figure 6.32 shows a probe that can be used to measure the permeability of dielectric films and

bulks (Steinhauer *et al.* 2000). The probe has a sharp-tipped center conductor extending beyond the outer conductor. The tip is in soft contact with the sample, so the sample is not scratched when it is scanned beneath the probe. Because the rf fields are concentrated at the probe tip, the resonant frequency and the quality factor of the resonator are a function of the properties of the part of sample near the probe tip.

On the basis of resonant-perturbation theory, we can estimate the frequency shift of the probe as a function of the fields near the probe tip. We define ε_{r1} and ε_{r2} as the permittivities of two samples. If $E1$ and $E2$ are the electric fields inside the two samples, the frequency shift of the microscope upon going from sample 1 to sample 2 is

$$\frac{\Delta f}{f} \approx \frac{\varepsilon_0(\varepsilon_{r2} - \varepsilon_{r1})}{4\,W} \cdot \int_{V_s} E_1 \cdot E_2 \, dV \qquad (6.107)$$

where W is the energy stored in the resonator, and the integral is over the volume V_s of the sample. We calculate an approximate W using the equation for the loaded Q of the resonator, $Q_L = \omega_0 W / P_l$, where ω_0 is the resonant frequency and P_l is the power loss in the resonator.

Usually, the calibration method is used. A bare LaAlO$_3$ (LAO) substrate can be used as sample 1 ($\varepsilon_{r1} = 24$, E_1) because its properties are well characterized and it is a common substrate for oxide dielectric thin films. In the measurement of a bulk dielectric sample, we calculate the frequency shift Δf upon the replacement of the LAO substrate by the sample of interest with permittivity ε_{r2}. In the measurement of thin films on a LAO substrate, we calculate the frequency shift Δf associated with replacing a thin top layer of the LAO substrate with a thin film of permittivity ε_{r2}.

6.5.3.3 Permeability measurement

Figure 6.33 shows a probe that can be used for imaging the magnetic permeability of metals (Lee *et al.* 2000). The magnetic permeability of the sample under test is derived from the changes of resonant frequency and quality factor due to the existence of the sample.

The coupling between the loop probe and a sample can be represented by an equivalent circuit shown in Figure 6.33(b). The loop probe is represented as an inductor L_0, and the sample under test is represented as a series combination of its effective inductance L_x and complex impedance $Z_x = R_x + X_x$. The coupling between the loop probe and the sample is represented by a mutual inductance M.

Since the microwave skin depth of the sample is much smaller than the thickness of the sample, the inductance of the sample can be modeled by an identical image of the loop probe:

$$L_x = L_0 \qquad (6.108)$$

If we assume the inner diameter (a) of the loop is equal to the wire thickness, the self-inductance of the loop probe can be roughly estimated:

$$L_0 \approx 1.25\mu_0 a. \qquad (6.109)$$

In the high-frequency limit, the surface impedance of the sample is given by

$$Z_x = \sqrt{j\mu_0\mu_r\omega\rho} \qquad (6.110)$$

where μ_r is the complex relative permeability of the material, ω is the microwave frequency, and ρ is the resistivity of the material, which is considered to be independent of μ_r. Since the loop and its image are roughly circular inductors, we can calculate M as the mutual inductance between two circular loops in the same plane. However, as this two-circular-loop model only approximately describes the geometry, we need to treat the value of M as a fitting parameter.

The microscope resonator can be taken as a transmission line that is capacitively coupled to the microwave source. The frequency shift and quality factor can be calculated using microwave transmission line theory. In the high-frequency limit, the value of ωL_0 is much greater than $|Z_x|$. According to the equivalent circuit shown in Figure 6.33(b), the load impedance presented by the probe and sample is

$$Z_{load} \approx j\omega L_0(1 - k^2) + k^2(R_x + jX_x) \qquad (6.111)$$

with

$$k = \frac{M}{\sqrt{L_0 L_x}} \qquad (6.112)$$

where the coupling coefficient k is a geometrical factor. The frequency shift is produced by the imaginary part of Z_{Load}, while the real part of Z_{Load} determines the quality factor of the probe. From Eqs. (6.110) and (6.111), we can obtain the magnetic permeability of the sample from the resonant frequency and the quality factor of the probe.

If an external magnetic field (H_{ext}) is applied uniformly parallel to the sample surface and the plane of the loop probe, as shown in Figure 6.33(a), we can study the ferromagnetic resonance of the sample. Detailed discussion on ferromagnetic resonance can be found in Chapter 8.

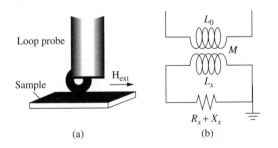

Figure 6.33 Interaction between a loop probe and a magnetized metal. (a) Measurement configuration and (b) equivalent circuit. Modified from Lee, S. C. Vlahacos, C. P. Feenstra, B. J. *et al.* (2000). "Magnetic permeability imaging of metals with a scanning near-field microwave microscope", *Applied Physics Letters*, **77** (26), 4404–4406

6.5.4 Metallic waveguide cavity

Metallic waveguide cavities can also be used in developing sensing probes for near-field microwave microscopes (Gutmann and Borrego 1987). In a probe developed from a one-port cylindrical-waveguide cavity, a slide-screw tuner is often used to ensure that the cavity is critically coupled to external circuit. The cavity interacts with a planar sample through a circular aperture at the center of the end surface. When the cavity operates in either the TM_{014} or TM_{022} modes, the evanescent field in the interaction region is dominantly electric. Though it is difficult to laterally confine the evanescent electric field, this kind of probe has high depth resolution.

As shown in Figure 6.34, to increase the lateral resolution, an aperture-coupled rectangular-waveguide resonator with an inductive-iris/stripline-fed coupling element of similar dimensions can be used (Gutmann and Borrego 1987). The critical coupling to the measurement instrument is achieved by adjusting the tuning screw with the rectangular-waveguide cavity. The spatial extent of the dominantly magnetic, evanescent field at the sample is controlled by the dimensions of the thin-conducting coupling element and the separation between the coupling element and the sample under test. Usually, the samples under test have conducting patterns.

Microwave scanning microscopy using evanescent wave coupling from one-port critically coupled cavities to planar samples have demonstrated lateral resolutions on the order of a few mils (0.002λ) and depth resolution on the order of a few microns (0.0001λ) with conventional X-band instrumentation (Gutmann and Borrego 1987).

6.5.5 Dielectric resonator

A long and narrow slot on the convex surface of a dielectric resonator can be used as a sensing probe (Abu-Teir *et al.* 2001). The probe developed by Abu-Teir *et al.* operates at 25 to 30 GHz and has a spatial resolution of 1 to 10 μm. The probe has a low impedance of about 20 Ω, and can be used for characterization of metallic layers with high conductivity, in particular, for thickness mapping, which can be done through the local measurement of the sheet resistance R_{sh} provided the film thickness is smaller than the skin depth:

$$R_{sh} = \rho/d \qquad (6.113)$$

where ρ is the resistivity and d is the film thickness. By increasing the dielectric constant of the resonator, the sensitivity can be further increased, and thicker films can be characterized.

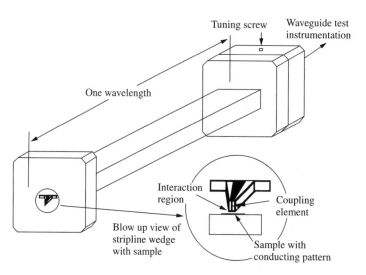

Figure 6.34   Rectangular-waveguide cavity. Source: Gutmann, R. J. Borrego, J. M. (1987). "Microwave scanning microscopy for planar structure diagnostics", *IEEE MTT-S Digest*, 281–284. © 2003 IEEE

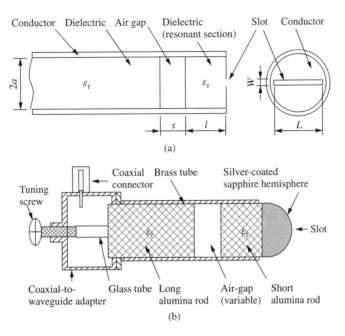

Figure 6.35 Probe design. (a) Schematic drawing and (b) actual design (Abu-Teir *et al.* 2001). Source: Abu-Teir, M. Golosovsky, M. Davidov, D. Frenkel, A. and Goldberger, H. (2001). "Near-field scanning microwave probe based on a dielectric resonator", *Review of Scientific Instruments*, **72** (4), 2073–2079

This probe combines a dielectric resonator with a small aperture, and can be taken as a modification of waveguide probe discussed in Section 3.6. Figure 6.35(a) shows a resonant slot in the end wall of a circular waveguide. The slot resonates when its length is approximately equal to the half wavelength. In the design of a sensing probe, we are interested in the slot radiation from the circular waveguide filled with dielectric material into free space. The design of probe mainly consists of two steps. In the first step, we model the unmatched dielectrically filled slot antenna and determine the radiation impedance (Z_{sl}) around the resonance frequency of the slot. Usually, the radiation impedance is quite high ($Z_{sl} \approx 1000\,\Omega$), and the frequency bandwidth is very narrow (less than 1 %). In the second step, we design a low-Q resonator as an impedance transformer. This allows good impedance matching in a frequency band comparable to the natural bandwidth of the slot. As shown in Figure 6.35, we can use a long waveguide filled with dielectric material as a feeding structure and a short segment of the waveguide filled with dielectric material as a resonator. It should be noted that in the air gap between the short

and long segments of the dielectrics, the dominant mode is evanescent, so the energy coupling between the feeding waveguide and the resonator can be adjusted by changing the length of the air gap.

Figure 6.35(b) schematically shows the structure of an actual probe. The basic design shown in Figure 6.35(a) is modified mainly in one respect – the flat front end of the probe is replaced by a spherical one. The spherical probe shape is advantageous for imaging purpose and alleviates sample approach, while there is no dramatic change in microwave performance. As shown in Figure 6.35(b), a short alumina rod is mounted at the end of a brass tube, and a sapphire hemisphere is attached to this alumina rod using a thin adhesive layer. Since sapphire and alumina have close dielectric constants, this assembly operates as a single dielectric resonator. The hemisphere is silver coated, leaving a long narrow slot at the center, and the length of the slot is chosen to ensure the working frequency. The brass tube with the resonator at the end is attached to a specially designed coax-to-cylindrical waveguide adapter. A tuning screw is mounted at the cover plate of the adapter along its symmetry axis. A long alumina rod is attached to the tuning screw through a glass tube transducer. To allow the motion of the dielectric rod inside the brass tube, the inner diameter of the brass tube slightly exceeds the diameter of the long alumina rod.

It is clear that the reflectivity increases with the increase of the film thickness d. In order to extract quantitative information about the intrinsic properties of the film under test from the reflectivity measurement, it is necessary to build a model for the probe–sample interaction. However, such models are still at the initial stages of development (Abu-Teir *et al.* 2001).

REFERENCES

Abu-Teir, M. Golosovsky, M. Davidov, D. Frenkel, A. and Goldberger, H. (2001). "Near-field scanning microwave probe based on a dielectric resonator", *Review of Scientific Instruments*, **72** (4), 2073–2079.

Barlow, H. E. M. (1962). "An improved resonant-cavity method of measuring high-frequency losses in materials", *Proceedings of the IEE-B*, **109** (23), 848–852.

Bonn, D. A. Morgan, D. C. and Hargy, W. H. (1991). "Split-ring for measuring microwave surface resistance of oxide superconductors", *Review of Scientific Instruments*, **62**(7), 1819–1823.

Chen, L. F. Ong, C. K. and Tan, B. T. G. (1996). "A resonant cavity for high-accuracy measurement of microwave dielectric properties", *Measurement Science and Technology*, **7**, 1255–1259.

Chen, L. F. Ong, C. K. and Tan, B. T. G. (1999). "Amendment of cavity perturbation method for permittivity measurement of extremely low-loss dielectrics", *IEEE Transactions on Instrumentation and Measurement*, **48**(6), 1031–1037.

Collin, R. E. (1992). *Foundations for Microwave Engineering*, McGraw-Hill, New York.

Derov, J. S. Drehman, A. J. Suscavage, M. J. Andrews, R. J. Cohen, E. Ianno, N. and Thompson, D. (1992). "Multiple frequency surface resistance measurement technique using a multimode TE_{01n} cylindrical cavity on a TlBaCaCuO superconducting film", *IEEE Microwave and Guided Wave Letters*, **2**, 452–453.

Dew-Hughes, D. (1997). "Microwave properties and applications of high temperature superconductors", in *Microwave Physics and Techniques*, H. Groll and I. Nedkov, Eds., Kluwer Academic Publishers, Dordrecht, 83–114.

Edgcombe, C. J. and Waldram, J. R. (1994). "Use of dielectric resonators in the measurement of surface impedance of thin films", *Research Review 94, IRC in Superconductivity*, University of Cambridge, Cambridge, 50.

Gao, C. Wei, T. Duewer, F. Lu, Y. L. and Xiang, X. D. (1997). "High spatial resolution quantitative microwave impedance microscopy by a scanning tip microwave near-field microscope", *Applied Physics Letters*, **71**(13), 1872–1874.

Gao, C. and Xiang, X. D. (1998). "Quantative microwave near-field microscopy of dielectric properties", *Review of Scientific Instruments*, **69**(11), 3846–3851.

Gutmann, R. J. and Borrego, J. M. (1987). "Microwave scanning microscopy for planar structure diagnostics", *IEEE MTT-S Digest*, **1**, 281–284.

Harrington, R. F. (1961). *Time-Harmonic Electromagnetic Field*, McGraw-Hill, New York.

Holstein, W. L. Parisi, L. A. Shen, Z. Y. Wilker, C. Brenner, M. S. and Martens, J. S. (1993). "Surface resistance of large-area $Tl_2Ba_2CaCu_2O_9$", *Journal of Superconductivity*, **6**, 191–200.

Jow, J. Hawley, M. C. Final, M. C. and Asmussen, J. Jr. (1989). "Microwave heating and dielectric diagnosis technique in a single-mode resonant cavity", *Review of Scientific Instruments*, **60** (1), 96–103.

Kajfez, D. and Guillon, P. (1986). *Dielectric Resonators*, Artech House, Dedham, MA.

Kaye, G. W. C. and Laby, T. H. (1995). *Tables of Physical and Chemical Constants*, 16th edition, Longman Group Limited, Essex.

Kobayashi, Y. and Katoh, M. (1985). "Microwave measurement of dielectric properties of low-loss materials by the dielectric rod resonator method", *IEEE Transactions on Microwave Theory and Techniques*, **33** (7), 586–592.

Kobayashi, Y. and Tanaka, S. (1980). "Resonant modes of a dielectric rod resonator short-circuited at both ends by parallel conducting plates", *IEEE Transactions on Microwave Theory and Techniques*, **28** (10), 1077–1095.

Kourov, D. N. and Shcherbakov, A. S. (1996). "Measurement of the magnitude of surface reactance", *Review of Scientific Instruments*, **67**, 274–278.

Krupka, J. and Mazierska, J. (1998). "Improvement of accuracy in measurements of the surface resistance of superconductors using dielectric resonator", *IEEE Transactions on Applied Superconductivity*, **8** (4), 164–167.

Kumar, A. (1976). "Measurement of the complex permittivity of lossy fluids at 9 GHz", *Review of Scientific Instruments*, **47** (2), 244–246.

Kumar, A. and Smith, D. G. (1976). "The measurement of the permittivity of sheet materials at microwave frequencies using an evanescent waveguide technique", *IEEE Transactions on Instrumentation and Measurement*, **25** (3), 190–193.

Landau, L. D. and Lifshitz, E. M. (1960). *Electrodynamics of Continuous Media*, Pergamon, Oxford.

Lee, S. C. Vlahacos, C. P. Feenstra, B. J. Schwartz, A. Steinhauer, D. E. Wellstood, F. C. and Anlage, S. M. (2000). "Magnetic permeability imaging of metals with a scanning near-field microwave microscope", *Applied Physics Letters*, **77** (26), 4404–4406.

Low, B. L. Xu, S. Y. Ong, C. K. Wang, X. B. and Shen, Z. X. (1997). "Substrate temperature dependence of texture quality in YBCO thin films fabricated by on-axis pulsed laser ablation", *Superconductor Science and Technology*, **10** (1), 41–46.

Lu, J. Ren, X. Y. and Zhang, Q. S. (1994). "A TE$_{011+\delta}$ mode sapphire resonator probe for accurate characterization of microwave surface resistance of HTS thin films", *ICCS '94*, IEEE, Singapore, 959–963.

Luo, Z. X. Yang, K. Lu, J. Tang, Z. X. and Zhang, Q. S. (1997). "Measurement system for accurate characterization of microwave surface resistance of high-Tc superconductors thin films", *Physica C*, **282–287**, 2537–2538.

Maier Jr., L. R. and Slater, J. C. (1952). "Field strength measurements in resonant cavities", *Journal of Applied Physics*, **23** (1), 68–77.

Mazierska, J. (1997). "Dielectric resonator as a possible standard for characterization of high temperature superconducting films for microwave applications", *Journal of Superconductivity*, **10** (2), 73–84.

Mazierska, J. and Grabovickic, R. (1998). "Circulating power, RF magnetic field, and RF current density of shielded dielectric resonators for power handling analysis of high-temperature superconducting thin films of arbitrary thickness", *IEEE Transactions on Applied Superconductivity*, **8** (4), 178–187.

Meng, B. Booske, J. and Cooper, R. (1995). "A system to measure complex permittivity of low loss ceramics at microwave frequencies and over large temperature ranges", *Review of Scientific Instruments*, **66** (2), 1068–1071.

Ni, E. H. and Ni, Y. Q. (1997). "Extra-cavity perturbation method for the measurement of dielectric resonator materials", *Review of Scientific Instruments*, **68** (6), 2524–2528.

Ong, C. K. Chen, L. F. Lu, J. Xu, S. Y. Rao, X. S. and Tan, B. T. G. (1999). "Mirror-image calibrator for resonant perturbation method in surface resistance measurement of high-T$_c$ superconducting thin films", *Review of Scientific Instruments*, **70** (7), 3092–3096.

Ormeno, R. J. Morgen, D. C. Broun, D. M. Lee, S. F. and Waldram, J. R. (1997). "Sapphire resonator for the measurement of surface impedance of high temperature superconducting thin films", *Review of Scientific Instruments*, **68**, 2121–2126.

Rosner, B. T. and van der Weide, D. W. (2002). "High-frequency near-field microscopy", *Review of Scientific Instruments*, **73** (7), 2505–2525.

Shen, Z. Y. Wilker, C. Pang, P. Holstein, W. L. Face, D. and Kountz, D. J. (1992). "High T$_c$ superconductor-

sapphire microwave resonator with extremely high Q-values up to 90 K", *IEEE Transactions on Microwave Theory and Techniques*, **40**, 2424–2432.

Shrom, A. J. (1998). *A Scanning Microwave Microscope*, Physics Department of Physics, Princeton University, Princeton, NJ.

Shu, Y. and Wong, T. Y. (1995). "Perturbation of dielectric resonator for material measurement", *Electronics Letters*, **31** (9), 704–705.

Steinhauer, D. E. Vlahacos, C. P. Dutta, S. K. Wellstood, F. C. and Anlage, S. M. (1997). "Surface resistance imaging with a scanning near field microwave microscope", *Applied Physics Letters*, **71** (12), 1736–1738.

Steinhauer, D. E. Vlahacos, C. P. Dutta, S. K. Feenstra, B. J. Wellstood, F. C. and Anlage, S. M. (1998). "Quantitative imaging of sheet resistance with a scanning near-field microwave microscope", *Applied Physics Letters*, **72** (7), 861–863.

Steinhauer, D. E. Vlahacos, C. P. Wellstood, F. C. Anlage, S. M. Canedy, C. Ramesh, R. Stanishevsky, A. and Melngailis, J. (2000). "Quantitative imaging of dielectric permittivity and tunability with a near-field scanning microwave microscope", *Review of Scientific Instruments*, **71** (7), 2751–2758.

Sucher, M. and Fox, J. (1963). *Handbook of Microwave Measurements*, 3rd edition, Vol. 2, Polytechnic Press of the Polytechnic Institute of Brooklyn, New York.

Tanabe, E. and Joines, W. T. (1976). "A non-destructive method for measuring the complex permittivity of dielectric materials at microwave frequencies using an open transmission line resonator", *IEEE Transactions on Instrumentation and Measurement*, **25** (3), 222–226.

Tian, B. Q. and Tinga, W. R. (1993). "Single-frequency relative Q measurements using perturbation theory", *IEEE Transactions on Microwave Theory and Techniques*, **41**(11), 1922–1927.

Tian, B. Q. and Tinga, W. R. (1995). "Linearity condition between a cavity's Q-factor and its input resonant resistance", *IEEE Transactions on Microwave Theory and Techniques*, **43** (3), 691–692.

Waldron, R. A. (1969). *Theory of Guided Electromagnetic Waves*, Van Nostrand Reinhold Company, London, Chapter VI.

Wang, H. M. and Ni, E. H. (1997). "Permittivity measurement of dielectric resonator materials at C-band", *Microwave Journal*, **40** (5), 296–302.

Xu, D. M. Liu, L. P. and Jiang, Z. Y. (1987). "Measurement of the dielectric properties of biological substances using an improved open-ended coaxial line resonator method", *IEEE Transactions on Microwave Theory and Techniques*, **35** (12), 1424–1428.

7

Planar-circuit Methods

This chapter discusses the applications of microwave planar circuits in the measurement of the electromagnetic properties of materials, including thin films, sheet samples, and substrate samples. After a brief introduction on the applications of planar circuits in materials property characterization, we discuss the methods developed from three typical types of planar transmission lines: stripline, microstrip, and coplanar line. We then discuss permeance meters, which can be used for permeability measurement of magnetic thin films. At the end of this chapter, we discuss the near-field microwave microscopes developed using planar circuits based on resonant perturbation.

7.1 INTRODUCTION

As discussed in previous chapters, most of the microwave methods for materials property characterizations are based on transmission lines and the resonant structures developed from transmission lines, and, correspondingly, materials property characterization methods generally fall into nonresonant methods and resonant methods. In the previous chapters, discussions are focused on three types of transmission lines: coaxial line, waveguide, and free space. In materials property characterization, three kinds of planar circuits, including stripline, microstrip, and coplanar line, are also widely used. Generally speaking, a planar transmission line consists of dielectric substrates, and conductors, including conducting strips and grounding conductors. Besides their geometrical structures, the electromagnetic properties of the

substrate and the conductors affect the properties of the planar transmission lines. Therefore, planar transmission lines can be used to characterize dielectric materials and conductors. Similar to the methods based on other types of transmission lines, the methods based on planar transmission lines also include nonresonant methods and resonant methods.

7.1.1 Nonresonant methods

Similar to the methods developed from coaxial line, waveguide and free space, planar-circuit non-resonant methods also include reflection methods and transmission/reflection methods. The basic principles of reflection methods and transmission methods have been discussed in Chapters 3 and 4. In the following discussion, we concentrate on the problems in the development of planar-circuit methods for the characterization of materials properties.

7.1.1.1 Reflection method

In a reflection method, the electromagnetic properties of the sample are obtained from the reflection properties of a segment of the transmission line. However, as different types of transmission lines can be used, the samples under test may be arranged in different ways. For example, for an open-end coaxial probe, the sample is attached to the open end of the probe, while for a planar-circuit reflection method, the sample under test usually serves as the substrate or part of the substrate.

Microwave Electronics: Measurement and Materials Characterization   L. F. Chen, C. K. Ong, C. P. Neo, V. V. Varadan and V. K. Varadan
© 2004 John Wiley & Sons, Ltd   ISBN: 0-470-84492-2

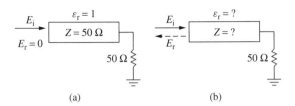

(a)   (b)

Figure 7.1 Transmission-line analogy of reflection method. (a) Transmission line with matched impedance and (b) transmission line with unmatched impedance

Figure 7.1 shows the basic principle of a planar-circuit reflection method. Consider a transmission line fabricated on a theoretical material having a dielectric constant of one, and the width of the 50 Ω planar transmission line can be calculated theoretically. In this case, a matched condition results in no reflection at the input point. When a transmission line with the same dimensions is fabricated on a material with a dielectric constant that is not equal to one, the impedance of the transmission line is no longer 50 Ω, but some value determined by the dielectric constant of the material. In experiments, the impedance of the transmission line can be determined by measuring the reflections caused by this new impedance, and calculations can then be made determining the corresponding value of the dielectric constant for that impedance.

To understand the principle of a transmission line used for the measurement of dielectric constant, the mathematical relationships between the parameters that can be measured experimentally and the materials properties of the sample should be built. The relationship between the impedance $Z(\varepsilon_r)$ and reflectivity (Γ) is given by

$$\Gamma = \frac{Z_0 - Z(\varepsilon_r)}{Z_0 + Z(\varepsilon_r)} \qquad (7.1)$$

where Z_0 is the termination impedance, and for the present case, it is 50 Ω. So the characteristic impedance $Z(\varepsilon_r)$ of the planar circuit can be calculated from the measured reflectivity Γ. The dielectric permittivity can then be obtained from the relationship between the characteristic impedance and the structural parameters. As discussed in Chapter 2, for different types of planar circuits, the expressions for characteristic impedance are different. For a microstrip line with

$w/h \geq 1$, its characteristic impedance is given by (Laverghetta 1997)

$$Z(\varepsilon_r) = \frac{120\pi/\sqrt{\varepsilon_{\text{eff}}}}{(w/h) + 1.393 + 0.667 \ln[(w/h) + 1.444]} \qquad (7.2)$$

with

$$\varepsilon_{\text{eff}} = \frac{\varepsilon_r + 1}{2} + \frac{\varepsilon_r - 1}{2} \left(1 + 12\frac{h}{w}\right)^{\frac{1}{2}} \qquad (7.3)$$

where w is the width of the microstrip and h is the thickness of the dielectric material. From Eq. (7.2) the dielectric properties of the substrate material can be calculated from the characteristic impedance of the microstrip line. The calculations of the characteristic impedances of different planar transmission lines are discussed in Chapter 2.

In Figure 7.1, the termination impedance is 50 Ω. In principle, any value of termination impedance can be used. Figure 7.2 shows two cases with open and short loads respectively. However, as there are technical difficulties to realize good open and short terminations, the configuration shown in Figure 7.1 is more widely used.

7.1.1.2 Transmission/reflection method

Planar-circuit transmission/reflection methods are also based on Nicolson–Ross–Weir algorithm, which has been discussed in Chapter 4. For a planar transmission line, the characteristic impedance and propagation constant are functions of the dielectric and magnetic properties of the substrate. As shown in Figure 7.3, if the substrate of a segment of a transmission line is replaced by the material under test, the characteristic

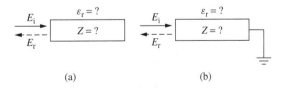

(a)   (b)

Figure 7.2 Reflection methods with different termination impedance. (a) Open ($L_0 = \infty$) and (b) short ($L_0 = 0$)

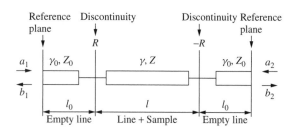

Figure 7.3 Schematic diagram for the transmission/reflection method

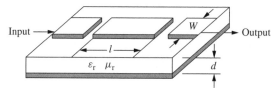

Figure 7.4 A microstrip resonator for material property characterization. The sample under test is the substrate of the resonator

impedance of the transmission line is changed from Z_0 to Z, and the propagation constant is changed from γ_0 to γ. There is reflectance at the impedance and propagation constant discontinuities, from which the electromagnetic properties of the material under test can be derived. In the calculation of dielectric permittivity and magnetic permeability, the relationship between the characteristic impedance, electromagnetic properties of the substrate and structure parameters are needed.

As shown in Figure 7.3, in actual measurements, the reference planes where calibrations are made and the sample surfaces may not be at the same positions. So in the calculation of materials properties, transformations are needed to obtain the scattering parameters of the segment of transmission line filled with the sample.

7.1.2 Resonant methods

Similar to their coaxial and waveguide counterparts, planar-circuit resonant methods include resonator methods and resonant-perturbation methods. Usually, resonant methods have higher accuracy and sensitivity than nonresonant ones, but resonant methods can only provide results at a single frequency or several discrete frequencies. In the following, we discuss two examples of resonator method and resonant-perturbation method.

7.1.2.1 Resonator method

In a resonator method, the sample under test is used as a substrate of a planar resonator. The

dielectric constant of the sample is calculated from the resonant frequency of the resonator, and the loss tangent of the material is calculated from the quality factor of the resonator. Figure 7.4 shows a straight-ribbon microstrip resonator. The measurement frequency is mainly determined by length of the microstrip, and the conductivity of the strip and ground conductors affects the sensitivity of loss tangent measurement.

7.1.2.2 Resonant-perturbation method

In a resonant-perturbation method, part of the substrate of a planar resonator is replaced by the sample under study. From the changes of the resonant frequency and quality factor, the properties of the sample can be derived (Abdulnour *et al.* 1995). Figure 7.5 shows a measurement fixture for resonant-perturbation method. The position A-A is close to the open end of the resonator, so the electric field dominates at the position A-A. If a sample is placed at position A-A, its dielectric properties can be measured. The position B-B is near the center of the resonator, so the magnetic field dominates at the position B-B. If a sample

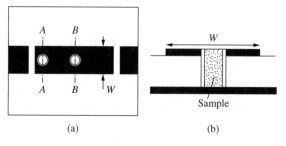

Figure 7.5 Perturbation to the substrate of a microstrip resonator. (a) Top view and (b) cross-section view at positions A-A and B-B

is placed at the position B-B, its magnetic properties can be measured. Usually, calibration with a standard sample is needed to obtain absolute values of materials properties.

7.2 STRIPLINE METHODS

In a stripline structure for the characterization of materials properties, the sample under test can be inserted above and below the center strip through the open sides. A stripline structure does not have low cutoff frequency, so broadband measurements can be performed. More discussions on the transmission properties of stripline structure can be found in Chapter 2. There are varieties of stripline structures, and, correspondingly, various stripline measurement techniques have been developed for different applications.

It deserves to be noted that the electric and magnetic fields of the transverse electromagnetic (TEM) wave propagating in a stripline structure is essentially unidirectional, and so the anisotropic effects of a sample can be characterized by rotating the sample. More discussion on the measurement of anisotropic samples can be found in Chapter 8.

7.2.1 Nonresonant methods

The transmission/reflection methods based on Nicolson–Ross–Weir algorithm have been developed using symmetrical and asymmetrical stripline structures. In these methods, the sample under study fills the space between the strip and the ground planes fully or partially. The samples are usually of rectangular shape, so the sample fabrication is relatively easy.

7.2.1.1 Symmetrical stripline method

Figure 7.6 shows a measurement fixture proposed by Barry for permittivity and permeability measurements (Barry 1986). The fixture mainly consists of three portions: coaxial-line portion, stripline portion, and another coaxial-line portion. There are two transitions between coaxial line and stripline. The stripline portion of the fixture is designed for a characteristic impedance

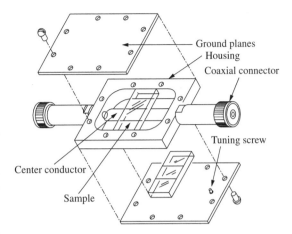

Figure 7.6 Stripline measurement fixture (Barry 1986). Source: Barry, W. (1986). "A broad-band, automated, stripline technique for the simultaneous measurement of complex permittivity and permeability", *IEEE Transactions on Microwave Theory and Techniques*, **34** (1), 80–84. © 2003 IEEE

$Z_0 = 50\,\Omega$. The samples under test should fit above and below the center conductor inside the housing securely. The total thickness of the sample must be less than $\lambda_m/2$ in order to avoid dimensional resonances. It should be noted that although Figure 7.6 indicates grooves in the samples for a better fit around the center conductor, samples without grooves could be measured equally well. Therefore, the grooves in samples may be deleted in order to save time when preparing samples.

By comparing the fixture shown in Figure 7.6 with the equivalent circuit shown in Figure 7.3, the empty line in a stripline measurement fixture includes the coaxial line and the stripline without sample, therefore, it is crucial to obtain the best possible transition from the stripline to coaxial connectors. In actual measurements, the transition between the stripline and coaxial line could not be perfect. The S-parameters of the imperfect transition regions between coaxial connectors and the front and back terminal planes of the sample region can be found using short circuits (Hanson *et al.* 1993). The S-parameters of the sample region can be subsequently de-embedded from the S-parameters measured at the coaxial connectors. This leads to substantially improved accuracy in determining the *S*-parameters of the segment of the stripline filled with sample, resulting in an enhanced accuracy in the determination of the complex permittivity and permeability of the sample.

7.2.1.2 Asymmetrical stripline method

In order to increase the sensitivity of stripline method, Salahun *et al.* proposed an asymmetrical stripline structure as shown in Figure 7.7(a). Using this structure, permeability and permittivity measurements of multilayered materials and thin films are allowed over a broad frequency band (Salahun *et al.* 2001).

Consider a general case of an asymmetrical stripline structure as shown in Figure 7.7(b). Four layers compose the cross section of the measurement fixture. This configuration enables us to take different locations of the material into account. The material may be laid on the ground plane as shown in Figure 7.7(a), or the sample may be laid on the strip, as shown in Figure 7.7(b), to increase the cell sensitivity. An air gap between the material and its support is taken into account to represent the experimental conditions. Because of the heterogeneity of the cross section, a TEM wave cannot be propagated. However, for low frequencies, longitudinal components of the microwave fields can be neglected compared with transversal ones. So the hypothesis of a quasi-TEM mode can be used.

Because of the heterogeneity of the loaded cross section, no analytical expression is available for the calculation of the electromagnetic properties of the material from the measured parameters. Therefore, two steps are necessary for the determination of electromagnetic properties of the sample. The first step is the direct problem. The effective constants of the transverse section are calculated from the complex permeability and permittivity of the sample. The second step is the inverse problem. The complex permeability and the permittivity of the material are extracted by matching theoretical and measurement results. Detailed discussion on the calculation algorithm is made in (Salahun *et al.* 2001).

7.2.2 Resonant methods

A stripline resonator is a segment of a stripline. It can be either open or short circuited at both ends (λ/2-resonator) or short circuited at one end and open at the other end (λ/4-resonator). If the substrate of a stripline is homogeneous and lossless, the stripline can support a pure TEM propagation mode. Dispersion does not exist unless the permittivity of the substrate has frequency dependence. In materials property characterization, the sample under test is inserted into the space between the grounds, and the properties of the sample are derived from the resonant frequency and quality factor of the resonator loaded with the sample.

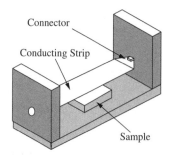 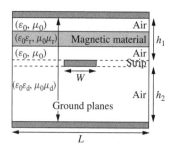

Figure 7.7 An asymmetrical stripline structure. (a) Schematic drawing of an asymmetrical without the upper ground plane. The sample is laid on the ground plane and (b) cross section of a general case of asymmetrical stripline containing a sample. The sample covers the stripline (Salahun *et al.* 2001). Source: Salahun, E. Queffelec, P. Le Floc'h, M. and Gelin, P. (2001). "A broadband permeameter for '*in situ*' measurements of rectangular samples", *IEEE Transactions on Magnetics*, **37** (4), 2743–2745. © 2003 IEEE

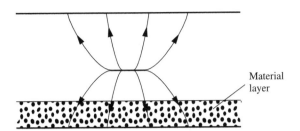

Figure 7.8 End view of the electric field in a stripline resonator. Reprinted with permission from *Industrial Microwave Sensors*, by Nyfors, E. and Vainikainen, P. (1989) Artech House Inc., Norwood, MA, USA. www.artechhouse.com

In a stripline transmission structure, there may exist one or more center conductors between the two planar ground plates. In the following, we discuss the applications of resonators made from striplines with one and two center conductors, respectively.

7.2.2.1 One-conductor stripline resonator method

Generally speaking, a stripline resonator has a higher quality factor than a microstrip resonator, and can be used for on-line measurement of sheet samples and material layers. A one-conductor stripline resonator usually has a symmetric electric field, and if the ground plates are large enough, the radiation loss is almost zero. However, as shown in Figure 7.8, any asymmetric loading will upset the balance in field distribution. The stripline resonator method is suitable for noncontact measurement of a dielectric layer on a conducting surface, in which case there are two possibilities – measurement of the permittivity of a layer of constant thickness, or measurement of the thickness of a layer of constant permittivity. This method is often used in industry for thickness measurement or moisture measurement (Nyfors and Vainikainen 1989). If two independent measurements, for example S_{11} and S_{21}, can be made, with proper calibration, two parameters of materials properties, for example, moisture and dry mass per area, may be obtained simultaneously.

Figure 7.9 shows a configuration for measuring the mass per unit area of powder samples. It is assumed that small variations in density are

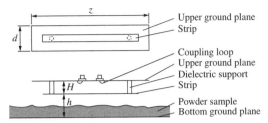

Figure 7.9 One-conductor stripline $\lambda/2$ resonator for the measurement of the mass per unit area of powder samples. Reproduced from Fischer, M. Vainikainen, P. and Nyfors, E. (1995). "Design aspects of stripline resonator for industrial applications", *Journal of Microwave Power and Electromagnetic Energy*, **30** (4), 246–257, by permission of International Microwave Power Institute

compensated by the variations in permittivity (Fischer *et al.* 1995). The measurement accuracy depends on the electric field in the space between the lower ground plane and the strip. A broad strip gives a uniform field and a good accuracy. A narrow strip emphasizes the upper layers of the material to be measured, but gives a better spatial resolution in the cross direction. To avoid the radiation from the stripline resonator, the distance between the two ground planes should not exceed 0.3λ. There is no strict requirement on the size of the upper ground plane. A small upper ground plane, only slightly larger than the strip, suffices to give reliable results (Nyfors and Vainikainen 1989).

7.2.2.2 Two-conductor stripline resonator method

Figure 7.10 shows a two-conductor $\lambda/2$ stripline resonator, and the sample under test is put between the two center conductors (Fischer *et al.* 1995). This structure is convenient for the measurement of sheet samples that can be transported between the two center conductors. For example, this structure can be used in the control of the paper thickness.

A resonator with two center conductors has either one or two first-order resonant modes. A $\lambda/2$-resonator has two degenerate resonant modes, even and odd, but a $\lambda/4$-resonator has only the even resonant mode. The electric potentials of the center conductors are equal for the even mode, but with opposite signs for the odd mode. The electric field patterns for a $\lambda/2$-resonator with two

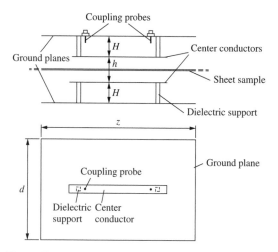

Figure 7.10   Two-conductor stripline λ/2 resonator for the measurement of sheet samples. Reproduced from Fischer, M. Vainikainen, P. and Nyfors, E. (1995). "Design aspects of stripline resonator for industrial applications", *Journal of Microwave Power and Electromagnetic Energy*, **30** (4), 246–257, by permission of International Microwave Power Institute

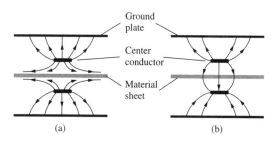

Figure 7.11   End views of the electric field strength patterns of a two-conductor stripline λ/2 resonator. (a) Even mode and (b) odd mode. Modified from Fischer, M. Vainikainen, P. and Nyfors, E. (1995). "Design aspects of stripline resonator for industrial applications", *Journal of Microwave Power and Electromagnetic Energy*, **30** (4), 246–257, by permission of International Microwave Power Institute

center conductors are presented in Figure 7.11. The electric field of the even mode is parallel to the material layer in the middle of the center conductors, and the electric field of the odd mode is perpendicular to the material layer. This orthogonality gives different properties for the two modes, because the field inside the material layer depends on the direction of the external field.

Theoretically, the two modes are degenerate, but, in practice, the resonant frequencies differ by about 10 % because of different fringing of the fields at the ends.

As shown in Figure 7.12, the radiation of the two-conductor stripline resonator can be practically eliminated by bending the edges of the ground planes 90° toward the plane of the sample (Nyfors and Vainikainen 1989). The edges may reach the levels of the strips without altering the normal performance of the resonator.

By taking the sheet sample as an ellipsoid and locating it in such a way that the external field is parallel to one axis of the ellipsoid, we have (Fischer *et al.* 1995)

$$
\frac{\Delta f}{f_0} + j\frac{1}{2}\Delta\left(\frac{1}{Q}\right)
$$

$$
= \frac{(\varepsilon_r' - 1) + A(\varepsilon_r' - 1)^2 + A\varepsilon_r''^2 - j\varepsilon_r''}{[1 + A(\varepsilon_r' - 1)]^2 + (A\varepsilon_r'')^2}
$$

$$
\cdot \frac{\displaystyle\int_{V_s} |E_0|^2 dV_s}{2\displaystyle\int_V |E_0|^2 dV} \tag{7.4}
$$

where Δf is the change in the resonant frequency due to material perturbation, f_0 is the resonant frequency of the empty resonator, Q is the quality factor of the resonator, A is the depolarization factor of the sample placed in the resonator, E_0 is the external electric field strength, V is the volume of the resonator, and V_s is the volume of the dielectric object. In Eq. (7.4), the unperturbed resonator is air-filled. The integral term can be considered as a filling factor that depends only on the location of the dielectric object and its volume. The depolarization factor can be accurately calculated only for ellipsoids. For a thin sheet of material $A \approx 1$ when the external field is perpendicular to the sheet of material and $A \approx 0$ when the external field is parallel to the sheet of material.

For a two-conductor stripline resonator, the sheet sample is transported through the resonator in the middle between the center conductors. The electric field is parallel to the material layer for the even mode and perpendicular to the layer for the

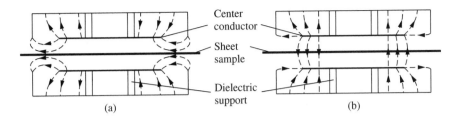

Figure 7.12 Electrical fields of a two-conductor stripline with bent ground planes. (a) Even mode and (b) odd mode. Modified from Fischer, M. Vainikainen, P. and Nyfors, E. (1995). "Design aspects of stripline resonator for industrial applications", *Journal of Microwave Power and Electromagnetic Energy*, **30** (4), 246–257, by permission of International Microwave Power Institute

odd mode, as shown in Figure 7.11. So for even mode ($A \approx 0$), Eq. (7.4) becomes

$$\frac{\Delta f}{f_0} + j\frac{1}{2}\Delta\left(\frac{1}{Q}\right)$$
$$= -[(\varepsilon_r' - 1) - j\varepsilon_r''] \cdot \frac{\displaystyle\int_{V_s} |E_0|^2 dV_s}{2\displaystyle\int_V |E_0|^2 dV} \qquad (7.5)$$

and for odd mode, Eq. (7.4) becomes

$$\frac{\Delta f}{f_0} + j\frac{1}{2}\Delta\left(\frac{1}{Q}\right)$$
$$= -\left[1 - \frac{\varepsilon_r'}{\varepsilon_r'^2 + \varepsilon_r''^2} - j\frac{\varepsilon_r''}{\varepsilon_r'^2 + \varepsilon_r''^2}\right]$$
$$\cdot \frac{\displaystyle\int_{V_s} |E_0|^2 dV_s}{2\displaystyle\int_V |E_0|^2 dV} \qquad (7.6)$$

For the even mode, the changes in f_0 and $1/Q$ depend linearly on ε_r' and ε_r''. But for the odd mode, the change in f_0 is rapidly saturated when ε_r' becomes large (Fischer *et al.* 1995). Thus, the odd resonant mode is suitable for the measurement of materials with low permittivity only.

It has been shown that the ratio $\varepsilon_r''/(\varepsilon_r' - 1)$ is independent of the density or thickness of the material layer for most natural materials. The ratio depends on the water content and is thus used for the measurement of moisture content. For the even mode, we can get from Eq. (7.5)

$$\frac{\Delta(1/Q)}{\Delta f/f_0} = \frac{2\varepsilon_r''}{\varepsilon_r' - 1} \qquad (7.7)$$

For the odd mode, we can get from Eq. (7.6)

$$\frac{\Delta(1/Q)}{\Delta f/f_0} = \frac{2\varepsilon_r''}{\varepsilon_r'(\varepsilon_r' - 1) + \varepsilon_r''^2} \qquad (7.8)$$

Equation (7.7) indicates that the moisture of a sheet sample, independent of its density, can be measured by using the ratio of the change of the quality factor and the change of the resonant frequency of the even mode.

7.2.2.3 Resonant-perturbation method

This method was first proposed by Waldron (1964), and improved by Weil *et al.* (2000). In this method, the sample under test is introduced into a stripline resonator, and the properties of the sample are derived from the changes of the resonant frequency and quality factor of the stripline resonator due to the presence of the sample. Because of the small specimen sizes used and the correspondingly low cavity filling factors, this method works well for materials that possess a wide range of dielectric and magnetic losses.

Compared to other measurement methods, the stripline cavity-perturbation method possesses some significant advantages and disadvantages (Weil *et al.* 2000). One obvious advantage is that the stripline unit is capable of multifrequency operation over a wide frequency range, using up to 10 or more harmonics of the fundamental resonance of the unit. Another advantage is that the resonator has a uniaxial or very nearly uniaxial field structure. So anisotropic materials can be characterized by orienting the sample under test either parallel or normal to the electric or magnetic fields of the resonator. Such measurements are not possible in the coaxial air line, owing to its radial field configuration. Other advantages include the capability to measure small samples of high-loss magnetic films, as well as the reduced cost of sample preparation (rectangular shape versus very accurately machined toroid). Besides, the samples under test can be readily introduced through the open sides of the structure. The disadvantages of this method are also obvious. This method is incapable of satisfactorily resolving the highly dispersive permeability properties of ferrites in the VHF/UHF frequency range due to the cavity's inadequate low-frequency limit and its poor frequency resolution. Furthermore, this method relies on small-perturbation theory for the derivation of material parameters. Since such techniques are known to be very susceptible to measurement error, this constitutes another potential disadvantage.

Figure 7.13 illustrates the basic structure of a stripline resonator for resonant-perturbation method (Weil *et al.* 2000). It mainly consists of a center-strip conductor mounted equidistantly between two ground planes and terminated by two end plates. A propagating TEM mode is excited

within the stripline structure using either coupling loops mounted in one of the end plates or a monopole probe mounted in the ground planes and adjustable in the axial position. The fundamental resonance is achieved when the resonator length corresponds to one half-guide wavelength of the exciting frequency. Additional resonances may occur at harmonic frequencies of the fundamental, assuming that the length remains unchanged. Two-port resonators, containing two coupling loops or monopole probes mounted on both sides of the center strip, allow for transmission factor measurements, which usually provide more accurate measurements than one-port resonators.

According to the perturbation theory, the dielectric permittivity can be derived only when the magnetic field intensity is zero, and similarly, the magnetic permeability can be derived only when the electric field intensity is zero. Therefore, complex dielectric permittivity measurements are performed by placing the specimen at an axial magnetic field node of the resonator. For the case of fundamental and odd-harmonic resonances, the magnetic field node occurs at the axial midpoint of the structure, as shown in Figure 7.13. For even-harmonic resonances, the test specimen needs to be axially moved to a new magnetic field node location. Similarly, measurements of the complex magnetic permeability are conducted by placing the specimen under test at an axial electric field node, located at the end plates of the resonator.

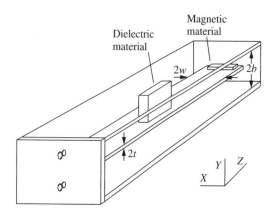

Figure 7.13 Stripline cavity showing the required sample locations for the measurement of complex permittivity (axial midpoint) and complex permeability (adjacent to end plate) (Weil *et al.* 2000). Source: Weil, C. M. Jones, C. A. Kantor, Y. and Grosvenor, Jr. J. H. (2000). "On RF material characterization in the stripline cavity", *IEEE Transactions on Microwave Theory and Techniques*, **48** (2), 266–275. © 2003 IEEE

Permittivity measurement

As shown in Figure 7.13, for dielectric-property measurements, the rectangular sample under test needs to have a width along the y-axis that covers the full distance $(b - t)$ between center strip and ground plane in order to minimize dielectric depolarization errors. The slab is located in the region of maximum and uniaxial electric field, specifically at the midpoint of the center strip ($x = 0$), and with the smallest dimension oriented along the x-axis in order to minimize electric field nonuniformity across the specimen. In the following discussions, the main dimensions of the resonators are resonator length l_0, ground plane separation $2b$, strip width $2w$, and strip thickness $2t$. The main dimensions of the dielectric sample are x-direction $2y$, y-direction $(b - t)$ and z-direction $2l_1$.

The dielectric properties of the sample can be calculated from the resonant properties of the stripline resonator with and without the sample, and the dimensions of the stripline resonator and sample (Weil *et al.* 2000):

$$\varepsilon_r' = 1 + \frac{f_0 - f_L}{f_0} \cdot \frac{bl_0}{2Ayl_1} \quad (7.9)$$

$$\varepsilon_r'' = \left(\frac{1}{Q_L} - \frac{1}{Q_0} \right) \cdot \frac{bl_0}{4Ayl_1} \quad (7.10)$$

where f_0 and Q_0 are the resonant frequency and quality factor of the empty cavity respectively, and f_L and Q_L are the resonant frequency and quality factor of the cavity loaded with the sample respectively. The geometrical factor A can be obtained by calibration, and a more detailed discussion on A can be found in (Weil *et al.* 2000; Jones 1999, Jones *et al.* 1998).

Permeability measurement

The way in which permeability measurements for ferrite specimens are usually conducted is illustrated in more detail in Figure 7.14. The need to ensure the magnetic field uniformity across the specimen and the critical need to minimize the magnetic depolarization error dictates the use of long and thin rectangular specimen shapes. The single specimen is located in a region of maximum and uniaxial magnetic fields; that is, on the center strip and flush against the cavity end plate. The sample under test has the following dimensions: $2y$ in the x-direction, s in the y-direction, and l_1 in the z-direction.

According to resonant-perturbation theory, the magnetic permeability of the sample can be calculated by (Weil *et al.* 2000)

$$\mu_r' = 1 + \frac{f_0 - f_L}{f_0} \cdot \frac{b(b - t)l_0}{Bysl_1} \quad (7.11)$$

$$\mu_r'' = \left(\frac{1}{Q_L} - \frac{1}{Q} \right) \cdot \frac{b(b - t)l_0}{Bysl_1} \quad (7.12)$$

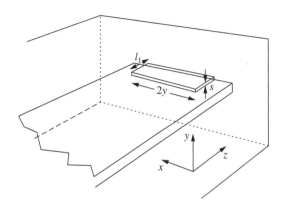

Figure 7.14 Placement of magnetic specimen for the measurement of the x-component of complex permeability (Weil *et al.* 2000). Source: Weil, C. M. Jones, C. A. Kantor, Y. and Grosvenor, Jr. J. H. (2000). "On RF material characterization in the stripline cavity", *IEEE Transactions on Microwave Theory and Techniques*, **48** (2), 266–275. © 2003 IEEE

where f_0 and Q_0 are the resonant frequency and quality factor of the empty cavity respectively, and f_L and Q_L are the resonant frequency and quality factor of the cavity loaded with the sample respectively. The geometrical factor B can be obtained by calibration, and a more detailed discussion on B can be found in (Weil *et al.* 2000; Jones 1999; Jones *et al.* 1998).

It should be noted that the permeability value measured by the orientation shown in Figure 7.14 is the component in x-direction. If the sample is magnetically anisotropic, other orientations of the sample are needed to measure the other two components of the permeability tensor (Jones *et al.* 1998).

7.3 MICROSTRIP METHODS

Among various types of planar circuits, microstrip is the most widely used in microwave electronics. Microstrip circuits are also widely used in the characterization of the electromagnetic properties of materials, especially substrate materials and thin films. The methods developed on the basis of microstrip circuits also fall into nonresonant methods and resonant methods.

7.3.1 Nonresonant methods

Nonresonant microstrip circuit methods mainly include transmission-line method and transmission/reflection method.

7.3.1.1 Transmission-line method

As shown in Figure 7.15, in a transmission-line method, the sample under test is used as a substrate for the development of a segment of microstrip transmission line, and the electromagnetic properties of the substrate sample are obtained from the transmission and reflection properties of the microstrip circuit (Hinojosa *et al.* 2001; Hinojosa 2001). An obvious advantage of this method is that there is no air gap between the strip and the substrate, and the materials properties obtained using this method can be directly used in the development of microstrip circuits. However, this method requires the complicated procedure of sample fabrication.

This method requires the propagation to be the quasi-TEM dominant mode. The calculation of materials properties is based on the *S*-parameter measurements at the microstrip access planes (P_1 and P_2 in Figure 7.15). The fixture for the measurement of microstrip cells is shown in Figure 7.16. It is necessary that the test fixture can be used for different sizes of microstrip. Usually, it employs a line-reflect-match (LRM) and/or line-reflect-line (LRL) calibration, and the two

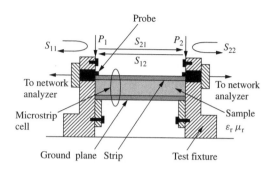

Figure 7.16   The measurement fixture with a microstrip cell (Hinojosa *et al.* 2001). Modified from Hinojosa, J. Faucon, L. Queffelec, P. and Huret, F. (2001). "S-parameter broadband measurements of microstrip lines and extraction of the substrate intrinsic properties", *Microwave and Optical Technology Letters*, **30** (1), 65–69, by permission of John Wiley & Sons

measurement reference planes (P_1 and P_2) are at the two probe outputs.

For a microstrip line on a substrate exhibiting both dielectric and magnetic properties in quasi-TEM mode, there are simple formulas for the characteristic impedance Z_c, propagation constant γ, and total effective loss tangent $\tan \delta_{\text{eff}}$:

$$Z_c = Z_0 \sqrt{\frac{\mu_{\text{reff}}}{\varepsilon_{\text{reff}}}} \tag{7.13}$$

$$\gamma = \omega \sqrt{\varepsilon_0 \mu_0} \cdot \sqrt{\varepsilon_{\text{reff}} \mu_{\text{reff}}} \tag{7.14}$$

$$\tan \delta_{\text{eff}} = \frac{1 - (\varepsilon'_{\text{reff}})^{-1}}{1 - (\varepsilon'_r)^{-1}} \tan \delta_d + \frac{1 - \mu'_{\text{reff}}}{1 - \mu'_r} \tan \delta_m \tag{7.15}$$

where Z_0 is the characteristic impedance when $\varepsilon_r = \mu_r = 1$, $\varepsilon_{\text{reff}}$ and μ_{reff} are the effective relative permittivity and permeability respectively, and $\tan \delta_e$ and $\tan \delta_m$ are the effective electric and magnetic loss tangents respectively.

The calculation of materials properties from *S*-parameter measurement mainly includes two steps: direct problem and inverse problem. The direct problem computes the *S*-parameters at the access planes of the microstrip cell under test propagating only the quasi-TEM mode, according to the complex substrate properties (ε_r, μ_r), the cell dimensions and the frequency. From the given complex ε_r and μ_r values, a given frequency point, and knowing the microstrip-cell structure, we can calculate the effective permittivity and

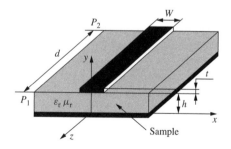

Figure 7.15   Microstrip cell in which the sample under test is the substrate. Reproduced from Hinojosa, J. Faucon, L. Queffelec, P. and Huret, F. (2001a). "S-parameter broadband measurements of microstrip lines and extraction of the substrate intrinsic properties", *Microwave and Optical Technology Letters*, **30** (1), 65–69, by permission of John Wiley & Sons

permeability, and then the complex propagation constant and the complex characteristic impedance of the microstrip cell from analytical equations. After that, the S-parameters can be computed by the reflection/transmission method. The inverse problem is based on an iterative technique derived from the gradient method. It simultaneously carries out the ε_r and μ_r computation and the convergence between measured values of S-parameters and those values computed by the analytical equations (the direct problem) through the successive increment of ε_r and μ_r initial values.

The transmission-line method works well on the condition that the transition effect of coax-to-microstrip is relatively small. This means that the approximate permittivity of the substrate sample should be known before the measurement, so that the characteristic impedance of the test section can be designed in the vicinity of $50\,\Omega$. Besides, this method gives accurate results when the electrical length of lines is long. Lee and Nam proposed a two-microstrip-line method for accurate measurement of a substrate dielectric constant (Lee and Nam 1996). This technique can be used for microstrip lines with arbitrary width. The measurement errors caused by the coax-to-microstrip transitions and the connection uncertainty can be minimized by multiple measurements and by choosing the measurement data set with minimum error cost.

7.3.1.2 Transmission/reflection method

As discussed in Chapter 4, transmission/reflection measurement systems are often built using coaxial air lines or rectangular waveguides. These transmission lines impose strict restrictions on the dimensions of the samples since the test material has to entirely fill the cross section of the cell. This makes sample machining difficult. Moreover, for thin films, this implies the propagation of the electromagnetic wave along the sample thickness, thus lowering the measurement sensitivity. As a consequence, these measurement systems are inadequate for thin-film measurements. On the basis of microstrip circuit, Queffelec *et al.* developed a broadband characterization method, which can be applied to thick samples (Queffelec *et al.* 1994) as well as to thin films (Queffelec *et al.* 1998).

The working principle of this method is similar to that of the stripline methods discussed in Section 7.2.1. As shown in Figure 7.17, a strong interaction between the wave and the sample under test is obtained when the thin film is laid on the microstrip substrate close to the central conductor, and this configuration is sensitive enough for the measurement of thin-film samples.

The determination of complex permittivity and permeability requires the knowledge of two measurable complex parameters. Those are the reflection (S_{11}) and transmission (S_{21}) coefficients measured with a network analyzer in the microstrip cell loaded with the thin film and its support. In

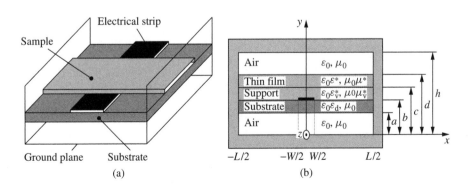

Figure 7.17 Test fixture loaded with a thin-film sample. (a) Three-dimensional view and (b) cross section (Queffelec *et al.* 1998). Source: Queffelec, P. Le Floc'h, M. and Gelin, P. (1998). "Broad-band characterization of magnetic and dielectric thin films using a microstrip line", *IEEE Transactions on Instrumentation and Measurement*, **47** (4), 956–963. © 2003 IEEE

this configuration, the thin film occupies a part of the cross section of the microstrip line, and a full-wave analysis is needed to accurately determine the material properties. The calculation of permittivity and permeability from the measured *S*-parameters needs the use of a numerical optimization procedure (inverse problem) combined with the full-wave analysis of the cell (direct problem).

Direct problem

In the direct problem, the electromagnetic analysis of the cell is made. On the basis of the mode-matching method, the parameters of the transmission line as functions of the permittivity and permeability of the sample under test can be determined. The first step is the determination of the dispersion characteristics of the microstrip line. As shown in Figure 7.17(b), the cross section of the cell is composed of a shielded microstrip structure, which constitutes the sample holder, a thin film, and its support. Full-wave analysis is required for the modal study of the microstrip line.

The second step in the direct problem is the calculation of the parameters imposing continuity conditions on electromagnetic fields at the edge of the cell discontinuities, as shown in Figure 7.18. The use of the orthogonality of modes enables us to determine the coupling coefficients ρ_n, t_n, R_n, and T_n between the modes from the continuity conditions. The parameters of the microstrip cell are given by (Queffelec *et al.* 1998)

$$S_{11} = S_{22} = \rho_1 \exp(-2j\gamma_0 l_0) \quad (7.16)$$

$$S_{21} = S_{12} = t_1 \exp(-2j\gamma_0 l_0) \quad (7.17)$$

where ρ_1 and t_1 represent the reflection and transmission coefficients of the fundamental mode through the cell discontinuities, l_0 is the length of the empty cell, and γ_0 is the propagation constant of the dominant mode. The parameters of the measurement fixture depend on permittivity and permeability of the sample under test through the coefficients ρ_1 and t_1.

Inverse problem

In the inverse problem, the permittivity and permeability of the sample are determined by

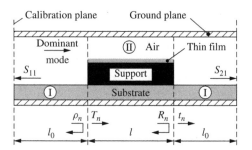

Figure 7.18 Measurement cell discontinuities. Region I: empty regions. Region II: region partly filled with the sample under test. ρ_n, t_n, R_n, and T_n are the coupling coefficients between the modes (Queffelec *et al.* 1998). Source: Queffelec, P. Le Floc'h, M. and Gelin, P. (1998). "Broad-band characterization of magnetic and dielectric thin films using a microstrip line", *IEEE Transactions on Instrumentation and Measurement*, **47** (4), 956–963. © 2003 IEEE

matching the calculated and measured values of the *S*-parameters using a numerical optimization procedure. The optimization problem involves minimizing an objective function, which can be expressed as a sum of squared functions as follows:

$$E(X) = \sum_{i=1}^{2} \sum_{j=1}^{2} \left| \frac{S_{ij}(X)|_{\text{theoretical}}}{S_{ij}|_{\text{measured}}} - 1 \right|^2 \quad (7.18)$$

where $X = (\varepsilon', \varepsilon'', \mu' \, \mu'')$. Suitable optimization technique is required for a fast location of the global minimum of $E(X)$ by avoiding local minima.

7.3.2 Resonant methods

Similar to the resonant methods using other transmission lines, microstrip resonant methods include resonator methods and resonant-perturbation methods. In a resonator method, the material under test is used as the substrate or a layer of the multilayer substrate, and the properties of the sample are calculated from the resonant frequency and quality factor of the resonator. In a resonant-perturbation method, the sample under test is brought close to a microstrip resonator, and the properties of the sample are derived from the change of resonant frequency and quality factor of the resonator due to the presence of the sample. Actually, most of the stripline resonant methods have their microstrip counterparts, and vice versa. In the following text, we discuss the several microstrip resonant methods often used.

7.3.2.1 Ring-resonator method

As discussed in Chapter 2, in a ring resonator, there is no open end, so the problems due to the end effects in straight-ribbon microstrip resonators are avoided. Usually, a ring resonator has higher quality factor than a straight-ribbon microstrip resonator. Ring resonator is often used in the characterization of dielectric sheet materials. There are two types of ring-resonator methods. In one method, the sample under test is used as the substrate for the fabrication of a ring resonator. In another method, a microstrip ring resonator is used as a measurement fixture, and the sample under test covers the microstrip ring circuit, so the sample can be taken as a layer of the multilayer substrate of the ring resonator.

Single-layer substrate ring resonator

As shown in Figure 7.19, the substrate of the microstrip ring resonator is the dielectric sample under test. The relationship between the resonant frequency f_n and the effective permittivity $\varepsilon_{\mathrm{reff}}(f_n)$ is given by

$$\pi D = \frac{nc}{f_n \sqrt{\varepsilon_{\mathrm{reff}}(f_n)}} \quad (n = 1, 2, 3, \ldots) \quad (7.19)$$

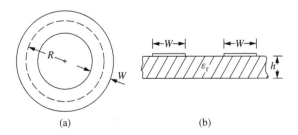

(a)   (b)

Figure 7.19 Microstrip ring resonator whose substrate is the sample under test. (a) Top view and (b) cross section

where D is the mean diameter of the ring ($D = 2R$), and n is the mode number.

The resonant frequencies f_n at different orders of harmonics can be measured directly, and the effective permittivity of the microstrip structure at that frequency can be calculated using Eq. (7.19). The dielectric constant of the sample can be calculated from the effective permittivity of the microstrip structure and the width of the microstrip and the thickness of the sample. In the calculation, we assume that the resonant frequency of the fundamental mode is known. Besides, the loss tangent of the sample can also be calculated from the quality factor of the resonator, provided that the surface resistance of the microstrip conductor is known.

Multilayer substrate ring resonator

Figure 7.20 shows a general configuration of a ring resonator for the measurement of sheet samples. The measurement fixture is the microstrip ring resonator fabricated on the substrate with thickness H, and the sample under test tightly covers the fixture. The polytetrafluoro-ethylene (PTFE) block situated above the sample is used to get rid of the air gap between the microstrip circuit and the sample by firm pressing.

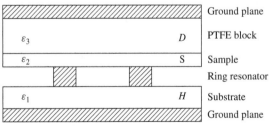

Figure 7.20 Cross section of a ring microstrip resonator. H denotes substrate thickness, S, test material thickness, and D, PTFE thickness. Not to scale. ε_1, ε_2, and ε_3 are the dielectric constants of the substrate, sample, and PTFE block respectively. Source: Bernard, P. A. and Gautray, J. M. (1991). "Measurement of dielectric constant using a microstrip ring resonator", *IEEE Transactions on Microwave Theory and Techniques*, **39** (3), 592–595. © 2003 IEEE

For a given measurement fixture, ε_1 is known. When no sample is loaded to fixture, the structure can be represented by $\varepsilon_2 = \varepsilon_3 = 1$. When the fixture is loaded with a sample, ε_3 is known, and ε_2 is the parameter to be measured. The resonant frequencies (F_1 and F_0) of the ring resonator with and without the sample under test can be measured. The effective permittivity $\varepsilon_{reff,0}$ for unloaded case can be calculated from the structure and substrate properties of the fixture, and the effective permittivity $\varepsilon_{reff,1}$ for the loaded case can be calculated based on the fact that the diameter of the ring resonator does not change due to sample loading (Bernard and Gautray 1991):

$$\pi D = \frac{nc}{F_0 \sqrt{\varepsilon_{reff,0}}} = \frac{nc}{F_1 \sqrt{\varepsilon_{reff,1}}} \qquad (7.20)$$

Equation (7.20) can be rewritten as:

$$\varepsilon_{reff,1} = \varepsilon_{reff,0} \left(\frac{F_0}{F_1} \right)^2 \qquad (7.21)$$

It should be noted that the resonances corresponding to F_0 and F_1 should have the same mode number (n).

Once the loaded effective permittivity ($\varepsilon_{reff,1}$) is obtained, the dielectric constant of the sample can be derived from $\varepsilon_{reff,1}$. In the calculation of the dielectric constant of the sample from $\varepsilon_{reff,1}$, numerical calculation is needed, and full-wave simulation is often used. The loss tangent of the sample can also be obtained if the surface resistance of the conductors, loss tangent values of the substrate, and the PTFE block are known.

7.3.2.2 Straight-ribbon resonator method

As shown in Figure 7.4, in a microstrip straight-ribbon resonator method, the sample under test is used as a substrate for the microstrip circuit, and the dielectric properties of the sample are determined from the resonant properties and structural parameters of the microstrip straight ribbon resonator. If the length of the ribbon is l, the relationship between the resonant frequency and the effective dielectric constant ε_{eff} of the microstrip can be calculated from the resonant frequency f:

$$\varepsilon_{eff} = \left(\frac{nc}{2lf} \right)^2 \qquad (7.22)$$

where c is the speed of light, and n is the order of resonance. The effective dielectric constant is a composite of the dielectric constants above and below the signal line. From ε_{eff}, the dielectric constant of the substrate can be calculated.

The gaps and the coupling to the rest of the lines add a loading that can be represented by an incremental change in the length of the resonant section. So the effective length of the resonator is a bit longer than its physical length, and Eq. (7.22) needs to be modified as

$$\varepsilon_{eff} = \left(\frac{nc}{2(l + l_c)f} \right)^2 \qquad (7.23)$$

where l_c is the contribution of the capacitance of the two open ends. It is clear that to obtain accurate value of dielectric constant of the substrate, the effect of l_c should be taken into consideration.

A method has been developed that allows dielectric constants to be determined from resonant frequency measurements made on two or more series resonant structures (Rudy *et al.* 1998). The resonant frequency of a segment of microstrip occurs at frequencies where the length of the segment is approximately equal to an integral number of half wavelengths. Since the incremental length is a property of the gap and coupling, it is relatively independent of the length of the resonant segment. The effect of the gap loading can be eliminated by measuring the resonant frequencies of two resonant segments with different lengths. Consider two microstrip resonators with different lengths, each coupled to identical external circuitry through identical gaps, as shown in Figure 7.21. Their resonance conditions are given by

$$l_1 + l_c = \frac{n_1}{2} \cdot \frac{c}{f_1 \sqrt{\varepsilon_{eff}}} \qquad (7.24)$$

$$l_2 + l_c = \frac{n_2}{2} \cdot \frac{c}{f_2 \sqrt{\varepsilon_{eff}}} \qquad (7.25)$$

where l_1 and l_2 are the physical lengths of the two ribbons respectively, l_c is the length correction, n_1 and n_2 are the orders of the resonances of the two resonators respectively, f_1 and f_2 are the resonant frequencies of the two resonators respectively, and c is the speed of light. Eliminating the item of l_c

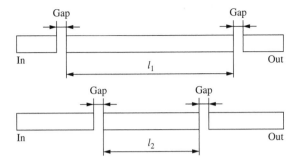

Figure 7.21 Straight-ribbon resonators with different lengths

in Eqs. (7.24) and (7.25) gives

$$\varepsilon_{\text{eff}} = \left[\frac{c(n_1 f_2 - n_2 f_1)}{2 f_1 f_2 (l_2 - l_1)} \right]^2 \qquad (7.26)$$

As l_c has a small dependence on frequency, it is desirable that resonant frequencies f_1 and f_2 are close together. It is common to make the lengths l_1 and l_2 integral multiples of each other and to use higher-order resonances to calculate the effective dielectric constant.

In a microstrip circuit, the dielectric media above the circuit usually is air and the dielectric below the circuit is the substrate material. With air as the dielectric above the circuit, the dielectric constant ε_r of the substrate and the effective dielectric constant ε_{eff} of the microstrip are related by a filling factor that weighs the amount of the field in air and the amount of field in the dielectric substrate. Analytical modeling of microstrip configurations is difficult and the results are generally complex, so semiempirical relations are typically used. For simple calculations, Eq. (7.3) can be used. However, for more precise determinations, a more complex set of expressions is needed. The substrate dielectric constant from the measured effective dielectric constant of a microstrip configuration can be determined by (Hammerstad and Jensen 1980)

$$\varepsilon_r = \frac{2\varepsilon_{\text{eff}} + f(u) - 1 - C}{f(u) + 1 - C} \qquad (7.27)$$

with

$$u = \frac{w}{h} \qquad (7.28)$$

$$a(u) = 1 + \frac{1}{49} \ln \left(\frac{u^4 + (u/52)^2}{u^4 + 0.432} \right)$$

$$+ \frac{1}{18.7} \ln \left(1 + \left(\frac{u}{18.1} \right)^3 \right) \qquad (7.29)$$

$$b(\varepsilon_r) = 0.564 \left(\frac{\varepsilon_r - 0.9}{\varepsilon_r + 3} \right) \qquad (7.30)$$

$$f(u) = \left(1 + \frac{10}{u} \right)^{a(u)b(\varepsilon_r)} \qquad (7.31)$$

$$C = \frac{t}{2.3 h \sqrt{u}} \qquad (7.32)$$

where h is the thickness of the dielectric substrate, w and t are the width and thickness of the metal strip. C is a small correction for the finite thickness of the circuit trace and frequently is neglected in the calculation of the dielectric constant, but it is important in the calculation of losses. The uncertainty of Eq. (7.27) is within a fraction of 1 %.

The loss tangent of the sample can be obtained from quality factor of the resonator, structural parameters, and surface resistance of the conducting strips. Detailed discussion on the calculation of loss tangent can be found in (Rudy *et al.* 1998).

The fringing fields of a $\lambda/2$ resonator can also be used in the characterization of dielectric materials (Nyfors and Vainikainen 1989). Figure 7.22(a) shows a microstrip resonator sensor for industrial applications. At the two open ends of the resonator, there are strong fringing fields. Meanwhile, as shown in Figure 7.22(b), in a microstrip circuit, the electric field is concentrated between the strip and ground plane, but a weak fringing field exists beyond the dielectric substrate. If the microstrip resonator is held against a dielectric slab, the fringing fields will penetrate the surface of the slab, thus changing the effective permittivity of the microstrip and hence the resonant frequency of the resonator. The dielectric constant of the slab can be obtained from the change of the resonant frequency, but it is difficult to measure the loss tangent of low-loss materials using this technique. In practice, the sensor is often coated with a protective dielectric layer to prevent wear and corrosion. This will decrease the sensitivity of the

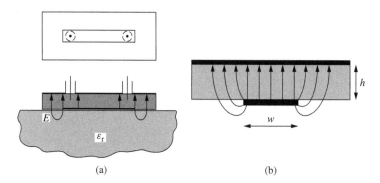

(a)   (b)

Figure 7.22   Microstrip resonator sensor. (a) Measurement configuration showing the fringing fields near the open ends and (b) electric field distribution at a cross section (Nyfors and Vainikainen 1989). Reprinted with permission from *Industrial Microwave Sensors*, by Nyfors, E. and Vainikainen, P. Artech House Inc., Norwood, MA, USA. www.artechhouse.com

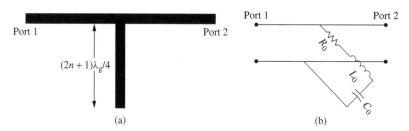

(a)   (b)

Figure 7.23   A T-resonator. (a) Structure and (b) equivalent circuit. Modified from Carroll, J. Li, M. and Chang, K. (1995). "New technique to measure transmission line attenuation", *IEEE Transactions on Microwave Theory and Techniques*, **43** (1), 219–222. © 2003 IEEE

sensor, but permits the measurement of substances with higher losses.

7.3.2.3 T-resonator method

The $\lambda/2$ resonator method has some inherent problems. First, the size of the resonator and required feed lines take up a large amount of substrate area. Secondly, the coupling gaps as shown in Figure 7.4 cause measurement inaccuracies at high frequencies due to radiation losses. The two gaps are actually four open radiating microstrip ends, whose large radiation losses at high frequencies are undesirably included in the measured transmission-line attenuation coefficient. To save space, reduce radiation, and improve transmission-line characterization, T-resonator structures are often used (Carroll *et al.* 1995). The T-resonator method has proven to be effective in the characterization of

many ceramic and organic materials (Amey and Horowitz 1997). The resonant frequency is used to determine the dielectric constant of the sample, and the measured total quality factor (Q) is used to determine the attenuation and the loss tangent of the material.

As shown in Figure 7.23, a T-resonator is actually a quarter-wave stub resonator. The T-pattern is an open-end transmission-line stub that resonates approximately at odd-integer multiples of its quarter-wavelength frequency. The T-resonator reduces the amount of space needed for the test structure by at least 50 % over the $\lambda/2$ resonator. This structure is directly coupled to the measurement transmission line so that the inaccuracies caused by the coupling gaps of a $\lambda/2$ resonator are avoided. There is only one open end in a T-resonator, which eliminates most of the radiation and discontinuity losses found in the $\lambda/2$ resonator.

The effective dielectric constant of the transmission line can be determined through the resonant frequency and the known length of the quarter-wave stub. The effective dielectric constant at the frequency of each of the stub's quarter-wave resonances is given by (Carroll *et al.* 1995)

$$\varepsilon_{\text{reff}} = \left(\frac{nc}{4f_n(l + l_c)}\right)^2 \quad (n = 1, 3, 5, \ldots)$$
(7.33)

where c is the speed of light in free space. Equation (7.33) indicates that the fringing field at the open end effectively lengthens the physical length of the resonator (l) by an extra length (l_c), which can be determined by empirical equations (Kirschning *et al.* 1981). From the effective dielectric constant of the microstrip, the dielectric constant of the sample can be determined.

The effect of the fringing field at the open end of the T-resonator can also be eliminated using the method discussed in straight-ribbon resonators. As shown in Figure 7.24, we can fabricate two T-resonators with different stub lengths, and Eq. (7.26) can be modified for T-resonators.

The loss tangent of the sample can be derived from the quality factor of the resonator. The measured quality factor Q includes the effects of conductor, dielectric, and radiation losses and can be expressed as

$$\frac{1}{Q} = \frac{1}{Q_c} + \frac{1}{Q_d} + \frac{1}{Q_r}$$
(7.34)

where Q_c is the quality factor due to conductor losses, Q_d is the quality factor due to dielectric losses, and Q_r is the quality factor due to radiation losses. The radiation losses may be neglected since they are very small. Q_d can then be expressed as

$$Q_d = \frac{Q Q_c}{Q_c - Q}$$
(7.35)

Q represents the measured value of quality Factor while Q_c is calculated from the measured physical properties (line width, line thickness, roughness, etc.) as well as the impedance, conductivity, and calculated dielectric constant. Loss tangent can then be derived from Q_d, the material dielectric constant (ε_r), as well as the effective material dielectric constant ($\varepsilon_{\text{reff}}$) using the following relationship (Amey and Horowitz 1997):

$$\tan \delta = \frac{\varepsilon_{\text{reff}}(\varepsilon_r - 1)}{Q_d \varepsilon_r(\varepsilon_{\text{reff}} - 1)}$$
(7.36)

7.3.2.4 Cross-resonator method

Figure 7.25(a) shows the circuit layout of a microstrip cross-resonator for the characterization of dielectric materials. It consists of the power feeding line L_3, a three-quarter-wavelength long resonator L_1 and two inductive stubs L_2. The aperture in this resonant probe is the open end of the line L_1. The equivalent circuit of the structure is shown in Figure 7.25(b). The two stubs form a short to the three-quarter-wavelength long resonator L_1, labeled *stub short*. At the position about $\lambda_g/4$ from the open end, there is an equivalent short. Between the stub short and the equivalent short is a $\lambda_g/2$ resonator. The position of the equivalent short is related to the loading conditions at the open end. As shown in Figure 7.25(b), when a dielectric sample is brought close to the open end of L_1, the loading, including the capacitance C and resistance R, of the open end will be changed. The capacitance C is related to the dielectric constant of the sample, and the resistance R is related to the loss tangent of the sample. The change of capacitance C causes the shift of the position of the equivalent short, and thus changes the resonant frequency. The change of resistance changes the overall equivalent resistance of the resonator, and thus changes the quality factor of the resonator. According to resonant-perturbation

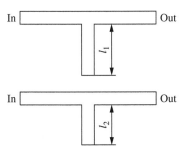

Figure 7.24 Microstrip T-resonators with different stub lengths

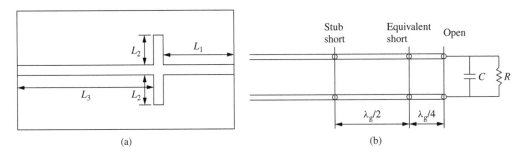

Figure 7.25   A cross microstrip resonator probe. (a) Microstrip circuit and (b) equivalent circuit. Source: Wang, M. S. Bothra, S. Borrego, J. M. and Kristal, K. W. (1990). "High spatial resolution dielectric constant uniformity measurements using microstrip resonant probes", *IEEE MTT-S Digest*, 1121–1124. © 2003 IEEE

theory, the dielectric properties of the sample can be derived from the change of the resonant properties of the resonator due to the presence of the sample.

The operation of the resonant probe is based on the fact that when a piece of dielectric material is brought in close proximity to the open aperture of the microwave resonator, the resonant frequency of the resonator is perturbed by the presence of the material. The shift in the resonant frequency depends upon the resonator, the aperture of the resonator, and the dielectric properties of the material, as given by (Wang *et al.* 1990)

$$\frac{\Delta f}{f_0} = \frac{V_a(\varepsilon_a - 1)}{V_c \varepsilon_c} \tag{7.37}$$

where f_0 is the resonant frequency before perturbation, Δf is the change of resonant frequency due to the perturbation, V_c represents the volume of the cavity and V_a represents the volume occupied by the evanescent fields at the aperture, ε_a and ε_c are the dielectric constant of the material in the aperture and in the resonator respectively. If the probe is properly calibrated, the dielectric constant of a dielectric material or its spatial variation can be obtained by measuring the change in resonant frequency of the cavity as the aperture scans a dielectric material. In experiments, the microstrip probe is often mounted inside a shielding box to eliminate the possible disturbance from the outside during measurements. The dielectric sample under test is placed on an *X-Y-Z* stage, and the sample is in soft contact with the aperture of the resonator. As the width of microstrip is much smaller than

the microwave wavelength, high spatial resolution can be obtained.

7.3.2.5 Two-section microstrip method

Consider a microstrip structure consisting of two identical, parallel metal strips placed closely together, as shown in Figure 7.26. On each metal strip, one end is open, and the other end, called *neighboring end*, is connected to a transmission line, usually coaxial line. As shown in Figure 7.27, there is a damping pole in the frequency response of this two-section microstrip structure. As the position and amplitude of the pole are related to the dielectric constant and loss tangent of the substrate material, the dielectric properties of the substrate can be obtained from the frequency response of the structure. Compared to a conventional microstrip resonator method, this technique has higher sensitivity, and the measurement fixture is much smaller (Belyaev *et al.* 1997; 1995).

The two sections of the microstrip structure are coupled both electrically and magnetically. The transmission coefficient can be calculated by the formula

$$T = \frac{(P_e - P_o)^2}{(1 + P_e^2)(1 + P_o^2)} \tag{7.38}$$

with

$$P_e = (Z/Z_e) \tan(k_e l) \tag{7.39}$$

$$P_o = (Z/Z_o) \tan(k_o l) \tag{7.40}$$

$$k_e = (\omega/c)\sqrt{\varepsilon_e} \tag{7.41}$$

$$k_o = (\omega/c)\sqrt{\varepsilon_o} \tag{7.42}$$

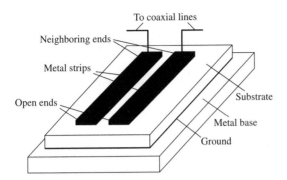

Figure 7.26 Two-section microstrip structure. Modified from Belyaev, B. A. Leksikov, A. A. and Tyurnev, V. V. (1995). "Microstrip technique for measuring microwave dielectric constant of solids", *Instruments and Experimental Techniques*, **38** (5), 646–650, by permission of Plenum Publishing Corporation

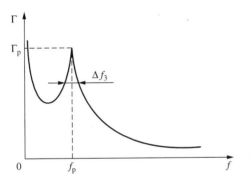

Figure 7.27 Typical frequency dependence of the reflection coefficient of a two-section microstrip structure shown in Figure 7.26

where ω is the angular frequency, c is the speed of light, l is the length of the microstrip section, Z is the impedance of the transmission line, Z_e, Z_o and ε_e, ε_o are the impedance of the microstrip lines and the effective dielectric constants of the substrate for even and odd modes respectively.

As shown in Figure 7.27, in the frequency response of the structure, there are passbands with high transmission and low reflection, and damping poles with deep minimum transmission and high reflection. The passbands are observed around frequencies that are multiples of the half-wave resonance in the microstrip sections, and the damping poles at frequencies where the inductive and capacitive couplings compensate for each other. As a rule, at a substrate dielectric constant $\varepsilon_r \geq 2$, the frequency response of the structure has only one damping pole, and its frequency is always lower than that of the half-wave resonance.

We define a relative sensitivity η of a sensor:

$$\eta = \frac{\Delta f_p}{\Delta f_3} \tag{7.43}$$

where Δf_p is the change of the damping frequency due to the change of one parameter of the two-section structure, and Δf_3 is measured at a level of 3 dB of loss at the damping pole frequency, as shown in Figure 7.27. Generally speaking, the relative measurement sensitivity increases with the increase of separation (S) between the two sections, but decreases with the increase of the strip width (w) and the dielectric constant of the substrate. In the determination of the dimensions of the microstrip structure, many practical aspects should be taken into consideration (Belyaev *et al.* 1997, 1995).

Two methods have been developed on the basis of the above principle. In one method, the sample under test is used as the substrate of the microstrip structure, and we call it *substrate method*. In the other method, the sample under test is loaded to the open ends of the structure, causing a perturbation on the frequency response of the structure, and we call it *perturbation method*.

Substrate method

In the substrate method, though the dimensions of a microstrip structure may be arbitrary, it is advisable to follow certain guidelines in order to obtain the maximum sensitivity. First, the gap S between the strips should be as large as possible, but no more than one-third of the strip length l. Otherwise, other damping poles may appear around the first one, which makes measurements difficult. Second, to get a higher quality factor for the microstrip, for most cases the substrate thickness h should be larger than 1 mm. Besides, the strip width w should not be less than the substrate thickness, otherwise the quality factor of the microstrip would be low and that would result in a small pole amplitude, and hence low measurement sensitivity.

For a given configuration, the dielectric constant ε_r' of the sample can be obtained from the damping frequency f_p and the structural parameters according to the following equation (Belyaev *et al.* 1995):

$$f_p = \frac{0.4c\sqrt{\varepsilon_r'}(h/S + 5/\varepsilon_r')}{2l \cdot (\varepsilon_r' + 4)} \qquad (7.44)$$

The uncertainty for the above equation is within 10 %.

Quasi-static analysis indicates that the amplitude L_p of the damping pole of the microstrip structure is given by

$$L_p = L_0 + 20\log_{10} Q_0 \qquad (7.45)$$

where L_0 is a constant that can be calculated for a specific microstrip structure with known frequency response. Equation (7.45) indicates that the inherent Q_0 of the microstrip in the test structure can be derived from the damping pole amplitude.

Similar to the traditional resonant techniques, the loss tangent of the dielectric substrate can be calculated from the inherent resonator Q_0 and the quality factor Q_m of the fixture:

$$\tan\delta = \frac{Q_m - Q_0}{Q_m Q_0} \qquad (7.46)$$

Q_m can be derived from the pole amplitude of a reference sample with known loss tangent. It should be noted that the amplitude of the damping pole depends both on the loss tangent of the dielectric and on the quality of substrate surface and metal coating. Therefore, measurements of the absolute value of $\tan\delta$ of low-loss materials are difficult. However, this technique can sensitively detect the changes in the dielectric loss, for example, due to temperature variations, even in low-loss materials with $\tan\delta < 10^{-3}$.

Perturbation method

The two-section microstrip method can be used to measure the dielectric constant ε_r and the loss tangent for samples of arbitrary shape and small size, using perturbation method. The sample under study is placed on the sensor either between metallic strips or directly on them, close to their

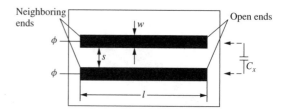

Figure 7.28 Two-section microstrip structure for perturbation method. Modified from Belyaev, B. A. Leksikov, A. A. Tyurnev, V. V. and Shikhov, Y. G. (1997). "A microstrip sensor for measuring microwave dielectric constants of solids", *Instruments and Experimental Techniques*, **40** (3), 395–398, by permission of МАИК Hayka/Interperiodica Publishiing

open ends at the antinode of the microwave electric field. As shown in Figure 7.28, placing a sample on the open ends of the microstrip segments is equivalent to introducing an additional capacitive coupling C_x between the strips.

In this case, the damping pole frequency f_p can be expressed as a function of the linear characteristics of the microstrip lines forming the two-section structure: the inductance L_1, the mutual inductance L_{12}, the capacitance C_1, and the mutual capacitance C_{12}. By assuming that $L_1 \gg L_{12}$ and $C_1 \gg C_{12}$, the damping pole frequency of the structure is given by (Belyaev *et al.* 1997)

$$f_p = f_0 \cdot \sqrt{\frac{3C_{12}/C_1}{L_{12}/L_1 - 2C_{12}/C_1}} \qquad (7.47)$$

where f_0 is the half-wave resonance frequency of the microstrip structure:

$$f_0 = \left(4\pi l \cdot \sqrt{L_1(C_1 + C_{12})}\right)^{-1}. \qquad (7.48)$$

Equation (7.47) indicates that, as the linear mutual strip capacitance C_{12} increases when the fixture is loaded with a sample, the damping pole frequency f_p increases as well. Furthermore, the smaller the value of $(L_{12}/L_1) - (2C_{12}/C_1)$, the sharper the function $f_p(C_{12})$.

It should be noted that the equivalent lumped-element parameters cannot be easily expressed using closed-form equations. In the design of the measurement fixture, and the derivation of materials properties, numerical calculations are often needed.

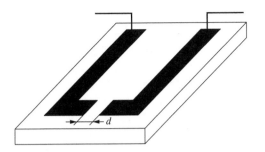

Figure 7.29 Optimized microstrip fixture. Modified from Belyaev, B. A. Leksikov, A. A. Tyurnev, V. V. and Shikhov, Y. G. (1997). "A microstrip sensor for measuring microwave dielectric constants of solids", *Instruments and Experimental Techniques*, **40** (3), 395–398, by permission of МАИК Hayka/Interperiodica Publishiing

It has been shown that the fixture shown in Figure 7.29 has high sensitivity (Belyaev *et al.* 1997). The optimal strip width in the region where transmission lines are connected to the microstrip line is approximately equal to the substrate thickness, and the gap between the strips in this region is determined by the dielectric constant of the substrate. The length of the opposite segments of the coupled lines on which the sample under study is placed and the gap between the strips in this region are selected so that the sample fully covers the gap.

7.4 COPLANAR-LINE METHODS

Similar to stripline and microstrip lines, coplanar lines have also been used in the development of measurement fixtures for materials property characterization in both nonresonant and resonant methods. Many types of coplanar structures, such as coplanar waveguide, two-strip coplanar line, and three-strip coplanar line, have been used in materials property characterization. In the following discussion, we concentrate on the applications of coplanar waveguides in the characterization of materials properties.

7.4.1 Nonresonant methods

Coplanar transmission lines have several advantages as sensors that stem from their single-sided configuration (Stuchly and Bassey 1998; Hinojosa

et al. 2002a,b). Generally speaking, nonresonant methods using coaxial lines, rectangular waveguides and coaxial lines are not suitable for the characterization of film samples, while the microstrip and coplanar lines allow characterizing film samples by using the samples as the substrates. It is possible to optimize the shape of the microstrip or coplanar cell in order to propagate the dominant quasi-TEM mode and to perform accurate *S*-parameter measurements in a certain frequency range.

Hinojosa *et al.* proposed a broadband electromagnetic characterization method of isotropic sheet samples using coplanar lines as sample cells (Hinojosa *et al.* 2002a, b), as shown in Figure 7.30. The propagation modes in a coplanar transmission line are hybrid because some of the electric and magnetic fields are fringed out into the air space above the conductive strip. In the extraction of the substrate properties, only the quasi-TEM dominant mode is considered. The values of the dielectric and magnetic constants entering into the characteristic impedance and propagation constant are less than the electromagnetic properties $(\varepsilon_\mathrm{r}, \mu_\mathrm{r})$ of the substrate and they are denoted as effective $(\varepsilon_\mathrm{reff}, \mu_\mathrm{reff})$.

As the coplanar quasi-TEM mode dispersion is low, the propagation analysis method based on the analytical relationships derived from conformal mapping techniques can be used. The characteristic impedance Z_c, propagation constant γ, and effective total loss tangent $\tan \delta_\mathrm{eff}$ for a coplanar line on a substrate exhibiting both dielectric and magnetic properties are given by

$$Z_\mathrm{c} = Z_0 \sqrt{\frac{\mu_\mathrm{reff}}{\varepsilon_\mathrm{reff}}} \qquad (7.49)$$

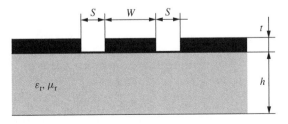

Figure 7.30 Coplanar cell for the characterization of sheet materials

$$\gamma = \omega\sqrt{\varepsilon_0\mu_0} \cdot \sqrt{\varepsilon_{\text{reff}}\mu_{\text{reff}}} \tag{7.50}$$

$$\tan\delta_{\text{eff}} = \frac{1 - (\varepsilon_{\text{reff}})^{-1}}{1 - (\varepsilon_{\text{r}})^{-1}}\tan\delta_{\text{d}} + \frac{1 - \mu_{\text{reff}}}{1 - \mu_{\text{r}}}\tan\delta_{\text{m}} \tag{7.51}$$

where Z_0 is the characteristic impedance when $\varepsilon_{\text{r}} = \mu_{\text{r}} = 1$. The dispersions of the effective electromagnetic properties $(\varepsilon_{\text{reff}}, \mu_{\text{reff}})$ can be expressed by (Hinojosa *et al.* 2002a; Hasnain *et al.* 1986)

$$\varepsilon_{\text{reff}}(f)$$
$$= \left[\sqrt{\varepsilon_{\text{reff,stat}}} + \frac{\sqrt{\varepsilon_{\text{r}}} - \sqrt{\varepsilon_{\text{reff,stat}}}}{1 + a \cdot (f/f_{\text{TE}})^{-b}}\right]^2 \tag{7.52}$$

$$\mu_{\text{reff}}(f)$$
$$= \left[\sqrt{(\mu_{\text{reff,stat}})^{-1}} + \frac{\sqrt{(\mu_{\text{r}})^{-1}} - \sqrt{(\mu_{\text{reff,stat}})^{-1}}}{1 + a \cdot (f/f_{\text{TM}})^{-b}}\right]^{-2} \tag{7.53}$$

with the effective electromagnetic properties at quasi-static limit given by

$$\varepsilon_{\text{reff,stat}} = 1 + (\varepsilon_{\text{r}}' - 1) \cdot \frac{K(k_1)K(k')}{2K(k_1')K(k)} \tag{7.54}$$

$$\mu_{\text{reff,stat}} = \left[1 + \left(\frac{1}{\mu_{\text{r}}'} - 1\right) \cdot \frac{K(k_1)K(k')}{2K(k_1')K(k)}\right]^{-1} \tag{7.55}$$

where the parameters k, k', k_1, and k_1' are determined by coplanar structures, and $K(k), K(k')$,

$K(k_1)$ and $K(K_1')$ are the complete elliptical integrals of first order of modulus k and k_1 and complementary modulus k' and k_1' (Hilberg 1969; Stuchly and Bassey 1998). The cutoff frequencies for the lowest-order TE mode (f_{TE}) and the lowest-order TM mode (f_{TM}), and the parameters a and b, are defined in (Hinojosa 2002a; Hasnain 1986).

Figure 7.31 shows a measurement fixture for this method. The two reference planes (P_1, P_2) are the outputs of the two probes of the measurement fixture. The calculation of the substrate properties $(\varepsilon_{\text{r}}, \mu_{\text{r}})$ is based on the *S*-parameter measurements made at the coplanar access planes. It requires an electromagnetic analysis of the coplanar cell (the direct problem) together with an optimization procedure (the inverse problem). The direct problem is computing the *S*-parameters at the access planes of the coplanar cell under test propagating only the quasi-TEM mode, according to the complex properties of the substrate $(\varepsilon_{\text{r}}, \mu_{\text{r}})$, the cell dimensions and the frequency. For a coplanar cell with known structure, given complex ε_{r} and μ_{r} values, at a given frequency point, we can compute the effective permittivity and permeability, and then the complex propagation constant (γ) and the complex characteristic impedance (Z_{c}) from the analytical relationships defined in Eqs. (7.49) to (7.55). The *S*-parameters can then be computed based on the reflection/transmission theory. In the

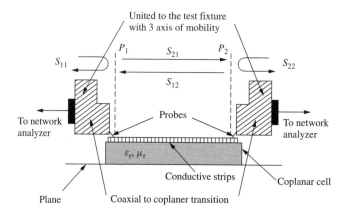

Figure 7.31 An on-coplanar measurement fixture with a coplanar cell (Hinojosa *et al.* 2002a). Reproduced from Hinojosa, J. Lmimouni, K. Lepilliet, S. and Dambrine, G. (2002a). "Very high broadband electromagnetic characterization method of film-shaped materials using coplanar", *Microwave and Optical Technology Letters*, **33** (5), 352–355, by permission of John Wiley & Sons

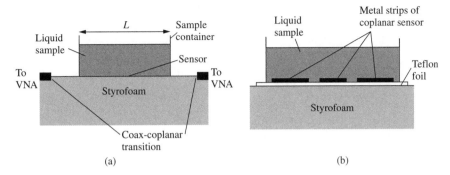

Figure 7.32 The experimental setup for the measurement of liquid samples. (a) Side view and (b) axial view

optimization procedure (the inverse problem), iterative technique derived from the gradient method is often used. It simultaneously carries out the ε_r and μ_r computation and the convergence between measured (S_{11}, S_{21}) values and computed (S_{11}, S_{21}) values by the direct problem through the successive increment of the initial values of ε_r and μ_r.

Figure 7.32 shows a configuration for the measurement of the dielectric properties of liquid samples (Stuchly and Bassey 1998). In this configuration, the coplanar sensor is fabricated on a very thin dielectric sheet with low dielectric constant, such as Teflon, and the contribution of the dielectric sheet to the impedance of the coplanar line can be neglected. The coplanar sensor is placed on a piece of Styrofoam whose dielectric permittivity is close to unity. The liquid sample is put inside a container. From the S-parameters, the dielectric constant of the liquid sample can be derived with a method similar to the one discussed above.

7.4.2 Resonant methods

In a coplanar resonator method, the sample under test is used as the substrate or part of the substrate of a coplanar resonator. From the resonant properties of the resonator, the electromagnetic properties of the sample can be derived. In principle, any types of resonant structure can be used, while in practical measurements, we should consider measurement sensitivity and accuracy.

Peterson and Drayton proposed a coplanar T-resonator method (Peterson and Drayton 2002). The basic principle of this method is similar to the one for microstrip T-resonator discussed earlier. As

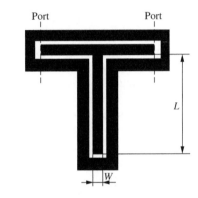

Figure 7.33 Layout of a CPW T-resonator

shown in Figure 7.33, at odd quarter wavelengths, a standing-wave distribution exists along the open-ended T-resonator stub and presents well-defined resonances. Under low-loss conditions, the resonant frequency can be used to extract the effective dielectric constant according to

$$\varepsilon_{\text{eff},n} = \left(\frac{n \cdot c}{4 \cdot L \cdot f_n} \right)^2 \qquad (7.56)$$

where n is the resonance index ($n = 1, 3, 5, \ldots$), c is the speed of light in vacuum and L is the effective physical length of the resonating stub. Most of the conclusions for microstrip T-resonators can be extended to coplanar ones.

7.5 PERMEANCE METERS FOR MAGNETIC THIN FILMS

Measurement of permeability of magnetic thin films at microwave frequencies is becoming more and more important owing to the increasing working frequencies of electronic circuits. The fixture for permeability measurement is often called *permeance meters*. Most of the permeance meters are based on the transmission measurement of a two-port system consisting of a driving coil and a pickup coil. In this section, we first discuss the working principle of permeance meters, and then discuss three examples of permeance meters.

7.5.1 Working principle

As shown in Figure 7.34, a typical permeance meter mainly consists of a driving coil and a pickup coil. The coupling between the driving coil and the pickup coil is related to the magnetic properties of the core material. When a magnetic sample is inserted into the driving coil and the pickup coil, the voltage output at the pickup coil will be changed, and the change of voltage output of the pickup coil is mainly determined by the magnetic permeability of the sample.

If we neglect the contribution of the dielectric permittivity of the sample, the relationship between the magnetic permeability of the sample and the change of output voltage is given by

$$(\mu_r - 1) \cdot \frac{S_s}{S_0} = \frac{V_s}{V_0} - 1 \qquad (7.57)$$

where μ_r is the magnetic permeability of the sample, S_0 and S_s are the areas of the cross sections of the pickup coil and the sample respectively, V_0 and V_s are the voltage outputs of the pickup coil without and with sample respectively. To measure magnetic permeability at microwave frequencies,

it is necessary to develop suitable driving coil and pickup coil working at microwave frequencies. In the following, we discuss three methods for the measurement of magnetic thin films, including two-coil method, single-coil method, and electrical impedance method.

7.5.2 Two-coil method

In the two-coil method, the measurement circuit shown in Figure 7.34 is realized at microwave frequencies using planar circuits. The measurement fixture mainly consists of a driving coil and a pickup coil.

7.5.2.1 Driving coil

Figure 7.35(a) shows the principle for permeability measurement, and it is actually a counterpart at microwave frequencies of Figure 7.34. Figure 7.35(b) shows a measurement fixture developed on the basis of Figure 7.35(a). In this configuration, the driving coil is a pair of parallel driving plates that generate propagating quasi-TEM mode electromagnetic field, and a voltage is induced in the pickup coil because of the change of magnetic flux at the area of the pickup coil.

The measurement frequency range and measurement reliability of a permeance meter are closely related to its driving coil. Two types of driving coils are often used: traveling-wave type (Yamaguchi *et al.* 1997, 1996) and standing-wave type (Yabukami *et al.* 2000, 1999). The driving coil shown in Figure 7.35(b) is terminated by a 50-Ω terminator, and it works in traveling-wave state. Figure 7.36 shows a standing-wave driving coil. The pair of parallel plates is shorted by a short plate. Usually, a standing-wave driving coil results in a higher sensitivity than a traveling-wave one.

In the design of driving coils, we should consider the impedance matching between external circuit (50 Ω) and the measurement fixture. The length of the driving coils should be less than the half wavelength of the highest working frequency, so that there is no LC resonance in the driving coil. Meanwhile, we should also consider effects of the pickup coil on the driving coil.

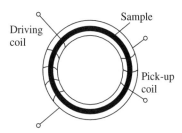

Figure 7.34 Permeability measurement at ac frequencies

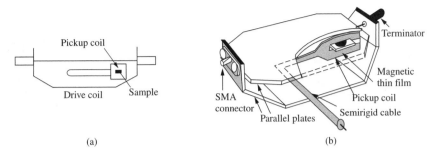

Figure 7.35   Transmission-type permeance meter. (a) Principle drawing and (b) structure (Yamaguchi *et al.* 1997). Modified from Yamaguchi, M. Yabukami, S. and Arai, K. I. (1997). "A new 1 MHz-2 GHz permeance meter for metallic thin films", *IEEE Transactions on Magnetics*, **33** (5), 3619–3621. © 2003 IEEE

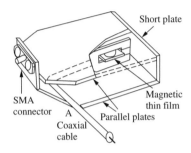

Figure 7.36   Reflection-type permeance meter (Yabukami *et al.* 2000). Source: Yabukami, S. Takezawa, M. Uo, T. Yamaguchi, M. Arai, K. I. Miyazawa, Y. Watanabe, M. Itagaki, A. and Ando H. (2000). "An evaluation of permeability for striped thin films in the gigahertz range", *Journal of Applied Physics*, **87** (9), 5998–6000

7.5.2.2 Pickup coil

The pickup coil is another key component in a permeance meter, and the sensitivity of a permeance meter is closely related to its pickup coil. In the design of a pickup coil, we should consider the coupling between the driving coil and the pickup coil, and we should also consider the wave resonance of the pickup coil and the effect of electric field on the pickup coil. To decrease the noise level, the ground plane of the pickup coil should be connected to the ground plate of the driving coil.

Figure 7.37 shows two types of pickup coils often used: microstrip coil and shielded loop coil (Yamaguchi *et al.* 1996). Figure 7.37(a) shows a single turn of microstrip line. The characteristic impedance of the microstrip is 50 Ω to ensure the

impedance matching. The end of the microstrip line is connected to a 50-Ω terminator, so there is no resonance in the loop. This kind of pickup coil is easy for fabrication, but it is sensitive to the electric field.

Figure 7.37(b) shows a shielded pickup coil made from a stripline. Similar to microstrip pickup coil, its characteristic impedance is 50 Ω. There is a narrow gap on the shields at both sides. The center conductor forms a half-turn pattern and is terminated on the ground planes of the opposite half-loop at the gap. The output voltage of the coil is proportional to the average value of the magnetic field through the loop aperture. The coil is insensitive to the electric field because the inner conductor is shielded electrically by the ground planes.

For different types of pickup coils, different measurement procedures should be followed. From Eq. (7.57), we have

$$(\mu_{\mathrm{r}} - 1) \cdot t_{\mathrm{m}} = \left(\frac{V_{\mathrm{s}}}{V_0} - 1 \right) \frac{S_0}{d_{\mathrm{m}}} \qquad (7.58)$$

where d_{m} and t_{m} are the width and thickness of the thin film respectively. The presence of the sample changes the electric field distribution. If a shielded pickup coil is used, as the shielded pickup coil is insensitive to the electric field, the value of V_0 in Eq. (7.58) is the voltage output without sample. Whereas, if a microstrip pickup coil is used, to obtain the reference voltage V_0, the sample should be saturated by applying a strong dc magnetic field along the easy axis so that its permeability is close to unity.

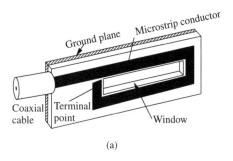
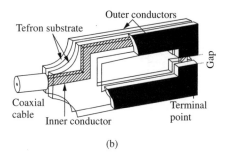

(a)

(b)

Figure 7.37   Two types of pickup coils. (a) Microstrip pickup coil and (b) shielded pickup coil (Yamaguchi *et al.* 1996). Modified from Yamaguchi, M. Yabukami, S. and Arai, K. L. (1996). "A new permeance meter based on both lumped elements/transmission line theories", *IEEE Transactions on Magnetics*, **32** (5), 4941–4943. © 2003 IEEE

To obtain more accurate results, Eq. (7.58) can be further modified (Yamaguchi *et al.* 1997) as

$$(\mu_r - 1) \cdot t_m = \frac{S_0}{d_m} \cdot \left[\frac{V_s(Z_s + 50)}{V_0(Z_0 + 50)} - 1 \right] \quad (7.59)$$

where Z_s is the impedance of the pickup coil during the main measurement and Z_0 is the impedance of the pickup coil during the reference measurement.

7.5.3 Single-coil method

In a single-coil method, the driving coil and the pickup coil is the same coil, and in this case, we usually analyze the input impedance of the coil. As shown in Figure 7.38, in a single-coil method for the measurement of permeability of magnetic thin films, the coil is essentially a short-ended transmission line (Pain *et al.* 1999; Ledieu *et al.* 2003). The input impedance of the loop is measured with and without the sample. The permeability is derived from the difference of the input impedance of the short-ended microstrip line with and without the sample under test.

The width w and the inner height h of the microstrip are chosen to ensure a line with characteristic impedance $Z_0 = 50\,\Omega$. The length of the microstrip line l is shorter than a quarter of wavelength in vacuum in the measurement frequency range to limit the electromagnetic resonance. The phase reference plane corresponding to the beginning of the microstrip line is set so that the reflection coefficient $S_{11} = -1$. This can be done by

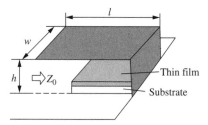

Figure 7.38   A short-ended microstrip line for the measurement of the permeability of magnetic thin films

shorting the microstrip line to the ground plane with a conducting choke. The input impedance Z_m of the cell is then given by

$$Z_m = Z_c \cdot \frac{1 + S_{11}}{1 - S_{11}} \quad (7.60)$$

where Z_c is the characteristic impedance of the microstrip. The impedance measurement is then made in two steps: the first one with only the substrate and the second one with the thin film deposited on the substrate.

The calculation of the permeability of a magnetic thin film is based on the magnetic flux perturbation induced by the insertion of the film into the coil. The change of magnetic flux causes a change in the voltage at the coil entry, which can also be expressed as the impedance variation ΔZ of the coil:

$$\Delta Z = j2\pi f \mu_0 K \cdot (\mu_r - 1) \cdot t_s \quad (7.61)$$

where f is the operating frequency, K is a constant that can be determined by a standard sample with known permeability, μ_r is the relative permeability of the sample under test, and t_s is the sample thickness. So the relative permeability is given by

$$\mu_r = 1 + \frac{\Delta Z}{j\mu_0 K t_s 2\pi f} \qquad (7.62)$$

The impedance analysis with equivalent circuits is adequate for the extraction of the permeability values. At low frequencies (typically lower than 1 GHz) the strip can be modeled as a coil with an inductance L (L-model) as shown in Figure 7.39(a). Z_a is an impedance correction due to the substrate. In this case, the change of impedance due to the magnetic thin film is

$$\Delta Z = Z_F - Z_S \qquad (7.63)$$

where Z_S and Z_F are the measured impedances of the cell loaded with the substrate and the film deposited on substrate respectively. But for higher frequencies, the measured impedance should be analyzed by a more appropriate equivalent RLC circuit (RLC model) as shown in Figure 7.39(b) (Pain *et al.* 1999):

$$Z_m = \frac{\left(R + \frac{1}{j2\pi f C}\right)(Z_a + \Delta Z + j2\pi f L)}{R + \frac{1}{j2\pi f C} + j2\pi f L + Z_a + \Delta Z} \qquad (7.64)$$

R, L, and C are determined from the best fit of the measured impedance of the cell loaded with the substrate. ΔZ is deduced from this equation. A first measurement with only the substrate inserted into the coil ($\Delta Z = 0$) yields Z_a. A second measurement with the magnetic film yields ΔZ.

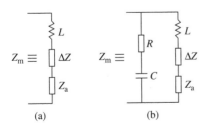

(a) (b)

Figure 7.39 Equivalent electric circuits for the measured impedance Z_m. (a) L-model and (b) RLC model. Z_a is the impedance correction corresponding to the behavior of the cell loaded with the substrate only

The permeability spectrum is then obtained from Eq. (7.62).

This method permits a large domain of investigation. For ferromagnetic thin films with in-plane uniaxial anisotropy, the permeability measurement can be done in any direction. The measurement can be easily performed under an external static field. With ferromagnetic thin films deposited on flexible substrates, measurements under stress are also possible.

Cross measurements of thin-film permeability between the two-coil method and the single-coil method have been made, and the results from these two methods agree very well (Yamaguchi *et al.* 2002). Finally, it should be indicated that the single-coil method can also be realized in other types of transmission structures, such as stripline (Moraitakis *et al.* 2000).

7.5.4 Electrical impedance method

Permeability measurements of narrow magnetic thin films are needed for evaluating high frequency-characteristics of magnetic write head or micromagnetic devices. Yabukami *et al.* proposed a method for the measurement of the permeability of striped conducting films (Yabukami *et al.* 2001). In this method, the permeability of a striped film is obtained by measuring its electrical impedance. In principle, this method uses the sample under study as the driving coil and the pickup coil. This method has high sensitivity, it is reported that a very narrow film ($\mu_r' = 10$, $0.4 \, \mu m$ wide) can be measured using this method.

Figure 7.40 shows a setup for the measurement of a striped magnetic thin film. The striped film is connected as the termination of a microstrip line. One end of the striped film is electrically connected to the microstrip conductor and the other end of the film is connected to the ground plane using bonding wires. High-frequency current then passes through the magnetic film. The relationship between the reflection coefficient and the impedance of the striped film is given by

$$Z_s = R + j2\pi f L_s = Z_c \frac{1 + S_{11}}{1 - S_{11}} \qquad (7.65)$$

where R is the resistance, L_s is the inductance of the film, f is the frequency, and Z_c is the

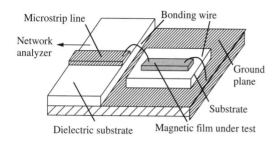

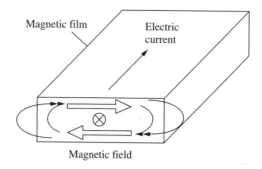

Figure 7.40 Configuration for the measurement of striped films (Yabukami *et al.* 2001). Source: Yabukami, S. Uo, T. Yamaguchi, M. Arai, K. I. Takezawa, M. (2001). "High sensitivity permeability measurements of striped films obtained by input impedance", *IEEE Transactions on Magnetics*, **37** (4), 2776–2778. © 2003 IEEE

Figure 7.41 Model of striped film and field distribution. Source: Yabukami, S. Uo, T. Yamaguchi, M. Arai, K. I. Takezawa, M. (2001). "High sensitivity permeability measurements of striped films obtained by input impedance", *IEEE Transactions on Magnetics*, **37** (4), 2776–2778. © 2003 IEEE

characteristic impedance of the transmission line. Usually, the resistance is fairly smaller than reactance, and it can be ignored in the estimation of inductance. The inductance of the film is approximately given by

$$L_s = 200l \cdot \left[\ln\left(\frac{2l}{w+l}\right) + 0.25049 \right.$$
$$\left. + \frac{w+t}{3l} + \frac{\mu_0 \mu_r}{4} \right] \tag{7.66}$$

where l is film length, t is film thickness, and w is film width.

Figure 7.41 shows a model of a striped film and the magnetic field distribution near the film. High-frequency current flows inside the film, and thus a closed magnetic flux loop is generated. The permeability obtained in this method is the one along the direction of film surface, but in the calculation we should consider the leakage flux from the edge of the film, correct orientation of magnetic moments in the film, and the effect of vertical (depth) permeability.

The measurement procedure of this method consists of two steps. The first step is the measurement of the impedance of the film Z_s, and the second step is the measurement of the background impedance Z_0, when a strong dc field (500 Oe) is applied to the longitudinal direction of the film. The effect of the contact resistance between the film and the bonding wire or the outer

inductance of the striped film can be eliminated by using the difference between Z_s and Z_0:

$$Z_s - Z_0 = Z_{sc} - Z_{0c} \tag{7.67}$$

where Z_{sc} and Z_{0c} are the calculated impedances, which are a function of the relative permeability of the film:

$$Z_{sc} = \frac{k_s \rho l}{2w} \coth\left(\frac{k_s t}{2}\right) \tag{7.68}$$

$$Z_{0c} = \frac{k_s \rho l}{2w} \coth\left(\frac{k_0 t}{2}\right) \tag{7.69}$$

$$k_s = \frac{(1+j)}{\sqrt{\rho/(\pi f \mu_0 \mu_r)}} \tag{7.70}$$

$$k_0 = \frac{(1+j)}{\sqrt{\rho/(\pi f \mu_0 \mu_{r0})}} \tag{7.71}$$

where ρ is the resistivity of the film, t is film thickness, l is the film length, and w is the film width. μ_r is the relative permeability of the magnetic film, and μ_{r0} is the relative permeability of the striped film under strong dc field (500 Oe). The relative permeability of the magnetic film μ_r can be calculated from Eqs. (7.67)–(7.71) using the Newton–Raphson method (Yabukami *et al.* 2001).

7.6 PLANAR NEAR-FIELD MICROWAVE MICROSCOPES

The basic principles for near-field microwave microscopes have been discussed in Chapter 6. In this section, we concentrate on the near-field microwave microscopes developed from planar circuits.

7.6.1 Working principle

Planar near-field microwave microscopes are based on resonant-perturbation theory. As shown in Figure 7.42, when a sample is brought close to the open end of a planar resonator, the electromagnetic field at the open end will be perturbed, and thus the resonant properties of the resonator will be changed. From the change of the resonant frequency f and quality factor Q, the electromagnetic properties of the sample can be determined. Generally speaking, dielectrics, semiconductors, conductors, and magnetic materials can be characterized by this technique.

To increase the measurement resolution, the open end of the resonator is tapered, becoming a tip, and the open end of the resonator can be taken as an electric dipole. Near its resonance frequency, the resonator can be modeled by a series

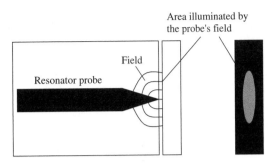

Figure 7.42 Interaction between a near-field probe and a sample. Modified from Tabib-Azar, M. and Akinwande, D. (2000). "Real-time imaging of semiconductor space charge regions using high-spatial resolution evanescent microwave microscope", *Review of Scientific Instruments*, **71** (3), 1460–1465

LCR circuit (Tabib-Azar *et al*. 1999a), as shown in Figure 7.43(a). When a sample is placed near the tip, it is coupled to the LCR circuit through a coupling capacitor. For simplicity, insulators and conductors are modeled in Figures 7.43(b) and (c) respectively. In these figures, R_0, L_0, and C_0 are the intrinsic circuit parameters of the stripline resonator; C_c is the coupling capacitance of the air gap between the tip and the sample; L_s, C_s, and R_s represent the microwave properties of the sample. The resonance frequency of planar

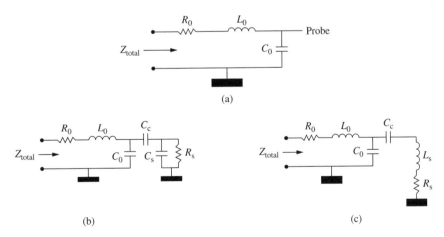

Figure 7.43 Equivalent circuit models of the probe for a near-field microwave microscope. (a) Series LCR models of the evanescent microwave probe, (b) circuit model in the presence of an insulating sample, and (c) circuit model in presence of a conducting sample. Source: Tabib-Azar, M. Su, D. P. Pohar, A. LeClair, A. Ponchak, G. (1999a). "0.4 μm spatial resolution with 1 GHz ($\lambda = 30$ cm) evanescent microwave probe", *Review of Scientific Instruments*, **70** (3), 1725–1729

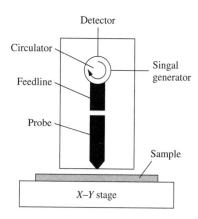

Figure 7.44 Measurement system of a near-field microwave microscope

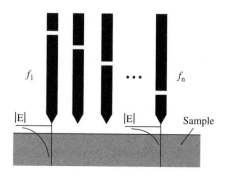

Figure 7.45 Probes with different operational frequencies

resonator shifts in the presence of a sample near the probe tip. It is clear that the shifts in resonance frequency f and quality factor Q mainly depend on the electromagnetic properties of the sample, the probe-sample distance, and effective area of the tip.

Figure 7.44 shows a typical measurement setup, and it is similar to the system discussed in Chapter 6. It consists of a microwave resonator coupled to a feed-line and connected to a circulator. The circulator is also connected to a signal generator and a microwave crystal detector. The detector output is a dc voltage proportional to the magnitude of the reflected wave. The probe is mounted vertically over an X-Y stage. A more detailed description of the measurement setup can be found in (Tabib-Azar *et al.* 1999c).

Owing to the penetration effect of the electromagnetic wave, electromagnetic waves with lower frequencies can travel deeper into bulk materials. Sometimes, near-field microwave microscopes with probes operating at different frequencies are developed to study the depth distributions of materials properties, as shown in Figure 7.45. Such microwave microscopes can be used for nondestructive testing, impurity testing, and biomedical checking.

7.6.2 Electric and magnetic dipole probes

Two types of probes are often used in near-field microwave microscopes: electric dipole probes and magnetic dipole probes. The discussions above are based on the assumption that the electric dipole probe is used, and the interaction between the probe and the sample is made through the coupling capacitor. For a magnetic dipole probe, the interaction between the probe and the sample is made through the coupling inductor of the probe. Electric probes have relatively high impedances and they are better suited to the characterization of insulators and semiconductors, whereas magnetic probes have lower impedances and are better suited to the characterization of high-conductivity materials.

Figure 7.46 shows three types of electric dipole probes. A tapered strip can be used as an electric dipole probe, as shown in Figure 7.46(a). To improve the resolution, a metal needle can be attached to the tip of the tapered strip, as shown in Figure 7.46(b). Another type of electric dipole probe is an open loop between the tip of the strip and the grounding plate, as shown in Figure 7.46(c) and (d). The open ends of the loop form a capacitor, and the resolution of the probe is thus greatly improved. The change in the reflection coefficient of the resonator when a material object is placed near the probe is calculated by treating the probe as an electrically short and insulated antenna (i.e., the length of the dipole is much shorter than the microwave wavelength) in which the insulation is provided by the air gap between the probe and the sample. In the case when the probe length T is larger than the distance b between the probe and the sample, the input admittance of this electrically short and insulated

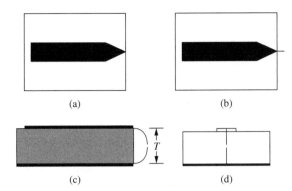

Figure 7.46 Three types of electric dipole probes. (a) Top view of a tapered strip, (b) top view of a needle dipole, (c) side view of an open loop dipole, and (d) front view of the open loop dipole whose side view is shown in (c)

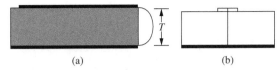

Figure 7.47 Closed loop magnetic dipole probe. (a) Side view and (b) front view

antenna is approximately given by (Tabib-Azar *et al.* 1993)

$$Y_L = \frac{\sigma_s \pi T}{\ln(T/b) - 1} \left(\frac{\gamma^2}{p_s^2 + (1 + \gamma)^2} \right)$$

$$+ j\omega \frac{\varepsilon_i \pi T}{\ln(b/c)} \left(\frac{p_s^2 + (1 + \gamma)}{p_s^2 + (1 + \gamma)^2} \right) \quad (7.72)$$

where σ_s is the conductivity of the sample, ε_i is the permittivity of the insulator ($\varepsilon_i = \varepsilon_{ri}\varepsilon_0$), c is the radius of the wire and

$$\gamma = \frac{\varepsilon_{ri}}{\varepsilon_s} \cdot \frac{\ln(T/b) - 1}{\ln(b/c)} \quad (7.73)$$

$$p_s = \frac{\sigma_s}{2\pi f \varepsilon_s} \quad (7.74)$$

where ε_s is the permittivity of the sample ($\varepsilon_s = \varepsilon_{rs}\varepsilon_0$) and f is the frequency of the microwave signal.

Figure 7.47 shows a magnetic dipole probe, which is a closed loop connecting the tip of the strip to the grounding plate. The method of images can be applied to calculate the loop's impedance placed in front of a dissipative medium like a real conductor or a semiconductor. In the case of ideal conductors, the electric field of the image loop cancels the electric field of the original loop at the surface of the conductor. In the case of dielectric and semiconductor samples, the electric field produced by the image loop

modifies the electric field at the surface of the sample. The phase and amplitude of the electric and magnetic fields, produced by the image loop, can be calculated from the reflection coefficients of the corresponding fields. The reflection coefficient for the electric field is given by (Tabib-Azar *et al.* 1993)

$$\Gamma_E = \frac{E_r}{E_i} = \frac{\sqrt{1/\varepsilon_0} - \sqrt{j\omega/(\sigma_s + j\omega\varepsilon_s)}}{\sqrt{1/\varepsilon_0} + \sqrt{j\omega/(\sigma_s + j\omega\varepsilon_s)}} \quad (7.75)$$

where σ_s and ε_s are the conductivity and the permittivity of the sample respectively. The reflection coefficient for the magnetic field contains the same information as Eq. (7.75), which is derived for nonmagnetic materials in which the relative magnetic permeability is unity.

The dielectric permittivity of the substrate of a probe also affects the performances of the probe. Higher permittivity substrates result in a smaller line width of the strip for 50-Ω characteristic impedance and have higher spatial resolutions. A more detailed discussion on this aspect can be found in (Tabib-Azar *et al.* 1999b).

7.6.3 Probes made from different types of planar transmission lines

In the above discussions, we assume that the probes are made from microstrip lines. Actually, all types of the planar transmission lines can be used for the development of near-field microwave microscope probes. In the following text, we discuss the probes made from stripline, microstrip, and coplanar line, and we focus on electric dipole probes whose tips are connected with a needle.

Figure 7.48 shows a probe developed from a stripline. Stripline supports strict TEM mode, and the electric fields near the tip are symmetrical about the center plane. Stripline probes usually

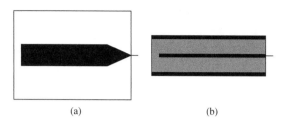

Figure 7.48 Stripline needle electric dipole probe. (a) Center plane and (b) side view

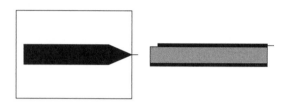

Figure 7.49 Microstrip needle electric dipole probe. (a) Top view and (b) side view

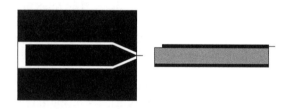

Figure 7.50 Coplanar needle electric dipole probe. (a) Top view and (b) side view

have high accuracy; however, the fabrication is complicated. In the measurement procedure, to increase the resolution, the adjustment of the coupling between the feed-line and the resonator is often needed, while it is not easy to make such an adjustment for a stripline probe.

Microstrip probes are most widely used in planar microwave microscopes. As shown in Figure 7.49, the fabrication of microstrip probes is easy, and such probes usually have high accuracy.

Microwave microscope probes can also be made from coplanar lines (Tabib-Azar and Akinwande 2000). As shown in Figure 7.50, the fabrication of a coplanar probe is easy, and it is also not difficult to adjust the couplings between the feed-line and the resonator. However, its accuracy and resolution may not be as high as those of stripline and microstrip ones.

REFERENCES

Abdulnour, J. Akyel, C. and Wu, K. (1995). "A generic approach for permittivity measurement of dielectric materials using discontinuity in a rectangular waveguide or a microstrip line", *IEEE Transactions on Microwave Theory and Techniques*, **43** (5), 1060–1066.

Amey, D. I. and Horowitz, S. J. (1997). "Tests characterize high-frequency material properties", *Microwave & RF*, **36** (8), 68–80.

Barry, W. (1986). "A broad-band, automated, stripline technique for the simultaneous measurement of complex permittivity and permeability", *IEEE Transactions on Microwave Theory and Techniques*, **34** (1), 80–84.

Belyaev, B. A. Leksikov, A. A. and Tyurnev, V. V. (1995). "Microstrip technique for measuring microwave dielectric constant of solids", *Instruments and Experimental Techniques*, **38** (5), 646–650.

Belyaev, B. A. Leksikov, A. A. Tyurnev, V. V. and Shikhov Y. G. (1997). "A microstrip sensor for measuring microwave dielectric constants of solids", *Instruments and Experimental Techniques*, **40** (3), 395–398.

Bernard, P. A. and Gautray, J. M. (1991). "Measurement of dielectric constant using a microstrip ring resonator", *IEEE Transactions on Microwave Theory and Techniques*, **39** (3), 592–595.

Carroll, J. Li, M. and Chang, K. (1995). "New technique to measure transmission line attenuation", *IEEE Transactions on Microwave Theory and Techniques*, **43** (1), 219–222.

Fischer, M. Vainikainen, P. and Nyfors, E. (1995). "Design aspects of stripline resonator for industrial applications", *Journal of Microwave Power and Electromagnetic Energy*, **30** (4), 246–257.

Hammerstad, E. and Jensen, O. (1980). "Accurate models for microstrip computer-aided design", *IEEE MTT-S International Microwave Symposium Digest*, 407–409.

Hanson, G. W. Grimm, J. M. and Nyquist, D. P. (1993). "An improved de-embedding technique for the measurement of the complex constitutive parameters of materials using a stripline field applicator", *IEEE Transactions on Instrumentation and Measurement*, **42** (3), 740–745.

Hasnain, G. Dienes, A. and Whinnery, J. R. (1986). "Dispersion of picosecond pulses in coplanar transmission lines", *IEEE Transactions on Microwave Theory and Techniques*, **34** (6), 738–741.

Hilberg, W. (1969). "From approximations to exact relations for characteristic impedances", *IEEE Transactions on Microwave Theory and Techniques*, **17** (5), 259–265.

Hinojosa, J. (2001). "S-parameter broad-band measurements on-microstrip and fast extraction of the substrate intrinsic properties", *IEEE Microwave and Wireless Components Letters*, **11** (7), 305–307.

Hinojosa, J. Faucon, L. Queffelec, P. and Huret, F. (2001). "S-parameter broadband measurements of microstrip lines and extraction of the substrate intrinsic properties", *Microwave and Optical Technology Letters*, **30** (1), 65–69.

Hinojosa, J. Lmimouni, K. and Dambrine, G. (2002b). "Fast electromagnetic characterization method of thin planar materials using coplanar line up to V-band", *Electronics Letters*, **38** (8), 373–374.

Hinojosa, J. Lmimouni, K. Lepilliet, S. and Dambrine, G. (2002a). "Very high broadband electromagnetic characterization method of film-shaped materials using coplanar", *Microwave and Optical Technology Letters*, **33** (5), 352–355.

Jones, C. A. (1999). "Permittivity and permeability measurements using stripline resonator cavities – a comparison", *IEEE Transactions on Instrumentation and Measurement*, **48** (4), 843–848.

Jones, C. A. Kantor, Y. Grosvenor, J. H. and Janezic, M. D. (1998). *Stripline Resonator for Electromagnetic Measurements of Materials*, NIST Technical Note 1505, NIST, Boulder, CO.

Kirschning, M. Jansen, R. H. and Koster, N. H. (1981). "Accurate model for open end effect of microstrip lines", *Electronics Letters*, **17** (3), 123–125.

Laverghetta, T. S. (1997). "A swept measurement for wireless material dielectric constant", *Microwave Journal*, **40** (9), 96–108.

Ledieu, M. Schoenstein, F. Le Gallou, J. H. Valls, O. Queste, S. Duverger, F. and Acher, O. (2003). "Microwave permeability spectra of ferromagnetic thin films over a wide range of temperatures", *Journal of Applied Physics*, **93** (10), 7202–7204.

Lee, M. Q. and Nam, S. K. (1996). "An accurate broadband measurement of substrate dielectric constant", *IEEE Microwave and Guided Wave Letters*, **6** (4), 168–170.

Moraitakis, E. Kompotiatis, L. Pissas, M. and Niarchos, D. (2000). "Permeability measurements of permalloy films with a broad band stripline technique", *Journal of Magnetism and Magnetic Materials*, **222**, 168–174.

Nyfors, E. and Vainikainen, P. (1989) *Industrial Microwave Sensors*, Artech House, Norwood, MA.

Pain, D. Ledieu, M. Acher, O. Adenot, A. L. and Duverger, F. (1999). "An improved permeameter for thin film measurements up to 6 GHz", *Journal of Applied Physics*, **85** (8), 5151–5153.

Peterson, A. L. and Drayton, R. F. (2002). "A CPW T-resonator technique for electrical characterization of microwave substrates", *IEEE Microwave and Wireless Components Letters*, **12** (3), 90–92.

Queffelec, P. Gelin, P. Gieraltowski, J. and Loaec, J. (1994). "A microstrip device for the broad band simultaneous measurement of complex permeability and permittivity", *IEEE Transactions on Magnetics*, **30** (2), 224–231.

Queffelec, P. Le Floc'h, M. and Gelin, P. (1998). "Broad-band characterization of magnetic and dielectric thin films using a microstrip line", *IEEE Transactions on Instrumentation and Measurement*, **47** (4), 956–963.

Rudy, D. A. Mendelsohn, J. P. and Muniz, P. J. (1998). "Measurement of RF dielectric properties with series resonant microstrip elements", *Microwave Journal*, **41** (3), 22–41.

Salahun, E. Queffelec, P. Le Floc'h, M. and Gelin, P. (2001). "A broadband permeameter for 'in situ' measurements of rectangular samples", *IEEE Transactions on Magnetics*, **37** (4), 2743–2745.

Stuchly, S. S. and Bassey, C. E. (1998). "Microwave coplanar sensors for dielectric measurements", *Measurement Science and Technology*, **9**, 1324–1329.

Tabib-Azar, M. and Akinwande, D. (2000). "Real-time imaging of semiconductor space charge regions using high-spatial resolution evanescent microwave microscope", *Review of Scientific Instruments*, **71** (3), 1460–1465.

Tabib-Azar, M. Akinwande, D. Ponchak, G. and LeClair, S. R. (1999c). "Novel physical sensors using evanescent microwave probes", *Review of Scientific Instruments*, **70** (8), 3381–3386.

Tabib-Azar, M. Pathak, P. S. Ponchak, G. and LeClair, S. (1999b). "Nondestructive superresolution imaging of defects and nonuniformities in metals, semiconductors, dielectrics, and plants using evanescent microwaves", *Review of Scientific Instruments*, **70** (6), 2783–2792.

Tabib-Azar, M. Shoemaker, N. S. and Harris, S. (1993). "Non-destructive characterization of materials by evanescent microwaves", *Measurement Science & Technology*, **4** (5), 583–590.

Tabib-Azar, M. Su, D. P. Pohar, A. LeClair, A. and Ponchak, G. (1999a). "0.4 μm spatial resolution with 1 GHz ($\lambda = 30$ cm) evanescent microwave probe", *Review of Scientific Instruments*, **70** (3), 1725–1729.

Waldron, R. A. (1964). "Theory of a strip-line cavity for measurement of dielectric constants and gyromagnetic-resonance line-widths", *IEEE Transactions on Microwave Theory and Techniques*, **12**, 123–131.

Wang, M. S. Bothra, S. Borrego, J. M. and Kristal, K. W. (1990). "High spatial resolution dielectric constant uniformity measurements using microstrip resonant probes", *IEEE MTT-S Digest*, **3**, 1121–1124.

Weil, C. M. Jones, C. A. Kantor, Y. and Grosvenor Jr., J. H. (2000). "On RF material characterization in the stripline cavity", *IEEE Transactions on Microwave Theory and Techniques*, **48** (2), 266–275.

Yabukami, S. Takezawa, M. Uo, T. Yamaguchi, M. Arai, K. I. Miyazawa, Y. Watanabe, M. Itagaki, A.

and Ando, H. (2000). "An evaluation of permeability for striped thin films in the gigahertz range", *Journal of Applied Physics*, **87** (9), 5998–6000.

Yabukami, S. Uo, T. Yamaguchi, M. Arai, K. I. and Takezawa, M. (2001). "High sensitivity permeability measurements of striped films obtained by input impedance", *IEEE Transactions on Magnetics*, **37** (4), 2776–2778.

Yabukami, S. Yamaguchi, M. and Arai, K. I. (1999). "Noise analysis of a 1 MHz-3 GHz magnetic thin film permeance meter", *Journal of Applied Physics*, **85** (8), 5148–5150.

Yamaguchi, M. Acher, O. Miyazawa, Y. Arai, K. I. and Ledieu, M. (2002). "Cross measurements of thin-film permeability up to UHF range", *Journal of Magnetism and Magnetic Materials*, **242–245**, 970–972.

Yamaguchi, M. Yabukami, S. and Arai, K. L. (1996). "A new permeance meter based on both lumped elements/transmission line theories", *IEEE Transactions on Magnetics*, **32** (5), 4941–4943.

Yamaguchi, M. Yabukami, S. and Arai, K. I. (1997). "A new 1 MHz-2 GHz permeance meter for metallic thin films", *IEEE Transactions on Magnetics*, **33** (5), 3619–3621.

8

Measurement of Permittivity and Permeability Tensors

Permittivity and permeability tensors are two important parameters describing the electromagnetic properties of anisotropic materials. The first section discusses the permittivity tensors of various anisotropic dielectric materials and the permeability tensors of magnetic materials with different magnetization directions. The second and third sections discuss the measurements of permittivity and permeability tensors respectively. In the final section, the methods for the study of ferromagnetic resonance are discussed.

8.1 INTRODUCTION

Anisotropic materials have found important applications in microwave engineering. Among the applications are substrates for microwave integrated circuits, antenna radomes, and electromagnetic wave absorbers. Besides, isotropic substrates may exhibit anisotropic properties due to the rotation of the principal axes in some application situations. Therefore, it is important to study the anisotropy of electromagnetic materials.

Though the electromagnetic theory of anisotropic materials is well established (Kong 1975), experimental techniques are always needed for the determination of the constitutive parameters of these materials. The determination of the constitutive parameters of anisotropic materials is a typical inverse problem, and the problem is complicated by the fact that, in general, such materials are both lossy and dispersive. First, we must ascertain

what measurement data are sufficient and necessary for unequivocal determination of the constitutive parameters, and, second, we should develop experimental setups that are suitable for collecting the needed data. In the past decades, a lot of resources have been spent for the characterization of anisotropic materials, and as a result, many methods have been developed for different applications.

8.1.1 Anisotropic dielectric materials

For an anisotropic dielectric material, the relationship between the electric displacement tensor $[D]$ and the electric field tensor $[E]$ is given by

$$\begin{bmatrix} D_1 \\ D_2 \\ D_3 \end{bmatrix} = \begin{bmatrix} \varepsilon_{11} & \varepsilon_{12} & \varepsilon_{13} \\ \varepsilon_{21} & \varepsilon_{22} & \varepsilon_{23} \\ \varepsilon_{31} & \varepsilon_{32} & \varepsilon_{33} \end{bmatrix} \begin{bmatrix} E_1 \\ E_2 \\ E_3 \end{bmatrix} \quad (8.1)$$

For a physically real material, its permittivity tensor $[\varepsilon]$ is Hermitian, defined by $\varepsilon_{ij} = \varepsilon_{ji}{}^{*}$. So the matrix $[\varepsilon]$ can be diagonalized by using the principal coordinates x, y, and z:

$$\begin{bmatrix} D_x \\ D_y \\ D_z \end{bmatrix} = \begin{bmatrix} \varepsilon_{xx} & 0 & 0 \\ 0 & \varepsilon_{yy} & 0 \\ 0 & 0 & \varepsilon_{zz} \end{bmatrix} \begin{bmatrix} E_x \\ E_y \\ E_z \end{bmatrix} \quad (8.2)$$

where ε_{xx}, ε_{yy}, and ε_{zz} are the principal permittivity components.

The electric-energy density U can then be expressed as

$$U = \frac{1}{2} \left(\frac{D_x^2}{\varepsilon_{xx}} + \frac{D_y^2}{\varepsilon_{yy}} + \frac{D_z^2}{\varepsilon_{zz}} \right) \quad (8.3)$$

Microwave Electronics: Measurement and Materials Characterization   L. F. Chen, C. K. Ong, C. P. Neo, V. V. Varadan and V. K. Varadan
© 2004 John Wiley & Sons, Ltd   ISBN: 0-470-84492-2

By defining quantities X, Y, and Z along the three principal spatial axes

$$X = \frac{D_x}{\sqrt{2\varepsilon_0 U}}, \tag{8.4}$$

$$Y = \frac{D_y}{\sqrt{2\varepsilon_0 U}}, \tag{8.5}$$

$$Z = \frac{D_z}{\sqrt{2\varepsilon_0 U}}, \tag{8.6}$$

Eq. (8.3) can be rewritten as

$$\frac{X^2}{\varepsilon_1} + \frac{Y^2}{\varepsilon_2} + \frac{Z^2}{\varepsilon_3} = 1 \tag{8.7}$$

with

$$\varepsilon_1 = \varepsilon_{xx}/\varepsilon_0, \tag{8.8}$$

$$\varepsilon_2 = \varepsilon_{yy}/\varepsilon_0, \tag{8.9}$$

$$\varepsilon_3 = \varepsilon_{zz}/\varepsilon_0. \tag{8.10}$$

From Eqs. (8.2) and (8.8) to (8.10), the relative permittivity tensor in the principal axes can be expressed as

$$[\varepsilon_r] = \begin{bmatrix} \varepsilon_1 & 0 & 0 \\ 0 & \varepsilon_2 & 0 \\ 0 & 0 & \varepsilon_3 \end{bmatrix} \tag{8.11}$$

Equation (8.7) describes an ellipsoid known as *index ellipsoid*, as shown in Figure 8.1. The semi-axes of the index ellipsoid of a dielectric are equal to the square roots of the relative permittivities in the three corresponding principal directions.

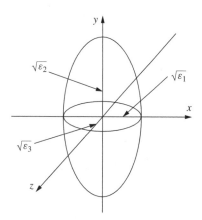

Figure 8.1   The index ellipsoid of a dielectric

For a crystal, the shape of the ellipsoid depends on its crystalline structural properties. Generally speaking, crystals can be classified into three types (Anderson 1964). Type I crystals have cubic symmetry and have three equivalent directions: $\varepsilon_1 = \varepsilon_2 = \varepsilon_3$. Their corresponding ellipsoids are spheres. Such crystals are isotropic, and their dielectric permittivity can be described by a complex number. Silicon and gallium arsenide are typical examples of cubic crystals. The measurement of dielectric permittivity of isotropic materials has been discussed in previous chapters.

Type II materials mainly include trigonal, tetragonal, and hexagonal crystals, all of which have one axis of symmetry. This axis is one of the principal axes of the permittivity ellipsoid. The symmetry of the crystal about this axis results in corresponding symmetry of the ellipsoid. Crystals of this type are called *uniaxial*. Quartz is a typical example of uniaxial dielectric, which crystallizes in the trigonal trapezohedral class with three crystal axes usually labeled a, b, and c. For α-phase SiO_2 crystals, the c-axis is of threefold symmetry: a rotation of such a crystal through $120°$ about the c-axis leaves the crystal in an identical configuration. Other examples of uniaxial dielectric materials include calcium carbonate, lithium niobate, and cadmium sulfide.

By assuming that the symmetric axis of the dielectric ellipsoid is in the z-direction ($\varepsilon_1 = \varepsilon_2$), the relative permittivity tensor of a uniaxial dielectric can be expressed in the form of

$$[\varepsilon_r] = \begin{bmatrix} \varepsilon_1 & 0 & 0 \\ 0 & \varepsilon_1 & 0 \\ 0 & 0 & \varepsilon_3 \end{bmatrix} \tag{8.12}$$

This type is relatively simple but is a very important group of anisotropic materials. In the second section of this chapter, we mainly discuss the measurement of permittivity tensors of uniaxial dielectric materials.

Type III materials are usually biaxial crystals, which have no axes of symmetry; and usually they have the orthorhombic, monoclinic, and triclinic structures. All the three principal axes of the index ellipsoid are different. The form of permittivity tensor is in the form of Eq. (8.11). The methods for the measurement of uniaxial dielectric materials can be extended for the measurement of biaxial materials.

8.1.2 Anisotropic magnetic materials

Similar to the relationship between the electric displacement tensor $[D]$ and electric field tensor $[E]$ for an anisotropic dielectric material, the relationship between inductance tensor $[B]$ and magnetic field tensor $[H]$ for an anisotropic magnetic material can be expressed in the form of

$$\begin{bmatrix} B_1 \\ B_2 \\ B_3 \end{bmatrix} = \begin{bmatrix} \mu_{11} & \mu_{12} & \mu_{13} \\ \mu_{21} & \mu_{22} & \mu_{23} \\ \mu_{31} & \mu_{32} & \mu_{33} \end{bmatrix} \begin{bmatrix} H_1 \\ H_2 \\ H_3 \end{bmatrix} \quad (8.13)$$

The permeability tensor $[\mu]$ can be also diagonalized by using the principal coordinates x, y, and z

$$\begin{bmatrix} B_x \\ B_y \\ B_z \end{bmatrix} = \begin{bmatrix} \mu_{xx} & 0 & 0 \\ 0 & \mu_{yy} & 0 \\ 0 & 0 & \mu_{zz} \end{bmatrix} \begin{bmatrix} H_x \\ H_y \\ H_z \end{bmatrix} \quad (8.14)$$

Similarly, the relative permeability tensor in the principal axes can be expressed in the following format:

$$[\mu_r] = \begin{bmatrix} \mu_1 & 0 & 0 \\ 0 & \mu_2 & 0 \\ 0 & 0 & \mu_3 \end{bmatrix} \quad (8.15)$$

with

$$\mu_1 = \mu_{xx}/\mu_0, \quad (8.16)$$

$$\mu_2 = \mu_{yy}/\mu_0, \quad (8.17)$$

$$\mu_3 = \mu_{zz}/\mu_0. \quad (8.18)$$

According to the relationships between μ_1, μ_2, and μ_3, we can also classify magnetic materials into three general types: isotropic, uniaxial, and biaxial.

When a magnetic material is magnetized by an external dc magnetic field, the relative permeability tensor $[\mu_r]$ is usually different from Eq. (8.15). When the external dc magnetic field H_0 is in the z-direction, the relative permeability tensor is in the form of

$$[\mu_r] = \begin{bmatrix} \mu & -j\kappa & 0 \\ j\kappa & \mu & 0 \\ 0 & 0 & \mu_z \end{bmatrix} \quad (8.19)$$

If the magnetization is saturated, $\mu_z = 1$. When the external magnetic field H_0 is applied in the x-direction, the relative permeability is in the form of

$$[\mu_r] = \begin{bmatrix} \mu_x & 0 & 0 \\ 0 & \mu & -j\kappa \\ 0 & j\kappa & \mu \end{bmatrix} \quad (8.20)$$

If the magnetization is saturated, $\mu_x = 1$. When the external magnetic field H_0 is applied in y-direction, the relative permeability tensor is in the form of

$$[\mu_r] = \begin{bmatrix} \mu & 0 & -j\kappa \\ 0 & \mu_y & 0 \\ j\kappa & 0 & \mu \end{bmatrix} \quad (8.21)$$

If the magnetization is saturate, $\mu_y = 1$. In the following discussions, we concentrate on the case with the external dc magnetic field applied in z-direction, so the relative permeability tensor is in the form given by Eq. (8.19). Meanwhile, in most cases, we assume that the magnetization is saturated, so the measurement of the permeability tensor becomes the measurement of two parameters μ and κ.

The components in permeability tensor are usually frequency-dependent. For a magnetic material with external magnetic field H_0 in z-direction, according to Eq. (8.19), the relationship between microwave magnetic inductance $[b]$ and microwave magnetic field $[h]$ is given by

$$\begin{bmatrix} b_x \\ b_y \\ b_z \end{bmatrix} = \mu_0 \begin{bmatrix} \mu & -j\kappa & 0 \\ j\kappa & \mu & 0 \\ 0 & 0 & 1 \end{bmatrix} \begin{bmatrix} h_x \\ h_y \\ h_z \end{bmatrix} \quad (8.22)$$

with

$$\mu = 1 + \frac{\omega_m \omega_0}{\omega_0^2 - \omega^2} \quad (8.23)$$

$$\kappa = -\frac{\omega_m \omega_0}{\omega_0^2 - \omega^2} \quad (8.24)$$

$$\omega_0 = \gamma H_0 \quad (8.25)$$

$$\omega_m = \gamma 4\pi M_0 \quad (8.26)$$

where ω_0 is the angular ferromagnetic resonance frequency, M_0 is the static magnetization due to external dc magnetic field. γ is the gyromagnetic ratio, and for most of the ferrites, $\gamma = 2.8 \times 10^6$ Hz/Oe $= (1/4\pi) \times 2.8 \times 10^3$ Hz/(A/m).

In a general case, the permeability of a magnetized magnetic material is a tensor. To some special types of magnetic fields, if we do not consider the loss factor, permeability tensor becomes a scalar. These fields can be obtained by finding the eigenvalues and eigenvectors of $[\mu]$ of the following equation:

$$\begin{bmatrix} \mu_0\mu & -j\mu_0\kappa & 0 \\ j\mu_0\kappa & \mu_0\mu & 0 \\ 0 & 0 & \mu_0 \end{bmatrix} \begin{bmatrix} h_x \\ h_y \\ h_z \end{bmatrix} = \lambda \begin{bmatrix} h_x \\ h_y \\ h_z \end{bmatrix} \tag{8.27}$$

where λ is the eigenvalue to be determined. Equation (8.27) can be transformed as

$$\begin{vmatrix} \mu_0\mu - \lambda & -j\mu_0\kappa & 0 \\ j\mu_0\kappa & \mu_0\mu - \lambda & 0 \\ 0 & 0 & \mu_0 - \lambda \end{vmatrix} = 0 \tag{8.28}$$

The three eigenvalues of Eq. (8.28) are

$$\lambda_1 = \mu_0 \tag{8.29}$$

$$\lambda_2 = \mu_0(\mu - \kappa) \tag{8.30}$$

$$\lambda_3 = \mu_0(\mu + \kappa) \tag{8.31}$$

The above eigenvalues are scalars, and their corresponding eigenvectors are

$$h_1 = h_z = \begin{bmatrix} 0 \\ 0 \\ h_0 \end{bmatrix} \tag{8.32}$$

$$h_2 = h_+ = \frac{1}{\sqrt{2}} \begin{bmatrix} h_0 \\ -jh_0 \\ 0 \end{bmatrix} \tag{8.33}$$

$$h_3 = h_- = \frac{1}{\sqrt{2}} \begin{bmatrix} h_0 \\ jh_0 \\ 0 \end{bmatrix} \tag{8.34}$$

The three microwave magnetic fields described by Eqs. (8.32) to (8.34) are usually called *z-direction linearly polarized magnetic field, right-hand (positive) circular-polarization magnetic field, and left-hand (negative) circular-polarization magnetic field* respectively. For these microwave magnetic fields, the magnetic material has a scalar permeability. For example, for a magnetic field h_z that is linearly polarized in z-direction, the relationship between $[b]$ and $[h]$ is given by

$$b_z = \mu_0 h_z \tag{8.35}$$

So the permeability is a scalar, which equals to the corresponding eigenvalue ($\lambda_1 = \mu_0$).

The concepts of right-hand polarization and left-hand polarization are relative to the direction of the external bias magnetic field H_0 in the z-direction. The permeability values, μ_+ and μ_-, corresponding to the right-hand and left-hand circular-polarization fields are eigenvalues, λ_1 and λ_2 respectively:

$$\mu_+ = \mu - \kappa = 1 + \frac{\omega_m}{\omega_0 - \omega} \tag{8.36}$$

$$\mu_- = \mu + \kappa = 1 + \frac{\omega_m}{\omega_0 + \omega} \tag{8.37}$$

Equation (8.36) indicates that μ_+ has resonant characteristics. When the microwave angular frequency approaches ω_0, the value of μ_+ will be approaching infinity. This phenomenon is called *ferromagnetic resonance*. The value of μ_- does not have resonant characteristics. The frequency dependences of μ_+ and μ_- for an ideal magnetic material with no magnetic loss are shown in Figure 8.2(a).

In the above discussion, the energy dissipation of magnetic materials is not taken into consideration. For actual magnetic materials, there exists magnetic losses, and so μ_+ and μ_- have their imaginary parts, representing the magnetic loss factors. Owing to the energy dissipation, the real part of μ_+ cannot be infinity. Figure 8.2(b) shows typical curves for complex μ_+ and μ_- of actual magnetic materials.

Ferromagnetic resonance is important for the study of ferromagnetism and the applications of ferromagnetic materials. In Sections 8.3 and 8.4, the origin of ferromagnetic resonance and the methods for the measurement of ferromagnetic resonance are further discussed.

8.2 MEASUREMENT OF PERMITTIVITY TENSORS

In the measurement of permittivity tensors, usually the electric field is applied to the sample along the direction of one of the main axes of the sample. Following the classification in Chapter 2, the methods for the measurement of the permittivity tensor of anisotropic materials generally fall into

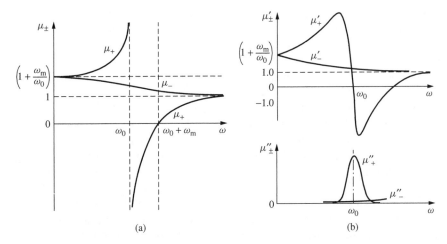

Figure 8.2 The frequency dependences of μ_+ and μ_-. (a) Ideal magnetic material with no magnetic loss and (b) magnetic material with magnetic loss

nonresonant and resonant methods. A nonresonant method can cover a certain frequency range, while a resonant method usually has higher accuracy.

In this section, we concentrate only on the measurement of permittivity tensors. The methods that can be used to measure both permittivity and permeability tensors are discussed in the Section 8.3.

8.2.1 Nonresonant methods

Nonresonant methods mainly include reflection and transmission/reflection methods. In principle, any type of transmission lines can be used for the measurement of anisotropic materials. As higher-order

modes are often involved, numerical techniques are often used in the calculation of tensor components.

8.2.1.1 Reflection methods

The principles for reflection methods have been discussed in Chapter 3. In the following text, we discuss the reflection methods developed from three types of transmission lines for the measurement of permittivity tensors: coaxial line, waveguide, and free space.

Coaxial-line method

Figure 8.3 shows a coaxial cell proposed for the measurement of uniaxial materials using reflection

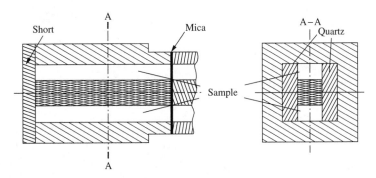

Figure 8.3 Structure of the coaxial cell for the measurement of uniaxial materials. Source: Parneix, J. P. Legrand, C. and Toutain, S. (1982). "Automatic permittivity measurements in a wide frequency range: application to anisotropic fluids", *IEEE Transactions on Microwave Theory and Techniques*, **30** (11), 2015–2017. © 2003 IEEE

method (Parneix *et al.* 1982). It is a short-circuited rectangular hybrid structure with the same geometrical sizes as the standard 7-mm coaxial connector, so that the measurement cell can be connected to a 7-mm coaxial connector. The input plane is a piece of mica sheet. The gap between the internal and the external conductors is partially filled with quartz that makes it possible, firstly, to get a homogeneous electric field inside the sample under test and, secondly, to limit the volume of sample. Quartz is chosen because of its very low thermal coefficient and low dielectric losses. What is more, it can be easily machined with high accuracy.

The essence of this method is that, in the measurement cell, the electric field lines are parallel to each other, and there are two measurement configurations: either the direction of the electric field is parallel to the optical axis (ε_3) or perpendicular to it (ε_1). So the components ε_1 and ε_3 of permittivity tensor in the form of Eq. (8.12) can be measured independently. Besides, the coaxial measurement cell has two additional features. First, a small sample volume is required so that samples with small quantities can be tested. Second, a wide frequency range of measurement can be realized by optimizing the geometrical sizes of the cell.

The calculation of materials properties mainly consists of two steps. In the first step, a fictitious permittivity (ε_f) is calculated using the way as discussed in Chapter 3, assuming the cell to be coaxial line of the same length and filled homogeneously with the sample. In the second step, the permittivity (ε_s) of the sample is obtained using a linear relationship

$$\varepsilon_f = \varepsilon_f' - j\varepsilon_f'' = (1-\theta)\varepsilon_q + \theta\varepsilon_s \qquad (8.38)$$

where ε_q is the permittivity of quartz and θ is a filling factor. As quartz is extremely low loss, its imaginary part of permittivity can be ignored. From Eq. (8.38), the real and imaginary parts of the permittivity of the sample can be obtained:

$$\varepsilon_s' = \frac{\varepsilon_f' - (1-\theta)\varepsilon_q}{\theta} \qquad (8.39)$$

$$\varepsilon_s'' = \frac{\varepsilon_f''}{\theta} \qquad (8.40)$$

In the calculation of the filling factor θ, a quasi-TEM propagation-mode approximation is assumed

in spite of the hybrid character of the measurement cell. By using a numerical method, ε_f can be calculated as a function of ε_s' by solving Laplace's equation in a cross-section plane of the cell. The relationship between ε_f and ε_s' is approximately linear, and the slope of the straight line gives the value of θ.

In the working range of the coaxial cell, the quasi-TEM approximation can give consistent results for low-loss dielectric materials with ε_s' values up to 7. Such a cell can be used to characterize both bulk and liquid materials. The measurement accuracy is close to that of a usual reflection method: the overall uncertainty is less than 2% for ε_s' and 5% for ε_s'' (Parneix *et al.* 1982).

Waveguide method

As discussed in Chapter 3, when a piece of dielectric sample covers an open-end waveguide probe with a flat plane, from the reflection at the interface between the open end of the waveguide and the sample, the dielectric permittivity of the sample can be calculated. Chang *et al.* extended this method for the characterization of anisotropic materials (Chang

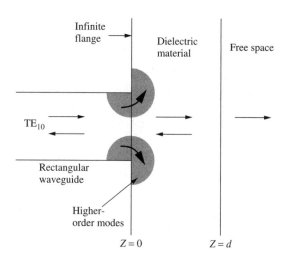

Figure 8.4 Nondestructive measurement using an open-end waveguide. Source: Chang, C. W. Chen, K. M. Qian, J. (1996). "Nondestructive measurements of complex tensor permittivity of anisotropic materials using a waveguide probe system", *IEEE Transactions on Microwave Theory and Techniques*, **44** (7), 1081–1090. © 2003 IEEE

et al. 1996). Figure 8.4 shows a measurement configuration in which a flanged open-end rectangular waveguide is placed against a layer of unknown anisotropic material with thickness d and backed by free space.

The essence of this method is that the total electric field in the probe aperture includes not only a dominant TE_{10} mode but also higher-order modes. Because of the higher-order modes, the electric fields in the aperture generally have components in any directions, and so the reflection from the aperture is related to all the components in the permittivity tensor of the material covering the aperture. To determine the components of permittivity tensor, independent reflection measurements are needed. For example, three independent measurements are required to determine the three principal permittivity components in Eq. (8.11).

Numerical calculation is needed for the determination of permittivity tensor. The algorithm consists of the direct problem and the inverse problem. The direct problem can be solved by matching the boundary conditions at discontinuity interfaces to constitute two coupled integral equations for the unknown aperture electric field. The electromagnetic field excited in the waveguide can be expressed on the basis of the Hertzian potentials, while the electromagnetic fields in the material layer and the free space backing the material layer can be derived on the basis of the concept of eigenmodes of the spectrum-domain transverse electromagnetic fields. Two coupled electric field integral equations (EFIEs) for the electric field in the aperture can then be derived. The EFIEs can be converted to a matrix equation using the method of moments, and the aperture electric field can then be determined by solving the matrix equation. Once the aperture electric field is determined, relevant quantities such as the reflection coefficient of the incident dominant wave and the probe input admittance can be expressed as functions of the electromagnetic properties of the material. If the reflection coefficient of the incident wave can be experimentally measured, the electromagnetic properties of the material can be determined using Newton-Raphson method in the inverse problem (Chang *et al.* 1996).

For a layer of anisotropic material with a diagonal form of the complex tensor permittivity in the form of Eq. (8.11), three independent measurements are required to determine the three unknown principal permittivity components ε_1, ε_2, and ε_3. The three measurements can be obtained by three different orientations of the probe with respect to a principal axis of the material, for example, 0, 45, and 90 degrees, or 0, 30, and 90 degrees. Actually, independent measurements can be obtained in various configurations. For example, the sample under test can be backed by a metal, and an air gap can be left between the sample and the backing metal. More configurations of reflection measurements can be found in Chapter 3.

To accurately determine the tensor permittivity components, especially the one in the z-direction, the sample cannot be very thin. Similar to the conventional reflection probe method, the air gap between the probe and the sample causes serious error to the measurement results. One possible solution is to introduce a layer of air with known thickness between the waveguide probe and the sample under test. Finally, it should be noted that, in principle, the waveguide probe system is capable of simultaneously determining both permittivity tensor and permeability tensor of anisotropic materials provided that sufficient independent measurements can be made.

Free-space method

Rubber sheets with various kinds of fillers have many applications, such as shielding materials and wave absorbers. These materials usually have anisotropy that comes from the rolling method of manufacturing, because the fillers tend to be aligned along the rolling direction. Hashimoto and Shimizu proposed a free-space reflection method for the measurement of the permittivity tensors of rubber sheets manufactured by rolling process (Hashimoto and Shimizu 1986). In this method, the permittivity tensor is determined from the reflection coefficient measured at normal incidence, and the procedure mainly consists of two steps. First, the frequency characteristics of the reflection coefficients of a rubber sheet backed by metal plate are measured by changing the

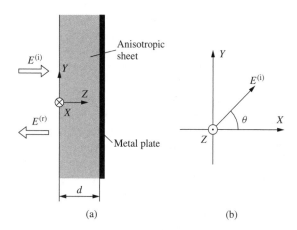

(a)   (b)

Figure 8.5 Measurement of permittivity tensor of sheet samples. (a) Lossy anisotropic sheet backed by a metal plate and (b) direction of incident electric field

angle between the rolling direction and the incident electric field. Next, the elements of the permittivity tensor are determined by the least-squares method from the reflection coefficients of the samples measured in the first step. After the tensor, including off-diagonal elements, is obtained, the principal directions of the tensor can be calculated from the measured elements.

Figure 8.5 shows a sheet sample backed by a metal plate, and the sample is anisotropic only in the $X-Y$ plane. The permittivity tensor of such a sheet is in the form of

$$[\varepsilon] = \begin{bmatrix} \varepsilon_{11} & \varepsilon_{12} & 0 \\ \varepsilon_{21} & \varepsilon_{22} & 0 \\ 0 & 0 & \varepsilon_{33} \end{bmatrix} \qquad (8.41)$$

where $\varepsilon_{12} = \varepsilon_{21}$, and the subscripts 1, 2, and 3 denote the directions of X-, Y- and Z-axes, respectively.

In order to measure the permittivity tensor of the anisotropic sheet, the reflection coefficients from the sheet backed by a metal plate must be analyzed. The coordinate system for analysis and the direction of the incident electric field are shown in Figure 8.5, in which the direction of the X-axis is usually taken in the direction of rolling. By substituting Eq. (8.41) into Maxwell's equations and eliminating the transverse components of the magnetic field, the following differential equation can be obtained for the rubber-sheet region (Hashimoto and Shimizu 1986):

$$-\frac{\partial}{\partial z^2} E_t = \omega^2 \mu_0 [\varepsilon_t] E_t \qquad (8.42)$$

where $[\varepsilon_t]$ is the dielectric permittivity tensor that has elements ε_{ij} $(i, j = 1, 2)$, and the subscript t denotes the transverse components.

The transverse electric field can be expressed by

$$E_t = \begin{bmatrix} A_1 \\ A_2 \end{bmatrix} \exp(-\gamma z) \exp(j\omega t) \qquad (8.43)$$

where A_1 and A_2 are amplitude coefficients of E_x and E_y respectively, and γ is the propagation constant. The reflection coefficients in the directions of X- and Y-axes, Γ_x and Γ_y, can be calculated by applying the boundary conditions that are the continuation of tangential field components. From Γ_x and Γ_y, the reflection coefficient Γ of the direction of angle θ can be calculated:

$$\Gamma = (\Gamma_x' \cos\theta + \Gamma_y' \sin\theta) + j(\Gamma_x'' \cos\theta + \Gamma_y'' \sin\theta) \qquad (8.44)$$

with

$$\Gamma_x = \Gamma_x' + j\Gamma_x'', \qquad (8.45)$$

$$\Gamma_y = \Gamma_y' + j\Gamma_y''. \qquad (8.46)$$

The elements of the tensor $[\varepsilon_t]$ can be calculated by the least-squares method, and the objective function F is defined by

$$F(\varepsilon_1, \varepsilon_2, \ldots, \varepsilon_6) = \sum_{v=1}^{N} (\hat{X}_v - X_v)^2 \qquad (8.47)$$

with

$$\hat{X}_v = 20 \log_{10} |\Gamma_{\text{calc}}|_v, \qquad (8.48)$$

$$X_v = 20 \log_{10} |\Gamma_{\text{meas}}|_v, \qquad (8.49)$$

where ε_i $(i = 1, 2, \ldots, 6)$ are the six real or imaginary parts of the three tensor elements (ε_{11}, ε_{12}, and ε_{22}) to be determined, and N is the number of data. The value of $|\Gamma_{\text{calc}}|_v$ is the reflection coefficient calculated theoretically according to the guess values of materials properties. $|\Gamma_{\text{meas}}|_v$, which is the absolute value of the reflection coefficient in the same direction as the direction of incident field, is measured by changing the angle θ between the direction of electric field and the X-axis. So the determination

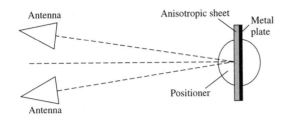

Figure 8.6 Top view of a free-space measurement setup

of materials properties becomes finding suitable ε_i ($i = 1, 2, \ldots, 6$) values, which result in minimum F value defined by Eq. (8.47).

Figure 8.6 shows a free-space measurement system for reflectivity measurement. Two antennas are used for transmitting and receiving microwave signals respectively. The anisotropic sheet backed by a metal plate is placed on a positioner. To decrease the measurement noise, the measurement is often conducted in an anechoic chamber. To reduce the effect of the sample edge to the measurement results, the samples under test are usually in a circular shape, as shown in Figure 8.7. The different angles between the electric field of the incident wave and the X-axis can be achieved by rotating the sample around its

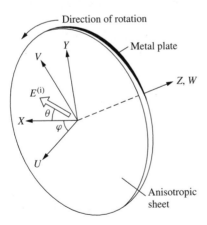

Figure 8.7 Schematic view of a circular sample. The directions of X, Y, and Z are defined in Figure 8.5, and U, V, and W denote the principal directions of the sample. Source: Hashimoto, O. Shimizu, Y. (1986). "Reflecting characteristics of anisotropic rubber sheets and measurement of complex permittivity tensor", *IEEE Transactions on Microwave Theory and Techniques*, **34** (11), 1202–1207. © 2003 IEEE

axis. After the complex permittivity tensors including off-diagonal elements are obtained, the principal directions of the tensor can be calculated from the measured permittivity tensor.

8.2.1.2 Transmission/reflection method

The basic principle of transmission/reflection method has been discussed in Chapter 4. Here, we discuss the circular-waveguide method and the coaxial discontinuity method for the measurement of permittivity tensors of uniaxial materials.

Circular-waveguide method

Pradoux *et al.* studied the electromagnetic wave propagation in circular waveguides filled with anisotropic media and proposed a method for the determination of the permittivity tensor of uniaxial dielectrics (Pradoux *et al.* 1973a, b).

Figure 8.8 shows two coordinate systems for a cylindrical uniaxial sample: the optical coordinates (UVW) and geometrical coordinates (XYZ). The W-axis in the optical coordinate system is along the Z-axis in the geometrical system. The U-axis is the main optical axis and is perpendicular to the Z-axis. The angle between the U-axis and the X-axis is φ. In the optical coordinates (UVW), its permittivity tensor is in the form of

$$[\varepsilon_r] = \begin{bmatrix} \varepsilon_1 & 0 & 0 \\ 0 & \varepsilon_2 & 0 \\ 0 & 0 & \varepsilon_2 \end{bmatrix} \qquad (8.50)$$

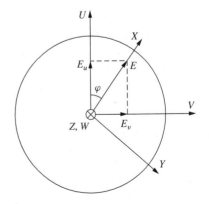

Figure 8.8 Two coordinate systems for a cylindrical uniaxial sample

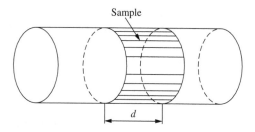

Figure 8.9 Circular-waveguide measurement cell loaded with a uniaxial sample

While in the geometrical coordinates (*XYZ*), the permittivity tensor is in the form of

$$[\varepsilon_r] = \begin{bmatrix} \varepsilon_1 + \eta \sin^2 \varphi & \eta \sin \varphi \cos \varphi & 0 \\ \eta \sin \varphi \cos \varphi & \varepsilon_2 + \eta \cos^2 \varphi & 0 \\ 0 & 0 & \varepsilon_2 \end{bmatrix}$$

(8.51)

with $\eta = \varepsilon_2 - \varepsilon_1$.

Figure 8.9 shows a circular waveguide filled with an anisotropic substance with one of its optical axes along the direction of propagation (*Z*-axis), and its *UVW* coordinates are defined in Figure 8.8. An incident TE wave can be split into two waves, one with its electric field along the direction of the *U*-axis, and the other along the *V*-axis, which is perpendicular to the *U*-axis. The wave with electric field parallel to *U*-axis propagates as a TE mode that is very close to the incident mode that propagates in an isotropic substance with dielectric constant ε_1. The electric field of this wave is $E_u = E \cos \varphi$, undergoes a phase change θ_1 when it emerges from the sample. The wave with electric field parallel to *V*-axis propagates as a TE mode that is very close to the incident mode that propagates in an isotropic substance of dielectric constant ε_2. The electric field of this wave is $E_v = E \sin \varphi$, and it undergoes a phase change θ_2. When they emerge from the sample, the two electric fields are combined together, taking into account their phase difference.

After the wave propagates through the sample, the wave polarization is changed from linear polarization to elliptic polarization with ellipticity β, and the major axis of the elliptic wave makes an angle α with the *U*-axis. From the rotation and the ellipticity of the wave, which has propagated through the sample, taking into account the multiple reflections, the two elements of the

permittivity tensor of the uniaxial sample can be obtained. Detailed discussion on this method and its calculation algorithms can be found in references (Pardoux *et al.* 1973a,b).

Coaxial discontinuity method

As shown in Figure 8.10, in this method, a segment of the central conductor of a coaxial line is moved away, and the cylindrical sample under test is inserted inside the outer conductor of coaxial line and between the two ends of the central conductors. As two independent complex parameters (reflection and transmission) can be measured, two independent material property parameters can be derived. The coaxial discontinuity method has been discussed in Chapter 4 for the characterization of both permittivity and permeability of isotropic materials.

This method can also be used in characterizing the permittivity tensor of uniaxial dielectric materials (Belhadj–Tahar and Fourrier–Lamer 1991). In the measurement, the main axis of the sample is along the axis of the coaxial line ($Y_0 - Y_0$). In the region between the two ends of the central conductors (T_1 and T_2), the electric field has two components in the directions parallel and perpendicular to the coaxial axis, respectively. Therefore, from the transmission and reflection parameters of the region filled with the sample, the two

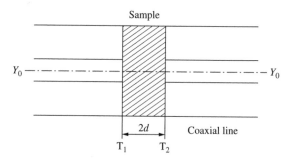

Figure 8.10 Coaxial discontinuity method for materials property characterization. Modified from Belhadj-Tahar, N. E. and Fourrier-Lamer, A. (1991). "Broad-band simultaneous measurement of the complex permittivity tensor for uniaxial materials using a coaxial discontinuity", *IEEE Transactions on Microwave Theory and Techniques*, **39** (10), 1718–1724. © 2003 IEEE

permittivity-tensor components can be determined. As there are no closed-form equations available for materials property deduction, numerical calculations are required, and often the mode-matching method is used.

8.2.2 Resonator methods

Because of their high accuracy and sensitivity, resonant methods are more suitable for low-loss materials. The resonant methods for the characterization of permittivity tensors also include resonator method and resonant-perturbation method. Here, we discuss the resonator methods for the characterization of permittivity tensors, and the resonant-perturbation methods for the characterization of permittivity tensors are discussed in Section 8.2.3.

As discussed in Chapter 5, in a resonator method, the sample under test serves as a dielectric resonator in the measurement circuit, and the dielectric properties of the sample are determined from the resonant properties of the dielectric resonator (Kobayashi and Tanaka 1980). To measure the two complex components of the permittivity tensor of a uniaxial sample using a dielectric resonator method, usually two modes with electric fields perpendicular to each other are needed. In the following text, we discuss three examples of dielectric resonator methods for the measurement of permittivity tensors of uniaxial dielectrics.

8.2.2.1 Shielded resonator methods

Figure 8.11 shows a cylindrical dielectric resonator enclosed by a closed metal shield. The main dielectric axis of the sample is along the cylindrical axis of the sample. The sample with radius b is at the center of a cylindrical metal shield with radius a. The sample and the metal shield have the same height, so there is no air gap between the open ends of the sample and the endplates of the metal shield.

In the measurements of isotropic materials, one of the TE_{0nl} modes (usually the TE_{011} mode) is used. The same circularly symmetric modes can also be used for the measurement of the permittivity-tensor component (ε_1) that is perpendicular to the anisotropy axis of the uniaxially anisotropic sample.

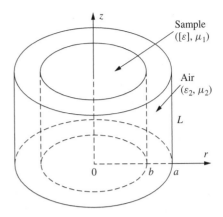

Figure 8.11 Configuration of the dielectric resonator

The determination of the parallel permittivity-tensor component (ε_3) requires one of the hybrid or TM modes. Usually, the HE_{111} and the HE_{211} modes are used because they can be easily excited. Since the permittivity component (ε_1) perpendicular to the anisotropy axis has been evaluated by solving the characteristic equation for the TE_{0nl} mode family, the permittivity component (ε_3) parallel to the anisotropy axis can be obtained from the resonant frequency of one hybrid mode by solving the characteristic equation for the hybrid mode (Geyer and Krupka 1995; Krupka *et al.* 1994a).

The dielectric loss tangent for the permittivity-tensor component perpendicular to the anisotropy axis can be calculated according to the method used in characterizing the loss tangent of isotropic materials:

$$\tan \delta = p_e \left(\frac{1}{Q_u} - \frac{1}{Q_c} \right) \tag{8.52}$$

where Q_u is the unloaded quality factor of the TE_{0nl} mode of the resonant system with the dielectric sample, Q_c is the quality factor depending on conductor losses of the metal shield, and p_e is the electric-energy filling factor of the dielectric sample. The electric-energy filling factor p_e, whose value in most cases is close to 1, can be evaluated with exact formulas that require prior knowledge of the permittivity (Geyer and Krupka 1995). Evaluation of Q_c requires additional knowledge of the surface resistances of endplates and sidewall of the metal shield. Evaluation of the dielectric loss tangent

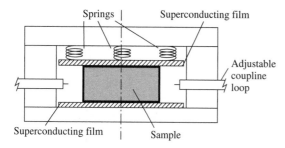

Springs | Superconducting film

Adjustable coupline loop

Superconducting film | Sample

Figure 8.12 Resonant setup for measurement of permittivity tensor of uniaxial dielectrics at cryogenic temperatures. Modified from Geyer, R. G. and Krupka, J. (1995). "Microwave dielectric properties of anisotropic materials at cryogenic temperatures", *IEEE Transactions on Instrumentation and Measurement*, **44** (2), 329–931. © 2003 IEEE

for the permittivity-tensor component parallel to the anisotropy axis is also theoretically possible. However, in this case, measurement error would be much greater because of the appearance of hybrid mode splitting. This splitting is caused by unavoidable imperfections in the axial symmetry of the resonant sample under test.

8.2.2.2 Sandwiched resonator method

Geyer and Krupka developed a method of measuring the permittivity tensor using a circularly symmetric mode and a hybrid mode of the dielectric resonator formed by shorting the two ends of a uniaxially anisotropic sample with two pieces of high-temperature superconducting thin films (Geyer and Krupka 1995; Krupka *et al.* 1994a). As shown in Figure 8.12, the dielectric sample under test constitutes a cylindrical dielectric resonator situated between two parallel superconducting films, and the cylindrical sample possesses uniaxial anisotropy along the cylinder axis. Copper–beryllium springs are used to hold the sample tightly between the endplates of the shield. If the sample faces are optically finished, there are essentially no air gaps between the sample and the endplate. The general resonant properties of such a sandwiched dielectric resonator has been discussed in Chapter 2. In Chapter 5, this configuration has been used for the measurement of the dielectric permittivity of isotropic dielectric samples and surface resistance of superconducting films.

When this configuration is used in the measurement of permittivity tensor of a uniaxial dielectric sample, the components of the permittivity tensor can be obtained in a way similar to that used in the shielded resonator method discussed above. The permittivity-tensor component (ε_1) that is perpendicular to the anisotropy axis of uniaxially anisotropic material can be obtained from the resonant properties of a circularly symmetric mode. Then with the knowledge of ε_1, the permittivity-tensor component (ε_3) that is parallel to the anisotropy axis can be obtained from the resonant properties of the hybrid mode. The use of superconducting films is to increase the measurement sensitivity of the loss tangent. However, as superconducting films are used, this configuration is only suitable for measurements at cryogenic temperatures.

8.2.2.3 Whispering-gallery resonator method

The properties of whispering-gallery mode resonators have been discussed in Chapter 2. The use of whispering-gallery modes in dielectric resonator specimens is one of the most accurate ways for determining the complex permittivity of ultra-low-loss dielectric materials, which may be either isotropic or uniaxially anisotropic (Krupka *et al.* 1999, 1994b).

When whispering-gallery modes are employed, the radiation losses are negligibly small, especially for the modes with large azimuthal mode number. Therefore, they can be used alternatively with or without a metallic shield for high-resolution dielectric loss tangent measurements of ultra-low-loss materials. Figure 8.13 shows two typical configurations of whispering-gallery resonators for the characterization of the properties of dielectric samples. In Figure 8.13(a), a crystal sample is enclosed in a metal shield, while in Figure 8.13(b) a ceramic sample is excited in an open space.

The relationship between resonant frequencies, dimensions of the resonant structure, and permittivity of the sample under test can be calculated with a radial mode-matching technique (Krupka *et al.* 1999, 1994b). In this method, the resonant structure is first subdivided into cylindrical regions having dielectric inhomogeneity only along the axial

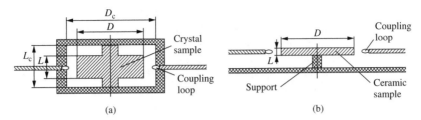

Figure 8.13 Two types of whispering-gallery resonators. (a) Shielded resonator and (b) unshielded resonator. Source: Krupka, J. Derzakowski, Abramowicz, A. Tobar, M. E. and Baker-Jarvis, J. (1999). "Use of whispering-gallery modes for complex permittivity determinations of ultra-low-loss dielectric materials", *IEEE Transactions on Microwave Theory and Techniques*, **47** (6), 752–759. © 2003 IEEE

direction. The electromagnetic field components are then expanded into series modal expansions separately in each region. Application of the boundary conditions at the boundaries between adjacent regions leads to a system of matrix equations with respect to the field expansion coefficients that has nonzero solutions only when the determinant vanishes. The resonant frequencies are the values making the determinant of the square matrix vanish. If the resonant structure has a plane of symmetry, the system of equations splits into two separate systems for the modes symmetrical and antisymmetrical, with respect to the plane of symmetry. Since for whispering-gallery modes the electromagnetic fields are very well confined to the dielectric sample, one can use subdivision into symmetrical and antisymmetrical modes, even for structures that physically do not exhibit a planar symmetry.

In order to find the principal permittivity-tensor components of a uniaxially anisotropic dielectric material, a cylindrical specimen whose cylindrical axis is along one principal direction of anisotropy should be used. The resonant frequencies of two whispering-gallery modes of the specimen that exhibit quasi-TE (H-mode) and quasi-TM (E-mode) structures, respectively, must then be identified and measured. After that, the following two nonlinear determinant equations should be solved (Krupka *et al.* 1999)

$$F_1(f^{(H)}, \varepsilon_1, \varepsilon_3) = 0 \qquad (8.53)$$

$$F_2(f^{(E)}, \varepsilon_1, \varepsilon_3) = 0 \qquad (8.54)$$

where $f^{(H)}$ and $f^{(E)}$ are the measured resonant frequencies for the quasi-TE and the quasi-TM whispering-gallery modes, ε_1 and ε_3 are the real

parts of the permittivity-tensor components perpendicular and parallel to the anisotropy axis, respectively. The eigenvalue equations represented by F_1 and F_2 result from the application of the mode-matching technique.

Once permittivity components are evaluated from Eqs. (8.53) and (8.54), dielectric loss tangents can be computed as solutions to the following two linear equations:

$$Q_{(E)}^{-1} = p_{e1}^{(E)} \tan \delta_1 + p_{e3}^{(E)} \tan \delta_3 + R_s/G^{(E)} \quad (8.55)$$

$$Q_{(H)}^{-1} = p_{e1}^{(H)} \tan \delta_1 + p_{e3}^{(H)} \tan \delta_3 + R_s/G^{(H)} \quad (8.56)$$

where $\tan \delta_1$ and $\tan \delta_3$ are the dielectric loss tangents perpendicular and parallel to the anisotropy axis; $p_{e1}^{(H)}$, $p_{e3}^{(H)}$, $p_{e1}^{(E)}$, $p_{e3}^{(E)}$ are the electric-energy filling factors perpendicular (subscript 1) and parallel (subscript 3) to the anisotropy axis of the resonant structure, for quasi-TM whispering-gallery modes (superscript E) and quasi-TE whispering-gallery modes (superscript H); and $G^{(E)}$ and $G^{(H)}$ are the geometric factors for quasi-TM and quasi-TE whispering-gallery modes. The electric-energy filling factors may be determined from the incremental frequency rule:

$$p_{e1}^{(H)} = 2 \left| \frac{\partial f^{(H)}}{\partial \varepsilon_1} \right| \frac{\varepsilon_1}{f^{(H)}} \qquad (8.57)$$

$$p_{e3}^{(H)} = 2 \left| \frac{\partial f^{(H)}}{\partial \varepsilon_3} \right| \frac{\varepsilon_3}{f^{(H)}} \qquad (8.58)$$

$$p_{e1}^{(E)} = 2 \left| \frac{\partial f^{(E)}}{\partial \varepsilon_1} \right| \frac{\varepsilon_1}{f^{(E)}} \qquad (8.59)$$

$$p_{e3}^{(E)} = 2 \left| \frac{\partial f^{(E)}}{\partial \varepsilon_3} \right| \frac{\varepsilon_3}{f^{(E)}} \qquad (8.60)$$

Using resonant modes at different frequencies, complex permittivity measurements may be performed over a relatively wide frequency range. As the size of a whispering-gallery resonator can have a much larger size than a conventional dielectric resonator, the whispering-gallery modes may be conveniently applied to dielectric measurements at frequencies as high as 100 GHz. However, there exists some difficulty in employing whispering-gallery modes for measuring complex permittivity. One of the difficult aspects is the identification of the resonant modes. The identification is based on resonance-frequency computations with initially assumed permittivity values. This difficulty especially occurs when the initially assumed permittivity differs substantially from the true intrinsic permittivity. Another difficulty is the mode degeneracy, which may seriously affect the accuracy of the measurement results.

8.2.3 Resonant-perturbation method

The permittivity tensor of uniaxial dielectric materials can also be characterized by resonant-perturbation method using two degenerate TE_{112} modes in a cylindrical bimodal cavity (Chen *et al.* 1999). Compared to resonator methods, this method has the following advantages. First, this method does not require an estimate of the dielectric properties of samples, and the amount of sample needed is small. Second, the measurement frequency is determined by the resonant frequency of the cavity that can be designed for a selected frequency. For a specially designed cavity, the two components in the permittivity tensor of a uniaxial sample can be measured at the same frequency. Third, as the conventional cavity-perturbation method is modified by using the frequency-retuning method discussed in Chapter 6, the uncertainties caused by the increase of the total stored energy in the cavity are eliminated.

8.2.3.1 Working principle

In this method, two resonant modes with electric fields perpendicular to each other are used to measure the two complex components in the permittivity tensor of a uniaxially anisotropic dielectric. To avoid the errors caused by the possible coupling between the two modes, the two complex components are measured separately and independently. In the measurement of each component, only the corresponding mode resonates, and the other mode is eliminated.

The measurement of each component is based on resonant perturbation theory, which has been discussed in Chapter 6. When a uniaxially anisotropic sample is introduced at the position of maximum electric field, its complex component ($\varepsilon_i = \varepsilon_i' - j\varepsilon_i''$) in the direction of the electric field can be obtained from the changes of the resonant frequency f and the quality factor Q of the cavity:

$$\frac{f_{ia} - f_{ib}}{f_{ib}} = A_i(\varepsilon_i' - 1)\frac{v}{V} \qquad (8.61)$$

$$\frac{1}{Q_{ib}} - \frac{1}{Q_{ia}} = B_i \varepsilon_i'' \frac{v}{V} \qquad (8.62)$$

where the subscript "i" indicates the component in the direction of the electric field; and subscripts "a" and "b" indicates the states before and after the introduction of the sample respectively; v and V are the volumes of the sample and the cavity respectively. The two parameters A_i and B_i are usually obtained by calibrating the cavity using a standard sample with known complex permittivity. As the introduction of the sample alters the coupling condition between the cavity and the transmission line, unloaded quality factors are used in Eq. (8.62). In the conventional cavity-perturbation method, because there is always a decrease in the resonant frequency due to sample insertion, Q_{ia} and Q_{ib} are usually measured at different frequencies, f_{ia} and f_{ib}. However, as indicated in Chapter 6, Eq. (8.62) is based on several assumptions, among which the most fundamental one is that the field configuration and the total stored energy in the cavity do not change because of the introduction of the sample (Sucher and Fox 1963; Waldron 1970). However, when a dielectric sample is placed at the position of maximum electric field in a resonant cavity, the resonant frequency of the cavity is decreased, and the total stored energy in the cavity is increased (Jow *et al.* 1989; Tian and Tinga 1993, 1995; Maier and Slater 1952). Therefore, in a strict sense, it is not appropriate to deduce ε_i'' from Q_{ia} and Q_{ib} measured at f_{ia} and f_{ib}.

The frequency-retuning method discussed in Chapter 6 can be used to modify the conventional cavity-perturbation method. For a resonant cavity with a dielectric sample at the position of maximum electric field, when the resonant frequency of the cavity is retuned to the value before the sample is inserted by introducing an additional perturbation, the total stored energy in the cavity is also retuned to its original value. If the additional perturbation is purely reactive, the energy dissipation in the cavity does not change because of frequency retuning. So the difference between the quality factor before sample insertion (Q_{ia}) and the quality factor after sample insertion and frequency retuning (Q_{ib}) reflects the increase of the energy dissipation due to sample insertion. Therefore, Q_{ia} and Q_{ib} can be used in Eq. (8.62) to calculate the value of ε_i''. In experiments, frequency retuning can be achieved by cavity shape perturbation. By installing a metal plunger on the cavity wall,

cavity shape perturbation can be made by adjusting the plunger's insertion depth in the cavity.

8.2.3.2 Bimode cavity

To measure the permittivity tensor of a uniaxially anisotropic material using cavity-perturbation method, a resonant cavity with the following features is needed. First, the cavity has two perpendicular modes working at the same frequency; second, the resonant frequency of the cavity is adjustable so that frequency retuning is possible; and third, the coupling coefficients between the resonant cavity and the external transmission lines are adjustable so that suitable coupling conditions can be achieved for different samples.

Figure 8.14 shows a cylindrical cavity coupled to two waveguides (labeled 1) through coupling irises (labeled 3). Near each coupling iris, there is a plunger (labeled 2) for adjusting the coupling

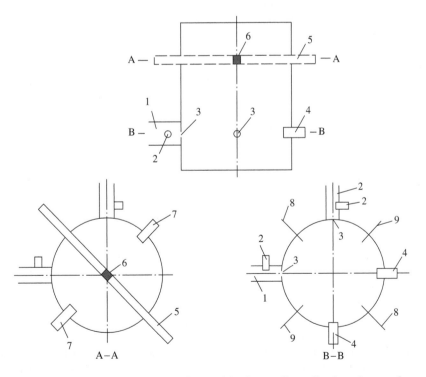

Figure 8.14 Structure of the bimodal cavity. 1. Waveguide, 2. coupling adjusting plunger, 3. coupling iris, 4. plunger, 5. quartz tube, 6. sample, 7. retuning plunger, 8 and 9. mode-selection needle. Source: Chen, L. F. Ong, C. K. and Tan, B. T. G. (1999). "Cavity perturbation technique for the measurement of permittivity tensor of uniaxially anisotropic dielectrics", *IEEE Transactions on Instrumentation and Measurement*, **48** (6), 1023–1030. © 2003 IEEE

condition by varying its insertion depth into the waveguide. There is a quartz tube (labeled 5) crossing the cavity, and the sample (labeled 6) in the quartz tube is located at the position of maximum electric field.

Because of the following considerations, the cavity is selected to work at the TE_{112} mode. First, as each TE_{11n} mode has two degenerate modes with electric fields perpendicular to each other, it can be used to measure two complex components in the permittivity tensor of a uniaxially anisotropic material. Second, the change of the electric field distribution of the TE_{11n} mode is less near the central part of the cavity, so the requirement for the accuracy of sample location is less stringent. And third, as shown in Figure 8.14, by selecting TE_{112} mode instead of TE_{111} mode, the coupling area (lower part of the cavity) and the perturbation area (upper part of the cavity) are separated, so that the field distribution in the cavity hardly changes because of sample insertion and frequency retuning. Four brass plungers (two labeled 4 and two labeled 7) are used to adjust the two degenerate TE_{112} modes so that they have the same resonant frequency.

The two perpendicular degenerate TE_{112} modes are used to measure the two components ε_1 and ε_3 separately. In the measurement of each component, only the corresponding mode resonates, and the other mode is eliminated. Using the similar technique shown in Figure 6.7, mode selection can be made by using two thin needles. As shown in Figure 8.15, the two thin metal needles eliminate the mode whose electric field is parallel to the two needles, while the mode whose electric field is perpendicular to the needles remains unperturbed. The two needles labeled 8 in Figure 8.14 are parallel to the axis of the quartz tube, and they are used to select the mode whose electric field is normal to the axis of the quartz tube. The two needles labeled 9 in Figure 8.14 are normal to the axis of the quartz tube, and they are used to select the mode whose electric field is parallel to the axis of the quartz tube.

Two brass plungers (labeled 7 in Figure 8.14) are also used to retune the resonant frequencies of the two perpendicular TE_{112} modes. For the mode with electric field perpendicular to the plungers, shown in Figure 8.16(a), the plungers are at the places of maximum magnetic field, so the resonant frequency of the cavity increases if the plungers are moved in. For the mode with the electric field parallel to the tuning plungers, shown in Figure 8.16(b), the plungers are situated at the positions where the electric fields are dominant, so the resonant frequency of the cavity increases if the two plungers are moved out. In frequency retuning, the two plungers are adjusted simultaneously to preserve the symmetry of the cavity.

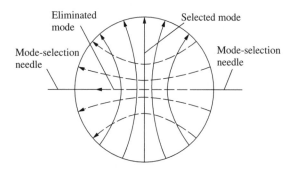

Figure 8.15 Mode selection. The mode whose electric field is in the direction of the mode-selection needles is eliminated. Modified from Chen, L. F. Ong, C. K. and Tan, B. T. G. (1999). "Cavity perturbation technique for the measurement of permittivity tensor of uniaxially anisotropic dielectrics", *IEEE Transactions on Instrumentation and Measurement*, **48** (6), 1023–1030. © 2003 IEEE

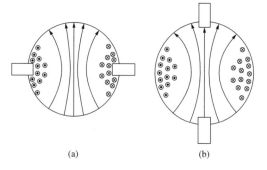

Figure 8.16 Frequency retuning. (a) The plungers are at the places of maximum magnetic field and (b) the plungers are at the places of maximum electric field. Modified from Chen, L. F. Ong, C. K. and Tan, B. T. G. (1999). "Cavity perturbation technique for the measurement of permittivity tensor of uniaxially anisotropic dielectrics", *IEEE Transactions on Instrumentation and Measurement*, **48** (6), 1023–1030. © 2003 IEEE

8.2.3.3 Measurement procedure

To measure the two complex components in the permittivity tensor, the perturbations of the sample to the two modes are measured separately. The measurements are carried out in the following four steps.

First, the degeneration of TE_{112} mode: Install the quartz tube, set the four metal plungers (labeled 4 and 7 in Figure 8.12) and the two coupling adjustment plungers (labeled 2 in Figure 8.14) to their proper positions, and move out all the mode-selection needles. Two resonant peaks corresponding to the two TE_{112} modes will be observed. Adjust the four metal plungers till the two resonant peaks overlap.

Second, the measurement of f_{ia} and Q_{ia} ($i = 1, 3$): Select the mode with electric field normal to the axis of the quartz tube by moving in the two mode-selection needles (labeled 8 in Figure 8.14). Measure the resonant frequency f_1 and the quality factor Q_{1a}, and move out the mode-selection needles after the measurement. Then, select the mode with electric field parallel to the axis of the quartz tube by moving in the two mode-selection needles (labeled 9 in Figure 8.14). Measure its resonant frequency f_{3a} and the quality factor Q_{3a}, and move out the mode-selection needles after the measurement. Fine adjustment is needed till f_{1a} is equal to f_{3a} in the measurement accuracy.

Third, the measurement of f_{1b} and f_{3b}: Insert the sample into the quartz tube and place the sample at the position of maximum electric field. Select resonant modes according to the method described in the second step, and measure the resonant frequencies of the two modes after sample insertion, f_{1b} and f_{3b}.

Fourth, the measurement of Q_{1b} and Q_{3b}: Select resonant modes according to the method described in the second step, retune the resonant frequency of the cavity to f_{ia} ($f_{1a} = f_{3a}$) by adjusting the retuning plungers (labeled 7 in Figure 8.14), and measure the quality factors Q_{1b} and Q_{3b}.

After obtaining the values of f_{ia}, f_{ib}, Q_{ia}, and Q_{ib} ($i = 1, 3$), the complex components ε_i can be calculated according to Eqs. (8.61) and (8.62). It should be pointed out that, for Eqs. (8.61) and (8.62), the two degenerate modes have their corresponding parameters A_i and B_i ($i = 1, 3$), due to

the presence of the quartz tube, metal plungers, and mode-selection needles. So the two modes should be calibrated separately before measurements.

This method has been used to study the orientation dependence of the permittivity of α-SiO$_2$ crystals (Chen *et al.* 1999). As shown in Figure 8.17, in a α-SiO$_2$ cylindrical sample, there is an angle θ between the cylindrical z-axis of the sample and the c-axis in its crystal structure. As α-quartz crystals with different crystalline orientations are widely used in electrical engineering, it is important to investigate the dielectric permittivity $\varepsilon(\theta)$ of quartz samples with different θ values. The five samples had different θ values: $0°, 30°, 45°, 60°$, and $90°$ respectively. All the samples have the same dimensions: $\Phi 2.9\,\text{mm} \times 2.9\,\text{mm}$, and all the quartz samples have been fired at $800\,°\text{C}$ for $4\,\text{h}$, and slowly cooled to room temperature to avoid possible defects and also to keep their piezoelectric structures.

For the sample shown in Figure 8.17, there are two dielectric permittivities with a special meaning: the dielectric permittivity when the electric field is in the plane perpendicular to z-direction, $\varepsilon_\perp$, and the one when the electric field is in the z-direction, $\varepsilon_\parallel$. Because of the piezoelectric effect, the permittivity tensor of an α-SiO$_2$ crystal is dependent on the cutting angle θ. The relationship between $\varepsilon_\perp$ and θ may be represented by (Cady 1946)

$$\begin{aligned}\varepsilon_\perp(\theta) &= \varepsilon_\perp' - j\varepsilon_\perp'' \\ &= \varepsilon_1 + (\varepsilon_3 - \varepsilon_1)\sin^2\theta \\ &= \varepsilon_1 \cos^2\theta + \varepsilon_3 \sin^2\theta \end{aligned} \quad (8.63)$$

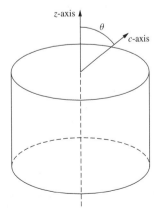

Figure 8.17 Schematic diagram of a α-SiO$_2$ sample

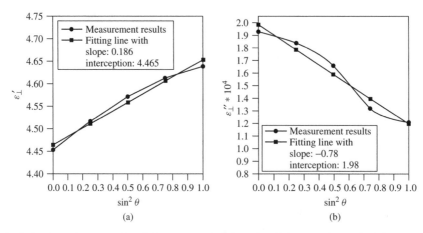

Figure 8.18 Relationships between the dielectric permittivity of α-SiO$_2$ samples and their crystal orientations. (a) Real part and (b) imaginary part. Source: Chen, L. F. Ong, C. K. and Tan, B. T. G. (1999). "Cavity perturbation technique for the measurement of permittivity tensor of uniaxially anisotropic dielectrics", *IEEE Transactions on Instrumentation and Measurement*, **48** (6), 1023–1030. © 2003 IEEE

where ε_1 and ε_3 are the permittivities in the directions normal and parallel to the c-axis of the crystal respectively.

The measurement results are shown in Figure 8.18. If a straight-line fit is used, $\varepsilon_\perp$ can be expressed as

$$\varepsilon'_\perp(\theta) = 4.465\cos^2\theta + 4.651\sin^2\theta \quad (8.64)$$

$$\varepsilon''_\perp(\theta) = 1.98 \times 10^{-4}\cos^2\theta$$
$$+ 1.20 \times 10^{-4}\sin^2\theta \quad (8.65)$$

From Eqs. (8.63) to (8.65), ε_1 and ε_3 can be obtained:

$$\varepsilon_1 = 4.465 - j1.98 \times 10^{-4} \quad (8.66)$$

$$\varepsilon_3 = 4.651 - j1.20 \times 10^{-4} \quad (8.67)$$

8.3 MEASUREMENT OF PERMEABILITY TENSORS

The methods for the measurement of permeability tensors can also be classified as the nonresonant methods and the resonant methods. For magnetic materials, there is a special phenomenon called *Faraday rotation*, based on which both nonresonant and resonant methods for the measurement of permeability tensors can be developed. The four parts of this section discuss nonresonant methods,

Faraday rotation methods, resonator methods, and resonant-perturbation methods, respectively.

It should be noted that in the study of ferromagnetic resonance, people are more interested in the relationships between the absorption of microwave energy, microwave frequency, and external dc magnetic field, and such information can be obtained from the permeability tensor. Most of the methods for the measurement of permeability tensor can be used for the study of ferromagnetic resonance. This section concentrates on the methods for the measurement of permeability tensors, and the methods specially designed for the measurement of ferromagnetic resonance are discussed in Section 8.4.

8.3.1 Nonresonant methods

The nonresonant methods for the measurement of permeability tensors are mainly the transmission/reflection methods, and the transmission lines often used are coaxial line and waveguide. In this part, we discuss coaxial air-line method, partially filled waveguide method, and waveguide-junction method.

8.3.1.1 Coaxial air-line method

Coaxial air-line method is widely used in characterizing the dielectric permittivity and magnetic permeability of isotropic materials. For the TEM mode inside a coaxial line, the electric field is in the radius direction, while the magnetic field is in the circular direction. So coaxial line can be used in measuring the permittivity component in the radius direction and permeability component in the circular direction. This method has been used in characterizing artificial high-impedance anisotropic magnetic materials (Jacquart and Acher 1996; Acher *et al.* 1994).

For some microwave applications, it is a significant advance to manufacture composite materials that combine the high saturation magnetization and the high permeability of ferromagnetic thin wires or films with small permittivity. However, composites manufactured by dispersing ferromagnetic particles in an insulating binder have intrinsic impedances lower than unity. As shown in Figure 8.19, one way of improving the intrinsic impedance is to fabricate coaxial samples by winding thin ferromagnetic wires or ferromagnetic films deposited on thin flexible substrates. This geometry can be approximated as consisting of coaxial alternated ferromagnetic and insulating cylinders, and the differences between strictly coaxial and spiral-wound geometries are of little consequence.

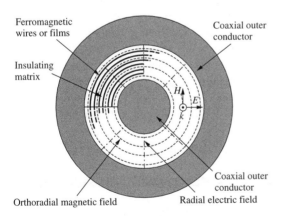

Figure 8.19 Sketch of a strongly anisotropic composite made of orientated conducting wires or films adapted to coaxial-line geometry. The propagation vector is *k*, and *E* and *H* represent the electromagnetic fields in TEM mode. Source: Jacquart, P. M. Acher, O. (1996). "Permeability measurement on composites made of oriented metallic wires from 0.1 to 18 GHz", *IEEE Transactions on Microwave Theory and Techniques*, **44** (11), 2116–2120. © 2003 IEEE

In such a composite, the component of the permittivity tensor parallel to the laminations is expected to be so large that a wave with electric field parallel to the laminations can not penetrate the composite.

The strongly anisotropic composites made from wiring thin wires or films can be characterized using coaxial air-line transmission/reflection method. Using this method, the two most important tensor components of such composites can be characterized: the permittivity along the radius direction and the permeability along the circular direction.

8.3.1.2 Partially filled waveguide method

A waveguide method has been developed for measuring the complex permeability tensor components and complex scalar permittivity of magnetized ferrites (Queffelec *et al.* 2000, 1999; Clerjon *et al.* 1999). The principle behind this method is using the anisotropy of the magnetic material to achieve the nonreciprocity of the measurement fixture loaded with the sample in order to have the same number of measurable parameters (the *S*-parameters of the fixture) as the number of the characteristics to be determined.

The configuration of the measurement fixture used in this method is identical to the one used for the realization of nonreciprocal waveguide devices, such as isolators and switches, except that there is no absorbing material in contact with the ferrite, as shown in Figure 8.20. To increase the cell sensitivity, the gyrotropic effect is intensified by setting the ferrite in a circularly polarized field area of the guide with a dielectric support. The waveguide is placed between the poles of an electromagnet to magnetize the sample.

When a uniform dc magnetic field H_0 is applied along the narrow side of the waveguide (*y*-axis of a Cartesian coordinate system), the field distributions of TE_{10} mode in the rectangular waveguide are given by

$$E_y = -j\omega\mu\frac{a}{\pi}H_{10}\sin\left(\frac{\pi x}{a}\right)e^{\mp jk_z z} \quad (8.68)$$

$$H_x = \pm jk_z\frac{a}{\pi}H_{10}\sin\left(\frac{\pi x}{a}\right)e^{\mp jk_z z} \quad (8.69)$$

$$H_z = H_{10}\cos\left(\frac{\pi x}{a}\right)e^{\mp jk_z z} \quad (8.70)$$

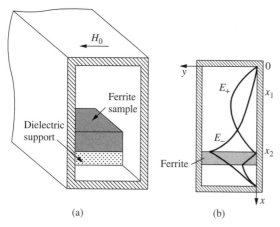

(a)   (b)

Figure 8.20 Measurement fixture. (a) Three-dimensional view and (b) cross section. E_+ and E_- represent the energy of microwave signals propagating along $+z$- and $-z$-directions respectively. Modified from Queffelec. P. Le Floc'h, M. Gelin, P. (1999). "Nonreciprocal cell for the broad-band measurement of tensorial permeability of magnetized ferrites: Direct problem", *IEEE Transactions on Microwave Theory and Techniques*, **47** (4), 390–397. © 2003 IEEE

where a is the width of the wide side, and μ is the permeability of the medium within the waveguide. The upper sign in the $\pm$ or $\mp$ symbol represents the wave in the $+z$-direction, while the lower sign represents the wave in the $-z$-direction. H_{10} is the amplitude of the components H_z.

From Eqs. (8.76) and (8.77), we can get

$$\frac{H_x}{H_z} = \pm jk_z \frac{a}{\pi} \tan\left(\frac{\pi x}{a}\right). \qquad (8.71)$$

At the position $x = x_1$ satisfying

$$\tan\left(\frac{\pi x_1}{a}\right) = \frac{\pi}{ak_z} = \frac{\lambda_g}{2a}, \qquad (8.72)$$

Eq. (8.71) becomes

$$\frac{H_x}{H_z} = \pm j \qquad (8.73)$$

Equation (8.73) indicates that at position $x = x_1$, the microwave magnetic field in the rectangular waveguide is of circular polarization. Relative to the direction of $\boldsymbol{H_0}$, the wave propagating in the $+z$-direction is of left-hand circular polarization, while the wave propagating in the $-z$-direction is of

right-hand circular polarization. It can also be found that at position $x = x_2 = a - x_1$, the microwave magnetic field is also of circular polarization, but the polarization direction is opposite to the polarization direction of the wave at position $x = x_1$.

As shown in Figure 8.20(b), when a magnetic material is placed at the position $x = x_2$, the field displacement occurs along the wide side of the guide (x-axis). This can be understood in the following way. We assume that the operating frequency is higher than the ferromagnetic-resonance frequency of the magnetic material ($\omega > \omega_0 = \gamma H_0$). The microwave magnetic field propagating in the $+z$-direction is of right-hand circular polarization. Figure 8.2(b) indicates that $\mu'_+ < 1$, so the electromagnetic field is "expelled" out of the magnetic material, and most of the electromagnetic energy of TE_{10} mode is concentrated in the space filled with air. The wave propagating in $-z$-direction is of left-hand circular polarization. Figure 8.2(b) indicates that $\mu'_- > 1$, so the electromagnetic energy is concentrated on the space filled with the magnetic sample.

The above discussions show that, in the measurement fixture, the sample has different effects on the forward modes (wave propagated in the positive direction) and the backward modes (wave propagated in the negative direction), so all the scattering matrix elements of the cell will be different. Thus, sufficient information is available to define specimen properties from scattering parameters.

Figure 8.21 shows the cross section along the z-direction of the measurement fixture loaded with

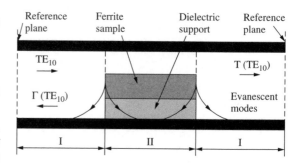

Figure 8.21 Discontinuities of the measurement fixture. Waveguide I: empty regions. Waveguide II: region partially filled with the ferrite

the sample under test. As the magnetic field is applied in *y*-direction, the permeability is in the form of Eq. (8.21). When the magnetization is saturate, $\mu_y = 1$. If the sample has isotropic permittivity ε_r, the electromagnetic properties of the sample can be expressed using three parameters: μ, κ, and ε_r. Therefore, three independent measurement parameters are needed for the determination of μ, κ, and ε_r.

The determination of material properties include two calculation procedures: direct problem and inverse problem. The direct problem is the electromagnetic analysis of the measurement cell that calculates the *S*-parameters as functions of the scalar permittivity, the permeability tensor, and dimensions of the ferrite. The electromagnetic analysis requires the modes determination in the waveguide and the use of the orthogonality conditions between the modes, and it is usually based on the mode-matching method applied to the waveguide discontinuities shown in Figure 8.21. The electromagnetic analysis for this structure is quite complicated, and detailed discussion on this issue can be found in (Queffelec *et al.* 2000, 1999).

The inverse problem is the computation of μ, κ, and ε_r of a given ferrite from the measured values of the *S*-parameters. It should be noted that the permeability component μ_y cannot be determined, as it does not appear in the *S*-parameters calculation, the modes being independent of this variable. The electromagnetic constants of the sample under test are determined by matching the calculated and the measured values of the *S*-parameters using a numerical optimization procedure, and the objective function for the optimization can be expressed as a sum of squared functions:

$$E(x) = \sum_{i=1}^{2} \sum_{j=1}^{2} |S_{ij,\text{theoretical}}(x) - S_{ij,\text{measured}}|^2$$

$$(8.74)$$

where $x = (\mu', \mu'', \kappa', \kappa'', \varepsilon_r', \varepsilon_r'')$. The optimization algorithm should lead to a fast location of the global minimum by avoiding local minima. The most significant input parameters for optimization algorithms are the number of variables, initial estimate of the solution, and expression of the objective function.

8.3.1.3 Waveguide-junction method

Damaskos *et al.* proposed a method for the determination of the constitutive parameters of an anisotropic material, by inserting a slab of the material in a rectangular waveguide (Damaskos *et al.* 1984). If both permittivity and permeability tensors have zero off-diagonal elements (biaxial material), the six diagonal elements can be determined by measuring amplitude and phase of reflection and transmission coefficients. If the material is nondispersive, two sets of measurements of TE$_{10}$ excitation at two different frequencies are sufficient. In the more general case of a lossy and dispersive material, two sets of measurements at the same frequency under TE$_{10}$ and TE$_{20}$ excitations are needed.

For a biaxial medium, its relative permittivity and permeability tensors $[\varepsilon_r]$ and $[\mu_r]$ are diagonal matrices represented by Eqs. (8.11) and (8.15) in a rectangular Cartesian reference system (x, y, z). The six constitutive parameters ε_i and μ_i $(i = 1, 2, 3)$ in $[\varepsilon_r]$ and $[\mu_r]$ are dimensionless numbers and are, in general, complex and frequency-dependent.

Consider a metallic rectangular waveguide oriented along the *z*-axis, with horizontal walls of width (a) parallel to the *x*-axis and vertical walls of height $(b \leq a)$ parallel to the *y*-axis. The waveguide is filled with a biaxial material in the range $0 \leq z \leq L$, and the principal axes of the biaxial material coincide with the (x, y, z) axes of the waveguide. The reflection coefficient R_m and transmission coefficient T_m for TE$_{m0}$ mode are given by (Damaskos *et al.* 1984)

$$\frac{1}{T_m} = \cos \beta_m L + \frac{\text{j}}{2} \left(u + \frac{1}{u} \sin \beta_m L \right) \qquad (8.75)$$

$$\frac{R_m}{T_m} = -\frac{\text{j}}{2} \left(u - \frac{1}{u} \sin \beta_m L \right) \qquad (8.76)$$

with

$$u = \frac{\mu_1 \beta_m}{\mu_3^2 \beta_{0m}} \qquad (8.77)$$

$$\beta_m = k_0 \sqrt{\frac{\mu_1}{\mu_3}} \sqrt{\varepsilon_2 \mu_3 - \left(\frac{m\pi}{k_0 a} \right)^2} \qquad (8.78)$$

$$\beta_{0m} = \sqrt{k_0^2 - \left(\frac{m\pi}{a} \right)^2} \qquad (8.79)$$

When L is given and the amplitude and phase of R_m and T_m have been measured, the inverse problem is to find u and β_m from Eqs. (8.75) and (8.76). The constitutive parameters of the sample can be found from u and β_m. In general, the parameters ε_2, μ_1, and μ_3 are complex, and therefore u and β_m are also complex.

From Eqs. (8.75) and (8.76), we can get

$$u = \frac{j \sin \beta_m L}{\dfrac{1}{T_m} + \dfrac{R_m}{T_m} - \cos \beta_m L}$$

$$= \frac{\dfrac{1}{T_m} - \dfrac{R_m}{T_m} - \cos \beta_m L}{j \sin \beta_m L} \quad (8.80)$$

$$\cos \beta_m L = \frac{T_m^2 - R_m^2 + 1}{2T_m} \quad (8.81)$$

In the following discussion, the real and imaginary parts of the right side of Eq. (8.81) are separated by

$$\frac{T_m^2 - R_m^2 + 1}{2T_m} = \alpha' + j\alpha'', \quad (8.82)$$

and the real and imaginary parts of β_m are defined by

$$\beta_m = \beta_m' - j\beta_m'' \quad (8.83)$$

where $\beta_m' > 0$ and $\beta_m'' \geq 0$.

On the basis of the above definitions, following relationships can be obtained from Eq. (8.81):

$$\cos \beta_m' L \cosh \beta_m'' L = \alpha' \quad (8.84)$$

$$\sin \beta_m' L \sinh \beta_m'' L = \alpha'' \quad (8.85)$$

For given values of α' and α'', the inverse problem becomes finding β_m' and β_m'' from Eqs. (8.84) and (8.85). Equations (8.84) and (8.85) also imply that $\cos(\beta_m' L)$ has the sign of α' and $\sin(\beta_m' L)$ has the sign of α''. Once β_m is known, u can be calculated from Eq. (8.80).

Eqs. (8.84) and (8.85) can be modified as

$$\cos \beta_m' L = \sqrt{\xi} \, \text{sign}(\alpha') \quad (8.86)$$

$$\sin \beta_m' L = \sqrt{1 - \xi} \, \text{sign}(\alpha'') \quad (8.87)$$

with

$$\xi = \frac{1}{2}\left[\alpha'^2 + \alpha''^2 + 1 - \sqrt{(\alpha'^2 + \alpha''^2 + 1)^2 - 4\alpha'^2}\right] \quad (8.88)$$

where $\text{sign}(\alpha')$ and $\text{sign}(\alpha'')$ mean the signs of α' and α'' respectively. From Eqs. (8.86) and (8.87), we can get

$$\beta_m' = \tilde{\beta}_m' + \frac{2\pi n}{L} \quad (n = 0, 1, 2, \ldots) \quad (8.89)$$

where $0 < \tilde{\beta}_m' L \leq 2\pi$, and $\tilde{\beta}_m'$ can be directly obtained from Eqs. (8.86) and (8.87).

From Eqs. (8.84) and (8.85), we can also get

$$\sinh \beta_m'' L = \sqrt{\alpha'^2 + \alpha''^2 - \xi} \quad (8.90)$$

from which β_m'' can be uniquely determined. So β_m'' can be determined aside from the choice of the integer n in Eq. (8.89). If the length L of the sample is sufficiently small, then $n = 0$; however, in many practical cases the length L cannot be arbitrarily chosen; then n can be uniquely determined from nondispersive media by carrying out measurements at two different frequencies (Damaskos *et al.* 1984).

If the medium is dispersive, the measured data at different frequencies cannot be mixed together, and the only way to obtain a second, independent measurement of R_m and T_m is to excite the waveguide with a different mode. Thus, we conduct two separate measurements at the same frequency ω, the first with the TE_{10} mode yielding R_1 and T_1, the second with the TE_{20} mode yielding R_2 and T_2. According to Eq. (8.78), β_1 and β_2 are given by

$$\beta_1 = \sqrt{\frac{\mu_1}{\mu_3}}\sqrt{k_0^2 \varepsilon_2 \mu_3 - \left(\frac{\pi}{a}\right)^2} \quad (8.91)$$

$$\beta_2 = \sqrt{\frac{\mu_1}{\mu_3}}\sqrt{k_0^2 \varepsilon_2 \mu_3 - 4\left(\frac{\pi}{a}\right)^2} \quad (8.92)$$

According to Eq. (8.77), β_1 and β_2 are given by

$$\beta_1 = \frac{\mu_3^2}{\mu_1}\beta_{01}\mu_1 \quad (8.93)$$

$$\beta_2 = \frac{\mu_3^2}{\mu_1}\beta_{02}\mu_2 \quad (8.94)$$

where μ_1 and μ_2 are the values of μ measured with TE_{10} and TE_{20} excitation respectively, and

$$\beta_{01} = \sqrt{k_0^2 - \left(\frac{\pi}{a}\right)^2} \qquad (8.95)$$

$$\beta_{02} = \sqrt{k_0^2 - 4\left(\frac{\pi}{a}\right)^2} \qquad (8.96)$$

Equating the ratio of Eqs. (8.91) and (8.92) and the ratio of Eqs. (8.93) and (8.94) and solving the product $\varepsilon_2\mu_3$ results in (Damaskos *et al.* 1984)

$$\varepsilon_2\mu_3 = \frac{1 - 4\left(\dfrac{\beta_{01}\mu_1}{\beta_{02}\mu_2}\right)^2}{1 - \left(\dfrac{\beta_{01}\mu_1}{\beta_{02}\mu_2}\right)^2}\left(\frac{\pi}{k_0 a}\right)^2 \qquad (8.97)$$

Equating Eq. (8.91) to Eq. (8.93) results in

$$\frac{\mu_1^3}{\mu_3^5} = \frac{\beta_{01}^2\mu_1^2}{k_0^2\varepsilon_2\mu_3 - (\pi/a)^2} \qquad (8.98)$$

If the sample is sufficiently thin ($\beta_1 L < 2\pi$), then $\beta_1 = \tilde{\beta}_1$. From Eqs. (8.91) and (8.92), we can get

$$\frac{\mu_1}{\mu_3} = \frac{\tilde{\beta}_1^2}{k_0^2\varepsilon_2\mu_3 - (\pi/a)^2} \qquad (8.99)$$

Therefore, μ_1 and μ_3 can be obtained from Eqs. (8.98) and (8.99), and ε_2 can be obtained from Eq. (8.97). Similar measurements with a rotated sample yield ε_1, ε_3, and μ_2 (Damaskos *et al.* 1984). This method is also applicable to the simpler case of a lossless and nondispersive sample.

Figure 8.22 shows a waveguide junction that is convenient for exciting TE_{10} and TE_{20} modes (Damaskos *et al.* 1984). The TE_{10} mode is excited by the horizontal arms that are standard waveguides beyond the tapered transitions. The TE_{20} mode is excited by the branch arms that also are standard waveguides. Both modes can propagate in the central part of the junction, which is of double width. Beyond the transition region, the horizontal arms are below cutoff for TE_{20} modes. With reasonable precision in the fabrication of the junction, TE_{10} mode will not be coupled into the branch arms. Therefore, a circuit through the branch arms can be used to measure reflection and transmission coefficients due to the TE_{20} mode, while a

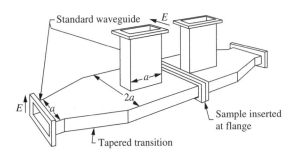

Figure 8.22 Sketch of TE_{10}–TE_{20} measurement fixture (Damaskos *et al.* 1984). Source: Damaskos, N. J. Mack, R. B. Maffett, A. L. Parmon, W. and Uslenghi, P. L. E. (1984). "The inverse problem for biaxial materials", *IEEE Transactions on Microwave Theory and Techniques*, **32** (4), 400–405. © 2003 IEEE

circuit through the side arms can be used to measure reflection and transmission coefficients for the TE_{10} mode. It is clear that impedance matching is required at the branch arms and at the tapered transitions.

Equations (8.81) and (8.91) to (8.99) require measurements of four complex quantities to determine three unknown complex components of the constitutive parameters from each orientation of the sample. The additional measured quantity greatly simplifies the mathematical steps that are required to solve the inverse problem. If three orientations of the sample are used to determine the six unknown components, the measurements provide ample redundancies that aid in immediately evaluating the quality of the measurement (Damaskos *et al.* 1984).

8.3.2 Faraday rotation methods

Consider the propagation of a linearly polarized wave in a ferrite biased by a dc magnetic field. If the direction of propagation is parallel to that of the dc magnetic field, Faraday effect can be observed, whereas, if the direction of propagation is transverse to the dc magnetic field, birefringence occurs (Pozar 1998). In this section, after introducing the effect of Faraday rotation, we discuss the circular waveguide and cavity methods for the measurement of permeability tensors based on Faraday effect.

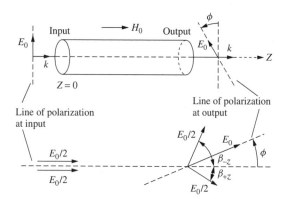

Figure 8.23 Faraday rotation of a ferrite magnetized in z-direction

8.3.2.1 Faraday rotation

As shown in Figure 8.23, a linearly polarized wave propagating in a magnetized ferrite can be decomposed into two circularly polarized modes, and the unequal propagation constants of these modes lead to Faraday rotation.

We assume that the electric field of a linearly polarized wave at position $z = 0$ is in the x-direction; so it can be expressed as

$$E|_{z=0} = \hat{x} E_0 = \frac{E_0}{2}(\hat{x} - j\hat{y}) + \frac{E_0}{2}(\hat{x} + j\hat{y})$$
(8.100)

The first term in the right side of Eq. (8.100) represents a right-handed circularly polarized wave (positive polarization), and it propagates in the z-direction with propagation constant β_+; the second term represents a left-handed circularly polarized wave (negative polarization), and it propagates in the z-direction with propagation constant β_-. So the total field E propagates as

$$E = \frac{E_0}{2}(\hat{x} - j\hat{y})e^{-j\beta_+ z} + \frac{E_0}{2}(\hat{x} + j\hat{y})e^{-j\beta_- z}$$

$$= \frac{E_0}{2}\hat{x}(e^{-j\beta_+ z} + e^{-j\beta_- z})$$

$$- j\frac{E_0}{2}\hat{y}(e^{-j\beta_+ z} - e^{-j\beta_- z})$$
(8.101)

$$= E_0 \left[\hat{x} \cos\left(\frac{\beta_+ - \beta_-}{2}\right) z \right.$$

$$\left. - \hat{y} \sin\left(\frac{\beta_+ - \beta_-}{2}\right) z \right] e^{-j(\beta_+ + \beta_-)z/2}$$

Eq. (8.101) shows that the wave in the magnetized ferrite is still a linearly polarized wave, but its polarization rotates as the wave propagates along the z-axis. At a given point along the z-axis, the polarization angle ϕ measured from the x-axis is given by

$$\phi = -\left(\frac{\beta_+ - \beta_-}{2}\right) z$$
(8.102)

In the above discussion, the loss of the ferrite is neglected. However, an actual ferrite medium has finite losses, which have a significant influence on the wave propagation. The values of relative permeability for the two circularly polarized waves can be expressed as

$$\mu_{r\pm} = \mu'_{r\pm} - j\mu''_{r\pm} = 1 + \chi' \pm \kappa' - j(\chi'' \pm \kappa'')$$
(8.103)

with

$$\chi = \chi' - j\chi'' = \chi_{xx} = \chi_{yy}$$
(8.104)

$$j\kappa = j(\kappa' - j\kappa'') = \chi_{xy} = -\chi_{yx}$$
(8.105)

The propagation constants for the two circularly polarized waves can be expressed as

$$\gamma_\pm = \alpha_\pm + j\beta_\pm$$
(8.106)

with

$$\beta_\pm = \frac{\omega\sqrt{\mu_0 \varepsilon_0 \varepsilon_r}}{\sqrt{2}}\left[1 + \chi' \pm \kappa' \right.$$

$$\left. + \sqrt{(1 + \chi' \pm \kappa')^2 + (\chi'' \pm \kappa'')^2}\right]^{1/2}$$
(8.107)

$$\alpha_\pm = \omega^2 \mu_0 \varepsilon_0 \varepsilon_r \frac{\chi'' \pm \kappa''}{2\beta_\pm} = \omega^2 \mu_0 \varepsilon_0 \varepsilon_r \frac{\mu''_{r\pm}}{2\beta_\pm}$$
(8.108)

where ε_r is the dielectric constant of the ferrite.

It should be noted that Faraday rotation is different in the two regions above and below the frequency of ferromagnetic resonance (Collin 1992). If the measurement frequency is lower than the frequency of ferromagnetic resonance ($\omega < \omega_0$), then $\beta_+ > \beta_-$; while if $\omega > \omega_0$, then $\beta_+ < \beta_-$.

Figure 8.24 shows a fixture for the measurement of Faraday rotation (Hogan 1953). In the fixture, two rectangular waveguides are separated by a circular waveguide with nonreflective transitions at each end of the circular section. One rectangular

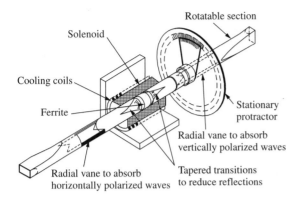

Figure 8.24 Test chamber for rotation measurement (Hogan 1953). Source: Hogan, C. L. (1953). *Reviews of Modern Physics*, **25** (1), 253–263, the American Physical Society

waveguide can be rotated about the longitudinal axis of the system. The dominant TE_{10} mode is excited in one rectangular waveguide, and by means of the smooth transition, it goes over into the dominant TE_{11} mode in the circular waveguide. The rectangular waveguide on the opposite end accepts only the component of the polarization that coincides with the TE_{10} mode in that waveguide, the other component being reflected at the transition. Absorbing vanes, inserted in the circular section, absorb the reflected components. The circular waveguide is placed in a solenoid to establish an axial magnetic field along its length. The ferrite cylinders to be measured are placed at the midsection of the circular guide.

Using the fixture shown in Figure 8.24, insertion loss can be measured by determining the power transmitted under identical conditions with the ferrite cylinder removed, and the ellipticity of the transmitted wave can be determined by measuring the power transmitted when the rectangular guide on the detector side is rotated both to positions of maximum and minimum transmissions.

8.3.2.2 Circularly polarized propagation method

Mullen and Carlson proposed a method making use of measurements performed on a ferrite rod placed along the axis of a circular waveguide (Mullen and Carlson 1956). The sample under test is excited successively with positive and negative circular

polarizations, and for each polarization, the phase and attenuation of the sample are measured. Using a perturbation treatment, the real and imaginary parts of the tensor components can be obtained from phase and attenuation of the positive and negative circular polarizations.

The theoretic model of this method is a circular waveguide containing a concentric ferrite rod, which is longitudinally magnetized, as shown in Figure 8.25. The general solution for this configuration can be found in (Lax and Button 1962). In actual measurements, the rods are often mounted in a polyfoam cylinder and inserted in the waveguide, which is inside a solenoid so that the magnetization field can be applied. The diagonal element χ and off-diagonal element κ can be calculated from the phase and attenuation information of the propagation constants of the negatively and positively polarized waves (Mullen and Carlson 1956):

$$\beta_{\mp} = \frac{\beta_0}{2}\left[1 - \frac{A \cdot s}{S}(\chi' \pm \kappa') + \frac{k_0^2}{\beta_0^2}\chi_e'\right] \quad (8.109)$$

$$\alpha_{\mp} = \frac{\beta_0}{2}\left[\frac{A \cdot s}{S}(\chi'' \pm \kappa'') + \frac{k_0^2}{\beta_0^2}\chi_e''\right] \quad (8.110)$$

where β_0 is the phase constant for the empty waveguide, s is the cross section of the ferrite rod, S is the cross-section area of the waveguide, the constant "A" is 1.04 for circular waveguide, k_0 is the wave number of free space, and χ_e is the dielectric susceptibility of the sample.

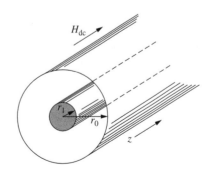

Figure 8.25 A cylindrical waveguide of radius r_0 containing a concentric ferrite rod of radius r_1 (Lax and Button 1962, p400). Source: Lax, B. and Button, K. J. (1962). *Microwave Ferrites and Ferrimagnetics*, New York: McGraw-Hill Book Company Inc.

Equations (8.109) and (8.110) indicate that this method cannot separate out the dielectric contributions to the phase and attenuation shifts. If the dielectric properties of the sample are known or can be estimated, the permeability tensor can be derived using Eqs. (8.109) and (8.110). It should be indicated that in the above discussion, the effect of demagnetization of the sample is not taken into consideration. If the ratio of the length/diameter of the rod is larger than 10:1, the demagnetization effect can be practically neglected, so this method is suitable for long and thin rod samples.

8.3.2.3 Reflection method

As discussed earlier, when an electromagnetic wave propagates along a ferrite cylinder, the plane of polarization rotates in proportion to its path length. The direction of the rotation depends simply upon the direction of dc magnetic field applied to the ferrite rod along its longitudinal axis. As shown in Figure 8.26, if the wave is reflected back through the ferrite sample, it does not return to its original direction of polarization, so Faraday rotation is nonreciprocal. Actually, the polarized plane tilts to the same side whether it is the polarized plane of the incident wave or the reflected one. In Figure 8.26, the rotation angle from $z = 0$ to $z = l$ is assumed to be ϕ, and the wave is reflected at the position $z = l$. After the wave returns back to position $z = 0$, the total rotation angle of the wave is 2ϕ.

On the basis of nonreciprocal property of Faraday rotation, Ito *et al.* proposed a reflection method for the measurement of the permeability tensor of ferrite in the microwave frequencies (Ito *et al.* 1984). In this method, a circular cylindrical waveguide containing a longitudinally magnetized

ferrite rod is used. A plane polarized TE_{11} wave is incident upon the section in which the ferrite cylinder is placed. One end of the circular waveguide is short circuited by a movable short-circuit plunger, which holds the ferrite cylinder through a hole drilled at the center. The angle of Faraday rotation and the position of the minimum voltage point of the elliptically polarized standing wave are measured far from the ferrite section.

Figure 8.27(a) schematically illustrates the measurement configuration. The ferrite rod is placed in the region of $0 \leq z \leq l$ and the end is terminated with a short circuit. An external dc magnetic field H_0 is applied to the ferrite along the z-axis. The incident wave is a TE_{11} plane polarized wave with polarization directed to the X-axis. If the reflection at the front end ($z = 0$) of the ferrite is negligibly small, the electric field E along the central axis in the region of $z < 0$ can be expressed by

$$E = E_+ + E_- \tag{8.111}$$

$$E_\pm = \frac{1}{2} V_\pm(z)(\hat{x} \mp j\hat{y})e^{j\omega t} \tag{8.112}$$

$$V_\pm = e^{-j\beta_a z} - e^{-2\gamma_\pm l + j\beta_a z} \tag{8.113}$$

where β_a is a phase constant in the empty waveguide region, $\gamma_\pm$, the propagation constants of right- and left-handed polarized waves and ω is the angular frequency.

The output voltage of the detector probe scanning the field in the cross section of the waveguide has a shape like a dumbbell and along the axis it has a standing-wave pattern as usual, as shown in Figure 8.27(a). Suppose that the dumbbell at the maximum ($z = z_{max}$) of the standing wave is observed. As shown in Figure 8.27(b), it may be characterized by a rotation angle if ϕ_1 and two extremes A_1 and B_1, where the subscript "1" stands for the value of A, B, or ϕ at the maximum. If the field intensity A_2 is measured when the probe is shifted to the node ($z = z_{min}$) and directed to the angle ϕ_1, the propagation constants $\gamma_\pm$ can be determined from the following equations (Ito *et al.* 1984):

$$\gamma_\pm = \alpha_\pm + j\beta_\pm \tag{8.114}$$

$$\alpha_\pm l = \frac{A_2}{A_1} \mp \frac{B_1}{A_1} \equiv \eta_\pm \tag{8.115}$$

$$\beta_\pm l = \beta_a z_{min} \mp \phi_1(1 + \alpha_\pm l) \equiv \theta_\pm \tag{8.116}$$

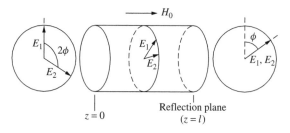

$z = 0$ Reflection plane $(z = l)$

Figure 8.26   Nonreciprocity of Faraday rotation

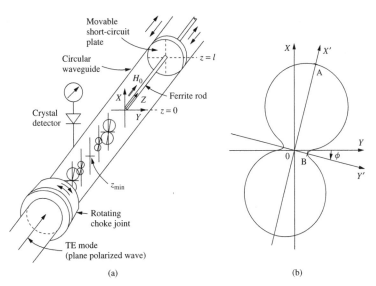

(a)   (b)

Figure 8.27 Reflection method based on Faraday rotation. (a) Schematic illustration of the measurement circuit and (b) locus of an output of the detector on the guide wall (Ito *et al.* 1984). Source: Ito, A. Ohkawa, S. and Yamamoto, H. (1984). "On a measurement method of the tensor permeability of ferrite using Faraday rotation", *IEEE Transactions on Instrumentation and Measurement*, **33** (1), 26–31. © 2003 IEEE

Equations (8.115) and (8.116) suggest that if $\eta_\pm$ and $\theta_\pm$ are plotted with various lengths l, $\alpha_\pm$ and $\beta_\pm$ can be obtained from the slopes of the expected straight lines.

When the specimen is demagnetized, θ_+ and θ_- (also η_+ and η_-) are reduced to the same value, there is no difference between γ_+ and γ_-, so $\gamma_\pm$ can be replaced by the notation γ_{dem}. If the special permeability of the demagnetized specimen μ_{dem} is known, the specific permeability $\mu_\pm$ and the relative permittivity ε_r can be obtained by (Ito *et al.* 1984)

$$\mu_\pm = \frac{\xi \cdot \left(1 + \dfrac{\mu_{dem} - 1}{\mu_{dem} + 1}\right) + (\gamma_\pm - \gamma_{dem})}{\xi \cdot \left(1 - \dfrac{\mu_{dem} - 1}{\mu_{dem} + 1}\right) - (\gamma_\pm - \gamma_{dem})}$$

(8.117)

$$\varepsilon_r = \frac{1 + \dfrac{1}{\xi} \cdot \left(\dfrac{\beta_a}{k_0}\right)^2 \cdot (\gamma_{dem} - j\beta_a) + \left(\dfrac{\beta_a}{k_0}\right)^2 \cdot \dfrac{1 - \mu_{dem}}{1 + \mu_{dem}}}{1 - \dfrac{1}{\xi} \cdot \left(\dfrac{\beta_a}{k_0}\right)^2 \cdot (\gamma_{dem} - j\beta_a) - \left(\dfrac{\beta_a}{k_0}\right)^2 \cdot \dfrac{1 - \mu_{dem}}{1 + \mu_{dem}}}$$

(8.118)

with

$$\xi = \frac{\chi^2}{2(\chi^2 - 1)J_1^2(\chi)} \cdot \left(\frac{d}{D}\right)^2 \beta_a$$

(8.119)

$$k_0 = \omega\sqrt{\varepsilon_0\mu_0},$$

(8.120)

where J_n is the Bessel function of the nth order, χ the first root of $J_0(\chi) - J_2(\chi) = 0$, k_0 is the wave number in free space, d and D are diameters of the ferrite rod or the waveguide, respectively. Practically, the value of μ_{dem} may be taken as unity, so Eqs. (8.117) and (8.118) can be simplified:

$$\mu_\pm = \frac{\xi + \gamma_\pm - \gamma_{dem}}{\xi - \gamma_\pm + \gamma_{dem}}$$

(8.121)

$$\varepsilon_r = \frac{1 + \dfrac{1}{\xi} \cdot \left(\dfrac{\beta_a}{k_0}\right)^2 \cdot (\gamma_{dem} - j\beta_a)}{1 - \dfrac{1}{\xi} \cdot \left(\dfrac{\beta_a}{k_0}\right)^2 \cdot (\gamma_{dem} - j\beta_a)}$$

(8.122)

It should be noted that in the above discussion, the possible higher-order modes and the reflection from the tip of the ferrite are neglected. The effects of these factors are discussed in (Ito *et al.* 1984).

8.3.2.4 Bimodal-cavity method

In the microwave Faraday rotation methods discussed above, a section of cylindrical waveguide is filled with the sample to be investigated, and the rotation of the plane of polarization of the microwaves on passage through the sample is measured. The rotation angle can be expressed as

$$\theta = 4\pi^2 (l/\lambda)(\xi/\omega) \tag{8.123}$$

where ξ is the parameter characterizing the rotation, l is the length of the sample, and λ is the wavelength in the sample. Equation (8.123) is based on the assumption that the sample is matched to the guide so that internal reflections may be neglected. If the magnetic medium is isotropic or is at least cylindrically symmetric, the rotation may be understood in terms of decomposition into circularly polarized microwave fields. If the medium has axial symmetry along the waveguide, the two circularly polarized components propagate independently, the phase velocity of each wave can be considered separately. The rotation parameter ξ is given by

$$\xi = \frac{1}{2}\omega(\chi_- - \chi_+) \tag{8.124}$$

For ferrites and ferromagnetic materials, the rotation parameter ξ can be calculated from

$$\xi/(\gamma H_0) = -M_0/H_0 \tag{8.125}$$

As the initial susceptibilities of ferrites and ferromagnetic materials usually have a value on the order of 10^2, their angles of Faraday rotation can be accurately measured. For paramagnetic samples, the rotation parameter ξ can be calculated from

$$\xi/(\gamma H_0) = -\chi_0 \tag{8.126}$$

where χ_0 is the static magnetic susceptibility. As the static magnetic susceptibilities of paramagnetic materials are on the order of 10^{-4} or smaller at room temperature, the rotation angles cannot be accurately measured using conventional Faraday rotation methods. The rotation angle of an antiferromagnetic sample would be smaller than that of a paramagnetic sample by a factor α (Portis and Teaney 1959):

$$\alpha = \omega^2/(\gamma^2 H_E H_A) \tag{8.127}$$

where H_E and H_A are the exchange and anisotropy fields. For a typical antiferromagnetic sample with 1-cm length in a field of 10,000 oersteds, the rotation angle is usually less than 10^{-5} radian. This angle is too small to be measured directly with any accuracy.

To increase the measurement sensitivity, we can make use of the fact that multiple reflections of the electromagnetic wave back and forth through a sample result in successive, cumulative rotations in the same sense, as the sense of Faraday rotation depends only on the direction of the external dc magnetic field, not on the direction of propagation. In the propagation method discussed in Section 8.3.2.2, the rotation angle is measured directly for just one passage through the sample. In the reflection method, the microwave signal reflected back through the sample is further rotated in the same direction. As long as there is no appreciable absorption of microwave energy, we can have the microwave signal multiply reflected through the sample. For a sample biased far from resonance, the cumulative absorption during the multiple passages is little. When such a sample is placed in a microwave cavity, the number of passages of the microwave signal through the sample is proportional to the quality factor of the cavity. Putting in representative numbers, one expects an enhancement of the rotation angle by a factor of 10^3.

The rotation angle can be detected using a bimodal cavity, which is resonant in two degenerate TE_{111} modes, as shown in Figure 8.28. Microwave power is coupled into the cavity through a coupling iris on the top from a rectangular waveguide propagating in a TE_{10} mode, and microwave power is coupled out through the end of the cavity into a second TE_{10} rectangular waveguide. For optimum coupling, the second waveguide is shorted a half-guide wavelength below the coupling window. If the input and output waveguides are parallel and the cavity is perfectly cylindrical, there can be no coupling into the output waveguide. This can be explained by the symmetry of the structure. The magnetic fields in the input waveguide and the cavity are symmetrical with respect to the plane of the paper, while the magnetic field in the output waveguide would be antisymmetrical with respect to the plane of the paper. So there should be no coupling between the input and output waveguides.

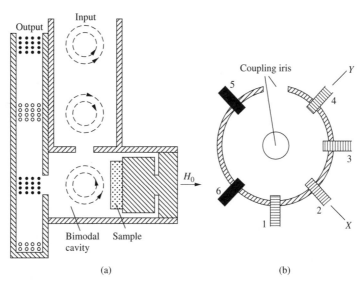

Figure 8.28 Structure of a bimodal cavity for the measurement of Faraday rotation (Portis and Teaney 1959). (a) Section through the bimodal cavity. Microwave power is coupled into the cavity through an iris in the top. The dotted circles represent the microwave magnetic field and (b) transverse section showing tuning plugs and coupling irises. Plugs 1 through 4 are metallic and vary the capacitance across the cavity. Plugs 5 and 6 are absorbers used to adjust the losses in the x and y modes. Source: Portis, A. M. and Teaney, D. (1959). *Physical Review*, **116** (4), 838–845, the American Physical Society

As shown in Figure 8.26(a), a magnetic sample is mounted opposite the output iris, and an external dc magnetic field is applied along the axis of the cavity. The sample produces Faraday rotation about the axis of the cavity, and the microwave magnetic field in the cavity is no longer symmetrical with respect to the plane of the paper, so there will be coupling to the output waveguide. The power coupled to the output waveguide is related to the angle of Faraday rotation.

It is desirable that no power is coupled between the input and the output waveguides in the presence of the unmagnetized sample so that any subsequent rotation induced can be detected with maximum sensitivity. Figure 8.26(b) shows a section perpendicular to the axis of the cavity, showing the arrangement of capacitive and resistive tuning plugs. Plugs 1, 2, 3, and 4 are metallic and may be adjusted so that the two cavity modes are degenerate. It can be arranged so that either plug 2 or 4 is well out of the cavity so that a quite sensitive final adjustment may be made with that plug. However, the cavity degeneracy alone does not insure that the input and output waveguides will be decoupled after

a nonmagnetized sample is loaded in the cavity. Consider two normal modes as having microwave magnetic fields oriented along the $-x$- and $+y$-axes. These modes are driven equally by a vertical microwave magnetic field. However, if the two modes have different loss rates, the amplitude of their response will be different. Then, the cavity magnetic field will not be vertical and there will be coupling to the output waveguide. This coupling may be avoided by artificially introducing additional losses into one of the modes. This can be accomplished by adjusting the resistive plugs 5 and 6 made from graphite-deposited resistance cards.

Further discussion on bimodal method for Faraday rotation measurement can be found in (Lax and Button 1962). The analysis of the Faraday rotation in terms of the susceptibility components for paramagnetic and antiferromagnetic cases can be found in (Portis and Teaney 1958, 1959).

8.3.3 Resonator methods

The permeability tensors of magnetic materials can also be characterized by using resonant methods, including resonator methods and

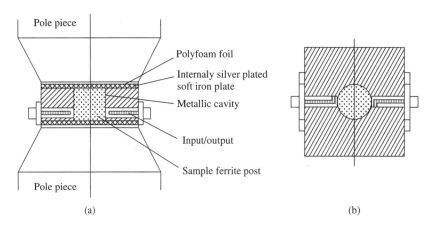

(a)   (b)

Figure 8.29   Measurement configuration for TE$_{111}$ resonator method. (a) Section view along z-axis and (b) section view perpendicular to z-axis showing the sample-filled resonator working at TE$_{111}$ mode and the input/output couplings (Ogasawara *et al.* 1976). Source: Ogasawara, N. Fuse, T. Inui, T. and Saito, I. (1976). "Highly sensitive procedures for measuring permeabilities ($\mu_\pm$) for circularly polarized fields in microwave ferrites", *IEEE Transactions on Magnetics*, **12** (3), 256–259. © 2003 IEEE

resonant-perturbation methods. In the design of resonant methods for the measurement of permeability tensors, one of the important considerations is the identification of the direction of the microwave magnetic field, so that the corresponding components of the permeability tensors can be measured. Resonant methods can work only at single or several discrete frequencies. As resonant methods are suitable for the measurement of low-loss materials, these methods are only applicable at the regions far from ferromagnetic resonance.

In this section, we discuss the resonator methods often used in the characterization of permeability tensors, including TE$_{111}$ resonator method, shielded cylindrical resonator method, ring resonator method, and whispering-gallery resonator method. The resonant-perturbation methods for the measurement of permeability tensors are discussed in Section 8.3.4.

8.3.3.1 TE$_{111}$ resonator method

Ogasawara *et al.* proposed a resonator method for the measurement of $\mu'_\pm - j\mu''_\pm$ of ferrites based on the resonance splitting of counterrotating TE$_{11}$ modes in the cylindrical metallic cavity resonator filled with axially magnetized ferrite sample under study (Ogasawara *et al.* 1976). As shown in

Figure 8.29, the cylindrical sample fills the metallic cavity, which is excited in the TE$_{111}$ mode by a pair of coupling apertures on diagonally opposite sides. The two end plates of the cavity are made of a silver-plated soft iron, so the cavity resonator can snugly fit the pole pieces.

In experiments, the change of reciprocal quality factor ($1/Q$) and the resonant frequency (f) while varying the biasing magnetic field are measured. At first, the biasing magnetic field is raised to a level well above the ferromagnetic resonance magnetic field, and one resonant peak can be observed, as shown in Figure 8.30(a), which is from the resonance of the TE$_{111}$ mode of a

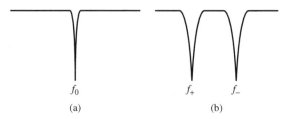

f_0 f_+ f_-

(a)   (b)

Figure 8.30   Resonance curves of the positive and negative polarities. (a) Resonance at biasing field far above the ferromagnetic resonance field. The two resonance curves coincide perfectly and (b) two resonance peaks corresponding to the waves with positive and negative polarities

dielectric-filled resonator. When the biasing field is lowered gradually, the resonance splits and the quality factor of the resonator, Q_s, decreases. As shown in Figure 8.30(b), the lower frequency resonance corresponds to the circularly polarized TE_{111} mode of the positive sense, and the higher frequency resonance corresponds to the mode of the negative sense.

The real and imaginary parts of $\mu_{\pm}$ can be derived from the changes of the reciprocal quality factor $(1/Q)$ and the resonant frequency ratios, both relative to those at the highest biasing field (Ogasawara *et al.* 1976):

$$\frac{f_{\pm}}{f_0} = (\mu'_{\pm})^{-\frac{1}{2}} \qquad (8.128)$$

$$\frac{1}{Q_{\pm}} - \frac{1}{Q_0} = \frac{\mu''_{\pm}}{\mu'_{\pm}} \qquad (8.129)$$

where f_0 and Q_0 are the resonant frequency and quality factor of the resonator when the sample is magnetized by a field well above the resonant magnetic field, respectively; $f_{\pm}$ and $Q_{\pm}$ are the resonant frequency and quality factor of positive and negative polarized modes of the resonator respectively. It should be noted that if the magnetic field is close to the ferromagnetic resonance field, the TE_{111} resonance could not be established because of large magnetic loss.

8.3.3.2 Shielded cylindrical resonator method

Figure 8.31 shows a shielded cylindrical dielectric resonator. In this configuration, the sample and the metal shield have the same height. In experiments, such configuration is usually realized by sandwiching the cylindrical sample by two parallel

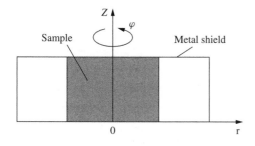

Figure 8.31 Shielded cylindrical dielectric resonator

conducting plates, and there is no air gap between the sample and the conducting plates. As discussed in Section 8.2.2, this configuration can be used in characterizing permittivity tensors. Here, we discuss the application of this configuration in the measurement of the permeability tensor.

When the external magnetic field is applied along the z-direction, the permeability tensor is in the form of Eq. (8.19). To fully characterize the three complex components μ, κ, and μ_z, three independent measurements are required. In the method proposed by Krupka (1989), two cylindrical parallel-plate resonators made of the same materials but with different dimensions are used. Suitable dimensions of the two resonators are chosen so that the resonant frequencies for the HE_{111} mode of the small resonator and the H_{011} mode of the large one are approximately the same. The main purpose of using two resonators is to minimize the errors caused by the frequency dependence of the tensor components. From the resonant frequency and unloaded quality factor of the HE_{111}^+, HE_{111}^-, and H_{011} modes, the real and imaginary parts of μ, κ, and μ_z can be determined. In the determination of these parameters, the surface resistance of conducting plates, the permittivity, and scalar permeability are needed and can be measured using the methods discussed in earlier chapters.

8.3.3.3 Ring resonator method

Krupka and Geyer proposed a ring resonator method for determining the real and imaginary parts of all three permeability tensor components using only one cylindrical ferrite sample (Krupka and Geyer 1996; Krupka 1991). The method is based on the simultaneous utilization of the $HE_{111}^{\pm}$ and H_{111} modes, which occur in dielectric ring resonators containing the cylindrical ferrite sample under test. Employing the corresponding eigenvalue equations for these modes, the material parameters can be computed from the measured resonant frequencies and quality factors of the three modes of the dielectric ring resonator with and without the ferrite sample. To minimize frequency variations in the measurement system, two dielectric ring resonators, having the same height and internal diameter but different external diameters, are needed. The

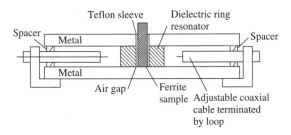

Figure 8.32 Resonant structure for the measurement of permeability tensor components of a cylindrical ferrite sample (Krupka *et al.* 1996). Source: Krupka, J. and Geyer, G. (1996). "Complex permeability of demagnetized microwave ferrites near and above gyromagnetic resonance", *IEEE Transactions on Magnetics*, **32** (3), 1924–1933. © 2003 IEEE

external diameters of the two ring resonators are chosen so as to obtain close resonant frequencies of the HE_{111} mode (the resonator with smaller diameter) and of the H_{111} mode (the resonator with larger diameter) for the two resonators containing typical nonmagnetized ferrite samples.

The measurement fixture is shown in Figure 8.32. As circularly polarized modes are used in measurement, the resonators are coupled to external circuits through four identical loop-terminated coaxial cables. The radial positions of the coaxial cables are adjustable so that suitable couplings can be achieved. As the computations are based on the exact eigenvalue equations of the resonant modes used in the measurement, the air gap that always exists between the ferrite sample and the dielectric resonator can be rigorously taken into consideration (Krupka 1991).

8.3.3.4 Whispering-gallery resonator method

The resonator methods discussed above are suitable for the measurement of permeability tensors at low microwave frequencies, usually less than 10 GHz. Although these techniques may be employed for the measurement of permeability tensors at higher frequencies, mainly due to two reasons, their measurement accuracies are decreased. Firstly, when the frequency increases, the metal loss and radiation loss increase, so the quality factor of metal cavities and dielectric

resonators decrease. Secondly, the sample dimensions become smaller with increasing frequency, so the uncertainties of sample dimensions have more serious effects on the measurement results. Therefore, the use of whispering-gallery modes for permeability tensor measurements at frequencies greater than 10 GHz becomes quite attractive since their quality factors depend principally on the intrinsic ferrite material losses, and the ferrite sample dimensions also remain reasonably large, even for frequencies up to 100 GHz. The basic properties of whispering-gallery resonators and their applications in materials property characterization have been discussed in Chapters 2 and 5.

The permeability tensor components of longitudinally magnetized ferrites can be determined from the resonant frequencies WGH_{n00} and WGE_{n00} as a function of the external dc magnetic field applied on the samples (Krupka *et al.* 1996). The measurement configuration is schematically shown in Figure 8.33. The sample has a circular disk shape, and the whispering-gallery modes are excited by adjustable coupling loops. The disk sample is magnetized along its axis, so the permeability tensor is in the form of Eq. (8.19).

Usually, a sample is only partially magnetized when the external static magnetic induction field is 0.05 to 0.3 T. The behavior of the WGH_{n00} resonant frequencies is very similar to that of the resonant frequencies of lower-order modes in a cavity or dielectric ring resonator partially filled with a magnetized ferrite sample (Krupka 1991). The split between WGE_{n00} resonant frequencies corresponds to modes polarized in opposite directions in a partially magnetized sample. The three components of the permeability tensor (μ, κ, μ_z) can be calculated from the resonant frequencies

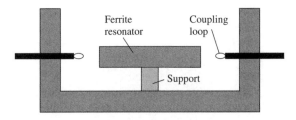

Figure 8.33 The structure of the resonator working at whispering-gallery modes for the measurement of permeability tensors

and quality factors of the WGH$_{n00}$ and the two WGE$_{n00}$ modes polarized in opposite directions.

For large external magnetic fields (usually above 0.4 T), the sample is saturated and the behavior of WGE$_{n00}$ resonant frequencies is the same as that expected for a sample that has a uniform longitudinal magnetization. The two components of the permeability tensor (μ, κ) may be determined from the resonant frequencies of two oppositely polarized WGH$_{n00}$ modes. The third component μ_z can be ascertained from the average of resonant frequencies of two WGE$_{n00}$ modes.

8.3.4 Resonant-perturbation methods

In a resonant-perturbation method for the measurement of permeability tensor, the small ferrite sample under test is introduced into a degenerate cavity in a region of maximum microwave magnetic field and zero electric field, and an external dc magnetic field is applied perpendicular to the microwave magnetic field. The magnetic permeability tensor of the sample is calculated from the perturbation effects of the sample on the resonance frequency and quality factor of the cavity. The prime requisite for a resonant-perturbation method for the study of permeability tensor is that the cavity is degenerate in two suitably chosen orthogonal linear modes. For a cavity with two orthogonal degenerate modes, after the sample is loaded and necessary adjustments are made, the resonant curves of the two modes overlap, so only one resonant peak can be observed. When an external dc magnetic field is applied, the single resonance peak splits into two, and the properties of the sample can be derived from the splitting of the resonance peak.

To understand what happens as the external magnetic field is varied, it is necessary to analyze the behavior of the electron spins in the sample (Artman and Tannenwald 1955). When a large dc magnetic field is applied to a ferromagnetic material, the unpaired electron spins tend to line up with the applied field, and, consequently, the resultant magnetic moment precesses about the steady magnetic field. The classical precessional motion would be quickly damped out by the crystal lattice unless energy is supplied by an oscillating magnetic field; the microwave magnetic field provides

this oscillating field. Maximum energy is transferred from the microwave field to the electron spin system when the frequency of the microwave field equals the precession frequency of the electron spins. This is the condition of ferromagnetic resonance, which is also called *gyromagnetic resonance*. Further discussions on ferromagnetic resonance are made in Section 8.4.

For a cavity excited in one linear mode, the precessing electron spins will couple to this mode and in turn couple to and excite the orthogonal mode. Because of this ferrite-cavity interaction, the description of the cavity properties in terms of two orthogonal linear modes is not convenient. The linearly polarized excitation can be considered as the sum of two modes circularly rotating in opposite directions. Before the steady magnetic field is applied, the cavity is degenerate in these two rotating modes. After the steady magnetic field is applied, the precessing spins in the ferrite affect these two modes differently. The mode rotating in the same direction as the precessing magnetic dipoles of the ferrite spins strongly interacts with the ferrite spins, so that the resonant frequency and quality factor of this mode are greatly changed. The other mode, rotating in a direction opposite to that of the precessing magnetic dipoles, is hardly affected.

In the following, we discuss the resonant properties of the resonant cavities perturbed by magnetic materials. Subsequently, we discuss the resonant-perturbation method for the measurement of permeability tensors, and we then discuss the geometrical effects on the measurement of permeability tensors. After that, several examples of exact-theory methods are given.

8.3.4.1 Resonant perturbation of magnetic samples

The general perturbation theory has been discussed in Chapter 6. In this part, we discuss the perturbation of magnetic materials to resonant cavities, including nondegenerate cavity and degenerate cavity.

Perturbation to a nondegenerate cavity

For a magnetic material with magnetic field applied in z-direction, its permeability tensor is in the form of Eq. (8.19). If the magnetization is saturated ($\mu_z = 1$), the susceptibility tensor $[\chi_m]$ of the magnetized sample can be expressed as

$$[\chi_m] = \begin{pmatrix} \chi & -j\kappa \\ j\kappa & \chi \end{pmatrix} \qquad (8.130)$$

For a nondegenerate cavity, the change in energy due to the magnetic sample depends on

$$\begin{pmatrix} H^* & 0 \end{pmatrix} \begin{pmatrix} \chi & -j\kappa \\ j\kappa & \chi \end{pmatrix} \begin{pmatrix} H \\ 0 \end{pmatrix} = \chi H^2 \qquad (8.131)$$

where H is the magnetic field of working mode. Equation (8.131) indicates that only the diagonal components χ of the tensor $[\chi_m]$ contributes to the perturbation. The general relationships between the change in resonant frequency and change in quality factor are given by (Artman and Tannenwald 1955)

$$\frac{\Delta\omega_r}{\omega_0} = -\frac{\chi' H^2 \Delta v}{2 \int_v H^2 \, d\tau} \qquad (8.132)$$

$$\Delta\left(\frac{1}{2Q}\right) = \frac{\chi'' H^2 \Delta v}{2 \int_v H^2 \, d\tau} \qquad (8.133)$$

with

$$\chi = \chi' - j\chi'' \qquad (8.134)$$

$$\Delta\omega = \Delta\omega_r + j\Delta\omega_i = \Delta\omega_r + j\Delta\left(\frac{\omega_0}{2Q}\right) \qquad (8.135)$$

where Δv is the volume of the sample and ω_0 is angular resonant frequency of the cavity before the sample is introduced.

It is clear that for a nondegenerate cavity, the information on κ cannot be obtained, so it cannot be used in the measurement of permeability tensor. But nondegenerate cavities can be used in the measurement of χ in the existence of κ. In the following text, we discuss two typical nondegenerate cavities: circular cavity and rectangular cavity.

Figure 8.34 shows a nondegenerate TE_{11n} cylindrical cavity with diameter a and length ln. In this configuration, the angular symmetry is disturbed by coupling through an iris located on the side of the cavity. The sample under study is placed near the center of one endplate, and the external

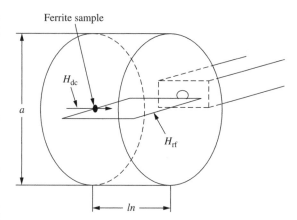

Figure 8.34 Nondegenerate TE_{11n} cylindrical cavity. Source: Artman, J. O. and Tannenwald, P. E. (1955). "Measurement of susceptibility tensor in ferrite", *Journal of Applied Physics*, **26** (9), 1124–1132

dc magnetic field is applied along the axis of the cavity. For a nondegenerate TE_{11n} cylindrical cavity, Eqs. (8.132) and (8.133) become (Artman and Tannenwald 1955)

$$\frac{\Delta\omega_r}{\omega_0} = -\frac{26.316\chi'}{13.560 + (\pi a/l)^2} \frac{\Delta v}{nl^3} \qquad (8.136)$$

$$\Delta\left(\frac{1}{2Q}\right) = \frac{26.316\chi''}{13.560 + (\pi a/l)^2} \frac{\Delta v}{nl^3} \qquad (8.137)$$

Figure 8.35 shows a rectangular nondegenerate cavity constructed from a closed-off section of rectangular waveguide resonating in a TE_{10n} mode. In the measurement of magnetic materials, the sample under test is usually placed at the point $x = a/2$, $y = b/2$, and $z = L$. The external dc magnetic field is applied along the z-direction, perpendicular to the microwave magnetic field. In this case, Eqs. (8.132) and (8.133) become (Artman and Tannenwald 1955)

$$\frac{\Delta\omega_r}{\omega_0} = -\chi' \frac{\Delta v}{V} \cdot 2\left[1 - \left(\frac{\lambda_0}{2a}\right)^2\right] \qquad (8.138)$$

$$\Delta\left(\frac{1}{2Q}\right) = \chi'' \frac{\Delta v}{V} \cdot 2\left[1 - \left(\frac{\lambda_0}{2a}\right)^2\right] \qquad (8.139)$$

where V is the volume of the cavity.

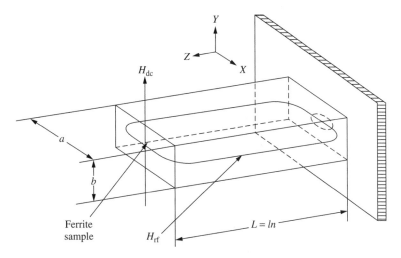

Figure 8.35 Rectangular TE_{10n} cavity. Modified from Artman, J. O. and Tannenwald, P. E. (1955). "Measurement of susceptibility tensor in ferrite", *Journal of Applied Physics*, **26** (9), 1124–1132

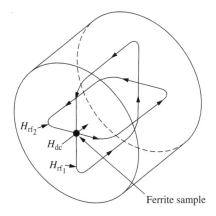

Figure 8.36 Orthogonal microwave magnetic field lines in a degenerate cavity. Modified from Artman, J. O. and Tannenwald, P. E. (1955). "Measurement of susceptibility tensor in ferrite", *Journal of Applied Physics*, **26** (9), 1124–1132

Perturbation to a degenerate cavity

Figure 8.36 shows a cavity that can support two orthogonal resonant modes. The ferrite is placed on the axis of the cavity in a region of zero microwave electric field. The axes are chosen in the orthogonal 1, 2 directions, and at the position where the ferrite is placed, the amplitudes of the microwave magnetic fields in these two directions are H_1, H_2 respectively. The resonant mode $(H_1 H_2)^T$ satisfies the following relation (Artman

and Tannenwald 1955):

$$
\begin{pmatrix} H_1^* & H_2^* \end{pmatrix} \begin{pmatrix} \chi & -j\kappa \\ j\kappa & \chi \end{pmatrix} \begin{pmatrix} H_1 \\ H_2 \end{pmatrix}
$$
$$
= \lambda \begin{pmatrix} H_1^* & H_2^* \end{pmatrix} \begin{pmatrix} H_1 \\ H_2 \end{pmatrix} \tag{8.140}
$$

Equation (8.140) can be rewritten as

$$
\begin{pmatrix} H_1^* & H_2^* \end{pmatrix} \begin{pmatrix} \chi - \lambda & -j\kappa \\ j\kappa & \chi - \lambda \end{pmatrix} \begin{pmatrix} H_1 \\ H_2 \end{pmatrix} = 0 \tag{8.141}
$$

where λ is the eigenvalue, which can be obtained from

$$
\begin{vmatrix} \chi - \lambda & -j\kappa \\ j\kappa & \chi - \lambda \end{vmatrix} = 0 \tag{8.142}
$$

Two solutions exist: $H_2 = +jH_1$ corresponding to $\lambda = \chi + \kappa$ and $H_2 = -jH_1$ corresponding to $\lambda = \chi - \kappa$. H_1 and H_2 are equal in magnitude, but due to the factor $(\pm j)$, they are in quadrature in time as well as in space. Therefore, the two natural modes

$$
H_\pm = \begin{pmatrix} H \\ \pm jH \end{pmatrix} \tag{8.143}
$$

are right- and left-hand circularly rotating modes.

Before the ferrite sample is introduced, the empty cavity resonates at the frequency ω_0, and the Maxwell's equations for the empty cavity are

$$
\nabla \times \boldsymbol{H}_0 = j\omega_0\varepsilon_0\boldsymbol{E}_0 \tag{8.144}
$$

$$
\nabla \times \boldsymbol{E}_0 = -j\omega_0\mu_0\boldsymbol{H}_0 \tag{8.145}
$$

The subscript 0 indicates the state of empty cavity. If a perturbation object with a small volume Δv is introduced into the cavity, the cavity will then resonate at a new frequency ω. The Maxwell's equations for the cavity containing the perturbation object are

$$\nabla \times \boldsymbol{H} = j\omega_0\varepsilon_0\boldsymbol{E} + \boldsymbol{J}_e \qquad (8.146)$$

$$\nabla \times \boldsymbol{E} = -j\omega_0\mu_0\boldsymbol{H} - \boldsymbol{J}_m \qquad (8.147)$$

where $\boldsymbol{J}_e$ and $\boldsymbol{J}_m$ only exist in the small volume Δv occupied by the perturbation object. If the perturbation object is a ferrite located in a region where the microwave magnetic field is circularly polarized, then

$$\boldsymbol{J}_m = j\omega\mu_0[\chi_m] \cdot \boldsymbol{H} \qquad (8.148)$$

$[\chi_m] \cdot \boldsymbol{H}$ represents the multiplication of the tensor $[\chi_m]$ into the vector $\boldsymbol{H}$. As $\boldsymbol{H}$ is circularly polarized, $[\chi_m]$ is the effective scalar $\chi \pm \kappa$. The relation between $\boldsymbol{J}_e$ and $\boldsymbol{E}$ is generally of a scalar nature

$$\boldsymbol{J}_e = \sigma\boldsymbol{E} \qquad (8.149)$$

where σ is the conductivity.

According to resonant perturbation theory, the shift of complex frequency of the cavity is related to the tensor components, the sample volume, and the microwave magnetic field configuration:

$$\frac{\Delta\omega_\pm}{\omega_0} = \frac{(\chi \pm \kappa)H^2\Delta v}{2\displaystyle\int_v H^2 d\tau} \qquad (8.150)$$

with

$$\chi = \chi' - j\chi'' \qquad (8.151)$$

$$\kappa = \kappa' - j\kappa'' \qquad (8.152)$$

$$\Delta\omega = \Delta\omega_r + j\Delta\omega_i = \Delta\omega_r + j\Delta\left(\frac{\omega_0}{2Q}\right) \qquad (8.153)$$

According to Eq. (1.150), the real and imaginary parts of the change of complex frequency due to the introduction of the sample are given by

$$\left(\frac{\Delta\omega_r}{\omega_0}\right)_\pm = \frac{(\chi' \pm \kappa')H^2\Delta v}{2\displaystyle\int_v H^2 d\tau} \qquad (8.154)$$

$$\Delta\left(\frac{1}{2Q}\right)_\pm = \frac{(\chi'' \pm \kappa'')H^2\Delta v}{2\displaystyle\int_v H^2 d\tau} \qquad (8.155)$$

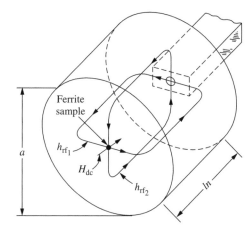

Figure 8.37 A degenerate TE_{11n} cylindrical cavity loaded with a magnetized ferrite sample. Source: Lax, B. and Button, K. J. (1962). *Microwave Ferrites and Ferrimagnetics*, New York: McGraw-Hill Book Company Inc.; Artman, J. O. and Tannenwald, P. E. (1955). "Measurement of susceptibility tensor in ferrite", *Journal of Applied Physics*, **26** (9), 1124–1132

Figure 8.37 shows a degenerate TE_{11n} cylindrical cavity with diameter a and length $L = ln$. When such a cavity is used for the measurement of susceptibility tensors, Eqs. (8.154) and (8.155) become (Artman and Tannenwald 1955)

$$\left(\frac{\Delta\omega_r}{\omega_0}\right)_\pm = -\frac{26.316}{13.560 + (\pi a/l)^2}(\chi' \pm \kappa')\frac{\Delta v}{nl^3} \qquad (8.156)$$

$$\Delta\left(\frac{1}{2Q}\right)_\pm = \frac{26.316}{13.560 + (\pi a/l)^2}(\chi'' \pm \kappa'')\frac{\Delta v}{nl^3} \qquad (8.157)$$

8.3.4.2 Measurement of permeability tensors

On the basis of the perturbation of magnetic materials to resonant cavities discussed above, resonant-perturbation method can be developed for the measurement of permeability tensors. In the design of resonant cavity, we should consider the field distributions of resonant modes, excitation of resonant modes, application of external dc magnetic field, and the way for fixing samples.

Figure 8.38 shows an example of TE_{112} mode degenerate cylindrical cavity for the characterization of permeability tensors. The sample under

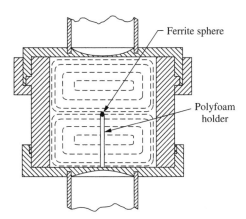

(a)　　　　　　　　　(b)

Figure 8.39 Typical resonant curves of a linearly polarized degenerate cavity containing a ferrite sample. (a) No external dc magnetic field is applied and (b) an external dc magnetic field is applied perpendicular to the microwave magnetic field

Figure 8.38 A typical TE_{112} mode cavity for the characterization of permittivity tensors (Spencer *et al.* 1956b). Source: Spencer, E. G. LeCraw, R. C. and Reggia, F. (1956b). "Measurement of microwave dielectric constants and tensor permeabilities of ferrite spheres", *Proceedings of the IRE*, **44** (6), 790–800. © 2003 IEEE

test is put at the center of the cavity, and the dc magnetic field is applied along the axis of the cylindrical cavity. The degenerate modes in the cavity for permeability tensor measurement can be in linear polarization or circular polarization. In the following, we discuss two methods for the excitation of resonant modes with linear and circular polarizations respectively.

Linearly polarized cavity

Figure 8.37 schematically shows a method for the excitation of degenerate TE_{11n} modes in linear polarization. In an actual structure, additional adjustment structures are needed to ensure that the cavity has two orthogonal degenerate modes after a sample is loaded, and the tuning structure shown in Figure 8.28(b) can be used. In the measurement, when the cavity is loaded with a sample and no steady magnetic field is applied, adjustments are made to ensure that the two modes are degenerate and perpendicular to each other, and it is preferable that the two modes have the same quality factor. After adjustments, only one resonant peak can be observed, as shown in Figure 8.39(a).

When an external dc magnetic field is applied perpendicularly to the microwave magnetic fields of the two degenerate modes, the magnetized sample

has different perturbations to the two modes, so the resonant peaks of the two modes split, as shown in Figure 8.39(b). Using Eqs. (8.154) and (8.155), the real and imaginary parts of χ and κ can be calculated from the changes of resonant frequencies and quality factors of the two modes (Artman and Tannenwald 1955).

However, it should be noted that, sometimes, the two resonant peaks shown in Figure 8.39(b) cannot totally split, so it may be difficult to accurately measure the perturbation effects of the sample to the two modes, especially the effects on quality factor. This problem can be solved by separately exciting the two circularly polarized modes, as will be discussed below.

Circularly polarized cavity

In a linearly polarized cavity, two circularly polarized modes coexist. The two circularly polarized modes required for the measurement of permeability tensor can be excited separately (Spencer *et al.* 1955, 1956a,b). Figure 8.40 shows typical resonant curves of a degenerate cavity in circular polarization, containing a magnetized sample with the steady magnetic field perpendicular to the microwave magnetic field. As shown in Figure 8.40(a), if the cavity is in an arbitrary elliptical polarization, two resonant peaks can be observed, corresponding to the positive and negative polarizations respectively. If the cavity is in a pure positive or negative circular polarization, only one resonant peak can be observed, as shown in Figure 8.40(b) and (c). By choosing positive and negative circular polarizations separately,

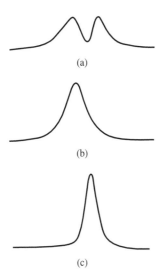

Figure 8.40 Typical resonant curves of a degenerate cavity. (a) Arbitrary elliptical polarization, (b) positive circular polarization, and (c) negative circular polarization. Modified from Spencer, E. G. LeCraw, R. C. and Reggia, F. (1956b). "Measurement of microwave dielectric constants and tensor permeabilities of ferrite spheres", *Proceedings of the IRE*, **44** (6), 790–800. © 2003 IEEE

using Eqs. (8.154) and (8.155), the permeability components χ and κ can be calculated from the changes of resonant frequencies and quality factors of the two modes due to the magnetized sample.

In experiments, circularly polarized resonant modes are often excited by two ways. Figure 8.41 shows a typical transmission-type TE_{11n}-mode cylindrical cavity in circular polarization for the measurement of permeability tensor (Baden Fuller 1987). In this configuration, at first the rectangular

waveguide is transformed to a circular waveguide using a rectangular to circular transition; then a polarized absorber is arranged to absorb any cross-polarized wave in the circular waveguide. After passing through the quarter-wave plate circular polarizer, the microwave in the circular waveguide becomes circularly polarized, and then the resonance with circular polarization in the cavity is excited through the coupling iris. The output waveguide system is the mirror of the input waveguide system, and the power coupled from the cavity to the output waveguide system is proportional to the response of the cavity.

In the measurement of permeability tensor, both the positive and negative polarizations are needed. In experiments, the direction of circular polarization can be achieved by changing the transmission direction of the cavity, or changing the direction of the dc magnetic field.

Figure 8.42 shows another way of exciting the circular-polarization resonant mode using two linear polarization waves. A TE_{112} mode cylindrical cavity is excited by two linearly polarized waves in space and time quadrature, which set up either a positive or negative circularly polarized resonant mode. By shifting the phase of one of the input waves by 180°, the positive or negative circularly polarized mode may be selected.

8.3.4.3 Geometrical effects and intrinsic properties

Many researchers have reported that, in the measurement of the permeability tensor using resonant-perturbation methods, the measurement results may be dependent on the geometrical factors of

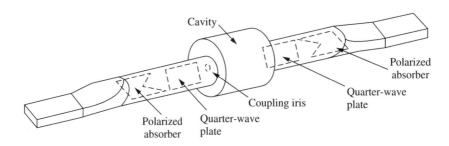

Figure 8.41 Circularly polarized cavity (Baden Fuller 1987). Modified from Baden Fulle, A. J. (1987). *Ferrites at Microwave Frequencies*, Peter Peregrinus Ltd., London, 235–255, by permission of Peter Peregrinus Ltd

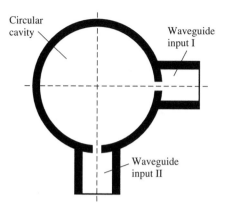

Figure 8.42 Excitation of a circular-polarization mode using two linear polarization modes

the samples. To obtain the intrinsic properties of the samples, it is necessary to consider the geometrical effects to the measurement results. Lax and Button have made systematic discussions on various geometrical effects. In the following text, we briefly analyze such effects, following the discussions made by Lax and Button (1962).

Wall effect

If a spherical ferrite sample is placed near the cavity wall, the measurement results of the permeability tensor of the sample are related to the distance between the sample and the wall. The conventional method in considering the wall effect is to introduce

the image of the sample on the other side of the wall. This renders the wall as an equipotential surface that may be removed from consideration and replaced by the dipole field shown in Figure 8.43(b). It should be noted that the dipole field of the sample and its image gives rise to slightly different demagnetizing conditions compared with the conditions arising from the linear rf field in the cavity. However, various refined electromagnetic treatments may be formulated to describe the wall effect quantitatively.

In experiments, the wall effects are often avoided by two methods. The first is to move the sample some distance from the wall until the correction becomes negligible. The second solution is to use a longer cavity having an rf magnetic field maximum at some point in the interior of the cavity. The problem of supporting the magnetic sample at some interior point of the cavity is usually solved by suspending the sample with a nylon string or attaching it to a polyfoam holder, as shown in Figure 8.38.

Size effect

In the measurement of permeability tensors, the measurement results are also related to the sizes of the samples. The major contribution to the size effect is the fact that the penetration of the electromagnetic field into a sample depends on the magnetic permeability and the dielectric permittivity, which are to be measured. The size

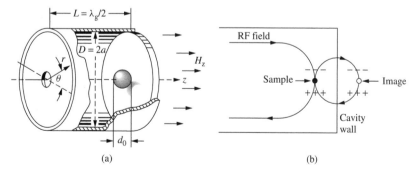

(a)   (b)

Figure 8.43 Cylindrical cavity loaded with magnetized spherical sample. (a) Three-dimensional view (Von Aulock and Rowen 1957). Source: Von Aulock, W. and Rowen, J. H. (1957). "Measurement of dielectric and magnetic properties of ferromagnetic materials at microwave frequencies", *The Bell System Technical Journal*, **36** (3), 427–448, by permission of American Telephone and Telegraph Co. and (b) sample and its image. Modified from Lax, B. and Button, K. J. (1962). *Microwave Ferrites and Ferrimagnetics*, New York: McGraw-Hill Book Company Inc.

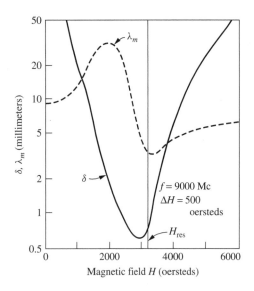

Figure 8.44 The calculated skin depth (solid line) and wavelength within the ferrite (dashed line) at 9 GHz and with a line width of 500 oersteds (Lax and Button 1962, p481). Source: Lax, B. and Button, K. J. (1962). *Microwave Ferrites and Ferrimagnetics*, New York: McGraw-Hill Book Company Inc.

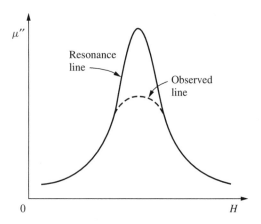

Figure 8.45 The size effect causes an apparent flattening of the resonance line because the skin depth becomes smaller in the resonance region

effect will be very obvious near the ferromagnetic resonance. As shown in Figure 8.44, in the range of ferromagnetic resonance, the permeability of the sample changes drastically, and so the skin depth and wavelength within the sample are also greatly changed. The changes of the skin depths and wavelength within the sample affect the resonant frequency and quality factor of the cavity loaded with the sample, from which the permeability values are obtained.

In the study of ferromagnetic resonance, we usually study the relationship between the externally applied dc magnetic field and the imaginary part of permeability. There are three major phenomena caused by the changes of the skin depth and wavelength within the sample: the broadening of the resonance line, the shift of the resonance peak to higher magnetic fields, and the frequency dependence of the line width. In Figure 8.45, the solid line shows the intrinsic resonance line of μ''. At the low-field side of the intrinsic resonance line, when the steady magnetic field H increases, μ'' increases, the skin depth decreases, and the electromagnetic field within the sample weakens. So

the maximum response cannot be reached and the curve is flattened, as shown by the dashed line, making the resonance appear to be broader because line width ΔH is taken at half-maximum. Besides, as shown in Figure 8.44, the skin depths are different at the corresponding points below and above resonance because of the reversal of the algebraic sign of μ'. This asymmetry causes the peak of the resonance curve to be shifted to higher fields. Furthermore, the size effect leads implicitly to a frequency dependence of the line width. For a sample with a given size, the electromagnetic field penetration will be smaller at higher frequencies. Therefore, the line broadening will increase when the operation frequency increases because of incomplete penetration.

Body resonance

In the measurement of permeability tensors, by changing the external dc magnetic field, resonance caused by the body of the sample can be observed, and such a resonance is usually called *body resonance*. As shown in Figure 8.44, when the external dc magnetic field is changed, both the skin depth and the wavelength within the sample experience drastic changes. If ferrite samples are on the order of a wavelength thick, multiple internal reflections may result in body resonance. There are two basic conditions for body resonance.

One is that the sample thickness must be equal to an integral number of half wavelengths, and the other is that the skin depth must be greater than the sample thickness to permit the propagation of multiple internal reflections.

It is necessary to compare the body resonance with the ferromagnetic resonance. In ferromagnetic resonance, the diagonal and off-diagonal elements of the permeability tensor exhibit anomalous dispersion, and microwave fields are excluded from the sample. In body resonance, nothing spectacular is happening to the components of the permeability tensor, but the fields in the sample become very large, giving rise to a large loss (Waldron 1960).

Shape effect

In the measurement of the magnetic properties of ferrites using cavity-perturbation methods, in order to observe the true resonance line shape, it is necessary to consider the effect of the sample shape. This may not be a serious problem at very high microwave frequencies, but the effects of incomplete magnetization of some sample shapes by the external dc magnetic field are often important at lower frequencies, for example, below 5 GHz (Lax and Button 1962).

Sphere is one of the most commonly used sample shapes. A spherical sample may approach the resonance condition while it is still incompletely magnetized. This effect will be serious when both the resonance line width and the value of $4\pi M_0$ of the material are relatively high. For a spherical sample, to sweep out the magnetic domains, the external steady magnetic field must be greater than the demagnetizing field $4\pi M_0/3$. So a ferrite having a saturation magnetization at room temperature of 3000 gauss requires an applied field of at least 1000 oersteds to be fully magnetized. However, if the measurements are to be made at 2.8 GHz, according to Kittel's formula, $\omega_r = \gamma H_0$, the peak of the resonance curve occurs at 1000 oersteds. Therefore, as the field intensity is increased toward resonance, the magnetization will be increasing. Since the rf susceptibility is proportional to magnetization, the resonance line will be asymmetrical: it will appear to be narrower below resonance. Furthermore, the low-field losses

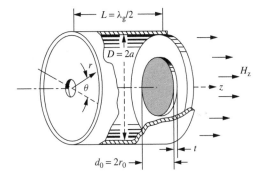

Figure 8.46 Degenerate TE_{111} cylindrical cavity with ferrite disk (Von Aulock and Rowen 1957). Source: Von Aulock, W. and Rowen, J. H. (1957). "Measurement of dielectric and magnetic properties of ferromagnetic materials at microwave frequencies", *The Bell System Technical Journal*, **36** (3), 427–448, by permission of American Telephone and Telegraph Co.

due to the existence of domains will contribute additional distortion to the resonance line.

The use of a thin disk rather than a small sphere as a cavity perturbation eliminates most of the difficulties discussed above because the intrinsic parameters χ and κ are measured directly. As shown in Figure 8.46, when a thin-disk sample is placed against the endplate of a cylindrical TE_{111} mode cavity, the intrinsic parameters χ and κ can be obtained from the splitting of the resonance frequency and the change in $(1/Q)$ (Von Aulock and Rowen 1957)

$$\frac{\Delta\omega_\pm}{\omega_0} = \frac{1}{4}\frac{\lambda_0^2 t}{L^3}(\chi' R_1 \pm \kappa' R_2) \qquad (8.158)$$

$$\Delta\left(\frac{1}{Q}\right)_\pm = \frac{1}{2}\frac{\lambda_0^2 t}{L^3}(\chi'' R_1 \pm \kappa'' R_2) \qquad (8.159)$$

where t is the thickness of the disc, L is the length of the cavity, ω_0 is the angular resonant frequency of the cavity before the magnetic field is applied, λ_0 is the wavelength in free space associated with ω_0, and R_1 and R_2 are functions of the geometry that take into account the fact that the microwave magnetic field is not constant over the face of the disk. Equations (8.158) and (8.159) are based on the assumption that the disk is thin enough to introduce only a first-order perturbation into the cavity field. Figure 8.47 shows a plot of R_1 and R_2 versus the ratio of disc diameter d_0 to

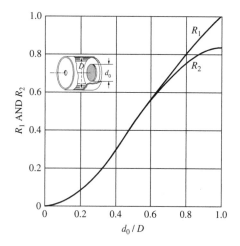

Figure 8.47 The functions of R_1 and R_2 versus the ratio of disk d_0 to cavity diameter D (Von Aulock and Rowen 1957). Source: Von Aulock, W. and Rowen, J. H. (1957). "Measurement of dielectric and magnetic properties of ferromagnetic materials at microwave frequencies", *The Bell System Technical Journal*, **36** (3), 427–448, by permission of American Telephone and Telegraph Co.

cavity diameter D. According to Figure 8.47, it is clear that the functions are closely equal for disc diameters less than $D/2$. This implies circular polarization of the magnetic field in this region, whereas the field becomes elliptically polarized as the sample diameter approaches the outer diameter of the cavity.

As the values of $\Delta\omega_\pm$ and $\Delta(1/Q)_\pm$ depend on the volume of the perturbation object, the perturbations caused by thin discs are at least an order of magnitude larger than those caused by small spheres. Therefore, accurate measurements of ferrite parameters can be made using thin discs in the region below saturation and below and above resonance. In particular, the loss parameters $\chi'' \pm \kappa''$ of low-loss ferrites can be determined accurately in these regions. However, in the resonance region, these values become very large, so it is difficult to accurately measure the resonant properties of the cavity, especially the quality factor. Therefore, the thin-disk method is not suitable for the measurement of resonance line width. The obvious advantage of the thin-disk method lies in its ability to explore the region below resonance where many ferrite devices operate.

It should be noted that the resonance field of a sample is related to the shape of the sample. According to Kittel's relations, the field in the ferrite at which resonance occurs for a given saturation magnetization and operating frequency is given by (Von Aulock and Rowen 1957)

$$H_{\text{res}} = \frac{\omega}{|\gamma|} \qquad \text{(for thin disk)} \qquad (8.160)$$

$$H_{\text{res}} = \frac{\omega}{|\gamma|} - \frac{M_z}{3} \qquad \text{(for small sphere)} \qquad (8.161)$$

$$H_{\text{res}} = \frac{\omega}{|\gamma|} - \frac{M_z}{2} \qquad \text{(for thin cylinder)} \qquad (8.162)$$

Equations (8.160) to (8.162) indicate that for some materials a small sphere or a thin cylinder may be at resonance even before it is saturated. Figure 8.48 is a plot of the real and imaginary parts of $\mu + \kappa$, for a theoretical Landau–Lifshitz line, as a function of H_0/H_r through ferromagnetic resonance. H_0 is the applied dc magnetic field,

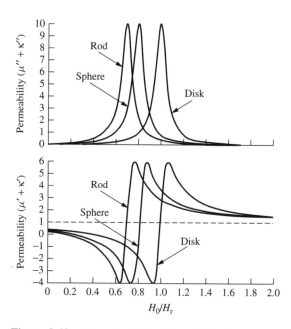

Figure 8.48 Theoretical circularly polarized wave permeability values for rod, sphere, and disk samples (Spencer *et al.* 1956a). Source: Spencer, E. G. Ault, L. A. and LeCraw, R. C. (1956a). "Intrinsic tensor permeabilities on ferrite rods, spheres, and disks", *Proceedings of the IRE*, **44** (10), 1311–1317. © 2003 IEEE

H_r is the resonance field. The results of disk measurements, labeled "disk" in the figure, are intrinsic. The peak of $\mu'' + \kappa''$, which is given by $4\pi M_0 / k H_r$ is normalized equal to 10 for convenience, and the damping constant k in the Landau–Lifshitz equation is chosen to be 0.06 (Spencer *et al.* 1956a).

8.3.4.4 *Exact-theory methods*

Cavity-perturbation methods may face difficulty. If the sample is small enough for the resonant-perturbation theory to be accurate, the frequency change may be too small to be accurately measured, whereas if a big sample is used so that the frequency change can be accurately measured, the approximations associated with the use of perturbation theory may become invalid. One way to overcome such difficulty is to use exact theory to calculate the resonant frequency of the cavity. An advantage of the exact-theory method is the convenience of measuring the comparatively large changes of resonant frequency and quality factor produced by samples. Furthermore, it should be noted that the exact calculations yield directly the intrinsic properties of the sample, that is, the dielectric constant and the magnetic permeability of the material from which the sample is made. On the other hand, perturbation methods yield a "demagnetized susceptibility," $(\mu - \mu_0)F(\mu)$, where $F(\mu)$ is a demagnetizing factor, so the geometrical effects should be considered in perturbation methods, as discussed in Section 8.3.4.3.

In an exact-theory method, it is necessary to draw up the relationships between resonant frequency and quality factor of the cavity loaded with the sample and the intrinsic properties of the sample. One of the requirements of exact-theory method is that the exact solutions of the electromagnetic field in the resonator loaded with the sample can be expressed in closed forms. Therefore, the exact-theory method has special requirements on the geometries of the cavity and the sample. For a sample with arbitrary shape, exact-theory method is difficult, but for a sample with symmetrical shape, such as a rod or a circular disk in a cylindrical cavity, or a rectangular disk or longitudinal slab in a rectangular cavity, the exact theory is available. To ensure that the exact-theory method is feasible, the samples are usually required to have either the same length or the same cross section as the cavity.

Thin rod

As shown in Figure 8.49, the most commonly used measurement configuration is a cylindrical cavity containing a cylindrical sample of full height operating on two circularly polarized, oppositely rotating TM_{110}^{+} and TM_{110}^{-} modes, and the rod sample is magnetized along its length. Such a measurement configuration can be taken as made from a circular waveguide containing a concentric ferrite sample. In the following discussion, we follow the analytical method by Bussy and Steinert (1958).

For a magnetic material magnetized by an external dc magnetic field in the z-direction, the relationship between the microwave magnetic field $[h]$ and the induction $[b]$ in right-circular cylindrical coordinates, is given by

$$\begin{bmatrix} b_r \\ b_\phi \\ b_z \end{bmatrix} = \begin{bmatrix} \mu & j\kappa & 0 \\ -j\kappa & \mu & 0 \\ 0 & 0 & \mu_z \end{bmatrix} \begin{bmatrix} h_r \\ h_\phi \\ h_z \end{bmatrix} \quad (8.163)$$

The signs of the off-diagonal terms specify that the time dependence is $\exp(-j\omega t)$. The antisymmetric matrix $[\mu]$ in Eq. (8.163) is identical with the

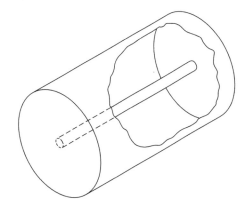

Figure 8.49 A cylindrical cavity containing a thin-rod sample (Baden Fuller 1987). Source: Baden Fulle, A. J. (1987). *Ferrites at Microwave Frequencies*, Peter Peregrinus Ltd., London, 235–255, by permission of Peter Peregrinus Ltd

matrix that represents the permeability tensor in the usual Cartesian coordinates. This may be shown by applying rotation operators to the matrix equation $[b] = [\mu][h]$ in Cartesian space and obtaining Eq. (8.163) as a result. For convenience in the later discussion, we introduce the inverse of the matrix $[\mu]$:

$$[\mu]^{-1} = \begin{bmatrix} M & jK & 0 \\ -jK & M & 0 \\ 0 & 0 & M_z \end{bmatrix} \qquad (8.164)$$

as its elements occur most naturally in the wave solutions to be obtained.

To obtain a solution for the cavity with a ferrite rod, the wave equation in the sample must be solved and the waves in it must match those in the remaining space of the cavity. For the simplicity of solution, by assuming that the waves are independent of the z-coordinate, two-dimensional modes of resonance for the cavity can be used. The two-dimensional wave equation is given by

$$(\nabla_p^2 + \omega^2 \varepsilon / M) E_z = 0 \qquad (8.165)$$

where the Laplacian in transverse (r, ϕ) space is given by

$$\nabla_p^2 = \frac{1}{r} \frac{\partial}{\partial r} r \frac{\partial}{\partial r} + \frac{1}{r^2} \frac{\partial^2}{\partial \phi^2}, \qquad (8.166)$$

where ω is the angular frequency, and the dielectric permittivity ε is assumed to be a scalar. Equation (8.165) is applicable for free space after replacing M by $M_0 = 1/\mu_0$ and ε by ε_0, where μ_0 and ε_0 are the constitutive parameters of free space.

The solutions of Eq. (8.165) are of the form

$$E_z = (Ae^{jn\phi} + Be^{-jn\phi})[FJ_n(\beta r) + GY_n(\beta r)] \qquad (8.167)$$

with

$$\beta = \omega(\varepsilon/M)^{1/2} \qquad (8.168)$$

where J_n and Y_n are Bessel functions of the first and second kinds, respectively, of integral order n; and A, B, F, and G are coefficients to be determined by the boundary conditions. The solutions of Eq. (8.165) for free space are the same as Eq. (8.167), provided that the cylinder functions

have the argument (kr) instead of (βr) with k given by

$$k = \omega(\varepsilon_0/M_0)^{1/2}. \qquad (8.169)$$

The waves given by Eq. (8.167) should be fit into the boundary conditions of a right-circular cylindrical cavity with radius b with a centered ferrite rod with radius a. Since the fields do not vary with z, the height of the structure is arbitrary. In this problem, since the $e^{+jn\phi}$ and $e^{-jn\phi}$ solutions have different resonant frequencies, either frequency can be selected for excitation. To fit the boundary conditions at $r = 0$ and $r = b$, Eq. (8.167) can be expressed as

$$E_z = De^{\pm jn\phi} J_n(\beta r), \qquad (r < a) \qquad (8.170)$$

$$E_z = Ee^{\pm jn\phi} C_n(kr), \qquad (r > a) \qquad (8.171)$$

with

$$C_n(kr) = J_n(kr) - Y_n(kr)[J_n(kb)/Y_n(kb)] \qquad (8.172)$$

Equation (8.172) indicates that $C_n(kr) = 0$ when $r = b$.

The microwave magnetic fields can be obtained from Eqs. (8.170) and (8.171) by applying Maxwell's curl equation in matrix form:

$$\begin{bmatrix} M & jK & 0 \\ -jK & M & 0 \\ 0 & 0 & M_z \end{bmatrix}$$

$$\times \begin{bmatrix} 0 & -D_z & r^{-1}D_\phi \\ D_z & 0 & -D_r \\ -r^{-1}D_\phi & r^{-1}D_r r & 0 \end{bmatrix}$$

$$\times \begin{bmatrix} E_r \\ E_\phi \\ E_z \end{bmatrix} = j\omega \begin{bmatrix} h_r \\ h_\phi \\ h_z \end{bmatrix} \qquad (8.173)$$

where the Ds are derivative operators.

From the boundary condition that E_z and h_ϕ should be continuous across the boundary at $r = a$, the ratio of the amplitudes D/E and the value of ω can be fixed, and the equation of resonance for the structure can be obtained:

$$\frac{M\beta a J_n'(\beta a)}{J_n(\beta a)} \pm nK = \frac{M_0 ka C_n'(ka)}{C_n(ka)} \qquad (8.174)$$

where the primes indicate derivatives with respect to the arguments. For $n > 0$, Eq. (8.174) represents two resonances, indicated by "+" and "−" signs. For $n = 0$, we have a nondegenerate TM_{0m0} mode, and in this mode, a small ferrite rod furnishes mostly a dielectric effect on the resonator. In order to evaluate the three components, ε, M and K, three measurements on a ferrite in three different cavity modes are needed. It is convenient to make one measurement in a TM_{0m0} resonator and two in a TM_{110} resonator (positive and negative circular polarization). To determine M_z defined in Eq. (8.164), a fourth measurement is often made in a TE_{011} cavity.

Figure 8.50 shows a cylindrical cavity containing a rod sample. The circularly polarized modes are excited through two adjacent irises by waves of equal magnitude and $90°$ out of phase. By this approach, both positive and negative modes can be excited. It should be noted that the symmetry in the mechanical structure of the resonator is important in tensor permeability measurements. In the mathematical solutions to Eq. (8.174), it is often assumed that the resonator is perfectly circular. Insufficient symmetry may result in a deleterious effect. To avoid the errors due to initial splitting of the degeneracy by geometrical perturbations, it is required that the symmetry group of the perturbed

TM_{110} resonator be C_3 or higher (Bussy and Steinert 1958). This means that with irises, loops, or probes around the periphery of the cylinder, there must be three or more identical irises, and so on, spaced at equal angles. A circular iris on the axis would be desirable, as it leaves the symmetry C_∞, but is not always practicable.

Besides the exact-theory method, approximation methods can also be used for thin-rod method. The permeability components μ and κ can also be calculated by

$$\left(\frac{\Delta f}{f}\right)_\pm = -\frac{2(\mu' \pm \kappa' - 1) \times 1.541 \cdot V_s}{(\mu' \pm \kappa' + 1)V_c}$$
(8.175)

$$\Delta\left(\frac{1}{Q}\right)_\pm = -\frac{2(\mu'' \pm \kappa'') \times 1.541 \cdot V_s}{(\mu' \pm \kappa' + 1)V_c}$$
(8.176)

where V_s and V_c are the volumes of the sample and the cavity respectively.

As mentioned above, the parallel component μ_z is usually determined on the basis of an independent measurement, for example, using cylindrical cavity working at TE_{011} mode. Figure 8.51 shows another method for measuring μ_z by placing the thin rod sample at the center of a rectangular cavity operating in the TE_{102} mode. The values of μ'_z

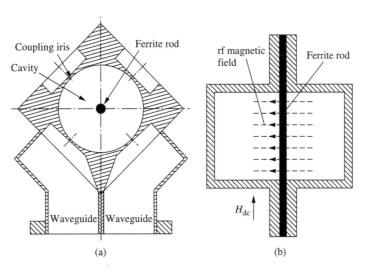

(a)   (b)

Figure 8.50   Configuration of a TM_{110} resonator with four irises and guides permitting excitation of a rotating field pattern. (a) Cross view and (b) top view. Modified from Bussy, H. E. and Steinert, L. A. (1958). "Exact solution for a gyromagnetic sample and measurements on ferrite", *IRE Transactions on Microwave Theory and Techniques*, **6** (1), 72–76. © 2003 IEEE

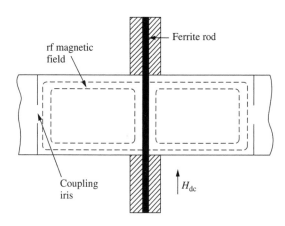

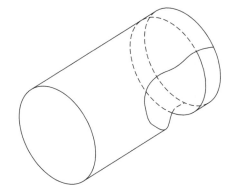

Figure 8.51 TE_{102} mode rectangular cavity for the measurement of μ'_z and μ''_z

Figure 8.52 Disk at one end of a cylindrical cavity (Baden Fuller 1987). Source: Baden Fulle, A. J. (1987). *Ferrites at Microwave Frequencies*, Peter Peregrinus Ltd., London, 235–255, by permission of Peter Peregrinus Ltd

and μ''_z can be calculated by

$$\frac{\Delta f}{f} = \frac{(\mu'_z - 1)V_s}{[1 + (l/2a)^2]V_c} \quad (8.177)$$

$$\Delta\left(\frac{1}{Q}\right) = \frac{2\mu''_z V_s}{[1 + (l/2a)^2]V_c} \quad (8.178)$$

where l is the length and a is the width of the cavity, V_s and V_c are the volumes of the sample and the cavity respectively.

Thin disk

Figure 8.52 shows a ferrite disk completely filling one end of a cylindrical cavity. This configuration can be analyzed as a short length of completely filled circular waveguide, coupled to a length of empty circular waveguide. As such, structure can be exactly analyzed in theory, no demagnetization factors enter into the expressions used for the microwave properties of the ferrite. Detailed discussions on this configuration can be found in (Baden Fuller 1987).

There are several other configurations of thin-disk method. A rectangular cavity operating in the TE_{10n}-mode can be used, and the thin ferrite slab completely fills one end of the rectangular cavity (Viswanathan and Murthy 1990). Another configuration often used is to mount a transversely magnetized ferrite slab in a rectangular waveguide cavity (Baden Fuller 1987).

Small sphere

Besides a cylindrical cavity coaxial loaded with thin rod and a cylindrical or rectangular cavity loaded with thin disk, a spherical cavity loaded with a small sphere sample can also be analyzed by using closed-form equations. Okada and Tanaka proposed a method for the measurement of the permeability tensor using a spherical cavity resonator loaded with a spherical ferrite sample under test (Okada and Tanaka 1990). For convenience, a spherical coordinate system is used as shown in Figure 8.53(a). The spherical TE_{011} mode has symmetry in ϕ direction, and a spherical cavity can support two degenerate TE_{011} modes. In Figure 8.53(b), the microwave magnetic field of one TE_{011} mode is shown by dotted line, labeled "TE_{011}," and that of the orthogonally degenerate TE_{011} mode (z-axis change to x-axis) is shown by broken line, labeled "degenerate TE_{011}." The external dc magnetic field H_{dc} is applied along the y-axis.

If the two orthogonally degenerate TE_{011} modes shown in Figure 8.53(b) are excited by a 90-degree phase difference, the microwave magnetic fields of the two modes constitute the circular polarized magnetic field around the center point. In the center point, the microwave electric field is almost zero, so if a small spherical ferrite sample is placed at this point and a biased H_{dc} field is impressed in the y-axis, the relative permeability $\mu_{r\pm}$ for right- and

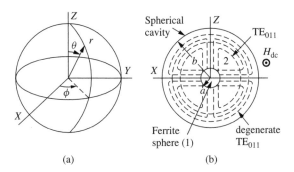

(a)   (b)

Figure 8.53 Spherical cavity for the measurement of permeability tensors. (a) Spherical coordinate and (b) distributions of magnetic fields of two degenerate TE$_{011}$ modes (Okada and Tanaka 1990). Source: Okada, F. and Tanaka, H. (1990). "Measurement of ferrites tensor permeability by spherical cavity resonator", *Digest of Conference on Precision Electromagnetic Measurements '90*, 230–231. © 2003 IEEE

left-hand circular polarized fields, the components of the tensor [μ_r], can be derived from the resonant properties of the spherical resonator loaded with the spherical magnetic sample.

The electromagnetic field E_ϕ, H_r, H_θ of the TE$_{011}$ mode in a ferrite sample region 1 ($r < a$) and outer region 2 ($a < r < b$) respectively can be obtained from the wave equation. Then, from the boundary conditions ($E_{\phi 1} = E_{\phi 2}$, $H_{\theta 1} = H_{\theta 2}$ at $r = a$ and $E_{\phi 2} = 0$ at $r = b$), the following equation can be derived (Okada and Tanaka 1990):

$$\frac{1}{\mu_{r\pm}} \cdot \frac{\frac{\partial}{\partial x}(xj_1(x))|_{x=k_f a}}{j_1(k_f a)}$$

$$= \frac{n_1(k_0 b)\frac{\partial}{\partial x}(xj_1(x))|_{x=k_0 a} - j_1(k_0 b)\frac{\partial}{\partial x}(xn_1(x))|_{x=k_0 a}}{n_1(k_0 b)j_1(k_0 a) - j_1(k_0 b)n_1(k_0 a)}$$

$$(8.179)$$

with

$$k_f = k_0 = \omega\sqrt{\varepsilon_r \mu_{r\pm}} \qquad (8.180)$$

$$k_0 = \omega\sqrt{\varepsilon_0 \mu_0} \qquad (8.181)$$

$$\omega = \omega_1 + j\omega_2 = \omega_1 + j\frac{\omega_1}{2Q_0} \qquad (8.182)$$

$$\mu_{r\pm} = \mu \mp \kappa \qquad (8.183)$$

where $\mu_{r\pm}$ and ε_r are relative permeability for circular-polarization wave and relative permittivity of ferrite, and μ and κ are the components of relative permeability tensor [μ_r]. The spherical Bessel functions $j_1(x)$ and $n_1(x)$ are given by the following trigonometric functions:

$$j_1(x) = \frac{1}{x}\left(\frac{\sin x}{x} - \cos x\right) \qquad (8.184)$$

$$n_1(x) = -\frac{1}{x}\left(\frac{\cos x}{x} + \sin x\right) \qquad (8.185)$$

If resonant frequency and quality factor are measured when the ferrite sample is inserted into the cavity, $\mu_{r\pm}$ can be computed from Eq. (8.179).

Figure 8.54 shows a spherical cavity constructed of two semispheres. A hollow thin dielectric rod is used to support the spherical ferrite sample. Couplings are made by using coaxial cable loops. 1 and 1' are the input and output ports for the TE$_{011}$ mode, while 2 and 2' are used for the orthogonally degenerate TE$_{011}$ mode. The pine labeled 3 is used to adjust the isolation between the two degenerate modes.

Finally, it deserves to be noted that the quality factor of a spherical cavity is usually about two times higher than that of a rectangular or circular cavity resonator, so spherical cavities usually

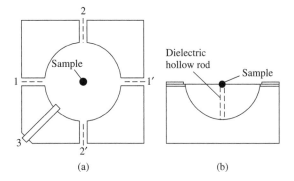

(a)   (b)

Figure 8.54 A spherical cavity with the top half moved. (a) Plane view and (b) side view (Okada and Tanaka 1990). Source: Okada, F. and Tanaka, H. (1990). "Measurement of ferrites tensor permeability by spherical cavity resonator", *Digest of Conference on Precision Electromagnetic Measurements '90*, 230–231. © 2003 IEEE

have higher accuracy than rectangular and circular cylindrical cavities in materials characterization.

8.4 MEASUREMENT OF FERROMAGNETIC RESONANCE

Ferromagnetic resonance is a special phenomenon of magnetic materials, and the measurement of ferromagnetic resonance is important for the study of ferromagnetism and applications of magnetic materials. The basic knowledge of ferromagnetic resonance has been discussed in previous sections. In this section, we further discuss the origin of ferromagnetic resonance and the parameters describing ferromagnetic resonance, and then we discuss the methods for the measurement of ferromagnetic resonance.

8.4.1 Origin of ferromagnetic resonance

Ferromagnetic resonance originates from the precession of atoms under a magnetic field (Viswanathan and Murthy 1990). For an atom possessing both magnetic moment μ and angular momentum J, the rate of change of angular momentum is equal to the torque acting on the system:

$$\hbar \frac{dJ}{dt} = \mu \times H \qquad (8.186)$$

where $\mu \times H$ is the torque acted on the atom in the magnetic field H. The relationship between the magnetic moment μ and the angular momentum J is

$$\mu = -\gamma \hbar J \qquad (8.187)$$

with the gyromagnetic ratio γ given by

$$\gamma = \frac{ge}{2mc} \qquad (8.188)$$

where g is the Lande's splitting factor, e and m are the electricity and mass of an electron, and c is the speed of light. The g-factor is related to many aspects, and $g = 2$ for free spin. From Eqs. (8.186) and (8.187), we obtain

$$\frac{d\mu}{dt} = -\gamma (\mu \times H) \qquad (8.189)$$

As the magnetization M of the medium is the sum of all the magnetic moments in a unit volume

$$M = \sum \mu_i \qquad (8.190)$$

From Eqs. (8.189) and (8.190), the following movement equation can be obtained:

$$\frac{dM}{dt} = -\gamma (M \times H) \qquad (8.191)$$

Equation (8.191) represents the precession of M under the magnetic field H. Figure 8.55 (a) shows the precession under dc magnetic field H_0.

To consider the losses of the system, Eq. (8.191) should be modified as

$$\frac{dM}{dt} = -\gamma (M \times H) + \frac{\lambda}{|M|^2}(M \times (M \times H)) \qquad (8.192)$$

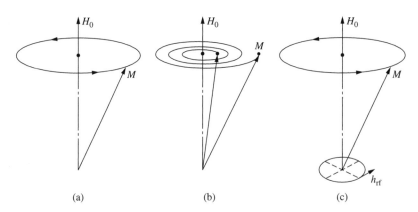

Figure 8.55 Precession of magnetization vector. (a) Precession without damping, (b) precession with damping, and (c) precession with damping and microwave field

where λ is the damping constant. Figure 8.55(b) shows a precession of M under external dc magnetic field H_0. Because of the damping effect, the angle of the precession decreases. Finally, the spin magnetic moment vector becomes parallel to H_0.

Figure 8.55(c) shows the case when a circularly polarized microwave field H_{rf} is applied to the magnetic moments. Depending on the direction of the rotation of H_{rf}, there are two results. If the electron and the microwave field are rotating at the same direction, the microwave field provides energy to the electron. The precession of magnetic moments continues, and when the frequency of precession of magnetic moments equals the frequency of the microwave field, the ferromagnetic resonance occurs. If the electron and the microwave field are rotating in opposite directions, the energy is alternately exchanged between the electron and magnetic field, so that the ferromagnetic resonance cannot take place. When a linearly polarized microwave field is applied to the sample, one of its two circular-polarization components contributes to the ferromagnetic resonance.

8.4.2 Measurement principle

In the study of ferromagnetic resonance, we usually measure the microwave energy dissipation parameters, such as the imaginary part of permeability or the magnetic power absorption P. Generally speaking, ferromagnetic resonance can be measured by two types of methods: sweeping magnetic field method and sweeping frequency method. In a sweeping magnetic field method, the operation frequency is fixed, and the relationships between the dissipation parameter and the external dc magnetic field are measured. As shown in Figure 8.56(a), two parameters can be determined from the ferromagnetic resonance curve: resonance magnetic field H_0 and resonance line width ΔH. The gyromagnetic ratio γ can be calculated from H_0 according to the following relationship:

$$\omega_0 = \gamma H_0. \tag{8.193}$$

In this method, ω_0 is the operating frequency of the measurement system. After γ is obtained, the g-factor can then be calculated from γ according to Eq. (8.188).

The line width ΔH is related to the gyromagnetic ratio γ and the relaxation time τ by

$$\Delta H = \frac{1}{\gamma \tau}. \tag{8.194}$$

The damping constant can then be calculated according to

$$\lambda = \frac{\gamma |M|}{\omega_0 \tau}. \tag{8.195}$$

In a sweeping frequency method, the external dc magnetic field is fixed, and the relationships between the energy dissipation parameter and the microwave frequency are measured. As shown in Figure 8.56(b), two parameters can be determined from the ferromagnetic resonance curve: resonance frequency ω_0 and resonance line width $\Delta \omega$. The gyromagnetic ratio γ can also be calculated from

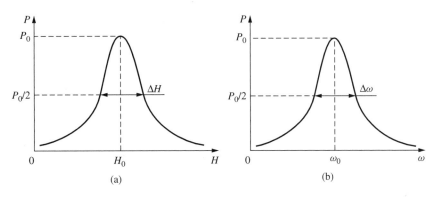

Figure 8.56 Two types of ferromagnetic resonance curves. (a) Sweeping magnetic field method and (b) sweeping frequency method

ω_0 using Eq. (8.193), and in this method, H_0 is the fixed external dc magnetic field. The relation between ΔH and $\Delta \omega$ measured using the two methods shown in Figure 8.56 is given by

$$\Delta \omega = \gamma \Delta H = \frac{1}{\tau}. \qquad (8.196)$$

In the measurement of ferromagnetic resonance curves, besides the power absorption method shown in Figure 8.56, cross-coupling method can also be used, as ferromagnetic resonance can also be characterized by the coupling between two orthogonal modes that are normally isolated when the sample is in off-resonance state. As discussed earlier, for a magnetic sample saturately magnetized in the z-direction, the rf magnetization $[m]$ is related to the rf field $[h]$ through a tensor susceptibility $[\chi]$:

$$\begin{bmatrix} m_x \\ m_y \end{bmatrix} = \begin{bmatrix} \chi_{xx} & \chi_{xy} \\ \chi_{yx} & \chi_{yy} \end{bmatrix} \begin{bmatrix} h_x \\ h_y \end{bmatrix} \qquad (8.197)$$

The coupling between the two orthogonal modes is related to the off-diagonal component χ_{xy} ($\chi_{xy} = -\chi_{yx}$). As the power coupled between the two modes is proportional to $|\chi_{xy}|^2$, the values of H_0 and ΔH are defined as shown in Figure 8.57, and the values of ω_0 and $\Delta \omega$ can also be similarly defined. It is clear that, the criteria for determining the resonance line width is

$$|\chi_{xy}|_{\Delta H} = |\chi_{xy}|_{max}/\sqrt{2} \qquad (8.198)$$

where $|\chi_{xy}|_{max}$ is the maximum value of $|\chi_{xy}|$ corresponding to H_0, and $|\chi_{xy}|_{\Delta H}$ is the value of $|\chi_{xy}|$ for the determination of ΔH.

The above discussion indicates that precise information on the magnetic state of the atom in a magnetic medium can be gathered from its ferromagnetic resonance curve, so this technique gives information about the process at the atomic level. Besides, the components of susceptibility tensor can be calculated from the parameters of ferromagnetic resonance.

The diagonal and off-diagonal components of the tensor susceptibility χ_{xx} and χ_{xy} are complex quantities and are functions of many parameters, including the dc magnetic field H_0, the rf frequency ω, the magnetization M, the gyromagnetic ratio γ, and the damping constant λ. From Eqs. (8.192) and (8.197), we can get

$$\chi_{xx} = \chi_{yy} = \frac{(\omega_0 + j\Delta\omega)\omega_m}{(\omega_0 + j\Delta\omega)^2 - \omega^2} \qquad (8.199)$$

$$\chi_{yx} = -\chi_{xy} = \frac{-j\omega_0\omega_m}{(\omega_0 + j\Delta\omega)^2 - \omega^2} \qquad (8.200)$$

with

$$\omega_m = \gamma 4\pi M_0 \qquad (8.201)$$

$$\omega_0 = \gamma H_0 \qquad (8.202)$$

where M_0 is the magnetization of the sample due to the external dc magnetic field H_0.

Typical methods for the measurement of permeability tensors have been discussed in Section 8.3, and in principle, these methods can be used in the

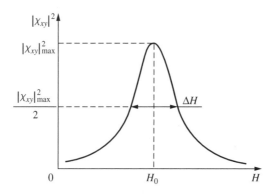

Figure 8.57 Ferromagnetic resonance curve in the form of $|\chi_{xy}|^2$

study of ferromagnetic resonance. However, if we only need to measure the ferromagnetic resonance curve, it is not necessary to characterize all the components in the permeability tensor. In the following text, we discuss three types of methods often used for the study of ferromagnetic resonance: cavity methods, waveguide methods, and planar-circuit methods.

8.4.3 Cavity methods

The usual method for measuring the microwave susceptibility and the ferromagnetic resonance curve of a ferrite depends upon the complex frequency perturbation of a resonant cavity, and the cavity-perturbation technique is used in a large number of experiments mainly because of its sensitivity and accuracy. However, the cavity perturbation has some disadvantages. It does not permit one to vary frequency easily, and the measurement procedure is quite complicated. The cavity-perturbation technique could be dispensed with for the routine measurement of the line width and the *g*-factor of microwave ferrites, especially when one is concerned with comparative data among a large number of ferrite samples. In such a situation, saving time is more important than obtaining very accurate data.

In the cavity methods that will be discussed below, the sample under study is placed within a resonant cavity. The energy dissipation of the sample is estimated from the energy dissipation of the cavity containing the sample. Though quality factor is an accurate indicator of the energy dissipation of the cavity, the measurement of quality factor requires complicated instruments, and furthermore, at the region of ferromagnetic resonance, the sample is usually quite lossy, so the measurement of quality factor may become difficult. In the following text, we discuss the principles of the reflection method and transmission method for the measurement of energy dissipation of a cavity loaded with a magnetic sample under test.

8.4.3.1 Reflection method

In this method, the energy dissipation of the sample under test is estimated from the reflection from the cavity containing the sample. As

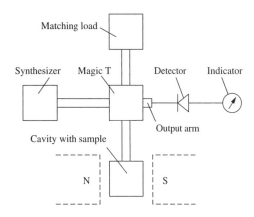

Figure 8.58 Reflection method for the measurement of ferromagnetic resonance using magic T

shown in Figure 8.58, the measurement circuit is a microwave bridge circuit built using a magic T or hybrid ring. The cavity containing the sample is connected to one arm of the magic T, and a matching load is connected to the opposite arm. Before the dc magnetic field is applied, the matching load is adjusted so that there is no output at the output arm. Then the external dc magnetic field is applied, and from the relationship between the applied dc magnetic field and the output at the output arm, the ferromagnetic resonance curve can be obtained.

In the above measurement procedure, it is assumed that the sample does not absorb microwave energy at zero dc magnetic field. To consider the loss of the sample at zero dc magnetic field, at first, an external dc magnetic field that is much higher than the resonance magnetic field ($H \gg H_0$) is applied, and the matching load is adjusted so that there is no output at the output arm. Then, the external dc magnetic field is decreased, and the relationship between the output at the output arm and the external dc magnetic field can be obtained. When the external dc magnetic field decreases to zero, the absorption of the sample at zero magnetic field can be obtained.

8.4.3.2 Transmission method

For a high-quality factor cavity loaded with a ferrite sample, most of the loss is due to the energy dissipation of the sample, and the transmission energy through the cavity at its resonance is related

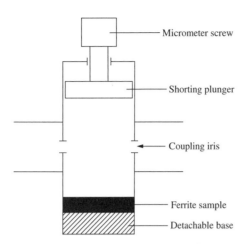

Figure 8.59 Schematic view of a transmission cavity. Modified from Viswanathan, B. and Murthy, V. R. K. (1990). *Ferrite Materials, Science and Technology*, Narosa Publishing House, New Delhi, 85–105, by permission of Narosa Publishing House

to the energy dissipation of the sample loaded in the cavity. If the input microwave power is fixed, from the relationship between the transmission power of the cavity and the applied dc magnetic field, the ferromagnetic resonance curve of the sample can be obtained. Figure 8.59 shows the basic structure of a transmission cavity. In the measurement procedure, the resonant frequency of the cavity may be changed because of the change of the external dc magnetic field, so that at each measurement point, it is desirable to adjust the shorting plunger to ensure that the cavity is at its resonant state.

8.4.4 Waveguide methods

In this type of method, the sample is loaded in a waveguide structure, and measurements can be made at a certain frequency range. In the following, we discuss three waveguide methods often used in the measurement of ferromagnetic resonance: frequency-variation method, cross-guide method, and pickup-coil method.

8.4.4.1 Frequency-variation method

In this method, the sample under test is put inside a shorted waveguide, which is placed between the two poles of the electromagnet, as shown in Figure 8.60(a). The electromagnet is driven by a dc current and an ac current, so the magnetic field H consists of a dc magnetic field H_{dc} and an ac magnetic field H_{ac}:

$$H = H_{dc} + H_{ac} \qquad (8.203)$$

Suitable value of the dc magnetic field H_{dc} and suitable amplitude of the ac magnetic field are chosen so that the ferromagnetic resonance curve of the sample can be observed in the measurement range.

As shown in Figure 8.60(b), two ferromagnetic resonance curves are measured corresponding to two microwave frequencies ω_1 and ω_2. When the operation frequency changes from ω_1 and ω_2, the absorption curve shifts ΔH along the H axis. The gyromagnetic ratio γ can be calculated by

$$\gamma = \frac{\Delta \omega}{\Delta H} \qquad (8.204)$$

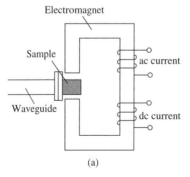

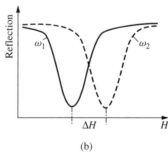

Figure 8.60 Frequency-variation method for the measurement of ferromagnetic resonance. (a) Measurement setup and (b) measurement of ΔH

with $\Delta\omega = (\omega_2 - \omega_1)$. This method is widely used by ferrite manufacturers.

8.4.4.2 Cross-guide method

Stinson proposed a method for the measurement of the resonance line width and the g-factor of a spherical ferrite sample in a cross-guide directional coupler (Stinson 1958). This method is simple and sufficiently accurate for most purposes. It is especially suitable for yielding reliable comparative data during the development of optimum manufacturing techniques.

As shown in Figure 8.61, this method uses a cross-guide directional coupler with a round, centered coupling hole in the common broad wall. The two waveguides of the cross-guide coupler are soldered together, and the wall thickness at the coupling hole is about half of the normal waveguide wall thickness. A suitable diameter of the hole is chosen so that the coupling is in the range of 40 to 50 dB. An access hole is made for inserting the sample into the coupling hole, and the fit of the cover plate on the access hole is not critical, as the leakage of power through it is not important in relative power measurements of this type as long as the leakage power remains constant during the measurements. The ferrite sample is

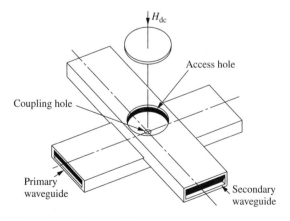

Figure 8.61 Cross-guide coupler for the measurement of ferromagnetic resonance (Stinson 1958). Source: Stinson, D. C. (1958). "Ferrite line width measurements in a cross-guide coupler", *IRE Transactions on Microwave Theory and Techniques*, **6**, 446–450. © 2003 IEEE

glued symmetrically in the coupling hole. Though the placement of the sample in the coupling hole has some effect on the level of the coupled power, the effect on the line width or the g-factor is not obvious.

The method for measuring the line widths of ferrite samples using a cross-guide coupler is developed from the theory of coupling through apertures containing ferrites. The theory for coupling through an aperture containing a ferrite assumes an aperture that is small compared to the guide wavelength and a ferrite sample that is small compared to the wavelength inside it (Stinson 1958, 1957). These assumptions can be easily realized, so the theoretical conclusions can be used in obtaining qualitative information concerning many ferrite-coupling problems.

The expressions for the coupling between waveguides contain terms for both the dielectric and magnetic susceptibilities. However, the dielectric susceptibility term could be removed by terminating the primary waveguide with a short located at a multiple of a half-guide wavelength from the aperture so as to annul the electric field at the aperture. Thus, the coupling expression for a collinear coupler contains only the diagonal magnetic susceptibility, while the coupling expression for a cross-guide coupler contains only the nondiagonal magnetic susceptibility. Also, for the cross-guide coupler, there is no coupled power in the absence of the external dc magnetic field. For the collinear coupler, there is coupled power in the absence of external dc magnetic field. Therefore, the cross-guide coupler is suitable for line-width measurements, and very small samples may be measured. In practice, garnet-structure ferrite spheres with 0.030-inch diameter are adequate for measurements at X-band frequencies (Stinson 1958).

The expression for the coupled power in dB in the cross-guide coupler for a centered aperture and with a short located in the primary waveguide so as to annul the electric field at the aperture is (Stinson 1958)

$$C = C_0 + 20\log_{10}|\chi_{xy}| \qquad (8.205)$$

with

$$C_0 = 20\log_{10}\left(\frac{2\pi d^3 F_H}{2ab\lambda_{\mathrm{g}}}\right) \qquad (8.206)$$

where χ_{xy} is the nondiagonal magnetic susceptibility of the ferrite sample in the coupling aperture; d is the diameter of the aperture; a and b are the wide and narrow dimensions of the rectangular waveguide respectively; λ_g is the guide wavelength; and F_H is a quantity that measures the attenuation due to finite aperture thickness. Equation (8.205) and all of the quantities involved are more completely defined in (Stinson 1958, 1957). According to Eq. (8.205), ferromagnetic resonance curve in the form of $|\chi_{xy}|^2$, as shown in Figure 8.57, can be obtained.

8.4.4.3 Pickup-coil method

The working principle of the pickup-coil is similar to that for the cross-guide method, but in the pickup-coil method, the secondary waveguide in the cross-guide method is replaced by a pickup coil (Masters *et al.* 1960). As shown in Figure 8.62, the sample is placed in the transverse, uniform microwave magnetic field at a distance $n\lambda_g/2$ from the short

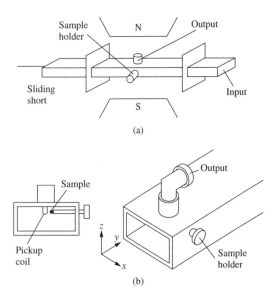

(a)

(b)

Figure 8.62 Sliding short method for the measurement of ferromagnetic line width. (a) Measurement configuration and (b) structure of the measurement fixture. Modified from Masters, J. I. Capone, B. R. Gianino, P. D. (1960). "Measurement technique for narrow line width ferromagnets", *IRE Transactions on Microwave Theory and Techniques*, **8**, 565–566. © 2003 IEEE

in a shorted waveguide. A nearby pickup coil, which senses the precession dipolar field of the magnetic sample, is oriented so that the plane of the loop and microwave magnetic field h are parallel, producing a negligible coupling to the transverse microwave field. When the dc magnetic field H is near resonance a voltage is induced in the loop.

As the spatial relationship of the loop and the sample is such that only the projection of the motion along the y-direction is effective, the induced voltage at the output port depends upon the absolute value of the off-diagonal component of $[\chi]$:

$$V_{\text{output}} \propto |\chi_{xy}|. \qquad (8.207)$$

Therefore, from the voltage at the output port, ferromagnetic resonance curve can be obtained.

In experiments, the microwave signal is fed to a section of waveguide terminated by a sliding short. The correct position of the sliding short can be found by substitution of an electric probe for the loop and adjusting for a null in the electric-probe output. The correct loop orientation is then found by rotation of the loop for minimum leakage of the microwave field with the external dc magnetic field $H = 0$, which is usually about 30 dB below the loop output when H is adjusted for resonance.

8.4.5 Planar-circuit methods

As discussed above, ferromagnetic resonance properties of magnetic materials can be measured using waveguide systems. However, because of the coherent limitation of the narrowband of waveguide systems, it is difficult to characterize the resonance line width ΔH in multioctaves, and ΔH values are sometimes reported at a single frequency point. In the following, we discuss two broadband planar-circuit methods for line width measurement: magnetostatic resonator method and slot-coplanar junction method, and these methods can cover wide frequency bands. Finally, we will discuss a microstrip resonator method, based on near-field microwave microscope techniques, which has high sensitivity and spatial resolution.

8.4.5.1 MSW-SER method

This method is suitable for film and sheet samples. The measurement fixture is a piece of microstrip transmission line, and the sample under test forms a magnetostatic wave straight-edge resonator (MSW-SER) (Chen *et al.* 1993). The resonance line width ΔH is calculated from the half resonance band ($\Delta \omega$) for unloaded quality factor of the MSW-SER. As the resonant frequency of a MSW-SER can be tuned via a bias magnetic field (Ishak and Chang 1986), using this method, measurements can be made over a wide frequency range.

Figure 8.63 shows a microstrip-line fixture loaded with a film sample. The impedance of the transmission line is $50\,\Omega$. The sample to be tested is placed on the top of the microstrip line. A proper spacer is inserted between the sample and the central conductor to adjust the coupling strength. The fixture loaded with sample is a typical two-port MSW-SER. Depending on the orientation of the external dc magnetic field H, two kinds of resonant modes can be excited. When the dc magnetic field is perpendicular to the film plane, forward volume wave is excited in the sample; whereas when the dc magnetic field is parallel to the film plane, surface wave is excited. Therefore, $\Delta H_\perp$ and $\Delta H_\parallel$ can be separately measured, corresponding to the cases when the external dc magnetic field is perpendicular and parallel to the sample plane, respectively.

The relationship between ΔH and $\Delta \omega$ can be obtained from the resonance condition of the MSW-SER. The resonant condition can be expressed in an implicit form

$$F(\omega, k, H) = 0 \tag{8.208}$$

where ω is the angular frequency, k is the wave number, and H is the external dc magnetic field. The ferromagnetic resonance line width (ΔH) of the magnetic film can be obtained from (Chen *et al.* 1993)

$$\Delta H = \Delta \omega / \gamma_k \tag{8.209}$$

with

$$\gamma_k = -\gamma \frac{\partial F / \partial \omega}{\partial F / \partial \omega_0} \tag{8.210}$$

$$\omega_0 = \gamma H \tag{8.211}$$

where γ is the gyromagnetic ratio, and γ_k is related to the orientation of external dc magnetic field and thus the resonance types of the MSW-SER. For forward volume wave resonators, γ_k is given by

$$\gamma_k = \frac{\gamma(\omega^2 + \omega_0^2)}{2\omega\omega_0} \tag{8.212}$$

and for surface-wave resonators, γ_k is given by

$$\gamma_k = \frac{\gamma(2\omega_0 + \omega_m)}{2\omega} \tag{8.213}$$

with

$$\omega_m = \gamma 4\pi M_s \tag{8.214}$$

8.4.5.2 Slot-coplanar junction method

Figure 8.64 schematically shows a slot-coplanar junction for the measurement of ferromagnetic resonance (Dorsey *et al.* 1992; Zhang *et al.* 1997). The planar fixture is designed such that a slot line is the input transmission line, with a coplanar waveguide (CPW) being the output transmission line. These collinearly aligned transmission lines are shorted in a junction region, which appears in the center. Figure 8.64(a) shows the slot line and CPW geometries, and the microwave magnetic field (h) patterns of their propagating modes. As the low-frequency CPW mode is approximately orthogonal to the slot-line propagating mode, the fixture has a good isolation between input and

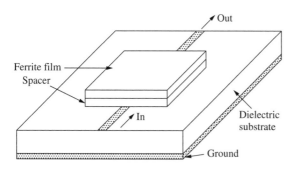

Figure 8.63 Microstrip-line fixture loaded with a film sample (Chen *et al.* 1993). Source: Chen, H. De Gasperis, P. and Marcelli, R. (1993). "Using microwave network analyzer and MSW-SER to measure linewidth spectrum in magnetic garnet film from 0.5 to 20 GHz", *IEEE Transactions on Magnetics*, **29** (6), 3013–3015. © 2003 IEEE

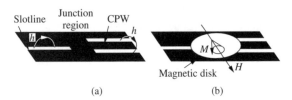

Figure 8.64   Slot-coplanar junction for ferromagnetic resonance measurement (Zhang *et al.* 1997). (a) The geometry of the slot line and CPW in the junction region and (b) sample placement. Modified from Zhang, S. Oliver, S. A. Israeloff, N. E. and Vittoria, C. (1997). "High-sensitivity ferromagnetic resonance measurements on micrometer-sized samples", *Applied Physics Letters*, **70** (20), 2756–2758

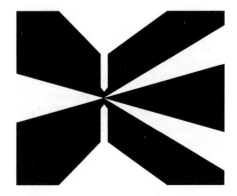

Figure 8.65   The junction region where a sample would be placed for testing

output ports, and the CPW effectively appears as a short to the input slot line.

The geometry of the junction region is crucial for fixture performance. The microwave magnetic field at the shorted end of a slot line is predicted to have a large component normal to the fixture plane because of the large electrical currents at the slot-line terminus. For a single shorted slot line in an otherwise continuous metal plane, this normal-oriented microwave magnetic field pattern is distributed over a large area around the slot-line terminus since the effective electrical length of the slot line extends a fraction of a wavelength into the metal. Placing the CPW close to the slot-line terminus effectively maintains the short, and greatly enhances the intensity of the normal-oriented microwave magnetic field since the electrical currents are restricted within the junction region. So it is expected that the shorter the junction region, that is, the slot line–to–CPW-terminus spacing, the higher the intensity of the normal-oriented microwave magnetic field in the junction region.

Zhang *et al.* developed a fixture suitable for measuring ferromagnetic resonance of micrometer–diameter magnetic disks (Zhang *et al.* 1997). Figure 8.65 shows the basic structure of the junction. The slot line enters from the left, while the CPW appears on the right. For measuring micrometer-diameter disks, the measurement fixture should have a junction of approximately 1-μm length. To meet the very small junction region dimension requirements, it is necessary to taper

the slot line and CPW slots down to a point at their termini.

The presence of intense microwave magnetic fields oriented normal to the fixture plane in the junction region is essential for the measurement of ferromagnetic resonance. Figure 8.64(b) shows the placement of a magnetic disk over the junction region, where it overlaps both transmission line termini. In this geometry, most of the microwave energy transmitted across the junction region is coupled through the sample, and so the transmission of the microwave energy between the input and output ports is highly dependent on the microwave absorption of the sample. By applying a magnetic field in the fixture plane, the ferromagnetic resonance of the sample can be observed.

8.4.5.3 Microstrip resonator method

It is desirable to study the distributions of the ferromagnetic resonance properties of magnetic films. The concept of near-field microwave microscope can be used for this purpose, and a microstrip resonator with a measurement hole on the ground plane can be used as the measurement sensor (Belyaev *et al.* 1997). The measurement hole serves as a localized microwave magnetic field source and a communication channel between the film and the resonator, and the magnetic absorption is obtained on the basis of resonant-perturbation theory, by measuring the change of quality factor of the resonator with sweeping external dc magnetic field applied along the film.

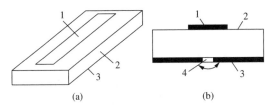

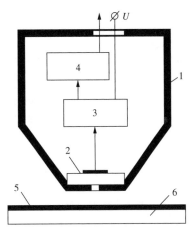

Figure 8.66 Schematic drawing of the microstrip resonator. (a) Three-dimensional view and (b) cross-section view. 1, metal strip; 2, dielectric substrate; 3, bottom metal; 4 measurement hole. The arrow shows the direction of the microwave magnetic field. Modified from Belyaev, B. A. Leksikov, A. A. Makievskii, I. Y. and Tyurnev, V. V. (1997). "Ferromagnetic resonance spectrometer", *Instruments and Experimental Techniques*, **40** (3), 390–394, by permission of МАИК Hayka/Interperiodica Publishiing

As shown in Figure 8.66, the measurement hole is under the middle of the microstrip line, where the microwave magnetic field antinode with the frequency of the first half-wavelength microstrip resonance is located. The microwave magnetic field near the measurement hole has two components: the normal component perpendicular to the grounding plane, and the tangent component parallel to the grounding plane and perpendicular to the microstrip line. The tangent field component is much larger than the normal component.

Figure 8.67 shows the basic structure of a measurement head for a ferromagnetic resonance microscope. The microstrip resonator is placed within a metal shield, and the head axis runs exactly through the center of the hole of the microstrip resonator. The microstrip resonator serves as the driving resonator of the transistor microwave oscillator operating in the autodyne mode. The output signal is proportional to the microwave power absorbed by film area located under the measurement hole. The prime advantage of this head is its high sensitivity. A ferromagnetic resonance spectrometer can be equipped with many replaceable measurement heads for different operating frequencies. Meanwhile, to investigate magnetic-film inhomogeneities, it is favorable to have replaceable measurement heads for the same microwave frequency but with measurement holes of different diameters, so that different degrees of measurement locality can be achieved.

Figure 8.67 Structure of a measurement head for ferromagnetic resonance microscope. 1, housing; 2, microstrip resonator; 3, microwave oscillator; 4, detector; 5, magnetic film; and 6 magnetic-film substrate. Modified from Belyaev, B. A. Leksikov, A. A. Makievskii, I. Y. and Tyurnev, V. V. (1997). "Ferromagnetic resonance spectrometer", *Instruments and Experimental Techniques*, **40** (3), 390–394, by permission of МАИК Hayka/Interperiodica Publishiing

In measurements, the sample under test is placed on a table that can move in two coordinates in the horizontal plane. A dc magnetic field H_{dc} is applied along the sample surface using Helmholtz coils, and the value of H_{dc} is chosen to ensure that ferromagnetic resonance can be observed in the measurement range. The dc magnetic field H_{dc} is modulated by an ac magnetic field H_{ac} produced by modulation coils. The amplitude of H_{ac} is usually set within a range of 0.05 to 5.0 Oe, depending on the line width of the ferromagnetic resonance of the film under study. If the measurement head can rotate by 90° along its axis, the tangent component of microwave magnetic field can be parallel or perpendicular to the external dc magnetic field.

REFERENCES

Acher, O. Vermeulen, J. L. Jacquart, P. M. Fontaine, J. M. and Baclet, P. (1994). "Permeability measurement on ferromagnetic thin films from 50 MHz up to 18 GHz", *Journal of Magnetism and Magnetic Materials*, **136**, 269–278.

Anderson, J. C. (1964). *Dielectrics*, Chapman & Hall, London, Chapter 9.

Artman, J. O. and Tannenwald, P. E. (1955). "Measurement of susceptibility tensor in ferrite", *Journal of Applied Physics*, **26** (9), 1124–1132.

Baden Fuller, A. J. (1987). *Ferrites at Microwave Frequencies*, Peter Peregrinus Ltd, London, 235–255.

Belhadj-Tahar, N. E. and Fourrier-Lamer, A. (1991). "Broad-band simultaneous measurement of the complex permittivity tensor for uniaxial materials using a coaxial discontinuity", *IEEE Transactions on Microwave Theory and Techniques*, **39** (10), 1718–1724.

Belyaev, B. A. Leksikov, A. A. Makievskii, I. Y. and Tyurnev, V. V. (1997). "Ferromagnetic resonance spectrometer", *Instruments and Experimental Techniques*, **40** (3), 390–394.

Bussy, H. E. and Steinert, L. A. (1958). "Exact solution for a gyromagnetic sample and measurements on ferrite", *IRE Transactions on Microwave Theory and Techniques*, **6** (1), 72–76.

Cady, W. G. (1946). *Piezoelectricity*, McGraw-Hill, London.

Chang, C. W. Chen, K. M. Qian, J. (1996). Nondestructive measurements of complex tensor permittivity of anisotropic materials using a waveguide probe system", *IEEE Transactions on Microwave Theory and Techniques*, **44** (7), 1081–1090.

Chen, H. De Gasperis, P. and Marcelli, R. (1993). "Using microwave network analyzer and MSW-SER to measure line width spectrum in magnetic garnet film from 0.5 to 20 GHz", *IEEE Transactions on Magnetics*, **29** (6), 3013–3015.

Chen, L. F. Ong, C. K. and Tan, B. T. G. (1999). "Cavity perturbation technique for the measurement of permittivity tensor of uniaxially anisotropic dielectrics", *IEEE Transactions on Instrumentation and Measurement*, **48** (6), 1023–1030.

Clerjon, S. Bayard, B. Vincent, D. and Noyel, G. (1999). "X-band characterization of anisotropic magnetic materials: application to ferrofluids", *IEEE Transactions on Magnetics*, **35** (1), 568–572.

Collin, R. E. (1992). *Foundations for Microwave Engineering*, 2nd edition, McGraw-Hill, New York.

Damaskos, N. J. Mack, R. B. Maffett, A. L. Parmon, W. and Uslenghi, P. L. E. (1984). "The inverse problem for biaxial materials", *IEEE Transactions on Microwave Theory and Techniques*, **32** (4), 400–405.

Dorsey, P. Oliver, S. A. Vittoria, C. Wittenauer, M. A. and Friedlander, F. J. (1992). "Novel technique for ferromagnetic resonance measurements", *IEEE Transactions on Magnetics*, **28** (5), 2450–2452.

Geyer, R. G. and Krupka, J. (1995). "Microwave dielectric properties of anisotropic materials at cryogenic temperatures", *IEEE Transactions on Instrumentation and Measurement*, **44** (2), 329–931.

Hashimoto, O. Shimizu, Y. (1986). "Reflecting characteristics of anisotropic rubber sheets and measurement of complex permittivity tensor", *IEEE Transactions on Microwave Theory and Techniques*, **34** (11), 1202–1207.

Hogan, C. L. (1953). "The ferromagnetic Faraday effect at microwave frequencies and its applications", *Reviews of Modern Physics*, **25** (1), 253–263.

Ishak, W. S. and Chang, K. W. (1986). "Tunable microwave resonator using magnetostatic wave in YIG films", *IEEE Transactions on Microwave Theory and Techniques*, **34** (12), 1383–1393.

Ito, A. Ohkawa, S. and Yamamoto, H. (1984). "On a measurement method of the tensor permeability of ferrite using Faraday rotation", *IEEE Transactions on Instrumentation and Measurement*, **33** (1), 26–31.

Jacquart, P. M. Acher, O. (1996). "Permeability measurement on composites made of oriented metallic wires from 0.1 to 18 GHz", *IEEE Transactions on Microwave Theory and Techniques*, **44** (11), 2116–2120.

Jow, J. Hawley, M. C. Final, M. C. (1989). "Microwave heating and dielectric diagnosis technique in a single-mode resonant cavity", *Review of Scientific Instruments*, **60** (1), 96–103.

Kobayashi, Y. and Tanaka, S. (1980). "Resonant modes of a dielectric rod resonator short-circuited at both ends by parallel conducting plates", *IEEE Transactions on Microwave Theory and Techniques*, **28** (10), 1077–1085.

Kong, J. A. (1975). *Theory of Electromagnetic Waves*, Wiley, New York.

Krupka, J. (1989). "Resonant modes in shielded cylindrical ferrite and single crystal dielectric resonator", *IEEE Transactions on Microwave Theory and Techniques*, **37** (4), 691–697.

Krupka, J. (1991). "Measurements of all complex permeability tensor components and the effective line widths microwave ferrites using dielectric ring resonators", *IEEE Transactions on Microwave Theory and Techniques*, **39** (7), 1148–1157.

Krupka, J. Blondy, P. Cros, D. Guillon, P. and Geyer, R. G. (1996). "Whispering-gallery modes and permeability tensor measurements in magnetized ferrite resonators", *IEEE Transactions on Microwave Theory and Techniques*, **44** (7), 1097–1102.

Krupka, J. Cros, D. Aubourg, M. and Guillon, P. (1994b). "Study of whispering gallery modes in anisotropic single crystal dielectric resonators", *IEEE Transactions on Microwave Theory and Techniques*, **42** (1), 56–61.

Krupka, J. Derzakowski, Abramowicz, A. Tobar, M. E. and Baker-Jarvis, J. (1999). "Use of whispering-gallery modes for complex permittivity determinations of ultra-low-loss dielectric materials", *IEEE Transactions on Microwave Theory and Techniques*, **47** (6), 752–759.

Krupka, J. and Geyer, G. (1996). "Complex permeability of demagnetized microwave ferrites near and above gyromagnetic resonance", *IEEE Transactions on Magnetics*, **32** (3), 1924–1933.

Krupka, J. Geyer, R. G. Kuhn, M. and Hinken, J. H. (1994a). "Dielectric properties of single crystals of Al_2O_3, $LaAlO_3$, $NdGaO_3$, $SrTiO_3$, and MgO at cryogenic temperatures", *IEEE Transactions on Microwave Theory and Techniques*, **42** (10), 1886–1890.

Lax, B. and Button, K. J. (1962). *Microwave Ferrites and Ferrimagnetics*, McGraw-Hill, New York.

Maier Jr., L. C. and Slater, J. C. (1952). "Field strength measurements in resonant cavities", *Journal of Applied Physics*, **23** (1), 68–77.

Masters, J. I. Capone, B. R. Gianino, P. D. (1960). "Measurement technique for narrow line width ferromagnets", *IRE Transactions on Microwave Theory and Techniques*, **8**, 565–566.

Mullen, E. B. and Carlson, E. R. (1956). "Permeability tensor values from waveguide measurements", *Proceedings of the IRE*, **44** (10), 1318–1323.

Ogasawara, N. Fuse, T. Inui, T. and Saito, I. (1976). "Highly sensitive procedures for measuring permeabilities $(\mu_{\pm})$ for circularly polarized fields in microwave ferrites", *IEEE Transactions on Magnetics*, **12** (3), 256–259.

Okada, F. and Tanaka, H. (1990). "Measurement of ferrites tensor permeability by spherical cavity resonator", *Conference on Precision Electromagnetic Measurements (CPEM) '90 Digest*, 230–231.

Parneix, J. P. Legrand, C. and Toutain, S. (1982). "Automatic permittivity measurements in a wide frequency range: application to anisotropic fluids", *IEEE Transactions on Microwave Theory and Techniques*, **30** (11), 2015–2017.

Portis, A. M. and Teaney, D. (1959). "Microwave Faraday rotation in antiferromagnetic MnF_2", *Physical Review*, **116** (4), 838–845.

Portis, A. M. and Teaney, D. (1958). "Microwave Faraday rotation: design and analysis of a bimodal cavity", *Journal of Applied Physics*, **29** (12), 1692–1698.

Pozar, D. M. (1998). *Microwave Engineering*, 2nd edition, John Wiley & Sons, New York.

Pradoux, D. Blanc, F. Fanguin, R. and Raoult, G. (1973a). "Dielectric constants of quartz. I. Theory", *Journal of Applied Physics*, **44** (12), 5222–5224.

Pradoux, D. Blanc, F. Fanguin, R. and Raoult, G. (1973b). "Dielectric constants of quartz. II. Experiment", *Journal of Applied Physics*, **44** (12), 5225–5226.

Queffelec. P. Le Floc'h, M. Gelin, P. (1999). "Nonreciprocal cell for the broad-band measurement of tensorial permeability of magnetized ferrites: Direct problem", *IEEE Transactions on Microwave Theory and Techniques*, **47** (4), 390–397.

Queffelec, P. Le Floc'h, M. Gelin, P. (2000). "New method for determining the permeability tensor of magnetized ferrites in a wide frequency range", *IEEE Transactions on Microwave Theory and Techniques*, **48** (8) 1344–1351.

Spencer, E. G. Ault, L. A. and LeCraw, R. C. (1956a). "Intrinsic tensor permeabilities on ferrite rods, spheres, and disks", *Proceedings of the IRE*, **44** (10), 1311–1317.

Spencer, E. G. LeCraw, R. C. and Reggia, F. (1955). "Circularly polarized cavities for measurement of tensor permeabilities", *Journal of Applied Physics*, **26** (3), 354–355.

Spencer, E. G. LeCraw, R. C. and Reggia, F. (1956b). "Measurement of microwave dielectric constants and tensor permeabilities of ferrite spheres", *Proceedings of the IRE*, **44** (6), 790–800.

Stinson, D. C. (1958). "Ferrite line width measurements in a cross-guide coupler", *IRE Transactions on Microwave Theory and Techniques*, **6**, 446–450.

Stinson, D. C. (1957). "Coupling through an aperture containing an isotropic ferrite", *IRE Transactions on Microwave Theory and Techniques*, **5**, 184–191.

Sucher, M. and Fox, J. (1963). *Handbook of Microwave Measurements*, Vol. 2, 3rd edition, Polytechnic Press of the Polytechnic Institute of Brooklyn, New York, 530–539.

Tian, B. Tinga, W. R. (1993). "Single-frequency relative Q measurements using perturbation theory", *IEEE Transactions on Microwave Theory and Techniques*, **41** (11), 1922–1927.

Tian, B. Tinga, W. R. (1995). "Linear condition between a cavity's Q-factor and its input resonant resistance", *IEEE Transactions on Microwave Theory and Techniques*, **43** (3), 691–692.

Viswanathan, B. and Murthy, V. R. K. (1990). *Ferrite Materials, Science and Technology*, Narosa Publishing House, New Delhi, 85–105.

Von Aulock, W. and Rowen, J. H. (1957). "Measurement of dielectric and magnetic properties of ferromagnetic materials at microwave frequencies", *The Bell System Technical Journal*, **36** (3), 427–448.

Waldron, R. A. (1960). "What is ferromagnetic resonance?" *British Journal of Applied Physics*, **11**, 69–73.

Waldron, R. A. (1970). *Theory of Guided Electromagnetic Waves*, Von Nostrand Reinhold Company, London, 292–334.

Zhang, S. Oliver, S. A. Israeloff, N. E. and Vittoria, C. (1997). "High-sensitivity ferromagnetic resonance measurements on micrometer-sized samples", *Applied Physics Letters*, **70** (20), 2756–2758.

9

Measurement of Ferroelectric Materials

Ferroelectric materials usually have high dielectric constants, and their dielectric properties are temperature- and electric-field-dependent. This chapter discusses the methods for the characterization of the dielectric properties of ferroelectric materials. After a brief introduction to the basic properties of ferroelectric materials, three types of measurement methods are discussed, including nonresonant methods, resonant methods, and planar-circuit methods. Special attention is paid to planar-circuit methods, which are suitable for ferroelectric thin films. The measurements of the responding time and the nonlinear properties of ferroelectric thin films are discussed in the final two sections.

9.1 INTRODUCTION

Ferroelectric materials come under a subclassification of pyroelectric materials that show a spontaneous polarization that depends on temperature and the orientation of which can be changed by applying an external dc electric field. Pyroelectric materials in turn come from a class of piezoelectric materials that show an electric polarization under mechanical stress. Therefore ferroelectric materials possess pyroelectric as well as piezoelectric properties (Lines 1977; Anderson 1964).

For many years, although ferroelectric materials are used in a wide variety of devices, it has always been that it is the other properties of the material such as pyroelectricity or piezoelectricity that are actually employed. This situation has been changed during the past few years. The applications of ferroelectric materials in random access memories are expected to replace magnetic core memories and magnetic bubble memories for many applications. Besides, ferroelectric materials can be used in fabricating capacitors for electronic industry because of their high dielectric constants, and this is important in the trend toward miniaturization and high functionability of electronic products. Furthermore, ferroelectric materials can be used for the development of tunable microwave devices, which have potential applications in satellite, terrestrial communications, and other microwave applications where the working frequencies are higher than the useful range of Si-based devices (Varadan *et al*. 1999, 1995; Teo *et al*. 2002, 2000). By using ferroelectric thin films, electrically tunable microwave integrated circuits can be developed. Therefore, it is very important to characterize the dielectric constant and tunability of ferroelectric thin films.

In this chapter, various methods for the measurement of the dielectric properties of ferroelectric materials will be discussed. As ferroelectric materials may be in composite state, polycrystalline ceramic state, bulk single crystal state, and thin-film state, different characterization methods are needed. Generally speaking, the methods for characterizing dielectric composites, ceramics, and bulk ferroelectric crystals include nonresonant methods and resonant methods, and people are interested in the dielectric relaxations of such materials, whereas ferroelectric thin films are usually characterized using planar-circuit methods, and almost all the dielectric properties, including the dielectric relaxations and electric tunability, are needed to be studied.

Microwave Electronics: Measurement and Materials Characterization   L. F. Chen, C. K. Ong, C. P. Neo, V. V. Varadan and V. K. Varadan
© 2004 John Wiley & Sons, Ltd   ISBN: 0-470-84492-2

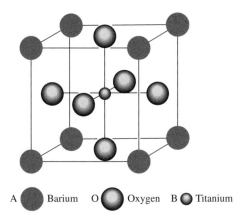

Figure 9.1 The ABO_3 perovskite structure. Reproduced from Jaffe, B. Cook, W. R. and Jaffe, H. (1971), *Piezoelectric Ceramics*, Academic Press, London, by permission of Elsevier

In principle, all the methods for characterizing the dielectric permittivity could be extended for the characterization of ferroelectric materials. However, in characterizing ferroelectric materials, the following aspects should be taken into consideration. First, the dielectric constants of ferroelectric materials are usually quite high. Second, it is necessary to consider the ways for applying the bias dc electric field. Third, it is also necessary to consider how to control the measurement temperature.

9.1.1 Perovskite structure

The principal structure for most ferroelectric materials is a perovskite structure, which is a simple cubic containing three different ions in the form ABO_3, as shown in Figure 9.1 (Jaffe *et al.* 1971). The A and B atoms represent +2 and +4 ions respectively, while the O atom is the oxygen ion (O^{2-}). This ABO_3 structure can be thought of as a face-centered cubic (FCC) lattice with the A atoms at the corners and the O atoms on the faces. The B atom is located at the center of the lattice, and there are minimum-energy positions off-centered from the original octahedron that can be occupied by the B atom. Therefore, when an electric field is applied, the B atom is shifted, causing the structure to be altered and dipoles to be created.

9.1.2 Hysteresis curve

As mentioned earlier, ferroelectric materials possess spontaneous electrical polarization, whose direction can be changed by an external electric field. Figure 9.2 shows the hysteresis behavior of the polarization P versus electric field E. At zero electric field, the polarization P has two stable values $+P_r$ and $-P_r$, known as the *remnant polarization*. Reverse electrical field, called the *coercive field E_c*, must be applied to annihilate the existing polarization to achieve zero polarization. The name "ferroelectricity" is used because the *P-E* relations of these materials are very similar to the *B-H* relations of ferromagnetic materials. Analogous to a ferromagnetic material, a ferroelectric material possesses a Curie temperature, above which it becomes paraelectric. Furthermore, a ferroelectric material also has an internal domain structure, similar to that of a ferromagnetic material.

9.1.3 Temperature dependence

One important feature of ferroelectric materials is that many of their properties are temperature-dependent. For example, the hysteresis curve, dielectric constant, and crystal structure of a ferroelectric material may change with temperature.

Figure 9.3 shows an example of temperature dependence of hysteresis curves of ferroelectric materials. At low temperatures, the hysteresis

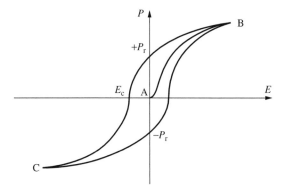

Figure 9.2 A typical *P-E* hysteresis loop for a ferroelectric. A: zero polarization at zero field; B and C: saturation polarization. P_r: remnant polarization; E_c: coercive field

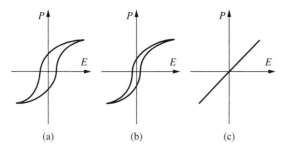

(a)   (b)   (c)

Figure 9.3   Typical temperature dependence of hysteresis curves. (a) $T \ll T_c$, (b) $T < T_c$, and (c) $T \geq T_c$

crystal structure is illustrated in Figure 9.5. Above the Curie point, BaTiO$_3$ has a cubic structure. On cooling through the Curie point at 120 °C, the cubic structure transforms into a tetragonal structure. A change from the tetragonal structure to an orthorhombic structure occurs at 5 °C, and a further change from the orthorhombic structure to a rhombohedral structure occurs at −90 °C. Owing to the variation of the crystal structure, a piece of BaTiO$_3$ crystal has different dielectric constants at different temperatures.

loop is relatively wide. This is because at lower temperatures the thermal motion energy is less and so a greater field is required to orientate the domains. When the temperature increases, the corresponding coercive field required for zero polarization decreases, resulting in a narrow hysteresis loop and smaller value of remnant polarization. If the temperature is increased beyond the Curie temperature T_c, which is defined as the temperature at which the material changes form a ferroelectric state to a paraelectric state, the hysteresis loop disappears.

The dielectric constant of a ferroelectric material is also dependent on temperature. Figure 9.4 shows the change of dielectric constant of BaTiO$_3$ near the Curie temperature T_c. The change of dielectric constant is mainly due to the phase transformations that BaTiO$_3$ undergoes. The variation of the

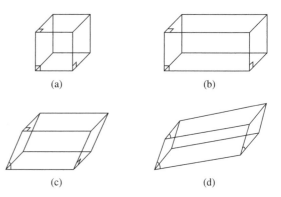

(a)   (b)

(c)   (d)

Figure 9.5   Phase transformations of BaTiO$_3$ at different temperatures. (a) $T > 120\,°C$, cubic, (b) $5\,°C < T < 120\,°C$, tetragonal, (c) $-90\,°C < T < 5\,°C$, orthorhombic, and (d) $T < -90\,°C$, rhombohedral

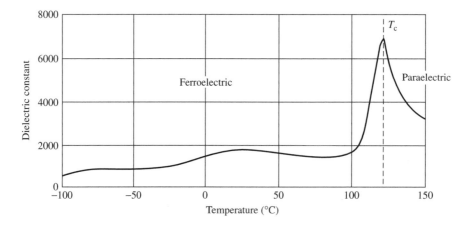

Figure 9.4   Temperature dependence of the dielectric constant of BaTiO$_3$ (Data source: http://www.novacap.com)

9.1.4 Electric field dependence

The dielectric constant of ferroelectric materials can be varied by applying an external dc electric field, and this property can be used in the development of electrically tunable electronic devices. The change of the dielectric constant because of the external electric field is often described by tunability, as defined below:

$$\xi = \frac{\varepsilon_{r0} - \varepsilon_{re}}{\varepsilon_{r0}} \times 100\,\% \qquad (9.1)$$

where ε_{r0} is the dielectric constant of the ferroelectric material without external dc electric field, and ε_{re} is the dielectric constant when the material is biased by an external dc electric field.

The dielectric loss of a ferroelectric material is also dependent on the applied dc electric field. Generally speaking, the dielectric loss decreases with the increase of the external dc electric field.

Experiments show that a ferroelectric material with higher loss tangent usually has larger tunability. As loss tangent of a material is an important factor affecting the performances of electric circuits, in the development of electrically tunable ferroelectric microwave devices, a figure of merit K, defined by $K = \xi/\tan\delta$, is often used to indicate the quality of ferroelectric materials. Usually, in the calculation of the figure of merit K, the loss tangent with no external dc electric field is used.

9.2 NONRESONANT METHODS

As discussed in Chapter 2, nonresonant methods can be generally classified into reflection methods and reflection/transmission methods. Both reflection methods and transmission/reflection methods have been used in characterizing ferroelectric materials. In principle, all types of transmission lines can be used in developing nonresonant methods for the characterization of ferroelectric materials. It should be noted that most of the nonresonant methods are used in studying the dielectric relaxation of ferroelectric composites and ferroelectric polycrystalline ceramics. In the following, we discuss several types of nonresonant methods often used in the study of ferroelectric materials.

9.2.1 Reflection methods

Figure 9.6 shows a shielded coaxial reflection method, whose working principle has been discussed in Chapter 3. In this method, the sample is placed at the open end of the central conductor of the coaxial line, and the coaxial line is shorted by a metal plunger. The plunger is movable for sample loading and unloading. If the movable plunger contacts the sample, the measurement configuration becomes the one shown in Figure 3.27(a).

Before measurement, calibration is required at the end of the coaxial line, and the standards used are usually short, open, and matched load. The dielectric properties of the ferroelectric sample are obtained from the reflection of the coaxial line, and the algorithms for deducting complex permittivity are given by (McNeal *et al.* 1998).

$$\varepsilon_r' = \frac{2\Gamma\sin\theta}{\omega C_0 Z_0(\Gamma^2 + 2\Gamma\cos\theta + 1)} - \frac{C_f}{C_0} \qquad (9.2)$$

$$\varepsilon_r'' = \frac{1 - \Gamma^2}{\omega C_0 Z_0(\Gamma^2 + 2\Gamma\cos\theta + 1)} \qquad (9.3)$$

where Γ and θ are the magnitude and phase angle of the reflection coefficient respectively, Z_0 is the characteristic impedance of the coaxial line, C_0 and C_f are determined by the structure of the measurement fixture, and can be obtained by

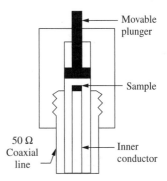

Figure 9.6 Shielded coaxial reflection method for the characterization of bulk ferroelectric materials (McNeal *et al.* 1998). Source: McNeal, M. P. Jang, S. J. and Newnham, R. E. (1998). "The effect of grain and particle size on the microwave properties of barium titanate (BaTiO₃)", *Journal of Applied Physics*, **83** (6), 3288–3297

calibrating the fixture using standard sample with known permittivity.

This method has been used to study the effects of the particle and grain sizes of ferroelectric powders, and the effects of the fabrication procedures and techniques to the dielectric properties of ferroelectric composites and ceramics (McNeal *et al.* 1998). As the transmission line used in this method is a coaxial line, this method can cover a wide frequency range. Although, similar to other non-resonant methods, the sensitivity and accuracy of this method are not very high, this method is sensitive and accurate enough for studying the dielectric relaxation of most of the ferroelectric composites and ceramics.

9.2.2 Transmission/reflection method

Figure 9.7 shows a coaxial discontinuity method that can be used for the study of the dielectric relaxation of ferroelectric composites and ceramics. In this configuration, a segment of the central conductor is moved away, and the sample is placed between the two open ends of the two sections of central conductors of the coaxial line. The bias dc voltage can be applied through the two sections of the central conductors of the coaxial line. This method can be taken as a special case of the coaxial discontinuity method discussed in Chapter 4.

The theoretical model of this method is similar to Figure 3.27(a) with the short end in Figure 3.27(a) being replaced by an electric wall and a magnetic wall. The dielectric permittivity of the sample can be obtained from the transmission and

reflection coefficients of the fixture. As two complex coefficients are used in deducting the dielectric properties, this method usually has higher accuracies than the reflection methods. Furthermore, similar to the coaxial discontinuity method discussed in Chapter 4, this method can be further extended to characterize uniaxially anisotropic dielectric samples.

9.3 RESONANT METHODS

Resonant methods, including dielectric resonator method and cavity perturbation method, have also been used in characterizing ferroelectric composites, polycrystalline ceramics, and single crystals. These methods have higher accuracy and sensitivity than nonresonant ones. As resonant methods can only be used in measuring dielectric properties at single or several discrete frequencies, they are not suitable for getting the dielectric spectrums of ferroelectric materials over wide frequency ranges but these methods are often used to characterize ferroelectric substrates or crystals for the development of microwave circuits and devices.

9.3.1 Dielectric resonator method

The working principle of dielectric resonator method has been discussed in Chapter 5. In the following, we discuss several configurations of resonators often used in the characterization of ferroelectric bulk materials.

9.3.1.1 Courtney resonator

Figure 9.8 shows a dielectric resonator configuration widely used in the characterization of dielectric materials. The sample under test is sandwiched between two metal plates, and the dielectric properties of the sample are determined from the resonant frequency and quality factor of the resonator. The two metal plates can be used as electrodes to apply the biasing dc electric field for tunability measurement. This method can be used in characterizing isotropic dielectric materials as discussed in Chapter 5 and can also be used to characterize anisotropic materials as discussed in Chapter 8.

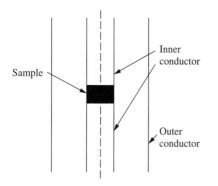

Figure 9.7   Coaxial discontinuity for measuring the dielectric properties of ferroelectric samples

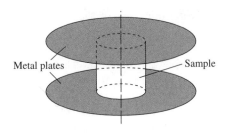

Figure 9.8 Configuration of a Courtney dielectric resonator

This method is often used to characterize ferroelectric composites and ceramics, which can be taken as isotropic materials, and it can also be used to characterize ferroelectric crystals, which are usually uniaxially anisotropic.

9.3.1.2 Disk resonator

Figure 9.9 shows a disk resonant configuration often used in the characterization of ferroelectric bulk materials (Vendik *et al.* 1995). The sample under test is in disk shape, with its top and bottom surfaces metallized. From the resonant frequency and quality factor of the resonator, the dielectric constant and loss tangent of the disk sample can be derived. Bias dc voltage can be applied through the two metallized top and bottom surfaces, so the electric field dependence of the dielectric properties can be studied.

As discussed in Chapter 2 and Chapter 5, the assumption of a perfect magnetic wall at the cylindrical surface is a fairly good approximation for a disk resonator. For the TM_{0n0} mode of the resonator, from the solutions of the Maxwell equations, the following relationship can be obtained (Vendik *et al.* 1995):

$$r = \frac{\kappa_{0n}\lambda_0}{2\pi\sqrt{\varepsilon_r}} \quad (9.4)$$

Figure 9.9 Disk resonator for characterization of ferroelectric material

where r is the radius of the disk, κ_{0n} is the nth root of the derivative of the zero-order Bessel function $J_0(z)$, λ_0 is the free-space wavelength that can be obtained from the resonant frequency of the disk resonator, and ε_r is the dielectric constant of the sample. Equation (9.4) indicates that the dielectric constant ε_r of the sample can be calculated from the resonant frequency of the resonator.

The loss tangent of the ferroelectric sample can be calculated from the resonant frequency and unloaded quality factor Q_u of the resonator, surface resistance of the metal, and the height h of the disk. For the first mode ($\kappa_{01} = 3.83$), it is given that (Vendik *et al.* 1995)

$$\frac{1}{Q_u} = \tan\delta + \frac{R_s}{120\pi} \cdot \frac{\lambda_0}{\pi h} \quad (9.5)$$

where Q_u is the unloaded quality factor of the disk resonator, $\tan\delta$ is the loss tangent of the ferroelectric disk sample, and R_s is the surface resistance of metal plates covering the disk.

9.3.1.3 Effects of air gap

As the dielectric constant of a ferroelectric material is usually very large, a small air gap between the ferroelectric sample and the metal plate may seriously affect the measurement results (Lacey *et al.* 1994). Lacey *et al.* studied the effect of air gap between a ferroelectric sample and metal plates using a single crystal square of strontium titanate, $SrTiO_3$ (STO), plate resonator with sizes ($10 \times 10 \times 1 \ mm^3$), as shown in Figure 9.10. A parallel-plate resonator metallized in this way can only exhibit TE_{mn} modes with a resonant frequency given by

$$f_{mn} = \frac{c}{4\sqrt{\varepsilon_r}}\sqrt{\left(\frac{m}{a}\right)^2 + \left(\frac{n}{b}\right)^2} \quad (9.6)$$

where a and b are the linear dimensions of the resonator and c is the free-space speed of light. In experiments, it is possible to measure the resonant frequencies of the first two or three modes (TE_{01}, TE_{11}, TE_{02}) and the value of dielectric constant ε_r can be calculated from Eq. (9.6). As the dielectric constant is mode independent, different resonant modes should result in close values of ε_r.

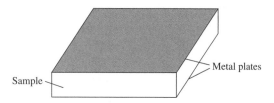

Figure 9.10 A square plate dielectric resonator with its top and bottom surfaces metallized

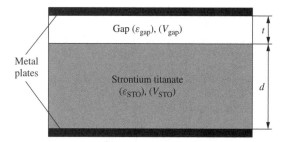

Figure 9.11 Maxwell–Wager two-layer condenser. Lines of d must be continuous across the dielectric–dielectric boundary

Effect of a gap on the dielectric constant can be modeled as a classic Maxwell–Wagner two-layer condenser as shown in Figure 9.11. The two-layer structure has an effective dielectric constant given by (Lacey *et al.* 1998)

$$\varepsilon_{\text{eff}} = \frac{d}{(t/\varepsilon_{\text{gap}}) + (d/\varepsilon_{\text{STO}})} \qquad (9.7)$$

where d is the thickness of the STO and t is the thickness of the gap. The values of $\varepsilon_{\text{STO}} = 300$, $\varepsilon_{\text{gap}} = 1$, $d = 1$ mm and $t = 2\,\mu$m yield $\varepsilon_{\text{eff}} = 187$. This has important implications because mode identification is based on the assumption that the permittivity of the sample lies between certain accepted values reported in the literature. Figure 9.12(a) illustrates the importance of this effect; ε_{eff} is normalized to the original estimation of 300.

The air gap also affects the bias voltage applied to the dielectric. In the structure shown in Figure 9.11, the total voltage V_{app} applied to the two-layer condenser can be expressed as (Lacey *et al.* 1998)

$$V_{\text{app}} = V_{\text{STO}} + V_{\text{gap}} \qquad (9.8)$$

$$V_{\text{STO}} = \frac{\varepsilon_{\text{gap}}}{\varepsilon_{\text{STO}}} \cdot \frac{d}{t} \cdot V_{\text{gap}} = \beta \cdot V_{\text{gap}} \qquad (9.9)$$

$$V_{\text{STO}} = \left(\frac{\beta}{1+\beta}\right) V_{\text{app}} \qquad (9.10)$$

where V_{STO} is the voltage applied to STO, V_{gap} is the voltage applied to the gap, and the parameter β is defined in Eq. (9.9). Figure 9.12(b) shows the relationships between the voltage applied to STO, the air-gap thickness and the temperature. When the temperature decreases, the dielectric constant of STO increases, so the effects of air gap become more obvious.

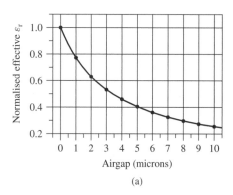

(a)

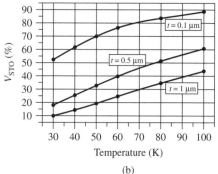

(b)

Figure 9.12 Effects of air gap on ferroelectric resonator (Lacy *et al.* 1998). (a) Normalized effective dielectric constant and (b) bias voltage. Modified from Lacey, D. Gallop, J. C. and Davis, L. E. (1998). "The effects of an air gap on the measurement of the dielectric constant of STO at cryogenic temperatures", *Measurement Science and Technology*, **9** (3), 536–539, by permission of IOP Publishing Ltd

The above discussion shows that in the characterization of ferroelectric samples, the gap thickness should be minimized. In experiments, though it is impractical to completely eliminate the gap, it is necessary to make an estimation of the air-gap thickness and its possible effects to measurement results.

9.3.2 Cavity-perturbation method

Figure 9.13 shows a TM_{010} mode cavity for characterizing the dielectric properties of ferroelectric materials using cavity perturbation method (McNeal *et al.* 1998). The sample, in the form of a long thin rod, is held symmetrically along the axis of the cavity in a hole drilled through a thin strip of mica. The bottom plate of the cavity is adjustable, and an adjustable plunger is used to adjust the position of the bottom plate to ensure that the cavity height equals the sample length. The upper plate of the cavity is removable for sample loading and unloading.

The sample in the cavity can be treated as a dipole having the same length as the sample (Parkash *et al.* 1979). Using the method of images, taking account of the effects of the polarization of the sample and its image dipoles on the net polarizing field in the sample, the effective

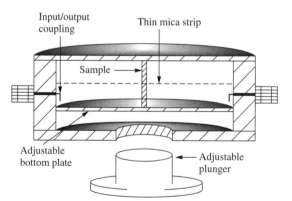

Figure 9.13   Cross section of TM_{010} cavity and sample used for cavity-perturbation measurements (McNeal *et al.* 1998). Source: McNeal, M. P. Jang, S. J. and Newnham, R. E. (1998). "The effect of grain and particle size on the microwave properties of barium titanate (BaTiO$_3$)", *Journal of Applied Physics*, **83** (6), 3288–3297

depolarizing factor is given by

$$N_e = N \frac{\pi h}{2H} \cot\left(\frac{\pi h}{2H}\right) \qquad (9.11)$$

where N is the depolarizing factor that depends on the axial ratio of the specimen, $2H$ is the height of the cavity, and $2h$ is the length of the sample. By considering the fields in both the unperturbed and perturbed cavity, the dielectric parameters can be obtained from (McNeal *et al.* 1998)

$$\varepsilon_r' = 1 + \frac{V_c \cdot \dfrac{\delta\omega}{\omega_0}\left[\dfrac{V_s}{2J_1^2(ka)} - N_e V_c \dfrac{\delta\omega}{\omega_0}\right] - N_e V_c^2 \left[\delta\left(\dfrac{1}{2Q}\right)\right]^2}{\left[\dfrac{V_s}{2J_1^2(ka)} - N_e V_c \dfrac{\delta\omega}{\omega_0}\right]^2 + \left[N_e V_c \delta\left(\dfrac{1}{2Q}\right)\right]^2} \qquad (9.12)$$

$$\varepsilon_r'' = \frac{V_c \cdot \delta\left(\dfrac{1}{2Q}\right)\left[\dfrac{V_s}{2J_1^2(ka)} - N_e V_c \dfrac{\delta\omega}{\omega_0}\right] + N_e V_c^2 \dfrac{\delta\omega}{\omega_0}\left[\delta\left(\dfrac{1}{2Q}\right)\right]^2}{\left[\dfrac{V_s}{2J_1^2(ka)} - N_e V_c \dfrac{\delta\omega}{\omega_0}\right]^2 + \left[N_e V_c \delta\left(\dfrac{1}{2Q}\right)\right]^2} \qquad (9.13)$$

where $\delta(1/2Q)$ represents $(1/2)[(1/Q_1) - (1/Q_0)]$, Q_1 and Q_0 are the quality factors of the filled and unfilled cavity respectively; ω_0 is the resonant cavity of the unperturbed cavity, $\delta\omega$ is the difference between resonant frequencies for the unperturbed and perturbed cavity, and $(1/2)J_1^2(ka)$ is 1.8522 for first-order mode excitation.

In the configuration shown in Figure 9.13, the sample and the cavity have the same height, so Eqs. (9.12) and (9.13) become (McNeal *et al.* 1998)

$$\varepsilon_r' = 1 + 0.539 \cdot \frac{V_c}{V_s} \cdot \frac{\delta\omega}{\omega_0} \qquad (9.14)$$

$$\varepsilon_r'' = \frac{0.539}{2\varepsilon_r'} \cdot \frac{V_c}{V_s} \cdot \left(\frac{1}{Q} - \frac{1}{Q_0}\right) \qquad (9.15)$$

Equations (9.14) and (9.15) indicate that the dielectric properties of the sample can be obtained from the changes in resonant frequency and quality factor due to the introduction of the sample. However, it should be indicated that this resonant-perturbation method is only suitable for bulk ferroelectric materials, such as ferroelectric composites and crystals. One disadvantage of using this method is that it is difficult to apply bias voltage.

9.3.3 Near-field microwave microscope method

Near-field scanning microwave microscopes are sensitive to linear permittivity as well as to nonlinear dielectric terms, which can be described as a function of an external dc electric field, so they can be used to image the permittivity and tunability of bulk and thin-film ferroelectric samples (Steinhauer *et al.* 2000). The working principle and measurement system for near-field microwave microscopes have been discussed in Chapter 6. The sensing probe is a key component for a microwave microscope, and a probe with a sharp-tipped center conductor extending beyond the outer conductor is often used because of its high spatial resolution. Figure 9.14 shows a probe used in characterizing ferroelectric thin films. A bias voltage V_b is applied to the thin-film sample under test through a bias T component, and the counter electrode beneath the thin film acts as the ground plane. In Figure 9.14, the bias voltage consists of a dc component and an ac component. The ac component of the bias voltage is mainly for

the study of the nonlinear dielectric properties of ferroelectric thin films.

9.4 PLANAR-CIRCUIT METHODS

Ferroelectric thin films are widely used in developing tunable microwave devices, such as tunable resonators, tunable filters, and phase shifters. The development of such devices requires accurate knowledge of the dielectric properties of the ferroelectric thin films. The dielectric properties of ferroelectric thin films are usually characterized by planar-circuit methods. The working principles for various planar-circuit methods have been discussed in Chapter 7.

The planar-circuit methods used in characterizing ferroelectric thin films mainly include transmission line method, resonator method, and capacitor method. Coplanar waveguide (CPW) method is a typical transmission line method, and other planar transmission lines, such as microstrip (Carroll *et al.* 1993), can also be used in characterizing ferroelectric thin films. In the resonator method, the thin film under test is fabricated into a planar resonator, and the dielectric properties of the thin film are derived from the resonant frequency and quality factor of the resonator. In the capacitor method, the thin film under test is fabricated into a capacitor, and the dielectric properties of the thin film are calculated from the capacitance and quality factor of the capacitor. In the choice of planar-circuit methods, it is desirable that the measurement arrangement is close to that of the actual applications of the films.

9.4.1 Coplanar waveguide method

Two types of CPWs are used in characterizing ferroelectric thin films: ungrounded waveguide and grounded waveguide. In the following, we concentrate on ungrounded CPW, and the conclusions obtained can be extended to grounded coplanar waveguide.

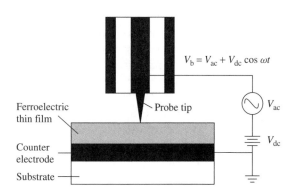

$$V_b = V_{ac} + V_{dc} \cos \omega t$$

Ferroelectric thin film

Probe tip

Counter electrode

Substrate

V_{ac}

V_{dc}

Figure 9.14 Measurement of a ferroelectric thin film using a scanning microwave microscope. The bias voltage V_b is applied through the central conductor of the sensing probe

9.4.1.1 Measurement principle

Figure 9.15 shows a shielded two-layered CPW, which can be used in characterizing the dielectric

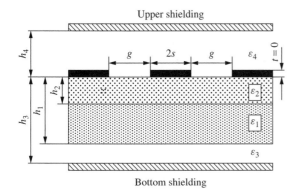

Figure 9.15 Two-layered substrate coplanar waveguide (Gevorgian *et al.* 1995). Source: Gevorgian, S. Linner, L. J. P. and Kollberg, E. L. (1995). "CAD models for shielded multi-layered CPW", *IEEE Transactions on Microwave Theory and Techniques*, **43** (4), 772–779. © 2003 IEEE

properties of ferroelectric thin films. The bottom layer is the substrate with thickness $(h_1 - h_2)$, on which the ferroelectric thin film with thickness h_2 is deposited. The dielectric permittivity values of the substrate and the ferroelectric thin film are ε_1 and ε_2 respectively. The coplanar structure is shielded by the bottom shielding and upper shielding whose distances to the circuit are h_3 and h_4, respectively.

The effective dielectric permittivity of the coplanar waveguide shown in Figure 9.15 is given by (Gevorgian *et al.* 1995)

$$\varepsilon_{r,\text{eff}} = 1 + q_1(\varepsilon_1 - 1) + q_2(\varepsilon_2 - \varepsilon_1) \quad (9.16)$$

with the filling factors q_1 and q_2 given by

$$q_1 = \frac{K(k_1)}{K(k_1')} \left[\frac{K(k_3)}{K(k_3')} \cdot \frac{K(k_4)}{K(k_4')} \right] \quad (9.17)$$

$$q_2 = \frac{K(k_2)}{K(k_2')} \left[\frac{K(k_3)}{K(k_3')} \cdot \frac{K(k_4)}{K(k_4')} \right] \quad (9.18)$$

with

$$k_i = \frac{\sinh\left(\dfrac{\pi s}{2h_i}\right)}{\sinh\left[\dfrac{\pi(s+g)}{2h_i}\right]} \quad (i = 1, 2) \quad (9.19)$$

$$k_i = \frac{\tanh\left(\dfrac{\pi s}{2h_i}\right)}{\tanh\left[\dfrac{\pi(s+g)}{2h_i}\right]} \quad (i = 3, 4) \quad (9.20)$$

$$k_i' = \sqrt{1 - k_i^2} \quad (i = 1, 2, 3, 4) \quad (9.21)$$

For unshielded coplanar waveguide, $h_3 = h_4 = \infty$. According to Eqs. (9.19)–(9.21)

$$k_0 = k_3 = k_4 = \frac{s}{s+g} \quad (9.22)$$

So the filling factors become

$$q_1 = \frac{1}{2} \cdot \frac{K(k_1)}{K(k_1')} \cdot \frac{K(k_0')}{K(k_0)} \quad (9.23)$$

$$q_2 = \frac{1}{2} \cdot \frac{K(k_2)}{K(k_2')} \cdot \frac{K(k_0')}{K(k_0)} \quad (9.24)$$

Once the effective permittivity is experimentally determined, the dielectric properties of the ferroelectric thin film can be derived according to Eq. (9.16).

Figure 9.16 shows the basic structure of a CPW fabricated from a ferroelectric thin film deposited on a dielectric substrate. At each end of the CPW, there is a taper discontinuity, with equal ratio of s:g, for circuit packaging and transition from CPW to a coaxial connector. To ensure good

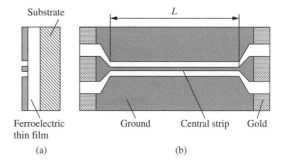

Figure 9.16 Two-layered coplanar waveguide. (a) Cross section and (b) plan view. Modified from *Physica C*, **308**, Chakalov, R. A. Ivanov, Z. G. Boikov, Y. A. Larsson, P. Carlsson, E. Gevorgian, S. and Claeson, T., "Fabrication and investigation of YBCO/BSTO thin film structures for voltage tunable devices", 279–288, (1998), with permission from Elsevier Science B.V.

electric contact, the ends of the circuit are coated with gold. The effective dielectric constant of the CPW can be determined from the absolute phase shift across the CPW that can be measured experimentally (Chakalov *et al.* 1998):

$$\varepsilon_{r,eff} = \frac{\varphi c_0}{2\pi f L} \quad (9.25)$$

where φ is the average phase of S_{12} and S_{21}, c_0 is the speed of light in free space, f is the microwave frequency, and L is the actual length of the transmission line. The effect of the taper discontinuities on the phase can be estimated by comparison between CPW lines with different lengths. So in actual experiments, the parameter L in Eq. (9.25) should be the effective length for an actual transmission line. After obtaining the effective dielectric constant of the CPW, the dielectric constant of the ferroelectric thin film can be extracted using Eq. (9.16).

In a first-order approximation, the effective loss tangent, $\tan \delta_{eff}$, of the CPW can be taken as the loss tangent of the ferroelectric thin film, $\tan \delta_{film}$. The effective loss tangent, $\tan \delta_{eff}$, of the CPW can be calculated from the insertion loss of the CPW line (Chakalov *et al.* 1998):

$$\tan \delta_{eff} = \frac{\alpha c_0}{1.89 L \pi f \sqrt{\varepsilon_{eff}}} \quad (9.26)$$

where the insertion loss α is the average magnitude of S_{12} and S_{21}.

More accurate results of $\tan \delta_{film}$ can be obtained from the measurements of identical CPW structures with and without the ferroelectric film. In this technique, two assumptions are made. First, the planar dimensions in the cases with and without ferroelectric film are the same. Second, the qualities of the conductors deposited directly on dielectric substrate and on the substrate covered by ferroelectric films are the same. Upon these assumptions, the difference of the losses between the two cases reveals the contribution of the ferroelectric film. But in the calculation, to obtain accurate results, the impedance difference in the cases with and without the ferroelectric thin film should be taken into consideration.

9.4.1.2 Packaging design

In experiments, the planar circuits developed from ferroelectric thin films are usually housed by metal packages. Figure 9.17 shows a package for coplanar circuit. The contact between the central conductor and the connector pin is maintained by the springs behind the dielectric platform. If the platform is thick enough, the effects of the springs are negligible.

To study the tunability of a ferroelectric film, it is necessary to apply a bias dc voltage to the ferroelectric film under study. The dc voltage can be applied across the two ground planes of the coplanar circuit (Lancaster *et al.* 1998). In this method, the two ground planes should be isolated from each other and from the metal package. The potential is dropped across the two gap widths. The advantage of this approach is that the bias voltage can be easily applied while the disadvantage is that the grounding of the CPW circuit is not good.

Another method of applying the bias voltage is to apply the positive bias voltage to the central conductor of the CPW circuit, and drop the potential across the gap between the central circuit and the ground planes. As shown in Figure 9.18, usually bias T components are used so that the bias voltage applied to the CPW cannot reach the microwave measurement instrument, for example network analyzer, while the microwave signals do not interfere with the dc source (Findikoglu *et al.* 1999).

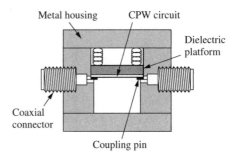

Figure 9.17   Brass package used to house the coplanar circuits (Lancaster *et al.* 1998). Source: Lancaster, M. J. Powell, J. and Porch, A. (1998). "Thin-film ferroelectric microwave devices", *Superconductor Science & Technology*, **11** (11), 1323–1334, IOP Publishing Ltd.

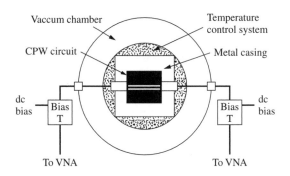

Figure 9.18 contains labels: Vaccum chamber, CPW circuit, dc bias, Bias T, To VNA, Temperature control system, Metal casing, Bias T, dc bias, To VNA

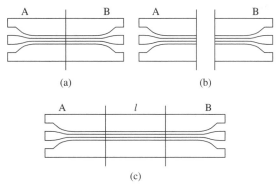

Figure 9.18 Measurement system for ferroelectric CPW. Modified from Findikoglu, A. T. Reagor, D.W. Rasmussen, K. O. Bishop, A. R. Gronbech-Jensen, N. Jia, Q. X. Fan, Y. Kwon, C. and Ostrovsky, L. A. (1999). "Electrodynamic properties of coplanar waveguides made from high-temperature superconducting $YBa_2Cu_3O_{7-\delta}$ electrodes on nonlinear dielectric SrTiO substrates", *Journal of Applied Physics*, **86** (3), 1558–1568

Figure 9.19 TRL calibration for CPW measurements. (a) "Thru" with reference planes directly connected. (b) "Reflect" with both reference planes opened and (c) "Line" with reference planes connected by a matching line. Source: Lue, H. T. and Tseng, T. Y. (2001). "Application of on-wafer TRL calibration on the measurement of microwave properties of $Ba_{0.5}Sr_{0.5}TiO_3$ thin films", *IEEE Transactions on Ultrasonics, Ferroelectrics, and Frequency Control*, **48** (6), 1640–1647. © 2003 IEEE

Temperature dependence is another important parameter in the study of ferroelectric materials. As shown in Figure 9.18, the temperature control system is used to control the measurement temperature so that samples can be measured at different temperatures. Usually, the temperature-control system is installed inside a vacuum chamber, so that temperature can be more easily controlled, and measurements at very high and very low temperature can be conducted in a room-temperature environment.

9.4.1.3 TRL calibration

In experiments, it is difficult to make a 50 Ω matching transmission line for measurement because the dielectric constants of ferroelectric thin films are not known before measurement. Because of the transmission mismatch, multiple reflections occur at the junctions, and this affects the measurement accuracy. In addition, tapers between the much wider electrical probing pads and the narrow transmission lines act as reactance in the circuits, leading to the distortion of the phase response. Taking these effects into account, Lue *et al.* proposed an on-wafer transmission-reflection-line (TRL) calibration technique for the measurement of the

dielectric constant and loss tangent of ferroelectric thin films (Lue and Tseng 2001).

As shown in Figure 9.19, the patterns for TRL calibration, "thru", "reflect", and "line", have the same transition regions A and B with tapers. Each "reflect" is terminated by open, and the "line" is made by connecting A and B with two segments of lines with different lengths, separately. As discussed in Chapter 2, the purpose of TRL calibration is to eliminate the unknown error box of transition regions A and B by calculating the "thru", "reflect", and "line"; then the embedded S-matrix of the device under test (DUT) can be converted to the de-embedded result, independent of A and B. The function of the calibration is equivalent to moving the reference plane from the probe pad to the desired reference plane. Because the transmission lines are identical at the two sides of the reference plane, the "line" is expected to be matched to the reference plane. Besides, the reference plane should be far away from the discontinuous region to prevent the disturbance of local mode. It is not necessary for the "reflect" to be exactly open, but it should be identical at both of the two ports. Usually, the "reflect" is simply opened to the air. When the phase shift through *l* is 0° or multiples of 180°, the error associated with the system becomes larger. The highest accuracy is obtained at the frequency at which the length *l* is one-quarter wavelength long or, equivalently, at a 90° phase shift. It is a common practice to limit a single line to between 20 and 160°; beyond this bandwidth, the calibration is replaced by another line, and such a procedure is called *split-band TRL calibration*. After TRL calibration, the transmission line is nonreflective and matched to the reference plane, and the effective dielectric constant and characteristic impedance can then be correctly extracted.

9.4.2 Coplanar resonator method

Figure 9.20 shows a coplanar resonator fabricated from a ferroelectric thin film deposited on a substrate (Lancaster *et al.* 1998). The dielectric constant of the ferroelectric thin film can be obtained from the resonant frequency of the resonator, and its loss tangent can be obtained from the quality factor of the resonator.

Similar to CPWs, to characterize ferroelectric thin films under bias voltage and different temperature, the coplanar resonators fabricated from ferroelectric thin films are usually packaged. Coplanar resonator can be packaged and excited in the way shown in Figure 9.17. Figure 9.21 shows another coupling approach. The couplings are made by coaxial coupling loops above the coplanar circuit,

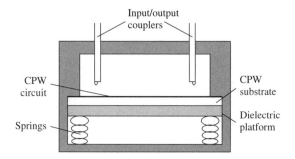

Figure 9.21   Coaxial loop coupling approach

and the couplings can be adjusted by adjusting the distance between the planar circuit and the coaxial loop couplers. The advantage of this approach is that samples can be easily loaded and unloaded, and the disadvantage of this method is that there may be direct couplings between the two coupling loops.

The ways of applying bias voltage and controlling temperature for this method are quite similar to those for CPW method. However, in applying the bias voltage to the central conductor of the CPW resonator, because the central conductor is not physically connected to the couplers, the bias voltage is often applied directly to the electric field node of the center conductor of the CPW structure. For a half-wavelength resonator, its electric field node is at the midpoint of the center conductor.

9.4.3 Capacitor method

In this method, the ferroelectric thin film under study is fabricated into a capacitor. The dielectric constant of thin film is calculated from the capacitance of the capacitor, and the loss factor of the thin film is estimated from the quality factor of the capacitor. The two important aspects of this method are the capacitor design and the capacitance measurement. As there are many types of capacitors (Vendik *et al.* 1999) and many methods for capacitance measurements, this method has many variations. In the following, we first discuss the capacitor design, and then discuss two methods for capacitance measurement at microwave frequencies: split-resonator method and fin-line resonator method.

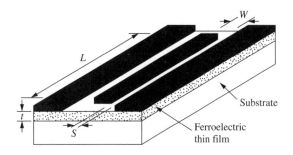

Figure 9.20   A ferroelectric coplanar resonator. Modified from Lancaster, M. J. Powell, J. and Porch, A. (1998). "Thin-film ferroelectric microwave devices", *Superconductor Science & Technology*, **11** (11), 1323–1334, IOP Publishing Ltd, by permission of IOP Publishing Ltd

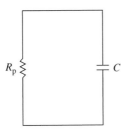

Figure 9.22 Equivalent of a capacitor. C is the capacitance of the capacitor and R_p is the equivalent parallel resistance of the capacitor

9.4.3.1 Capacitor design

As shown in Figure 9.22, an actual capacitor can be represented by a parallel RC circuit. Besides the capacitance C, the dissipation factor D and quality factor Q are also often used to describe a capacitor:

$$D = \frac{1}{Q} = \frac{1}{2\pi f C R_p} \qquad (9.27)$$

where f is the frequency and R_p is the equivalent parallel resistance of the capacitor. The capacitance is related to the dielectric constant of the dielectric material between the electrodes of the capacitor and the quality factor is related to the loss tangent of the dielectric material between the electrodes of the capacitor.

It should be noted that different results might be obtained if different types of capacitors are used. So we should choose a type of capacitor that is closest to the practical application of the ferroelectric thin film under test. Four types of capacitors often used in the characterization of ferroelectric thin films are shown in Figure 9.23. If a trilayer capacitor shown in Figure 9.23(a) is used, the dielectric constant can be obtained from (Chakalov *et al.* 1998)

$$C = \varepsilon_r \varepsilon_0 \frac{wl}{h} \qquad (9.28)$$

where C is the capacitance that can be measured by LCR meter, w and l are the width and length of the capacitor, and h is the thickness of the ferroelectric thin film. The advantage of trilayer capacitor is that the model for permittivity calculation is simple, but it is difficult to fabricate trilayer capacitors.

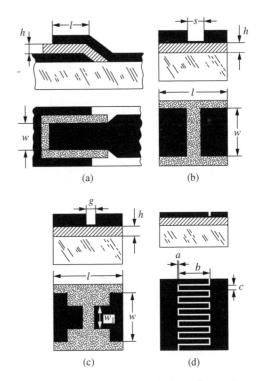

Figure 9.23 Cross sections and plane views of capacitors often used in the characterization of ferroelectric thin films. (a) Trilayer capacitor; (b) rectangular capacitor; (c) step capacitor; (d) interdigital capacitor. Reproduced from Vendik, O. G. Hollmann, E. K. Kozyrev, A. B. and Prudan, A. M. (1999). "Ferroelectric tuning of planar and bulk microwave devices", *Journal of Superconductivity*, **12** (2), 325–338, by permission of Plenum Publication Corporation

Figure 9.23 also shows three types of coplanar capacitors, and according to their shapes, these capacitors are often called rectangular capacitor, step capacitor, and interdigital capacitor. The calculation of the capacitance of these capacitors usually requires numerical methods. Generally speaking, interdigital capacitor has higher sensitivity than the others in the characterization of dielectric films. In the following, we discuss the design of interdigital capacitor for the characterization of ferroelectric thin films, following the method discussed in references (Gevorgian *et al.* 1996a, 1996b).

In the interdigital capacitor shown in Figure 9.24, the substrate of the capacitor consists of two layers: ferroelectric thin film with dielectric

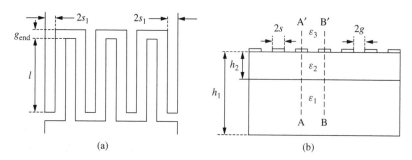

(a)   (b)

Figure 9.24   Interdigital capacitor. (a) Top view and (b) cross section. Reproduced from Gevorgian, S. Carlsson, E. *et al.* (1996a). "Modeling of thin-film HTS/ferroelectric interdigital capacitors", *IEE Proceedings – Microwave and Antenna Propagations*, **143** (5), 397–401, by permission of IEE

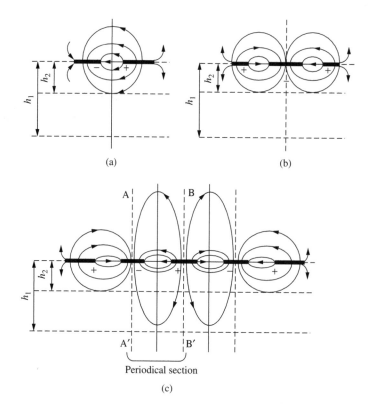

Figure 9.25   Potential distributions and schematic field distributions of interdigital capacitors with different finger numbers. (a) Two-finger capacitor, (b) three-finger capacitor, and (c) multifinger ($n > 3$) capacitor. Modified from Gevorgian, S. S. Martinsson, T. Linner, P. L. J. and Kollberg, E. L. (1996b). "CAD models for multi-layered substrate interdigital capacitors", *IEEE Transactions on Microwave Theory and Techniques*, **44** (6), 896–904. © 2003 IEEE

permittivity ε_2 and the substrate with dielectric permittivity ε_1 for the ferroelectric thin film. The total thickness of the substrate for the interdigital capacitor is h_1 and the thickness of the thin film is h_2. In most cases, the finger number n is larger than 3. As shown in Figure 9.25, because of the difference in the electric field distributions and symmetry of the capacitance structures, the capacitances of the two- and three-finger capacitors, C_2 and C_3, need to be analyzed separately.

According to Figure 9.25, a multifinger ($n > 3$) capacitor can be taken as a combination of a three-finger capacitor and ($n - 3$) periodical sections. Each periodical section is the part between two adjacent magnetic walls AA' and BB'. Therefore, the capacitance C of a multifinger ($n > 3$) capacitor consists of three parts (Gevorgian et al. 1996a, 1996b):

$$C = C_3 + C_n + C_{\text{end}} \qquad (9.29)$$

where C_3 is the capacitance of the three-finger capacitor, C_n is the capacitance of the ($n - 3$) periodical sections, and C_{end} is the correction term for the fringing fields at the ends of the strips.

As shown in Figure 9.25(c), the three-finger capacitor consists of three outer fingers of the multifinger capacitor, one and a half on each side. The capacitance of the three-finger capacitor is given by

$$C_3 = 4\varepsilon_0 \varepsilon_{e3} \frac{K(k_{03}')}{K(k_{03})} l \qquad (9.30)$$

where l is the finger length and $K(k)$ is the first kind of Basset function, which is the modified Bessel function of the second kind, and other parameters are given by

$$\varepsilon_{e3} = 1 + q_{13} \frac{\varepsilon_1 - 1}{2} + q_{23} \frac{\varepsilon_2 - \varepsilon_1}{2} \qquad (9.31)$$

$$q_{i3} = \frac{K(k_{i3})}{K(k_{i3}')} \cdot \frac{K(k_{03}')}{K(k_{03})} \quad (i = 1, 2) \qquad (9.32)$$

$$k_{03} = \frac{s}{s + 2g} \cdot \sqrt{\frac{1 - \left(\dfrac{s + 2g}{s + 2s_1 + 2g}\right)^2}{1 - \left(\dfrac{s}{s + 2s_1 + 2g}\right)^2}} \qquad (9.33)$$

$$k_{i3} = \frac{\sinh\left(\dfrac{\pi s}{2h_i}\right)}{\sinh\left[\dfrac{\pi(s + 2g)}{2h_i}\right]}$$

$$\times \frac{\sqrt{1 - \sinh^2\left[\dfrac{\pi(s + 2g)}{2h_i}\right] \Big/ \sinh^2\left[\dfrac{\pi(s + 2s_1 + 2g)}{2h_i}\right]}}{\sqrt{1 - \sinh^2\left(\dfrac{\pi s}{2h_i}\right) \Big/ \sinh^2\left[\dfrac{\pi(s + 2s_1 + 2g)}{2h_1}\right]}} \quad (i = 1, 2)$$

$$\qquad (9.34)$$

$$k_{i3}' = \sqrt{1 - k_{i3}^2} \qquad (9.35)$$

The capacitance of the ($n - 3$) periodical sections is given by

$$C_n = (n - 3)\varepsilon_0 \varepsilon_{en} \frac{K(k_0)}{K(k_0')} l \qquad (9.36)$$

where

$$k_0 = \frac{s}{s + g} \qquad (9.37)$$

$$k_0' = \sqrt{1 - k_0^2} \qquad (9.38)$$

$$\varepsilon_{en} = 1 + q_{1n} \frac{\varepsilon_1 - 1}{2} + q_{2n} \frac{\varepsilon_2 - \varepsilon_1}{2} \qquad (9.39)$$

$$q_{in} = \frac{K(k_{in})}{K(k_{in}')} \cdot \frac{K(k_0')}{K(k_0)} \quad (i = 1, 2) \qquad (9.40)$$

$$k_{in} = \frac{\sinh\left(\dfrac{\pi s}{2h_i}\right)}{\sinh\left[\dfrac{\pi(s + 2g)}{2h_i}\right]}$$

$$\times \sqrt{\frac{\cosh^2\left[\dfrac{\pi(s + 2g)}{2h_i}\right] + \sinh^2\left[\dfrac{\pi(s + g)}{2h_i}\right]}{\cosh^2\left(\dfrac{\pi s}{2h_i}\right) + \sinh^2\left[\dfrac{\pi(s + g)}{2h_i}\right]}}$$

$$\qquad (i = 1, 2) \quad (9.41)$$

For a thin ferroelectric thin film, $s/h_i \gg 1$, the last expression can be simplified to

$$k_i = \sqrt{2} \exp\left(-\frac{\pi g}{2h_i}\right) \qquad (9.42)$$

$$k_i' = \sqrt{1 - k_i^2} \qquad (9.43)$$

The capacitance of the finger end can be approximately calculated by

$$C_{\text{end}} = 4ns(2+\pi)\varepsilon_0\varepsilon_{\text{end}}\frac{K(k_{0,\text{end}})}{K(k'_{0,\text{end}})} \quad (9.44)$$

where

$$\varepsilon_{\text{end}} = 1 + q_{1,\text{end}}\frac{\varepsilon_1 - 1}{2} + q_{2,\text{end}}\frac{\varepsilon_2 - \varepsilon_1}{2} \quad (9.45)$$

$$q_{i,\text{end}} = \frac{K(k_{i,\text{end}})}{K(k'_{i,\text{end}})} \cdot \frac{K(k'_{0,\text{end}})}{K(k_{0,\text{end}})} \quad (i = 1, 2) \quad (9.46)$$

$$k_{0,\text{end}} = \frac{x}{x + 2g_{\text{end}}} \quad (9.47)$$

$$k_{i,\text{end}} = \frac{\sinh\left(\dfrac{\pi s}{2h_i}\right)}{\sinh\left[\dfrac{\pi(x + 2g_{\text{end}})}{2h_i}\right]} \quad (i = 1, 2) \quad (9.48)$$

Parameter x takes into account the fringing of the field near the finger end. The most reasonable value of x is $0.5s$ with s defined in Figure 9.24(b) (Gevorgian *et al.* 1996a, 1996b). For long fingers ($l/s \gg 1$), it is not necessary to take into account the correction of the fringing fields at the ends of the fingers. In this case, the accuracy of the computed capacitance is around 5–10 %.

As will be discussed in the following, the capacitance can be experimentally measured. For a given capacitor, its capacitance can be calculated according to the equations discussed above. By comparing the measurement results and the calculation results, the dielectric properties of the film can be obtained.

9.4.3.2 Planar split-resonator method

After a ferroelectric capacitor is designed and fabricated, its capacitance and quality factor of the capacitor should be experimentally characterized so that the dielectric properties of the ferroelectric thin film from which the capacitor is fabricated can be calculated. At low microwave frequencies, capacitors can be characterized using LCR meters. At high microwave frequencies, capacitors are usually integrated into resonant structures, and the capacitance and quality factor of

the capacitors are derived from the resonant properties of the resonators. Two types of resonant structures are often used: planar split resonator and fin-line resonator. This section concentrates on the planar split-resonator method, and the discussion on fin-line-resonator method can be found in Section 9.4.3.3.

Working principle

The concept of split resonator has been used in characterizing dielectric sheet samples, as discussed in Chapter 5. Figure 9.26 shows a half-wavelength resonator that splits into two parts at its center point, and the planar capacitor is placed between the two parts of the planar circuits. The resonant frequency and quality factor of the resonant structure, including the two parts of the planar resonator circuit and the planar capacitor, are related to the capacitance and quality factor of the planar capacitor.

For a planar resonator with effective dielectric constant $\varepsilon_{\text{r,eff}}$ and effective length L_{eff}, it can resonate at different modes with frequencies given by

$$f_n = \frac{c}{\sqrt{\varepsilon_{\text{r,eff}}}} \cdot \frac{n}{2L_{\text{eff}}} \quad (9.49)$$

where c is the speed of light in free space, and n is a positive integer, indicating the numbers of cycles of field/current variations along the length of the circuit. According to the values of n, resonant modes can be classified into odd and even modes.

Figure 9.27 shows the current distributions of the first and the second modes of a straight-ribbon resonator and a split resonator, which have the same effective length. For a straight-ribbon resonator at its first mode, as shown in

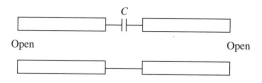

Figure 9.26 Circuit model of a split-resonator method loaded with a capacitor

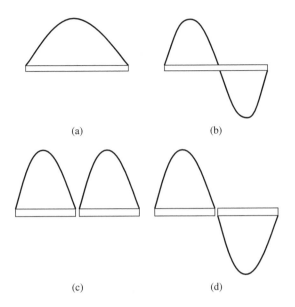

(a)   (b)

(c)   (d)

Figure 9.27 The current distributions of a straight-ribbon resonator and a split resonator at their first and second modes. (a) A straight-ribbon resonator at its first mode, (b) a straight ribbon at its second mode, (c) split resonator at its first mode, and (d) split resonator at its second mode

Figure 9.27(a), the electric current reaches the maximum value at the center of the resonator, and the current distributions along the resonant circuit are in the same direction; while for a straight-ribbon resonator at its even mode, as shown in Figure 9.27(b), there is a current node at the center of the resonator, and the current distributions are in opposite directions at the two sides of the current node. If there is a narrow gap at the center of the resonator, for an even mode, as there is no current at the center of the resonator, the narrow gap hardly affects the current distribution and resonant frequency of the resonator, as shown in Figure 9.26(d). However, for an odd mode, as the electric current reaches maximum value at the center, the gap at the center greatly changes the current distribution and the resonant frequency of the resonator. In an ideal case, if the capacitance between the two parts of the split resonator can be neglected, there is no current at the gap, as shown in Figure 9.27(c). For a split resonator, the difference between the first and the second mode is that, for the first mode, the currents at the two

parts have the same direction, while for the second mode, the currents at the two parts are in opposite direction. By comparing Figures 9.27(c) and (d), one can also find that, if the capacitance between the two parts of a split resonator is neglected, the first and second modes have the same resonant frequency and quality factor.

When a planar capacitor is connected in series between the two parts of the resonator as shown in Figure 9.26, the current distribution and resonant frequency for an even mode do not change. While if the resonator works at an odd mode, as shown in Figure 9.28, its current distribution and resonant frequency are greatly changed because of the capacitor. The capacitance of the capacitor can be calculated from (Galt *et al.* 1995)

$$C = -\frac{\tan(k_0 L_{\mathrm{eff}}/2)}{2\,\omega_0 Z_0} \qquad (9.50)$$

where k_0 is the resonant wave number, ω_0 is the resonant angular resonant frequency, and Z_0 is the characteristic impedance of the resonator.

However, as the effective length of a split resonator loaded with a capacitor cannot be easily determined, it is not convenient to use Eq. (9.50) for capacitance calculation. In experiments, capacitance is often calculated from the resonant frequencies of the first and second modes of the split resonator loaded by the capacitor (Kozyrev *et al.* 1998a):

$$C = -\frac{\tan(\pi f_1/f_2)}{4\pi f_1 Z_0} \qquad (9.51)$$

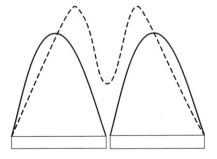

Figure 9.28 The current distributions of the split resonator at the first mode with and without the capacitor. The solid line represents the current distribution without the capacitor and the dashed line represents the current distribution with the capacitor

where f_1 and f_2 are the resonant frequencies of the first and the second modes respectively. After the value of the capacitance is obtained, the dielectric constant of the thin film can be obtained using Eq. (9.29).

The quality-factor component Q_c due to the loss in the ferroelectric capacitor is given by

$$\frac{1}{Q_c} = \frac{1}{Q_1} - \frac{1}{Q_2} \tag{9.52}$$

where Q_1 and Q_2 are the quality factors of the first and the second modes respectively. The resultant Q_c can be transferred to the loss tangent of the ferroelectric thin film from which the capacitor is fabricated. The loss tangent of the ferroelectric thin film can also be estimated in an approximate way (Kozyrev *et al.* 1998a):

$$\tan \delta = \frac{1}{\xi} \left(\frac{1}{Q_1} - \frac{1}{Q_2} \right) \tag{9.53}$$

with

$$\xi = \frac{2}{1 - 2\varphi / \sin 2\varphi} \tag{9.54}$$

$$\varphi = \pi \frac{f_1}{f_2} \tag{9.55}$$

Therefore, according to Eqs. (9.51)–(9.55), the capacitance of the ferroelectric capacitor and the loss tangent of the ferroelectric thin film can be obtained from the resonant frequencies and quality factors of the first and second modes of the split resonator integrated with the capacitor.

It should be noted that Eqs. (9.51)–(9.55) are based on the assumption that the dielectric properties of the ferroelectric sample are independent of the frequency. It should also be noted that, though Figures 9.27 and 9.28 refer to the first and the second modes, the above conclusions can be extended to higher-order modes. When the resonant mode becomes higher, it is more difficult for mode identification. Usually, the first six modes can be distinguished. If the dielectric properties of the ferroelectric thin film are independent of frequency, three groups of measurements based on Eqs. (9.51)–(9.55) can be made, and they should give close results.

In the following, we introduce two types of planar split resonators often used in characterizing ferroelectric thin films: microstrip and stripline split resonators.

Microstrip split resonator

Microstrip is the most widely used planar circuit in microwave electronics, and microstrip split resonator can be easily developed for the characterization of ferroelectric thin films. In this method, the ferroelectric thin film is fabricated into a planar capacitor, and the capacitor covers the gap of the split resonator, as shown in Figure 9.29 (Sok *et al.* 2000). To achieve good electrical contact, the metal strips of the microstrip split resonator and the electrodes of the planar capacitor can be coated with gold, and the planar capacitor can be soldered to the microstrip split resonator. In Figure 9.29, the ferroelectric thin film is fabricated into a step capacitor. Actually, other structures of planar capacitors, such as the ones shown in Figure 9.23, can also be used. The selection of capacitor is based on the sensitivity and accuracy requirements and the fabrication techniques available.

Applying bias dc voltage is an important aspect for the characterization of ferroelectric thin films. Usually, the bias dc voltage can be applied to the two parts of the microstrip circuit of the split resonator at their voltage nodes through rf chokes. Besides, Galt *et al.* used adjustable needles to apply the dc bias voltage to the capacitor at voltage

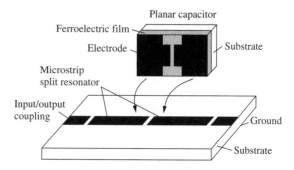

Figure 9.29 Microstrip split resonator for the measurement of ferroelectric thin films. The ferroelectric thin film under test is fabricated into the planar capacitor, which is connected at the gap between the two parts of the split resonator

nodes of the microstrip resonator, and this method of biasing is satisfactory for the characterization of ferroelectric thin films. Detailed discussions on this method can be found in (Galt *et al.* 1995).

Stripline split resonator

Besides the microstrip split resonator discussed above, other types of planar split resonators can also be used in characterizing planar capacitor. As shown in Figure 9.30, a planar capacitor made from the ferroelectric thin film under test is connected in series at the gap between the two parts of a suspended stripline split resonator. The stripline split resonator is coupled to the input and output waveguides through irises (Kozyrev *et al.* 1998a). The dc bias voltage could be applied to the ferroelectric capacitor through rf chokes or bias T components.

The advantage of this approach is that high-power microwave signals can be applied (Kozyrev *et al.* 2000, 1999), so this approach can be used to study the nonlinear properties of the ferroelectric thin films, as will be discussed in Section 9.6. The

disadvantage of this method is that the frequency range of the measurement is limited by the working band of the input/output waveguide. In the design of the measurement fixture, it is usually required that the measurement system can support one set of odd and even modes, so that Eqs. (9.51) and (9.55) can be applied to the computation of permittivity.

Nondestructive method

In the approaches shown in Figure 9.29 and Figure 9.30, the ferroelectric thin film under test is fabricated into a planar capacitor, and the capacitor is connected to the test fixture: a microstrip split resonator or a stripline split resonator. The disadvantage of these approaches is that they are destructive. Actually, the ferroelectric capacitor in Figure 9.26 can also be formed by covering the gap between the two parts of the split resonator with the ferroelectric thin film to be tested, and the dielectric properties of the ferroelectric thin film can be obtained in a similar way as discussed

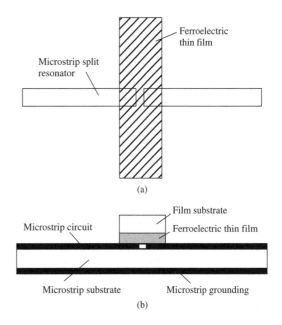

(a)

(b)

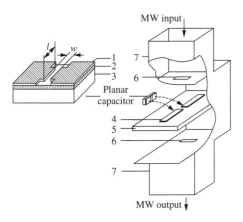

Figure 9.30 Schematic diagram of stripline resonator and ferroelectric film. (1) Electrode, (2) ferroelectric film, (3) capacitor substrate, (4) strip line, (5) suspended substrate, (6) coupling iris, and (7) waveguide. (Kozyrev *et al.* 2000). Source: Kozyrev, A. Ivanov, A. V. Samoilova, T. Soldatenkov, O. Astafiev, K. and Sengupta, L. C. (2000). "Nonlinear response and power handling capability of ferroelectric Ba$_x$Sr$_{1-x}$TiO$_3$ film capacitors and tunable microwave devices", *Journal of Applied Physics*, **88** (9), 5334–5342

Figure 9.31 Nondestructive testing of ferroelectric thin films using a microstrip split resonator. The gap between the two parts of the split resonator is covered by the ferroelectric thin film under test. (a) Top view and (b) cross section. The air gap between the thin film and the microstrip circuit is not shown in the figure

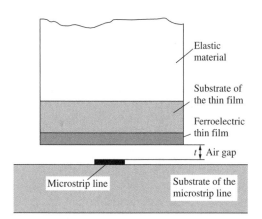

Figure 9.32 Schematic drawing of the air gap between the planar circuit and the ferroelectric thin film under test

above (Galt *et al.* 1995; Tan *et al.* 2004), as shown in Figure 9.31. This approach is nondestructive, and requires little or no sample preparation. It is applicable to both microstrip and stripline split resonators. In the following discussion, we concentrate on microstrip split resonators.

The approach shown in Figure 9.31 has an obvious disadvantage. When a piece of thin film covers the gap between the two parts of a split resonator, there is an inevitable air gap between the planar circuit and the thin film. This air gap greatly changes the resonant frequency of the resonant structure consisting of the planar split resonator and the ferroelectric thin film deposited on a dielectric substrate, and it also affects the actual bias voltage applied to the ferroelectric thin film. To improve the measurement repeatability, an elastic material, such as spring or rubber, is often used to fix the position of the ferroelectric thin film (Tan *et al.* 2004). To obtain accurate results, the effects of the air gap should be considered.

Figure 9.32 shows the air gap between a ferroelectric thin film and the circuit of a split resonator. The air gap increases the resonant frequency of the resonant structure. Owing to the existence of the air gap, the model for the calculation of capacitance discussed in Section 9.4.3.1 is not applicable. If the air gaps between the fixture circuit and the thin films are repeatable for different samples and different times of sample loadings, the dielectric constant of the sample under test can be calculated

from the resonant frequency of the resonant structure using a numerical way.

The dielectric constant of a ferroelectric thin film can be obtained in the following steps (Tan *et al.* 2004). First, determine the thickness of the air gap. Measure the frequency response of the split resonator when a substrate without ferroelectric thin film is loaded, then adjust the thickness of the air gap in the simulation code until the simulation results fit the measurement results. Second, establish the calibration curve between the resonant frequency of the measurement fixture and the dielectric constant of the sample loaded in the fixture. The resonant frequencies for different dielectric constant values are calculated using the simulation code. From the relationships between the dielectric constant and the resonant frequency, the calibration curve can be obtained. Figure 9.33 shows an example of calibration curve. The two lines in the figure correspond to the upper and lower limits of the thickness value of the air gap, which is determined from the uncertainties of frequency measurement for different times of loading. After the calibration curve is obtained, according to the measured resonant frequency of the fixture loaded with the sample under test, the dielectric constant of the sample and its uncertainties can be determined from the upper and lower limit lines in the calibration curve.

It should be noted that the air-gap thickness discussed above does not mean the exact thickness value of the air gap. Actually it considers all the uncertainty factors which affect the resonant frequency of the sample-loaded fixture, so it can be taken as a correction factor. It should also be noted that for different fixtures, different materials of substrates, different substrate thicknesses, or different film thicknesses, the calibration curves are different.

It should be indicated that when the air gap between the microstrip circuit and the thin film under study is small, the air gap has similar effects to the quality factors of the even and odd modes of the resonant structures. So Eq. (9.53) can still be used in calculating the loss tangent of the thin film under study.

The air gap also affects the results of tunability measurement, as the actual bias voltage applied to

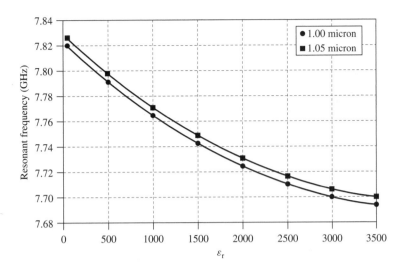

Figure 9.33 The relationship between the dielectric constant and the resonant frequency. The two lines correspond to the upper limit ($1.05\,\mu$m) and the lower limit ($1.00\,\mu$m) of the air gap, respectively

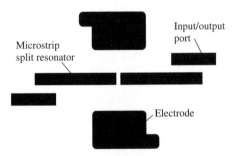

Figure 9.34 An approach for applying bias voltage. The ferroelectric thin film under study covers two electrodes, and the bias voltage is applied to the ferroelectric thin film through the two electrodes

the ferroelectric thin film is different from the dc voltage applied to the measurement fixture because of the presence of the air gap. The two parts of the split resonator can be used as the electrodes for the application of bias dc voltage. Figure 9.34 shows another way of applying bias voltage (Tan *et al.* 2004). The bias voltage is applied through the two additional metal pads as electrodes. The distance between the microstrip circuit and the two electrodes should be large enough so that the effects of the electrodes to the microstrip circuit are negligible. The advantage of this method is that the electric field is uniform at the gap area, so that more accurate information can be obtained for

the response of ferroelectric thin films to external dc electric field. The disadvantage of this method is that high voltage is required to establish the electric field that is strong enough for the study of the tunability of ferroelectric thin films.

As shown in Figure 9.35, the air gap between the electrode and the ferroelectric thin film can be represented by equivalent capacitance. To the external bias voltage, the total capacitance C is given by

$$\frac{1}{C} = \frac{1}{C_s} + \frac{1}{C_{g1}} + \frac{1}{C_{g2}} \qquad (9.56)$$

where C_s is the part of capacitance due to the ferroelectric sample, C_{g1} and C_{g2} are the equivalent

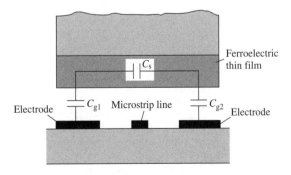

Figure 9.35 The equivalent circuit model for the air gap between the electrode and the ferroelectric thin film

capacitances because of the air gaps at the two electrodes. If the bias voltage applied to the two electrodes is V, the actual voltage applied to the sample V_s is given by

$$V_s = \frac{V}{1 + (C_s/C_{g1}) + (C_s/C_{g2})} \qquad (9.57)$$

If there is no air gap between the sample, C_{g1} and C_{g2} become infinity, and from Eq. (9.57), we can get $V_s = V$. However, in actual situations, the air gap is inevitable, so the V_s is always less than V. Therefore, it is important to decrease the air gap between the electrodes and the thin film under test, and the tunability obtained in experiments can be taken as a lower-limit value.

9.4.3.3 Fin-line-resonator method

In principle, any resonant structure can be used to characterize planar capacitors provided planar capacitors can be integrated into the structure at the positions with microwave electric currents. Kozyrev *et al.* proposed a fin-line resonator for the characterization of ferroelectric capacitors (Kozyrev *et al.* 2001). As shown in Figure 9.36, the symmetrical fin line is installed in a rectangular waveguide. The fin-line resonator

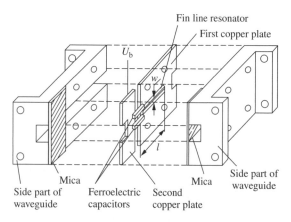

Figure 9.36   Schematic configuration of a fin-line resonator for the characterization of ferroelectric capacitor (Kozyrev *et al.* 2001). Copyright 2001 from 'Microwave properties of ferroelectric film planar varactors' by Kozyrey & Keis in Integrated Ferroelectrics, **34**. Reproduced by permission of Taylor & Francis, Inc., http://www.routledge_ny.com

consists of two copper plates and two ferroelectric capacitors. The first copper plate is a segment of short-ended fin line, and it is grounded owing to the contact with the side parts of the waveguide. The second plate is insulated from the waveguide by mica, and it is intended for controlling the bias dc voltage applied to the ferroelectric capacitors. The two planar ferroelectric capacitors under test are soldered at the ends of the fin-line resonator between the first and the second plates.

Similar to microstrip split-resonator method, the capacitance and quality factor of the planar capacitor can be derived from the resonant frequency and quality factor of the resonant structure. However, it should be noted that the resonant properties of the whole resonant structure are related to the capacitance and the quality factor of the two ferroelectric capacitors together. To characterize one capacitor, a standard capacitor with known capacitance and quality factor is needed.

9.4.4 Influence of biasing schemes

In the characterization of ferroelectric thin films, different measurement methods may result in different results. In the development of measurement methods, we should ensure that the sample is in the state we are interested in. In the measurement of ferroelectric thin films, the films under study are often used as one layer of multilayer substrates, on which planar circuits are developed. Besides the circuit layouts, the ways of applying bias voltage may also affect the measurement results.

Van Keuls *et al.* studied the influence of biasing schemes to the performances of ferroelectric microstrip ring resonator (Van Keuls *et al.* 1998). In microwave electronics, microstrip ring resonators are often used as materials characterization fixtures and critical resonant components of high-frequency devices, such as the stabilizing elements in oscillators. So the analysis of the properties of microstrip ring resonators is helpful for not only the characterization of ferroelectric thin films but also the development of tunable ferroelectric devices.

As discussed in Chapter 2, the resonant wavelength λ_g of a ring resonator is given by

$$2\pi R = n\lambda_g \quad (n = 1, 2, 3, \ldots) \qquad (9.58)$$

where R is the mean radius of the ring resonator, and n is the mode number. The unloaded quality factor Q_0 of a ring resonator is closely related to the microwave properties of the materials forming the resonant circuit (Tonkin and Hosking 1997):

$$Q_0 = \frac{\pi}{\alpha \lambda_g} = \frac{\pi f \sqrt{\varepsilon_{\text{eff}}}}{c\alpha} \qquad (9.59)$$

where α is the attenuation coefficient of the transmission line, c is the speed of light, f is the resonant frequency, ε_{eff} is the effective dielectric constant of the circuit. According to the model for multilayer substrate discussed in Section 9.4.1, the dielectric properties of the film can be obtained from the resonant frequency and quality factor of the resonator. Because of its geometry, ring resonators usually have higher quality factors than straight-ribbon resonators, so ring resonator methods usually have higher sensitivities and accuracies.

Figure 9.37 shows a typical scheme for applying bias dc voltage to a microstrip ring resonator (Van Keuls *et al.* 1998). The ring resonator is coupled through a microstrip line. Both the microstrip line and the ring resonator can be biased. Voltage V_L is applied to the microstrip transmission line through bias T components, while V_R is applied to the ring resonator through a gold wire bonded to the ring resonator. The gold wire is bonded to the ring at a virtual short circuit position, nearly opposite of the coupling gap, so does not to disturb the resonance.

There are three typical cases according to the relationships between V_L and V_R (Van Keuls *et al.*

1998). In the first case ($V_L = V_R$), maximum coupling between the ring resonator and the microstrip line is obtained. The dc electric field only exists between the resonant circuit and the ground plane, with most of the electric field concentrated in the dielectric substrate, so the dielectric constant of the ferroelectric thin film in the range between the ring and the line is close to its highest possible value. When the applied bias voltage increases, the resonant frequency increases, but the change of resonant frequency is small, and the coupling between the resonator and the microstrip line almost does not change. In the second case, the microstrip line is grounded, and the bias dc voltage is applied to the ring resonator ($V_L = 0$ and $V_R > 0$). When the bias dc voltage increases, the dielectric constant decreases, so the coupling between the ring resonator and microstrip line also decreases. Therefore, the coupling coefficient can be adjusted. Meanwhile, the increase of bias voltage will increase the resonant frequency of the ring resonator because of the decrease of the dielectric constant of ferroelectric thin film. The third case ($V_R > V_L$) could be the most desirable biasing scheme for practical microwave applications. Both the resonant frequency and the coupling coefficient can be adjusted more easily than the above two schemes.

9.5 RESPONDING TIME OF FERROELECTRIC THIN FILMS

In microelectronics, ferroelectric materials are used in fabricating tunable devices, electronic controllers, and switching modules. So, the responding time of the permittivity of ferroelectric materials on exposure to microwave signals and under the action of videofrequency voltage pulses needs to be investigated.

Kozyrev *et al.* proposed a method for studying the responding time of the permittivity of ferroelectric films (Kozyrev *et al.* 1998b). In this method, the ferroelectric film under study is fabricated into a planar ferroelectric capacitor, and the ferroelectric capacitor is loaded to a planar resonator excited by a weak microwave signal. A pulsed control voltage is applied to the ferroelectric

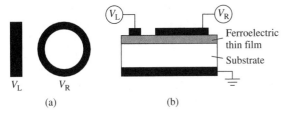

(a)   (b)

Figure 9.37   Biasing scheme for testing a tunable ring resonator. (a) Top view and (b) front view. Modified from Van Keuls, F. W. Romanofsky, R. R. Bohman, D. Y. and Miranda, F. A. (1998). "Influence of the biasing scheme on the performance of Au/SrTiO₃/LaAlO₃ thin film conductor/ferroelectric tunable ring resonators", *Integrated Ferroelectrics*, **22**, 363–372.

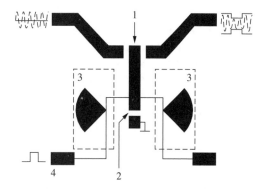

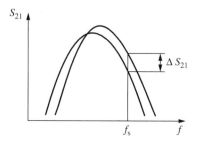

Figure 9.39   Change of the resonant characteristics of the microstrip resonator when a dc control voltage is applied to the ferroelectric capacitor. ΔS_{21} is the change of the transmission coefficient at a fixed microwave frequency (f_s)

Figure 9.38   Microstrip microwave circuit for the measurement of the responding time of planar ferroelectric capacitors (Kozyrev *et al.* 1998b). (1) Microstrip resonator, (2) connection point for planar capacitor, (3) rf chokes, and (4) circuits for applying the pulsed control voltage. Modified from Kozyrev, A. B. Soldatenkov, O. I. and Ivanov, A. V. (1998b). "Switching time of planar ferroelectric capacitors using strontium titanate and barium strontium titanate films", *Technical Physics Letters*, **24** (10), 755–757

capacitor. The responding time of the ferroelectric film to the pulsed control voltage is determined from the variation of the response of the resonator loaded with the ferroelectric capacitor to the weak microwave signal. Figure 9.38 shows an example of measurement circuit. The ferroelectric capacitor is loaded at the gap of the microstrip split resonator near its short-circuited end, and the other end of the resonator is open. The resonator is coupled to the external microwave circuit by capacitive couplings, which also protect the microwave circuit from the action of the control voltage. The pulsed control voltage is applied to the ferroelectric capacitor through rf chokes, which are connected to the microstrip resonator near the points of zero microwave electric field, so the influence of the circuits for applying the pulsed control voltage on the resonant properties of the resonator is small.

The relationship between the resonant frequency f_0 of the resonator and the capacitance of the capacitor is given by (Kozyrev *et al.* 1998b)

$$2\pi Z_0 f_0 C(U_b) = \tan\left(\frac{2\pi}{c}\sqrt{\varepsilon_{\text{eff}}} \cdot f_0 \cdot l_{\text{eff}}\right) \quad (9.60)$$

where $C(U_b)$ is the capacitance of the capacitor when a control voltage U_b is applied, Z_0 and ε_{eff} are the impedance and the effective permittivity of the microstrip resonator, l_{eff} is the effective length of the microstrip resonator, and c is the speed of light. When the control voltage pulse is applied to the capacitor, the capacitance of the capacitor will be changed, so the resonant frequency will also be changed. Therefore, at a fixed microwave frequency, the transmission coefficient S_{21} will also be changed, as shown in Figure 9.39. A comparison between the applied voltage pulse and the detected microwave response pulse can reveal the responding time of the resonator loaded with the ferroelectric capacitor under test.

Kozyrev *et al.* studied the responding time of $SrTiO_3$ and $(Ba, Sr)TiO_3$, and the results are shown in Figure 9.40. Figure 9.40(a) shows the trace of the control voltage pulse (pulse length $100\,\mu s$, rise time $t_f \approx 30\,ns$, and amplitude $U_m = 30\,V$) and Figure 9.40(b) shows the detected microwave signal pulse after passing through the resonator loaded with an $SrTiO_3$ capacitor. For the resonator loaded with a $(Ba, Sr)TiO_3$ capacitor, the envelope of the microwave signal, shown in Figure 9.40(c) indicates that there are two different mechanisms responsible for the change in the permittivity of the $(Ba, Sr)TiO_3$ film under the action of the control voltage pulse: a fast variation in the permittivity over a time less than the rise time of the control pulse ($\sim 30\,ns$) and a slow variation with a response time of the order of $20\,\mu s$. The slow variation in the amplitude of the detected microwave signal is less than 5 to 10 % of the total amplitude of the pulse.

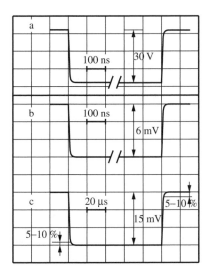

Figure 9.40 Traces showing (a) control pulse, (b) switching time of SrTiO₃ capacitor, and (c) switching time of (Ba, Sr)TiO₃ capacitor. The scales of the microwave responses of the SrTiO₃ and (Ba, Sr)TiO₃ elements are different to reveal the "slow" ($\sim$20 µs) relaxation time (Kozyrev *et al*. 1998). Source: Kozyrev, A. B. Soldatenkov, O. I. and Ivanov, A. V. (1998b). "Switching time of planar ferroelectric capacitors using strontium titanate and barium strontium titanate films", *Technical Physics Letters*, **24** (10), 755–757

9.6 NONLINEAR BEHAVIOR AND POWER-HANDLING CAPABILITY OF FERROELECTRIC FILMS

Most of the investigations on ferroelectric thin films have been devoted to the properties of these materials to weak microwave signals. In recent years, progress has been made on the study of the nonlinear response and power- handling capability (PHC) of ferroelectric films and devices (Kozyrev *et al*. 2000, 1998c). The research in this area is important because the nonlinear response of ferroelectric elements to a microwave signal is undesirable in passive tunable devices and should be minimized to reduce nonlinear distortion of the operating microwave signal. On the other hand, the nonlinear microwave properties can be utilized for the development of nonlinear active devices. Meanwhile, the heating of ferroelectric film due to the microwave energy dissipation also leads to variations in the properties of the films and,

consequently, in the performance of the microwave electronic devices.

Two approaches have been used in the study of the nonlinear behavior of ferroelectric films and devices. One is to measure the response of a resonator loaded with a ferroelectric capacitor to a pulsed microwave signal. By using pulsed signals, the effect of microwave electric field on the dielectric properties of the ferroelectric films can be separated from the effect of the overheating of the ferroelectric capacitor under the action of microwave power. So, the nonlinear parameters of the ferroelectric film capacitors can be extracted. The other method is to measure the intermodulation distortion (IMD) of a resonator loaded with a ferroelectric capacitor. This technique provides the information of the IMD products generated in a capacitor-loaded resonator under the action of two-tone incident microwave signals.

PHC is another important parameter for microwave electronic devices. PHC is defined as the level of incident microwave power up to which the device can function properly. The shift of the working frequency of devices employing nonlinear ferroelectric elements and the level of IMD products under elevated power are two parameters often used for the determination of PHC. The nonlinear properties of ferroelectric films and the heating of ferroelectric films by microwave power are the main phenomenon governing these parameters. In the following, we discuss the two methods often used in the study of the nonlinear properties of ferroelectric films: pulsed signal method and intermodulation method.

9.6.1 Pulsed signal method

The structure proposed by Kozyrev *et al*. shown in Figure 9.30 can be used for both small-signal study and large-signal study. The capacitance of ferroelectric capacitor and the loss tangent of the ferroelectric film can be calculated from Eqs. (9.51) and (9.53). For the study of high microwave power properties of ferroelectric film capacitors, it is necessary to separate the thermal and microwave electric field effects responsible for the nonlinearity of the ferroelectric capacitor. Kozyrev *et al*. used pulsed microwave signal of

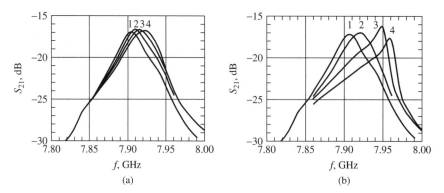

Figure 9.41   Frequency response of a resonator loaded with a (Ba, Sr)TiO$_3$ film capacitor to pulsed microwave power of various levels measured at (a) leading and (b) trailing fronts. Incident power levels P_{inc}: (1) +6 dBm, (2) +26 dBm, (3) +32 dBm, and (4) +35 dBm (Kozyrev *et al.* 2000). Source: Kozyrev, A. Ivanov, A. V. Samoilova, T. Soldatenkov, O. Astafiev, K. and Sengupta, L. C. (2000). "Nonlinear response and power handling capability of ferroelectric Ba$_x$Sr$_{1-x}$TiO$_3$ film capacitors and tunable microwave devices", *Journal of Applied Physics*, **88** (9), 5334–5342

pulse duration 10 μs and pulse duty factor 10^{-3}. The duration of the pulse leading and trailing fronts was $T_f = 10^{-7}$ s. The pulsed microwave power was varied between −10 and +50 dBm. The transmission S_{21} and reflection S_{11} coefficients were measured during the pulse duration, and the equipment used allowed the registration of time variations in the S_{21} and S_{11} coefficients with time resolution on the order of 10 ns (Kozyrev *et al.* 2000).

Figure 9.41 shows the frequency responses of a resonator loaded with a (Ba, Sr)TiO$_3$ capacitor at various values of incident pulsed power P_{inc}. At a low level of incident power, the resonant curves coincide with the small-signal measurements at zero dc bias voltage. The increase of microwave power leads to the following results. When the microwave power is not very high, the maximum of the transmission coefficient S_{21} shifts toward higher frequencies and the resonant curve is slightly widened. When P_{inc} is further increased, the resonant curve becomes asymmetrical, and this is a common characteristic of nonlinear resonators.

Owing to the thermal effect on the dielectric properties of the (Ba, Sr)TiO$_3$ film, the nonlinear effects are more drastic at the trailing front than those at the leading front. The deformation of the envelope of the transmitted microwave pulse under the elevated level of the incident microwave pulsed power of +35 dBm at the different frequencies

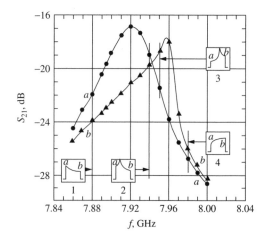

Figure 9.42   The deformation of the envelope of the transmitted microwave pulse (insets) at different frequencies for incident pulsed power +35 dBm. Curves (a) and (b) show the resonant curves measured at leading and trailing fronts of the microwave pulse, respectively (Kozyrev *et al.* 2000). Source: Kozyrev, A. Ivanov, A. V. Samoilova, T. Soldatenkov, O. Astafiev, K. and Sengupta, L. C. (2000). "Nonlinear response and power handling capability of ferroelectric Ba$_x$Sr$_{1-x}$TiO$_3$ film capacitors and tunable microwave devices", *Journal of Applied Physics*, **88** (9), 5334–5342

is illustrated in the insets shown in Figure 9.42. The resonator frequency response measured at the leading (a) and the trailing (b) fronts at $P_{inc} =$

+35 dBm are shown in the figure as well. At frequencies far below or far above the small-signal resonant frequency (insets 1 and 4), the output signal varies in time monotonically because of the increase in resonant frequency during the microwave pulse. For frequencies higher but close to the peak transmission, nonmonotonic variation in the transmitted power during the pulse takes place (insets 2 and 3) because of the effect of crossover between the measurement frequency and resonant frequency.

It should be noted that, in the calculation of the loss tangent of the ferroelectric film using Eq. (9.53), the quality factor should be measured. The measurement of quality factor of asymmetrical transmission curve is discussed in Chapter 2.

9.6.2 Intermodulation method

For a microwave electronic device, it is desirable to transmit high-power microwave signals without causing unacceptable signal degradation due to the generation of intermodulation distortion (IMD) products. IMD products are caused by the non-linear behavior of the device, and arise when the original signal is mixed with a second signal at a second frequency resulting in a third signal at a third frequency. Nonlinear behavior also generates harmonic signals at frequencies that are an integer multiple of the input signal. A summary of the different types of distortion caused by IMD and harmonic signal generation is shown in Figure 9.43. The net result is that the nonlinear distortion takes

energy from the desired signal at the frequency of interest, thus reducing the amplitude of this signal, and the energy is used to generate harmonic and IMD signals at frequencies that may interfere with the desired signal (Mueller and Miranda 2000).

The intermodulation distortion becomes obvious when the input to the system consists of two closely spaced frequencies, usually called two tones:

$$v_{\text{in}} = \cos \omega_1 t + \cos \omega_2 t \qquad (9.61)$$

Then, the output spectrum will consist of all the harmonics in the form of

$$\omega = m\omega_1 + n\omega_2 \qquad (9.62)$$

where m and n may be positive or negative integers. The order of a given product is defined as $j = |m| + |n|$, and the jth order harmonics is denoted v^j. The second-order harmonic products v^2 include $2\omega_1$, $2\omega_2$, $\omega_1 - \omega_2$ and $\omega_1 + \omega_2$. These frequencies are generally far away from the fundamentals ω_1 and ω_2, and so can be easily filtered, but for a mixer, the $\omega_1 - \omega_2$ product is usually the desired result. The third-order products v^3 include such as $3\omega_1$, $3\omega_2$, $2\omega_1 + \omega_2$ and $\omega_1 + 2\omega_2$ which can be filtered, and the products $2\omega_1 - \omega_2$ and $2\omega_2 - \omega_1$, which generally cannot be easily filtered. Such products that arise from mixing two input signals are called intermodulation, and the products $2\omega_1 - \omega_2$ and $2\omega_2 - \omega_1$ are especially important because they may set the dynamic range or bandwidth of the system (Pozar 1998).

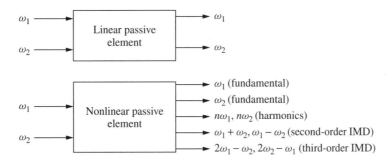

Figure 9.43 Intermodulation distortion caused by transmission of signals through nonlinear components. Reproduced from Mueller, C. H. and Miranda, F. A. (2000). "Tunable dielectric materials and devices for broadband wireless communications", *Handbook of Thin Film Devices*, edited by Francombe, M. H., volume 5: *Ferroelectric Film Devices*, Academic Press, by permission of Elsevier

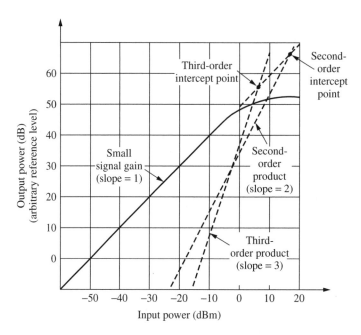

Figure 9.44   Intercept diagram for a nonlinear component (Pozar 1998). Reproduced from Pozar, D. M. (1998). *"Microwave Engineering"*, 2nd ed., John Wiley & Sons, Inc., New York, by permission of John Wiley & Sons

Figure 9.44 shows a graph of output power versus input power for a nonlinear system, and the intercept points between different output products are indicated. These intercept points show the second- or third-order intermodulation distortion. A plot of output signals (ω_1 and ω_2) power versus input power has a slope of unity for small-signal levels. As the input power increases, saturation sets in, causing clipping of the output waveform and signal distortion. This distortion manifests itself by diverting part of the input power to various harmonics. The curve for a second-order product has a slope of two. If the linear part of the small-signal gain curve is extended, it will intercept the second-order product power curve at the second-order intercept point. This point can be specified by either the input or the output power at the intersection, and is a measure of the amount of second-order inter-modulation distortion. The component would actually be operated well below this point. Similarly, the intercept point between small-signal gain and third-order product can also be obtained.

Figure 9.45 shows a microstrip resonator for IMD measurement of nonlinear planar capacitors (Kozyrev *et al.* 1998c). The planar capacitor under

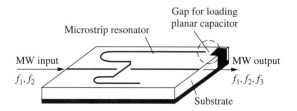

Figure 9.45   Ferroelectric microstrip resonator for IMD measurements (Kozyrev *et al.* 1998c). Modified from Kozyrev, A. B. Samoilova, T. B. Golovkov, A. A. Hollmann, E. K. Kalinikos, D. A. Loginov, V. E. Prudan, A. M. Soldatenkov, O. I. Galt, D. Mueller, C. H. Rivkin, T. V. and Koepf, G. A. (1998c). "Nonlinear behavior of thin film SrTiO capacitors at microwave frequencies", *Journal of Applied Physics*, **84** (6), 3326–3332

test is loaded at the gap near the short end of a λ/4 resonator, and the resonator is loosely coupled to the external circuits. Two sinusoidal microwave signals of the same power, P_{1inc}, with slightly different frequencies ($f_1 - f_2 \sim 1$ MHz) are applied to the resonator input. The resonator output contains IMD products at frequencies $f = \pm m f_1 \pm n f_2$ where m and n are positive integers. The properties that can be measured are the fundamental transfer function, P_{1out} as a function of P_{1inc}; the third-order IMD transfer function, P_{3out} at frequency $f_3 = 2f_1 - f_2$ (or $2f_2 - f_1$) as a function of P_{1inc}; the loaded quality factor, Q_L; and the resonant frequency. All measurements should be carried out under the condition that f_1, f_2 and f_3 are localized within the passband of the resonator. To eliminate the possibility of IMD phenomena arising from nonlinear behavior of contacts between the capacitor electrodes and the resonator microstrip line, identical measurements can be performed using linear, nontunable capacitor connected into the circuit. With the nontunable capacitors, usually no IMD products can be detected.

Figure 9.46 shows the typical intermodulation measurement results (Kozyrev *et al.* 1998c). With the increase of the input power, both the fundamental-order and third-order output powers increase, and the difference between the fundamental power and the third-order power decreases. Application of a dc bias voltage across the ferroelectric capacitor changes the properties of the resonator loaded with the ferroelectric capacitor. Figure 9.46(a) shows the fundamental and third-order transfer functions for the resonator loaded with capacitor at dc bias voltages of 0 and 100 V. Figure 9.46(b) illustrates the difference in output power levels for the fundamental and intermodulation products ($\Delta P = P_{3out} - P_{1out}$) as a function of dc bias voltage for incident power levels of 9 and 21 dBm. From Figure 9.46, it can be found that the level of the third-order intermodulation product is reduced by applying a dc electric field across the capacitor. Besides the external dc electric field, sometimes it is necessary to consider the microwave electric voltage across the capacitor in analyzing the tunability of ferroelectric capacitors. Usually, the microwave electric voltage is very small compared to the bias dc voltage; however, when the input power becomes very large, the effect of microwave electric voltage cannot be neglected.

The study of nonlinear properties is helpful for estimating the microwave power-handling capability of tunable microelectronic devices that incorporate nonlinear ferroelectric capacitors. The power handling capability of a device can be defined as the level of incident microwave power up to which the device functions properly and the deviations

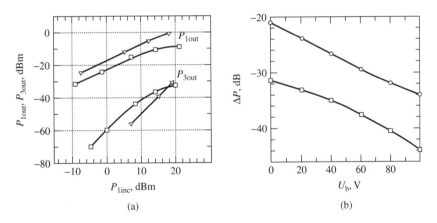

(a)   (b)

Figure 9.46 Effects of bias dc voltage to IMD products. (a) Output power for the fundamental- and third-order intermodulation distortion as a function of input power per tone. The square and triangle legends represent two different sets of data taken with a 0 and 100 V dc bias applied across the capacitor, respectively. (b) Difference between the output power levels of the fundamental- and third-order intermodulation distortion signals versus bias dc voltage at different incident power levels. The circle and square legends represent two different incident power $P_{1inc} = +21$ and $+9$ dBm, respectively (Kozyrev *et al.* 1998c). Source: Kozyrev, A. B. Samoilova, T. B. Golovkov, A. A. Hollmann, E. K. Kalinikos, D. A. Loginov, V. E. Prudan, A. M. Soldatenkov, O. I. Galt, D. Mueller, C. H. Rivkin, T. V. and Koepf, G. A. (1998c). "Nonlinear behavior of thin film SrTiO capacitors at microwave frequencies", *Journal of Applied Physics*, **84** (6), 3326–3332

of the operating parameters do not exceed acceptable levels. Usually, the allowable levels of the shift of the working frequency and the intermodulation product power at the output of the device are the essential parameters determining the power-handling capability of a device loaded with nonlinear components.

REFERENCES

Anderson, J. C. (1964). *Dielectrics*, Chapman & Hall, London.

Carroll, K. B. Pond, J. M. Chrisey, D. B. Horwitz, J. S. and Leuchtner, R. E. (1993). "Microwave measurement of the dielectric constant of $Sr_{0.5}Ba_{0.5}TiO_3$ ferroelectric thin films", *Applied Physics Letters*, **62** (15), 1845–1847.

Chakalov, R. A. Ivanov, Z. G. Boikov, Y. A. Larsson, P. Carlsson, E. Gevorgian, S. and Claeson, T. (1998). "Fabrication and investigation of YBCO/BSTO thin film structures for voltage tunable devices", *Physica C*, **308**, 279–288.

Findikoglu, A. T. Reagor, D. W. Rasmussen, K. O. Bishop, A. R. Gronbech-Jensen, N. Jia, Q. X. Fan, Y. Kwon, C. and Ostrovsky, L. A. (1999). "Electrodynamic properties of coplanar waveguides made from high-temperature superconducting $YBa_2Cu_3O_{7-\delta}$ electrodes on nonlinear dielectric SrTiO substrates", *Journal of Applied Physics*, **86** (3), 1558–1568.

Galt, D. Price, J. C. Beall, J. A. and Harvey, T. E. (1995). "Ferroelectric thin film characterization using superconducting microstrip resonators", *IEEE Transactions on Applied Superconductivity*, **5** (2), 2575–2578.

Gevorgian, S. Carlsson, E. Rudner, S. Wernlund, L. D. Wang, X. and Helmersson, U. (1996a). "Modeling of thin-film HTS/ferroelectric interdigital capacitors", *IEE Proceedings – Microwave and Antenna Propagations*, **143** (5), 397–401.

Gevorgian, S. Linner, L. J. P. and Kollberg, E. L. (1995). "CAD models for shielded multi-layered CPW", *IEEE Transactions on Microwave Theory and Techniques*, **43** (4), 772–779.

Gevorgian, S. S. Martinsson, T. Linner, P. L. J. and Kollberg, E. L. (1996b). "CAD models for multi-layered substrate interdigital capacitors", *IEEE Transactions on Microwave Theory and Techniques*, **44** (6), 896–904.

Jaffe, B. Cook, W. R. and Jaffe, H. (1971). *Piezoelectric Ceramics*, Academic Press, London.

Kozyrev, A. Ivanov, A. V. Samoilova, T. B. Soldatenkov, O. I. Sengupta, L. C. Rivkin, T. V. Carlson, C. M. Parilla, P. A. and Ginley, D. S. (1999). "Microwave properties of ferroelectric BSTO

varactors a high microwave power", *Integrated Ferroelectrics*, **24**, 297–307.

Kozyrev, A. Ivanov, A. V. Samoilova, T. Soldatenkov, O. Astafiev, K. and Sengupta, L. C. (2000). "Nonlinear response and power handling capability of ferroelectric $Ba_xSr_{1-x}TiO_3$ film capacitors and tunable microwave devices", *Journal of Applied Physics*, **88** (9), 5334–5342.

Kozyrev, A. Keis, V. Buslov, O. Ivanov, A. Soldatenkov, O. Loginov, V. Taricin, A. and Graul, J. (2001). "Microwave properties of ferroelectric film planar varactors", *Integrated Ferroelectrics*, **34**, 271–278.

Kozyrev, A. B Samoilova, T. B. and Golovkov, A. A. Hollmann, E. K. Kalinikos, D. A. Loginov, V. E. Prudan, A. M. Soldatenkov, O. I. Galt, D. Mueller, C. H. Rivkin, T. V. and Koepf, G. A. (1998c). "Nonlinear behavior of thin film SrTiO capacitors at microwave frequencies", *Journal of Applied Physics*, **84** (6), 3326–3332.

Kozyrev, A. B. Soldatenkov, O. I. Samoilova, T. B. Ivanov, A. V. Mueller, C. H. Rivkin, T. V. and Koepf, G. A. (1998a). "Response time and power handling capability of tunable microwave devices using ferroelectric films", *Integrated Ferroelectrics*, **22**, 329–340.

Kozyrev, A. B. Soldatenkov, O. I. and Ivanov, A. V. (1998b). "Switching time of planar ferroelectric capacitors using strontium titanate and barium strontium titanate films", *Technical Physics Letters*, **24** (10), 755–757.

Lacey, D. Gallop, J. C. and Davis, L. E. (1998). "The effects of an air gap on the measurement of the dielectric constant of STO at cryogenic temperatures", *Measurement Science and Technology*, **9** (3), 536–539.

Lancaster, M. J. Powell, J. and Porch, A. (1998). "Thin-film ferroelectric microwave devices", *Superconductor Science & Technology*, **11** (11), 1323–1334.

Lines, M. E. and Glass, A. M. (1977). *Principles and Applications of Ferroelectric and Related Materials*, Clarendon Press, Oxford.

Lue, H. T. and Tseng, T. Y. (2001). "Application of on-wafer TRL calibration on the measurement of microwave properties of $Ba_{0.5}Sr_{0.5}TiO_3$ thin films", *IEEE Transactions on Ultrasonics, Ferroelectrics, and Frequency Control*, **48** (6), 1640–1647.

McNeal, M. P. Jang, S. J. and Newnham, R. E. (1998). "The effect of grain and particle size on the microwave properties of barium titanate ($BaTiO_3$)", *Journal of Applied Physics*, **83** (6), 3288–3297.

Mueller, C. H. and Miranda, F. A. (2000). "Tunable dielectric materials and devices for broadband wireless communications", *Handbook of Thin Film Devices*, Volume 5: *Ferroelectric Film Devices*, M. H. Francombe, Ed., Academic Press, San Diego.

Parkash, A. Vaid, J. K. and Mansingh, A. (1979). "Measurement of dielectric parameters at microwave frequencies by cavity-perturbation technique", *IEEE*

Transactions on Microwave Theory and Techniques, **27** (9), 791–795.

Pozar, D. M. (1998). *Microwave Engineering*, 2nd edition, John Wiley & Sons, New York.

Sok, J. Y. Park, S. J. Lee, E. H. Hong, J. P. Kwak, J. S. and Kim, C. O. (2000). "Characterization of ferroelectric BaSrTiO3 thin films using a flip-chip technique at microwave frequency ranges", *Japanese Journal of Applied Physics*, **30**, 2752–2755.

Steinhauer, D. E. Vlahacos, C. P. Wellstood, F. C. Anlage, S. M. Canedy, C. Ramesh, R. Stanishevsky, A. and Melngailis, J. (2000). "Quantitative imaging of dielectric permittivity and tunability with a near-field scanning microwave microscope", *Review of Scientific Instruments*, **71** (7), 2751–2758.

Tan, C. Y. Chen, L. F. Chong, K. B. and Ong, C. K. (2004). "Nondestructive microwave permittivity characterization of ferroelectric thin film using microstrip dual resonator", *Review of Scientific Instruments*, **75** (1), 136–140.

Teo, P. T. Jose, K. A. Gan, Y. B. and Varadan, V. K. (2000). "Beam scanning of array using ferroelectric phase shifters", *Electronics Letters*, **36** (19), 1624–1626.

Teo, P. T. Jose, K. A. Wang, Y. J. Lee, C. K. and Varadan, V. K. (2002). "Linear scanning array with bulk ferroelectric-integrated feed network", *IEEE Transactions on Ultrasonics, Ferroelectrics, and Frequency Control*, **49** (5), 558–564.

Tonkin, B. A. and Hosking, M. W. (1997). "Automated microwave measurements of microstrip resonators at low temperatures", *IEEE Transactions on Applied Superconductivity*, **7** (2), 1865–1868.

Van Keuls, F. W. Romanofsky, R. R. Bohman, D. Y. and Miranda, F. A. (1998). "Influence of the biasing scheme on the performance of Au/SrTiO$_3$/LaAlO$_3$ thin film conductor/ferroelectric tunable ring resonators", *Integrated Ferroelectrics*, **22**, 363–372.

Varadan, V. K. Jose, K. A. and Varadan, V. V. (1999). "Design and development of electronically tunable microstrip antennas", *Smart Materials and Structures*, **8** (2), 238–242.

Varadan, V. K. Jose, K. A. Varadan, V. V. Hughes, R. and Kelly, J. F. (1995). "A novel microwave planar phase shifter", *Microwave Journal*, **38** (4), 244–254.

Vendik, O. G. Hollmann, E. K. Kozyrev, A. B. and Prudan, A. M. (1999). "Ferroelectric tuning of planar and bulk microwave devices", *Journal of Superconductivity*, **12** (2), 325–338.

Vendik, O. G. Kollberg, E. Gevorgian, S. S. Kozyrev, A. B. and Soldatenkov, O. I. (1995). "1 GHz tunable resonator on bulk single crystal SrTiO$_3$ plated with Yba$_2$Cu$_3$O$_{7-x}$ films", *Electronics Letters*, **31** (8), 654–656.

10

Microwave Measurement of Chiral Materials

"Chiral" literally means "handed" and comes from the Greek *cheir* for "hand" since our hands are nonsuperimposable on their mirror images. Its optical activity, which includes optical rotatory dispersion and circular dichroism, has been known for almost two centuries. It can be explained by the direct substitution of new constitutive relations, $D = \varepsilon E + \beta \varepsilon \nabla \times E$ and $B = \mu H + \beta \mu \nabla \times H$, $\vec{D} = \varepsilon \vec{E} + i\xi \vec{B}$ and $\vec{H} = i\xi \vec{E} + \vec{B}/\mu$, etc., into Maxwell's equations. Here, ε and μ are the complex dielectric permittivity and magnetic permeability respectively, while β is a complex chirality parameter that results directly from the handedness in the microstructure of the medium. Determination of these three complex constitutive parameters requires three measured complex S-parameters.

10.1 INTRODUCTION

"I call any geometrical figure, or group of points, chiral, and say it has chirality, if its image in a plane mirror, ideally realized, cannot be brought to coincide with itself." This is a celebrated definition of chirality stated by Lord Kelvin in 1904, in his Baltimore Lectures on "Molecular Dynamics and the Wave Theory of Light." Such materials have been known and studied since 1813, when Biot (1812) first explained observations of color changes for white linearly polarized light transmitted through quartz. This phenomenon became known as *optical rotation* or, alternatively, as *optical activity*. Cotton (1895) later showed that chiral

media can transform a linearly polarized wave into an elliptically polarized wave. The phenomenon of the rotation angle or activity changing sign as a function of frequency is later called *Cotton effect*, which can be observed in the region in which optically active absorption bands occur (Velluz *et al.* 1965; Crabbe and Syntex 1965). Since light is an electromagnetic wave, similar phenomena can be constructed at microwave and millimeter-wave frequencies by embedding appropriately sized chiral inclusions such as helices of one-handedness in a host medium. All these effects can be explained by substitution of new sets of constitutive equations, for instance, $\vec{D} = \varepsilon \vec{E} + \beta \varepsilon \nabla \times \vec{E}$ and $\vec{B} = \mu \vec{H} + \beta \mu \nabla \times \vec{H}$, $\vec{D} = \varepsilon \vec{E} + i\xi \vec{B}$ and $\vec{H} = i\xi \vec{E} + \vec{B}/\mu$, etc., into Maxwell's equations. Here, ε and μ are the usual permittivity and permeability, respectively, while β and ξ are the chirality parameters that result from the handedness or lack of inversion symmetry in the microstructure of the medium. It can be shown that left and right circularly polarized fields (LCP and RCP) are the eigenstates of polarization that propagate with a different phase velocity as well as attenuation (Bohren 1974).

Chiral materials have received considerable attention during recent years (Jaggard *et al.* 1979; Mariotte *et al.* 1995; Theron and Cloete 1996; Hui and Edward 1996) and might have a variety of potential applications in the field of microwaves, such as microwave absorbers, microwave antennas, and devices (Varadan *et al.* 1987; Lindell and Sihvola 1995). Lakhtakia *et al.* (1989) have given a fairly complete set of references on the subject.

Microwave Electronics: Measurement and Materials Characterization   L. F. Chen, C. K. Ong, C. P. Neo, V. V. Varadan and V. K. Varadan
© 2004 John Wiley & Sons, Ltd   ISBN: 0-470-84492-2

Bokut and Federov (1960), Jaggard *et al.* (1979), Silverman (1986) and Lakhtakia *et al.* (1986) have studied the reflection and refraction of plane waves at planar interfaces involving chiral media. The possibility of designing broadband antireflection coatings with chiral materials was addressed by Varadan *et al.* (1987). These researchers have shown that the introduction of chirality radically alters the scattering and absorption characteristics. In these papers, the authors have used assumed values of chirality parameter, permittivity, and permeability in their numerical results.

Winkler (1956) and Tinoco and Freeman (1960) have studied the rotation and absorption of electromagnetic waves in dielectric materials containing a distribution of large helices. Direct and quantitative measurements are made possible with the recent advances in microwave components and measurement techniques. Urry and Krivacic (1970) have measured the complex, frequency-dependent values of $(n_L - n_R)$ for suspensions of optically active molecules, where n_L and n_R are the refractive indices for left and right circularly polarized waves. LCP and RCP waves propagate with different velocities and attenuation in a chiral medium. Still, these differential measurements are unable to characterize completely the chiral medium. More recently, Guire *et al.* (1990) have studied experimentally the normal incidence reflection of linearly polarized waves of metal-backed chiral composite samples at microwave frequencies. The beginning of a systematic experiment work came from Umari *et al.* (1991) when they reported measurements of axial ratio, dichroism, and rotation of microwaves transmitted through chiral samples. However, in order to characterize completely the chiral composites, the chirality parameter, permittivity, and permeability have to be determined. The main objective of this chapter is to present the free-space and waveguide technique for complete characterization of chiral composites.

A free-space setup consisting sweep synthesizer, vector network analyzer, waveguides, connectors, and a pair of spot-focusing horn lens antennas has been used for measurements. The free-space through, reflect, line (TRL) calibration technique along with time-domain gating developed by Ghodgaonkar *et al.* (1989, 1990) has been implemented for reducing measurement errors. The antennas provide plane-wave illumination at the planar chiral composites. The transmitted wave becomes an elliptically polarized wave whose major axis is rotated from the direction of linear polarization of the incident wave. Three scattering parameters were obtained to solve for three complex electromagnetic parameters. A procedure using time-domain response has been implemented to remove the ambiguity in the inversion of equations involving exponential terms. Standard materials with known electromagnetic parameters such as Quartz were used to assess the accuracy of the experiments and the inversion algorithm.

Besides the free-space method, a waveguide-measurement method can be used for measuring the electromagnetic parameters of chiral materials. The propagation modes in a waveguide filled with chiral materials are complicated. It has been demonstrated (Hollinger *et al.* 1991; Pelet and Engheta 1990; Engheta and Pelet 1989, 1990) that there are no independent TE, TM, and TEM modes in waveguides filled with chiral materials. The propagation mode is the hybrid mode of the electrical field and the magnetic field. It is not easy to obtain relations between the measurable quantities and the electromagnetic parameters of chiral materials. Hollinger *et al.* (1992) measured the rotation angle and the axial ratio in a circular waveguide (CW). The results show that the high-order modes have little effect on the measurable quantities. As such, only the dominant mode in the CW is considered. The relations between the measurable quantities and the electromagnetic parameters are derived. The rotation angle, the axial ratio, and the complex reflection coefficients of short-circuited and open-circuited samples are measured in order to simultaneously determine the electromagnetic parameters of chiral materials.

10.2 FREE-SPACE METHOD

The free-space setup is shown in Figure 10.1. It consists of a sweep synthesizer, vector network analyzer, waveguides, connectors and a pair of spot-focusing horn lens antennae. The free-space TRL calibration technique along with time-domain gating developed by Ghodgaonkar *et al.* (1989,

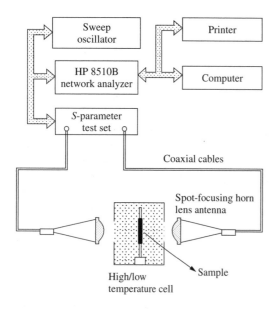

Figure 10.1   Free-space measurement system

the electromagnetic properties of chiral composites. This concentration is given by the percent of volume occupied by the metal inclusions in the 10 cm × 10 cm region that they lie within. The chiral sample consists of two layers, one layer contains only Eccogel, and the other layer contains the random orientation of the dispersed springs. A pure Eccogel sample has also been fabricated to determine its electromagnetic parameters. This is to facilitate the computation of electromagnetic properties of the chiral composite without the effect of the Eccogel layer by using the two-layer inversion method described in the next section. A more comprehensive report on how the samples are fabricated is found in (Guire 1990). A total of nine chiral samples at three concentrations containing left-handed only, right-handed only, and equichiral (racemic) mixtures of helices were prepared and characterized as shown in the following.

1990) has been implemented for reducing measurement errors. The antenna provide plane-wave illumination at the planar chiral composites. In this section, sample preparation, experimental procedure, calibration, computation of ε, μ, and β of the chiral composite samples, accuracy consideration and verification, and experimental results for chiral composites have been discussed.

10.2.1 Sample preparation

The chiral samples are prepared by embedding the required numbers of miniature metal helices in an epoxy host material and are usually about 15 cm (length) × 15 cm (breadth) × 1.2 cm (thick) in dimension. The matrix material, Eccogel 1365–90, which is manufactured by Emerson and Cumings, Inc., Canton, Massachusetts, is homogeneous, isotropic, and achiral. The metal helices have three full turns, and the helix parameters are radius, $a = 0.05842$ cm, pitch, $p = 0.05292$ cm, and are either left- or right-handed. Samples with metal volume concentration varying from 0.8 % (37 springs/cm^3) to 3.2 % (147 springs/cm^3) were fabricated in order to observe the effect of spring concentration on

10.2.2 Experimental procedure

A detailed experimental setup has been described in (Guire 1990; Umari 1991; Ghodgaonkar *et al.* 1990) and a schematic diagram is presented in Figure 10.1. Basically, it consists of a pair of spot-focusing horn lens antennas, a vector network analyzer, mode transitions, precision cables, and a computer. Planar samples of the chiral composite are placed at the common focal plane of the two antennas such that linearly polarized plane waves are normally incident on the samples. The use of spot-focusing lens antennas has minimized the diffraction effects at the edges of the samples. In addition, by implementing the TRL calibration (Ghodgaonkar *et al.* 1990), errors due to multiple reflections between the two antennas and mismatches at the various transitions are corrected. The TRL calibration is implemented in free space by establishing TRL standards. The reference planes for port 1 and port 2 are located at the focal planes of the transmitting and receiving antennas, respectively. The "through" standard is configured by keeping the distance between the two antennas equal to twice the focal distance. The "reflect" standards for port 1 and port 2 are obtained by placing a metal plate at the focal planes of the transmitting and receiving antennas, respectively. The "line" standard is achieved by separating the focal planes of the transmitting and receiving antennas by a distance equal to a quarter of the free-space wavelength at midband.

Moreover, time-domain gating is employed to remove the residual postcalibration errors such as source and load mismatches, etc., which appear as a ripple in the measured scattering parameters. The network analyzer makes measurements in the frequency domain and then computes the inverse Fourier transform to give the time-domain response. Then, gating is applied over the time-domain response, which includes the main reflection (or transmission) response and multiple reflections within the sample. The Fourier transform of the gated time-domain response gives the frequency-domain response, which is an approximate average of the ungated response.

In order to obtain three electromagnetic parameters, one reflection and two transmission measurements are needed. After conducting the free-space TRL calibration, the planar chiral sample is placed in the sample holder for reflection and transmission measurements. The reflected field S_{11co}, which is copolarized and one component of the elliptically polarized transmitted field, S_{21co}, are measured by keeping the transmit and receive antennas in the copolarized position. The other transmission measurement, $S_{21\alpha}$, is obtained by rotating the receiving antenna by an angle α from its copolarized position. The angle α is chosen to be $40°$ clockwise (CW) or $40°$ counterclockwise (CCW) since it leads to best accuracy in measurements for the setup used (Ro 1991).

The rotation angle can be measured more accurately by starting two cross-polarized antennas (Umari *et al.* 1991), which means that the direction of the electric field of the receiving antenna is perpendicular to that of the transmitting antenna, rotating the receiving antenna by $1°$ steps, and measuring the transmitted field at each angle. The direction of the minor axis is determined by a minimum in the transmitted field. This is the minor axis of the transmitted field polarization ellipse. Subsequently, the direction of the major axis can easily be determined by adding $90°$ to this angle.

The direction of the minor axis can be determined with greater accuracy than the direction of the major axis because a slight angle deviation from the direction of the axis will cause a greater relative change in the magnitude of the transmitted field for the minor axis than for the major axis (Umari *et al.* 1991). The measurement accuracy of the rotation angle is $±0.5°$.

The synthesized source is swept in the range 8 to 40 GHz by using mode transitions and different pairs of spot-focusing antennas in the free-space measurement system. Frequency responses of each sample are obtained for four different frequency bands: X-band (8 to 12.4 GHz), Ku-band (12.4 to 18 GHz), K-band (18 to 26.5 GHz), and A-band (26.5 to 40 GHz). The TRL calibration has to be repeated for each frequency band when the antennas are changed. Three amplitude and phase measurements are obtained for each frequency. A total of 400 data points were obtained in the 8 to 40 GHz range, and at each frequency 1000 signals were averaged. The averaging is done over time to decrease the influence of noise. This procedure was repeated for each of the nine samples.

10.2.3 Calibration

Measurement errors exist in any microwave measurement. Whether the measurement system is as simple as a power meter or as complex as a vector network analyzer, measurement ambiguities associated with the system will add uncertainty to the results. Measurement errors in network analysis can be separated into two categories, random and systematic errors. Random errors are nonrepeatable measurement variations due to noise, temperature, and other physical changes in the test setup. System errors include leakage and mismatch signals in the test setup, isolation characteristics between reference and test signal paths, and system frequency response. In most microwave measurements, systematic errors are the most significant sources of measurement uncertainty. However, in a stable measurement environment systematic errors are repeatable and can be measured by the network analyzer.

During measurement calibration, a series of known devices (standards) are connected. The systematic effects are determined as the difference between the measured and the known response of the standards. Once characterized, these errors can be mathematically related by solving a signal flow graph. The twelve-term error model, shown in Figure 10.2, includes all the significant systematic effects for the two-port case (Hewlett–Packard Product Note 8510-8 1987 and Hewlett–Packard Product Note 8510-5A 1988).

In Figure 10.2, directivity is defined as the ratio of power coupled into the coupled arm when the coupler is in the forward direction to the power available in the coupled arm when the coupler is in the reverse direction, using a Z_0 termination that will not reflect any signal. Source match is defined as the vector sum of signals appearing at the system test input because of the inability of the source to maintain constant power at the test device input, as well as adapter and cable mismatches and losses encountered by the reflected signal. Load match results from the imperfect termination of the device output by the test system. The frequency response (reflection or transmission tracking) is the vector sum of all test setup variations in magnitude and phase with frequency, including signal separation devices, test cables, adapter,

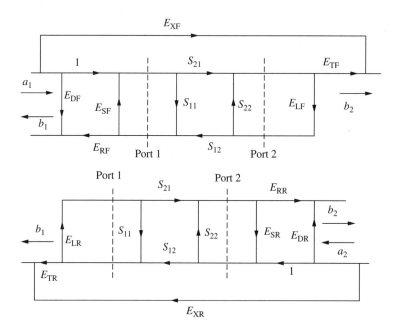

Forward Coefficients		Reverse Coefficients
E_{DF}	Directivity	E_{DR}
E_{SF}	Source Match	E_{SR}
E_{RF}	Reflection Tracking	E_{RR}
E_{XF}	Isolation	E_{XR}
E_{LF}	Load Match	E_{LR}
E_{TF}	Transmission Tracking	E_{TR}

Figure 10.2   Two-port twelve-term error model Source: Ro, Ruyen (1991), *Determination of the electromagnetic properties of chiral composites, using normal incidence measurements*, The Pennsylvania State University, Department of Engineering Science and Mechanics, Ph.D. Thesis

and variations in frequency response between the reference and the test channels. Isolation is due to leakage of energy between the system test and reference channels.

The process of mathematically removing these systematic effects is called *error correction*. Under ideal conditions, with perfectly known standards, systematic effects would be completely characterized and removed. Since this calibration is at the measurement plane, the errors caused by cables, adapters, and fixtures can be greatly reduced. In conventional two-port calibration, three known impedance references, such as open, short, and load (fixed or sliding), and a single transmission standard are required. The accuracy to which these standards are known establishes how well the systematic effects can be characterized.

In non-coaxial measurements, it is more difficult to build impedance standards that are easily characterized. In free space, for example, short circuits can be achieved by using metal plates, open circuits are frequency-dispersive, and it is difficult to build a high-quality purely resistive load. Because of these limitations, an alternative method for calibration in non-coaxial environments is needed that uses simple, realizable standards.

THRU-REFLECT-LINE is a two-port calibration that relies on transmission lines rather than a set of discrete impedance standards. The TRL calibration technique was initially developed by the United States National Bureau of Standards as a precision calibration method for a special-purpose network analyzer. Since that time,

Hewlett–Packard has adapted this method to the HP 8510B network analyzer. Specifically, since the only precision impedance reference required for TRL calibration is a transmission line, obtaining the required set of calibration will not be discussed here. For the interested reader, a detailed description can be found from (Franzen and Speciale 1975; Engen and Hoer 1979; Rytting 1987).

Eight of the twelve error terms are characterized using the basic TRL calibration and are shown in Figure 10.3 (Hewlett–Packard Product Note 8510-8 1987). Although this error model has a slightly different topology than the twelve-term model, the traditional error terms can be simply derived. For example, forward reflection tracking is simply the product of e_{10} and e_{01}. In the model, e_{11} and e_{22} serve as both the source and the load match terms. To solve for these eight unknown error terms, eight linearly independent equations are required. To compute the remaining four error terms, additional measurements are needed.

The basic TRL calibration process is shown in Figure 10.4 (Hewlett–Packard 1987). For the THRU step, the test ports are mated and then transmission frequency response and port match are measured in both directions (four measurements: S_{11T}, S_{21T}, S_{12T}, S_{22T}). For the REFLECT step, the same highly reflective device is connected to each port and its reflection coefficient is measured (two measurements: S_{11R}, S_{22R}). In the LINE step, a short transmission line is inserted and again the frequency response and port match are measured in each direction (S_{11L}, S_{21L}, S_{12L}, S_{22L}).

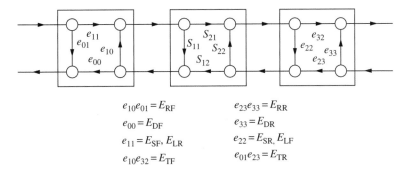

Figure 10.3 Eight-term TRL error model and generalized coefficients. Source: Ro, Ruyen (1991), *Determination of the electromagnetic properties of chiral composites, using normal incidence measurements*, The Pennsylvania State University, Department of Engineering Science and Mechanics, Ph.D. Thesis

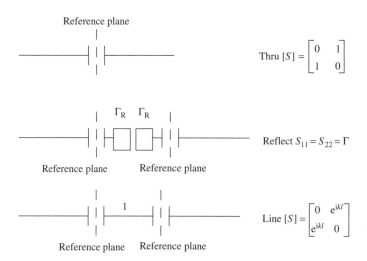

Figure 10.4 TRL procedure and assumed *S*-parameter values for each step. Source: Ro, Ruyen (1991), *Determination of the electromagnetic properties of chiral composites, using normal incidence measurements*, The Pennsylvania State University, Department of Engineering Science and Mechanics, Ph.D. Thesis

At this point, ten measurements have been made resulting in ten equations. However, the basic TRL error model, shown in Figure 10.3, has only eight unknowns. In the TRL solution, the complex reflection coefficient (Γ) of the REFLECT standard and the propagation constant (k) of the LINE are also determined.

Two additional steps are required to complete the calibration. Forward and reverse isolation is measured as the leakage from port 1 to port 2 with each port terminated. Up to this point, the solution for the error model assumes a perfectly balanced test system. The e_{11} and e_{22} terms represent both source and load match. However, in any switching test set, these terms are not equal. Additional correction is provided by measuring the ratio of the incident signals (a_1 and a_2) during the THRU and LINE steps. Once the impedance of the switch is measured, it is used to modify the e_{11} and e_{22} error terms. The term e_{11} is then modified to produce the forward source match (ESF) and the reverse load match (ELR). Now, all the 12 terms of the two-port error model are determined. Then, the actual *S*-parameters can be determined from the measured *S*-parameters and 12 error terms.

Recently, Ghodgaonkar *et al.* (1989, 1990) have implemented the TRL calibration for the free-space measurement system. The basic requirements and

operational procedures for implementing the HP 8510B's TRL calibration method in the free-space measurement system can be considered in four steps as follows: (1) Selecting standards appropriate for the application that meet the basic requirements of the TRL technique, (2) defining these standards for use with HP 8510B by modification of the internal calibration kit registers, (3) performing the calibration, and (4) checking the performance.

For the THRU standard, it can be either a zero-length or a nonzero length line. S_{21T} and S_{12T} are defined to be 1 at 0 degrees and S_{11T} and S_{22T} are defined to be zero for the zero-length THRU standard. For the nonzero length THRU standard, characteristic impedance Z_0 of the THRU and LINE must be the same and insertion phase or electrical length must be specified if the THRU is used to set the reference plane. For the REFLECT standard, it is required that its value be the same on both ports. It may be used to set the reference plane if the phase response of the REFLECT is well known and specified. For the LINE standard, insertion phase of the line must never be the same as that of the THRU (zero or nonzero length) and line length is 1/4 wavelength or 90 degrees relative to the THRU at the center frequency. S_{11L} and S_{22L} of the LINE are also defined to be zero. With this assumption, the system impedance is set to the characteristic impedance of the LINE. If Z_0 is known but not the desired value, the impedance of the LINE can be specified when defining the calibration standards.

For free space, the following expression can be used to select a LINE with 1/4 wavelength line at the center frequency.

Electrical Length (cm) = (LINE-THRU)

$$= \frac{15}{f_1 \text{ (GHz)} + f_2 \text{ (GHz)}},$$

where f_1 and f_2 are the minimum and maximum frequencies of the frequency range respectively. For example, $f_1 = 12.4$ GHz and $f_2 = 18$ GHz for Ku-band and the length of the LINE is equal to 0.49342 cm. Electrical length can be related to physical length when the effective permittivity of the LINE is known.

Once appropriate standards have been selected, they must be defined mathematically and entered into the calibration kit registers of the HP 8510B. Under the CAL menu, there are submenus, MODIFY CAL 1 and MODIFY CAL 2. Either register may be modified to accept user-specified definitions. Default values for the TRL standards may exist, but can be changed simply by entering a new value. Further information on defining calibration standards in the HP 8510B can be found in (Hewlett–Packard Product Note 8510-5A 1988). Although a variety of options and measurement conditions exist, there are two fundamental classes in which TRL calibration technique will be applied for the free-space measurement system. They are: (1) TRL with a zero-length THRU, reference plane set by THRU, and (2) TRL with a zero-length THRU, reference plane set by REFLECT. Under the MODIFY CAL KIT menu, there is TRL OPTION menu. In this menu, SET REF allows either the THRU or the REFLECT to set the location of the reference plane. In addition, CAL Z_0 relative to the impedance of the LINE standards is chosen for the free-space measurement system. In general, the default standard numbers for the THRU, REFLECT, and LINE classes are 15, 14, and 16 respectively. Each of the standards must be defined in terms of the coefficients required by

the Standard Definition Table. It can be achieved as follows. In DEFINE STANDARD menu, input number 14 (15 or 16), selecting the corresponding standard type REFLECT (THRU or LINE), and then selecting SPECIFY OFFSET. One can then define the REFLECT (THRU or LINE) standard in the SPECIFY OFFSET menu.

THRU is specified to have an offset delay of 0 ps and operates over the 12.4 to 18 GHz frequency range (Ku band for example). REFLECT is specified to have an offset delay of 0 ps and operates over the same frequency range. LINE is 0.49342 cm for Ku-band. The approximate offset delay for Ku-band is then 0.49342 cm/(2.9979 × 10^{10} cm/s) ≈ 16.45 ps. If the ratio of the frequency span to the start frequency is less than 8, one THRU/LINE pairs will be sufficient. For each frequency band, LINE operates in the same frequency range as THRU.

Once the standards are fully specified, the free-space TRL calibration can be performed. To begin the calibration process, select the CAL menu. Press TRL TWO-PORT. The TRL submenu will be displayed. The THRU, S11REFLECT, S22REFLECT, ISOLATION, and LINE calibrations can be performed in any convenient order. For the free-space measurement system, the reference planes for port 1 and port 2 are located at the focal planes of the transmitting and receiving antennas, respectively. The THRU standard is configured by keeping the distance between the two antennas equal to twice the focal distance. The LINE standard is achieved by separating the focal plane of the transmit and receive antennas by a distance equal to a quarter of free-space wavelength at the center of the band. The REFLECT standards for port 1 and port 2 are obtained by placing a metal plate at the focal planes of the transmitting and receiving antenna respectively. To measure the system cross talk, ISOLATION is measured with each port terminated. When systematic cross talk is sufficiently below the levels that are to be measured, it does not have to be characterized. Press ISOLATION, OMIT ISOLATION, and ISOLATION DONE.

Once a calibration has been generated, its performance should be checked before making device measurements. To check the accuracy that can be obtained using the new calibration kit, a device with a well-defined frequency response (preferable unlike any of the standard used) should be measured. When these devices are measured, the difference between the displayed results and the known values indicate the level of measurement accuracy. In free space, unfortunately, these verification devices do not exist. As a result, it will be difficult to state the absolute accuracy of such measurements.

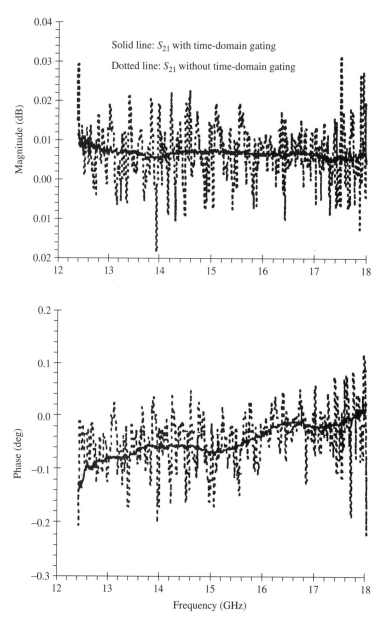

Figure 10.5   The measured transmission coefficients of the THRU configuration after performing the free-space TRL calibration, where the reference is set by THRU. Source: Ro, Ruyen (1991), *Determination of the electromagnetic properties of chiral composites, using normal incidence measurements*, The Pennsylvania State University, Department of Engineering Science and Mechanics, Ph.D. Thesis

However, there are some performance checks that can be made to provide some insight into the measurement integrity. Repeatability is the single largest factor that will limit the effectiveness of the calibration. The following figures plot the measured reflection and transmission coefficients of the THRU and REFLECT standards in the frequency range of 12.4 to 18 GHz after performing the free-space calibration, where the reference plane is set either by the THRU or REFLECT.

Figures 10.5 and 10.6 show the respective transmission and reflection coefficients of the THRU

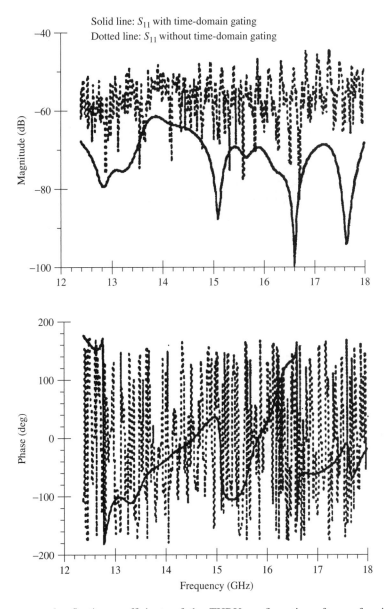

Figure 10.6 The measured reflection coefficients of the THRU configuration after performing the free-space TRL calibration, where the reference is set by THRU. Source: Ro, Ruyen (1991), *Determination of the electromagnetic properties of chiral composites, using normal incidence measurements*, The Pennsylvania State University, Department of Engineering Science and Mechanics, Ph.D. Thesis

standards after performing the free-space TRL calibration, where the reference plane is set by THRU.
The solid and dotted lines in these figures are the
measured quantities with and without respectively
the time-domain gating being applied. The detailed
description of the time-domain measurement will
be discussed in the next section. The vertical axes
of the upper and lower figures are the magnitude

in dB and phase in degree, respectively. The horizontal axes are the frequency range in GHz. Without time-domain gating, the absolute values of the
magnitude and phase of the transmission coefficient shown in Figure 10.5, are less than 0.03 dB
and 0.2°, respectively. After applying the time-
domain gating, the absolute values of the magnitude and phase of the transmission coefficients are

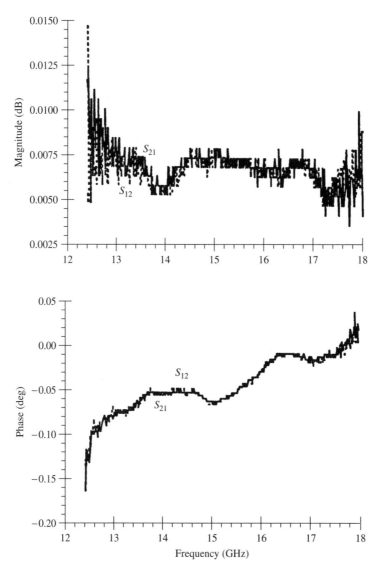

Figure 10.7 The measured forward and reverse transmission coefficients of the THRU configuration after
performing the free-space TRL calibration, where the reference is set by THRU. Source: Ro, Ruyen (1991),
Determination of the electromagnetic properties of chiral composites, using normal incidence measurements, The
Pennsylvania State University, Department of Engineering Science and Mechanics, Ph.D. Thesis

less than 0.015 dB and 0.1°, respectively. Since the THRU standard is set to be perfect, the difference between the measured transmission coefficients and the actual transmission coefficients (0 dB and 0°) is the error for the repeatability of the THRU standard. It can be seen in Figure 10.6 that the magnitudes of the reflection coefficients with and without time-domain gating are less

than −45 dB and −60 dB, respectively. However, the actual reflection of the THRU standard is assumed to be −∞ dB. Hence, one can say that the dynamic range of the reflection coefficients is around −60 dB after applying time-domain gating.

Figures 10.7 and 10.8 show the comparison between the forward and reverse transmission and reflection coefficients of the THRU

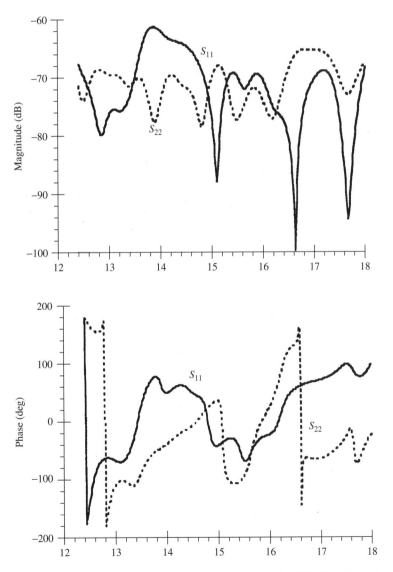

Figure 10.8 The measured forward and reverse reflection coefficients of the THRU configuration after performing the free-space TRL calibration, where the reference is set by THRU. Source: Ro, Ruyen (1991), *Determination of the electromagnetic properties of chiral composites, using normal incidence measurements*, The Pennsylvania State University, Department of Engineering Science and Mechanics, Ph.D. Thesis

standard, respectively. In Figure 10.7, the difference between forward transmission coefficients (S_{21}) and reverse transmission coefficients (S_{12}) is less than $-70\,$dB in magnitude and $0.03°$ in phase. Comparing with the error of the transmission coefficients, one can say S_{12} is equal to S_{21}, as it should according to the concept of reciprocity. It can be seen in Figure 10.8 that the difference between the magnitudes of the forward reflection coefficients (S_{11}) and reverse reflection coefficients (S_{22}) is less than $-65\,$dB, which is less than the value of the dynamic range of the reflection coefficients. Hence, it is possible to say that S_{11} is equal to S_{22} and their values are approximately zero. Under this circumstance, the phases of the reflection coefficients are unimportant.

Figures 10.9 and 10.10 show the respective reflection and transmission coefficients of the

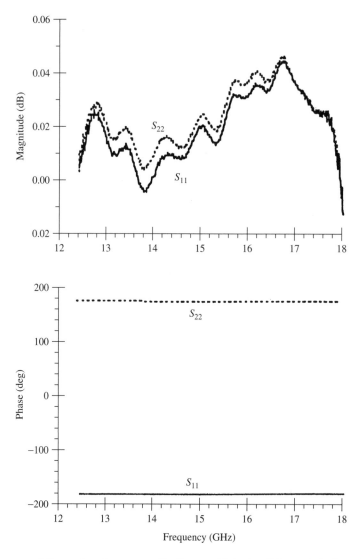

Figure 10.9   The measured forward and reverse reflection coefficients of the metal plate after performing the free-space TRL calibration, where the reference is set by THRU. Source: Ro, Ruyen (1991), *Determination of the electromagnetic properties of chiral composites, using normal incidence measurements*, The Pennsylvania State University, Department of Engineering Science and Mechanics, Ph.D. Thesis

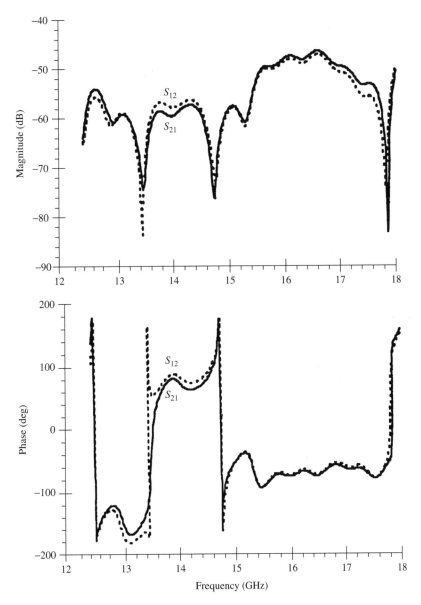

Figure 10.10   The measured forward and reverse transmission coefficients of the metal plate after performing the free-space TRL calibration, where the reference is set by THRU. Source: Ro, Ruyen (1991), *Determination of the electromagnetic properties of chiral composites, using normal incidence measurements*, The Pennsylvania State University, Department of Engineering Science and Mechanics, Ph.D. Thesis

metal plate after performing the free-space TRL calibration, where the reference plane is set by THRU. It can be observed in Figure 10.9 that the difference between the magnitudes of S_{11} and S_{22} is less than -63 dB and their absolute values are less than 0.05 dB; the difference between the phase of S_{11} and S_{22} is less than $3°$ and their values range from -178.5 to $-180°$ and from 178.5 to $180°$ for S_{11} and S_{22} respectively. Hence, one can say S_{11} is equal to S_{22}, which implies that the measurement system is well balanced and symmetric. Since the metal plate is a perfect conductor and the reference plane is set by THRU, the difference between the measured reflection coefficients and the actual reflection coefficients (0 dB and $\pm 180°$) is the measurement error of the metal plate. It can be seen in Figure 10.10 that the difference between the magnitudes of S_{12} and S_{21} is less than -55 dB and their values are less than -50 dB. The phases of S_{12} and S_{21} are not important under this circumstance. However, the actual transmission

coefficients of the metal plate equal to $-\infty$ dB. Hence, it is possible to say that the dynamic range of the transmission coefficients is around -50 dB.

Figures 10.11 and 10.12 plot the measured reflection coefficients of the metal plate and the transmission coefficients of the THRU standard, respectively after performing the free-space TRL calibration, where the reference plane is set by REFLECT. It can be seen in Figure 10.11 that the difference between S_{11} and S_{22} is less than

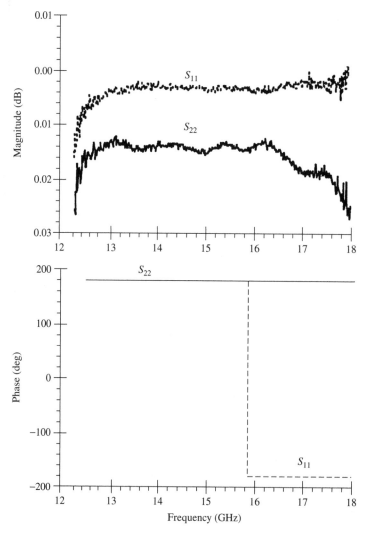

Figure 10.11 The measured reflection coefficients of the metal plate after performing the free-space TRL calibration, where the reference is set by REFLECT. Source: Ro, Ruyen (1991), *Determination of the electromagnetic properties of chiral composites, using normal incidence measurements*, The Pennsylvania State University, Department of Engineering Science and Mechanics, Ph.D. Thesis

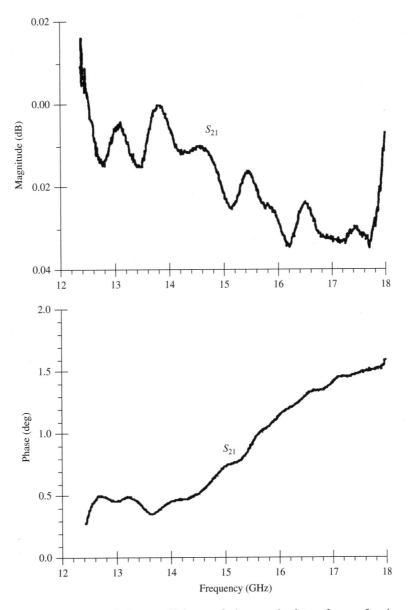

Figure 10.12 The measured transmission coefficients of the metal plate after performing the free-space TRL calibration, where the reference is set by REFLECT. Source: Ro, Ruyen (1991), *Determination of the electromagnetic properties of chiral composites, using normal incidence measurements*, The Pennsylvania State University, Department of Engineering Science and Mechanics, Ph.D. Thesis

-59 dB in magnitude and $\pm 2°$ in phase and the absolute values of the phases are from 179.2 to 180°. Since the REFLECT standard is chosen to define the reference plane, the difference between the measured reflection coefficients and the actual reflection coefficients (0 dB and $\pm 180°$) is the error for the repeatability of the REFLECT standard. It can be observed in Figure 10.12 that the magnitudes of the transmission coefficients of the THRU standard are less than 0.04 dB; the absolute values of the phases are less than 2°. Since the actual transmission coefficients of the

THRU standard are assumed to be 0 dB and 0°, the difference between the measured transmission coefficients and the actual transmission coefficients (0 dB and 0°) is the error of the transmission coefficients of the THRU standard.

Ghodgaonkar *et al.* (1989, 1990) have implemented the TRL calibration for the free-space measurement system. They reported that after the free-space TRL calibration and time-domain gating, the measurement errors in the reflection and transmission coefficients are mainly due to

(a) residual postcalibration errors resulting from imperfections in the calibration standards, the instrumentation and coaxial cables connecting the spot-focusing antennas to the test ports of the test set. In TRL calibration, the only error caused by the calibration standards is due to the wave impedance of the line standard (if it is different from the wave impedance in free space). This impedance error will cause normalization of reflection and transmission coefficients to the impedance of the line standard. The residual source and load mismatch errors due to imperfections in the calibration standards are minimized by the time-domain gating;

(b) small changes in the reference planes (as defined by THRU or REFLECT standard) between calibration and measurement due to small changes in the position of plates holding the samples.

Because of the combined effect of the sample holder and the residual postcalibration errors, the magnitude and phase errors in S_{11} are less than ± 0.055 dB and $\pm 2°$ respectively. And the amplitude and phase errors in S_{21} are less than ± 0.035 dB and $\pm 2°$ respectively.

10.2.4 Time-domain measurement

The relationship between the frequency-domain response and time-domain response of a network is described by the Fourier transform, and the response of a device may be completely specified in either domain. The network analyzer makes measurements in the frequency domain and then computes the inverse Fourier transform to give the time-domain response. This computation technique benefits from the wide dynamic range and the error correction of the frequency domain data. Detailed information on the time-domain measurement can be found in (Rytting 1985).

In the time domain, the horizontal axis represents the propagation delay through the device. In other words, this is the amount of time it takes for an impulse to travel through the device. In transmission measurements, the value displayed is the actual one-way travel time of the impulse; while for the reflection measurements, the horizontal axis shows the two-way travel time of the impulse. The electrical length is obtained by multiplying the time value by the speed of light in free space. To find the actual physical length of the device, multiply the electrical length by the relative propagation velocity of the transmission medium. This concept has been proposed in this study when dealing with the ambiguity of the inverse problem. The peak value of the time-domain response represents an average reflection or transmission coefficients for the reflection or transmission coefficients over the frequency range of the measurement.

There are two time domain modes available: Time Band Pass and Time Low Pass modes. The Time Band Pass mode is used for bandpass devices (devices that do not operate down to dc), and it allows any frequency-domain response to be transformed to the time domain. The Time Low Pass mode is used to make time-domain measurements on low-pass devices (devices that operate down to or near DC), and it simulates the traditional Time Domain Reflectometer (TDR) measurement, which gives the response of the device to a step or impulse stimulus. The Time Band Pass mode is selected in this work for the time-domain measurements.

The HP 8510B time domain has a feature called *Windowing* that is designed to enhance the time-domain measurements. Owing to the limited bandwidth and the measurement system, the transformation to the time domain gives the time-domain response of the device to the $\sin(x)/x$ stimulus. This nonideal impulse is the time-domain-resolving stimulus, which has two effects that limit the usefulness of the time-domain response. The finite impulse width caused by the band-limited nature of the test system limits the ability to resolve between two closely spaced responses. The impulse side lobes, caused by the abrupt cutoff at the stop frequency, limit the dynamic range of the time-domain measurement by hiding low-level response within the side lobes of adjacent higher-level responses.

For time bandpass measurements, the frequency-domain response has two cutoff points, at f_{start} and at f_{stop}. Therefore, in the time bandpass mode, the windowing function rolls off both the low end and the higher end of the frequency-domain response.

There are three windows available: Maximum, Normal, and Minimum. Selecting the Minimum window minimizes the filtering applied to the frequency domain data; Normal selects a general-purpose value for the roll-off; and Maximum applies maximum filtering to the frequency-domain data. The effect of windowing on the time- bandpass impulse response is a reduction in side lobes with the trade-off of increased impulse width. In this study, the Minimum window is always selected.

In order to use the time domain effectively, one must be aware of its limitations and ambiguities. Since the measured frequency domain is not continuous, it is sampled at discrete frequency points with the spacing of Δf Hz and each time-domain response is repeated every $1/\Delta f$ seconds. This length of time defines the range of the measurement. Time-domain response resolution is defined as the ability to resolve two closely spaced responses.

Response resolution depends upon the time domain mode, the frequency range, whether it is a reflection or a transmission measurement, and the relative propagation velocity of the signal path. The response resolution for the time bandpass using the minimum window can be expressed as follows:

$$\text{Response resolution} = 1.2/f_{\text{span}}$$

where f_{span} carries the units of Hz. For Ku-band with the frequency range 5.6 GHz, the response resolution is approximately equal to 214 and 107 ps for transmission and reflection measurements

respectively. It will be difficult or impossible to distinguish between equal magnitude responses separated by less than 214 ps for transmission measurements.

Time-domain range response is the ability to locate a single response in time. When only one response is present, this is how closely one can locate the peak of the response and thus the location of the discontinuity. Range resolution is related to the digital resolution of the time domain display, which uses the same number of points as the frequency domain. Therefore, the range resolution can be computed directly from the time span and the number of points selected. In this study, the time span is always 6 ns and the number of points is 401. Therefore, the marker can read the location of the response with range resolution of 15 ps (6 ns/400). In the time domain, one can center a response on the display and then zoom in on the response by simply narrowing the time span. This improves the range resolution by a factor equal to the reduction in the time span. Although the network analyzer transformation algorithm allows one to make the range resolution as small as one wishes, this does not imply that the actual physical location of the response may be measured with arbitrary accuracy. First, the nonuniform propagation velocity in the typical transmission medium will limit the ability to precisely locate the physical location of the discontinuity. In dispersive media such as waveguides, the nonlinear phase response will cause the time-domain impulse to be smeared in time and thus limit the ability to locate the actual peak of the response.

In the HP 8510B, the time-domain response can be obtained from the frequency-domain response as follows. Pressing DOMAIN presents the Domain menu, selecting TIME BAND PASS, pressing SPECIFY TIME, and then selecting IMPULSE and MINIMUM WINDOW. In the time domain, the STIMULUS hardkeys are redefined to the time value of the horizontal axis, and have no effect on the frequency-domain value. Figure 10.13 plots the time-domain responses of the transmission coefficient of the THRU standard and the reflection coefficient of the metal plate after performing the free-space TRL calibration, where the reference plane is set by THRU. The vertical

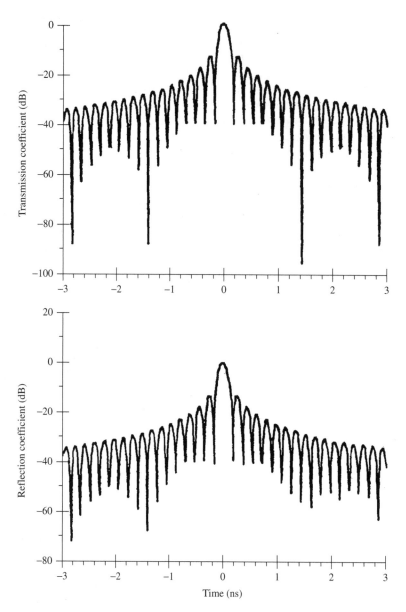

Figure 10.13 The measured time-domain responses of the transmission coefficients of the THRU standard (upper) and the reflection coefficients of the metal plate (lower). Source: Ro, Ruyen (1991), *Determination of the electromagnetic properties of chiral composites, using normal incidence measurements*, The Pennsylvania State University, Department of Engineering Science and Mechanics, Ph.D. Thesis

axes are the magnitude in dB and the horizontal axes represent the time span in nanosecond. It can be observed in Figure 10.13 that the propagation delay time for either THRU configuration or the metal plate is 0 ns and the peak value is around 0 dB. Since the reference plane is set by THRU, the propagation delay time for the THRU configuration should be 0 ns. The propagation delay time of the metal plate being 0 ns assures accuracy in the location of the reference plane. Figure 10.14 plots the time-domain responses of the transmission coefficients of Eccogel and Quartz plates.

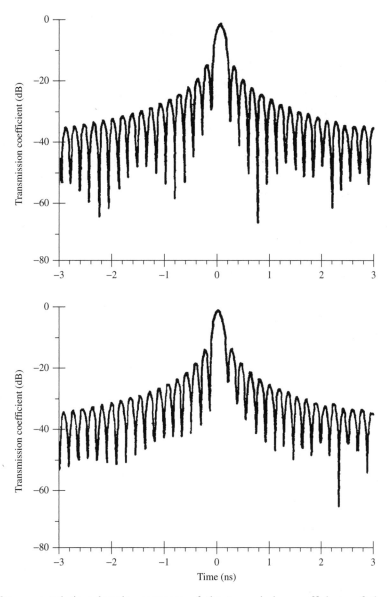

Figure 10.14 The measured time-domain responses of the transmission coefficients of the Eccogel (upper) and Quartz plate (lower). Source: Ro, Ruyen (1991), *Determination of the electromagnetic properties of chiral composites, using normal incidence measurements*, The Pennsylvania State University, Department of Engineering Science and Mechanics, Ph.D. Thesis

The thickness of the Eccogel and Quartz plate are 1.18364 and 0.65532 cm respectively. The propagation delay for the Eccogel and Quartz plates are 65 and 45 ps respectively. Hence, the approximate propagation velocities for the Eccogel and Quartz are around 1.8×10^8 m/s and 1.5×10^8 m/s respectively. Then it is possible to determine the

precise value of the real part of wave number from these data. Further discussion of the uniqueness of the solution will be illustrated when we deal with materials properties.

The network analyzer has a very powerful feature called *gating* that provides the flexibility to selectively remove unwanted reflection or transmission time-domain responses. In converting back to the frequency domain, the effects of the responses outside the gate are removed from the measurement. A gate is a time filter that is used to filter out specific time-domain responses. There are four gate shapes available: MAXIMUM, WIDE, NORMAL, and MINIMUM. Each of the four gate shapes has a different filter characteristic. The minimum gate has the least ripple and highest side-lobe levels, and the maximum gate has the largest ripples and lowest side-lobe levels. In this study, the maximum gate shape is always selected.

However, the gating operation can introduce errors in the measurement, which, depending upon the size and separation of the responses and how the gate is set up, can be relatively small or restrictively large. The first source of gating errors to consider is an offset gate. This can produce distortion of the gated response. In general, the distortion caused by edge effects for the maximum gate shape may be produced in both the upper and the lower 10 % of the band. The second source of gating error is the incomplete removal of adjacent responses. This is most significant when the responses that are to be removed are large in magnitude and close in time to those responses one wants to retain. In order to avoid the errors induced by incorrectly positioning the gate, it is important to center the gate at the peak of the time-domain response. Secondly, one must ensure that gate span is wide enough to include all responses of interest. Using the widest gate span possible will reduce the effect of the offset gate. The third gating guideline is to use the widest gate shape possible. This will minimize the effects of the offset gate as well as reduce effects of gate passband flatness and side lobes.

In the HP 8510B, the gate can be set up as follows. Pressing DOMAIN, selecting SPECIFY GATE, GATE SHAPE, and MAXIMUM. Then, selecting GATE CENTER by pressing MARKER, selecting MARKER to MAXIMUM, and then pressing DOMAIN, selecting CENTER, and pressing the MARKER key. The GATE SPAN is chosen as wide as possible to include all of the responses of interest.

Figure 10.15 plots the reflection coefficients (upper) and transmission coefficients (lower) of the Eccogel plate with and without the time-domain gating being applied. It can be seen in Figure 10.15 that ripples appear in the reflection and transmission coefficients, which result from the residual postcalibration errors, without applying time-domain gating. After applying time-domain gating, the gated frequency-domain responses are approximately the average of the ungated responses as shown in Figure 10.15. In this study, the gated reflection and transmission coefficients are used as the reflection and transmission characteristics of the measured material while determining material properties.

10.2.5 Computation of ε, μ, and β of the chiral composite samples

Out of several forms of the constitutive equations for chiral media (Sihvola and Lindell 1991), the set of constitutive equations, $\vec{D} = \varepsilon\vec{E} + \beta\varepsilon\nabla \times \vec{E}$ and $\vec{B} = \mu\vec{H} + \beta\mu\nabla \times \vec{H}$, is chosen. The coefficient β (in meters) is the chirality parameter, which is equal to zero for achiral or nonchiral as well as equichiral media. Following (Bohren 1974), the electromagnetic fields in achiral medium are expressed as a combination of the LCP, $\vec{Q}_L$, and RCP, $\vec{Q}_R$, waves by $\vec{E} = \vec{Q}_L + a_R\vec{Q}_R$ and $\vec{H} = a_L\vec{Q}_L + \vec{Q}_R$, where $a_R = -i\sqrt{\mu/\varepsilon}$ and $a_L = -i\sqrt{\varepsilon/\mu}$. These fields are found to propagate with the wave numbers $k_L = k/(1 - k\beta)$; $k_R = k/(1 + k\beta)$; $k = \omega\sqrt{\mu\varepsilon}$. In a chiral medium, the LCP and RCP fields propagate with different complex propagation constants. Since these constants must be positive, its denominator cannot become negative and hence $|k\beta|\langle 1$. This condition implies that the real and imaginary parts of the impedance are positive. Otherwise, the medium will become an active medium. This results in an upper bound on the chirality parameter, which is a function of frequency and other material properties. In many cases, β is expected to decrease with frequency.

For the case of a normal incident plane wave, the field representations for the planar two-layer

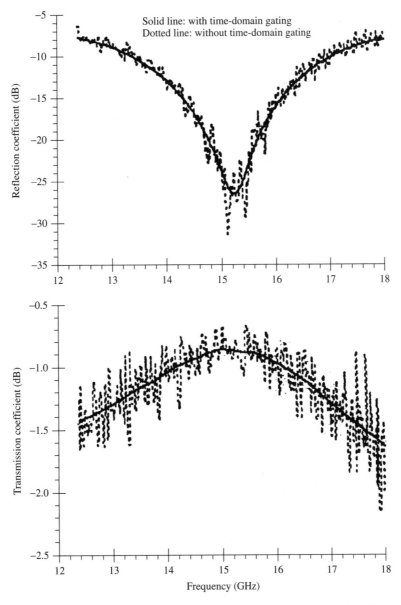

Figure 10.15 The measured reflection (upper) and transmission (lower) coefficients of the Eccogel plate with and without the time-domain gating being applied. Source: Ro, Ruyen (1991), *Determination of the electromagnetic properties of chiral composites, using normal incidence measurements*, The Pennsylvania State University, Department of Engineering Science and Mechanics, Ph.D. Thesis

chiral sample can be illustrated by Figure 10.16. Figure 10.16 is divided into four regions. Two of them, region 0 ($z \leq 0$) and region 3 ($z \geq d_2$), are free spaces that do not exhibit chirality. Region 2, the eccogel layer ($d_1 \leq z \leq d_2$), is also achiral material. The last one, region 1 ($0 \leq z \leq d_1$), is a chiral medium with electromagnetic properties, ε, μ and β. In the chiral medium, the LCP and RCP

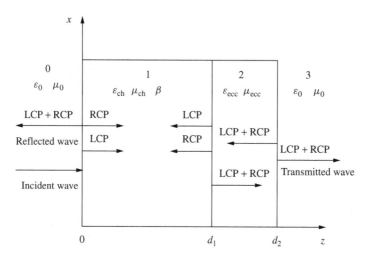

Figure 10.16 Field representations for the normally incident plane wave. Source: Ro, Ruyen (1991), *Determination of the electromagnetic properties of chiral composites, using normal incidence measurements*, The Pennsylvania State University, Department of Engineering Science and Mechanics, Ph.D. Thesis

waves propagate with different wave numbers, while for the achiral media, these wave numbers are identical.

In region 0, the normal plane wave is given by

$$\vec{E}_{inc} = (A_E \hat{e}_y + A_H \hat{e}_x)e^{ik_0 z} \qquad (10.1)$$

$$\vec{H}_{inc} = \frac{1}{\eta_0}(A_H \hat{e}_y - A_E \hat{e}_x)e^{ik_0 z} \qquad (10.2)$$

In the above expressions, $\eta_0 = k_0/\omega\mu_0$ and k_0 are the free-space impedance and wave number respectively. The coefficients $A_E \neq 0$, $A_H = 0$ refer to a TE-polarized incident plane wave, whereas $A_E = 0$, $A_H \neq 0$ denote an incident TM-polarized field. The coefficients $A_H = iA_E$ and $A_H = -iA_E$ represent LCP and RCP incident plane waves respectively.

In the chiral region, the field has to be expressed in terms of positive- and negative-traveling LCP and RCP plane waves. The representations of the fields in the chiral medium are given by

$$\vec{Q}_L = A_1(\hat{e}_y - i\hat{e}_x)e^{ik_L z} + A_2(\hat{e}_y + i\hat{e}_x)e^{-ik_L z} \qquad (10.3)$$

$$\vec{Q}_R = C_1(\hat{e}_y + i\hat{e}_x)e^{ik_R z} + C_2(\hat{e}_y - i\hat{e}_x)e^{-ik_R z} \qquad (10.4)$$

where A_1, A_2, C_1 and C_2 are the unknown coefficients to be determined. From (Bohren 1974), one

obtains the electric and magnetic fields in the chiral medium as

$$\vec{E}_{ch} = \vec{Q}_L + a_R \vec{Q}_R \qquad (10.5)$$

$$\vec{H}_{ch} = a_L \vec{Q}_L + \vec{Q}_R \qquad (10.6)$$

In the Eccogel (achiral) layer, the electric and magnetic fields are expressed in terms of circularly polarized waves as

$$\vec{E}_{ecc} = ((A_3 + C_3)\hat{e}_y + i(C_3 - A_3)\hat{e}_x)e^{ik_{ecc} z}$$
$$+ ((A_4 + C_4)\hat{e}_y + i(A_4 - C_4)\hat{e}_x)e^{-ik_{ecc} z} \qquad (10.7)$$

$$\vec{H}_{ecc} = \frac{1}{\eta_{ecc}}[((C_3 - A_3)\hat{e}_y - i(A_3 + C_3)\hat{e}_x)e^{ik_{ecc} z}$$
$$+ ((A_4 - C_4)\hat{e}_y + i(A_4 + C_4)\hat{e}_x)e^{-ik_{ecc} z}] \qquad (10.8)$$

in which η_{ecc} and k_{ecc} are the impedance and wave number of Eccogel respectively, and A_3, A_4, C_3 and C_4 are the unknown coefficients to be determined. The transmitted field in region 3 (free space), propagating only in the positive z-direction, is given by

$$\vec{E}_{tra} = ((T_L + T_R)\hat{e}_y + i(T_R - T_L)\hat{e}_x)e^{ik_0 z} \qquad (10.9)$$

$$\vec{H}_{tra} = \frac{1}{\eta_0}[((T_L - T_R)\hat{e}_y - i(T_L + T_R)\hat{e}_x)e^{ik_0 z}] \qquad (10.10)$$

The reflected fields in region 0 (free space) traveling along the negative z-direction has the form

$$\vec{E}_{\text{ref}} = ((R_{\text{L}} + R_{\text{R}})\hat{e}_y + i(R_{\text{L}} - R_{\text{R}})\hat{e}_x)e^{-ik_0 z} \tag{10.11}$$

$$\vec{H}_{\text{ref}} = \frac{1}{\eta_0}[((R_{\text{L}} - R_{\text{R}})\hat{e}_y + i(R_{\text{L}} + R_{\text{R}})\hat{e}_x)e^{-ik_0 z}] \tag{10.12}$$

In the above expressions, T_{L}, T_{R}, R_{L}, and R_{R} are the coefficients that need to be determined as solutions to the boundary value problem. The boundary conditions are:

At $z = 0$,

$$\hat{e}_z \times \lfloor \vec{E}_{\text{inc}} + \vec{E}_{\text{ref}} - \vec{E}_{\text{ch}} \rfloor = 0 \tag{10.13}$$

$$\hat{e}_z \times \lfloor \vec{H}_{\text{inc}} + \vec{H}_{\text{ref}} - \vec{H}_{\text{ch}} \rfloor = 0 \tag{10.14}$$

At $z = d_1$,

$$\hat{e}_z \times \lfloor \vec{E}_{\text{ecc}} - \vec{E}_{\text{ch}} \rfloor = 0 \tag{10.15}$$

$$\hat{e}_z \times \lfloor \vec{H}_{\text{ecc}} - \vec{H}_{\text{ch}} \rfloor = 0 \tag{10.16}$$

At $z = d_2$,

$$\hat{e}_z \times \lfloor \vec{E}_{\text{ecc}} - \vec{E}_{\text{tra}} \rfloor = 0 \tag{10.17}$$

$$\hat{e}_z \times \lfloor \vec{H}_{\text{ecc}} - \vec{H}_{\text{tra}} \rfloor = 0 \tag{10.18}$$

In Eqs. (10.13)–(10.18), we obtain a system of 12 equations for the unknowns A_j, C_j, ($j = 1$–4), $T_{\text{L}}, T_{\text{R}}, R_{\text{L}}$ and R_{R}. The unknowns can be obtained numerically. It can be shown that for a normally incident LCP plane wave, which implies that only waves having the polarization $\hat{e}_y - i\hat{e}_x$ can propagate in either the positive or negative z-direction, $A_2, A_4, C_1, C_3, T_{\text{R}}$, and R_{L} equal zero. Similarly, it can be shown that for a normally incident RCP plane wave, which implies that only waves having the polarization $\hat{e}_y + i\hat{e}_x$ can propagate in either the positive or negative z-direction, $A_1, A_3, C_2, C_4, T_{\text{L}}$ and R_{R} equal zero. The field representations for normally incident, circularly polarized waves are shown in Figure 10.17. There is no mode conversion, and LCP and RCP waves will propagate in each medium without interfering with each other. Therefore, the impedance method can be applied for a normally incident LCP or RCP wave.

According to the impedance method, the scalar field in the ith layer is given by

$$\Phi_i(z) = A_i^- e^{-k_i^- z} + A_i^+ e^{k_i^+ z} \tag{10.19}$$

where the $\pm$ superscripts denote a wave traveling in the positive and negative z-direction respectively. The wave numbers k_i^+ and k_i^- are equal to k_{L} or k_{R} according to

$$k_i^+ = k_{\text{L}} \text{ and } k_i^- = k_{\text{R}} \text{ (LCP incident)}$$

$$k_i^+ = k_{\text{R}} \text{ and } k_i^- = k_{\text{L}} \text{ (RCP incident)} \tag{10.20}$$

with $k_i^+ = k_i^- = k_i = \omega\sqrt{\mu_i \varepsilon_i}$ for the achiral region.

The generalized reflection coefficient for any medium is defined as the ratio of the back propagation field to the forward propagating field. Thus,

$$\Gamma_i(z) = \frac{A_i^-}{A_i^+} e^{-i(k_i^- + k_i^+)z}. \tag{10.21}$$

Substituting Eq. (10.21) into (10.19) gives

$$\Phi_i(z) = A_i^+ e^{k_i^+ z}(1 + \Gamma_i(z)) \tag{10.22}$$

The total field impedance is defined as

$$Z_i(z) = \eta_i \frac{1 + \Gamma_i(z)}{1 - \Gamma_i(z)} \tag{10.23}$$

where $\eta_i = \sqrt{\mu_i/\varepsilon_i}$ is the intrinsic impedance of each medium. In Eq. (10.23),

$$\Gamma_i(z) = \frac{Z_i(z) - \eta_i}{Z_i(z) + \eta_i} \tag{10.24}$$

For the first layer, the incident wave amplitude A_1^+ is known. For the last layer, region 3 is a free space, which extends to infinity; there is no back propagating wave in this medium and that implies $\Gamma_3 = 0$.

The reflection function at a point z' if the reflection function at a point z is known is given by

$$\Gamma_i(z') = \Gamma_i(z)e^{-i(k_i^- + k_i^+)(z' - z)} \tag{10.25}$$

Thus, the generalized reflection coefficient $\Gamma_i(z)$ of each layer can be obtained layer by layer from the last layer. Then, the reflection coefficient A_1^-/A_1^+ and the transmission coefficient, A_3^+/A_1^+

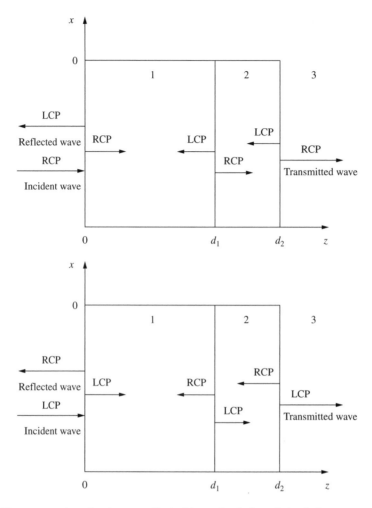

Figure 10.17   Field representations for the normally incident, circularly polarized plane wave. Source: Ro, Ruyen (1991), *Determination of the electromagnetic properties of chiral composites, using normal incidence measurements*, The Pennsylvania State University, Department of Engineering Science and Mechanics, Ph.D. Thesis

can be obtained. For normally incident, circularly polarized waves, one obtains (Ro 1991)

$$S_{11L} = S_{11R} = \frac{A + ABP + BCF + PCF}{1 + PB + ABCF + PACF} \tag{10.26}$$

$$S_{21L} = \frac{(1 + \Gamma_2(0))(1 + B)(1 + A)C^{1/2}}{(1 + PB + ABCF + PACF)e^{-ik_2 d_2}} \tag{10.27}$$

$$S_{21R} = S_{21L}e^{i(k_R - k_L)d} \tag{10.28}$$

where S_{11L} and S_{21L} (S_{11R} and S_{21R}) are reflection and transmission coefficients for LCP (RCP) waves respectively, and

$$A = \frac{\eta - \eta_0}{\eta + \eta_0} \tag{10.29}$$

$$B = \frac{\eta_2 - \eta}{\eta_2 + \eta} \tag{10.30}$$

$$C = e^{2ik_L d_1} \tag{10.31}$$

$$\Gamma_2(0) = \frac{\eta_0 - \eta_2}{\eta_0 + \eta_2} \tag{10.32}$$

$$P = \Gamma_2(0)e^{2ik_2 d_2} \tag{10.33}$$

$$F = e^{i(k_R - k_L)d_1} \tag{10.34}$$

where $\eta(=\sqrt{\mu/\varepsilon})$ is the impedance of the chiral composite, k_2 and η_2 are the respective wave number and the impedance of Eccogel, d_2 and d_1 are the thicknesses of Eccogel and chiral layer respectively.

For a normally incident, linearly polarized wave, one can obtain the reflection and transmission characteristics by solving the 12×12 matrix by Gaussian–Jordan elimination method. For normally incident, circularly polarized waves, the reflection and transmission characteristics are obtained from Eqs. (10.26)–(10.28). As the linearly polarized waves can be mathematically decomposed into coherent LCP and RCP waves of the same magnitude, the reflection and transmission characteristics of the different polarization of these waves are given by

$$S_{11\text{co}} = \frac{1}{2}(S_{11L} + S_{11R}) = S_{11L} = S_{11R} \quad (10.35)$$

$$S_{21R} = S_{21\text{co}} + iS_{21\text{cross}} \quad (10.36)$$

$$S_{21L} = S_{21\text{co}} - iS_{21\text{cross}} \quad (10.37)$$

$$S_{21\text{cross}} = \frac{S_{21\theta} - S_{21\text{co}} \cos\theta}{\sin\theta} \quad (10.38)$$

where $S_{11\text{co}}$ and $S_{21\text{co}}$ are the measured reflection and transmission fields with two copolarized antennas. $S_{21\text{cross}}$ and $S_{21\theta}$ are the measured transmission fields that are obtained by rotating the receiving antenna 90° and θ from its copolarized position respectively. It can be seen that reflection coefficients for LCP and RCP waves have the same values, meaning that the resultant reflected wave is a linearly polarized wave and is copolarized with the incident wave. On the other hand, the transmission coefficients for RCP and LCP waves are different. The resultant transmitted wave of these two circularly polarized waves is an elliptically polarized wave and its polarization direction is rotated from that of the incident wave. It can also be observed from Eq. (10.28) that the difference between k_R and k_L, which results from the chirality parameter, causes circular birefringence and circular dichroism between LCP and RCP waves. Without chirality, transmission coefficients for LCP and RCP waves are identical and the resultant transmission field is linearly polarized, as it should be for an achiral medium.

Once $S_{11\text{co}}$, $S_{21\text{co}}$ and $S_{21\theta}$ are obtained from experiments for a normally incident, linearly polarized wave, the reflection and transmission coefficients for a normally incident, circularly polarized wave can be computed from Eqs. (10.35)–(10.38). The electromagnetic properties of the chiral material then may be determined from these quantities. From Eqs. (10.29) and (10.30), one gets

$$B = \frac{(1+A)\eta_0 + (A-1)\eta_2}{(A-1)\eta_2 - (A+1)\eta_0} \quad (10.39)$$

Substituting Eq. (10.39) into Eq. (10.26) gives

$$C = \frac{(A - S_{11L})((A-1)\eta_2(P+1) + (1+A)\eta_0(P-1))}{F(AS_{11L} - 1)((A-1)\eta_2(P+1) + (1+A)\eta_0(1-P))} \quad (10.40)$$

Substituting Eqs. (10.39) and (10.40) into Eq. (10.36) and after some algebraic manipulations, the following quadratic equation is obtained.

$$aA^2 + bA + a = 0 \quad (10.41)$$

and A can be obtained as

$$A = \frac{-b \pm \sqrt{b^2 - 4a^2}}{2a} \quad (10.42)$$

where

$$a = S_{21L}^2 F(\eta_2^2(1+P)^2 - \eta_0^2(1-P)^2) \\ - 4S_{11L}\eta_2^2(1+\Gamma_2(0))^2 e^{2ik_2 d_2} \quad (10.43)$$

$$b = -2(\eta_2^2(1+P)^2 + \eta_0^2(1-P)^2)S_{21L}^2 F \\ + 4(S_{11L}^2 + 1)\eta_2^2(1+\Gamma_2(0))^2 e^{2ik_2 d_2} \quad (10.44)$$

The appropriate sign in Eq. (10.44) is chosen so that $|A| \leq 1$, as it should be from the definition given in Eq. (10.29). The impedance of the chiral material is then obtained as

$$\eta = \frac{A+1}{1-A}\eta_0 \quad (10.45)$$

and C can be computed from Eq. (10.40). Then the wave numbers for LCP and RCP waves, k_L

and k_R, can be calculated from C and F as shown in Eqs.(10.31) and (10.34) as follows:

$$2ik_L d_1 + i2n\pi = \ln(C) \tag{10.46}$$

$$i(k_R - k_L)d_1 \pm i2m\pi = \ln(F) \tag{10.47}$$

where m and $n = 0, 1, 2, \ldots$

Since the actual phase velocity cannot be found from experiment, the real parts of k_L and k_R, which relate to the respective wave speeds for LCP and RCP waves, have infinite numbers of solutions as shown in Eqs. (10.46) and (10.47). The approximate phase velocity concept is proposed in this study in order to properly choose the appropriate values for m and n. Once k_L, k_R, and η are determined uniquely, the macroscopic constitutive parameters, the permittivity ε, the permeability μ and the chirality parameter β can be calculated from the following equations:

$$k = \frac{2}{\dfrac{1}{k_R} + \dfrac{1}{k_L}} \tag{10.48}$$

$$\beta = \frac{1}{2}\left(\frac{1}{k_R} - \frac{1}{k_L}\right) \tag{10.49}$$

$$\mu = \frac{k\eta}{w} \tag{10.50}$$

$$\varepsilon = \frac{k}{w\eta} \tag{10.51}$$

10.2.6 Experimental results for chiral composites

First, the accuracy of the technique is established by obtaining the properties of pure Eccogel and quartz samples. The properties of pure Eccogel are required for the computation of electromagnetic parameters of chiral composite samples. Since the transmitted field is a linearly polarized wave, $S_{21cross}$ is zero for both quartz and pure Eccogel. The algorithm developed earlier is still applicable to nonchiral materials by setting $d_2 = 0$, $k_2 = k_0$, and $\eta_2 = \eta_0$. The electromagnetic properties of Eccogel and Quartz are obtained from S_{21co} and S_{11co}. Figure 10.18 is a plot of the relative permittivity and the relative permeability for quartz and

Eccogel. The properties were computed from S_{21co} and S_{11co}. Ghodgaonkar *et al.* (1989) have reported that the accuracy for the dielectric constant was better than $\pm 4\%$. For low-loss dielectric samples, the maximum error in dielectric loss tangent is around ± 0.06. Because the actual loss tangents of Teflon and glass are less than ± 0.06, the dielectric loss tangent cannot be measured accurately. It can be seen in Figure 10.2 that both the real parts of permittivity and permeability have accuracy within $\pm 2\%$. Both the imaginary parts of the relative permittivity and permeability cannot be measured accurately since both quartz and Eccogel are low-loss materials.

Figure 10.19 shows both the real and imaginary parts of the chirality parameter β of quartz and Eccogel. The chirality parameter β shown in Figure 10.19 was calculated using additional transmission measurement in the polarization direction $40°$ CW. It can be seen in Figure 10.19 that both the real and imaginary parts of the chirality parameter β are of the order of 10^{-6} or smaller in magnitude.

Since quartz and Eccogel are achiral media, both the real and imaginary parts of the chirality parameter should be zero. The measured chirality parameter for both Quartz and Eccogel is indicative of the error in the measurement of the chirality parameter. Both the permittivity and permeability can also be obtained from one reflection and two transmission measurements. When compared with the data shown in Figure 10.19, the difference between the real parts of the permittivity and permeability was found to be less than 1%, but the difference between the imaginary parts of the permittivity and permeability is 8%. The bigger difference in the imaginary parts of the permittivity and permeability is due to the fact that the loss tangent is less than 0.06, and the loss factor cannot be measured accurately.

The measurements are one reflection and two transmission measurements for a normally incident, linearly polarized plane wave. They are used for characterizing chiral composites. As the samples have local inhomogeneities, which are resulted from preparation error, the required S-parameters are measured at five different positions and averaged. This is done by rotating the sample in the

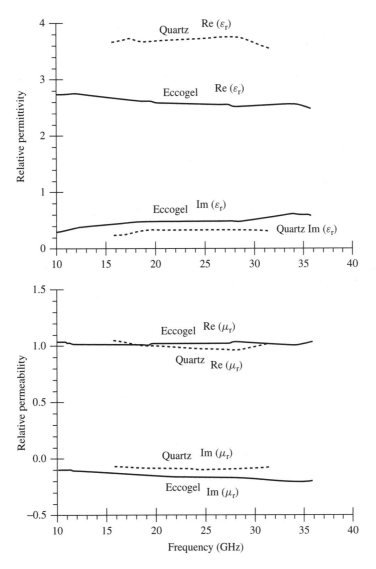

Figure 10.18   Electromagnetic properties (ε_r, μ_r) of Eccogel and quartz. Source: Ro, Ruyen (1991), *Determination of the electromagnetic properties of chiral composites, using normal incidence measurements*, The Pennsylvania State University, Department of Engineering Science and Mechanics, Ph.D. Thesis

sample holder or by moving it so that the spot focus of the antenna illuminates different positions of each sample in the five measurements.

It is known that for achiral materials, the imaginary part of k must be greater than or equal to zero such that the wave traveling through the medium will not gain any energy. The only constraint is $|k\beta| \leq 1$ for chiral materials. The following results show $|k\beta| \leq 1$ is of the order

10^{-1} or smaller in magnitude and is hence within the constraint.

Figures 10.20–10.22 are the plots of the electromagnetic properties ε, μ, and β of the left-handed chiral composite samples versus frequency. The properties shown in these figures were computed from one transmission measurement $S_{21\theta}$ at the polarization direction $40°$ CW, in addition to one reflection measurement S_{11co} and

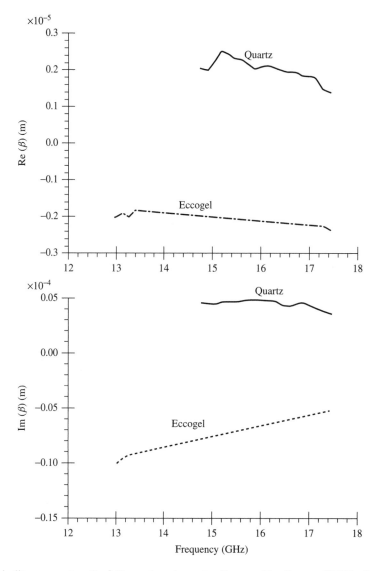

Figure 10.19   Chirality parameter β of Eccogel and quartz. Source: Ro, Ruyen (1991), *Determination of the electromagnetic properties of chiral composites, using normal incidence measurements*, The Pennsylvania State University, Department of Engineering Science and Mechanics, Ph.D. Thesis

one transmission measurement S_{21co}. The relative permittivities of the left-handed chiral samples are shown in Figure 10.20. It can be seen in Figure 10.20 that all the curves of the real parts of permittivities are concave. The frequency at which the minimum is observed for each concentration varies as the concentration changes. In general, the minimum for the lower-concentration sample is at a higher frequency than that for the

higher-concentration sample. It is also observed that the real part of the permittivity decreases as the concentration increases. The imaginary parts of the permittivities for all three samples have positive values at lower frequency and negative values at higher frequency. The frequency at which the imaginary part of the permittivity is equal to zero decreases as the concentration increases.

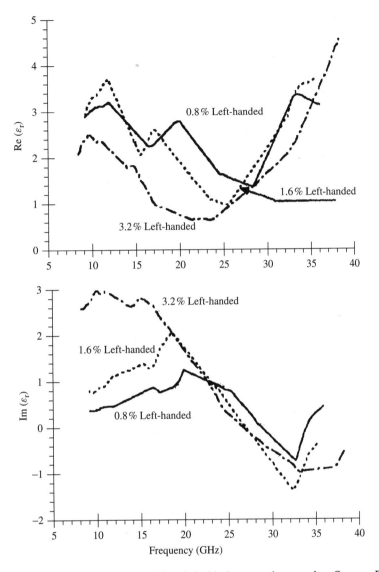

Figure 10.20 Relative permittivity ε_r of the left-handed chiral composite samples. Source: Ro, Ruyen (1991), *Determination of the electromagnetic properties of chiral composites, using normal incidence measurements*, The Pennsylvania State University, Department of Engineering Science and Mechanics, Ph.D. Thesis

Figure 10.21 shows the relative permeability of the left-handed chiral composite samples versus frequency. It can be seen that the real parts of the relative permeability have the same trend for all the three samples. Each curve of the real part of the relative permeability is convex and has a positive peak, whose value increases as the concentration increases. The peak observed for the lower-concentration sample is at higher frequency than that for the higher-concentration sample. It also can be observed from Figure 10.21 that each curve of the imaginary part of the relative permeability exhibits anomalous dispersion, first decreasing, then increasing, and then decreasing with frequency. It has the value zero at the lower frequency, then decreases to reach a negative maximum and drastically moves to a maximum peak, followed by a drop to zero at the higher

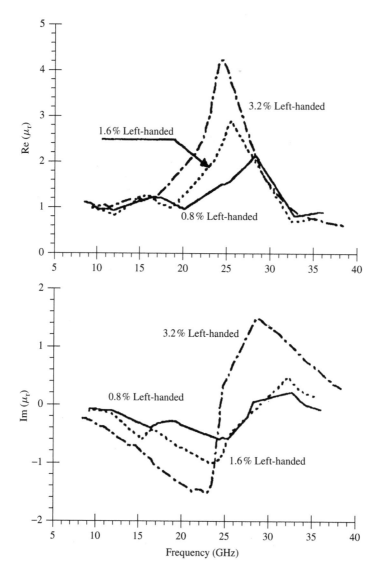

Figure 10.21   Relative permeability μ_r of the left-handed chiral composite samples. Source: Ro, Ruyen (1991), *Determination of the electromagnetic properties of chiral composites, using normal incidence measurements*, The Pennsylvania State University, Department of Engineering Science and Mechanics, Ph.D. Thesis

frequencies. It can be seen that both the negative and positive maxima occur at lower frequencies for the higher concentration than for the lower-concentration sample.

As shown in Figures 10.20 and 10.21, the curves of the real parts of the relative permittivity and permeability are concave and convex respectively. The curves of the imaginary parts of the relative permittivity and permeability are *S*-shaped, and their absolute values increase as the concentration increases in the frequency range of 8 to 40 GHz. These results show that the wavelength for either the LCP or the RCP wave in the chiral medium decreases as the concentration increases at the same frequency. With the wavelength inversely proportional to the frequency, one can expect that the cotton effect phenomenon, which occurred at some particular value of L/λ (Tinoco and Freeman

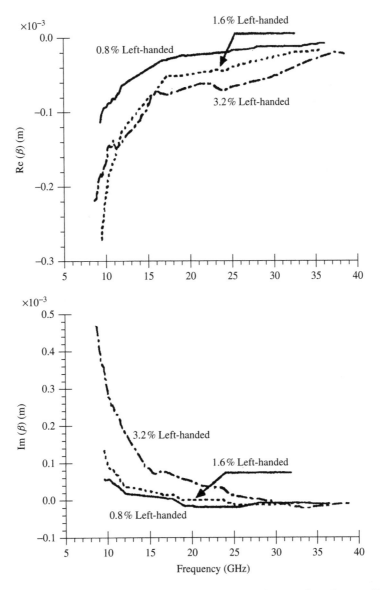

Figure 10.22   Chiral parameter β of the left-handed chiral composite samples. Source: Ro, Ruyen (1991), *Determination of the electromagnetic properties of chiral composites, using normal incidence measurements*, The Pennsylvania State University, Department of Engineering Science and Mechanics, Ph.D. Thesis

1960), will be observed at a lower frequency for a higher-concentration sample as compared to that for the lower-concentration sample.

Figure 10.22 plots the complex chirality parameter of the left-handed chiral samples. It can be seen that the real parts of the chirality parameters for all the three concentration samples have negative values in the whole frequency range.

In general, the absolute value of the real part of the chirality parameter increases as the concentration increases. The real part of the chirality parameter appears to approach to a negative maximum at lower frequency. As the maximum observed varies as the concentration changes, it resulted in the absolute value of the real part of the chirality parameter for the 3.2 % sample is somewhat smaller than the 1.6% sample at lower frequencies. The imaginary part of the chirality parameters for all the three concentration samples have the same trend, and their values are positive at lower frequencies and become negative at higher frequencies. In general, the absolute value of the imaginary part of the chirality parameter increases as the concentration increases.

The wave numbers for the LCP and RCP waves k_L and k_R in all three left-handed chiral composite samples can be calculated. It is seen that the real part of the wave number k_L increases as frequency increases over the whole frequency range. It also

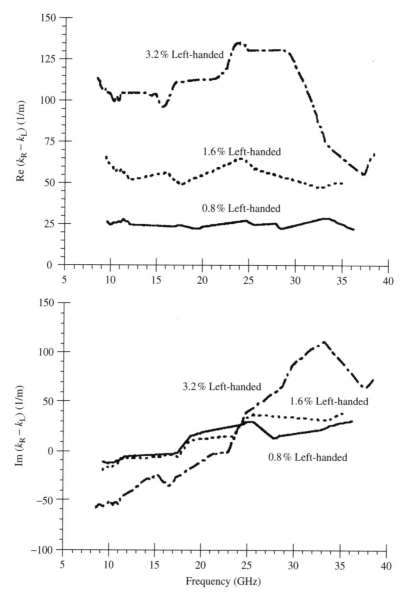

Figure 10.23 The difference between wave numbers k_R and k_L in the left-handed chiral samples. Source: Ro, Ruyen (1991), *Determination of the electromagnetic properties of chiral composites, using normal incidence measurements*, The Pennsylvania State University, Department of Engineering Science and Mechanics, Ph.D. Thesis

increases as the volume fraction of the chiral inclusion increases. Since the wave number is inversely proportional to the wave speed, it can be deduced that electromagnetic waves propagating through the chiral sample become slower as the concentration increases. The imaginary part of the wave number k_L increases as the concentration increases over the whole frequency range, which implies that LCP wave traveling through the left-handed chiral samples is absorbed more as

concentration increases. Similarly, the real part of the wave number k_R increases as frequency increases over the whole frequency range; the imaginary part of the wave number k_R increases as the concentration increases. The difference between the wave number k_L and k_R in the left-handed samples are plotted in Figure 10.23. The real part of $(k_L - k_R)$ has a positive value and increases as the concentration increases over the whole frequency range. This implies that LCP

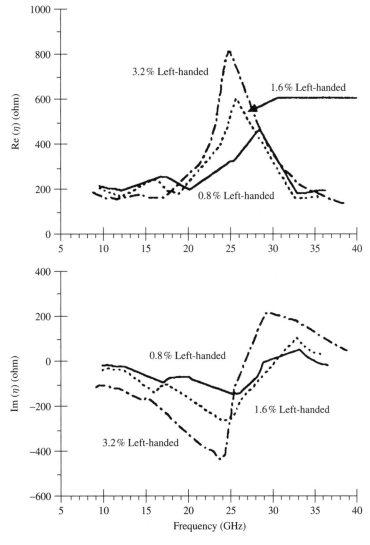

Figure 10.24 Impedance η of the left-handed chiral samples. Source: Ro, Ruyen (1991), *Determination of the electromagnetic properties of chiral composites, using normal incidence measurements*, The Pennsylvania State University, Department of Engineering Science and Mechanics, Ph.D. Thesis

waves propagate faster through the left-handed chiral samples than RCP waves. The imaginary part of $(k_L - k_R)$ increases as the concentration increases in the frequency range of 8 to 40 GHz. Since rotation angle and ellipticity are related to the real and imaginary parts of $(k_L - k_R)$ shown in Figure 10.23, they have the same characteristics.

Figure 10.24 presents the impedance η of the left-handed chiral samples versus frequency. The impedance η has the same characteristics as the relative permeability μ. For example, the real part of the impedance and permeability have a positive peak, which increases as the concentration increases; the imaginary part of the impedance and

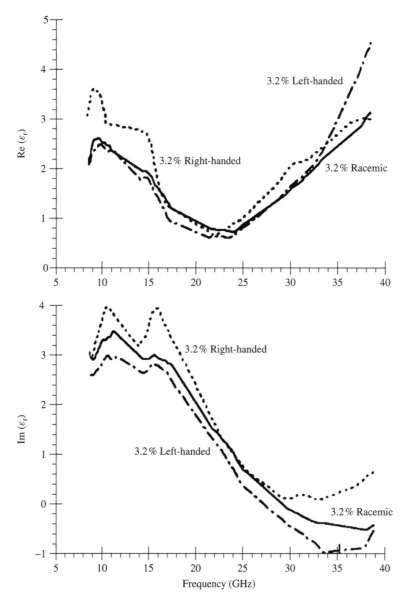

Figure 10.25 Relative permittivity ε_r of the 3.2% chiral composite samples. Source: Ro, Ruyen (1991), *Determination of the electromagnetic properties of chiral composites, using normal incidence measurements*, The Pennsylvania State University, Department of Engineering Science and Mechanics, Ph.D. Thesis

permeability have a negative value at lower frequencies and a positive value at higher frequencies. This means that left-handed chiral samples behave like capacitive and inductive media at lower and higher frequencies, respectively, in the frequency range of 8 to 40 GHz.

Figures 10.25–10.27 show the electromagnetic properties ε, μ, and β of the 3.2 % chiral composite samples. These plots allow us to compare ε, μ, and β at one concentration between left-handed, right-handed and racemic samples. It can be seen in Figures 10.25 and 10.26 that all the curves of the real part of the relative permittivity and permeability are concave and convex respectively for all samples. The curves of the imaginary parts of the relative permittivity and permeability are

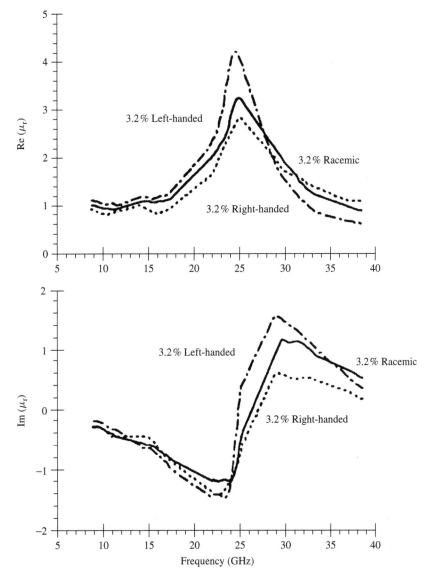

Figure 10.26   Relative permeability μ_r of the 3.2 % chiral composite samples. Source: Ro, Ruyen (1991), *Determination of the electromagnetic properties of chiral composites, using normal incidence measurements*, The Pennsylvania State University, Department of Engineering Science and Mechanics, Ph.D. Thesis

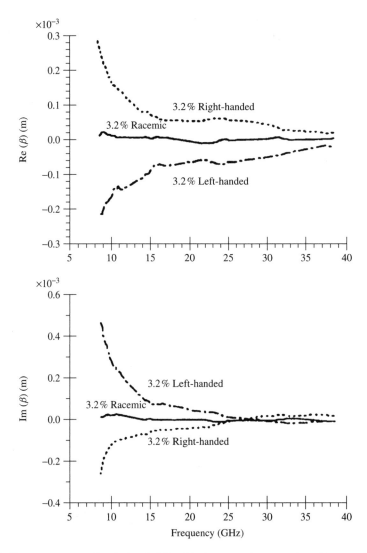

Figure 10.27 Chirality parameters β of the 3.2% chiral composite samples. Source: Ro, Ruyen (1991), *Determination of the electromagnetic properties of chiral composites, using normal incidence measurements*, The Pennsylvania State University, Department of Engineering Science and Mechanics, Ph.D. Thesis

S-shaped. Both the real and imaginary parts of the relative permittivity for all samples have almost the same value except at the lower and higher frequencies. At lower frequencies (8 to 16 GHz), there are two thickness resonance frequencies observed for the Eccogel plate with a thickness of 1.2 cm. The two peaks in the imaginary part of the relative permittivity of the 3.2% right-handed sample at lower frequencies shown in Figure 10.25 looks almost like the resonance effect

of Eccogel, which may be due to heterogeneity of the 3.2% right-handed sample. The discrepancy of the imaginary part among all samples at higher frequencies may be due to variations in sample preparation, which become more apparent at higher frequencies.

In Figure 10.26, the frequencies at which the maxima of the real parts of the relative permeability for all samples occurred are almost the same. The difference of the imaginary part of the relative permeability among all samples at higher frequencies may be due to the sample variation as explained above. From Figures 10.25 and 10.26, one may conclude that both the real and imaginary parts of the relative permittivity and permeability for all three-handed samples are almost the same in the frequency range of 8 to 40 GHz.

Figure 10.27 shows the real and imaginary parts of the chirality parameter of left-, right-handed and racemic samples. The left- and right-handed samples have almost the same values of β but are

of opposite sign. Similarly, the imaginary parts of the left- and right-handed samples change sign at the same frequency. On the other hand, racemic sample has a value of β nearly equal to zero for both real and imaginary parts and is of the order of 10^{-5}, which compares with the value of 10^{-6} for quartz, which is also achiral.

From Figures 10.25–10.27, one can conclude that the electric permittivity and magnetic permeability of the same concentration samples have almost the same values regardless of their

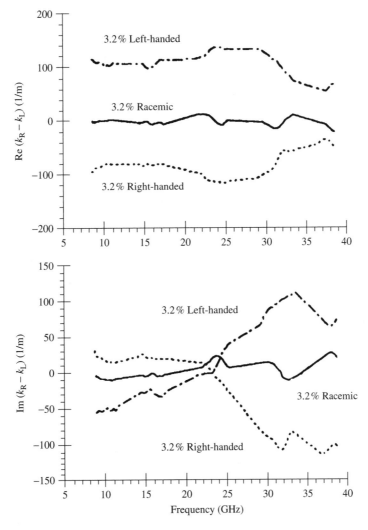

Figure 10.28 The difference between wave number k_R and k_L in the 3.2 % chiral samples. Source: Ro, Ruyen (1991), *Determination of the electromagnetic properties of chiral composites, using normal incidence measurements,* The Pennsylvania State University, Department of Engineering Science and Mechanics, Ph.D. Thesis

handedness, while both the real and imaginary parts of the chirality parameter are zero for the racemic sample and have the same values but with different signs for the left- and right-handed samples.

The difference between the wave numbers k_L and k_R in the 3.2 % chiral samples are plotted in Figure 10.28. The real part of $(k_L - k_R)$ has a positive value in the 3.2 % left-handed chiral sample and a negative value in the 3.2 % right-handed chiral sample, which implies that LCP waves traveling through a left-handed medium are faster than RCP waves, and vice versa. The imaginary part of $(k_L - k_R)$ in the 3.2 % left-handed chiral sample has a negative value at lower frequencies and a positive value at higher frequencies. For the right-handed chiral sample, the imaginary part of $(k_L - k_R)$ has a positive value at lower frequencies and a negative value at higher frequencies. Both the real and imaginary parts of $(k_L - k_R)$ are approximately equal to zero over the whole frequency range. The rotation and ellipticity have the same characteristics as the real and imaginary parts of $(k_L - k_R)$ respectively.

The impedance η of the 3.2 % chiral samples versus frequency has the same characteristics as the relative permeability μ. For example, the frequencies at which the maxima of the real parts of the relative permeability for all the handed samples occurred are almost the same; the imaginary part of the impedance has a negative value at lower frequencies and a positive value at higher frequencies. This means that all the handed samples behave like capacitive and inductive media at lower and higher frequencies respectively in the frequency range of 8 to 40 GHz.

10.3 WAVEGUIDE METHOD

Waveguide measurement method can be used for measuring the electromagnetic parameters of chiral materials. It has been demonstrated (Hollinger *et al.* 1991; Pelet and Engheta 1990; Engheta and Pelet 1989, 1990) that there are no independent TE, TM, and TEM modes in waveguides filled with chiral materials. Hollinger *et al.* (1992) measured the rotation angle and the axial ratio in a circular waveguide. The results show that the high-order

modes have little effect on the measurable quantities. As such, only the dominant mode in the CW is considered. The relations between the measurable quantities and the electromagnetic parameters are derived. The rotation angle, the axial ratio, and the complex reflection coefficients of short- and open-circuited samples are measured in order to determine the electromagnetic parameters of chiral materials.

10.3.1 Sample preparation

The chiral material samples are prepared by randomly embedding copper helices in low-loss paraffin wax that has the electromagnetic parameters $\varepsilon = 2.2 + 0i$ and $\mu = 1.0 + 0i$. Samples are right-handed, left-handed, and racemic containing an equal mix of right- and left-handed helices. The helices have a diameter of 0.521 mm, a pitch of 0.500 mm, a gauge diameter of 0.110 mm and three turns. These helices are distributed homogeneously and randomly in the paraffin wax. All the chiral samples have the same thickness of 4 mm and the same diameter of 25 mm. However, they have different concentrations of helices and handedness. The volume fractions of the helices are 0.4, 0.8, and 1.6 %, which are the percentage volumes occupied by the metal wires that constitute the helices (Liu *et al.* 1999).

10.3.2 Experimental procedure

The main part of the experimental setup used is shown in Figure 10.29. The mode launched in this setup is linearly polarized. The dominant mode in the empty rectangular waveguide (RW) is the TE_{10} mode; its polarization is parallel to the smaller side of the RW. In the empty CW, the dominant mode is the TE_{11} mode. The electrical field intensity is the highest near the center of the CW and the mode possesses an approximate plane-wave characteristic. Usually, only the TE_{11} mode is considered in the CW filled with nonchiral material or in the empty CW. It has been demonstrated (Hollinger *et al.* 1991) that the $HE_{\pm 11}$ mode is the dominant mode in the CW filled with chiral materials and that the transverse electrical field pattern of the mode is similar to that of the

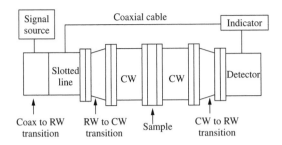

Figure 10.29   The main part of the experimental setup. Source: Liu, Z.L. Sun, G.C. Huang, Q.L. and Yao, K.L. (1999), "A circular waveguide method for measuring the electromagnetic parameters of chiral materials at microwave frequencies," *Meas. Sci. Technol.* **10**, 374–379, IOP Publishing Ltd, by permission of IOP Publishing Ltd

TE$_{ll}$ mode, especially near the center of the CW. Thus, in order to simplify the measurement and treatment of measured data, the HE$_{\pm11}$ mode is approximately treated as the TE$_{ll}$ mode.

The distribution of the electrical field in the longitudinal direction in the waveguide is measured by the slotted line, and the complex reflection coefficients are then calculated. The RW-to-CW transition and the detector can be rotated around the axis of the CW. Thus, the rotation angle and the axial ratio of the transmission field can be measured when electromagnetic waves travel through chiral samples.

Since the chiral composite samples will rotate the direction of the electrical field, the component that is parallel to the longer side of the RW will be reflected back to the sample. This will lead to multiple reflections and cause errors in the measurement. In the RW-to-CW transition and the CW-to-RW transition, there is a metal slice that is parallel to the longer side of the RW. The end facing the CW is cut to a taper in order to obtain impedance matching. The electrical field component perpendicular to the slice will experience little attenuation, while the component parallel to the slice will be highly attenuated. Hence, the slice can reduce the influence of multiple reflections.

Before measuring the complex reflection coefficients, a calibration is conducted by putting a metal plate in place of the chiral sample to act as the short circuit. Before measuring the rotation

angle and the axial ratio, a calibration is performed by connecting the two CWs directly to obtain the transmission standard.

10.3.3 Computation of ε, μ, and ξ of the chiral composite samples

For an electromagnetic wave in a nonchiral material in free space, the propagation constant k and wave impedance Z can be expressed as

$$k = w(\varepsilon\mu)^{1/2} \qquad (10.52)$$

$$Z = (\mu/\varepsilon)^{1/2} \qquad (10.53)$$

For an electromagnetic wave in a CW filled with a nonchiral material, the propagation constant k' and wave impedance Z' are

$$k' = [w^2 - k_c^2/\varepsilon\mu]^{1/2} \qquad (10.54)$$

$$Z' = w\mu/k' \qquad (10.55)$$

where $k_c = 1.841/R$ is a constant corresponding to the TE$_{ll}$ mode, 1.841 is the first root of the first-order Bessel function of the first kind and R is the radius of the CW. The waveguide angular frequency is defined as $w' = 2\pi c/\lambda_w$, where c is the velocity of light in vacuum and λ_w is the wavelength in the waveguide. The waveguide angular frequency can be expressed as

$$w' = [w^2 - k_c^2/\varepsilon\mu]^{1/2} \qquad (10.56)$$

Substituting Eq. (10.56) into Eq. (10.54) gives

$$k' = w'(\varepsilon\mu)^{1/2} \qquad (10.57)$$

which has the same form as the propagation constant for nonchiral materials in free space (shown in Eq. (10.52)). The wave impedance Z' in the CW filled with a nonchiral material is given by

$$Z' = w\mu/k' = Zw/w' \qquad (10.58)$$

For an electromagnetic wave in a chiral material in free space, there are only two eigenmodes of propagation. One is the right-circular polarization wave, the other is the left-circular polarization wave. The wave numbers of the RCP and LCP waves are expressed as k_+ and k_- (Jaggard *et al.* 1990).

$$k_{\pm} = \pm w\mu\xi + w(\varepsilon\mu + \mu^2\xi^2)^{1/2} \qquad (10.59)$$

The intrinsic impedance of chiral materials is

$$Z_c = [\mu/(\varepsilon + \mu\xi^2)]^{1/2} \qquad (10.60)$$

In the CW filled with a chiral material, the waveguide angular frequency w_w is related to ε, μ and ξ. Here, w_w can be expressed as

$$w_w = [w^2 - k_c^2/(\varepsilon\mu + \mu^2\xi^2)]^{1/2} \qquad (10.61)$$

Therefore, the propagation constants k_+ and k_- in a CW filled with a chiral material are expressed as

$$k_{w\pm} = \pm w_w\mu\xi + w_w(\varepsilon\mu + \mu^2\xi^2)^{1/2} \qquad (10.62)$$

The average propagation constants are given by

$$a = (k_{w+} + k_{w-})/2 = w_w(\varepsilon\mu + \mu^2\xi^2)^{1/2} \qquad (10.63)$$

$$b = (k_{w+} - k_{w-})/2 = w_w\mu\xi \qquad (10.64)$$

The wave impedance in a CW filled with a chiral material is expressed as

$$Z_w = Z_c w/w_w = w\mu/(w^2\varepsilon\mu + w^2\mu^2\xi^2 - k_c^2)^{1/2} \qquad (10.65)$$

By using the same method as that for propagation in free space (Sun *et al.* 1998), the complex reflection coefficients of the open-circuited sample Γ_1 and the short-circuited Γ_2 in the CW can be derived as

$$\Gamma_1 = \frac{i(1 - Z_w^2/Z_{w0}^2)\tan(ad)}{2Z_w/Z_{w0} - i(1 + Z_w^2/Z_{w0}^2)\tan(ad)} \qquad (10.66)$$

$$\Gamma_2 = \frac{iZ_w\tan(ad) + Z_{w0}}{iZ_w\tan(ad) - Z_{w0}} \qquad (10.67)$$

where $Z_{w0} = w\mu_0/k_0'$ and $k_0' = [w^2 - k_c^2/(\varepsilon_0\mu_0)]^{1/2}(\varepsilon_0\mu_0)^{1/2}$ are the wave impedance and propagation constant for the empty CW, respectively. d is the thickness of the chiral samples.

According to the definition of the rotation angle (θ) and the axial ratio (R_A) (Hollinger *et al.* 1992), one can also get the following relation:

$$b = \frac{1}{2d}\left[2\theta - i\tanh^{-1}\left(\frac{2R_A}{1 + R_A^2}\right)\right] \qquad (10.68)$$

On combining Eqs. (10.66) and (10.67), one can get

$$a = \frac{1}{2d}\arccos\left(\frac{\Gamma_1\Gamma_2^2 + \Gamma_2^2 - 2\Gamma_1\Gamma_2 - \Gamma_1 + 1}{2(\Gamma_1 - \Gamma_2)}\right) \qquad (10.69)$$

$$Z_w = Z_{w0}\left(\frac{\Gamma_1\Gamma_2 + \Gamma_1 + \Gamma_2 + 1}{\Gamma_1\Gamma_2 - 3\Gamma_1 + \Gamma_2 + 1}\right)^{1/2} \qquad (10.70)$$

From Eqs. (10.61–10.65), the following equations are derived:

$$\mu = \frac{Z_w a}{w} \qquad (10.71)$$

$$\xi = \frac{b(a^2 + k_c^2)^{1/2}}{a\mu w} \qquad (10.72)$$

$$\varepsilon = \frac{(a^2 - b^2)\mu\xi^2}{b^2} \qquad (10.73)$$

Using the setup shown in Figure 10.29, the parameters $\Gamma_1, \Gamma_2, \theta$ and R_A can be measured. So ε, μ and ξ can be calculated using Eqs. (10.68)–(10.73).

10.3.4 Experimental results for chiral composites

The measurements are carried out in the range 8.5 to 11.5 GHz. Experiments (Liu *et al.* 1999) show that the variance of the results for the same sample is about 6 % when the samples are rotated 120° and then measured. In order to improve the accuracy of the results, the samples are rotated 120° twice and measured each time, giving a total of three measurements. The result for each sample is the average of the three measurements.

Firstly, a paraffin wax sample is measured in order to test the accuracy of this measurement system. Figure 10.30 plots the permittivity and permeability of the paraffin wax sample. Both the

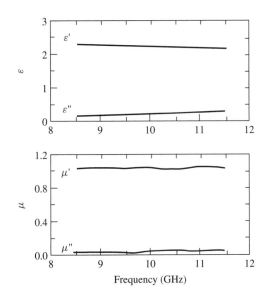

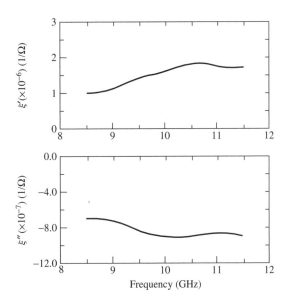

Figure 10.30 The complex permittivity and permeability of the paraffin wax sample. Source: Liu, Z.L. Sun, G.C. Huang, Q.L. and Yao, K.L. (1999), "A circular waveguide method for measuring the electromagnetic parameters of chiral materials at microwave frequencies," *Meas. Sci. Technol.* **10**, 374–379, IOP Publishing Ltd, by permission of IOP Publishing Ltd

Figure 10.31 The complex chirality parameter of the paraffin wax sample. Source: Liu, Z.L. Sun, G.C. Huang, Q.L. and Yao, K.L. (1999), "A circular waveguide method for measuring the electromagnetic parameters of chiral materials at microwave frequencies," *Meas. Sci. Technol.* **10**, 374–379, IOP Publishing Ltd, by permission of IOP publishing Ltd

measured values of the permittivity and permeability agree with the values given in section 10.3.1. Because it is a low-loss, nonmagnetic material, the imaginary parts of the complex permittivity and permeability of the paraffin wax sample should be zero. The measured values are indicative of the error of the measurement system in the measurement of the complex permittivity and permeability. It can be seen that the mean value of the imaginary part of the permeability is smaller than that of the imaginary part of the complex permittivity. This indicates that the accuracy of the measurement of the imaginary part of the complex permeability is higher than that of the complex permittivity.

Figure 10.31 is a plot of the complex chirality parameter of the paraffin wax sample. It can be seen that the real and imaginary parts of the chirality parameter of the paraffin wax are of the order of 10^{-6} and 10^{-7}, respectively. Since the paraffin wax is a nonchiral material, both the real and the imaginary part of the chirality

parameter should be zero. The measured value gives an indicative value of the error of the measurement system in the measurement of the chirality parameter. From the measured values of the electromagnetic parameters of the paraffin wax sample, it is obvious that the accuracy of the CW method is comparable to that of the free-space method (Varadan *et al.* 1994).

Figures 10.32–10.34 are plots of the electromagnetic parameters $\varepsilon, \mu,$ and ξ of the right-handed chiral samples versus frequency, respectively. Figure 10.32 shows the plots of the complex permittivity of the right-handed chiral samples. It can be seen that the real part of the permittivity increases as the concentration increases from 0.4 to 0.8%. However, an increase in the concentration to 1.6% leads to a decrease in the real part of the permittivity. The curves of the imaginary part of the permittivity of the 0.4 and 0.8% samples are flat, which indicates that less sensitive to frequency. At lower frequencies, the imaginary part of the permittivity increases as

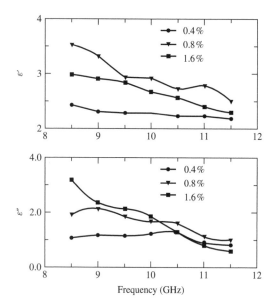

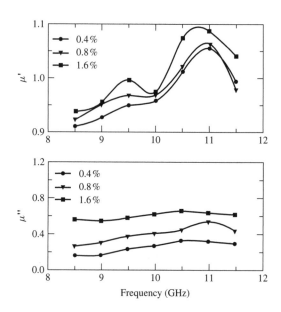

Figure 10.32 The complex permittivities of the right-handed chiral samples. Source: Liu, Z. L. Sun, G. C. Huang, Q. L. and Yao, K. L. (1999), "A circular waveguide method for measuring the electromagnetic parameters of chiral materials at microwave frequencies," *Meas. Sci. Technol.* **10**, 374–379, IOP Publishing Ltd, by permission of IOP Publishing Ltd

Figure 10.33 The complex permeabilities of the right-handed chiral samples. Source: Liu, Z.L. Sun, G.C. Huang, Q.L. and Yao, K.L. (1999), "A circular waveguide method for measuring the electromagnetic parameters of chiral materials at microwave frequencies," *Meas. Sci. Technol.* **10**, 374–379, IOP Publishing Ltd, by permission of IOP Publishing Ltd

the concentration increases. At higher frequencies, the imaginary part of the permittivity of the 1.6 % sample is smaller than those of the 0.4 and 0.8 % samples, owing to the higher frequency sensitivity.

Figure 10.33 shows the plot of the complex permeability of right-handed chiral samples. It can be seen that both the real and the imaginary part of the permeability increase as the concentration increases. It appears that the real part of the permeability has a peak in between 10.5 and 11.0 GHz.

Figure 10.34 plots the complex chirality parameter of right-handed chiral samples. It can be seen that the real and the imaginary parts of the chirality parameter of the right-handed chiral materials are positive and negative respectively. The values of the real and the imaginary parts of the chirality parameter are of the order of 10^{-4}, which are much greater than the values obtained from achiral materials. The absolute values of the real and the

imaginary parts of the chirality parameter increase as the concentration increases.

Figures 10.35–10.37 are plots of the electromagnetic parameters of the 0.8 % chiral samples. Figures 10.35 and 10.36 show the complex permittivity and permeability of the 0.8 % chiral samples respectively. It can be seen that the complex permittivities and permeabilities of right-handed, left-handed, and racemic samples have almost the same values. They also have the same trend as the frequency changes. It can be concluded that the complex permittivity and permeability are independent of the handedness of chiral inclusions.

Figure 10.37 is the plot of the complex chirality parameter of 0.8 % chiral samples. It can be seen that the right- and left-handed samples have almost the same absolute values of the chirality parameter. The real and the imaginary parts for a racemic sample are of the order of 10^{-6}, which is comparable to the values for nonchiral materials.

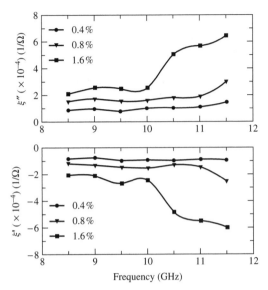

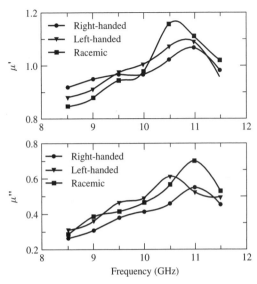

Figure 10.34   The complex chirality parameters of the right-handed chiral samples. Source: Liu, Z.L. Sun, G.C. Huang, Q.L. and Yao, K.L. (1999), "A circular waveguide method for measuring the electromagnetic parameters of chiral materials at microwave frequencies," *Meas. Sci. Technol.* **10**, 374–379, IOP Publishing Ltd, by permission of IOP Publishing Ltd

Figure 10.36   The complex permeabilities of the 0.8 % chiral samples. Source: Liu, Z.L. Sun, G.C. Huang, Q.L. and Yao, K.L. (1999), "A circular waveguide method for measuring the electromagnetic parameters of chiral materials at microwave frequencies," *Meas. Sci. Technol.* **10**, 374–379, IOP Publishing Ltd, by permission of IOP Publishing Ltd

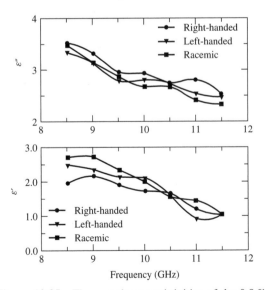

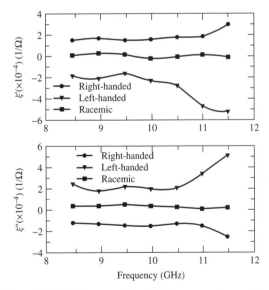

Figure 10.35   The complex permittivities of the 0.8 % chiral samples. Source: Liu, Z.L. Sun, G.C. Huang, Q.L. and Yao, K.L. (1999), "A circular waveguide method for measuring the electromagnetic parameters of chiral materials at microwave frequencies," *Meas. Sci. Technol.* **10**, 374–379, IOP Publishing Ltd, by permission of IOP Publishing Ltd

Figure 10.37   The complex chirality parameters of the 0.8 % chiral samples. Source: Liu, Z.L. Sun, G.C. Huang, Q.L. and Yao, K.L. (1999), "A circular waveguide method for measuring the electromagnetic parameters of chiral materials at microwave frequencies," *Meas. Sci. Technol.* **10**, 374–379, IOP Publishing Ltd, by permission of IOP Publishing Ltd

10.4 CONCLUDING REMARKS

The electromagnetic parameters of chiral materials can be simultaneously measured by using the CW short-circuit and open-circuit method. We have reported the principle and techniques of this method. The complex permittivity, permeability, and chirality parameters are calculated from the measured rotation angle, the axial ratio and the complex reflection coefficients of short-circuited and open-circuited samples. The dependence of the electromagnetic parameters on the concentration and handedness of the chiral inclusions has been studied. The results are comparable to those obtained by the use of free-space method (Ro 1991).

REFERENCES

Biot, J. B. 1812). "Memoire sur un Nouveau Genre d'Oscillation que les Molecules de la lumiere Eprouvent en Transversant Certains Cristaux", *Memoires de la Classe des Sciences Mathemathiques et Physiques de l'Institut Imperial de France, Part 1*, **1**, 1–372.

Bohren, C. F. (1974). "Light scattering by an optically active sphere", *Chemical Physics Letters*, **29**, 458–462.

Bokut, B. V. and Federov, F. I. (1960). "Reflection and refraction of light in an optically isotropic active media", *Optika i Spektroskopiya*, **9**, 334–336.

Cotton, A. (1895). "Absorption Inegale des Rayons Circulaires Droit et Gauche dans Certains Corps Actifs", *Comptes Rendus Hebdomadaires des Seances de l'Academie des Sciences*, **120**, 989–991.

Crabbe, P. and Syntex, S. A. (1965). *Optical Rotatory Dispersion and Circular Dichroism in Organic Chemistry*, Holden-Day, San Francisco.

Engen, G. F. and Hoer, C. A. (1979). "An improved technique for calibrating the dual six-port automatic network analyzer", *IEEE Transactions on Microwave Theory and Techniques*, **27** (12), 987–993.

Engheta, N. and Pelet, P. (1989). "Modes in chirowaveguides", *Optics Letters*, **14**, 593–595.

Engheta, N. and Pelet, P. (1990). "Mode orthogonality in chirowaveguides", *IEEE Transactions on Microwave Theory and Techniques*, **38**, 1631–1634.

Franzen, N. R. and Speciale, R. A. (1975). "A new procedure for system calibration and error removal in automated S-parameter Measurements", *Proceeding 5th European Microwave Conference*, Hamburg, 69–73.

Ghodgaonkar, D. K. Varadan V. V. and Varadan, V. K. (1989). "A free-space method for measurement of

dielectric constants and loss tangent at microwave frequencies", *IEEE Transactions on Instrumentation and Measurement*, **38**, 789–793.

Ghodgaonkar, D. K. Varadan V. V. and Varadan, V. K. (1990). "Free-space measurement of complex permittivity and complex permeability of magnetic materials at microwave frequencies", *IEEE Transactions on Instrumentation and Measurement*, **39**, 387–394.

Guire, T. Varadan, V. V. and Varadan, V. K. (1990). "Influence of chirality on the reflection of em waves by planar dielectric slabs", *IEEE Transactions on Electromagnetic Compatibility*, **32**, 300–304.

Hewlett-Parkard Product Note 8510-8 (1987). "Applying the HP 8510B TRL calibration for non-coaxial measurements", 5954–8382.

Hewlett-Parkard Product Note 8510-5A (1988). "Specifying calibration standards for the HP 8510 network analyzer", 5956–4352.

Hollinger, R. Varadan, V. V. and Varadan, V. K. (1991). "Eigenmodes in a circular waveguide containing an isotropic chiral material", *Radio Science*, **26**, 1335–1344.

Hollinger, R. Varadan, V. V. Ghodgaonkar, D. K. and Varadan, V. K. (1992). "Experimental characterization of isotropic chiral composites in circular waveguides", *Radio Science*, **27**, 161–168.

Hui, H. T. and Edward, K. N. (1996). "Modal expansion of dyadic Green's functions of the cylindrical chirowaveguide", *IEEE Transactions on Microwave and Guided Wave Letters*, **6**, 360–362.

Jaggard, D. L. Engheta, N. and Liu, J. (1990). "Chiroshield: a Salisbury/Dallenbach shield alternative", *Electronics Letters*, **26**, 1332–1334.

Jaggard, D. L. Mickelson, A. R., and Papas, C. H. (1979). "On electromagnetic waves in chiral media", *Applied Physics*, **18**, 211–216.

Lakhtakia, A. Varadan, V. V. and Varadan, V. K. (1986). "A Parametric study of microwave reflection characteristics of a planar achiral-chiral interface", *IEEE Transactions on Electromagnetic Compatibility*, **28**, 90–95.

Lakhtakia, A. Varadan, V. K. and Varadan, V. V. (1989). *Time-Harmonic Electromagnetic Fields in Chiral Media*, Lect. Note Ser. 35, Springer-Verlag, New York.

Lindell, I. V. and Sihvola, A. H. (1995). "Plane-wave reflection from uniaxial chiral interface and its application to polarization transformation", *IEEE Transactions on Antennas and Propagation*, **43**, 1397–1404.

Liu, Z. L. Sun, G. C. Huang, Q. L. and Yao, K. L. (1999). "A circular waveguide method for measuring the electromagnetic parameters of chiral materials at microwave frequencies", *Materials Science and Technology*, **10**, 374–379.

Mariotte, F. Guerin, F. Bannelier, P. and Bourgeade, A. (1995). "Numerical computations of the electromagnetic field scattered by complex chiral bodies",

Journal of Electromagnetic Waves and Applications, **9**, 1459–1485.

Pelet, P. and Engheta, N. (1990). "The theory of chirowaveguides", *IEEE Transactions on Antennas and Propagation*, **38**, 90–98.

Ro, R. (1991). *Determination of the Electromagnetic Properties of Chiral Composites, Using Normal Incidence Measurements*, Ph.D. Thesis., The Pennsylvania State University, Department of Engineering Science and Mechanics.

Rytting, D. (1987). "Advances in microwave error correction techniques", Hewlett-Packard, RF and MW Symp., 5954–8378.

Sihvola, A. H. and Lindell, I. V. (1991). "Bi-isotropic constitutive relations", *Microwave and Optical Technology Letters*, **4** (8), 295–297.

Silverman, M. P. (1986). "Reflection and refraction at the surface of a chiral medium: comparison of gyrotropic constitutive relations invariant and noninvariant under a duality transformation", *Journal of Optical Society of America*, **3**, 830–837.

Sun, G. C. Yao, K. L. Liu. Z. L. and Huang, Q. L. (1998). "A study on measuring the electromagnetic parameters of chiral materials", *Journal of Physics D: Applied Physics*, **31**, 2109–2111.

Theron, I. P. and Cloete, J. H. (1996). "The optical activity of an artificial non-magnetic uniaxial chiral crystal at microwave frequencies", *Journal of Electromagnetic Waves and Applications*, **10**, 539–561.

Tinoco, I. and Freeman, M. P. (1960), "The optical activity of oriented copper helices, II. Experimental", *J. Phys. Chem*, **61**, 1196–2000.

Umari, M. Varadan, V. V. and Varadan, V. K. (1991). "Rotation and dichroism associated with microwave propagation in chiral composite samples", *Radio Science*, **26**, 1327–1334.

Urry, D. W. and J. Krivacic (1970). "Differential scatter of left and right circularly polarized light by optically active particulate system", *Proceedings of National Academy of Sciences*, **65**, 845–852.

Varadan, V. V. Ro R. and Varadan V. K. (1994). "Measurement of the electromagnetic properties of chiral composites materials in the 8–40 GHz range", *Radio Sci.*, **29**, 9–22.

Varadan, V. K. Varadan, V. V. and Lakhtakia, A. (1987). "On the possibility of designing anti-reflection coatings using chiral composites", *Journal of Wave-Material Interaction*, **2**, 71–81.

Velluz, L. Legrand, M. and Grosjean, M. (1965). *Optical Circular Dichroism*, Verlag Chemie Academic Press, New York.

Winkler, M. H. (1956). "An experimental investigation of some models for optical activity", *Journal of Physical Chemistry*, **60**, 1656–1659.

11

Measurement of Microwave Electrical Transport Properties

This chapter discusses the measurement of electrical transport properties of materials using microwave Hall effect techniques. After a brief introduction of the Hall effects at different frequencies, the nonresonant and resonant methods for the measurement of microwave Hall effects are discussed. The nonresonant methods include transmission method, reflection method, and turnstile-junction method, and the resonant methods include hollow cavity method, dielectric resonator method, and planar resonator method. For a magnetic material, its Hall effect includes an ordinary Hall effect and an extraordinary Hall effect. The methods for the measurement of microwave electrical transport properties of magnetic materials are discussed in Section 11.4.

11.1 HALL EFFECT AND ELECTRICAL TRANSPORT PROPERTIES OF MATERIALS

The electrical transport properties of conductors, semiconductors, dielectrics, and magnetic materials have always been active research topics in solid-state physics and materials sciences, and are important for the development of electronic components and circuits. The electrical transport properties of a material include carrier density n, carrier mobility μ_H, and conductivity σ. The three parameters describing the electrical transport properties of materials (n, μ_H, and σ) are related as follows:

$$n = \frac{\sigma}{q\mu_H} \qquad (11.1)$$

where q is the charge of the free carriers and usually equals the electricity of an electron $e = 1.602 \times 10^{-19}$ Coulomb. Usually, the conductivity σ and the carrier mobility μ_H are measured experimentally, and the carrier density n is calculated from the conductivity σ and carrier mobility μ_H according to Eq. (11.1).

The measurement of the conductivity at direct current (dc) or alternative current (ac) frequencies is based on Ohm's law. At microwave frequencies, for a low-conductivity material, its conductivity σ is usually calculated from the imaginary part of the relative permittivity ε_r'' of the medium:

$$\sigma = \omega \varepsilon_r'' \varepsilon_0 \qquad (11.2)$$

where ω is the angular frequency. So the methods for the measurement of ε_r'' discussed in the previous chapters can be used in determining the conductivity of a low-conductivity sample. For a high-conductivity material, its conductivity can be calculated from its surface resistance R_s:

$$\sigma = \frac{\omega \mu}{2R_s^2} \qquad (11.3)$$

So, the methods for the measurement of surface resistance can be used in determining the high-frequency conductivity of high-conductivity materials.

The measurement of carrier mobility μ_H is based on the Hall effect. When an electric field E_x is applied to a material, it induces an electric current. If a static magnetic field B is then

Microwave Electronics: Measurement and Materials Characterization   L. F. Chen, C. K. Ong, C. P. Neo, V. V. Varadan and V. K. Varadan
© 2004 John Wiley & Sons, Ltd   ISBN: 0-470-84492-2

applied perpendicularly to the electric current, the charge carriers in the electric current will be deflected, resulting in an additional electric field E_H, perpendicular to both the electric current and the external magnetic field B. From the relationship between E_x and E_H, the mobility of the sample can be obtained as

$$\mu_H = k_\mu \frac{E_H}{E_x B} \qquad (11.4)$$

where k_μ is a parameter determined by the dimensions of the sample.

According to the nature of the external electric field applied to the sample, Hall effects can be generally classified into dc Hall effect, ac Hall effect, and microwave Hall effect.

11.1.1 Direct current Hall effect

As shown in Figure 11.1, the driving electric field in a dc Hall effect is a dc electric field, and correspondingly the Hall field E_H is also a dc field. The measurement of the dc Hall effect requires electrodes contacting the sample along the direction of the Hall field, and this method is usually known as the four-point probe method. If the two electrodes for the measurement of the Hall field are not symmetrical about the driving electric field, there will be a systematic error in the results of the Hall effect measurement. Such a systematic error can be eliminated by making two measurements when the driving electric currents are in opposite directions and by calculating the average value of the two measurements.

11.1.2 Alternate current Hall effect

As shown in Figure 11.2, when the external electric field is an ac electric field, the Hall field E_H is also an ac electric field, and the corresponding Hall effect is called the *ac Hall effect*. The measurement of ac Hall effect also requires electrodes contacting the sample along the direction of the ac Hall field. Similar to dc Hall effect, if the two electrodes are not symmetrical about the driving electric field, there will be systematic error in the Hall effect measurement. Such error can be eliminated by shifting the Hall

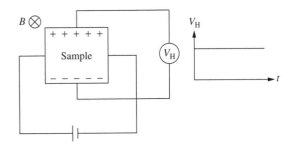

Figure 11.1   Measurement of dc Hall effect

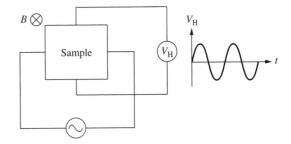

Figure 11.2   Measurement of ac Hall effect

voltage curve to make the maximum and minimum values symmetrical about the zero line.

Both the ac and dc Hall effect measurements require electrodes contacting the samples under test. Though this is feasible for bulk materials, the measurement of ac and dc Hall effects are less straightforward for powder samples, as electrode contact is very difficult for such samples. Even when the powder samples are pressed and molded into bulk samples, the measurement results may not be accurate because of charge polarization, grain boundaries, and particle–particle interactions.

11.1.3 Microwave Hall effect

If the driving field is at microwave frequency, the Hall field is also at microwave frequencies, and such a Hall effect is called microwave Hall effect. As the propagation of microwave signal does not require direct electrode contact, the measurement of microwave Hall effect does not require direct electrode contact with the sample under test, so both bulk and powder samples can be measured. Furthermore, as the clock speeds of electronic circuits have already stepped into the microwave frequency range, it becomes crucial to investigate the electrical transport properties of materials at microwave frequencies.

Figure 11.3 shows a microstrip measurement cell for the measurement of electrical transport properties of semiconductor layer samples (Druon *et al.* 1990). In this method, the sample is only lightly pressed on the fixture, and the fixture has capacitive metal-insulator-semiconductor contacts with the sample. The metal is the microstrip line, which acts as the electrode, with the insulator placed at the end of the line. The semiconductor sample is in physical contact with the insulated electrode. The sample size and the working frequency are chosen in order to get small values of the capacitive impedances with respect to the sheet resistance of the sample and to satisfy the conditions of measurement that enable the sample to be considered as a lumped circuit element. As the electrical contacts between the sample and the cell are capacitive, this method is nondestructive and requires no technological process for the samples. Using this method, the sheet resistance (R_s), the carrier density (n), and the mobility (μ) of epitaxial layers can be measured.

Figure 11.4 shows the basic structure of the measurement fixture. The measurement cell is placed inside the case with four connecting

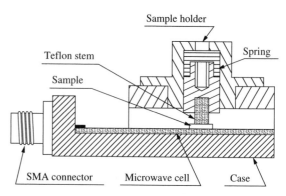

Figure 11.4 Cross-sectional view of the measurement fixture (Druon *et al.* 1990). Source: Druon, C. Tabourier, P. Bourzgui, N. and Wacrenier, J. M. (1990). "Novel microwave devices for nondestructive electrical characterization of semiconducting layers", *Review of Scientific Instruments*, **61** (11), 3431–3434

plugs and a device for pressing the sample on the measurement cell. The position of the sample on the microstrip lines can be reset with sufficient accuracy to perform the measurements. The magnetic field is applied perpendicularly to plane of the microstrip circuit.

Figure 11.5 shows the circuit for the measurement of electrical transport properties of layer samples using the fixture shown in Figures 11.3

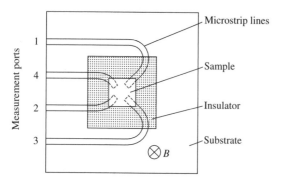

Figure 11.3 Top view of a microstrip measurement cell for the measurement of microwave Hall effect (Druon *et al.* 1990). Modified from Druon, C. Tabourier, P. Bourzgui, N. and Wacrenier, J. M. (1990). "Novel microwave devices for nondestructive electrical characterization of semiconducting layers", *Review of Scientific Instruments*, **61** (11), 3431–3434

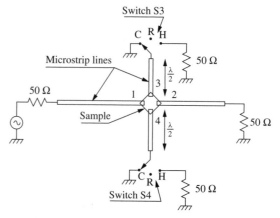

Figure 11.5 Measurement circuit for electrical transport properties of layer samples. Modified from Druon, C. Tabourier, P. Bourzgui, N. and Wacrenier, J. M. (1990). "Novel microwave devices for nondestructive electrical characterization of semiconducting layers", *Review of Scientific Instruments*, **61** (11), 3431–3434

and 11.4. The determination of R_s, μ, and n is performed in two steps (Druon *et al.* 1990): a low-frequency measurement and a high-frequency measurement. In the low-frequency step, measurements are performed at a frequency (usually several MHz) low enough for capacitive impedances to remain large with respect to R_s. The purpose of the low-frequency measurement is to determine the contact capacitances between the measurement cell and the sample under test. The capacitance values (typically 20 pF) are obtained from the measurement of the power transmitted through the sample between two chosen ports, the two others being grounded, as shown in Figure 11.6. These values provide control of the quality of the contacts. Furthermore, they are used in the high-frequency step to determine the correction factors needed when R_s and the capacitive impedances are of the same order of magnitude.

In the high-frequency step, measurements are made in an external dc magnetic field B, and the sheet resistance R_s, mobility μ, and carrier density n are determined. In the measurement, it is only necessary to measure the modulus of the transmission factor S_{ij}, and phase information is not needed. Meanwhile, the measurement cell is usually designed to be symmetrical, so that measurements can be easily made.

The sheet resistance measurement is performed between two opposite ports. In this case, microswitches S3 and S4, in Figure 11.5, which

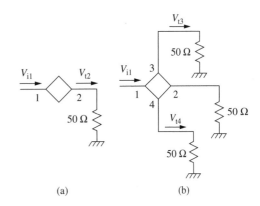

(a)   (b)

Figure 11.7 Equivalent circuits for the high-frequency step. (a) Measurement of R_s and $\Delta R/R$ (S3 and S4 on R shown in Figure 11.5) and (b) measurement of the Hall effect (S3 and S4 on H shown in Figure 11.5). Source: Druon, C. Tabourier, P. Bourzgui, N. and Wacrenier, J. M. (1990). "Novel microwave devices for nondestructive electrical characterization of semiconducting layers", *Review of Scientific Instruments*, **61** (11), 3431–3434

are located one-half wavelength away from the sample, enable the introduction of an open circuit on the other two electrodes, strictly at their point of contact with the sample. The equivalent circuit for this measurement is shown in Figure 11.7(a), and the value of R_s can be calculated by

$$R_s = C_R(1/|S_{21}| - 1) \qquad (11.5)$$

where C_R is a parameter related to the measurement fixture and sample shape and can be obtained by calibration.

The mobility and carrier density can be measured on the basis of Hall effect. The measurement of the Hall effect using the circuit and cell, shown in Figure 11.5 and Figure 11.3 respectively, is similar to the standard one performed with a dc, as shown in Figure 11.1. In this case, the Hall resistance Z_H is defined as

$$Z_H = V_H/I \qquad (11.6)$$

As shown in Figure 11.7(b), when the incoming current I is applied at port 1, the Hall voltage V_H is obtained between ports 3 and 4. The value of the Z_H can be determined from the measurement of the transmission factors $|S_{31}|$ and $|S_{41}|$. Then the

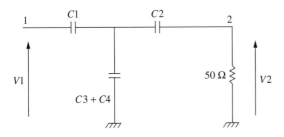

Figure 11.6 Equivalent circuit for the measurement of the contact capacitances in the low-frequency step (switches S3 and S4 on C shown in Figure 11.5). Source: Druon, C. Tabourier, P. Bourzgui, N. and Wacrenier, J. M. (1990). "Novel microwave devices for nondestructive electrical characterization of semiconducting layers", *Review of Scientific Instruments*, **61** (11), 3431–3434

Hall mobility can be obtained from (Druon *et al.* 1990)

$$\mu_H = C_H Z_H / (R_s B) \qquad (11.7)$$

where C_H is a factor that can be determined by calibration.

The carrier concentration of the sample can then be obtained from

$$n = 1/(R_s q \mu_H h_e) \qquad (11.8)$$

where h_e is the electrical thickness of the sample layer, and q is the charge of the free carriers. It should be noted that the electrical thickness h_e of the sample layer is different from the geometrical thickness h of the sample, as the existence of interfacial charged defects leads to a correction of h (Druon *et al.* 1990).

It should be indicated that, in the measurement circuit shown in Figure 11.5, the Hall effect measurement is only carried out when $\mu_H^2 B^2 \ll 1$, and Eq. (11.8) is also based on this assumption. If this is not the case, magnetoresistance measurement is often made. The equivalent circuit for magnetoresistance measurement is shown in Figure 11.7(a), and the value of mobility can be obtained from (Druon *et al.* 1990)

$$\mu_H = C_M (\Delta R/R)^{1/2} / B \qquad (11.9)$$

where C_M is a factor that can be obtained by calibration. With known μ_H and R_s, the carrier concentration n can also be obtained according to Eq. (11.8). It should be noted that in the magnetoresistance method, it is assumed that the drift mobility is equal to the Hall mobility. This magnetoresistance measurement method has been extended for nondestructive mapping of large samples from the measurement of the magnetoresistance effect, and the distribution of the electrical transport properties of semiconductor wafers can thus be obtained (Belbounaguia *et al.* 1994).

Besides the effects discussed above, the electrical transport properties also show themselves in two other important phenomena: Faraday rotation and coupling between two orthogonal resonant modes. On the basis of these phenomena, nonresonant and resonant methods have been developed for the measurement of microwave Hall effect.

11.2 NONRESONANT METHODS FOR THE MEASUREMENT OF MICROWAVE HALL EFFECT

Most of the nonresonant methods for the measurement of microwave Hall effect are based on Faraday rotation. In this section, after a brief discussion of the phenomenon of Faraday rotation, three types of nonresonant methods will be discussed: transmission method, reflection, and turnstile-junction method.

11.2.1 Faraday rotation

As indicated earlier, in a Hall effect, the external dc magnetic field is perpendicular to the driving electric field. The external dc magnetic field may be parallel or perpendicular to the direction of microwave propagation, and so there are two types of geometries for the study of microwave Hall effect: Cotton–Mouton geometry and Faraday geometry. Let θ be the angle between the direction of microwave propagation and the direction of the external dc magnetic field. In a Cotton–Mouton geometry, $\theta = \pi/2$, which means that the dc magnetic field transverses the direction of microwave propagation. Engineer and Nag analyzed the propagation of electromagnetic waves in a rectangular waveguide in the presence of a transverse magnetic field (Engineer and Nag 1965). In a Faraday geometry, $\theta = 0$, which means that the dc magnetic field is along the direction of microwave propagation. In most of the microwave Hall effect measurements, Faraday geometry is used. In the following discussion, we concentrate on Faraday geometry.

For a gyrotropic dielectric medium under an external dc magnetic field B in the z-direction, its dielectric permittivity ε, resistivity ρ, and conductivity σ are tensors in the following format:

$$[a] = \begin{bmatrix} a_{xx} & a_{xy} & 0 \\ -a_{xy} & a_{yy} & 0 \\ 0 & 0 & a_{zz} \end{bmatrix} \qquad (11.10)$$

where $[a]$ may be $[\varepsilon]$, $[\rho]$, or $[\sigma]$. The component a_{xy} reflects the Hall effect.

As shown in Figure 11.8, a linearly polarized electromagnetic wave propagating in the direction of the external magnetic field is decomposed into

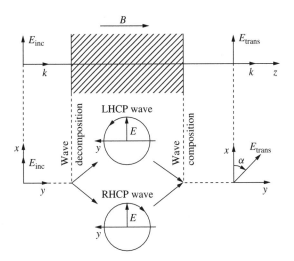

Figure 11.8 Schematic view of the Faraday rotation in a lossless gyrotropic dielectric medium. Modified from Musil, J. and Zacek, F. (1986). *Microwave Measurements of Complex Permittivity by Free-space Methods and their Applications*, Amsterdam, with permission from Elsevier

two circularly polarized waves: left-handed circularly polarized (LHCP) wave and right-handed circularly polarized (RHCP) wave. As the permittivity ε is in the form of Eq. (11.10), the LHCP wave and the RHCP wave propagate with different velocities, and the different propagation velocities of these two waves result in the fact that the polarization plane of the resultant wave passing through the medium rotates. The magnitude of the angle of rotation of the polarization plane depends on the path that the wave passes through and the materials property parameters.

When a linearly polarized electromagnetic wave passes through a material under a dc magnetic field, the polarization plane will be rotated a degree α given by

$$\alpha = \frac{1}{2} \cdot \frac{\omega}{c} \cdot (n_L - n_R) \cdot l \qquad (11.11)$$

where ω is the angular frequency, c is the speed of light in free space, and l is the length of the path passed by the electromagnetic wave. The parameters n_L and n_R are the indexes of refraction of the material for the LHCP wave and RHCP wave, respectively, and they are related to the Hall

mobility of the sample (μ_H) and the external dc magnetic field (B).

Musil and Zacek made detailed discussions on the relationship between the angle of the Faraday rotation α and the Hall mobility of the sample (Musil and Zacek 1986). Here, we only give the equations for the weak magnetic field ($\mu_H B \ll 1$) and low-frequency ($\omega\tau < 1$) case, where τ is the relaxation time of the sample. For low-loss materials with $\omega\varepsilon_r > \sigma$, where ε_r is the dielectric constant and σ is the conductivity of the material, the angle of rotation α is given by

$$\alpha = -\frac{1}{2}\sqrt{\frac{\mu_0}{\varepsilon_0\varepsilon_r}} \cdot \sigma\mu_H Bl \qquad (11.12)$$

For high-loss materials with $\omega\varepsilon_r < \sigma_0$, α is given by

$$\alpha = -\frac{1}{2}\sqrt{\frac{\mu_0\omega\sigma}{2}} \cdot \mu_H Bl \qquad (11.13)$$

The above discussion indicates that the Hall mobility of a material can be measured using the Faraday rotation method. In the following, we discuss three types of methods for measuring the rotation of the polarization plane after a linear polarized electromagnetic wave passes through or is reflected back from a medium under study: transmission method, reflection method, and turnstile-junction method.

11.2.2 Transmission method

In a transmission method, the sample under study is inserted in a segment of transmission line, and the rotation angle of the polarization plane is measured after the electromagnetic wave passes through the sample. In the following, we discuss the methods developed from the three types of transmission lines: waveguide, dielectric waveguide, and free space.

11.2.2.1 Waveguide method

Figure 11.9 shows a waveguide system for the measurement of Faraday rotation angle α (Musil and Zacek 1986). Two transitions between rectangular and circular waveguides are used. The sample holder is a segment of a circular waveguide. A

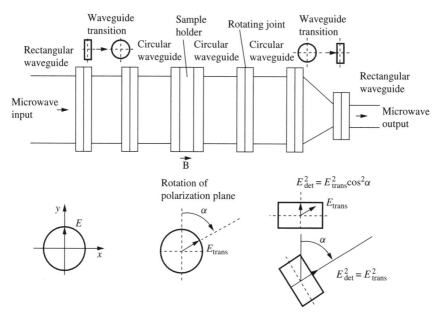

Figure 11.9   Principle of the waveguide method for the measurement of the angle of rotation of the polarization plane. Modified from Musil, J. and Zacek, F. (1986). *Microwave Measurements of Complex Permittivity by Free-space Methods and their Applications*, Amsterdam, with permission from Elsevier

wave linearly polarized in the y-direction passes through the sample placed between two circular waveguides. If the external dc magnetic field B is in the direction of the wave propagation, the polarization plane of the wave passing through the specimen is rotated by an angle α.

The rotation of the polarization plane manifests itself by the decrease in a signal received by a detector located in a rectangular waveguide. With the square-law detection, the detected signal is proportional to $\cos^2 \alpha$:

$$U_{det} \propto E_{trans}^2 \cdot \cos^2 \alpha$$

Therefore, the rotation angle α can be found by rotating the rectangular waveguide by means of a circular rotary joint into a position where a maximum signal can be obtained. Accuracy of the determination of the angle α is determined mainly by the perfection of the mechanical construction of the rotary joint and the technique used for measuring the angle of rotation. In actual experiments, measurement of the angle that gives the minimum received signal would give more accurate results ($\alpha_{real} = \pi/2 - \alpha_{meas}$).

In the above discussion, the loss of the specimen is neglected. A wave after passing a lossy specimen is elliptically polarized. In this case, the properties of a semiconductor can be evaluated from the ellipticity of the emerging wave. The ellipticity of the wave, defined as the ratio of the minor to the major semiaxis of the ellipse, may be measured by rotating the rectangular waveguide with the help of a circular rotary joint and measuring minimum and maximum signals at the detector.

This method has high requirements on the waveguide transitions. To increase the measurement accuracy and sensitivity, it is necessary to insert a damping plate into the circular part of a TE_{10} rectangular to TE_{11} circular mode transition in such a way that its face is parallel to the longer transverse dimension of the rectangular waveguide. This plate absorbs a wave reflected from the rectangular waveguide and thus prevents undesirable interference with the incident wave. A similar structure is shown in Figure 8.41, where such a damping plate is labeled "polarized absorber".

Figure 11.10 shows a crossed waveguide arrangement for the measurement of the Hall conductivity of a two-dimensional sample at

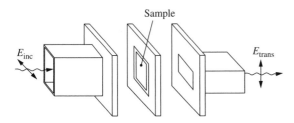

Figure 11.10 Crossed waveguide arrangement (Meisels and Kuchar 1987). Modified from Meisels, R. and Kuchar, F. (1987). "The microwave Hall effect of a two-dimensional electron gas", *Zeitschrift fur Physik B – Condensed matter*, **67**, 443–447, by permission of Springer-Verlag GmbH & Co. KG.

microwave frequencies (Meisels and Kuchar 1987). In this arrangement, the two rectangular waveguides are crossed by rotating them by 90° around their long axis against each other. It can be taken as a special case of the arrangement shown in Figure 11.9.

Under an external dc magnetic field in the z-direction, according to Eq. (11.10), the conductivity of a two-dimensional sample in the structure is a tensor in the form of

$$[\sigma] = \begin{bmatrix} \sigma_{xx} & \sigma_{xy} \\ -\sigma_{xy} & \sigma_{xx} \end{bmatrix} \qquad (11.14)$$

The component σ_{xy} is the Hall conductivity. The field pattern in this structure is more complicated than the case of a pure plane wave, while the essential conclusion remains unchanged. The relationship between the input power and the transmitted power is given by (Meisels and Kuchar 1987)

$$P_{\text{trans}} = \left| \frac{\sigma_{xy} Z^*}{2} \cdot [(1 + \sigma_{xx} Z^*/2)^2 \right.$$

$$\left. + (\sigma_{xy} Z^*/2)^2]^{-1} \right|^2 P_{\text{inc}} \qquad (11.15)$$

where $Z^* = \mu \mu_0 \omega / k_z$ is the impedance of the waveguide. For $\sigma_{xx} Z^* \ll 1$ and $(\sigma_{xy} Z^*) \ll 1$, Eq. (11.15) can be simplified as

$$P_{\text{trans}} \approx \left| \frac{\sigma_{xy} Z^*}{2} \cdot (1 - \sigma_{xx} Z^*) \right|^2 P_{\text{inc}} \qquad (11.16)$$

With $\sigma_{xx} Z^* \ll 1$, Eq. (11.16) can be further approximated:

$$P_{\text{trans}} \approx \left| \frac{\sigma_{xy} Z^*}{2} \right|^2 P_{\text{inc}} \qquad (11.17)$$

Equation (11.17) can be rewritten as

$$\left| \frac{E_{\text{trans}}}{E_{\text{inc}}} \right| = \left(\frac{P_{\text{tr}}}{P_{\text{inc}}} \right)^{\frac{1}{2}} \approx \left| \frac{Z^*}{2} \cdot \sigma_{xy} \right| \qquad (11.18)$$

So, from the relationship between the transmitted electric field or power and the incident electric field or power, we can estimate the Hall conductivity σ_{xy} of the two-dimensional sample.

11.2.2.2 Dielectric waveguide method

Dielectric waveguides can also be used in characterizing the Hall mobility of materials using the transmission method, and the measurement arrangement is shown in Figure 11.11. The sample under test is inserted between the open ends of two segments of dielectric waveguides. The specimen is in direct contact with the dielectric waveguides. This method is nondestructive and does not require special preparation on specimens. The only requirement on a specimen is that it should form a plane-parallel plate. In a dielectric waveguide transmission line, the electromagnetic field is focused on the range of the dielectric waveguide; specimens whose transverse dimensions D are practically equal to the transverse dimensions

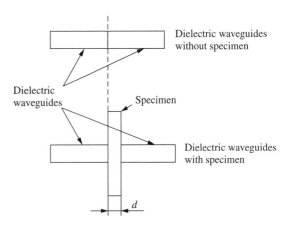

Figure 11.11 Transmission method using dielectric waveguide

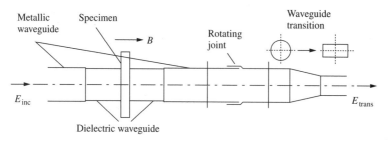

Figure 11.12 is made up of labels: Metallic waveguide, Specimen, Rotating joint, Waveguide transition, E_{inc}, B, E_{trans}, Dielectric waveguide.

Figure 11.12 Dielectric waveguide circuit for the measurement of the angle of rotation of the polarization plane (Musil and Zacek 1986, p187). Modified from Musil, J. and Zacek, F. (1986). *Microwave Measurements of Complex Permittivity by Free-space Methods and their Applications*, Amsterdam, by permission of Springer-Verlag GmbH & Co. KG.

of waveguide can also be measured. The properties of the dielectric waveguide and the application of materials property characterization using dielectric waveguides have been discussed in Chapters 2 and 4.

The measurement circuit shown in Figure 11.9 for the measurement of the angle of rotation of the polarization plane can be constructed using dielectric waveguides (Musil and Zacek 1986). As shown in Figure 11.12, the sample is placed between two segments of dielectric waveguides. There are two transitions between the metallic and dielectric waveguides and two transitions between the circular and rectangular waveguides. Similar to the arrangement shown in Figure 11.9, the measurement accuracy and the sensitivity of the measurement system shown in Figure 11.12 are closely related to the quality of the four transitions and the rotating joint.

11.2.2.3 Free-space transmission method

The transmission measurement system can also be built in a free-space approach. When a free-space approach is used, the sample under test is placed between two antennas that are usually corrected by dielectric lenses. The properties of the free-space propagation and its application in materials property characterization have been discussed in Chapters 2 and 4. Usually, the free-space method has higher accuracy at higher frequencies.

Figure 11.13 shows another free-space measurement system for the measurement of transmission tensor (Parks *et al.* 1997; Mittleman *et al.* 1997; Spielman *et al.* 1994). This system can achieve high accuracy at a millimeter wave frequency range. The input wave is polarized in the *x*-direction by the first polarizer. The components of the transmission tensor are selected by orientating

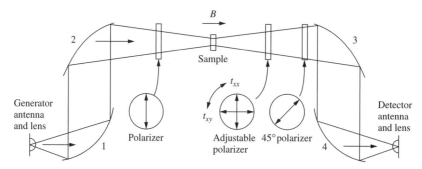

Figure 11.13 Free-space measurement system for the measurement of complex transmission tensors in the presence of a magnetic field. The items labeled as 1, 2, 3, and 4 are reflectors (Parks *et al.* 1997). Modified from Spielman, S. Parks, B. Orenstein, J. Nemeth, D. T. Ludwig, F. Clarke, J. Merchant, P. and Lew, D. T. (1997). "High-frequency Hall effect in the normal state of YBa$_2$Cu$_3$O$_7$", *Physical Review B*, **56** (1), 115–117

the second polarizer either parallel (t_{xx}) or perpendicular (t_{xy}) to the first. The third polarizer at 45° ensures that the polarization state of the signal finally reaching the detector is not affected by the orientation of the polarization. After obtaining the diagonal t_{xx} and off-diagonal t_{xy} components, the angle of rotation α can be determined. The effect of linear birefringence can be removed by reversing the direction of the dc magnetic field B.

11.2.3 Reflection method

Besides transmission methods, the electrical transport properties of materials can also be measured by reflection methods. In the following, we discuss two reflection methods: adjustable-short method and cross-slit probe method.

11.2.3.1 Adjustable-short method

Baturina *et al.* analyzed a waveguide method for the determination of the Hall mobility and conductivity of thin layer samples (Baturina *et al.* 1996), and Coue *et al.* also discussed a similar method (Coue *et al.* 1994). This method is based on the measurements of magnetic field dependences of the derivative of the reflection coefficient with respect to the magnetic field. This method does not require careful calibration of the microwave system and does not require the accurate measurements of the absolute values of the reflection coefficient and phase. In the general case of arbitrary microwave frequency, the expressions for the derivative of the reflection are complicated, but in some specific cases, simple analytical expressions are possible.

Working principle

Figure 11.14 shows a waveguide structure that consists of a rectangular waveguide with an adjustable short. Assume that only the basic mode TE_{10} propagates in the rectangular waveguide and that the reflection is determined by the diagonal element of the conductivity tensor σ_{yy} only. It is because the dominant TE_{10} wave mode inside a rectangular waveguide possesses the only component E_y of the electric field that

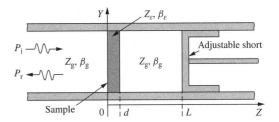

Figure 11.14 Electrodynamic scheme of the waveguide cell (Baturina *et al.* 1996). Source: Baturina, T. I. Borodovski, P. A. Studenikin, S. A. (1996). "Microwave waveguide method for the measurement of electron mobility and conductivity in GaAs/AlGaAs heterostructures", *Applied Physics A*, **63**, 293–298, by permission of Springer-Verlag GmbH & Co. KG.

the absorption of the microwave radiation is conditioned only by electron motion parallel to E_y. The diagonal element σ_{yy} of the conductivity tensor of the layer sample, subjected to an alternating electric field and static magnetic field, is given by Baturina *et al.* (1996)

$$\sigma \equiv \sigma_{yy} = \sigma_0 \frac{1 - i\omega\tau}{(1 - i\omega\tau)^2 + \mu_H^2 B^2} = \sigma_R + i\sigma_I \tag{11.19}$$

with

$$\sigma_R = \sigma_0 \frac{1 + \mu_H^2 B^2 + \omega^2\tau^2}{(1 - \omega^2\tau^2 + \mu_H^2 B^2)^2 + 4\omega^2\tau^2} \tag{11.20}$$

$$\sigma_I = \sigma_0 \frac{\omega\tau \, (1 + \mu_H^2 B^2 + \omega^2\tau^2)}{(1 - \omega^2\tau^2 + \mu_H^2 B^2)^2 + 4\omega^2\tau^2} \tag{11.21}$$

where $\sigma_0 = en\mu_H$ is the conductivity at zero magnetic field, $\mu_H = e\tau m^*$ is the carrier mobility, τ is the momentum relaxation time, m^* is the effective mass of carriers, n is the sheet carrier concentration, and $\omega = 2\pi f$ is the microwave angular frequency.

For the case of the layer sample in the geometry shown in Figure 11.14, the reflection coefficient is given by (Baturina *et al.* 1996)

$$\Gamma = \frac{1 - \overline{\sigma} - ia}{1 + \overline{\sigma} + ia} \tag{11.22}$$

with the normalized complex conductivity of the sample given by

$$\overline{\sigma} = \overline{\sigma}_R + i\overline{\sigma}_I = \sigma_{yy}(\omega, B) \frac{1}{\sqrt{1 - \lambda/\lambda_c}} \tag{11.23}$$

where λ is the wavelength in the free space, and λ_c is the cutoff wavelength in the waveguide. The normalized waveguide reactive conductivity, including the sample substrate, is given by

$$a = \frac{\left[\sqrt{\varepsilon}\tan(\beta_\varepsilon d)\tan(\beta_g l)/\sqrt{1 - \lambda/\lambda_c}\right] - 1}{\tan(\beta_g l) + \sqrt{\varepsilon}\tan(\beta_\varepsilon d)/\sqrt{1 - \lambda/\lambda_c}}$$

(11.24)

where ε is the dielectric constant of the substrate, the length of the shorted waveguide segment is $l = L - d$, and the propagation coefficients are given by

$$\beta_g = \frac{2\pi\sqrt{1 - \lambda/\lambda_c}}{\lambda}$$

(11.25)

$$\beta_\varepsilon = \frac{2\pi\sqrt{\varepsilon}\cdot\sqrt{1 - \lambda/\lambda_c}}{\lambda}$$

(11.26)

According to Eq. (11.22), the modulus of the reflection coefficient is given by

$$|\Gamma| = \sqrt{\frac{(1 - \overline{\sigma}_R)^2 + (\overline{\sigma}_I + a)^2}{(1 + \overline{\sigma}_R)^2 + (\overline{\sigma}_I + a)^2}}$$

(11.27)

Generally speaking, the derivative of the modulus of the reflection coefficient with respect to the magnetic field $d|\Gamma|/dB$ deduced from Eq. (11.27) is quite complicated. In the following, only several special cases are discussed.

Low-frequency limit ($\omega\tau \ll 1$)

At low frequencies, if the length of the shorted waveguide segment l after the sample is tuned to produce the minimum of the reflected power at $B = 0$, the reactive conductivity a of the microwave circuit at the plane of the sheet sample is equal to zero. In this condition, the reflection coefficient and its derivative are given by (Baturina 1996):

$$|\Gamma| = \left|\frac{1 + \mu_H^2 B^2 - \overline{\sigma}_0}{1 + \mu_H^2 B^2 + \overline{\sigma}_0}\right|$$

(11.28)

$$\frac{d|\Gamma|}{dB}$$

$$= 4\mu_H^2\overline{\sigma}_0 B\frac{1 + \mu_H^2 B^2 - \overline{\sigma}_0}{|1 + \mu_H^2 B^2 - \overline{\sigma}_0|(1 + \mu_H^2 B^2 + \overline{\sigma}_0)^2}$$

(11.29)

with

$$\overline{\sigma}_0 = \sigma_0\frac{1}{\sqrt{1 - \lambda/\lambda_c}}$$

(11.30)

When Eq. (11.29) has a maximum, we can obtain

$$1 + \overline{\sigma}_0 = 3\mu_H^2 B_{m1}^2$$

(11.31)

where B_{m1} is the magnetic field at which the derivative has a maximum in the case of $\omega\tau \ll 1$ and $a = 0$.

For the ideal case with $a = 0$ and $\omega\tau = 0$ exactly, the derivative $d|\Gamma|/dB$ has a jump from a negative to a positive value at the magnetic field B_{01} satisfying the following condition:

$$\overline{\sigma}_0 = 1 + \mu_H^2 B_{01}^2.$$

(11.32)

When a is small but not zero, the value of B_{01} can still be considered as the zero point of the derivative. Equation (11.32) can be fulfilled only for samples with sufficiently high conductivity $\overline{\sigma}_0 \geqslant 0$. The conductivity and carrier mobility of the sample can be determined by

$$\mu_H = \frac{\sqrt{2}}{\sqrt{3B_{m1}^2 - B_{01}^2}}$$

(11.33)

$$\sigma_0 = \frac{\sqrt{1 - \lambda/\lambda_c}}{120\pi}\cdot\frac{3B_{m1}^2 + B_{01}^2}{3B_{m1}^2 - B_{01}^2}$$

(11.34)

Figure 11.15(a) shows the $d|\Gamma|/dB$ curves corresponding to $\omega\tau = 0.1$ (solid line) and $\omega\tau = 0$ (dashed line), with the other parameters: $\overline{\sigma}_0 = 1.1$, $\mu_H = 1.13\,\mathrm{m}^2/\mathrm{Vs}$, and $m^* = 0.068m_e$. It is clear that the effect of $\omega\tau \neq 0$ substantially modifies curves and has minor influence on the positions of B_{01} and B_{m1}.

For samples with low conductivity ($\overline{\sigma}_0 < 1$), Eq. (11.31) is still applicable. In order to determine $\overline{\sigma}_0$, additional measurements are needed. This requires the measurements of $d|\Gamma|/dB$ under the condition $|a| \gg \overline{\sigma}_0$. This condition can be achieved by tuning the position of the short for the maximum of the reflected microwave power, because when the length of the terminated waveguide segment is changed, the reactive conductivity

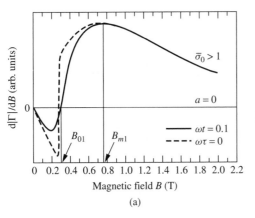
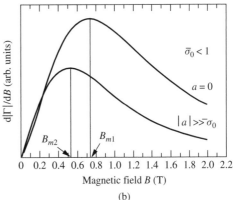

(a)

(b)

Figure 11.15 Calculated curves of the derivative of the modulus of the reflection coefficient with respect to the magnetic field against the magnetic field in the case $\omega\tau \ll 1$. (a) $\overline{\sigma}_0 = 1.1$ and (b) $\overline{\sigma}_0 = 0.9$ (Baturina *et al.* 1996). Source: Baturina, T. I. Borodovski, P. A. Studenikin, S. A. (1996). "Microwave waveguide method for the measurement of electron mobility and conductivity in GaAs/AlGaAs heterostructures", *Applied Physics A*, **63**, 293–298, by permission of Springer-Verlag GmbH & Co. KG.

of the waveguide is subsequently changed. Under this condition, Eq. (11.27) can be approximated as

$$|\Gamma| \approx 1 - \frac{2\overline{\sigma}_0}{a^2(1 + \mu_H^2 B^2)} \quad (11.35)$$

$$\frac{d|\Gamma|}{dB} \approx \frac{4\mu_H\overline{\sigma}_0}{a^2} \cdot \frac{\mu_H B}{(1 + \mu_H^2 B^2)^2} \quad (11.36)$$

Equation (11.36) has a maximum at B_{m2} given by

$$3\mu_H^2 B_{m2}^2 = 1 \quad (11.37)$$

Therefore, in this case ($\overline{\sigma}_0 < 1$), two measurements should be made at two positions of the adjustable short corresponding to the maximum ($|a| \gg \overline{\sigma}_0$) and minimum ($a = 0$) of the reflected power at $B = 0$. Figure 11.15(b) shows the curves for this case with initial parameters $\omega\tau = 0.1$, $\overline{\sigma}_0 = 0.9$, $\mu = 1.13 \ \text{m}^2/\text{Vs}$, and $m^* = 0.068m_e$.

High-frequency regime

For high frequencies (and/or high mobilities), the conductivity becomes complex and the expression for $d|\Gamma|/dB$ becomes complicated (Baturina *et al.* 1996). However, simple expressions are available for the two limit cases $|a| \gg \overline{\sigma}_0$ and $a = 0$. In experiments, by measuring the magnetic field dependences of the derivative $d|\Gamma|/dB$ at different

positions of the adjustable short, the carrier mobility and conductivity of a layer sample can be determined. Similar to the case in low-frequency regime, this method uses only relative measurements and does not need absolute value of the reflection coefficient and phase of the reflection signal. A more detailed discussion for high-frequency regime can be found in (Baturina *et al.* 1996).

Experimental setup

In experiments, two types of magnetic fields are needed. As shown in Figure 11.16(a), the measurement cell is placed between two magnetic poles, and the dc magnetic field is along the z-axis of the rectangular waveguide. The sample-loading cell is shown in detail in Figure 11.16(b). The layer sample on a substrate is clamped between two plates, with the layer sample facing the incident wave. The sample completely covers the waveguide cross section. The waveguide section behind the sample is terminated by adjustable short, which enables one to adjust the value of the reactive conductivity in the plane of the layer sample. For the modulation of the magnetic field, a two-section coil is placed near the sample.

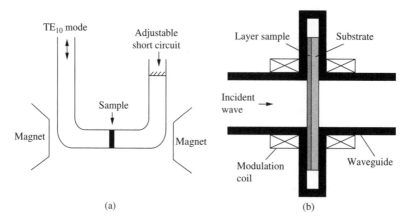

(a)   (b)

Figure 11.16   Experimental set. (a) Application of dc magnetic field and (b) application of the modulating magnetic field

11.2.3.2 Cross-slit probe method

Lann *et al.* proposed a near-field microwave scanning probe, which allows local polarimetric measurements with a subwavelength spatial resolution (Lann *et al.* 1999). The probe is a symmetrical transmitting/receiving antenna formed by two orthogonal slits at the end plate of a circular waveguide. If a linearly polarized incident wave passes through one slit, and the wave reflected from the sample mounted in the near field of the probe generally has two components perpendicular to each other, the reflected wave component with the incident polarization is received by the same slit, while the orthogonal polarization component is received by the orthogonal slit. From the cross-polarization, the information about the Hall mobility of the sample can be obtained. So, such a probe can be used in mapping the Hall mobilities in semiconductor wafers, and, in principle, its spatial resolution is determined by the common area of the two slits.

Figure 11.17(a) shows the basic structure of a cross-slit probe. The fundamental working principle for a slit aperture as a transmitting/receiving antenna has been discussed in Section 3.6. The probe is fabricated on the surface of a ball-shaped polymethyl-methacrylate (PMMA) insert protruding from the end of a circular waveguide (Lann *et al.* 1999). A layer of thick silver film is deposited on the insert through a mask formed by two perpendicularly placed thin varnished wires.

The length of each slit is approximately $\lambda/2$, while the width w is determined by the mask and can be very small.

Figure 11.17(b) shows the measurement setup. The circular waveguide with the probe is attached to the circular output of an orthomode transducer. A linearly polarized incident wave is launched

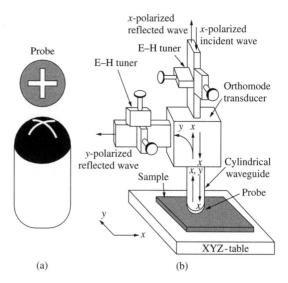

(a)   (b)

Figure 11.17   A cross-slit near-field microwave microscope (Lann *et al.* 1999). (a) A cross-slit probe and (b) measurement configuration. Source: Lann, A. F. Golosovsky, M. Davidov, D. and Frenkel, A. (1999). "Microwave near-field polarimetry", *Applied Physics Letters*, **75** (5), 603–605

through the rectangular input port of the orthomode transducer and comes out through the slit along the longer dimension of this input (regular slit). This wave is reflected from the sample mounted in close proximity to the probe. The reflected wave with the same polarization returns through the same slit and comes out through the same rectangular port of the orthomode transducer. If there is a change of polarization upon reflection from the sample, the orthogonal polarization component of the reflected wave passes through the orthogonal slit (cross-slit) to the other rectangular port of the orthomode transducer. An impedance matching component (E–H tuner) is inserted into each of the two rectangular ports of the orthomode transducer. The E–H tuner in the input port is tuned in such a way that in the absence of the sample there is no reflection. The E–H tuner in the orthogonal port is tuned so that maximum signal with orthogonal polarization can be obtained.

In experiments, there may be a cross-coupling between the two ports owing to incomplete isolation of the orthomode transducer and owing to deviation of the slits from orthogonality. In order to minimize this effect, one may measure the magnitude and phase of the cross-coupling signals in the absence of the sample and in the presence of a homogeneous metallic sample and determine the full cross-coupling matrix. In the analysis of the measurement results of samples under test, the cross-coupling signals can be vectorially subtracted (Lann *et al.* 1999).

The signal transmitted from a slit can be approximated as a plane wave. Assuming an x-polarized plane wave normally incident on a flat surface and neglecting the transmitted wave, the relation between the fields in the incident and reflected waves is given by

$$E_{\text{refl}} = \frac{R - Z_0}{R + Z_0} E_{\text{inc}} \qquad (11.38)$$

where R is the surface impedance tensor of the sample and Z_0 is the impedance of free space. According to Eq. (11.10), when an external dc magnetic field is the z direction, R can be expressed in the form of

$$R = \begin{bmatrix} R_{xx} & R_{xy} \\ -R_{xy} & R_{yy} \end{bmatrix} \qquad (11.39)$$

where the component R_{xy} reflects the Hall effect. For thin samples with the thickness d much smaller than the skin depth δ, the surface impedance should be replaced by the sheet resistance tensor $R = \rho/d$; for dielectrics, it should be replaced by $(\mu/\varepsilon)^{1/2}$ (Lann *et al.* 1999).

From Eqs. (11.38) and (11.39), the components along the x and y directions of the reflected wave can be obtained

$$E^x_{\text{refl}} = \frac{R^* - Z_0}{R^* + Z_0} E^x_{\text{inc}} \qquad (11.40)$$

$$E^y_{\text{refl}} = -\frac{2Z_0 R_{xy}}{(R^* + Z_0)(R_{yy} + Z_0)} E^x_{\text{inc}} \qquad (11.41)$$

with

$$R^* = R_{xx} + \frac{R^2_{xy}}{Z_0 + R_{yy}} \qquad (11.42)$$

For highly conductive samples ($R^* \ll Z_0$), Eqs. (11.40) and (11.41) can be simplified

$$E^x_{\text{refl}} \propto (1 - 2R^*/Z_0) \qquad (11.43)$$

$$E^y_{\text{refl}} \propto R_{xy}/Z_0 \qquad (11.44)$$

Equation (11.44) indicates that the orthogonal polarization component of the reflected wave is proportional to the off-diagonal element of the surface impedance tensor, which is related to the Hall mobility of the sample.

11.2.4 Turnstile-junction method

The angle of rotation α of the polarization plane of an elliptically polarized wave can be determined on the basis of the measurement of the amplitudes of two perpendicular components of the wave and measurement of the phase difference between them (Musil and Zacek 1986). As shown in Figure 11.18, an elliptically polarized wave can be decomposed into two perpendicular components oscillating in the x and y directions with amplitudes a and b, mutually phase-shifted by an angle φ:

$$E_x = a \cos(\omega t) \qquad (11.45)$$

$$E_y = b \cos(\omega t + \varphi) \qquad (11.46)$$

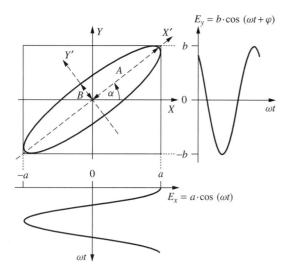

$E_y = b \cdot \cos (\omega t + \varphi)$

$E_x = a \cdot \cos (\omega t)$

Figure 11.18 Principle of the formation of polarization ellipse. Modified from Musil, J. and Zacek, F. (1986). *Microwave Measurements of Complex Permittivity by Free-space Methods and their Applications*, Amsterdam, with permission from Elsevier

From Eqs. (11.45) and (11.46), an ellipse equation in the $x - y$ coordinate system can be obtained

$$\left(\frac{E_x}{a}\right)^2 - 2\frac{E_x E_y}{ab}\cos\varphi + \left(\frac{E_y}{b}\right)^2 = \sin^2\varphi \tag{11.47}$$

This ellipse is inscribed within a rectangle with sides $2a$ and $2b$. The ellipse is expressed in terms of maximum amplitudes a and b of two perpendicular components E_x and E_y, respectively, and by means of the phase shift φ between these components. If parameters a, b, and φ are known, the angle of rotation of the ellipse α and its semi-axes A and B may be expressed by (Musil and Zacek 1986)

$$\tan(2\alpha) = \frac{2ab}{a^2 - b^2}\cos\varphi = \tan(2\gamma)\cos\varphi \tag{11.48}$$

$$\sin(2\beta) = \sin(2\gamma)\sin\varphi \tag{11.49}$$

with

$$\tan\gamma \equiv \frac{b}{a} \qquad \left(0 < \gamma < \frac{\pi}{2}\right) \tag{11.50}$$

$$\tan\beta \equiv \frac{B}{A} \qquad \left(0 < |\beta| < \frac{\pi}{4}\right). \tag{11.51}$$

According to Eqs. (11.48) to (11.51), from the amplitudes of the field components in the x- and y-directions and the phase shift φ between these components, both the angle of rotation of the polarization plane α and the ellipticity $e = B/A$ of the ellipse can be determined, from which the electrical transport properties of the sample can be derived.

Figure 11.19 shows the structure of a turnstile junction, which is a six-port device formed by joining one circular waveguide perpendicularly to two orthogonal rectangular waveguides. Using a turnstile junction, an elliptically polarized wave can be decomposed into two perpendicular components. The properties of a turnstile junction can be described by its scattering matrix $[S]$:

$$[S] = \frac{1}{2}\begin{bmatrix} 0 & 1 & 0 & 1 & \sqrt{2} & 0 \\ 1 & 0 & 1 & 0 & 0 & \sqrt{2} \\ 0 & 1 & 0 & 1 & -\sqrt{2} & 0 \\ 1 & 0 & 1 & 0 & 0 & -\sqrt{2} \\ \sqrt{2} & 0 & -\sqrt{2} & 0 & 0 & 0 \\ 0 & \sqrt{2} & 0 & -\sqrt{2} & 0 & 0 \end{bmatrix} \tag{11.52}$$

When an elliptically polarized wave enters the circular waveguide port with its E_x component along the arm 6 and E_y component along the arm 5, the wave components at the four rectangular

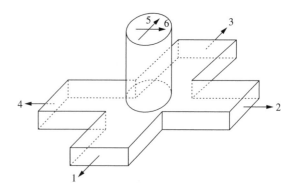

Figure 11.19 Schematic sketch of turnstile junction. Reprinted from Musil, J. and Zacek, F. (1986). *Microwave Measurements of Complex Permittivity by Free-space Methods and their Applications*, Amsterdam, with permission from Elsevier

waveguide arms are given by

$$E_1 = \frac{b}{\sqrt{2}} \cos(\omega t + \varphi) \quad (11.53)$$

$$E_2 = \frac{a}{\sqrt{2}} \cos(\omega t) \quad (11.54)$$

$$E_3 = -\frac{b}{\sqrt{2}} \cos(\omega t + \varphi) \quad (11.55)$$

$$E_4 = -\frac{a}{\sqrt{2}} \cos(\omega t) \quad (11.56)$$

Equations (11.53) to (11.56) indicate that a wave entering a turnstile junction is divided into different rectangular waveguide ports. The components of a wave entering the circular waveguide in direction 5 emerge from the arms 1 and 3, while the components of a wave entering the junction in direction 6 emerge from the arms 2 and 4. The signal in the arm 3 is in an opposite phase to the signal in the arm 1, and the signal in the arm 4 is in an opposite phase to the signal in the arm 2. In order to achieve high measurement accuracy, high isolation between the neighboring rectangular waveguides and between directions 5 and 6 in the circular waveguide is required.

For the determination of the rotation angle α of the polarization plane and the ellipticity e, it is sufficient to use signals from the arms 1 and 2, and usually the arms 3 and 4 are terminated by matching loads. More discussion on the turnstile-junction method for the measurement of the electrical transport properties of materials can be found in (Musil and Zacek 1986).

11.3 RESONANT METHODS FOR THE MEASUREMENT OF MICROWAVE HALL EFFECT

Besides transmission and reflection systems as discussed above, microwave Hall effects can also be observed in resonant structures. For a resonator with two orthogonal degenerate modes, when a sample is placed at the electric field antinodes of the two modes, and an external dc magnetic field is applied perpendicular to the microwave electric fields, the coupling between the two modes due to the dc magnetic field is related to the Hall mobility

of the sample. Such a bimodal resonator is usually called microwave Hall effect (MHE) resonator.

In the following, we first discuss the coupling between the two orthogonal resonant modes of a bimodal cavity due to microwave Hall effect and the typical circuit for the measurement of microwave Hall effect, and then discuss several typical MHE resonators.

11.3.1 Coupling between two orthogonal resonant modes

Figure 11.20 schematically shows the origin of resonant microwave Hall effect. When a linearly polarized driving microwave signal excites one resonant mode, the carriers in the sample move accordingly. If an external dc magnetic field is applied to the sample in the direction perpendicular to the direction of the electric field of the driving microwave signal, the carriers in the sample move in ellipsoid orbits. The resultant motion of carriers has a component perpendicular to the electric field of the driving microwave signal and the external magnetic field, and this component excites another resonant mode, called the MHE mode, perpendicular to the driving microwave signal and the external dc magnetic field. In this sense, microwave Hall effect can be regarded as the coupling between the two orthogonal modes in a

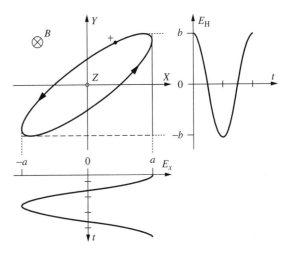

Figure 11.20 The origin of resonant microwave Hall effect

bimodal cavity due to the application of an external dc magnetic field. From the relationships between the input and output signals, the microwave Hall mobility can be deduced.

To observe resonant microwave Hall effect, an MHE resonator is needed. The basic requirement for an MHE resonator is that the resonator has two perpendicular and isolated resonant modes with the same resonant frequency. Meanwhile, in the resonator, there exists a region where the electric fields of the two perpendicular modes coexist, and in the measurement of microwave Hall effect, the sample is placed in this region.

Figure 11.21 shows a basic system for the measurement of microwave Hall effect. A dc magnetic field B is applied to the sample in the direction perpendicular to the electric fields of the two resonant modes. A network analyzer is used to conduct microwave measurements. As the primary and secondary modes of the MHE resonator are perpendicular to each other, in an ideal case, no microwave signals can be transmitted between the two modes. However, in actual experiments, after the sample is loaded, in the absence of

a dc magnetic field, some microwave signal, at times greater than the signal associated with the Hall effect, could be detected from the secondary mode owing to the nonideality of the MHE resonator. Imperfections in the geometry of the resonator, impurities in the cavity material, and the aberrations caused by the coupling and tuning structures of the MHE resonator contribute to this nonideal mode coupling. This nonideal transmitted signal reduces the sensitivity of the measurement of the microwave Hall effect.

The canceling channel composed of a variable attenuator and a phase shifter is used to reduce this nonideal transmitted microwave signal. A portion of the primary mode microwave signal is taken by a directional coupler (labeled 2) and sent to the canceling channel. If the canceling channel is tuned properly, at the second directional coupler (labeled 5) the nonideal transmitted signal and the signal from the canceling channel can have the same amplitude but a phase difference of 180°, so the nonideal transmitted signal can be cancelled out, and no microwave signal can be detected when no external dc magnetic field is applied. When an external dc magnetic field is applied, the microwave signal at the secondary mode causes the microwave Hall effect, and it can then be sensitively detected.

In microwave measurements, the coupling between the primary and secondary modes is described by S_{21}. Figure 11.22 shows a typical change of S_{21} due to the microwave Hall effect. When the canceling channel is not in operation, there is a resonant peak in the S_{21} curve corresponding to the resonant peak in the S_{11} curve. When the phase shifter and attenuator are suitably adjusted, there is a sharp absorption peak in the S_{21} curve (solid S_{21} curve). When an external dc magnetic field is applied, the value of S_{21} increases (dashed S_{21} curve). The change of the transmitted signal (ΔS_{21}) corresponds to the Hall effect of the sample.

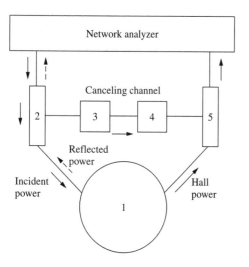

Figure 11.21   Basic measurement system for microwave Hall effect. 1. MHE resonator, 2. directional coupler, 3. attenuator, 4. phase shifter, and 5. directional coupler. Modified from Na, B. K. Vannice, M. A. and Walters, A. B. (1992). "Measurement of the effect of pretreatment and adsorption on the electrical properties of ZnO powders using a microwave-Hall-effect technique", *Physical Review B*, **46** (19), 12266–12277

11.3.2 Hall effect of materials in MHE cavity

Hollow metallic MHE cavities are widely used in the characterization of the electrical transport properties of materials. As shown in Figure 11.23, the sample under test is placed within the MHE

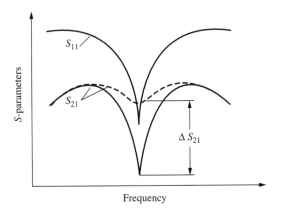

Figure 11.22   Measurement of microwave Hall effect. The trace on the top is the S_{11} curve indicating the resonant frequency of the MHE resonator. The solid S_{21} curve represents the state in the absence of a dc magnetic field, and the dashed S_{21} curve represents the state when an external magnetic field is applied

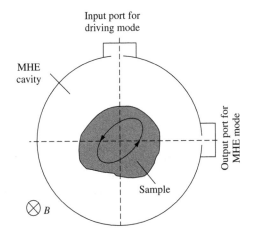

Figure 11.23   Microwave Hall effect of a sample in an MHE cavity

cavity. Under the effects of the driving mode and the external dc magnetic field, the carriers in the sample move in elliptic orbits, and the component of the motion perpendicular to the driving mode and the external dc magnetic field result in the MHE mode.

Hollow metallic MHE cavity is suitable for low-conductivity samples, and the sample may be in the form of powders, bulks, or liquids. Generally speaking, MHE methods are superior to dc and

ac Hall effect methods in the study of electrical transport properties of powder samples because it requires no electrode contact. The properties of powders may be affected by their sizes, preparation methods, degree of purity, and the prevailing conditions (such as gas adsorption), and MHE methods are useful for research on how these aspects affect the properties of powders.

Now we describe resonant microwave Hall effect using macroscopic parameters. The resistivity tensor $[\rho]$ of a material relates the electric field E_i with the current density J_j through

$$E_i = \sum_j \rho_{ij} J_j \qquad (11.57)$$

where $i = x,\ y,\ z;\ j = x,\ y,\ z;$ and ρ_{ij} is the component of the resistivity tensor. When a dc magnetic field is applied along the z direction, an isotropic medium has a resistivity tensor in the form of (Buschow 1991)

$$[\rho] = \begin{bmatrix} \rho_a & -\rho_H & 0 \\ \rho_H & \rho_a & 0 \\ 0 & 0 & \rho_b \end{bmatrix} \qquad (11.58)$$

where ρ_a and ρ_b are the resistivities for the electrical currents applied perpendicular and parallel to the direction of magnetization, and ρ_H is the Hall resistivity.

When a sample is placed in an MHE cavity, in the presence of a dc magnetic field perpendicular to both of the two electric fields of the two orthogonal modes, the resistivity of the sample is a tensor in the form of Eq. (11.58). The Hall resistance ρ_H causes the coupling between the two orthogonal modes. The power transmission between the two orthogonal modes due to the application of the dc magnetic field perpendicular to the electric fields, embodies the microwave Hall effect.

MHE cavity techniques are often combined with the cavity-perturbation method to characterize the electrical transport properties of samples. From the changes in the resonant frequency f and in the quality factor Q of the cavity due to the introduction of a small sample, and the change of the power transmission between the two orthogonal modes due to the application of a dc magnetic field, the electrical transport properties of the sample can be deduced. MHE cavity techniques have been developed over decades. After Portis and Teaney first introduced the bimodal cavity for measurements of small samples (Portis and Teaney 1958), MHE cavity techniques have been used in various fields. Using MHE cavity techniques, Sayed *et al.* studied low-mobility semiconductors and insulators (Sayed and Westgate 1975a, b); Ong *et al.* studied the electron transport in TTF-TCNQ (Ong and Portis 1977; Ong *et al.* 1981); Caverly *et al.* measured the microwave Hall mobility of copper phthalocyanine single crystals (Caverly and Westgate 1982); Na *et al.* studied the electrical properties of ZnO powders (Na *et al.* 1993, 1992); and Chen *et al.* studied the electrical transport properties of magnetic powders (Chen *et al.* 1998).

11.3.2.1 Measurement of microwave Hall mobility

Figure 11.24 shows a TE_{11n} MHE cylindrical cavity and the sample is placed at the antinode of the electric field. For a specially designed cavity, the two modes can be adjusted to be orthogonal, and degenerate in frequency, and of equal amplitude (i.e., $E_a = E_b$). Before a dc magnetic field B is applied, the output mode is not excited, that is, $E_2 = 0$. When an external dc magnetic field is applied, the relationship between the output electric field E_2 and input electric field E_1 is a measure of the microwave Hall mobility μ_H (Na *et al.* 1991):

$$\left(\frac{E_2}{E_1}\right) = K_\mu B \left(\frac{Q_0 - Q_L}{Q_0}\right) \mu_H \qquad (11.59)$$

where Q_0 and Q_L are the quality factors of the cavity before and after the introduction of the sample respectively; B is the dc magnetic field; K_μ is a constant determined by the structural parameters of the MHE cavity. In experiments, K_μ can be obtained by calibration.

By taking the bimodal cavity as a two-port network, Eq. (11.59) can be rewritten as

$$\mu_H = \frac{1}{K_\mu B}\left(\frac{Q_0}{Q_0 - Q_L}\right)(S_{21}) \qquad (11.60)$$

The subscripts 1 and 2 indicate the input port and output port respectively (shown in Figure 11.24).

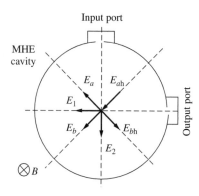

Figure 11.24   The origin of the microwave Hall effect in a bimodal cavity. $E_1 = E_a + E_b$, $E_2 = E_{ah} + E_{bh}$, where E_{ah} is the Hall field of mode a, and E_{bh} is Hall field of mode b

Equation (11.60) shows that the mobility of the sample can be calculated from the change of quality factor due to the introduction of the sample, and the scattering coefficient S_{21} due to the application of the external dc magnetic field B.

11.3.2.2 Determination of the sign of charge carriers

From the phase relationship between the output and input microwave fields, the sign of the charge carriers can be determined, but the phase information is ignored in the above discussion. However, the sign of the carriers can still be determined by calibration or by introducing a small imbalance into the cavity as described below.

In an actual MHE cavity, because of the structural asymmetries and manufacturing errors, the two degenerate modes in the bimodal cavity are not strictly orthogonal. Even if the two modes are adjusted to be "orthogonal," and the canceling channel is used, there may be power transmission between the two modes when no dc magnetic field is applied. When a magnetic field is applied, the total output microwave field E_2' is a vector sum of two parts: $E_2' = E_{20} + E_2$, where E_{20} corresponds to the power transmission due to the structural asymmetries and manufacturing errors, and E_2 corresponds to the power transmission due to the Hall effect. E_{20} does not contain any information about the Hall effect, and does not change with the application of the dc magnetic field.

E_{20} and E_2 can be in either the same phase or the opposite phase depending on the phase of E_2 that is dependent on the sign of the charge carriers of the sample under study. If E_{20} and E_2 are in the same phase, the S_{21} value increases after the application of the magnetic field; and if E_{20} and E_2 are in the opposite phase, the S_{21} value decreases after the application of the magnetic field. As such, the sign of the charge carriers can be determined by calibration. For example, if we know that S_{21} increases after a dc magnetic field is applied for positive carriers, the decrease of S_{21} indicates that the charge carriers are negative.

Besides the calibration method, Caverly and Westgate used another approach by introducing a small imbalance into the cavity by adjusting the amplitude of one of the two modes (E_a and E_b in Figure 11.24) to be a little smaller than that of the other one (Caverly and Westgate 1982). The imbalance between the two modes causes the value of E_2 to either increase or decrease, depending on the choice of the reduced mode, direction of the dc magnetic field, and the sign of the carriers. Thus, the sign of the charge carriers can be determined from the change of the output microwave field (Sayed and Westgate 1975b; Na *et al.* 1992). Actually, this method increases the coupling between the two modes before the dc magnetic field is applied.

11.3.2.3 Measurement of microwave conductivity

As discussed in Chapter 1, the conductivity σ of a material can be determined from the imaginary part of the complex permittivity:

$$\varepsilon_r'' = \frac{\sigma}{\omega \varepsilon_0} = \frac{\sigma}{2\pi f \varepsilon_0} \quad (11.61)$$

where $\varepsilon_0 = 8.85 \times 10^{-12}$ F/m is the permittivity of the vacuum. So the microwave conductivity can be measured by the cavity-perturbation method.

On the basis of resonant perturbation theory, the conductivity σ of the sample can be measured from the change of the quality factors due to the introduction of the sample (Na *et al.* 1993; Liu *et al.* 1994; Jow *et al.* 1989). For low-conductivity

samples, the conductivity can be obtained from (Na *et al.* 1993, 1992)

$$\sigma = 2.78 \times 10^{-13} f \left(\frac{Q_0 - Q_L}{Q_0 Q_L} \right) \frac{1}{\alpha} \quad (11.62)$$

and for intermediate conductivity samples, the conductivity can be obtained from (Na *et al.* 1993, 1992)

$$\sigma = 1.11 \times 10^{-12} f \left(\frac{\alpha}{N} \right)^2 \left(\frac{Q_0 Q_L}{Q_0 - Q_L} \right) \frac{1}{\alpha} \quad (11.63)$$

where α is the filling factor and N is a depolarization factor. For a TE_{112} cylindrical cavity,

$$\frac{\alpha}{N} = \frac{\Delta f}{f}, \quad (11.64)$$

$$\alpha = 2 \frac{V_s}{V_c} \quad (11.65)$$

where V_s and V_c are the volumes of the sample and the cavity respectively. The unit of the conductivity σ calculated using Eqs. (11.62) and (11.63) is $(ohm-cm)^{-1}$. After obtaining σ and μ_H, the density of carriers n can be obtained according to Eq. (11.1).

11.3.2.4 Typical MHE cavities

MHE cavities are the key components in the measurements of microwave Hall effect. In the design of an MHE cavity, two requirements should be satisfied. First, the cavity should support two isolated and perpendicular degenerate modes. Second, to increase the measurement sensitivity, at the place for sample loading, the electric fields of the two modes should have their maximum values. Usually, cavities with simple structures are preferred, such as circular cylindrical cavities and rectangular cavities.

Circular cylindrical MHE cavity

In a circular cylindrical MHE cavity, usually a TE_{11n} mode is used. As discussed in Chapter 2, a circular TE_{11n} mode has two perpendicular degenerate modes, and the electric field distributions near the maximum values are relatively uniform. These

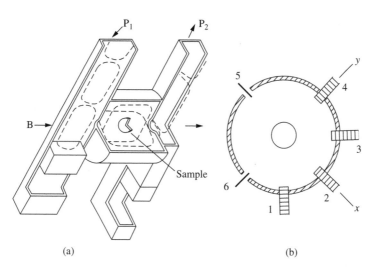

(a)   (b)

Figure 11.25   A circular cylindrical TE$_{111}$ bimodal cavity. (a) Schematic diagram of bimodal cavity with input and output coupled modes and matching waveguide terminations (Eley and Lockhart 1983). Source: Eley, D. D. and Lockhart, N. C. (1983). "Microwave Hall effects in low-mobility materials", *J. Phys. E: Sci. Instrum.* **16**, 47–51 and (b) section view perpendicular to the axis of the bimodal cavity showing coupling iris and tuning plugs. Four capacitive plugs (labeled 1, 2, 3, and 4) and two resistive plugs (labeled 5 and 6) are employed to balance the cavity (Portis and Teaney 1958). Source: Portis, A. M. and Teaney, D. (1958) "Microwave Faraday rotation: design and analysis of a bimodal cavity", *Journal of Applied Physics*, **29** (12), 1692–1698

properties are advantageous for the design of MHE cavities.

Figure 11.25(a) schematically shows the basic structure of a bimodal circular cylindrical cavity, and the dotted lines represent the microwave magnetic field (Portis and Teaney 1958; Eley and Lockhart 1983). The cavity is excited in a cylindrical TE$_{111}$ mode through an iris in the narrow face of the waveguide on the left. This waveguide is terminated an odd number of quarter wavelengths below the entrance iris in order to conveniently match the cavity to the line. The exit iris couples into the broad face of a second waveguide. For matched conditions, this waveguide is terminated an even number of quarter wavelengths below the iris. If the waveguides are parallel and the cavity is perfectly cylindrical, there is no coupling into the second waveguide. This can be explained by the symmetry of the arrangement. The magnetic fields in the first waveguide and in the cavity are symmetrical with respect to the plane of the paper. The propagating mode in the second waveguide is, however, antisymmetrical with respect to the plane of the paper, and so there

is no coupling from the cavity to this mode. If the cavity contains a sample under a dc magnetic field, the microwave magnetic field in the cavity will no longer be symmetrical and there will be coupling to the second waveguide.

Figure 11.25(b) is a sectional view perpendicular to the axis of the cavity showing the arrangement of capacitive and resistive tuning plugs. Plugs 1, 2, 3, and 4 are metallic and may be adjusted so that the two modes are degenerate. One can arrange them so that either plug 2 or plug 4 is well out of the cavity, so that a quite sensitive final adjustment may be made with that plug. However, the cavity degeneracy alone does not insure that the two waveguides are decoupled. Consider the normal modes as having microwave magnetic fields oriented along the $-x$ and $+y$ axes. These two modes are driven equally by a vertical microwave magnetic field. If the two modes have different quality factors, the amplitudes of their responses are different. Then, the cavity magnetic field will not be vertical and there will be coupling to the second waveguide. This coupling may be avoided by artificially introducing additional losses into

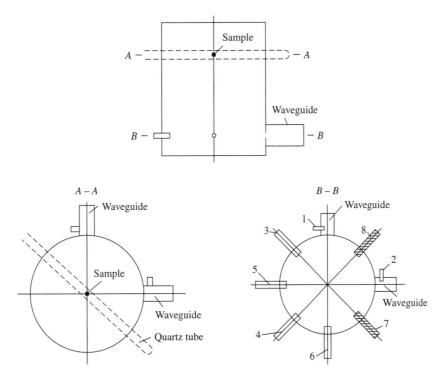

Figure 11.26   The schematic drawing of the TE_{112} bimodal resonant cavity. 1, 2: coupling screw, 3, 4, 5, 6: capacitive tuning bar, 7, 8: resistive tuning bar. Source: Chen, L. F. Ong, C. K. and Tan, B. T. G. (1998). "Study of electrical transport properties of fine magnetic particles using microwave Hall effect techniques", *IEEE Transactions on Magnetics*, **34** (1), 272–277. © 2003 IEEE

one of the modes. This can be accomplished by adjusting the resistive plugs 5 and 6, which are made from graphite-deposited resistance card (Portis and Teaney 1958).

The TE_{112} mode is also often used in the measurement of microwave Hall effect. Figure 11.26 shows a circular cylindrical MHE cavity working at TE_{112} mode. The TE_{112} mode gives more space for installing tuning bars. In Figure 11.26, the perturbation area is at the upper part while the coupling structure is at the lower part of the cavity. So the dominant field at the coupling part can be kept almost unchanged after the sample is introduced, ensuring that the coupling mechanism and the field pattern of the mode remain the same. In this structure, the coupling screws (labeled 1 and 2) are used to adjust the coupling conditions between the waveguides and the cavity. Four capacitive tuning bars (labeled 3, 4, 5, and 6) are metallic and may be adjusted so that the two TE_{112} modes are degenerate

and orthogonal. Two resistive tuning bars (labeled 7 and 8) are graphite-deposited, and may be adjusted so that the amplitudes or the quality factors of the two modes are equal. In the determination of the sign of the carriers, the two resistive tuning bars may be used to introduce an imbalance to the bimodal cavity.

Rectangular MHE cavity

As discussed in Section 2.3.4, for a rectangular cavity with a square section, its TE_{101} and TE_{011} modes are degenerate. These two degenerate modes can be used in the measurement of the microwave Hall effect (Cross *et al.* 1987). As shown in Figure 11.27, the MHE cavity is excited by electric dipoles, and its fundamental modes are TE_{101} and TE_{011}. The sample under test is held at the electric field antinode at the center of the cavity using a sample holder with a low dielectric

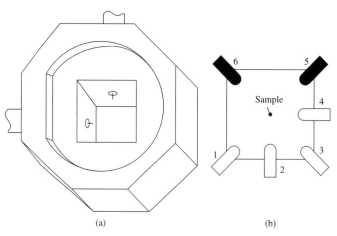

(a)

(b)

Figure 11.27 $TE_{101/011}$ bimodal cavity for the measurement of microwave Hall effect (Cross *et al.* 1987). (a) Schematic view without the top and bottom plates and (b) positions of the four capacitive and two resistive tuning stubs. Modified from Cross, T. E. De Cogan, D. Al Zoubi, A. Y. (1987). "Microwave Hall effect studies of low mobility materials", *New Materials and Their Application*, Warwick, Inst. Phys. Conf. Ser. No. 89, 301–306, by permission of IOP Publishing Ltd

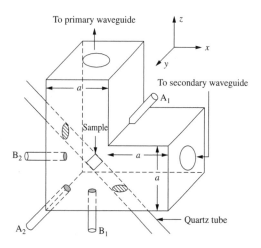

Figure 11.28 Schematic drawing of a bimodal rectangular cavity and sample tube positioning. A and B represent cavity-tuning screws (Liu *et al.* 1994). Modified from Liu, C. C. Na, B. K. Walters, A. B. Vannice, M. A. (1994). "Microwave absorption measurements of the electrical conductivity of small particles", *Catalysis Letters*, **26**, 9–24, by permission of J.C. Baltzer, Science Publishers

constant and a low loss tangent. Four capacitive stubs (labeled 1, 2, 3, and 4) and two resistive stubs (labeled 5 and 6) are used for reactive and resistive tuning of the cavity to produce a maximum degeneracy of the two TE modes.

Similar to the case of circular cylindrical MHE cavities, in order to separate the sample-loading region and the coupling region, higher-order modes can be used. Figure 11.28 shows a bimodal TE_{102} rectangular cavity (Liu *et al.* 1994). The two TE_{102} modes share half of each of their length. In the region shared by two modes, the electric field distributions are the same as those of a TE_{101} and TE_{011} bimodal cavity. The sample under test is loaded at the antinodes of electric fields of the two modes. The cavity is coupled to external waveguides through two coupling irises, and the couplings can be adjusted using two screws (not shown in the figure) near the coupling irises. The cavity has four cavity tuning screws. The two perpendicular TE_{102} modes are built by adjusting the tuning screws A_1 and A_2, and the resonant frequencies of the two modes are equalized by adjusting the tuning screws B_1 and B_2.

11.3.3 Hall effect of endplate of MHE cavity

As discussed above, in the measurement of microwave Hall effect, the material under test is usually placed within an MHE cavity and subjected to an external dc magnetic field. Owing to the microwave Hall effect of the material, there is a power transfer between the input mode and the output orthogonal mode. The measurement sensitivity is limited by a signal that appears when an external magnetic field is applied to the empty MHE cavity. This unwanted signal is superimposed on the signal from the sample and may become dominant for materials of low mobility and high resistivity. It has been suggested that this empty cavity signal is due to the Hall effect in the cavity end walls (Trukham 1966), and it has been shown that this signal is indeed consistent with that produced by the Hall effect in the end walls (Fletcher 1976).

Al Zoubi *et al.* proposed a theoretical model that describes the empty cavity signal in a rectangular MHE cavity with a square section operating at TE_{011} and TE_{101} modes (Al Zoubi 1990, 1991a, b). As shown in Figure 11.29, the square-section rectangular bimodal cavity of cross-sectional side a and length d supports two independent orthogonal modes. The cavity is iris-coupled to the matched coaxial probes so that, in principle, there is complete isolation between the input and the output. In the case of perfect symmetry, only the dominant TE_{011} mode coupled to the input is excited. Owing to the Hall effect in the end walls, the application of an external magnetic field B along the cavity axis causes the mode I (TE_{011}) to couple to the orthogonal mode II (TE_{101}). This results in a transmission of a portion of the incident power P_i to the output P_o, which is usually referred to as the empty cavity signal.

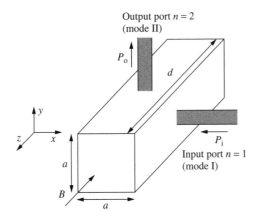

Figure 11.29 A bimodal rectangular cavity with a square section

If the cavity is critically coupled at both the input and the output ports, and tuned for equal resonant frequency and quality factor using special tuning stubs, then the Hall mobility μ_H of the end wall of the cavity can be found:

$$\mu_H = \left(\frac{R}{B}\right) \cdot S_{21} \qquad (11.66)$$

where R is a cavity geometry coefficient given by

$$R = \left(\frac{\pi^2}{8}\right) \cdot \left[2 + \left(\frac{d}{a}\right) + 3 \cdot \left(\frac{d}{a}\right)^3\right] \qquad (11.67)$$

Eq. (11.66) can be rewritten as

$$S_{21} = B \cdot \left(\frac{\mu_H}{R}\right) \qquad (11.68)$$

Equation (11.68) indicates that the unwanted empty cavity signal is dependent upon the cavity dimensions as well as the end-wall mobility. The cavity geometry coefficient R, which is related to the ratio (d/a), should be made as large as possible when the cavity is used in the measurement of low mobility samples. The empty signal can be reduced by mimicking cross-grids on the end plates of the cavity. The empty signal may also be reduced by painting one of the cavity end walls with a metal that exhibits a *p*-type mobility (e.g., Zn, Cd, or In), and such an MHE cavity is usually called a hybrid cavity.

Eq. (11.66) implies that if a conducting plate is placed at the end wall of the bimodal cavity, its mobility can be calculated from the value of S_{21}, the external magnetic field B and the cavity geometry constant R. So the theoretical analysis presented above can be employed to determine the microwave Hall mobility of metals and superconductors. Superconductors exhibit special microwave properties. This suggests that superconductors may have different electrical transport properties at microwave frequencies from their electrical transport properties at dc or ac frequencies, and understanding such properties is helpful for understanding the mechanism of superconductivity (Dorsey 1992; Harris *et al.* 1993; Bhattacharya *et al.* 1994).

In the measurement of the microwave Hall effects of high-conductivity samples, the end plates of the cavity could be made from the material under study, or a layer of the material under study, several skin depths thick, could be plated on the endplates of the MHE cavity. The skin depth for most metals is about 1 μm at 10 GHz.

11.3.4 Dielectric MHE resonator

As discussed above, the theoretical model developed by Al Zoubi *et al.* can be extended to the measurement of microwave Hall mobility of high-conductivity materials. To increase the measurement sensitivity and to measure small samples at low microwave frequencies, an MHE dielectric resonator with two degenerate and orthogonal modes could be used instead of a hollow metallic MHE cavity.

Figure 11.30 shows the working principle of a bimodal rectangular dielectric resonator for the measurement of the electrical transport properties of high-conductivity samples. For a specially designed dielectric resonator with two orthogonal modes with the same resonant frequency and the same quality factor, if the open end of the dielectric resonator is connected to a high-conductivity sample, under the application of an external dc magnetic field B in the direction perpendicular to the sample surface, the electrical transport properties of the sample can be derived from the coupling between the two degenerate modes of the dielectric resonator. To build an actual measurement structure following the principle shown in Figure 11.30, methods should be found to provide suitable couplings for the input and

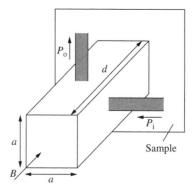

Figure 11.30 A bimodal rectangular dielectric resonator for the measurement of microwave Hall effect of high-conductivity samples

the output ports, and suitable tuning methods are needed to ensure the degeneracy of the two working modes.

Abu-Teir *et al.* proposed a polarization-sensitive scanning microwave microscope with a cross-slit aperture based on a bimodal dielectric resonator (Abu-Teir *et al.* 2002). The microscope operates in the reflection mode and has a subwavelength spatial resolution. It allows contactless mapping of the conductivity tensor, including magnetic field−induced terms such as the Hall effect.

As shown in Figure 11.31, the probe is an open bimodal dielectric resonator with a cross-slit aperture, and each mode can radiate only through one of the slits. The key part of the probe is an oriented sapphire rod tightly mounted in the cylindrical metal waveguide. One face of

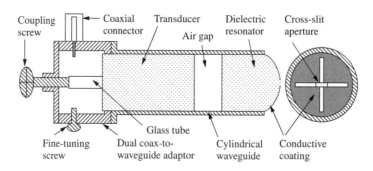

Figure 11.31 Structure of the bimodal dielectric probe (Abu-Teir *et al.* 2002). Source: Abu-Teir, M. Sakran, F. Golosovsky, M. and Davidov, D. (2002). "Local contactless measurement of the ordinary and extraordinary Hall effect using near-field microwave microscopy", *Applied Physics Letters*, **80** (10), 1776–1778

the rod is planar and uncoated, and another face is convex and metal coated, leaving two narrow orthogonal slits crossing in the center. The rod and the slits are resonant elements with one common resonant frequency. The length of the slit is about $\lambda/[2(\varepsilon_r)^{1/2}]$, where ε_r is the dielectric constant of the rod, while the width w may be as small as $0.5\,\mu m$ (Abu-Teir *et al.* 2002). Such a probe can be taken as a modification of the cross-slit probe discussed in Section 11.2.3.2 and the dielectric resonator probe discussed in Section 6.5.5.

The probe mainly consists of a circular waveguide, a dielectric resonator, and a transducer integrated with a specially designed dual-channel coaxial-to-waveguide adapter. The coupling to the resonator is achieved by the transducer consisting of a long dielectric rod attached to a coupling screw via a thin-wall glass tube. An air gap between the transducer and the resonator is a segment of waveguide beyond cut off in which the fields decay exponentially. By changing the gap, one can achieve critical coupling at the resonant frequency, whereupon almost all the input energy is radiated out through the corresponding slit. Owing to the circular symmetry, critical coupling is achieved simultaneously for both the TE_{11} modes.

The measurement setup is shown in Figure 11.32 (Abu-Teir *et al.* 2002). Both the resonant modes are adjusted into the critical coupling condition (minimum of S_{11}, S_{22}) using the coupling screw. Fine-tuning is achieved by additional screws mounted at the broad faces of the coaxial-to-waveguide adapter and at the adapter cover plate. Then, the sample is mounted very close to the probe and the S matrix is measured once again. The magnetic field is provided by an external source or by a small coil mounted on the circular waveguide. A polarized microwave signal at resonant frequency, launched through port 1, exits through the corresponding slit, and hits the sample. A reflected wave with the same polarization comes back through port 1, while a reflected wave with a 90°-rotated polarization comes back through port 2. The procedure is repeated with ports 1 and 2 interchanged. The magnitude and phase of the reflected waves yield the full conductivity tensor of the sample.

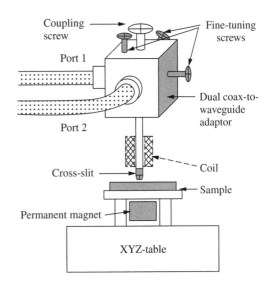

Figure 11.32 Measurement setup of a microwave microscope with a bimodal dielectric probe (Abu-Teir *et al.* 2002). Modified from Abu-Teir, M. Sakran, F. Golosovsky, M. and Davidov, D. (2002). "Local contactless measurement of the ordinary and extraordinary Hall effect using near-field microwave microscopy", *Applied Physics Letters*, **80** (10), 1776–1778

The terms S_{11} and S_{22} carry information on the sample resistance in two perpendicular directions, while the off-diagonal terms S_{12} and S_{21} yield the Hall effect. Since the Hall effect is nonreciprocal ($S_{12} = -S_{21}$), by taking the difference $(S_{12} - S_{21})/2$, it can be separated from the reciprocal background contributions such as cross talk between the two resonant modes and the birefringence in the sample. Quantitative measurements of the Hall effect are possible via an appropriate calibration procedure. For free-standing films, the transmission-line formalism can be applied, which yields (Abu-Teir *et al.* 2002)

$$S_{12} = -2\sigma_{xy}Z_0 d \text{ for } d \ll 1/Z_0\sigma_{xx} \text{ (thin film)}$$
$$(11.69)$$

$$S_{12} = -2\rho_{xy}/Z_0\delta \text{ for } d \gg \delta \text{ (thick film)} \quad (11.70)$$

$$S_{12} = -2\rho_{xy}/Z_0 d \text{ for } \delta \gg d \gg 1/Z_0\sigma_{xx}$$
$$\text{(intermediate thickness)} \quad (11.71)$$

where Z_0 is the transmission-line impedance, d is the layer thickness, and δ is the layer skin depth.

For a layer on a substrate, Eqs. (11.69) and (11.71) should be modified using the impedance transformation (Collin 1992). In the thin-film limit, S_{12} yields the Hall conductance $\delta_{xy}d$, while for the film with intermediate thickness it yields the Hall resistance ρ_{xy}/d. So, S_{12} depends on the film thickness nonmonotonously. When Eqs. (11.69) and (11.71) are used for the description of results, Z_0 can be replaced by the effective probe impedance Z_p, which can be found by calibrating the probe using the samples with known ρ_{xx} and ρ_{xy}.

11.3.5 Planar MHE resonator

Planar bimodal resonators have been used in developing filters (Curtis and Fiedziuszko 1991; Hong and Lancaster 1995), and they can also be used in characterizing the electrical transport properties of high-conductivity materials by fabricating the sample under test into a bimodal planar resonator.

Figure 11.33 shows two typical microstrip bimodal resonators that can be used for the characterization of electrical transport properties: circular disk resonators and square patch resonators. In the design and fabrication of planar bimodal resonators, the degeneracy of the two modes of the resonator should be ensured, and the direct coupling between the two ports should be minimized. The external dc magnetic field is applied perpendicular to the plane of the microstrip circuit. In experiments, special adjustment structures are needed for correcting the errors in the fabrication procedure and tuning the resonant frequencies and quality factors of the two modes.

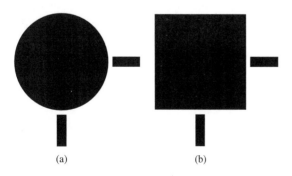

(a)   (b)

Figure 11.33 Two microstrip bimodal resonators. (a) Bimodal circular disk resonator and (b) bimodal square patch resonator

11.4 MICROWAVE ELECTRICAL TRANSPORT PROPERTIES OF MAGNETIC MATERIALS

The electrical transport properties of magnetic materials are important in magnetoelectronics. For a magnetic material, in addition to the ordinary Hall electric field, the charge carriers produce an electric field called the extraordinary Hall field, that arises from the asymmetric scattering in a magnetic material (Viswanathan and Murthy 1990; Bergmana 1979; Craik 1975; Lavine 1958). Usually, the extraordinary Hall effect is much greater than the ordinary Hall effect. In the following, we first discuss the ordinary and extraordinary Hall effects of magnetic materials, and then discuss the bimodal cavity method for bulk and powder samples and the bimodal dielectric probe method for film samples.

11.4.1 Ordinary and extraordinary Hall effect

The Hall resistivity ρ_H of a magnetic material is composed of an ordinary part and an extraordinary part (Lavine 1959; Wohlforth 1982):

$$\rho_H = \mu_0(R_0 H + R_1 M_1) \qquad (11.72)$$

where $\mu_0 = 4\pi \times 10^{-7}$ Henry/m, R_0 is the ordinary Hall coefficient, R_1 is the extraordinary Hall coefficient, H is the applied magnetic intensity, and M_1 is the magnetization in the sample. In the measurement of Hall effect, $J_y = 0$, $J_z = 0$, and E_y is also called the Hall electric field E_H. According to Eq. (11.57), the Hall electric field E_H is given by

$$E_H = \mu_0(R_0 H + R_1 M_1)J_x \qquad (11.73)$$

On the basis of the following relationship

$$J_x = \sigma E_x = \mu_H n e E_x, \qquad (11.74)$$

where σ is the conductivity of the material, n is the density of carriers, and μ_H is the Hall mobility of the material, Eq. (11.73) can be rewritten as

$$\frac{E_H}{E_x} = \mu_0(R_0 H + R_1 M_1)ne\mu_H \qquad (11.75)$$

In Eq. (11.75), M_1 changes with the variation of the applied magnetic field, and R_1 is related to the average resistivity ρ:

$$R_1 = a\rho + b\rho^2, \qquad (11.76)$$

where a and b are constants, and

$$\rho = \frac{1}{3}\rho_b + \frac{2}{3}\rho_a, \qquad (11.77)$$

with ρ_a and ρ_b defined in Eq. (11.58). It should be noted that the value of ρ also changes with the applied magnetic field due to the phenomenon of magnetoresistance (Karplus and Luttinger 1954; Kondo 1962).

When the level of magnetization of the magnetic material is well beyond saturation, $R_1 M_1$ can be regarded as a constant. By considering the derivative of Eq. (11.75) with respect to H, one can obtain the relationship of the change of (E_H/E_x), the change of H, and the value of the Hall mobility μ_H:

$$\Delta\left(\frac{E_H}{E_x}\right) = \mu_0 n e R_0 (\Delta H)\mu_H \qquad (11.78)$$

As $B = \mu_0(H + M)$, and M changes little beyond saturation, the change of magnetic intensity ΔH is proportional to the change of magnetic field ΔB. So Eq. (11.78) can also be written as

$$\Delta\left(\frac{E_H}{E_x}\right) = K_\mu \Delta B \mu_H \qquad (11.79)$$

where K_μ is a constant. Equation (11.79) indicates that the Hall mobility of a magnetic material can be deduced from the change of the Hall field and the change of the magnetic field provided that the magnetization is saturated.

By analogy to the argument leading to Eq. (11.59), we can obtain

$$\Delta\left(\frac{E_2}{E_1}\right) = K_\mu \Delta B\left(\frac{Q_0 - Q_L}{Q_0}\right)\mu_H \qquad (11.80)$$

Similarly, Eq. (11.80) works well provided that the magnetization is saturated. By taking the bimodal cavity as a two-port network, Eq. (11.80) can be rewritten as

$$\mu_H = \frac{1}{K_\mu \Delta B}\left(\frac{Q_0}{Q_0 - Q_L}\right)(\Delta S_{21}) \qquad (11.81)$$

The subscripts 1 and 2 indicate the input and output ports respectively. Equation (11.81) shows that an MHE resonator can be calibrated by a standard sample. If K_μ is known, the mobility of the sample can be calculated from the change of quality factor due to the introduction of the sample, and the change of scattering coefficient ΔS_{21} due to the variation of the magnetic field ΔB, provided that the magnetization is beyond saturation.

11.4.2 Bimodal cavity method

The bimodal cavity method is suitable for the measurement of bulk and powder samples. It has been used in characterizing the electrical transport properties of ferrite (Fe_3O_4) powders at room temperature (Chen *et al.* 1998). Figure 11.34 shows the relationship between the transmission coefficient (S_{21}) and the dc magnetic field B applied to the Fe_3O_4 powder sample. It is clear that the Hall effect of the Fe_3O_4 sample is a combination of two parts: ordinary Hall effect and extraordinary Hall effect. In the initial part, where

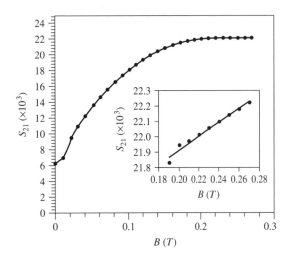

Figure 11.34 The relationship between S_{21} and the external dc magnetic field B for the Fe_3O_4 powder sample

the magnetization is not saturated, the curve is very steep, indicating a very large Hall coefficient (a combination of the ordinary and the extraordinary coefficients). For Fe_3O_4, it has been reported that the extraordinary Hall coefficient is about 100 times larger than the ordinary one and so obscures the latter, making it very difficult to separate the ordinary Hall effect from the extraordinary before the saturation of magnetization. At levels above technical saturation, M remains a constant while the magnetic field increases, so the slope of the curve of S_{21} as a function of the applied dc magnetic field B is related only to the ordinary Hall effect. This portion of the curve is enlarged, and inserted in the lower part of Figure 11.34. The enlarged portion can be used to resolve the slope of the curve above technical saturation to obtain the ordinary Hall mobility using Eq. (11.81).

The ordinary Hall mobility of Fe_3O_4 powders is estimated to be 0.49 cm²/volt-s. It should be indicated that, if a stronger dc magnetic field is applied, the measurement accuracy and sensitivity could be improved. The difficulty in ensuring that the magnetization is saturated over the region of the measurement causes an uncertainty in the mobility measurement. However, it is safe to regard the value of the mobility obtained by the above method as an upper limit. An analysis made by Lavine shows that this effect does not alter the measured results more than ±1% (Lavine 1959, 1958).

The sign of the Hall carriers Fe_3O_4 was found to be negative. This is in accord with Verwey's mode, which shows that the inverted spinel structure of Fe_3O_4 has one Fe^{2+} and Fe^{3+} per Fe_3O_4 group located in octahedral sites, and this may be

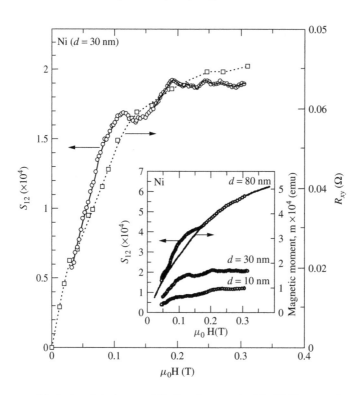

Figure 11.35 Microwave (circles) and dc (squares) Hall resistances in thin Ni films on glass substrates (Abu-Teir *et al.* 2002). The inset shows that the microwave Hall effect increases with increasing film thickness. The continuous line (inset) shows the magnetization of a small piece of the same film (~3 × 3 mm²) measured by SQUID magnetometry. Source: Abu-Teir, M. Sakran, F. Golosovsky, M. and Davidov, D. (2002). "Local contactless measurement of the ordinary and extraordinary Hall effect using near-field microwave microscopy", *Applied Physics Letters*, **80** (10), 1776–1778

alternatively described as two Fe^{3+} ions plus an electron. The electrical properties of Fe_3O_4 are due to the continuous interchange of the extra electrons.

The conductivity of Fe_3O_4 was estimated to be $\sigma = 227$ (ohm-cm)$^{-1}$ using Eq. (11.63). Among the ferrites, Fe_3O_4 has the highest room temperature conductivity. This is due to the free exchange of electrons between equivalent ions with different valence in the octahedral sites. Using the values of σ and μ, the density of the carriers can be calculated from Eq. (11.1): $n = 2.89 \times 10^{21}$ (cm^3)$^{-1}$. Because of the uncertainty in the Hall mobility as mentioned above, the value of n is better regarded as a lower limit on the charge carrier density.

11.4.3 Bimodal dielectric probe method

The microwave microscope with bimodal dielectric probe proposed by Abu-Teir *et al.*, as shown in Figure 11.31 and Figure 11.32, can be used for contactless measurement of the ordinary and extraordinary Hall effects of magnetic films (Abu-Teir *et al.* 2002). Owing to the high spatial resolution of the probe, the results obtained are localized to a certain position, while the results from dc measurements are global.

Figure 11.35 shows an example of the Hall effect in thin Ni films, which are evaporated on glass by an e-beam. The results of local microwave measurements agree well with the results of global dc measurements, and the small differences may come from sample inhomogeneity. As predicted by Eq. (11.69), the Hall effect in thin Ni films is proportional to the film thickness. Since, in the fields below 0.3 T, the ordinary Hall effect in Ni is negligible, Figure 11.35 mainly shows the extraordinary Hall effect. This effect yields the magnetization, which is consistent with the results of superconducting quantum interference device (SQUID) magnetometry on small pieces of the same Ni films, shown in the inset. It should be noted that the extraordinary Hall effect in the 10-nm-thick Ni film can be easily measured using the bimodal dielectric probe, while such a film is at the edge of the sensitivity of SQUID magnetometry (10^{-5} emu) (Abu-Teir *et al.* 2002).

The bimodal dielectric probe method is actually a microwave polar Kerr effect study in the near-field zone. Compared to the magneto-optic Kerr effect routinely used for contactless measurement of thin-film magnetization, this method has several advantages (Abu-Teir *et al.* 2002). Firstly, it has deeper penetration into the conducting layers ($\sim 1\,\mu m$ vs $\sim 20\,nm$ for optical waves); secondly, it tracks conductivity at the ns timescale, which is exactly the timescale of most applications; and thirdly, the scattering effect is negligible.

REFERENCES

Abu-Teir, M. Sakran, F. Golosovsky, M. and Davidov, D. (2002). "Local contactless measurement of the ordinary and extraordinary Hall effect using near-field microwave microscopy", *Applied Physics Letters*, **80** (10), 1776–1778.

Al Zoubi, A. Y. (1990). "Theoretical model of a microwave Hall-effect empty-cavity signal", *IEE Proceedings-H*, **137** (1), 78–80.

Al Zoubi, A. Y. (1991a). "Microwave Hall effect in a TE11p cylindrical cavity", *IEEE Transactions on Microwave Theory and Techniques*, **39** (11), 1899–1901.

Al Zoubi, A. Y. (1991b). "Determination of Hall mobility of metals using a microwave resonator", *Measurement Science and Technology*, **2**, 1165–1168.

Baturina, T. I. Borodovski, P. A. and Studenikin, S. A. (1996). "Microwave waveguide method for the measurement of electron mobility and conductivity in GaAs/AlGaAs heterostructures", *Applied Physics A*, **63**, 293–298.

Belbounaguia, N. Druon, C. Tabourier, P. and Wacrenier, J. M. (1994). "Nondestructive mapping of GaAs wafers from measurement of magnetoresistance effect using a novel microwave device", *IEEE Transactions on Instrumentation and Measurement*, **43** (1), 24–29.

Bergmana, G. (1979). "The anomalous Hall effect", *Physics Today*, **32** (8), 25–30.

Bhattacharya, S. Higgins, M. J. and Ramakrishnan, T. V. (1994). "Anomalies in free flux-flow Hall effect", *Physical Review Letters*, **73** (12), 1699–1702.

Buschow, K. H. J. (1991). *Handbook of Magnetic Materials*, Vol. 6, North-Holland, Amsterdam.

Caverly, R. H. and Westgate, C. R. (1982). "Low temperature microwave Hall mobility measurements on copper phthalocyanine single crystals", *Journal of Applied Physics*, **53**, 7410–7413.

Chen, L. F. Ong, C. K. and Tan, B. T. G. (1998). "Study of electrical transport properties of fine magnetic particles using microwave Hall effect techniques", *IEEE Transactions on Magnetics*, **34** (1), 272–277.

Collin, R. E. (1992). *Foundations for Microwave Engineering*, 2nd edition, McGraw-Hill, New York.

Coue, E. Chausse, J. P. Robert, H. and Barbarin, F. (1994). "Conductivity and mobility contactless measurements of semiconducting layers by microwave absorption at 35 GHz", *Journal de Physique III*, **4** (4), 707–718.

Craik, D. J. (1975). *"Magnetic oxides"*, John Wiley & Sons Ltd, London, UK.

Cross, T. E. De Cogan, D. and Al Zoubi, A. Y. (1987). "Microwave Hall effect studies of low mobility materials", *New Materials and Their Application*, Warwick, Inst. Phys. Conf. Ser. No. 89, 301–306.

Curtis, J. A. and Fiedziuszko, S. J. (1991). "Miniature dual mode microstrip filters", *IEEE MTT-S Digest*, vol. 2, 443–446.

Dorsey, A. T. (1992). "Hall effect near the vortex-glass transition in high-temperature superconductors", *Physical Review Letters*, **68** (5), 694–697.

Druon, C. Tabourier, P. Bourzgui, N. and Wacrenier, J. M. (1990). "Novel microwave devices for nondestructive electrical characterization of semiconducting layers", *Review of Scientific Instruments*, **61** (11), 3431–3434.

Eley, D. D. and Lockhart, N. C. (1983). "Microwave Hall effects in low-mobility materials", *Journal of Physics E: Scientific Instruments*, **16**, 47–51.

Engineer, M. H. and Nag, B. R. (1965). "Propagation of electromagnetic waves in rectangular guides filled with semiconductor in the presence of a transverse magnetic field", *IEEE Transactions on Microwave Theory and Techniques*, **13** (5), 641–646.

Fletcher, J. R. (1976). "An improved bimodal cavity for microwave Hall effect measurement", *Journal of Physics E: Scientific Instruments*, **9**, 481–483.

Harris, J. M. Ong, N. P. and Yan, Y. F. (1993). "Hall effect of vortices parallel to CuO_2 layers and the origin of the negative Hall anomaly in $YBa_2Cu_3O_{7-\delta}$", *Physical Review Letters*, **71** (9), 1455–1458.

Hong, J. S. and Lancaster, M. J. (1995). "Bandpass characteristics of new dual-mode microstrip square loop resonators", *Electronics Letters*, **31** (11), 891–892.

Jow, J. Hawley, M. C. Finzel, M. C. and Asmussen Jr., J. (1989). "Microwave heating and dielectric diagnosis techniques in a single-mode resonant cavity", *Review of Scientific Instruments*, **60** (1), 96–103.

Karplus, R. and Luttinger, J. M. (1954). "Hall effect in ferromagnetics", *Physical Review*, **95** (5), 1154–1160.

Kondo, J. (1962). "Anomalous Hall effect and magnetoresistance of ferromagnetic metals", *Progress of Theoretical Physics*, **27** (4), 772–792.

Lann, A. F. Golosovsky, M. Davidov, D. and Frenkel, A. (1999). "Microwave near-field polarimetry", *Applied Physics Letters*, **75** (5), 603–605.

Lavine, J. M. (1958). "Alternate current apparatus for measuring the ordinary Hall coefficient of ferromagnetic metals and semiconductors", *Review of Scientific Instruments*, **29** (11), 970–976.

Lavine, J. M. (1959). "Ordinary Hall effect in Fe_3O_4 and $(NiO)_{0.75}(FeO)_{0.25}(Fe_2O_3)$ at room temperature", *Physical Review*, **114** (2), 482–488.

Liu, C. C. Na, B. K. Walters, A. B. and Vannice, M. A. (1994). "Microwave absorption measurements of the electrical conductivity of small particles", *Catalysis Letters*, **26**, 9–24.

Meisels, R. and Kuchar, F. (1987). "The microwave Hall effect of a two-dimensional electron gas", *Zeitschrift fur Physik B – Condensed matter*, **67**, 443–447.

Mittleman, D. M. Cunningham, J. Nuss, M. C. and Geva, M. (1997). "Noncontact semiconductor wafer characterization with the terahertz Hall effect", *Applied Physics Letters*, **71** (1), 16–18.

Musil, J. and Zacek, F. (1986). *Microwave Measurements of Complex Permittivity by Free-Space Methods and their Applications*, Elsevier, Amsterdam.

Na, B. K. Vannice, M. A. and Walters, A. B. (1992). "Measurement of the effect of pretreatment and adsorption on the electrical properties of ZnO powders using a microwave-Hall-effect technique", *Physical Review B*, **46** (19), 12266–12277.

Na, B. K. Kelly, S. L. Vannice, M. A. and Walters, A. B. (1991). "Development of the microwave Hall effect technique using an ESR spectrometer and a network analyser", *Measurement Science and Technology*, **2**, 770–779.

Na, B. K. Walters, A. B. and Vannice, M. A. (1993). "Studies of gas adsorption on ZnO using ESR, FTIR spectroscopy, and MHE (Microwave Hall Effect) Measurements", *Journal of Catalysis*, **140**, 585–600.

Ong, N. P. Bauhofer, W. and Wei, C. J. (1981). "Microwave Hall measurements in the intermediate conductivity regime using a bimodal cavity", *Review of Scientific Instruments*, **52** (9), 1367–1375.

Ong, N. P. and Portis, A. M. (1977). "Microwave Hall effect in a quasi-one-dimensional system: tetracyanoquinodimethanide (TTF-TCNQ)", *Physical Review B*, **15** (4), 1782–1789.

Parks, B. Spielman, S. and Orenstein, J. (1997). "High-frequency Hall effect in the normal state of $YBa_2Cu_3O_7$", *Physical Review B*, **56** (1), 115–117.

Portis, A. M. and Teaney, D. (1958). "Microwave Faraday rotation: design and analysis of a bimodal cavity", *Journal of Applied Physics*, **29** (12), 1692–1698.

Sayed, M. M. and Westgate, C. R. (1975a). "Microwave Hall measurement techniques on low mobility semiconductors and insulators. I. Analysis", *Review of Scientific Instruments*, **46** (8), 1074–1079.

Sayed, M. M. and Westgate, C. R. (1975b). "Microwave Hall measurement techniques on low mobility semiconductors and insulators. II. Experimental procedures", *Review of Scientific Instruments*, **46** (8), 1080–1085.

Spielman, S. Parks, B. and Orenstein, J. Nemeth, D. T. Ludwig, F. Clarke, J. Merchant, P. and Lew, D. T. (1994). "Observation of the quasiparticle Hall effect in superconducting $YBa_2Cu_3O_{7-\delta}$", *Physical Review Letters*, **73** (11), 1537–1540.

Trukham, E. M. (1966). "A microwave method of measurement of the Hall effect with the aid of a degenerate cavity", *Radio Engineering and Electronic Physics*, **19** (9), 1097–1103.

Viswanathan, B. and Murthy, V. R. K. (1990). "*Ferrite Materials, Science and Technology*", Springer-Verlag, Berlin and Narosa Publishing House, New Delhi.

Wohlforth, E. P. (1982). *Ferromagnetic Materials*, vol. 3, Chapter 9, North Holland, Amsterdam.

12

Measurement of Dielectric Properties of Materials at High Temperatures

Measurements of the dielectric properties of materials at high temperatures are important for industrial, scientific, and medical applications. The methods for the measurements at high temperatures can be generally classified into nonresonant methods and resonant methods. In the second, third, and fourth sections, three types of nonresonant methods are discussed, including coaxial-line method, waveguide method, and free-space method. In the fifth and sixth sections, two types of resonant methods are discussed, including cavity-perturbation method and dielectric-loaded cavity method.

12.1 INTRODUCTION

Considerable efforts have been spent in the measurement of dielectric properties of materials at high temperatures. The conventional measurement methods for measurements of materials properties at room temperatures have been modified to circumvent the various practical limitations of measurements at high temperatures, and many new approaches are being developed to obtain data at higher temperatures, of greater accuracy and for a larger variety of materials.

In this chapter, various methods for the measurement of dielectric properties of materials at high temperatures are discussed. In this section, the significance of measurements of dielectric properties of materials at high temperatures is discussed, and then the problems and solutions for dielectric property measurements at high temperatures are raised,

and finally a brief review on the methods often used for the measurements of high-temperature dielectric properties is made.

12.1.1 Dielectric properties of materials at high temperatures

The temperature dependence of the electromagnetic properties of materials is important for academic research, microwave propagation, and microwave processing. In this chapter, we concentrate on the study of dielectric properties of materials at high temperatures; the high-temperature measurement techniques can be extended to the measurement of other properties, such as permeability, of materials at high temperatures.

12.1.1.1 Academic research

The research on temperature dependence of the electromagnetic properties of materials has been very active in physics and materials sciences; especially, the dielectric properties of materials at high temperatures are under intense investigation. For example, as discussed in Chapter 9, along with the increase of temperature, ferroelectric materials may experience phase transitions, and subsequently, their dielectric properties will change, especially near the transition temperatures (Mouhsen et al. 2001). Experimental characterization of dielectric properties at various temperatures is helpful in understanding the underlying science of such materials.

Microwave Electronics: Measurement and Materials Characterization L. F. Chen, C. K. Ong, C. P. Neo, V. V. Varadan and V. K. Varadan
© 2004 John Wiley & Sons, Ltd ISBN: 0-470-84492-2

For ordinary polycrystalline ceramics, the dielectric constant at microwave frequencies generally increases with the increase of temperature, and the temperature dependence of the loss tangents of polycrystalline ceramics usually shows a gradual increase with the increase of temperature, followed by a rapid rise around the range 800–1000 °C. The high-temperature loss tangents can be over an order of magnitude larger than their corresponding room-temperature values. The mechanism giving rise to the loss at room temperature and over the temperature region of gradual increase appears to be different from the one controlling the region of rapid rise. The rapid increase in loss tangents at high temperatures may be attributed to the additional loss associated with increasing electrical conductivity when the temperature is approaching the softening temperatures of residual amorphous glassy phases (Ho 1988). For synthesizing materials with the desired temperature dependence of dielectric properties, further and more systematic investigations on the mechanisms governing the temperature dependence of dielectric properties of materials are needed.

12.1.1.2 Microwave propagation

Sophisticated knowledge of the dielectric properties of materials and their variation with temperature is important for various fields of microwave engineering, especially in the design of components for microwave propagations, including microwave communications. Antenna window is a typical example with temperature stability requirements, as an antenna window is often required to work in different temperatures, and the knowledge of the dielectric properties of materials and their variation with temperature is crucial in the design of antenna windows with special applications; for example, the antenna windows used in space shuttles (Basset and Bomar 1973). Besides, knowledge of the dielectric properties of materials at high temperatures is also important for the development of rf and microwave windows for high-energy, high-current, charged-particle storage loops and accelerators (Hutcheon *et al.* 1991).

For some engineering materials, though the general information about the properties of these materials may be well known, an engineered optimum design requires detailed knowledge of the rf and microwave properties, which is often not available. For example, alumina materials are widely used in various fields of microwave engineering. For two samples with adequately low dielectric loss at room temperature, the loss of one of them may increase much more rapidly with temperature. Therefore, the measurement of electromagnetic properties at high temperatures is important for the applications of these materials at high temperatures.

12.1.1.3 Microwave processing

Microwave processing uses the energy of microwave fields for various purposes, such as heating materials, promoting chemical reactions, and retorting minerals (Thuery and Grant 1992; Metaxas and Meredith 1983). Two microwave frequencies are generally used for microwave processing: 915 MHz with wavelength 32 cm or about 12.5 in., and 2450 MHz with wavelength 12 cm or about 5 in. These two frequencies are allocated by the US Federal Communications Commission for industrial, scientific, and medical use, and they are often called the ISM frequencies. However, it deserves to be noted that the low ISM frequency in the United Kingdom is 896 MHz, and currently, most of the effort in microwave processing is being devoted to 2.45 GHz because it provides a suitable compromise between power deposition and penetration depth. The designing of microwave processing systems requires a knowledge of the temperature dependence of the dielectric properties of the materials to be processed, as the rapid increase in tanδ above a certain critical value observed in a number of materials can lead to difficulties in process control unless the rapid change in material behavior can be predicted (Binner *et al.* 1994).

Microwave heating

The use of microwave energy for obtaining heating effects offers a number of potential advantages over conventional radiant heating techniques (Arai *et al.* 1992; Sutton 1989; Metaxas and Meredith

1983). The increasing use of microwave heating in the polymer/rubber industries and the processing of ceramic materials has resulted in the need to characterize the relevant material properties over a wide range of temperatures. The ability of a material to extract energy from the microwave field depends upon its dielectric properties; however, these properties change as the material heats. So there is a need for obtaining the microwave dielectric properties at elevated temperatures (Bringhurst *et al.* 1992; Craven *et al.* 1996). In general, the effective loss factor of many ceramics increases with temperature, with the rate of increase depending upon the type of ceramic and the operating frequency. When a material exhibiting this positive slope in the relationship between the loss tangent and the temperature is heated using microwave energy, a "runaway effect" or uncontrolled rise in temperature often occurs. Damage to the material is highly probable unless steps are taken to avoid such a cumulative effect. If microwave processing is to be applied to these materials, then a thorough understanding of the dielectric properties, at least up to their sintering temperatures, is required.

Promoting of chemical reactions

Microwave energy can also be used in promoting chemical reactions that require heat (Jow *et al.* 1987). Many molecules contain polar groups that undergo molecular rotation due to thermal Brownian motion. Incident microwave radiation interacts with the polar groups in the molecules so that the normal random orientation of the dipoles becomes ordered. The molecules then relax to their normal random orientation. Since energy is required to hold the dipoles in place, the relaxation of the dipole is accompanied by transfer of thermal energy to the material. The relaxation is described by an exponential decay function with a characteristic relaxation time. When the frequency of the incident radiation is equal to the reciprocal of the relaxation time, the system is said to be in resonance. This resonant frequency, which is defined as the molecular resonant frequency, is a characteristic of the absorbing material.

Microwave processing for promoting chemical reactions has several obvious advantages. Firstly,

selective and controlled heating can be made because of the absorption of microwave energy by polar groups. Secondly, the thermal degradation can be decreased due to rapid uniform bulk heating. Thirdly, the control of material temperature time profile and cure cycle can be improved. These advantages may cause microwave-cured materials to have superior mechanical characteristics when compared to conventionally cured materials. Furthermore, the heat-transfer mechanism is also different between microwave curing and conventional thermal curing. The microwave energy directly heats the polymer; however, in the conventional thermal case, the mold is heated first and the heat is subsequently transferred into the epoxy via thermal conduction. So rapid cure and high efficiency of energy utilization can be obtained using microwave heating instead of conventional thermal process.

Microwave retorting

The temperature dependence of dielectric properties helps in evaluating the potential of microwave retorting techniques (Iskander and DuBow 1983). Researchers at the US Bureau of Mines have centered on a group of minerals identified as strategic to the United States and are involved in finding improved ways to process and recover these minerals by microwave processing (Holderfield and Salesman 1992). Such improvements would ensure that the United States does not have to depend on other nations for supply of these natural resources. However, the benefits cannot be exploited without knowledge of how these minerals are affected by microwave energy. The physical properties of a material that describe how that material is affected by microwaves are mainly its dielectric constant and loss factor. As part of this research effort, the U.S. Bureau of Mines has developed several methods for measuring the dielectric properties of minerals over a wide frequency and temperature range.

12.1.2 Problems in measurements at high temperatures

There are many practical reasons why high temperatures cause measurement problems (Tinga 1992;

Tinga and Xi 1993a). Though the measurement principles of any method remain the same as for room temperature operations, many technical difficulties arise when the sample temperature is raised. The measurement accuracy typically decreases as temperature increases because of the increasing parameter uncertainties. Stating and understanding these problems makes it easier to find ways of minimizing them. The three main difficulties in the measurement of the dielectric properties of materials at high temperatures include the intrinsic properties of samples, thermal expansion, and heating and temperature measurement of samples. It does not seem likely that we can obtain a single optimum solution to all these problems so that the permittivity of an arbitrary material can be measured at an arbitrary high temperature and at an arbitrary frequency. In the design of measurement methods, it is necessary to consider their specific applications.

12.1.2.1 Temperature dependence of materials' intrinsic properties

As the dielectric properties of materials are temperature dependent, the propagation constant and wavelength in a material will change when the temperature changes. So the measurement accuracy will also change with the change in temperature. Furthermore, at certain temperature ranges, the dielectric loss or electrical conductivity may become very high; so, it is very difficult to make optimum matching and tuning of the system. These problems are, more or less, common to all the dielectric measurement techniques since all the measurement techniques involve the wavelength and propagation constant of the microwave signals.

12.1.2.2 Thermal expansion

The constraints associated with operations at high temperatures mean that only a small number of conventional dielectric measurement techniques are suitable for use at high temperatures. Thermal expansion distorts and changes the geometry of both the sample and the measurement fixture, and the fitting between them. To minimize the

effects of thermal expansion, the measurement system must be thermally isolated from heating, for example by water cooling; or the measurement time must be short enough to prevent significant thermal conduction from the hot sample to the cool measurement system; or there should be no physical contact between the hot sample and the measurement fixture. For loaded waveguide techniques, the sample is heated through the apparatus and its measurement temperature range is limited. For free-space measurements, only the sample needs to be heated. As there is no physical contact between the sample and the measurement system in free-space measurements, the effects of thermal expansion are not severe. A variety of materials can be accurately characterized at high temperature using free-space techniques.

12.1.2.3 Heating of samples

Heating technique is an important aspect in the measurement of materials properties at high temperatures. The required sample size and the unavoidable heat losses, such as thermal conduction loss due to a required sample contact with a metal part, set a practical upper temperature limit for the measurement. The often used heating techniques include conventional oven heating, electrical heating, infrared heating, laser heating, solar energy heating, and microwave heating.

As discussed above, a beneficial measurement improvement would be the use of a noncontact measurement method to minimize the thermal expansion and heat conduction problems. Free-space techniques have been successfully and widely used in high-temperature measurements. However, one disadvantage of free-space techniques is that the samples should be electromagnetically large, so the samples usually have large thermal inertia, and the heating-testing time cycle may be long. A reduction in effective sample size results in lower energy requirements to reach a given temperature and reduces testing time. Small sample size naturally leads one to consider resonant techniques with the sample perturbing the resonance in one way or another. Certain resonator configurations allow suspension of the sample in a more or less noncontacting manner thereby reducing the heat loss. However, it is necessary to note that, for high-temperature noncontact dielectric measurements, the radiation heat loss will ultimately set the achievable upper temperature limit. To minimize the heat radiation losses in noncontact methods, a high quality, microwave transparent, high-temperature heat insulation becomes necessary at temperatures exceeding 1200 °C.

Using methods with very short test cycles is another way to counter the problem of heat loss over time. Measuring the sample's microwave properties dynamically, while heating the sample, minimizes the duration of the heating and testing cycle. This suggests the use of microwaves for both heating and characterizing a sample. A major advantage of electromagnetic sample heating is the inherently higher heating rate and the reduced thermal gradient due to the localized volumetric heating. However, one of the major difficulties is the tuning system needed to keep the microwave power source tuned and matched to the changing cavity resonance and impedance.

Finally, it should be indicated that the heat loss could also be used as an advantage in the following way. First, a small sample is heated to its maximum required temperature in an external furnace in close proximity to a microwave measurement fixture. After quickly pushing the sample from the furnace into the measurement fixture, the microwave properties of the sample are rapidly measured while the sample is cooling back to room temperature. But it should be noted that the thermally induced material hysteresis effects would affect the values of the cooling curve data.

12.1.2.4 Temperature measurement

Temperature measurement is also an important problem in the measurement of dielectric properties of materials at high temperatures. As the sample under test is in a microwave environment, commonly used thermocouples significantly perturb the heating and testing fields. Moreover, microwave current could be coupled into the thermocouple-detector circuit and mixed with the true temperature change signal, thereby further corrupting the results. Infrared radiation thermometer can be used as no contact is needed, and measurements can be made at a certain distance away from the

sample. Optical fiber thermometer is also widely used in high-temperature measurements, as optical fiber probe is microwave transparent, and has little effect on the microwave environment. Using an optical fiber thermometer, temperature can be measured with the optical fiber probe contacting the sample.

In temperature measurements, it is necessary to consider the thermal gradient caused by heating techniques, so the temperatures at different positions of a sample are often tested. In some cases, the results of temperature measurement are used in controlling the heating system, so the speed and accuracy of temperature measurements are required. If the sample under measurement is heated by microwave energy, owing to the high heating rate of microwave power, the temperature of the sample should be measured quickly and accurately to ensure stable and controllable heating.

12.1.3 Overviews of the methods for measurements at high temperatures

The methods for the measurement of materials properties at high temperatures can be generally classified into nonresonant methods and resonant methods. The use of these two categories of techniques is complementary. For example, the coaxial probe requires a minimum of sample preparation, is inherently broadband, can cope with a wide range of permittivity values, and may be operated under isothermal conditions. However, its use is difficult with samples that cannot be prepared with the required surface finish. In contrast, the resonant-cavity-based technique can be used with a wider range of samples, such as powders, and has a high inherent accuracy. However, in a cavity-perturbation technique, during measurement the sample is removed from the furnace and this will inevitably lead to some cooling. Furthermore, because of the necessity to satisfy the perturbation conditions, the cavity is limited in the range of permittivity values that may be measured (Binner *et al.* 1994).

12.1.3.1 Nonresonant methods

Nonresonant methods are useful in showing the frequency relaxation behavior of the materials under test. Nonresonant methods mainly include reflection method and transmission/reflection method, and the reflection method is more widely used. Three typical types of transmission lines are often used including coaxial line, waveguide, and free-space. In the second, third, and fourth sections of this chapter, the applications of these transmission lines in high-temperature measurements are discussed.

12.1.3.2 Resonant methods

Many microwave engineering applications remain narrow band thereby making the resonant methods attractive for obtaining accurate microwave behavior of materials at high temperature. The resonant methods used in the measurement of dielectric properties of materials at high temperatures are mainly cavity-perturbation method and dielectric-loaded cavity method, which are discussed in the fifth and sixth section respectively.

In resonant methods, microwave energy can also be used to heat the sample under test. Usually, the heating microwave power and the testing microwave signal work at different modes and have different frequencies. However, in some cases, one resonant mode can work as both the heating mode and the testing mode. In the active loop method discussed in the final part of this chapter, the sample under test is heated by the microwave power excited from the oscillation of the loop consisting of the cavity loaded with the sample.

12.2 COAXIAL-LINE METHODS

Nonresonant methods can be classified into refection methods and transmission/reflection methods.

In the measurement of dielectric properties of materials at high temperatures, reflection methods are more widely used (Blackham 1992). Reflection methods include short-reflection method and open-reflection method, and the open-reflection methods are more widely used in the measurement of dielectric properties of materials at high temperatures. In this section, we concentrate on the measurements of samples with infinite thickness, and the techniques and conclusions can be extended to the measurements of sheet samples with finite thickness.

The advantages of the open-ended coaxial-probe method mainly include the relative ease of the measurement procedure, the broadband measurement capabilities, and the possible on-line measurement. As shown in Figure 12.1(a), an open-ended coaxial probe is basically a truncated section of a coaxial line, with an optional extension of a ground plane. The sample under test is placed flush with the probe, and, in principle, the values of the complex permittivity are determined from the input impedance of the probe. The measurement technique basically requires that the sample possesses a single flat and smooth surface. Its diameter should be at least two times larger than the probe diameter. Most theoretical models assume that the sample is semi-infinitely thick. In experiments, it is required that the presence of any object behind the sample does not affect the measurement results, therefore the sample thickness should allow the magnitude of the electric field at the far end of the sample to be at least two orders smaller than that at the probe/sample interface. For samples with thin thickness, which do not satisfy this condition, a multiplayer dielectric model with an

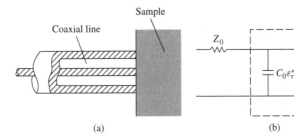

(a)   (b)

Figure 12.1 Measurement of dielectric properties using an open-ended coaxial-line probe (a) Measurement configuration and (b) equivalent circuit

air termination could be applied. One of the critical elements in dielectric properties measurements, using an open-ended coaxial probe, is making a good contact between the probe and the material under test. It has been shown that air gaps on the order of fractions of a millimeter strongly influence the measurement results.

12.2.1 Measurement of permittivity using open-ended coaxial probe

As shown in Figure 12.1, in the measurement of materials properties using an open-ended coaxial probe, the complex permittivity of the sample is determined from the measurement of reflection from the open end of the probe. The design of a suitable coaxial-line probe involves various trade-offs such as operating frequencies, range of measurement temperatures, range of complex permittivities of the materials to be tested, possible propagation of higher-order modes, adequate line length for cooling purposes, and the selection of a low thermal-expansion material for the fabrication of the probe (Andrade *et al.* 1992; Arai *et al.* 1992). A detailed discussion on the theory for open-ended coaxial probes can be found in Chapter 3. Here we introduce the basic working principle for the measurement of dielectric permeability using an open-ended coaxial probe.

As shown in Figure 12.1(b), there is an impedance discontinuity at the interface between the open-end of the coaxial line and the sample, and the impedance discontinuity results in a reflection at the interface. From the complex reflection coefficient Γ seen at the end of the line, the complex relative dielectric permittivity ε_r can be obtained (Salsman 1991):

$$\varepsilon_r = \frac{1 - \Gamma}{j\omega Z_0 C_0 (1 + \Gamma)} - \frac{C_f}{C_0} \qquad (12.1)$$

where ω is the angular frequency, C_0 and C_f are the fringing capacitances associated with the coaxial-line probe, and Z_0 is the characteristic impedance of the coaxial line. To obtain the permittivity value of the sample under study, the values of C_0 and C_f should be determined before the measurements, and this can be done by calibration.

Open-ended coaxial probe is widely used in measuring the dielectric properties of liquid and semisolid sample. When a coaxial probe is used to measure solid samples, the roughness of the sample surface will affect the measurement results (Arai *et al.* 1995b), so the measurement surfaces of samples are usually polished. Furthermore, the air gap between the probe and the sample greatly affects the measurement results, and the air gap also decreases the measurement sensitivity. Many efforts have been made to eliminate the air gap between the probe and the sample. It should also be indicated that relatively small errors in the measurement of S_{11} of either the material under test or the standard material result in significant errors in the complex permittivity results, and the resulting errors are much greater in ε_r'' values than those in the ε_r' values. This is due to open-ended coaxial probes' inherent problem of not being suitable for measuring the loss factor accurately, in particularly low-loss materials. However, the broadband capabilities of this probe and the simplicity of use continue to make this measurement technique attractive and frequently used.

12.2.2 Problems related to high-temperature measurements

Besides the general problems in measurements at high temperatures discussed in Section 12.1.2, there are special problems related to the applications of coaxial probes for measurement of dielectric properties of materials at high temperatures. We discuss the measurement temperature ranges of normal coaxial probes, heating techniques often used in coaxial probes, and the thermal elongation of coaxial line.

12.2.2.1 Measurement temperature ranges of conventional coaxial probes

The differential thermal expansion, which is inherent in conventional coaxial probes, can lead to erroneous results in the measurement of the complex permittivity at high temperatures. This is mainly due to both the air gaps between the probe and the material under test, and changes in the characteristic impedance of the coaxial line, because of the differential thermal expansion of the inner and outer conductors in metal probes. Most of the commercially available coaxial probes cannot measure samples at temperatures higher than 250 °C. Even the use of metals with small thermal-expansion coefficients, such as kovar, has resulted in limited success in providing reliable results for temperatures up to 800 °C. At higher temperatures, the differential thermal expansion between the inner and outer conductors of the kovar probe significantly increases and, hence, affects the accuracy of the measurements. None of the available metal probes can be used to provide accurate results at temperatures higher than 1000 °C.

12.2.2.2 Heating techniques for coaxial probes

Figure 12.2 shows a coaxial probe and its heating system for measurements at high temperatures (Arai *et al.* 1992). The probe is constructed of heat

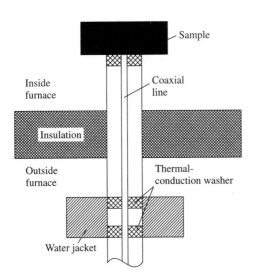

Figure 12.2 Schematic representation of high-temperature coaxial dielectric probe. Modified from Arai, M. Binner, J. G. P. Carr, G. E. and Cross, T. E. (1992). "High temperature dielectric measurements on ceramics", *Materials Research Society Symposium Proceedings*, **269**, 611–616, by permission of Materials Research Society

resistant metal alloy, and the thin outer conductor of the coaxial line reduces heat conduction. Ceramic discs near the open end act as both the support for the center conductor and a thermal path to allow cooling of the center conductor, and they also prevent sample debris from falling into the open end of the probe. The end of the probe is inserted into a conventional furnace and a length of high-temperature coaxial line then passes through the walls of the furnace. Outside the furnace, a water jacket is used to ensure that the conventional microwave parts remain at room temperature. The thermal-conduction washers direct the heat conducting down through the inner conductor to the water jacket, and the washers are made of dielectric materials with high thermal conductivity, such as aluminum nitride (AlN) and beryllium oxide (BeO). Using this technique, it is possible to measure the dielectric properties at a wide range of temperatures. This heating technique is widely used in various coaxial probes for measurements at high temperatures.

Besides the conventional oven method discussed above, sample heating can also be done by optical means using laser or image oven (Mouhsen *et al.* 2001). A conventional oven has the disadvantage that the measurement cell and the connection between the cell and the coaxial line that provides the wave propagation are at a high temperature. To solve this problem, an optical heating source, for example, a CO_2 laser emitting infrared beam, can be used. The power of the laser is variable, and can be controlled by a computer. The laser beam is widened so that the heating of the sample on the rear face (the face in contact with the central conductor) is as uniform as possible. To get a homogeneous distribution of the laser power on the sample, the beam is spread over all of the rear face by using an infrared mirror system. In these conditions, the sample could be brought to high temperatures in a time long enough to avoid fractures, but short by comparison to the heating period of a conventional oven. Moreover, the measuring cell is heated only by the thermal transfer due to its contact with the sample, and it remains at a relatively low temperature, which allows the use of coaxial lines and connections in contact with the measuring cell.

12.2.2.3 Thermal elongation of coaxial line

Besides the air gap between the sample and the open end of the probe, another source of critical errors emerges when the probe is adapted for high-temperature systems: the thermal elongation of the probe causes phase shift errors (Arai *et al.* 1995a). Though this problem can be minimized by fabricating the open-ended coaxial probe from invar to limit dilatations, theoretical analysis for correcting the errors owing to thermal elongation of high-temperature coaxial probe is needed.

As shown in Figure 12.3, the thermal elongation of a coaxial line moves the aperture plane of the probe from a reference plane where calibration is initially performed at room temperature to a new plane at high temperature, causing critical phase shift errors. The extra length added to the probe at a given temperature causing the shift in the aperture plane position can be considered as a lossless transmission line. Denoting the scattering parameters for this extra line as S_{11}, S_{21} $(=S_{12})$ and S_{22}, the relationship between the reflection coefficient with respect to the initial calibration plane Γ_c, and that with respect to the actual aperture measurement plane Γ_m is given by (Arai *et al.* 1995a)

$$\Gamma_c = S_{11} + \frac{S_{21}^2 \Gamma_m}{1 - S_{22}\Gamma_m} \qquad (12.2)$$

Using the condition for a lossless line $S_{11} = S_{22} = 0$ and substituting the transmission coefficient z for S_{21}, Eq. (12.2) becomes

$$\Gamma_c = z^2 \Gamma_m = \exp(-2\gamma L)\Gamma_m \qquad (12.3)$$

where γ is the propagation constant and L is the distance between the two planes, which is the elongation of the coaxial probe. Equation (12.3) can be rewritten as

$$\Gamma_m = |\Gamma_c| \exp(j\phi_c + 2\gamma L) \qquad (12.4)$$

where $|\Gamma_c|$ and ϕ_c are the magnitude and phase of Γ_c respectively. Equation (12.4) can be further rewritten as

$$\Gamma_m = |\Gamma_c| \exp\left(j\phi_c + \frac{2\omega L}{c}\right) \qquad (12.5)$$

where the item $2\omega L/c$ $(=\Delta\phi)$ is the extra phase owing to the extra line added. This is a function of both temperature and frequency and can be measured with either an open circuit or a short circuit, from which an experimental model of L as a function of temperature can be established.

After analyzing the problems related to measurements at high temperatures using coaxial probes, in the following text we discuss three typical techniques for high-temperature measurements: correction of phase shift, spring-loaded coaxial probe, and metallized ceramic coaxial probe.

12.2.3 Correction of phase shift

Arai *et al.* developed a coaxial probe system for the measurement of dielectric properties at high temperatures as shown in Figure 12.4 (Arai *et al.* 1996). The sample is placed in contact with the open end of the probe, and the sample is heated using a conventional furnace via a ceramic worktube. The probe is fabricated from a piece of 7-mm thin-walled coaxial airline made of a heat-resistant alloy. A neutral gas, normally nitrogen with an addition of 5 % hydrogen, is introduced into the tube to prevent the probe from being oxidized. A water jacket is mounted beneath the worktube to protect the conventional microwave components from heat damage. Heat conducting down through the inner conductor is directed to and absorbed by the water jacket via aluminum nitride (AlN) washers.

As the system is cycled over a wide temperature range, the probe will undergo thermal expansion. As discussed earlier, the thermal expansion results

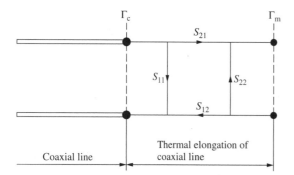

Figure 12.3   The thermal elongation of coaxial line

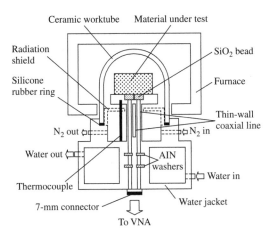

Figure 12.4 High-temperature coaxial-probe system (Arai *et al.* 1996). Reproduced from Arai, M. Binner, J. G. P. and Cross, T. E. (1996). "Comparison of techniques for measuring high-temperature microwave complex permittivity: measurements on an alumina/zircona system", *Journal of Microwave Power and Electromagnetic Energy*, **31** (1), 12–18, by permission of International Microwave Power Institute

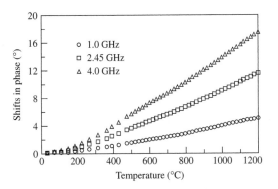

Figure 12.5 The temperature dependence of the shifts in phase in reflection measurement using a coaxial probe (Arai *et al.* 1996). Reproduced from Arai, M. Binner, J. G. P. and Cross, T. E. (1996). "Comparison of techniques for measuring high-temperature microwave complex permittivity: measurements on an alumina/zircona system", *Journal of Microwave Power and Electromagnetic Energy*, **31** (1), 12–18, by permission of International Microwave Power Institute

in an increase in the effective length of the probe, and thus there is an increase in phase shifts of the reflection coefficient, as shown in Figure 12.5.

The phase shift can be fitted by a polynomial expression and incorporated into the expression for

reflection coefficient (Arai *et al.* 1996):

$$
\Gamma_m = |\Gamma_c| \exp \left\{ j \left[\phi_c + \frac{2\omega}{c_0} \right. \right.
$$
$$
\left. \left. \times \, (a_0 + a_1 T + a_2 T^2 + a_3 T^3 + \cdots) \right] \right\}
$$
(12.6)

where Γ_m is the reflection coefficient with respect to the actual sample-probe contact plane, and $|\Gamma_c|$ and ϕ_c are the magnitude and phase of the reflection coefficient Γ_c with respect to the initial calibration plane. ω is the angular frequency ($2\pi f$), c_0 is the speed of light in free space and ($a_0 + a_1 T + a_2 T^2 + a_3 T^3 + \cdots$) describes the thermal expansion of the probe as a function of temperature, T, in which the polynomial coefficients, a_0, a_1, etc., can be determined experimentally. Normally, a polynomial expression consisting of four lower-order terms can simulate the expansion sufficiently.

To verify the applicability of the phase correction method, complex permittivity of air is measured with and without phase shift correction. Figure 12.6 indicates that the increase in phase angle is effectively offset by the method discussed above. The error associated with this technique is

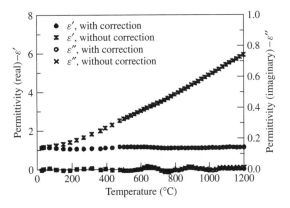

Figure 12.6 Permittivity values of air at different temperatures with and without phase shift corrections. Measurements are made at 2.45 GHz (Arai *et al.* 1996). Reproduced from Arai, M. Binner, J. G. P. and Cross, T. E. (1996). "Comparison of techniques for measuring high-temperature microwave complex permittivity: measurements on an alumina/zircona system", *Journal of Microwave Power and Electromagnetic Energy*, **31** (1), 12–18, by permission of International Microwave Power Institute

dependent upon the complex permittivity of the material and the frequency of measurement, and it is estimated to be within ±8 % for the samples reported in (Arai *et al.* 1996).

12.2.4 Spring-loaded coaxial probe

The differential thermal expansion is inherent in metal probes. In the heating procedure, the outer conductor is heated up faster than the inner conductor, and will therefore expand more than the inner conductor, causing a measurement air gap between the inner conductor and the material under test. In the cooling down procedure, the outer conductor is cooled faster than the inner conductor, and therefore shrinks faster, causing an air gap between the outer conductor and the material under test. The air gaps in both cases may cause severe errors to the measurement results, especially when the sample is heated or cooled quickly.

As shown in Figure 12.7, Gershon *et al.* built a stainless-steel coaxial probe with an air dielectric,

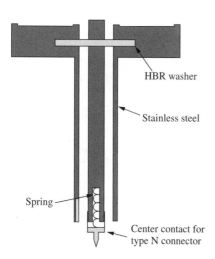

Figure 12.7 An open-ended coaxial probe for high-temperature measurements. The spring-loaded inner conductor forces the inner and outer conductors to maintain intimate contact with the sample under test (Gershon *et al.* 1999). Modified from Gershon, D. L. Calame, J. P. Carmel, Y. Antonsen, T. M. and Hutchen, R. M. (1999). "Open-ended coaxial probe for high-temperature and broad-band dielectric measurements", *IEEE Transactions on Microwave Theory and Techniques*, **47** (9), 1640–1648. © 2003 IEEE

and such a probe can maintain intimate contact with the sample under test at temperatures up to 1200 °C (Gershon *et al.* 1999). The spring-loaded inner conductor is centered by a boron nitride (HBR) washer near the probe head and by the center contact of the type-N connector at the other end. The inner diameter of the washer is large enough to provide a slip fit for the inner conductor at 1200 °C. The spring is positioned inside the inner conductor, which is slip fitted into the center contact of type-N connector. The center contact assists in electrically shielding the spring. In order to prevent damage to the connector and other microwave components during a high-temperature operation, a water jacket, which is positioned near the connector, removes heat from the outer conductor. Similar to the system shown in Figure 12.4, the probe is housed inside a mullite tube with a flowing neutral gas, minimizing the oxidation of the probe and the sample

12.2.5 Metallized ceramic coaxial probe

To solve the thermal expansion problem in coaxial probes, Bringhurst and Iskander developed a metallized ceramic coaxial probe for high-temperature complex permittivity measurements (Bringhurst and Iskander 1996). On the basis of carrying out the network analysis calibration procedure up to 1000 °C, the probe provides accurate dielectric measurements over a broad frequency range (500 MHz to 3 GHz) and for temperatures up to 1000 °C.

The probe is manufactured by using an alumina rod and an alumina tube for the inner and outer conductors of the coaxial probe respectively. The dimensions of the rod and the tube are chosen so that the characteristic impedance of the probe is 50 Ω. In such a probe, the metallization should be of sufficient thickness so that skin-depth problems are avoided at lower frequencies. The tube and rod are metallized with about 3 mil of moly-manganese, and a 0.5-mil protective coating of nickel plating is applied upon the moly-manganese layer. To connect the probe to the standard accessories of a network analysis system, a GR-N-type connector is chosen, because of its broadband performance in the desired frequency range. The

GR half of the connector is modified so that it could facilitate the connection of the open-ended coaxial probe (Bringhurst and Iskander 1996).

The thermal expansion of the metallized ceramic is proved to be minimal by calibrating the probe with a short. So, by using metallized-ceramic coaxial dielectric probes, the differential thermal expansion problem can be solved. Therefore, using such a coaxial probe with small thermal expansions coefficients significantly improves the quality of the calibration procedure and certainly improves the accuracy of the dielectric measurement results (Bringhurst *et al.* 1997; Andrade *et al.* 1992). Meanwhile, as the heat conduction of a metallized ceramic probe is small, the water jacket for cooling is usually not needed.

12.3 WAVEGUIDE METHODS

The conventional waveguide methods have been extended for the measurement of materials properties at high temperatures. The problems related to high-temperature measurement using waveguide methods mainly include how to heat the sample, how to keep the measurement system at room temperature, and how to correct the possible geometrical change of the measurement fixture due to the change of measurement temperature. In the following sections, we discuss two kinds of waveguide methods for high-temperature measurements: open-ended waveguide method and dual-waveguide method.

12.3.1 Open-ended waveguide method

Chevalier *et al.* developed an open-end waveguide method for permittivity measurement at temperatures up to $1000\,^{\circ}\mathrm{C}$ (Chevalier *et al.* 1992). The test apparatus consists of an open-ended flanged circular waveguide propagating TE_{11} mode, which radiates into the material under test placed in front of the flange. The complex permittivity is deduced from the reflection coefficient. A solar source is used for sample heating so that the complex permittivity can be measurement as a function of temperature.

As shown in Figure 12.8, the open-ended flanged circular waveguide propagating the dominant TE_{11} mode radiates into a layer of dielectric

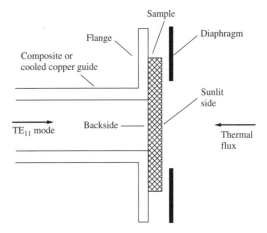

Figure 12.8 Open-ended circular waveguide method for the measurement of dielectric permittivity at high temperatures. The sample under study is heated by solar flux (Chevalier *et al.* 1992). Modified from Chevalier, B. Chatard-Moulin, M. Astier, J. P. and Guillon, P. Y. (1992). "High temperature complex permittivity measurements of composite materials using an open-ended waveguide", *Journal of Electromagnetic Waves and Applications*, **6** (9), 1259–1275, by permission of VSP Brill

material with finite thickness backed by free space. A solar furnace is used to heat the sample. Such a source allows us to raise the temperature of the sample in a definite area without directly heating the measurement device or perturbing the electromagnetic waves. A solar furnace operates by concentrating the energy received by a large sunlit surface onto a relatively small surface by a concentrator. The concentrator is usually a focusing mirror, and the solar energy is concentrated on the focus of the mirror. The flux strength at the focus can be controlled by adjusting a regulator (or flux attenuator) between the large sunlit surface and the concentrator. More information on the solar furnace can be found in (Chevalier *et al.* 1992).

To protect the measurement instrument, the normal copper waveguide should be kept at room temperatures. If the waveguide system is cooled, for example, by water cooling, there may exist thermal gradients in the sample. The sample temperature can be homogenized by using a metallized composite waveguide (Chevalier 1992). The function, design, and fabrication of a metallized composite are similar to those of the metallized

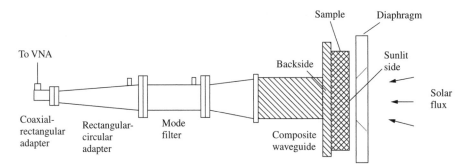

Figure 12.9   Open-ended waveguide system for the measurement of dielectric properties at high temperatures (Chevalier 1992). Modified from Chevalier, B. Chatard-Moulin, M. Astier, J. P. and Guillon, P. Y. (1992). "High temperature complex permittivity measurements of composite materials using an open-ended waveguide", *Journal of Electromagnetic Waves and Applications*, **6** (9), 1259–1275, by permission of VSP Brill

ceramic coaxial probe discussed in Section 12.2.5. In Figure 12.9, a metallized composite guide associated with a heatproof mode filter is used. Such a system does not need to be cooled, and the thermal gradients in the sample are less important. Moreover, such a device has a very small dilation factor versus temperature.

12.3.2 Dual-waveguide method

Among the room-temperature measurement techniques, those using rectangular waveguide lend themselves quite naturally to modification for high-temperature measurements. As discussed earlier, when extending any room-temperature technique to high-temperature techniques, the effects caused by temperature changes in the experimental apparatus must be taken into full consideration. Thermal expansion is the most apparent of these effects, although thermally caused changes in the waveguide electrical properties may also need attention if the waveguide loss characteristics change significantly upon heating. Batt *et al.* developed a variety of microwave measurement techniques for high temperatures, and central to these techniques is the development of a "dual-waveguide" test fixture (Batt *et al.* 1992). These techniques can be used for a variety of material types and sample geometries such as slab, post, and sheet. In the following discussion, we concentrate on the dual-waveguide method for the measurement of slab samples.

In the implementation of a dual-waveguide, two adjacent and symmetric waveguides are machined

from a single block of alloy, one serving as the test waveguide, the other as the reference waveguide. The dual-waveguide is surrounded by a programmable clam-shell furnace. The reference waveguide provides data, which is used to compensate for the electrical and dimensional thermal changes in the test section. With the use of the reference and test waveguide arms, thermal effects can be compensated for. This allows room-temperature transmission line measurement techniques to be modified for high-temperature measurements (Batt *et al.* 1992).

The dual-waveguide technique relies on the use of a reference waveguide to monitor electrical and dimensional changes during the heating process. Reference waveguide data is used to calculate corrections for the raw data generated in the arm of the waveguide containing the test sample. This modification procedure can be applied to the reflection and the reflection/transmission methods. A standard twelve-term network analyzer calibration procedure is employed when a reflection/transmission method is used, while a three-term procedure is used for a reflection method. For a standard two-port test, at chosen temperatures, data is collected for all four *S*-parameters in the reference and test waveguide sections. Two coaxial switches are used so that data can be generated in both the waveguide sections in rapid succession.

Figure 12.10 shows the dual-waveguide configuration developed by Batt *et al.* The waveguide assembly has vertical symmetry except for

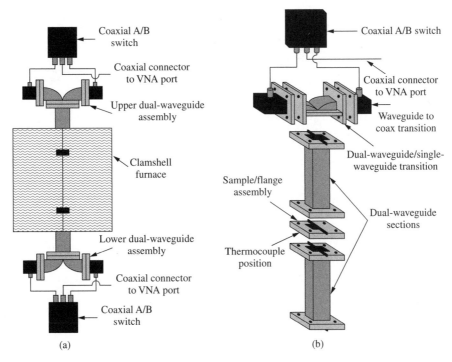

Figure 12.10   Measurement configurations for dual-waveguide method (Batt *et al.* 1992). (a) Dual-waveguide/ furnace assembly and (b) dual-waveguide assembly. Reproduced from Batt, J. A. Rukus, R. and Gilden, M. (1992). "General purpose high temperature microwave measurement of electromagnetic properties", *Materials Research Society Symposium Proceedings*, **269**, 553–559, by permission of Materials Research Society

the flange/sample region. Samples in the form of transverse slabs are relatively easy to produce and analyze. Vertical mounting facilitates slab positioning by the use of small projecting supports from the waveguide wall, and the slab is held in place by gravity. The upper and lower waveguide sections are machined from single blocks of inconel alloy. The septum between the waveguide sections is chosen to match the dual-waveguide/single-waveguide transition. A slot is machined in the interface wall for the placement of an inconel key. This step significantly improved the isolation between the adjacent waveguides at the junction interfaces. Short waveguide sections of stainless steel (not shown in Figure 12.10) are positioned between the inconel sections and the transition assemblies. The top section, which has a tendency to heat because of convection as well as conduction and radiative heat transfer, has a thin brass plate located at the inconel/stainless steel joint region. Air blown on the plate during the heating cycle provides adequate cooling of the upper waveguide region. The lower waveguide assembly is attached to a large aluminum plate, which serves as a heat sink.

The selection of the two flange/sample configurations for slab samples shown in Figure 12.11 is related to the choice of measurement technique. The choice of the reference side of the waveguide is between a short and a thru. Since the purpose of the reference guide data is primarily to provide phase corrections to the sample test data, the S-parameter test data determines the flange choice options for the reference guide. In the most general case where all four S-parameters are required, a shorting plate is needed as shown in Figure 12.11(a). Reflection coefficient data can be measured directly, whereas the transmission coefficient phase changes can be determined by summing the phase changes in S_{11} and S_{22} data. In the flange region, asymmetry exists between test and reference waveguide arms. Phase changes in

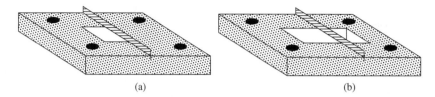

(a)   (b)

Figure 12.11   Two flange/sample configurations for slab samples. (a) Short and (b) through (Batt *et al.* 1992). Reproduced from Batt, J. A. Rukus, R. and Gilden, M. (1992). "General purpose high temperature microwave measurement of electromagnetic properties", *Materials Research Society Symposium Proceedings*, **269**, 553–559, by permission of Materials Research Society

this region must be calculated from the knowledge of thermal expansion behavior of the sample and flange. If only transmission data is required for the test arm, one may measure the transmission coefficient directly, and correct the transmission phase using the thru sample/flange configuration shown in Figure 12.11(b).

It should be noted that errors arise from the air gaps caused by the differential expansion between the waveguide and the sample. Owing to the air gap, the measured dielectric constant and loss tangent are less than their actual values. A detailed discussion on this problem can be found in Chapter 4.

12.4 FREE-SPACE METHODS

In the free space method, the antennas focus microwave energy at the measurement plane and the sample is fixed at the common focal plane between the two antennas. Since the focused beam is very nearly a plane wave, it is possible to have a precise reference for calibration, which increases the accuracy of the method. Since the sample is at the focal plane of the antenna and is not in contact with the applicator, it can easily be adapted for measurements at high or low temperatures and hostile environments (Jose *et al.* 2000). Discussions on the basic working principles of free-space methods can be found in Section 4.5.

In 1989, Ghodgaonkar *et al.* (1989, 1990) used the free-space measurement system developed by HVS Technologies (1986) for the measurement of dielectric properties in the frequency range

14.5 to 17.5 GHz. This compact measurement system proved to be ideal for the characterization of nonplanar surfaces (see e.g., Varadan *et al.* 2000) and for measuring the properties of materials at different temperatures (Varadan *et al.* 1991). The free space method presented here is capable of measuring complex permittivity, permeability, and chirality parameters of materials from 5.8 to 110 GHz as a function of temperature.

The test setup was designed to keep the samples in different temperatures while measuring the propagation parameters. Figure 12.12 shows the

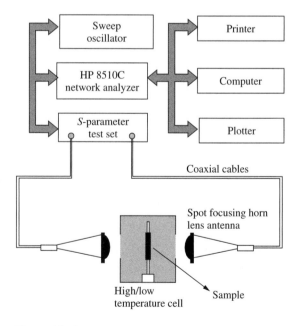

Figure 12.12   Free-space measurement system

configuration of the measurement system using spot focused antennas. The spot focusing horn lens system is a combination of two equal plano-convex dielectric lenses mounted back to back in a conical antenna. At the focal point, the electromagnetic wave has the properties of a plane wave. As the lens dimensions determine the width and depth of the focus, the accuracy in evaluation of permittivity from the experimental data is determined by the design of an optimum focused beam at the focal plane. The measured focal length of the antenna is 30.5 cm with a spot size of 4.37×3.2 cm at 9.1 GHz (Varadan *et al.* 2000). The HP 8510XF single connection, single sweep Network Analyzer is used for the measurement of the complex reflection (S_{11}) and transmission coefficients (S_{21}). The focused plane wave beam from the horn lens antenna system is incident on the sample positioned at the common focal plane of the two antennas. The sample is placed in the sample holder inside an environmental chamber, one for low temperature (down to $-40\,^\circ$C) and another one for high-temperature measurements (up to $1500\,^\circ$C). For high-temperature measurements, a furnace was specially designed and fabricated along with a ceramic sample holder. The sidewalls of the furnace are lined with a material, which is thermally insulating but is virtually transparent to microwaves. The low-temperature cell was also designed and fabricated by HVS Technologies. This consists of an aluminum sample holder, and the insulating materials used are transparent to microwaves. A chiller unit, along with a heat exchanger placed inside the cell, was used to control the temperature inside the low-temperature cell. Dry nitrogen was also pumped inside the cell to prevent any moisture from condensing on the sample. Free space TRL calibration (Ghodgaonkar *et al.* 1990) along with time domain gating was used for accurate measurements. Three sets of antennas were used to cover the range 5.8 to 110 GHz, one pair each for 5.8 to 8.2 GHz, 8.2 to 12.4 GHz, and 12.4 to 110 GHz. The coaxial to waveguide adapters and rectangular to circular waveguide transitions were changed for different waveguide bands from 12.4 to 110 GHz as required.

12.4.1 Computation of ε_r^*

The complex permittivity can be computed either by reflection or transmission measurements. In the reflection method, a perfectly conducting plate is placed behind the sample being measured and the permittivity is determined from the measured reflection coefficient. In the transmission method, the dielectric sample is measured in free space. In this case, the complex permittivity of the unknown sample is determined from the measured transmission coefficient. If the sample size is greater than three times the E-plane 3-dB beamwidth of the focused antennas, the diffraction effects at the edges of the sample can be neglected. At 10 GHz, the measured 3-dB beamwidth in E- and H-planes of the spot focused antennas are 3.20 cm and 4.37 cm.

Computation of complex permittivity from the measured reflection and transmission coefficients is presented by Ghodgaonkar *et al.* (1989). It is assumed that magnetic permeability of the sample is $\mu_r = 1 + j0$. It can be shown that the S_{11} and S_{21} parameters are related to the reflection and transmission coefficients by the following equation

$$S_{21} = \frac{T(1 - \Gamma^2)}{1 - \Gamma^2 T^2} \qquad (12.7)$$

where Γ and T can be written as

$$\Gamma = \frac{Z_{sn} - 1}{Z_{sn} + 1} \qquad (12.8)$$

$$T = e^{ikd} \qquad (12.9)$$

Z_{sn} and k are normalized characteristic impedance and propagation constants of the sample respectively and are related by

$$Z_{sn} = \frac{1}{\sqrt{\varepsilon_r}} \qquad (12.10)$$

$$k = k_0 \sqrt{\varepsilon_r} \qquad (12.11)$$

where $k_0 = 2\pi/\lambda_0$, λ_0 is the wavelength in free space.

The complex permittivity ε_r^* is obtained by an iterative technique with an initial estimate. From

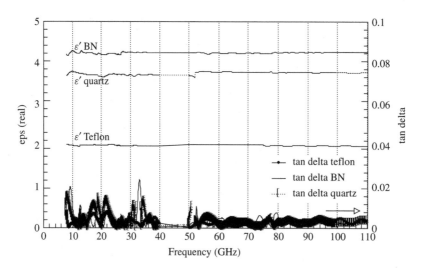

Figure 12.13 Measured permittivity of Teflon ($t = 13.31$ mm), quartz ($t = 6.45$ mm), boron nitride ($t = 12.82$ mm): reported values: Polytetrafluroethylene (Teflon-DuPont) 2.08, 0.0006 at 10 GHz (Von Hippel 1961); quartz, Dynasil 4000 3.82, 0.0004 at 35.1 GHz (Ho 1980); boron nitride 4.17, 0.009 at 15 GHz (Bussey 1969). Sample size is $6'' \times 6''$

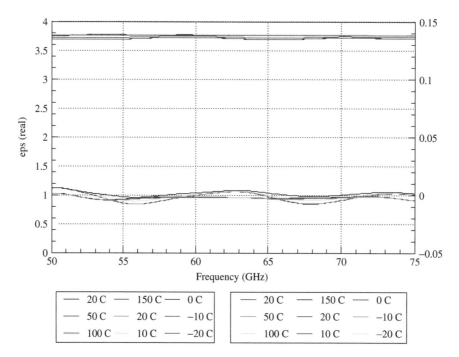

Figure 12.14 Measured change in permittivity of quartz, $t = 6.5270$ mm, due to the change in temperature from -20 to $150\,°$C, sample size is $6'' \times 6''$

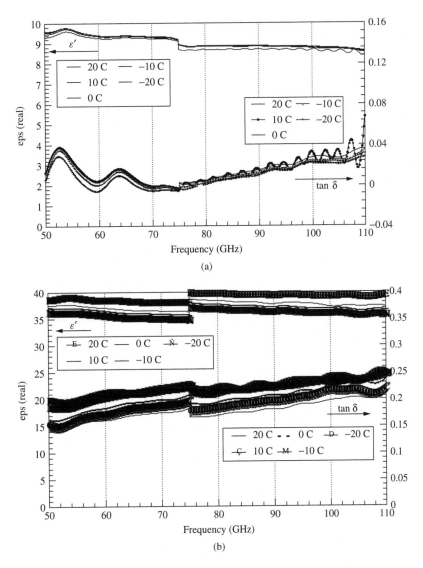

Figure 12.15 (a) Measured permittivity of metal-backed 2×2 sample alumina, $t = 0.6604$ mm. Sample size is $2'' \times 2''$ and (b) permittivity of BST-50, $2'' \times 2''$ sample, $t = 2.5654$ mm. Sample size is $2'' \times 2''$

Eq. (12.8) and the definitions that follow, it is possible to obtain S_{21} in terms of ε_r^*. However, the exact solution for ε_r^* is not straightforward due to the multiple roots for Eq. (12.9). Hence, an iterative technique using an initial guess for the ε_r^* is used. The initial guess is obtained from the estimated value of the phase velocity in the sample computed from the difference in phase at the measurement plane without and with the sample.

Samples of different dielectric constant, thickness, and size were tested and compared with that of the published data (see Jose *et al.* 2000). Measurements have been done for samples of size $6'' \times 6''$ and $2'' \times 2''$ both with and without metal backing for different temperatures. For the low temperature measurements, precautions were taken to avoid any condensation of moisture in the environmental chamber or on the sample. TRL calibration is first performed with an empty sample holder; this also calibrates out the presence of the insulating material from the measurements.

Figure 12.13 presents typical results of measured permittivity of $6'' \times 6''$ samples of Teflon, quartz, and boron nitride at room temperature. Measured values of dielectric constant agree well with those reported in (Von Hippel 1995); Ho (1980); and ASTM Designation D2520-86. There is some discrepancy in the loss tangent of very low-loss materials like Teflon at 10 GHz. This may be due to measurement and computation errors in wideband measurement of S_{11} and S_{21} for low-loss materials. The dielectric constant data reported in (Ho 1980) and ASTM Designation D2520-86 were mainly measured by resonant cavity or transmission line methods, which are known to have the best accuracy for single frequency measurements.

The measurement of the change in permittivity due to variation in temperature is very important for substrates used in real time systems. Jose *et al.* (2000) measured the permittivity of the samples of size $6'' \times 6''$ as well as those of $2'' \times 2''$ in the low temperature cell. Figure 12.14 presents the measured permittivity of a $6'' \times 6''$ quartz sample from $-20\,^\circ\text{C}$ to $150\,^\circ\text{C}$ while Figure 12.15(a) presents the results for the alumina sample and 12.15(b) for a tunable ferroelectric substrate (BST) of size $2'' \times 2''$ used for antenna applications. These samples were measured for two frequency bands and the edges of the bands are clearly seen at 75 GHz, in both the figures. While the permittivity of Teflon and alumina are not expected to change significantly over this range of temperature, BST will change considerably as the temperature drops below the Curie point for this sample which is close to $10\,^\circ\text{C}$.

The free-space technique described in this section gives accurate results of dielectric properties of materials. This *in situ* wideband characterization is possible for different temperatures, which is needed for many practical applications. It is already established that the free space setup is useful for accurate evaluation of dielectric properties of planar and nonplanar materials for a wide range of temperatures and frequencies. Results for a number of samples agree with published data (please contact HVS Technologies if interested)

and demonstrate the usefulness of spot-focused free space system for wideband *in situ* characterization and evaluation of materials. The noncontact nature of the method allows measurements of solids and liquids in high-low-temperature environments. The TRL calibration and error correction technique can be very easily applied to real time industrial applications ensuring accuracy of the measurement data and subsequent inversion.

12.5 CAVITY-PERTURBATION METHODS

Resonant methods for characterizing the electromagnetic properties of materials at high temperatures mainly include cavity-perturbation methods and dielectric-loaded cavity methods. In a resonant method, usually only the sample under test is heated, while the cavity itself is kept at the room temperature; so in most cases, no corrections are needed. Besides, in a resonant method, the cavity for materials property testing can also be used as the microwave applicator for heating samples under test.

In this section, cavity-perturbation methods will be discussed. We first give a brief discussion on the cavity-perturbation methods for measurements of materials properties, and then introduce several typical cavities often used in high-temperature measurements, including TE mode rectangular cavities and TM mode cylindrical cavities.

12.5.1 Cavity-perturbation methods for high-temperature measurements

The working principle for cavity-perturbation methods has been discussed in Chapter 5. In the following section, we give a brief introduction on the cavity-perturbation theory based on which the cavity-perturbation methods are developed, and then discuss the heating techniques often used in the cavity-perturbation methods for high-temperature measurements.

12.5.1.1 Cavity-perturbation theory

The shift of the resonant frequency of a cavity due to the introduction of a small dielectric material may be expressed as (Hutcheon *et al.* 1991)

$$\frac{f_2 - f_1}{f_1} + j \cdot \left(\frac{1}{2Q_2} - \frac{1}{2Q_1} \right)$$

$$= -\alpha \cdot \frac{\chi_e}{1 + \chi_e F_{sh}} \cdot \frac{V_s}{V_c} \qquad (12.12)$$

where $\chi_e = \varepsilon_r - 1$ is the relative complex susceptibility; f_1 and f_2 are the resonant frequencies of the cavity before and after the sample is introduced; Q_1 and Q_2 are the unloaded quality factors of the cavity before and after the sample is introduced; V_s and V_c are the volumes of the sample and the cavity; F_{sh} is a dimensionless sample shape factor; and α is a dimensionless real constant depending only on the shape of the cavity and the shape of the unperturbed fields. Equation (12.12) is based on the following three major assumptions. Firstly, the sample is located in a region of uniform field. Secondly, the sample shape is an ellipsoid of rotation. Thirdly, the stored energy in the sample is small compared to that stored in the rest of the cavity. Equation (12.12) is applicable to nonellipsoidal shapes if empirical values of α and F_{sh} are used. This approximation has been checked with full complex numerical calculations and found to be accurate to $\pm 3\%$ for sample length to diameter ratios down to 3.5:1 (Hutcheon *et al.* 1991).

In actual measurements, more straightforward equations are often used:

$$\frac{f_1 - f_2}{f_1} = A(\varepsilon_r' - 1)\frac{V_s}{V_c} \qquad (12.13)$$

$$\frac{1}{Q_2} - \frac{1}{Q_1} = B\varepsilon_r'' \frac{V_s}{V_c} \qquad (12.14)$$

where A and B are constants depending on the shapes of the cavity and the sample. In experiments, A and B are usually obtained by calibrating the cavity with standard dielectric samples with known permittivity values.

12.5.1.2 Heating techniques

Heating techniques should be fully considered in the design of resonant cavity for high-temperature measurements using cavity-perturbation method. With some modification, the heating techniques used in nonresonant methods can be used in resonant methods, and these techniques mainly include conventional ovens, infrared radiation, laser, and solar energy. In these techniques, the sample is heated through heat conduction or heat radiation. As the sample is heated from outside to the inside, the temperature distribution within the sample may be not uniform.

In a cavity-perturbation method, the cavity itself can be used as a microwave applicator, so the sample under test can be heated by the microwave energy (Tinga and Xi 1993a). A comparison between a conventional heating scheme and microwave heating scheme is shown in Figure 12.16. In a microwave heating scheme, the heat is produced within the sample. Under suitable control conditions, the sample can be heated uniformly and quickly. In the following discussion, both the conventional heating scheme and microwave heating scheme will be used.

In a cavity-perturbation method or a dielectric-loaded cavity method, the sample under test is usually held by an amorphous silica holder, because

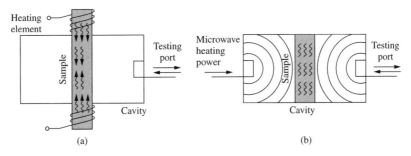

Figure 12.16 Two types of heating schemes (Tinga and Xi 1993a) (a) Conventional heating scheme and (b) microwave heating scheme. Reproduced from Tinga, W. and Xi, X. (1993a). "Design of a new high-temperature dielectrometer system", *Journal of Microwave Power and Electromagnetic Energy*, **28** (2), 93–103, by permission of International Microwave Power Institute

of the ability of fused silica to withstand high temperatures, and also because of the small variation of its dielectric properties with temperature (including low loss). In the use of amorphous silica holders, two effects should be taken into consideration: first, the gradual nucleation and recrystallization to cristoballite above 1000 °C, and second, the chemical and physical interaction with the sample, particularly if the sample contains mobile or reactive metal ions (Hutcheon *et al.* 1992a). The first effect results in heavy frosting of the surface of the holder. Though there is very little change in the dielectric properties due to this effect, embrittlement and cracking of the holder will eventually occur. Regarding the second effect, the chemical interaction with the holder is a function of the number of physical contact points; so the reactive powders, packed in the holder, may produce intolerable reaction, while pressed pellets of reactive material can often be measured with insignificant contamination of the holder.

12.5.2 TE$_{10n}$ mode rectangular cavity

Rectangular TE$_{10n}$ mode (n odd) cavities are often used in characterization of the permittivity of dielectric samples. Figure 12.17 shows a TE$_{10n}$ (n odd) mode rectangular cavity with width a, height b and length L, and a rectangular or cylindrical sample is partially inserted at one antinode of electric field. As shown in Figure 12.17(a), for a cylindrical sample with radius r and height b', its dielectric permittivity can be calculated from

(Andrade *et al.* 1992)

$$\varepsilon_r' = 1 + \frac{f_1 - f_2}{f_1} \cdot \frac{aL}{2\pi r^2} \cdot \frac{b}{b'} \quad (12.15)$$

$$\varepsilon_r'' = \left(\frac{1}{Q_2} - \frac{1}{Q_1}\right) \cdot \frac{aL}{4\pi r^2} \cdot \frac{b}{b'} \quad (12.16)$$

For a rectangular sample with width a', height b' and length L', as shown in Figure 12.13(b), its dielectric permittivity is given by (Andrade *et al.* 1992)

$$\varepsilon_r' = 1 + \frac{f_1 - f_2}{f_1} \cdot \frac{b}{b'}$$
$$\cdot \frac{2aL}{\left\{a' + \dfrac{a}{\pi} \sin\left[\dfrac{(a - a') \cdot \pi}{a}\right]\right\} \left\{L' + \dfrac{L}{3\pi} \sin\left[\dfrac{3(L - L') \cdot \pi}{L}\right]\right\}} \quad (12.17)$$

$$\varepsilon_r'' = \left(\frac{1}{Q_2} - \frac{1}{Q_1}\right) \cdot \frac{b}{b'}$$
$$\cdot \frac{2aL}{\left\{a' + \dfrac{a}{\pi} \sin\left[\dfrac{(a - a') \cdot \pi}{a}\right]\right\} \left\{L' + \dfrac{L}{3\pi} \sin\left[\dfrac{3(L - L') \cdot \pi}{L}\right]\right\}} \quad (12.18)$$

In Eqs. (12.15)–(12.18), f_1 and Q_1 are the resonant frequency and quality factor of the cavity before the sample is inserted, and f_2 and Q_2 are the resonant frequency and quality factor of the

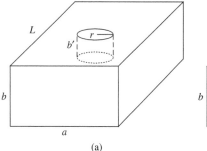
(a)

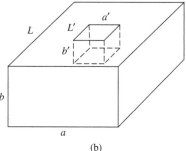
(b)

Figure 12.17 Rectangular cavity with partially inserted (a) cylindrical sample and (b) rectangular sample (Andrade *et al.* 1992). Reproduced from Andrade, O. M. Iskander, M. F. and Bringhurst, S. (1992). "High temperature broadband dielectric properties measurement techniques", *Materials Research Society Symposium Proceedings*, **269**, 527–539, by permission of Materials Research Society

cavity after the sample is inserted. It should be noted that these equations are based on resonant-perturbation theory, so the assumptions based on which the resonant-perturbation theory is developed should be fulfilled. One basic requirement is that the sample should be much smaller than the cavity.

Bringhurst *et al.* developed two TE_{10n} rectangular cavities to measure the complex permittivity of ceramic rod samples with circular and rectangular cross sections up to $1000\,^{\circ}C$ (Bringhurst *et al.* 1992). The S-band cavity (2.45 GHz) works at the TE_{103} mode, and the X-band cavity (10 GHz) operates in the TE_{107}-mode. The S-band rectangular cavity is constructed from a section of WR-340 brass waveguide. As shown in Figure 12.18, the two electric dipole couplers are installed in the cavity's broad wall on opposite sides at positions of maximum electric field, which in this case is at a distance of 1/6 of the length of the cavity from both ends. The lengths of the couplers are adjustable to obtain optimum couplings for different measurements. Two small holes at the center of the cavity are for sample insertion, and their diameter is chosen such that the circular waveguides thus formed are well below cutoff at the resonant frequency of the cavity.

For measurements at high temperatures, the sample is held in an adequate sample holder, which is to be heated in a conventional furnace prior to insertion into the cavity. Usually a thin-walled

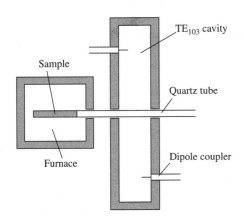

Figure 12.18 A TE_{103} mode rectangular cavity for the measurements of dielectric permittivity at high-temperature

fused-silica cylindrical tube is selected as a sample holder and a stop plate inside the silica tube is used to accurately calculate the sample length. This arrangement serves two purposes: to reduce the volume of the sample inside the cavity when losses are so high that a large reduction in the quality factor would affect the accuracy of the measurement, and also to allow for monitoring the sample temperature with a thermocouple located at the tip of the sample protruding from the cavity (Bringhurst *et al.* 1992).

During the measurement, the tube is quickly moved from the furnace into the cavity when the desired temperature is reached. Initially, the empty sample holder is calibrated as a function of temperature (both resonant frequency and quality factor), and these values are used as the "empty cavity" parameters in the perturbation expressions. Since the cooling rates of the sample and the tube are very fast after removal from the furnace and insertion into the cavity, special insulating blocks are placed around the tube part that protrudes from the furnace to minimize this effect. Also, to avoid the heating of the cavity during the measurement, fast data acquisition is needed when the sample is positioned in the cavity.

As shown in Figure 12.19, Ho developed a cavity-perturbation measurement system consisting of a rectangular reflection-type cavity operating in the TE_{107} mode near 35 GHz (Ho 1988). A small cylindrical sample of the material to be tested is inserted into the cavity through holes that center on the broad sides of the cavity. In this position, the sample is aligned along the direction of maximum field intensity. The dielectric constant of the sample is calculated from the observed shift in resonant frequency and the loss tangent from the change in the quality factor of the cavity.

To achieve a high temperature without significant increase in the temperatures of the cavity and the associated microwave measurement components, small tungsten heating coils are positioned around that portion of the sample that protrudes from the cavity. Cooling channels are attached to the body of the cavity. Constant-temperature recirculating water is used to maintain the cavity walls at near room temperature. The cavity system is placed inside a vacuum enclosure with viewing

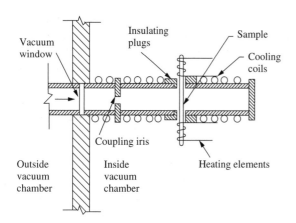

Figure 12.19 A TE$_{107}$ mode rectangular cavity for determining the dielectric properties of small rod-shaped samples by cavity-perturbation measurement method (Ho 1988). Modified from Ho, W. W. (1988). "High-temperature dielectric properties of polycrystalline ceramics", *Materials Research Society Symposium Proceedings*, **124**, 137–148, by permissions of Materials Research Society

ports, which allow observation of samples and measurement of sample temperatures. The vacuum enclosure can be filled with a neutral gas mixture to prevent oxidation of the sample and the tungsten heating element at high temperatures.

12.5.3 TM mode cylindrical cavity

The cylindrical cavities for the characterization of dielectric properties of materials usually operate at TM modes, and TM modes often used include two categories: TM$_{0n0}$ modes and TM$_{01n}$ modes. In the following section, we discuss the applications of these two categories of TM modes in the measurement of dielectric properties of materials at high temperatures.

12.5.3.1 TM$_{0n0}$ mode cavity

Owing to their field and current distributions, TM$_{0n0}$ mode cylindrical cavities are widely used in the characterization of dielectric materials, and the TM$_{010}$ mode is the most widely used. Figure 12.20 shows the field distributions in a TM$_{010}$ mode cylindrical cavity. The sample is placed at a point near the axis of the cavity. As the field distribution

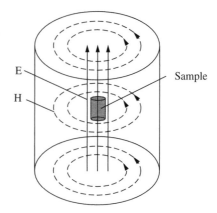

Figure 12.20 Field distributions of a TE$_{010}$ mode cylindrical cavity

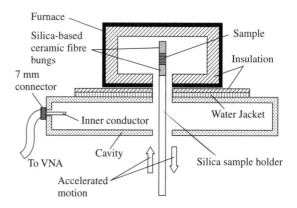

Figure 12.21 TM$_{0n0}$ circular cylindrical cavity system for dielectric property measurement at high temperatures (Arai *et al.* 1996). Reproduced from Arai, M. Binner, J. G. P. and Cross, T. E. (1996). "Comparison of techniques for measuring high-temperature microwave complex permittivity: measurements on an alumina/zircona system", *Journal of Microwave Power and Electromagnetic Energy*, **31** (1), 12–18, by permission of International Microwave Power Institute

does not change with z-position, this method does not have strict requirements on sample positioning.

Figure 12.21 shows a TM$_{0n0}$ circular cylindrical cavity system developed by Arai *et al.* (1996). The cavity is positioned below the furnace in order to avoid hot air entering the cavity and causing dimensional changes that can lead to an additional error (Binner *et al.* 1994). A cylindrical solid sample or a powder sample held in a silica

(SiO$_2$) sample holder using silica-based ceramic-fiber bungs is rapidly moved from the furnace to the cavity, measured and moved back again into the furnace. Repeating this procedure allows measurements over a range of temperatures. Usually, several TM$_{0n0}$ modes can be used before other resonant modes appear, so dielectric properties at several frequencies can be obtained.

It should be noted that this technique inevitably leads to a sample cooling when the sample under test is moved from the furnace to the cavity for measurement, and the actual sample temperature needs to be estimated using appropriate cooling curves. This may introduce some errors unless the curves are accurately designed to simulate the cooling of the sample. Another course of errors is attributable to volume changes of the sample with temperature, and this effect can be estimated using volume expansion coefficients.

Figure 12.22 shows a modified TM$_{010}$ mode cavity (Hutcheon *et al.* 1992a, b). In this structure, the electric field is uniform in a larger area near the axis of the cavity, and the electric field is more concentrated near the central space, so this cavity has higher accuracy and sensitivity than a conventional cylindrical TM$_{010}$ mode cavity (Hutcheon *et al.* 1992c). Besides, the two openings at both ends of the cavity make it easy to load and unload samples.

12.5.3.2 TM$_{01n}$ mode cavity

The TM$_{01n}$ modes used in materials property characterization mainly include TM$_{012}$ mode (Jow *et al.* 1989, 1987; Ali *et al.* 2000) and TM$_{013}$ mode (Tian and Tinga 1993). In these modes, the cavities are in single-mode states. Usually the single mode in a cavity is used to characterize the dielectric properties of the sample, and meanwhile the energy of this mode is also used in heating the sample under test. In the following discussion, we concentrate on TM$_{012}$ mode.

Resonant perturbation of a TM$_{012}$ mode cavity

Figure 12.23 shows a diagram of the field patterns in a TM$_{012}$ mode cavity loaded with a dielectric sample. The cylindrical shape of the sample is located at the center of the cavity, and this is the position of the maximum TM$_{012}$ mode electric field. The perturbation equations for a small material rod located at the center of a TM$_{012}$ mode cavity are given as follows (Jow *et al.* 1989):

$$\frac{f_0 - f_s}{f_0} = (\varepsilon_r' - 1)ABG \cdot \frac{V_s}{V_c} \quad (12.19)$$

$$\frac{1}{Q_s} - \frac{1}{Q_0} = 2\varepsilon_r''ABG \cdot \frac{V_s}{V_c} \quad (12.20)$$

with

$$A = J_0^2(2.405R_s/R_c) + J_1^2(2.405R_s/R_c) \quad (12.21)$$

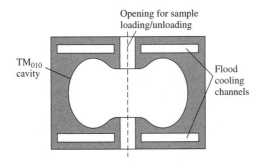

Figure 12.22   Modified TM$_{010}$ mode cylindrical cavity for the measurement of dielectric properties at high temperatures

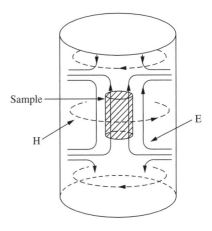

Figure 12.23   Field distribution of TM$_{012}$ mode resonant cavity. Source: Jow, J. Hawley, M. C. and Finzel, M. C. and Asmussen, Jr. J. (1989). "Microwave heating and dielectric diagnosis technique in a single-mode resonant cavity", *Review of Scientific Instruments*, **60** (1), 96–103, by permission of American Institute of Physics

$$B = 1 + [L_c/(2\pi L_s)] \cdot \sin(2\pi L_s/L_c)$$
$$\cdot \cos(4\pi H/L_s) \qquad (12.22)$$

$$G = 0.2178 \cdot [c_0/(f_0 R_c)] \qquad (12.23)$$

where L_c, R_c, V_c, f_0, and Q_0 are the length, the radius, the volume, the resonant frequency, and the quality factor of the empty cavity respectively; H, R_s, and V_s are the height, the radius, and the volume of the sample of the material under test respectively; f_s and Q_s are the resonant frequency and the quality factor of the loaded cavity respectively; $J_0(x)$ and $J_1(x)$ are the zero-order and the first-order cylindrical Bessel functions respectively, and c_0 is the velocity of light in free space.

Equations (12.19)–(12.23) indicate that when the assumptions for resonant-perturbation theory are satisfied, the dielectric properties of a sample can be calculated from the changes of the resonant frequency and quality factor due to the introduction of the sample and the geometrical parameters of the cavity and the sample.

As shown in Figure 12.24, Jow *et al.* developed a TM_{012} mode cavity with tunable resonant frequency for processing and characterizing materials (Jow *et al.* 1989). The cavity is made of a cylindrical brass tube, covered with three transverse brass circular plates. The shorting plate inside the cavity is adjustable to provide a variable cavity length L_c. One sample insertion hole is drilled through the center of the top plates. The sample can be loaded through this hole using a cotton thread without removing the top plates of the cavity. The removable and shorting bottom plate is designed to easily place and process large disk materials. Silver fingers are soldered around both the internal short plates to provide good electrical contact with the cavity wall. A semirigid 50 Ω impedance copper coaxial probe serves as a field excitation probe, and the coupling of this coupling probe can be adjusted by adjusting the inner conductor depth L_p. A microcoaxial probe is used as an E-field diagnostic probe. The inner conductor depth of the diagnostic probe is carefully chosen so that the electric field is detectable and is essentially undisturbed.

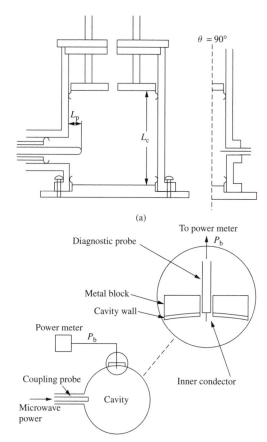

Figure 12.24 A TM_{012} mode cylindrical cavity for processing and diagnosing dielectric samples (Jow *et al.* 1989). (a) Cross-sectional view (the $\theta = 0°$ plane passes through the coupling probe) and (b) top view of the diagnostic probe. Modified from Jow, J. Hawley, M. C. Finzel, M. C. and Asmussen, Jr. J. (1989). "Microwave heating and dielectric diagnosis technique in a single-mode resonant cavity", *Review of Scientific Instruments*, **60** (1), 96–103, by permission of American Institute of Physics

Swept-frequency method and single-frequency method

To obtain complex permittivity using cavity-perturbation technique, the resonant frequencies and quality factors before and after the introduction of the sample should be measured. Two methods have been used to measure the changes of resonant frequency and quality factor due to sample insertion: swept-frequency method and single-frequency method. The swept-frequency method measures conventional resonant frequency shift and quality factor changes from the resonance curve with sweeping frequency of a cavity at a fixed cavity length. The single-frequency method measures the changes of the cavity length and the power reflection or transmission at a single frequency. It is also found that the error in a single-frequency method can be smaller than that in the swept-frequency method (Tian and Tinga 1993; Jow *et al.* 1989).

Both swept- and single-frequency methods can be applied to on-line diagnosing and processing of materials. However, most of the incident input power in the swept-frequency method is reflected, while, using the single-frequency method, the microwave power can be more efficiently transferred into the materials under process and test. Besides, power and cavity length measurements in the single-frequency method are faster and more accurate. Furthermore, the microwave power input can also be easily controlled using the single-frequency method. Therefore, compared with the swept-frequency method, the single-frequency method is more suitable to simultaneously heat the material and diagnose the heating process.

Single-frequency methods for materials' property diagnosis

When a single-frequency method is used to simultaneously process and diagnose materials, the temperature of the material should be controlled by adjusting the microwave power, and meanwhile, the dielectric properties of the material under process should be monitored. The techniques of temperature control in the methods discussed in Section 12.6 can be used in this method, with suitable modifications. In the following section, we discuss two single-frequency methods for the measurement of materials properties: transmission method and reflection method.

Transmission method

In the single-frequency method developed by Jow *et al.* (1989), the cavity is kept at the same resonant frequency by adjusting the cavity length. The shift of resonant frequency due to sample insertion is obtained from the change of the length of the cavity, and change of the quality factor is obtained from the change of the ratio between the power detected from the cavity and the power incident to the cavity.

The measurement circuit is shown in Figure 12.25. The microwave source is protected by an isolator (not shown in the figure) and incorporates an amplifier. Another isolator (not shown in the figure) is used to protect the amplifier from power reflected from the cavity. The incident power P_i from the source is decoupled by a directional coupler and measured by a power

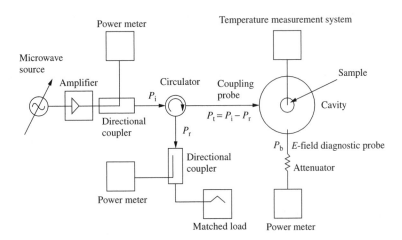

Figure 12.25 Microwave processing and diagnostic system circuit (Jow *et al.* 1989). Modified from Jow, J. Hawley, M. C. Finzel, M. C. and Asmussen, Jr. J. (1989). "Microwave heating and dielectric diagnosis technique in a single-mode resonant cavity", *Review of Scientific Instruments*, **60** (1), 96–103, by permission of American Institute of Physics

meter. The incident microwave energy is fed into a tunable cavity by the coupling probe. The reflected power P_r from the cavity is guided by a circulator, attenuated down by an attenuator and measured by another power meter. The power P_b detected by the diagnostic probe and measured by a power meter is proportional to the storage of microwave energy in the cavity.

The cavity shown in Figure 12.24 is used in the circuit shown in Figure 12.25. Continuous microwave input power at 2.45 GHz is fed into the cavity, and the length of the cavity is tuned to achieve TM_{012}-mode resonance at this frequency, indicated by the minimum reflection power. This resonant frequency (2.45 GHz) is the frequency f_0 in Eq. (12.19).

The coupling between the cavity and the external circuit is determined by the depth of the coupling probe L_p in Figure 12.24(a). A critically coupled structure is established when the reflected power P_r is negligible compared with the incident power P_i. The total dissipated power P_t in the microwave system is the difference between the measured incident and reflected powers. The quality factor of the cavity is proportional to the power ratio (P_b/P_c) at the resonant frequency f_0 (Jow *et al.* 1989):

$$Q_c \propto P_b/P_c \propto P_b/P_t \qquad (12.24)$$

where P_c is the dissipated power in the cavity, and P_b is the power detected by the diagnostic probe as shown in Figure 12.25. So the ratio of the quality factor of the cavity with and without the sample can be determined by the power ratio measurements as described by

$$Q_s/Q_0 = (P_{bs}/P_{ts})/(P_{b0}/P_{t0}) \qquad (12.25)$$

where the subscript "0" indicates the state without sample, and the subscript "s" indicates the state with sample. Equation (12.25) indicates that only the initial measurements of the quality factor Q_0 and the power ratio of P_{b0} to P_{t0} for the unloaded resonant cavity are required to determine the quality factor Q_s from the power ratio of P_{bs} to P_{ts} for the loaded resonant cavity.

The measurement procedure mainly consists of four steps (Jow *et al.* 1989). In the first step, the microwave system operates at low-power swept

frequency state, and the cavity is adjusted to critical coupling state. The resonant frequency f_0 and the quality factor Q_0 of the empty TM_{012} mode cavity are measured using the swept-frequency method. In the second step, the swept-frequency power input is switched to a single-frequency power input at frequency f_0, and the values of P_{b0}, P_{i0}, P_{r0}, and L_{c0} are measured. In the third step, the sample is loaded to the cavity at a position of maximum electric field (about the center of the cavity). Owing to the presence of the sample, the cavity is not in a resonant state. By adjusting the cavity length from L_{c0} to a new length L_{cs}, where the minimum reflection power is obtained, the resonant state is built again. The critical coupling is then restructured by adjusting the coupling probe depth to the point where the reflected power could be neglected compared to the incident power. Measurements of P_{bs}, P_{is}, P_{rs}, and L_{cs} are then made. The quality factor of the loaded cavity is determined according to Eq. (12.25). In the fourth step, the material is removed out of the cavity, and the system is switched into a swept-frequency state. The resonant frequency f_s of the empty cavity at the new cavity length L_{cs} is then determined using the swept-frequency method.

For a given cavity, the calibration curves of the resonant frequency shift versus the new cavity length and the cavity quality factor versus the power ratio can be obtained by calibration (Jow *et al.* 1989). The measured values of the cavity length and the power ratio can be directly related to the resonant frequency shift and the cavity quality factor from these calibration curves. Generally speaking, the cavity length and the power ratio decrease when the dielectric constant and dielectric loss factor of the sample increase. To monitor the properties of the sample on-line during the single-frequency microwave heating, continuous measurements of the cavity length and the power ratio are required.

Reflection method

Tian and Tinga developed a single-frequency reflection method for the measurement of relative quality factor (Tian and Tinga 1993), and this method is based on the fact that the quality factor of a resonator is directly or inversely proportional

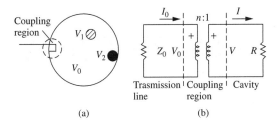

(a)   (b)

Figure 12.26 Cavity detuning and retuning. (a) Cavity containing a coupling device and two disturbances V_1 and V_2 and (b) equivalent circuit of the cavity at resonance. Modified from Tian, B. Q. and Tinga, W. R. (1993). "Single-frequency relative Q measurements using perturbation theory", *IEEE Transactions on Microwave Theory and Techniques*, **41** (11), 1922–1927. © 2003 IEEE

to the normalized input resonant resistance if a moderate perturbation condition is satisfied.

Figure 12.26 shows a resonant cavity, magnetically coupled to an external transmission line with characteristic impedance Z_0. The quality factor Q of the cavity can be expressed as (Tian and Tinga 1993)

$$Q = C_m/\overline{R} \qquad (12.26)$$

with

$$\overline{R} = n^2 R/Z_0 \qquad (12.27)$$

where R is the resonant resistance of the cavity, $\overline{R}$ is the normalized resistance looking into the cavity from the outside, n is the current transformer ratio of the coupling mechanism, and C_m is a constant for a cavity satisfying the perturbation assumptions. For a cavity electrically coupled to an external transmission line, by the principle of duality, we have (Tian and Tinga 1995, 1993)

$$Q = C_e \overline{R} \qquad (12.28)$$

where C_e is also a constant for a cavity satisfying the perturbation assumptions. It should be noted that the use of an ideal transformer model representing the coupling device in no way limits the generality of the above conclusions.

According to transmission line theory, the normalized resistance $\overline{R}$ is related to the reflection coefficient $|\Gamma|$ by

$$\overline{R} = \frac{1 \pm |\Gamma|}{1 \mp |\Gamma|} \qquad (12.29)$$

where the choice of the sign depends on whether the cavity is overcoupled or undercoupled, respectively. Applying Eqs. (12.26) – (12.29), the relative quality factor can be determined through the measurement of the reflection coefficient. Only if an absolute value of quality factor needs to be found must a value for C_m or C_e be determined. The value of C_m or C_e can be obtained by comparing the results obtained from the single frequency method and the standard swept-frequency method.

As discussed in Chapter 6, there are two typical cases in perturbation applications. One is detuning, where a dielectric disturbance, V_1 as in Figure 12.26(a), causes a resonant frequency shift. The other is retuning, where, after a first disturbance, the resonant frequency is retuned via a second disturbance, say V_2, to the original unperturbed resonant frequency. In retuning, the second disturbance of volume V_2 to pull the disturbed resonant frequency back to its undisturbed value can be done by adjusting a short circuit plunger or a tuning stub installed in a cavity. To maintain a fixed resonant frequency, the addition of the energy variations caused by the two disturbances must be zero. In other words, the energy in the twice-disturbed cavity is equal to that of the undisturbed cavity, and the Q measured in the retuned case will be more accurate than that measured in the detuned case.

The application of the single-frequency relative quality factor measurement in the determination of the dielectric loss factor is based on the fact that the dielectric loss factor can be derived from the drop in quality factor resulting from the loss of the sample inserted in the cavity (Tian and Tinga 1993):

$$\varepsilon_r'' = \frac{\varepsilon_r'}{F}\left(\frac{1}{Q_s} - \frac{1}{Q_0}\right) \qquad (12.30)$$

where Q_0 and Q_s are the quality factors of the cavity before and after the sample is loaded respectively, and F is the filling factor. Equation (12.30) can be rewritten as

$$\varepsilon_r'' = \frac{\varepsilon_r'}{F Q_0}\left(\frac{Q_0}{Q_s} - 1\right) \qquad (12.31)$$

Item (Q_0/Q_s) in Eq. (12.31) can be experimentally measured. By calibrating the value of (FQ_0), instead of F, using a standard sample, the absolute value of ε_r'' can be determined from Q_0/Q. Throughout the process, neither C_m (C_e) nor the absolute quality factor need to be determined.

12.6 DIELECTRIC-LOADED CAVITY METHOD

Besides the cavity-perturbation method, the dielectric-loaded cavity method is another type of resonant method, which can be extended to measure the dielectric properties of materials at high temperatures. The resonant properties of a dielectric-loaded cavity are determined by both the cavity and the dielectric material. The effects of the dielectric material to the resonant properties of the dielectric-loaded cavity may be great, and such a resonant structure cannot be analyzed using the resonant-perturbation theory. The dielectric-loaded cavity can be taken as a case between a perturbed cavity and a dielectric resonator, and accurate analysis of such a cavity is complicated, and usually numerical methods are needed. In this section, we discuss two types of dielectric-loaded cavities: coaxial reentrant cavity and open resonator.

In a dielectric-loaded cavity method, microwave energy can be used in heating the sample under test. A dielectric-loaded cavity may operate in dual-mode approach or single-mode approach. In the dual-mode approach, two resonant modes are used for heating and characterizing the sample respectively. In the single-mode approach, one resonant mode is used for both heating and characterizing the sample. Besides, in the oscillation-loop method discussed in the final part of this section, the microwave energy for sample heating is generated by the oscillation loop including the cavity loaded with the dielectric sample under test.

12.6.1 Coaxial reentrant cavity

Figure 12.27 shows two types of coaxial reentrant cavities that are often used in materials property characterization: single-post reentrant coaxial cavity and double-post reentrant coaxial cavity. In a single-post coaxial cavity, the dielectric sample is placed at the gap region between the open end of the central conductor and one

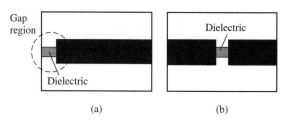

Figure 12.27 Two types of coaxial reentrant cavities loaded with dielectric samples. (a) Single-post coaxial cavity and (b) double-post coaxial cavity

end-wall of the cavity (Xi *et al.* 1992b). In a double-post coaxial cavity, the sample under test is placed between the two open-ends of the two central conductors (Xi *et al.* 1994). The single-post structure has the advantage of easier sample loading than the double-post structure. In the following discussion, we concentrate on single-post structure.

A single-post reentrant coaxial cavity, as shown in Figure 12.27(a), supports a series of quasi-TEM modes whose field patterns closely match the TEM modes in the region far from the gap, and in the gap region, are similar to the TM mode pattern. These TM modes provide a strong E_z component so that high measurement sensitivity and high heating rate are attainable for a sample present in the gap. Based on this structure, Tinga *et al.* developed a coaxial reentrant cavity both as a microwave heater and a dielectrometer with the capability of measuring the complex permittivity of a small sample from 300 to 1600 °C (Tinga 1992; Xi and Tinga 1992c). By virtue of the selective and localized microwave heating, a small sample can be rapidly heated to its melting point, while the cavity remains at room temperature, thereby greatly reducing the expansion errors that occur in a conventional sample-heating scheme.

12.6.1.1 Cavity structure

Figure 12.28 shows the gap regions of three single-post coaxial reentrant cavities:

(a) without insertion hole or a sample holder,

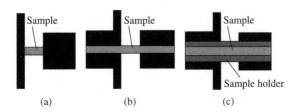

(a)   (b)   (c)

Figure 12.28 The gap regions of three cavities: (a) without an insertion hole or a sample holder, (b) with an insertion hole but without a sample holder, and (c) with both an insertion hole and a sample holder

(b) with insertion hole but without sample holder, and

(c) with both insertion hole and sample holder

In the design of measurement fixtures, the structure (c) has obvious advantages. Using a hollow center conductor and a metallic support in the endplate, one can insert a cylindrical sample through the insertion hole and push it through the focused-field gap into the hollow center conductor, without dismantling the cavity. By making the diameter of the hole much smaller than the wavelength in the sample, the hole can only support evanescent modes. Thus, the fields are strongly attenuated inside the hole, and the physical discontinuities at the sample ends are virtually invisible to the cavity fields (Tinga and Xi 1993a). The use of the sample holder makes it possible that various types of samples can be processed and tested, including powders and liquids. More discussions on the effects of the hole and sample holder can be found in (Xi and Tinga 1992a).

Figure 12.29 shows the basic structure of the reentrant coaxial cavity developed by Tinga *et al.* In this cavity, the gap structure shown in Figure 12.28(c) is used. As this cavity is expected to operate at dual-mode approach, the dimensions of the cavity should be chosen such that two quasi-TEM modes resonate simultaneously at the testing and heating frequencies (Tinga and Xi 1993a).

Figure 12.30 shows the resonant frequency shift Δf_0 of three cavities with different gap structures shown in Figure 12.28. Except for the gap regions, the three cavities have identical dimensions: $r_s = 0.24$ cm, $r_1 = 1.23$ cm, $r_2 = 4.51$ cm, $L =$

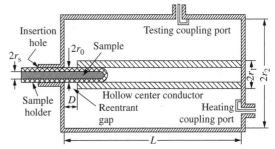

Figure 12.29 Cross section of a single-post coaxial reentrant cavity used as a heating and testing chamber (Xi and Tinga 1992a). Modified from Xi, W. G. and Tinga, W. R. (1992a). "Field analysis of new coaxial dielectrometer", *IEEE Transactions on Microwave Theory and Techniques*, **40** (10), 1927–1934. © 2003 IEEE

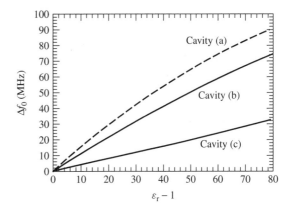

Figure 12.30 Resonant frequency shift produced by a sample at about 3 GHz (Xi and Tinga 1992a). Cavities (a), (b), and (c) respond to the three structures shown in Figure 12.28. For cavity (a), $r_0 = 0$. For cavity (b), $r_0 = r_s$. For cavity (c), $r_0 = 0.37$ cm, and the dielectric constant of the sample holder is 3.78. The structural parameters are defined in Figure 12.29. Modified from Xi, W. G. and Tinga, W. R. (1992a). "Field analysis of new coaxial dielectrometer", *IEEE Transactions on Microwave Theory and Techniques*, **40** (10), 1927–1934. © 2003 IEEE

20.5 cm, and $D = 0.3$ cm (Xi and Tinga 1992a). The above structural parameters are defined in Figure 12.29. The slope of Δf_0 versus ε_r determines the sensitivity in dielectric determinations, and it also commands the measurable range of dielectric properties because a larger detuning of a resonator tends to bring about more difficulties such as impedance mismatch in practical measurements. Therefore, a coaxial cavity with sample insertion holes and holder can extend the measurable range of dielectric properties. Moreover, it presents a better linearity of frequency shift as a function of dielectric constant. This improvement in linearity is not only the consequence of a reduction of the gap field but also the outcome of an increased effective hole depth at a higher dielectric constant, which offsets the saturation exhibited for a normal detuning curve, such as the curve for cavity (a) (Xi and Tinga 1992a).

Convenient, rapid and repeatable measurements can be made with this structure, which remains at or near room temperature during the entire test cycle time. If necessary, the hollow center conductor can be cooled using liquid to keep the cavity temperature constant. To allow a wider complex permittivity range, the gap width can be increased. The loss factor of the sample can be determined from the cavity quality factor. In addition, an accurate, theoretically derived calibration of frequency shift versus permittivity is available, from a mode-matching model of

this reentrant structure, making calibration against reference materials unnecessary (Xi and Tinga 1992a). Error analysis for this method can be found in (Tinga and Xi 1993b).

12.6.1.2 Control and measurement circuit for dual-mode method

In a cavity for dual-mode method, both the heating mode and testing mode resonate, and these two modes are driven by two separate microwave sources. Figure 12.31 shows a block diagram of the measurement system. In this setup, the cavity is connected to three channels. A 2- to 4-GHz reflectometer is used as the testing channel, in which the frequency shift and change of quality factor produced by an inserted sample are measured for dielectric determinations. The heating channel is mainly a microwave power source at 915 MHz, tunable over a 30-MHz band and capable of providing high microwave power. An electronic tuner is used to automatically adjust the power source frequency to the resonant frequency of the cavity, thus ensuring effective power delivery. An optical fiber thermometer (OFT) makes up the third

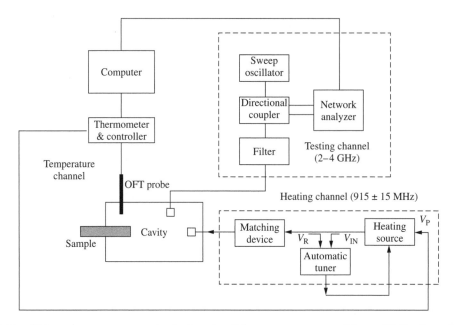

Figure 12.31 High-temperature automatic dual-mode dielectrometer system (Tinga 1992). Reproduced from Tinga, W. R. (1992). "Rapid high-temperature measurement of microwave dielectric properties", *Materials Research Society Symposium Proceedings*, **269**, 505–516, by permission of Materials Research Society

channel for measuring and controlling the temperature of the sample. Besides, a computer is used for automatic instrument setting, data acquisition and processing, and testing and heating control (Xi and Tinga 1992c; Tinga 1992).

12.6.1.3 Single-mode method

On the basis of single-frequency quality factor measurement techniques discussed in Section 12.5.3, the dielectric-loaded cavity can work at single-mode state (Tinga 1992). In a single-mode cavity method, the dielectric sample is heated while, simultaneously, its permittivity is measured with a single frequency microwave source. This eliminates most of the otherwise required precision instruments and makes the method suitable for on-line monitoring of microwave material properties when those materials are being heated with microwave power. The trade-off of this method is that electronic or mechanical adjustment is required to keep the resonator tuned and matched to the power source frequency. The frequency tuning can be made by the movable end wall, and the coupling matching can be accomplished by adjusting the coupling structure, for example, a movable coupling loop.

12.6.2 Open-resonator method

The open-resonator method is the resonant method corresponding to the free-space method discussed in Section 12.4. The open-resonator method for measuring dielectric materials at high temperatures offers certain distinct advantages. The sample under test does not need to contact the cavity wall, and the cavity has high quality factor though it is an open structure. Figure 12.32 shows two structures for dielectric measurement using open resonators. The spherical-mirror resonator is convenient for flat sheet specimens (Cullen *et al.* 1972). The applications of using spherical-mirror resonators for materials property characterization have been discussed in Chapter 5. In this section, we concentrate on the barrel-resonator method, and follow the analyzing approach in (Nagenthieram and Cullen 1974).

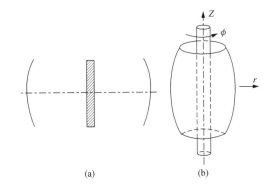

The inside surface of a barrel resonator is constructed in the shape of a barrel, and both the top and the bottom of the barrel are open. A barrel resonator is convenient for rod specimens. A long cylindrical rod of material may be introduced into the barrel through the open ends, and the change in resonant frequency and quality factor of the barrel can be measured and related to the dielectric properties of the rod. Regarding the measurement procedure, the barrel resonator is usually coupled to the source and detector waveguides by small holes, as used in conventional closed resonators, and the resonant modes can be identified by a small bead perturbation method.

The resonant frequency of an empty barrel resonator can be calculated by two distinct methods (Nagenthieram and Cullen 1974). The first method synthesizes the expected Hermite function form of axial field variation from circular waveguide modes using a Fourier integral method. According to this method, the resonant frequency of an empty barrel cavity is given by

$$f_{mnq} = \frac{c}{2\pi r_0} v_{mn} \ (\text{or} \ v'_{mn})$$

$$+ (2q+1)\frac{c}{2\pi r_0}\sqrt{\frac{r_0}{4R_0 - r_0}} = f_c \quad (12.32)$$

with

$$J_m(v_{mn}) = 0 \ \text{for} \ E \ \text{modes} \quad (12.33)$$

$$J'_m(v'_{mn}) = 0 \ \text{for} \ H \ \text{modes} \quad (12.34)$$

where $m, n,$ and q are the azimuthal, radial, and axial mode numbers, respectively; R_0 is the radius of curvature of the barrel in a plane $\phi =$ constant; r_0 is the radius of the barrel in the plane $z = 0$; and c is the speed of light in vacuum. The second method first obtains an approximate solution to the wave equation in prolate spheroidal coordinates, which is then used in a variational formula to get a more accurate result for the resonant frequency. This method gives the resonant frequency of an empty barrel cavity expressed as

$$f_{mnq} = \frac{c}{2\pi r_0} v_{mn} \text{ (or } v'_{mn})$$

$$+ \left(q + \frac{1}{2}\right) \frac{c}{2\pi r_0} \sqrt{\frac{r_0}{R_0 - r_0}} = f_v.$$

(12.35)

Measurement results were found to be in equally good agreement with both theories and within the estimated accuracy of 0.15 %, which is the same for both theories. However, the variational method is potentially the more powerful, as it is capable of giving higher accuracy if more accurate field expressions are available (Nagenthieram and Cullen 1974).

For the measurement of dielectric permittivity, it is necessary to calculate the resonant frequency of a dielectric-loaded resonator, as shown in Figure 12.32(b). We assume $n_1 = (\varepsilon_r)^{1/2}$ is the refractive index of the dielectric material with respect to air, and the radius of the dielectric rod is t. The equation of resonance in this case can be determined by matching field components across the dielectric-air interface (Nagenthieram and Cullen 1974):

$$\begin{aligned} &J_0(x)J_1(y)Y_0(z) \\ &\quad -J_0(x)Y_1(y)J_0(z) \\ &= n_1 \cdot [J_0(y)J_1(x)Y_0(z) \\ &\quad -J_1(x)Y_0(y)J_0(z)] \end{aligned} \quad \text{for } E_{0n0} \text{ modes} \quad (12.36)$$

$$\begin{aligned} &J_0(y)J_1(x)Y_1(z) \\ &\quad -J_1(x)Y_0(y)J_1(z) \\ &= n_1 \cdot [J_0(x)J_1(y)Y_1(z) \\ &\quad -J_0(x)Y_1(y)J_1(z)] \end{aligned} \quad \text{for } H_{0n0} \text{ modes} \quad (12.37)$$

with

$$x = n_1 kt - \frac{1}{2} \tan^{-1}\left(\frac{2t}{n_1 kw_0^2}\right) \quad (12.38)$$

$$y = kt - \frac{1}{2} \tan^{-1}\left(\frac{2t}{n_1 kw_0^2}\right) \quad (12.39)$$

$$z = kr_0 - \frac{1}{2} \tan^{-1}\left[\frac{2t}{kw_0^2} \cdot \left(r_0 - t + \frac{t}{n_1^2}\right)\right] \quad (12.40)$$

$$kw_0^2 = 2\sqrt{\left(R_0 - r_0 + t - \frac{t}{n_1^2}\right)\left(r_0 - t + \frac{t}{n_1^2}\right)} \quad (12.41)$$

where w_0 is the scale width of the resonant (beam) mode in the z direction. The preceding transcendental equations may be used to calculate n_1, and hence ε_r from known values of R_0, r_0, t and the measured resonant frequency f of the barrel cavity loaded with the dielectric rod.

As the same approximate theory is used to predict the resonant frequencies of the empty and dielectric-loaded resonators, the error in the resonant frequencies due to the approximation in the theory is probably almost the same in both cases. Therefore, it is not necessary to mechanically measure r_0. The value of r_0 can be determined by using the corresponding resonant frequency of the empty resonator in Eq. (12.35), and the errors arising from the approximate theory will tend to cancel, therefore giving a more accurate result for n_1. Besides, it needs to be noted that a large value of R_0 could make the inaccuracy in the theory negligible.

In a barrel-cavity method for measurements at high temperatures, the rod sample may be heated in a conventional furnace mounted above the open end of the barrel. The sample may then be rapidly transferred to the barrel where the dielectric properties can be monitored as the sample cools (Arai *et al.* 1992). Besides, we can also load the rod sample to the barrel cavity first, and then heat the sample using IR or laser methods. In order to reduce heat transfer to the cavity, the inside surface of the cavity is usually highly polished. Meanwhile, the cavity itself can be maintained at a constant temperature by water-cooling.

12.6.3 Oscillation method

Most of the microwave dielectric measurement methods discussed in this book are passive. In a passive method, the microwave source and the sample under test are independent of each other. In contrast, an active microwave dielectric measurement method incorporates the dielectric under test into the microwave oscillation mechanism, so that the properties of this dielectric are among the factors that determine the oscillation parameters such as the oscillation frequency and the power level. The use of an active oscillation loop for dielectric measurement was formally proposed by Ajmera *et al.* (1974). Tian and Tinga designed and analyzed a microwave oscillation loop formed by a dielectric-loaded cavity, amplifiers, and transmission lines for the dielectric-constant measurement (Tian and Tinga 1994). From the measured loop oscillation frequency, the cavity resonant frequency, and thereby the dielectric constant of the sample in the cavity can be determined. Furthermore, if a power amplifier is used in the oscillation loop, a power high enough to heat the sample can be achieved. In this case, the microwave generated in the loop serves not only the dielectric measurement but also the sample heating, resulting in a novel, active high-temperature dielectrometer. Because of its self-regulated frequency, no tuning and tracking system is needed, which normally is an indispensable part of a dielectrometer using microwave sample heating.

In this section, we first discuss the oscillation condition of an arbitrary two-port network having its input and output ports connected. Then, we discuss the experimental systems for the dielectric constant measurement at room temperature and high temperatures respectively.

12.6.3.1 Oscillation conditions of an active loop

An active oscillation loop consisting of an amplifier, transmission lines, and a cavity loaded with the unknown dielectric, as shown in Figure 12.33(a) is a typical active microwave dielectric measurement system. The oscillation frequency, f_L, of the loop is regulated by the dielectric constant to be measured, and, in return, the unknown dielectric constant can be deduced from the oscillation frequency f_L, which can be directly measured.

In Figure 12.33(b), the oscillation loop is represented by a two-port network having its input and output ports connected. The oscillation condition can be expressed in the form of *S*-parameters (Tian and Tinga 1994):

$$(1 - S_{21})(1 - S_{12}) = S_{11}S_{22} \qquad (12.42)$$

As the position of the reference plane is arbitrary, Eq. (12.42) is valid at any position in the loop that may contain many different components. In the special case of a unidirectional network such as a loop with a high gain amplifier, one of its transmission parameters, say S_{12}, is zero, so Eq. (12.42) becomes

$$1 - S_{21} = S_{11}S_{22} \qquad (12.43)$$

Moreover, when one of its ports, say port 2, is matched, S_{22} is also zero, and Eq. (12.43) can be further reduced to

$$S_{21} = 1 \qquad (12.44)$$

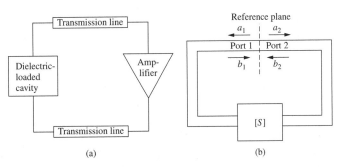

(a)   (b)

Figure 12.33 An oscillation loop for dielectric-constant measurement. (a) Oscillation loop consisting of a cavity, an amplifier, and transmission lines and (b) a network with *S*-parameter representation. Modified from Tian, B. Q. and Tinga, W. R. (1994). "A microwave oscillation loop for dielectric constant measurement", *IEEE Transactions on Microwave Theory and Techniques*, **42** (2), 169–176, © IEEE

Equation (12.44) indicates that the phase of the transmission parameter is equal to zero and its magnitude is equal to unity or its gain is zero, for the special cases when $S_{11} = S_{21} = 0$ or $S_{22} = S_{12} = 0$. In these cases, the oscillation constraint of an active loop can also be expressed as a phase equation and a gain equation:

$$\phi = 2n\pi \qquad (12.45)$$

$$G = 0 \qquad (12.46)$$

where ϕ is the phase shift of the loop in radians, G is the gain of the loop in dB and n is an integer. However, it should be noted that Eqs. (12.45) and (12.46) are applicable only for a unidirectional loop at a matched reference plane (Tian and Tinga 1994).

As the oscillation loop for dielectric constant measurement usually includes high gain amplifiers and a matched isolator at its port 2, the condition of $S_{22} = S_{12} = 0$ is well satisfied, so we can properly use Eq. (12.44) as its oscillation condition. Meanwhile, for the given reference plane in the following discussion, the loop phase shift is actually the phase ϕ_{21} of S_{21}, so it is not necessary to distinguish between the loop phase and the phase of S_{21}.

12.6.3.2 Circuit for measurements at room temperature

An oscillation loop for dielectric measurement should consist, at least, of a cavity, an amplifier, and transmission lines. In practice, the loop may need more components such as attenuators for adjusting the gain, directional couplers for coupling signals from the loop, and isolators for improving the impedance match to certain components. Usually, an attenuator can be included in an amplifier as a gain adjustment device, and a directional coupler or an isolator can be treated approximately as a section of transmission line. Therefore, from a theoretical modeling point of view, a cavity, an amplifier, and a transmission line are the basic elements.

The active loop for the dielectric measurement should be designed to operate in a cavity mode instead of any noncavity modes (Tian and Tinga

1994). Using such a loop for a dielectric constant measurement requires a quantitative relation between the loop oscillation frequency f_L and the cavity resonant frequency f_c, since it is f_L that will be measured, whereas it is f_c that will be used to determine the dielectric constant of the sample in the cavity. If high accuracy is not required, f_L can be taken as f_c. However, for high-accuracy measurement, the exact value of f_c should be used. The cavity resonant frequency f_c and the loop oscillation frequency f_L are related to each other through Eq. (12.44). According to the oscillation condition Eq. (12.44), the loop oscillation frequency f_L is the frequency that makes $\phi_{21} = 0$. Detailed discussions on the relationship between f_L and f_c can be found in (Tian and Tinga 1994).

Figure 12.34 shows a dielectric-constant measurement system using an active loop. In principle, the cavity can be of any type, for example, the reentrant coaxial cavity discussed earlier. The limiting level of the limiter, depending on its dc bias, can be adjusted to enable the amplifier to achieve its maximum output. A circulator provides a matched input as seen from the cavity

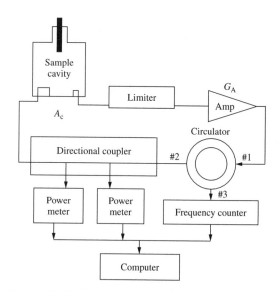

Figure 12.34 Oscillation loop for dielectric-constant measurement (Tian and Tinga 1994). Modified from Tian, B. Q. and Tinga, W. R. (1994). "A microwave oscillation loop for dielectric constant measurement", *IEEE Transactions on Microwave Theory and Techniques*, **42** (2), 169–176. © 2003 IEEE

while isolating the cavity from the amplifier and providing the frequency counter with a measurement signal through its port 3. A dual directional coupler forms a reflectometer to monitor the reflections from the cavity. Two power meters, a frequency counter, and a computer form a data acquisition system, which records the forward and the reflected power as well as the loop frequency for analysis by the computer (Tian and Tinga 1994). In experiments, f_L is initially measured for the empty cavity. After a sample is inserted into the cavity, f_L is measured again. Then the f_c values can be calculated from corresponding f_L values. The values of Δf_L and Δf_c are then used to calculate the dielectric constant.

Here we discuss how the gain condition affects the dynamic range of a sample's allowable dielectric loss factor. Considering the directional coupler and the circulator as part of the transmission line, the total gain around the loop should be the summation of the gains, in dB, of the cavity, the amplifiers and the attenuation in the rest of the loop. According to (12.46), the gain should equal zero when the loop oscillates:

$$A_c + A_L + G_A = 0 \qquad (12.47)$$

where A_c is the cavity attenuation (a reduction in gain) including sample loss, G_A is the amplifier gain, A_L is the attenuation in the rest of the loop made up of the limiter and transmission lines. Besides having to satisfy Eq. (12.47), in a practical case, a stable oscillation cannot be established unless the amplifier is saturated.

12.6.3.3 Circuits for measurements at high temperatures

The oscillation method can be extended to dielectric-constant measurement at high temperatures (Tian and Tinga 1994). By adding a power amplifier and a temperature measurement channel, as shown in Figure 12.35, this active loop can be used to both heat the sample and measure its dielectric constant. The two preamplifiers enable the loop oscillation to start from the system noise level and supply the power amplifier with the required input driving power. The attenuator between the two preamplifiers ensures a proper

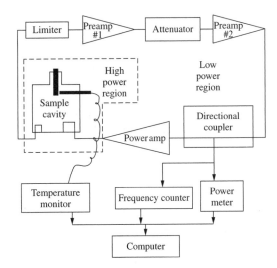

Figure 12.35 Oscillation system for the measurements of dielectric constant at high temperatures. In the system, the sample is heated in the high-power region, while the measurement components are all in the low-power region (Tian and Tinga 1994). Modified from Tian, B. Q. and Tinga, W. R. (1994). "A microwave oscillation loop for dielectric constant measurement", *IEEE Transactions on Microwave Theory and Techniques*, **42** (2), 169–176. © 2003 IEEE

input power level for the amplifier. The power amplifier boosts the power to a level high enough to heat up the sample. The microwave power, loop frequency, and the sample temperature are measured with a power meter, a frequency counter, and an optical fiber thermometer respectively.

As indicated in Figure 12.35, there are two different power regions in the active loop: a high-power region and a low-power region. Only the sample is located in the high-power region while the rest of the components are in the low-power region, avoiding high power–related problems such as overheating, performance deterioration, and damage to the measurement instruments.

In experiments, before the sample is introduced, it is necessary to adjust the empty cavity frequency and the output of the power amplifier to establish a proper oscillation in the loop, then measure the oscillation frequency, which serves as the reference for the subsequent frequency shift due to the introduction of the sample. The sample is inserted into the cavity, and then the microwave energy generated in the loop heats the sample. When the sample reaches the desired temperature, the temperature and frequency data will be recorded. The dielectric constant of the sample can be then calculated from the measured frequency shift due to the sample. In this way, the temperature dependence of the dielectric constant can be obtained in a short period, depending on the self-oscillating microwave power, which is used in sample heating.

REFERENCES

Ajmera, R. C. Batchelor, D. B. Moody, D. C. and Lashinsky, H. (1974). "Microwave measurements with active systems", *Proceedings of the IEEE*, **62** (1), 118–127.

Ali, I. A. Al-Amri, A. M. and Dawoud, M. M. (2000). "Dielectric properties of dates at 2.45 GHz determined with a tunable single-mode resonant cavity", *Journal of Microwave Power and Electromagnetic Energy*, **35** (4), 242–252.

Andrade, O. M. Iskander, M. F. and Bringhurst, S. (1992). "High temperature broadband dielectric properties measurement techniques", *Materials Research Society Symposium Proceedings*, **269**, 527–539.

Arai, M. Binner, J. G. P. Carr, G. E. and Cross, T. E. (1992). "High temperature dielectric measurements on ceramics", *Materials Research Society Symposium Proceedings*, **269**, 611–616.

Arai, M. Binner, J. G. P. and Cross, T. E. (1995a). "Correction of errors owing to high temperature coaxial probe for microwave permittivity measurement", *Electronics Letters*, **31** (4), 1138–1139.

Arai, M. Binner, J. G. P. and Cross, T. E. (1995b). "Estimating errors due to sample surface roughness in microwave complex permittivity measurements obtained using a coaxial probe", *Electronics Letters*, **31** (2), 115–117.

Arai, M. Binner, J. G. P. and Cross, T. E. (1996). "Comparison of techniques for measuring high-temperature microwave complex permittivity: measurements on an alumina/zircona system", *Journal of Microwave Power and Electromagnetic Energy*, **31** (1), 12–18.

ASTM Designation D2520-86 (1986). Standard test methods for complex permittivity of electrical insulating materials at microwave frequencies and temperature to 1650 °C, ASTM International, West Conshohocken, PA.

Basset, H. L. Bomar Jr., S. H. (1973). Complex Permittivity Measurements During High Temperature Recycling of Space Shuttle Antenna Window and Dielectric Heat Shield, NASA Technical Report, **CR-2302**, National Technical Information Service, Springfield, VA.

Batt, J. A. Rukus, R. and Gilden, M. (1992). "General purpose high temperature microwave measurement of electromagnetic properties", *Materials Research Society Symposium Proceedings*, **269**, 553–559.

Binner, J. G. P. Cross, T. E. Greenacre, N. R. and Naser-Moghadasi, M. (1994). "High temperature dielectric property measurements – an insight into microwave loss mechanism in engineering ceramics", *Materials Research Society Symposium Proceedings*, **347**, 247–252.

Blackham, D. (1992). "Calibration method for open-ended coaxial probe/vector network analyzer system", *Materials Research Society Symposium Proceedings*, **269**, 595–599.

Bringhurst, S. Andrade, O. M. and Iskander, M. F. (1992). "High-temperature dielectric properties measurements of ceramics", *Materials Research Society Symposium Proceedings*, **269**, 561–568.

Bringhurst, S. and Iskander, M. F. (1996). "Open-ended metallized ceramic caxial probe for high-temperature dielectric properties measurements", *IEEE Transactions on Microwave Theory and Techniques*, **44** (6), 926–935.

Bringhurst, S. Iskander, M. F. and White, M. J. (1997). "Thin-sample measurements and error analysis of high-temperature coaxial dielectric probes", *IEEE Transactions on Microwave Theory and Techniques*, **45** (12), 2073–2083.

Bussey, H. E. (1969). "Measurement of radio frequency properties of materials, a survey", *Proceedings of the IEEE*, **55**, 1046–1053.

Chevalier, B. Chatard-Moulin, M. Astier, J. P. and Guillon, P. Y. (1992). "High temperature complex permittivity measurements of composite materials using an open-ended waveguide", *Journal of Electromagnetic Waves and Applications*, **6** (9), 1259–1275.

Craven, M. P. Cross, T. E. and Binner, J. G. P. (1996). "Enhanced computer modelling for high temperature microwave processing of ceramic materials", *Materials Research Society Symposium Proceedings*, **430**, 351–356.

Cullen, A. L. Nagenthieram, P. and Williams, A. D. (1972). "Improvement in open-resonator permittivity measurement", *Electronics Letters*, **23** (8), 577–579.

Gershon, D. L. Calame, J. P. Carmel, Y. Antonsen, T. M. and Hutchen, R. M. (1999). "Open-ended coaxial probe for high-temperature and broad-band dielectric measurements", *IEEE Transactions on Microwave Theory and Techniques*, **47** (9), 1640–1648.

Ghodgaonkar, D. K. Varadan, V. V. and Varadan, V. K. (1989). "A free space measurement for dielectric constant and loss tangent at microwave frequencies", *IEEE Transactions on Instrumentation and Measurements*, **37** (3), 789–793.

Ghodgaonkar, D. K. Varadan, V. V. and Varadan, V. K. (1990). "Free space measurement for complex permittivity and complex permeability of magnetic materials at microwave frequencies", *IEEE Transactions on Instrumentation and Measurements*, **39** (5), 387–393.

Ho, W. W. (1980). High Temperature Millimeter Wave Characterization of the Dielectric Properties of Advanced Window Materials, Technical Report, Rockwell International Science Center, Thousand Oaks, CA.

Ho, W. W. (1988). "High-temperature dielectric properties of polycrystalline ceramics", *Materials Research Society Symposium Proceedings*, **124**, 137–148.

Holderfield, S. P. and Salesman, J. B. (1992). "Observed trends in the dielectric properties of minerals at elevated temperatures", *Materials Research Society Symposium Proceedings*, **269**, 589–594.

Hutcheon, R. de Jong, M. and Adams, F. (1992a). "A system for rapid measurements of RF and microwave properties up to 1400 °C, part 1: theoretical development of the cavity frequency-shift data analysis equations", *Journal of Microwave Power and Electromagnetic Energy*, **27** (2), 87–92.

Hutcheon, R. M. de Jong, M. S. Adams, F. P. Lucuta, P. G. McGregor, J. E. and Bahen, L. (1992c). "RF and microwave dielectric measurements to 1400 °C and dielectric loss mechanisms", *Materials Research Society Symposium Proceedings*, **269**, 541–551.

Hutcheon, R. M. de Jong, M. S. Lucuta, P. McGregor, J. E. Smith, B. H. and Adams, F. P. (1991). "Measurements of high-temperature RF and microwave properties of selected aluminas and ferrites used in accelerators", *IEEE Particle Accelerator Conference (1991): Accelerator Science and Technology*, **2**, 795–797.

Hutcheon, R. de Jong, M. Adams, F. Wood, G. McGregor, J. and Smith, B. (1992b). "A system for rapid measurements of RF and microwave properties up to 1400 °C, part 2: description of apparatus, data collection techniques and measurements on selected materials", *Journal of Microwave Power and Electromagnetic Energy*, **27** (2), 93–102.

HVS Technologies, State College, PA, Website: http://www.hvstech.com.

Iskander, M. F. and DuBow, J. B. (1983). "Time- and frequency-domain techniques for measuring the dielectric properties of rocks: a review", *Journal of Microwave Power*, **18** (1), 55–74.

Jose, K. A. Varadan, V. V. Hollinger, R. D. Tellakula, A. and Varadan, V. K. (2000). "Non contact broadband microwave material characterization at low and high temperatures", *Proceedings of the SPIE*, **4129**, 324–331.

Jow, J. Hawley, M. C. Finzel, M. and Asmussen Jr., J. (1987). "Microwave processing and diagnosis of chemically reacting materials in a single-mode cavity applicator", *IEEE Transactions on Microwave Theory and Techniques*, **35** (12), 1435–1443.

Jow, J. Hawley, M. C. Finzel, M. C. and Asmussen Jr., J. (1989). "Microwave heating and dielectric diagnosis technique in a single-mode resonant cavity", *Review of Scientific Instruments*, **60** (1), 96–103.

Metaxas, A. C. and Meredith, R. J. (1983). *Industrial Microwave Heating*, Peter Peregrinus Ltd, London.

Mouhsen, A. Achour, M. E. Miane, J. L. and Ravez, J. (2001). "Microwave dielectric relaxation of ferroelectric PLZT ceramics in the range of 300–900 K", *The European Physical Journal – Applied Physics*, **15**, 97–104.

Nagenthieram, P. and Cullen, A. L. (1974). "A microwave barrel resonator for permittivity measurements on dielectric rods", *Proceedings of the IEEE*, **62** (11), 1613–1614.

Salsman, J. B. (1991). "Technique for measuring the dielectric properties of minerals as a function of temperature and density at microwave heating frequencies", *Materials Research Society Symposium Proceedings*, **189**, 509–515.

Sutton, W. H. (1989). "Microwave processing of ceramic materials", *Ceramic Bulletin*, **68** (2), 376–386.

Thuery, J. and Grant, E. H. (1992). *Microwaves: Industrial, Scientific, and Medical Applications*, Artech House, Boston.

Tian, B. Q. and Tinga, W. R. (1993). "Single-frequency relative Q measurements using perturbation theory", *IEEE Transactions on Microwave Theory and Techniques*, **41** (11), 1922–1927.

Tian, B. Q. and Tinga, W. R. (1994). "A microwave oscillation loop for dielectric constant measurement", *IEEE Transactions on Microwave Theory and Techniques*, **42** (2), 169–176.

Tian, B. Q. and Tinga, W. R. (1995). "Linearity condition between a cavity's Q-factor and its input resistance", *IEEE Transactions on Microwave Theory and Techniques*, **43** (3), 691–692.

Tinga, W. R. (1992). "Rapid high-temperature measurement of microwave dielectric properties", *Materials Research Society Symposium Proceedings*, **269**, 505–516.

Tinga, W. and Xi, X. (1993a). "Design of a new high-temperature dielectrometer system", *Journal of Microwave Power and Electromagnetic Energy*, **28** (2), 93–103.

Tinga, W. and Xi, X. (1993b). "Error analysis and permittivity measurements with re-entrant high-temperature dielectrometer", *Journal of Microwave Power and Electromagnetic Energy*, **28** (2), 104–112.

Varadan, V. V. Hollinger, R. D. Ghodgaonkar, D. K. and Varadan, V. K. (1991). "Free space broadband measurement of high temperature complex dielectric properties at microwave frequencies", *IEEE Transactions on Instrumentation and Measurements*, **40** (5), 842–846.

Varadan, V. V. Jose, K. A. and Varadan, V. K. (2000). "In situ microwave characterization of non planar specimens", *IEEE Transactions on Microwave Theory and Techniques*, **48** (3), 388–394.

Von Hippel, A. R. (1995). *Dielectric Materials and Applications*, Artech House, Boston.

Xi, W. G. Tian, B. Q. and Tinga, W. R. (1994). "Numerical analysis of a movable dielectric gap in coaxial resonators for dielectric measurements", *IEEE Transactions on Instrumentation and Measurement*, **43** (3), 486–487.

Xi, W. G. and Tinga, W. R. (1992a). "Field analysis of new coaxial dielectrometer", *IEEE Transactions on Microwave Theory and Techniques*, **40** (10), 1927–1934.

Xi, W. G. and Tinga, W. R. (1992c). "Microwave heating and characterization of machinable ceramics", *Materials Research Society Symposium Proceedings*, **269**, 569–577.

Xi, W. G. Tinga, W. R. Voss, W. A. G. and Tian, B. T. (1992b). "New results for coaxial re-entrant cavity with partially dielectric filled gap", *IEEE Transactions on Microwave Theory and Techniques*, **40** (4), 747–753.

Index

Admittance
  characteristic, 46
  matrix, 119–121
Anisotropic material, 28
  dielectric, 323–325
  magnetic, 325–326
Antenna
  as transition, 83
  bandwidth, 85
  conical monopole, 156
  directive gain, 85
  hemispherical, 156
  power gain, 84
  spheroidal, 156
  with step transition, 156
Antiferromagnetic material, 5
Atomic polarization, 11, 12
Attenuation coefficient, 10

BCS theory, 4, 17
Beam
  focused, 86
  parallel 85
  width, 86
Bohr's model, 3

Calibration, 123–126. *See also* Free-space calibration,
  Measurement error, Network analyzer.
  one-port, 123
  two-port, 124–126
    full two-port, 124
    TRL, 125
Cavity perturbation, *see also* Resonant perturbation.
  extra-cavity, 265–267
  modification, 261–264
    calibration, 263
    frequency retuning, 262

  permeability measurement, 258
  permittivity measurement, 256
Cavity resonator
  coupled to one transmission line, 90
  coupled to two transmission lines, 91
  cylindrical, 100–103
    mode chart, 102
    TE resonant mode, 100
    TM resonant mode, 101
  rectangular, 97–99
    TE resonant mode, 98
    TM resonant mode, 99
  sample-loaded, 258
Chiral material, 25, 414–415
Chirality, 414, 434, 440, 445–450, 455–456.
Chirality measurement
  free space, 415–452
  waveguide, 452–457
  rotation angle, 414–415, 417, 448, 452–453.
Circular birefringence, 439
Circular dichroism, 439
Circular-radial resonator, 216–219
Coaxial air-line transmission/reflection methods,
  182–187
  enlarged circular coaxial line, 186
  enlarged square coaxial line, 186
  measurement uncertainty, 183
Coaxial line:
  attenuation, 57
  characteristic impedance, 57
  field distribution, 57
  shielded, 158
Coaxial resonators:
  capacitor-loaded, 94
  half-wave length, 93
  quarter-wave length, 94
Coaxial surface-wave:
  resonator, 228–231

Coaxial surface-wave: (*continued*)
    closed, 229
    open, 228
    transmission structure, 81–83
Coaxial-line probe, *see also* Coaxial-line reflection
  method.
    terminated into layered materials, 151–154
      sample backed by free space, 153, 154
      sample backed by metal, 152, 153
    modeling, 145–149
      capacitance, 145
      full wave simulation, 149
      radiation, 146
      rational function, 148
      virtual line, 147
    modification,
      conical tip, 157
      large, 149
      with elliptic aperture, 150
Coaxial-line reflection method, 144–161. *See also*
  Coaxial-line probe.
    flange effect, 149
Cohn resonator, 214
Cole–Cole diagram, 14
Composite material:
    dielectric-conductor, 31
    dielectric–dielectric, 29
Conductivity, 7
    complex, 18
    dielectric, 8
    ionic, 15
Conductor, 4, 11, 16
    perfect, 17
Constitutive parameter, 7
Constitutive relation, 6, 414, 434
Coplanar nonresonant method, 309
Coplanar resonant method, 311
Coplanar waveguide:
    attenuation factor, 62
    characteristic impedance, 61
    effective dielectric constant, 61
Cotton effect, 414, 444
Coupling:
    capacitive, 97
    coefficient, 91
    electric, 95
    magnetic, 95
    parallel-line, 97
    tap, 97
Courtney resonator, 209–214
    conducting plate, 212
    configuration, 209
    higher resonant mode, 214

    mode chart, 209
    usable range, 212
Cut-off frequency, 63

Dallenbach layer, 33
Debye theory, 13
Diamagnetic material, 4
Dielectric interface, 74
Dielectric material, 11
Dielectric microstrip, 78
Dielectric property at high temperature, 492–494
Dielectric property measurement at high temperature:
    cavity perturbation, 510–520
      heating technique, 511
      TE mode rectangular cavity, 512
      TM mode cylindrical cavity, 514
    coaxial re-entrant cavity, 520–523
      heating technique, 522
      measurement circuit, 522
      structure, 520
    coaxial-line, 497–503
      heating technique, 499
      phase shift correction, 500
      thermal elongation, 500
      spring-loaded, 502
    free-space, 199, 506–510
      measurement system, 506
    measurement problem, 494
    open-resonator, 523
      barrel, 523
      spherical-mirror, 523
    oscillation, 524–528
      circuit for high temperature, 527
      circuit for room temperature, 526
      oscillation condition, 525
    waveguide, 503–506
      dual-waveguide, 504
      open-ended, 503
Dielectric resonator, 34, 103–119
    isolated, 105
    shielded, 106
      asymmetrical parallel plate, 107
      cut-off magnetic-wall waveguide, 109
      closed metal shields, 110, 222–227
      parallel-plate, 106
Dielectric resonator perturbation, 267
Dielectric slab:
    grounded, 76
    ungrounded, 76
Dielectric waveguide, *see also* Surface wave
  transmission line.
    cylindrical, 79–81
    rectangular, 77–78

Dielectric waveguide transmission / reflection method, *see* Surface waveguide transmission / reflection method.
Dipolar polarization, 13, 14

Effective medium method, 29
Electrical transport property, 460. *See also* Hall effect.
  carrier density, 460
  conductivity, 460
  mobility, 461
  of magnetic material, 486
    bimodal cavity, 487
    bimodal dielectric probe, 489
Electronic polarization, 11, 12
Energy band
  conduction band, 3
  forbidden gap, 3
  valence band, 3
Extrinsic performance, 32

Fabry–Perot resonator, *see* Open resonator.
Faraday rotation, 464. *See also* MHE measurement, Permeability tensor measurement.
Ferrimagnetic material, 5
Ferroelectric material, 15, 382
Ferroelectricity, 15–16, 382–385
  Curie temperature, 15, 16
  electric field dependence, 385
  hysteresis curve, 383
  perovskite, 383
  temperature dependence, 383
  tunability, 385
Ferroelectricity measurement 385–412
  biasing scheme, 404
  capacitor, 394
    capacitor design, 395
    fin-line resonator, 404
    planar split-resonator, 398
  cavity perturbation, 388
  coplanar resonator, 394
  coplanar waveguide, 390
  dielectric resonator, 386
    Courtney resonator, 386
    disk resonator, 387
  near-field microscope, 390
  nonresonant, 385
    reflection, 385
    transmission/reflection, 386
  nonlinear behavior, 406
    inter-modulation, 409
    pulsed signal, 407
  responding time, 406
Ferromagnetic material, 5, 325–326, 346, 355.

Ferromagnetic resonance, 325–326, 370–371. *See also* Magnetic resonance.
Ferromagnetic resonance measurement, *see also* Permeability tensor measurement.
  cavity, 373
    reflection, 373
    transmission, 373
  planar-circuit, 376
    microstrip resonator, 378
    MSW-SER, 376
    slot-coplanar junction, 377
  principle, 371
  waveguide, 374
    cross-guide, 375
    frequency variation, 374
    pickup coil, 376
Fourier transform, 127
Free space, 83
Free-space calibration, 417–430. *See also* Calibration.
Free-space reflection method, 161–164
  bistatic reflection, 164
  far-field requirement, 161
  movable metal backing, 162
  short-circuited reflection, 162
Free-space transmission / reflection method, 195–200
  algorithm, 195
  high-temperature measurement, 199, 506–510
  TRL calibration, 197
  uncertainty, 198

Hall effect, *see also* Electrical transport property.
  ac, 461
  dc, 461
  extraordinary, 486
  microwave, 461
  ordinary, 486
Helix, 414, 416, 452
Hysteresis loop
  ferroelectric material, 15
  magnetic material, 19

Image guide,
  cylindrical, 81
  rectangular, 78
Impedance,
  characteristic, 46
  input, 48
  matching, 71
  matrix, 119–121
  wave, 10
Index ellipsoid, 324
Insulator, 3, 4, 11

Intrinsic property, 32
Isotropic material, 28

Left-handed material, 25–27
Linear material, 28
Loss tangent:
  dielectric, 8
  magnetic, 9

Macroscopic scale, 6–11
Magnetic material, 11, 19–24
  electrical transport property, 486–489
  hard, 22
  soft, 22
  thin film, 311
Magnetic moment, 4
Magnetic resonance, 22–24. *See also* Ferromagnetic
  resonance
  natural resonance, 22
  wall resonance, 24
Magnetization, 19
  easy direction, 20
  hard direction, 20
Matching,
  field, 71
  impedance, 71
Material perturbation, 253–255
  small object, 254
  whole medium, 254
Maxwell's equation, 7
Mean-field method, 29
Measurement error, 122–123. *See also* Calibration,
  Network analyzer.
  drift, 123
  random, 123, 417
  systematic error, 122, 417
Metamaterial, 24
MHE cavity, 475–484. *See also* Hall effect, MHE
  measurement, MHE resonator.
  circular cylindrical, 479
  endplate, 482
  materials in, 476
  orthogonal resonant modes, 475
  rectangular, 481
MHE measurement, 464–486
  Faraday rotation, 464
  reflection, 469
    adjustable short, 469
    cross-slit probe, 472
  resonance, 475. *See also* MHE cavity, MHE
    resonator.
    measurement circuit, 476
  transmission, 465

  dielectric waveguide, 468
  free-space, 468
  waveguide, 465
  turnstile junction, 473
MHE resonator, 484–486. *See also* Hall effect, MHE
  cavity
  cross-slit dielectric probe, 484
  planar resonator, 486
  rectangular dielectric resonator, 484
Microscopic scale, 2–6
Microstrip, 59–61
  attenuation, 61
  characteristic impedance, 60
  effective dielectric constant, 60
Microstrip nonresonant method, 298–300
  transmission line, 298
  transmission / reflection, 299
Microstrip resonant method, 300–309
  cross-resonator, 305
  ring resonator, 301
  straight-ribbon resonator, 302
  T-resonator, 304
  two-section microstrip, 306
Microwave propagation, 42–87
  high temperature, 493
Microwave processing, 493
Monolithic material, 29

Near-field microwave microscope, *see also*
  Ferroelectricity measurement.
  planar circuit, 317–320
    electric dipole probe, 318
    magnetic dipole probe, 319
    working principle, 317
  reflection, 170–172
    capacitive mode, 172
    inductive mode, 171
    rectangular slit, 170
  resonant perturbation, 278–286
    dielectric resonator, 284
    open-ended coaxial resonator, 280
    permeability measurement, 283
    permittivity measurement, 282
    sheet resistance measurement, 281
    tip-coaxial resonator, 279
    waveguide cavity, 284
    working principle, 278
Network analyzer, 121–126. *See also* Calibration,
  Measurement error.
Nonlinear material, 28
Nonresonant method, 38–40. *See also* Planar-circuit
  nonresonant method.

reflection, 38
transmission/reflection, 39

Open resonator, 115–119, 523. *See also* Open
  resonator method.
  barrel, 523
  concentric, 117
  confocal, 117
  coupling, 117
  parallel plane, 115
  spherical-mirror, 523
  stability requirement, 116
  stability diagram, 116
Open resonator method, 238–242. *See also* Open
  resonator.
  bi-concave resonator, 239
    frequency variation, 239
    length variation, 240
  plano-concave resonator, 241
Ordered magnetic material, 5

Paraelectric material, 15
Paramagnetic material, 5
Penetration depth, 18
Percolation, 31
Permeability,
  complex, 7, 10
  relative, 9
  tensor, 325
Permeability tensor measurement, 340–370. *See also*
  Ferromagnetic resonance measurement.
  coaxial air-line, 340
  Faraday rotation, 345
    bimodal cavity, 350
    circularly polarized propagation, 347
    reflection 348
  partially-filled waveguide, 341
  resonant perturbation, 355
    circularly polarized cavity, 359
    exact theory, 365
    geometrical effect, 360–365
    linearly polarized cavity, 359
  resonator, 351
    ring resonator, 353
    shielded cylindrical resonator, 353
    TE_{111} resonator, 352
    whispering-gallery resonator, 354
Permeance meter for magnetic thin film, 311–316
  electrical impedance, 315
  single-coil, 314
  two-coil, 312
  working principle, 312

Permittivity:
  complex, 7, 10
  relative, 8
  tensor, 323
Permittivity and permeability, measurement using
  reflection method, 164–168
  combination, 166
  different backing, 167
  different position, 165
  frequency-variation, 167
  time-domain, 168
  two-thickness, 164
Permittivity tensor measurement,
  326–340
  reflection, 327
    coaxial line, 327
    free space, 329
    waveguide, 328
  resonant perturbation, 336
  resonator, 333
    sandwiched resonator, 334
    shielded resonator, 333
    whispering-gallery resonator, 334
  transmission / reflection, 331
    circular waveguide, 331
    coaxial discontinuity, 332
Phase change coefficient, 10
Phase velocity, 63
Photonic band-gap material, 27–28
Piezoelectric material, 15, 382
Planar-circuit nonresonant method, 288–290. *See also*
  Coplanar nonresonant method, Microstrip
  nonresonant method, Stripline nonresonant method.
  reflection, 288
  transmission / reflection, 289
Planar circuit resonant method, 290–291. *See also*
  Coplanar resonant method, Microstrip resonant
  method, Stripline resonant method.
  resonant perturbation, 290
  resonator, 290
Planar-circuit resonator, 95–97
  circular, 96
  ring, 96
  straight ribbon, 96
Propagation coefficient, 10
Pyroelectric material, 15, 382

Quality factor for material, 9
Quality factor for resonator, 87
  loaded, 135
  measurement,
    134–139
    nonlinear phenomena, 136

Quality factor for material (*continued*)
    reflection, 135
      transmission, 136
   unloaded, 135

Reflection, 47, 142–144. *See also* Reflection method.
  open-circuited, 142
  short-circuited, 143
Reflection coefficient, 48
Reflection method, 38–39. *See also* Permittivity and
  permeability, measurement using reflection methods,
  Reflection.
  open, 38
  shorted, 39
  surface impedance measurement, 39
Reflectivity, 33
Refraction index, 238
Resonant frequency, 87
Resonant mode chart, 102
Resonant perturbation, 103, 250–256. *See also* Cavity
  perturbation, Dielectric resonator perturbation,
  Material perturbation.
  cavity shape, 103, 252
  material, 103, 253
  wall impedance, 255
Resonant perturbation method, 41–42. *See also*
  Resonant perturbation.
  permeability measurement, 41
  permittivity measurement, 41
  surface impedance measurement, 41
Resonator method, 40–41
  permittivity measurement, 3, 40
  surface resistance measurement, 41

Scattering parameter, 120
Semiconductor, 4, 11, 16
  extrinsic, 16
  intrinsic, 16
Sheet resonator, 219
Skin depth, 10
Smith chart, 51–56
  admittance, 54
  impedance, 52
Snell effect, 26
Split coaxial resonator, 233
Split cylinder cavity, 231
Split dielectric resonator, 236
Stripline, 58–59
  thick central conductor, 58
  thin central conductor, 58
Stripline nonresonant method, 291–292
  asymmetrical stripline, 292
  symmetrical stripline, 291

Stripline resonant method, 292–297
  resonant perturbation, 295
    permeability measurement, 297
    permittivity measurement, 296
  resonator, 293
    one-conductor, 293
    two-conductor, 293
Superconductor, 4, 11, 17, 18
  critical temperature, 17
  high-temperature, 17
  low-temperature, 17
Surface impedance, 10, 11, 18, 19
Surface impedance measurement, *see also* Surface
  reactance measurement, Surface resistance
  measurement.
  dielectric resonator, 243–247
    dual-mode resonator, 245
    two-resonator, 243
  reflection, 169
Surface reactance, 10
Surface reactance measurement, *see also* Surface
  impedance measurement
  resonant perturbation, 275–278
    material perturbation, 275
    wall-replacement, 275
Surface resistance, 10
Surface resistance measurement, *see also* Surface
  impedance measurement
  dielectric resonator, 242
  resonant perturbation, 269–275
Surface wave transmission line, 73–83. *See also*
  Coaxial surface-wave.
Surface waveguide transmission / reflection method,
  190–195
  circular dielectric waveguide, 190
  rectangular dielectric waveguide, 192

Telegrapher equations, 45
Time-domain technique, 127–134, 430–434
  gating, 131, 433
  windowing, 131, 430
    band pass mode, 430
    low pass mode, 430
Transition, 71–73
  antenna as, 83
  between coaxial line and microstrip line, 73
  between rectangular waveguide and circular
    waveguide, 71
  between rectangular waveguide and coaxial line, 72
Transmission line theory, 42–51
  TE wave, 43
  TEM wave, 44
  TM wave, 43

Transmission / reflection method, *see also* Coaxial
   air-line transmission / reflection method, Free-space
   transmission / reflection method, Surface waveguide
   transmission / reflection method, Waveguide
   transmission / reflection method.
   complex conductivity measurement, 203
   modification, 200–203
      antenna probe, 201
      coaxial discontinuity, 200
      cylindrical cavity between transmission lines, 200
      dual probe, 201
      dual-line probe, 201
   permittivity and permeability measurement, 39
   surface impedance measurement, 40
   theory
      effective parameter, 179
      nonlinear least-square, 180
      NRW algorithm, 177
      precision mode, 178
      working principle, 175

Velocity:
   group, 64
   phase, 63
   wave, 10

Wave:
   mixed, 51
   pure standing, 50
   pure traveling, 49
Wave number, 43, 433, 434, 437, 439, 446
Wave polarization
   circular, 347, 359, 371, 415,
      436–439, 464
   elliptical, 414, 415, 417, 439
      axial ratio, 415, 452–454
      ellipticity, 448, 452
      major axis, 415, 417
      minor axis, 417
   linear, 359, 415, 439, 464
Waveguide, 62–71
   circular, 68
   rectangular, 65
   transition, 69
Waveguide transmission / reflection method,
   187–190
Whispering gallery dielectric resonator, 112–115. *See
   also* Permeability tensor measurement, Permittivity
   tensor measurement.
   WGE mode, 114
   WGH mode, 114